Lecture Notes in Control and Information Sciences

Edited by A.V. Balakrishnan and M.Thoma

63

Analysis and Optimization of Systems

Proceedings of the Sixth International Conference on Analysis and Optimization of Systems
Nice, June 19–22, 1984

Part 2

Edited by
A. Bensoussan and J. L. Lions

Springer-Verlag
Berlin Heidelberg New York Tokyo 1984

Editors
A. Bensoussan
J. L. Lions

INRIA
Institut National de Recherche en Informatique et en Automatique
Domaine de Voluceau, Rocquencourt, B.P. 105
78153 Le Chesnay/France

Library of Congress Cataloging in Publication Data
International Conference on Analysis and Optimization of Systems
(6th : 1984 : Nice, France)
Analysis and optimization of systems.
(Lecture notes in control and information sciences ; 62–63)
"Organized by the Institut national de recherche en informatique et [en] automatique"
Foreword. English and French.
1. System analysis––Congresses.
2. Mathematical optimization––Congresses.
3. Automatic control––Congresses.
4. Biotechnology––Congresses.
I. Bensoussan, Alain.
II. Lions, Jacques Louis.
III. Institut national de recherche en informatique et en automatique (France).
IV. Title.
V. Series.
QA402.I533 1984 003 84-5601

ISBN 3-540-13552-9 Springer-Verlag Berlin Heidelberg New York Tokyo
ISBN 0-387-13552-9 Springer-Verlag New York Heidelberg Berlin Tokyo

Offsetprinting: Mercedes-Druck, Berlin
Binding: Lüderitz und Bauer, Berlin
2061/3020-543210

FOREWORD

This volume contains most of the 94 papers presented during the Sixth International Conference on Analysis and Optimization of Systems organized by the Institut National de Recherche en Informatique et Automatique.

The audience has increased by more than 50 % in comparison with the Fifth Conference. These papers, some invited and most of them submitted, were presented by speakers coming from 26 different countries. Most of the topics of System Theory are covered.

At the theoretical level, a trend towards algebraic and geometric methods was confirmed. Signal processing which was one of the main topics of the call for papers had a favourable result : two special sessions on non stationary models and on rupture detection were organized. In the field of applications, one can notice the increasing importance of the CACSD tools. Also, the progress of the biomedical and biotechnological engineering session is remarkable. It has justified the sponsorship of INSERM, for the first time.

In order to improve the coordination with the IEEE Control and Decision Conference, the Organizing Committee has decided to shift the date of the conference which from now on will be held in June. The conference took place near the new center of INRIA at Sophia Antipolis.

We would like to express our thanks to the Organisations which have given their sponsorship to this meeting : AFCET, IEEE, IFAC and INSERM.

We also would like to extend our gratitude to :

- the authors who have shown their interest in this conference,
- the numerous referees who have accepted the difficult task of selecting papers,
- the Chairpersons for having run with energy and efficiency the different sessions,
- our colleagues of the Organisation Committee,
- the Scientific Secretaries,
- Miss Bricheteau and the staff of the Public Relations Department for the difficult but successful job they have carried out in the organization of the Conference,
- Professor Thoma who has accepted to publish these proceedings in the Lecture Notes in Control and Information Sciences, and to the Publisher SPRINGER VERLAG.

A. BENSOUSSAN

J.L. LIONS

PREFACE

Ce volume contient la presque totalité des textes des 94 communications présentées lors de la Sixième Conférence Internationale sur l'Analyse et l'Optimisation des Systèmes, organisée par l'Institut National de Recherche en Informatique et Automatique.

Cette Conférence connaît une audience grandissante puisque le nombre de communications soumises a augmenté de plus de 50 % par rapport à sa dernière édition, confirmant ainsi une tendance antérieure. Ces communications, invitées ou pour la plupart soumises, émanent de 26 pays différents. La plupart des domaines de la "Théorie des Systèmes" y sont abordés.

Sur le plan théorique, on constate la confirmation d'une évolution vers les méthodes géométriques et algébriques. Le traitement du signal qui était l'un des thèmes principaux de l'appel aux communications a connu un succès certain : deux sessions spéciales sur les modèles non stationnaires et les détections de ruptures ont été organisées.

Du point de vue des applications, les communications présentées portent plus sur des outils généraux de CAO en Automatique que sur des applications spécifiques. Il faut cependant noter les progrès de la session présentant des applications au domaine du génie biomédical et des biotechnologies. Pour la première fois, la Conférence a reçu le patronage de l'INSERM.

La coordination avec la "Control and Decision Conference" de l'IEEE a conduit a déplacer les dates de la Conférence qui se tient désormais au mois de juin. La Conférence s'est déroulée à proximité du nouveau centre INRIA de Sophia-Antipolis.

Nous tenons à remercier les organismes qui ont accepté d'accorder leur patronage à cette manifestation : AFCET, IEEE, IFAC, INSERM.

Nos remerciements s'adressent également :

- aux auteurs qui ont manifesté leur intérêt pour cette conférence ;

- aux nombreux experts qui ont accepté la difficile tâche de sélectionner les communications,

- aux présidents de sessions qui ont accepté d'animer les débats,

- à nos collègues du Comité d'Organisation,

- aux Secrétaires Scientifiques,

- à Mademoiselle Bricheteau et ses collaboratrices du Service des Relations Extérieures qui ont largement participé à l'organisation de cette Conférence,

- à Monsieur le Professeur Thoma pour avoir accepté la publication de ce volume dans la série qu'il dirige, ainsi qu'à l'éditeur SPRINGER VERLAG.

A. BENSOUSSAN

J.L. LIONS

This conference is under the sponsorship of

Cette Conférence a reçu le patronage de

- Association Française pour la Cybernétique Economique et Technique (AFCET)
- Institute of Electrical and Electronics Engineers, Inc (I.E.E.E.) "Co-sponsorship"
- International Federation of Automatic Control (I.F.A.C.)
- Institut National de la Santé et de la Recherche Médicale (INSERM)

Organization Committee

Comité d'Organisation

K. J. ASTRÖM	Lund Institute of Technology, (Suède)
A. BENSOUSSAN	Université Paris-Dauphine / INRIA-Rocquencourt (France)
A. BENVENISTE	INRIA-Rennes (France)
P. BERNHARD	INRIA-Sophia-Antipolis (France)
P. FAURRE	SAGEM (France)
A. J. FOSSARD	ENSAE-Toulouse (France)
J. L. LIONS	Collège de France / INRIA (France)
A. G. J. Mac FARLANE	Cambridge University (GB)
M. THOMA	Technische Universität Hannover (RFA)
J. C. WILLEMS	Groningen University (Pays Bas)

Scientific Secretaries

Secrétaires Scientifiques

F. DELEBECQUE	INRIA-Rocquencourt (France)
J. HENRY	INRIA-Rocquencourt (France)

Conference Secretariat

Secrétariat de la Conférence

Th. BRICHETEAU	INRIA (France)
S. GOSSET	Service des Relations Extérieures

REFEREES

ABRAMATIC	J.F.	(FRANCE)
AEYELS	D.	(BELGIUM)
ALING		(THE NETHERLANDS)
ALMEIDA	L.B.	(PORTUGAL)
ASTROM	K.J.	(SWEDEN)
AUBIN	J.P.	(FRANCE)
BABARY	J.P.	(FRANCE)
BAILLIEUL	John	(U.S.A.)
BARAS	J.	(U.S.A.)
BARATCHART	L.	(FRANCE)
BARRAUD	M.	(FRANCE)
BASSEVILLE	Michelle	(FRANCE)
BENSOUSSAN	Alain	(FRANCE)
BENVENISTE	A.	(FRANCE)
BERNHARD	P.	(FRANCE)
BERNUSSOU	J.	(FRANCE)
BERTHOMIER	C.	(FRANCE)
BINDER		(FRANCE)
BISMUT	J.M.	(FRANCE)
BOIS VIEUX	J.F.	(FRANCE)
BONNANS	J.F.	(FRANCE)
BORNARD		(FRANCE)
BOSGRA	O.	(THE NETHERLANDS)
BREMAUD	P.	(FRANCE)
BRILLET	J.L.	(FRANCE)
BROCKETT	R.W.	(U.S.A.)
CALLIER	F.	(BELGIUM)
CARPENTIER		(FRANCE)
CHAPMAN		(G.B.)
CHAVENT	G.	(FRANCE)
CHENIN	P.	(FRANCE)
CHERRUAULT	Y.	(FRANCE)
CHERUY	Arlette	(FRANCE)
CHEVALIER	F.	(FRANCE)
CHOPLIN	J.	(FRANCE)
CHRETIEN		(FRANCE)
CLAASEN		(THE NETHERLANDS)
CLARA	F.	(FRANCE)
CLAUDE		(FRANCE)
CLERGEOT	H.	(FRANCE)
CLERGET		(FRANCE)
COHEN	Guy	(FRANCE)
COLLETER	P.	(FRANCE)
COMMAULT		(FRANCE)
COURVOISIER	J.P.	(FRANCE)

DAMLAMIAN	A.	(FRANCE)
DAVIS	M.H.A.	(G.B.)
DELEBECQUE	F.	(FRANCE)
DELFOUR	Michel	(FRANCE)
DELMAS	J.	(FRANCE)
DENHAM	M.	(G.B.)
DEPEYROT	Michel	(FRANCE)
DESCUSSE	M.	(FRANCE)
DESHAYES	J.	(FRANCE)
DESOER	C.A.	(U.S.A.)
DION	J.M.	(FRANCE)
DODU		(FRANCE)
DUBOIS	D.	(FRANCE)
DUPONT		(FRANCE)
EKELAND	I.	(FRANCE)
ESPIAU	B.	(FRANCE)
FAUGERAS	O.	(FRANCE)
FAVIER		(FRANCE)
FLIESS	M.	(FRANCE)
FORESTIER	J.P.	(FRANCE)
FOSSARD	A.	(FRANCE)
GAUTHIER		(FRANCE)
GAUVRIT		(FRANCE)
GERMAIN	F.	(FRANCE)
GLOWINSKI	Roland	(FRANCE)
GOMEZ	C.	(FRANCE)
GONDRAN		(FRANCE)
GOODWIN	G.C.	(AUSTRALIE)
GOURSAT	M.	(FRANCE)
GRENIER	Y.	(FRANCE)
GUEGEN	C.	(FRANCE)
HALME		(FINLAND)
HAUTUS	M.L.J.	(THE NETHERLANDS)
HAZEWINKEL	M.	(THE NETHERLANDS)
HENRY	J.	(FRANCE)
IRVING	E.	(FRANCE)
ISIDORI		(ITALY)
JACOB	G.	(FRANCE)
KERNEVEZ	J.P.	(FRANCE)
KOKOTOVIC	P.	(U.S.A.)
KOREZLIOGLIU	H.	(FRANCE)
KRENER	A.J.	(U.S.A.)
KUCERA	W.	(TCHEKOSLOVAKIA)
LACOUME	J.L.	(FRANCE)
LANDAU	I.D.	(FRANCE)
LAUB	A.	(U.S.A.)
LE LETTY	C.	(FRANCE)
LEDERER	P.	(FRANCE)
LEMARECHAL	C.	(FRANCE)
LEVINE	Jean	(FRANCE)
LIONS	P.L.	(FRANCE)
LJUNG	L.	(SWEDEN)
LOBRY	C.	(FRANCE)
LORINO	H.	(FRANCE)

MARMORAT	J.P.	(FRANCE)
MARROCCO	A.	(FRANCE)
MAURRAS	J.F.	(FRANCE)
MENALDI	J.L.	(FRANCE)
MICHEL		(FRANCE)
MIGNOT	F.	(FRANCE)
MINOUX		(FRANCE)
MIQUEL		(FRANCE)
MOALLA		(TUNISIA)
MORSE	A.S.	(U.S.A.)
MUNACK	A.	(F.R.G.)
MURON	O.	(FRANCE)
NAIN	Philippe	(FRANCE)
NEPOMIASTCHY	P.	(FRANCE)
NIJMEYER	H.	(THE NETHERLANDS)
OPPENHEIM	G.	(FRANCE)
ORTEGA		(FRANCE)
PARDOUX	E.	(FRANCE)
PAVE	A.	(FRANCE)
PICCI	G.	(ITALY)
PLATEN	R.	(G.D.R.)
POLAK	E.	(U.S.A.)
PRALY	L.	(FRANCE)
PROTH	J.M.	(FRANCE)
PUN		(FRANCE)
QUADRAT	J.P.	(FRANCE)
ROBIN	Maurice	(FRANCE)
ROFMAN	E.	(FRANCE)
ROUBELLAT		(FRANCE)
ROUCHALEAU	Y.	(FRANCE)
RUCKEBUSH	G.	(FRANCE)
SAGUEZ	Christian	(FRANCE)
SAMSON	C.	(FRANCE)
SENTIS	R.	(FRANCE)
SERMANGE	M.	(FRANCE)
SORINE	M.	(FRANCE)
STEER	S.	(FRANCE)
SULEM	Agnes	(FRANCE)
SZPIRGLAS	Jacques	(FRANCE)
TEMPELAAR	D.	(THE NETHERLANDS)
TITLI	A.	(FRANCE)
VAN DER SCHAFT	A.	(THE NETHERLANDS)
VAN DER WEIJDEN	A.	(THE NETHERLANDS)
VAN DOOREN	P.	(BELGIUM)
VAN SCHUPPEN		(THE NETHERLANDS)
VARAIYA	P.	(U.S.A.)
VIOT	M.	(FRANCE)
WEISS		(U.S.A.)
WILLEMS	J.C.	(THE NETHERLANDS)
WILLEMS	J.L.	(BELGIUM)
WILLSKY	A.S.	(U.S.A.)
WONHAM	M.W.	(CANADA)
YVON	J.P.	(FRANCE)
ZABCZYK	J.	(POLAND)
ZAMES	George	(CANADA)
ZOLESIO		(FRANCE)

TABLE OF CONTENTS / TABLE DES MATIERES

SESSION 11

NUMERICAL METHODS / METHODES NUMERIQUES

SESSION 12

STOCHASTIC CONTROL / CONTROLE STOCHASTIQUE

SESSION 13

LINEAR SYSTEMS II / SYSTEMES LINEAIRES II

SESSION 14

COMPUTER AIDED CONTROL SYSTEM DESIGN I / CAO EN AUTOMATIQUE I

SESSION 15

SIGNAL PROCESSING / TRAITEMENT DU SIGNAL

SESSION 16

NONLINEAR SYSTEMS I / SYSTEMES NON LINEAIRES I

SESSION 17

BIOTECHNOLOGICAL SYSTEMS AND BIOENGINEERING

GENIE BIOMEDICAL ET SYSTEMES BIOTECHNOLOGIQUES

SESSION 18

NONLINEAR SYSTEMS II / SYSTEMES NON LINEAIRES II

SESSION 19

COMPUTER AIDED CONTROL SYSTEM DESIGN II / CAO EN AUTOMATIQUE II

SESSION 20

PRODUCTION AUTOMATION / AUTOMATISATION DE LA PRODUCTION

ADDITIONAL INFORMATION CONCERNING
SOFTWARE DEMONSTRATIONS PRESENTED DURING THE MEETING

INFORMATION SUPPLEMENTAIRE CONCERNANT
LA PRESENTATION DE LOGICIELS AU COURS DE LA CONFERENCE

PART 1
(published as Lecture Notes in Control and Information Sciences, Vol. 62)

TABLE OF CONTENTS / TABLE DES MATIERES

SESSION 1

NON STATIONARY PROCESSES / PROCESSUS NON STATIONNAIRES

SESSION 2

STABILITY I / STABILITE I

SESSION 3

UTILITY SYSTEMS / RESEAUX DE SERVICE

SESSION 4

DETECTION OF CHANGES IN SYSTEMS / DETECTION DE CHANGEMENTS DANS LES SYSTEMES

SESSION 5

STABILITY II / STABILITE II

SESSION 6

LINEAR SYSTEMS I / SYSTEMES LINEAIRES I

SESSION 7

DISTRIBUTED PARAMETER SYSTEMS / SYSTEMES A PARAMETRES DISTRIBUES

SESSION 8

IDENTIFICATION AND DETECTION / IDENTIFICATION ET DETECTION

SESSION 9

DETERMINISTIC CONTROL / CONTROLE DETERMINISTE

SESSION 10

FILTERING / FILTRAGE

Session 11

NUMERICAL METHODS

MÉTHODES NUMÉRIQUES

OPTIMAL CONTROL OF SYSTEMS WITH MULTIPLE STEADY-STATES

E.J. Doedel+, M.C. Duban++, G. Joly++, and J.P. Kernevez++
+ Computer Science, Concordia University, Montreal, Canada
++ U.T.C., BP 233, 60206 Compiègne, France

ABSTRACT

This paper describes algorithms for the optimal control of multistate systems. As a test problem we use the reaction diffusion equation governing the steady-states of an enzyme system. The originality of such problems is that the state is not uniquely defined as a function of the control.

I - INTRODUCTION

We are interested in the problem of optimizing the activity of a biochemical system. This system is an artificial enzyme membrane (KERNEVEZ [1]) governed by the partial differential equation

$$s_t - s_{xx} + \sigma F(s) = o, \qquad o < x < 1,\ t > o \tag{1.1}$$

with the boundary conditions

$$s(o,t) = v_o \quad \text{and} \quad s(1,t) = v_1. \tag{1.2}$$

Here

$$F(s) = \frac{s}{1 + |s| + ks^2} \tag{1.3}$$

and v_o and v_1 are positive parameters.

The function $s(x,t)$ is the concentration of a substrate diffusing and reacting in the membrane. Each molecule of substrate which is consumed yields a molecule of product.

Suppose an industrial application where such a membrane functions at a steady-state, governed by :

$$\begin{cases} -s'' + \sigma F(s) = 0 & 0 < x < 1 \\ s(0) = v_0, \quad s(1) = v_1. \end{cases} \tag{1.4}$$

The activity of the membrane is by definition

$$act = \sigma \int_\Omega F(s)\, dx. \tag{1.5}$$

It is proportional to the number of molecules produced per unit of time.

The problem we address to is to find v_0 and v_1 such that this activity be maximum. In other words we wish to minimize

$$J(v,s) = -\sigma \int_\Omega F(s)\, dx \tag{1.6}$$

where the control variable $v = (v_0, v_1)$ belongs to

$$\mathcal{U}_{ad} = \{ v \mid v \geq 0 \} \tag{1.7}$$

and v and s are related by (1.4).

The originality of this problem lies in the non-uniqueness of solutions to (1.4). Already when $v_0 = v_1$ the plot of

$\|s\| = (\int_\Omega s^2(x)\, dx)^{1/2}$ (resp. act) as a function of v_0 looks like in figure (1.1) (resp. (1.2)). Incidentally the latter shows that our optimal control problem admits exactly one solution when $v_0 = v_1$. More precisely for $\sigma = 1200$ and $k = 0.1$ this optimum is obtained for $v_0 \simeq 67$ and $J \simeq -419$.

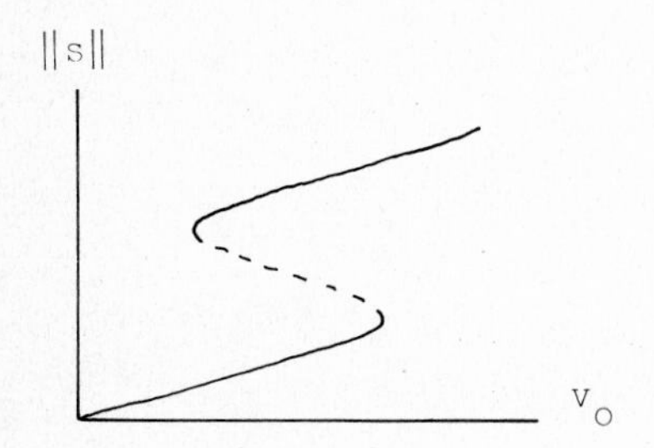

Figure 1 : the full (resp. dotted line represent stable (resp. unstable) steady-states.

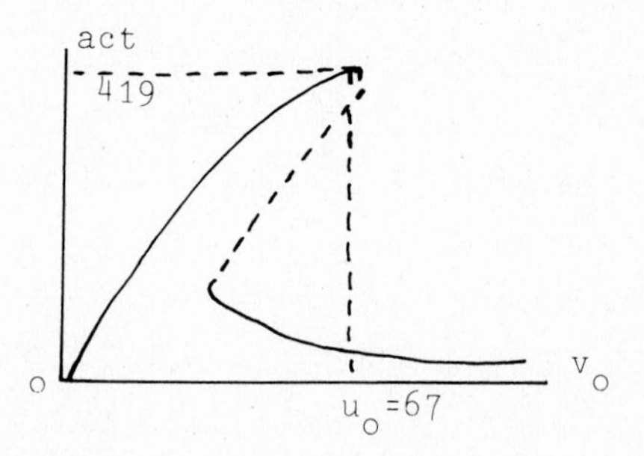

Figure 2 : existence of a maximum for the membrane activity.

A structure which occurs frequently in biological systems is a planar disk. Therefore it is of interest to generalize somewhat (1.4) by

$$\begin{cases} -\Delta s + \sigma F(s) = o & \text{in } \Omega \\ s/\Gamma = v \end{cases} \tag{1.8}$$

where Ω is the membrane (a bounded open set in $\mathbb{R}^2$!) and Γ its boundary.

Both the 1- and 2-dimensional cases can be discretized by the finite difference method, thus yielding the approximate problem :

to minimize $J(v,z)$ where $v \in \mathcal{U}_{ad}$ and v and z are related by $f(v,z) = o$

(1.9)

Here $v \in \mathbb{R}^m$ and $z \in \mathbb{R}^n$ represent the discretizations of the control and state variables, $J \in \mathbb{R}$ and $f \in \mathbb{R}^n$ the discretization of the cost function and state equations. We suppose the discretization fine enough for the problem (1.9) to inherit the essential property of its continuous counterpart, namely the state equation

$$f(v,z) = o \tag{1.10}$$

<u>does not uniquely</u> define z as a function of v :

$$f(v,z) = o \Leftrightarrow z = z_1(v) \ \underline{\text{or}} \ z = z_2(v) \ \underline{\text{or}} \ \ldots \tag{1.11}$$

Schematically in the classical framework the problem is to minimize

$$J(v,y) \qquad \text{(cost function)} \tag{1.12}$$

subject to the constraint

$$f(v,y) = o \qquad \text{(equation of state)} \tag{1.13}$$

where equation (1.13) can be solved for y as a function of v :

$$f(v,y) = o \qquad y = y(v), \tag{1.14}$$

whereas in the case of multistate systems (1.14) no more holds for every v but instead we have (1.11).

The theoretical aspect of closely related problems has been studied by J.L. LIONS [1,2]. For some preliminary discussions and results about the computational aspects we refer to BANKS, DUBAN and KERNEVEZ [1], KERNEVEZ, JOLY and SHARAN [1]and KERNEVEZ [2].
In this paper we begin by deriving the optimality system, in § 2.

The discussion of methods based upon the solution of the optimality system is given in § 3. Several types of penalty methods are given in respectively § 4, 5 and 6. Continuation methods of 2 types are explained in § 7 and 8. Of course, these methods can be applied to other systems than our biochemical system.

Remark 1.1 : In all the numerical applications the parameter values were

$$\sigma = 1200 \quad \text{and} \quad k = 0.1. \tag{1.15}$$

2 - SYSTEM OF NECESSARY CONDITIONS OF OPTIMALITY

It is well known that in the classical case (i.e. when (1.13) uniquely defines z as a function of v) we have, for an optimal pair v = u and z = y,

$$\begin{cases} f(u,y) = o \\ J_z(u,y) + p^T f_z(u,z) = o \\ g(v-u) \geq o \quad \forall v \in \mathcal{U}_{ad} \\ g = J_v(u,y) + p^T f_v(u,y). \end{cases} \tag{2.1}$$

Here $\mathcal{U}_{ad}$ denotes the closed convex subset of $\mathbb{R}^m$ where v lies. Indeed it can be shown (KERNEVEZ and LIONS [1]), that a similar system of optimality holds for the solution(s) of problem (1.6), (1.7), (1.8):

$$\begin{cases} -\Delta y + \sigma F(y) = o \\ -\Delta p + \sigma F'(y)(p-1) = o \\ p/\Gamma = o \\ \int_\Gamma \frac{\partial p}{\partial n}(v-u)\, d\Gamma \geq o \quad \forall u \in \mathcal{U}_{ad} \end{cases} \tag{2.2}$$

To establish (2.1) for problem (1.9), we employ the penalty method. With notations J,f,v,z for the cost function, the state equation, and the control and state variables, define

$$J_\varepsilon(v,z) = J(v,z) + \frac{1}{2\varepsilon} \|f(v,z)\|^2 \tag{2.3}$$

and calculate

$$\begin{cases} J_{\varepsilon z} = J_z + \frac{1}{\varepsilon} f^T f_z = J_z + p_\varepsilon^T f_z \\ J_{\varepsilon v} = J_v + p_\varepsilon^T f_v \end{cases} \tag{2.4}$$

where

$$p_\varepsilon = \frac{1}{\varepsilon} f \qquad (2.5)$$

Then the optimality conditions for u_ε, y_ε to minimize J_ε are

$$\begin{cases} f(u_\varepsilon, y_\varepsilon) = \varepsilon\, p_\varepsilon \\ p_\varepsilon^T f_z(u_\varepsilon, y_\varepsilon) + J_z(u_\varepsilon, y_\varepsilon) = o \\ [p_\varepsilon^T f_v(u_\varepsilon, y_\varepsilon) + J_v(u_\varepsilon, y_\varepsilon)]\,(v - u_\varepsilon) \geq o \quad \forall\, v \in \mathcal{U}_{ad}. \end{cases} \qquad (2.6)$$

Then we have to find, like in LIONS [1,2] a priori estimations on the adjoint variable p_ε. This is indeed a non trivial task, but once it is achieved, it is easy to pass to the limit as $\varepsilon \longrightarrow o$:

$$\begin{aligned} u_\varepsilon &\longrightarrow u \\ y_\varepsilon &\longrightarrow y \\ p_\varepsilon &\longrightarrow p, \end{aligned} \qquad (2.7)$$

u,y and p satisfying (2.1).

3 - SOLUTION OF THE OPTIMALITY SYSTEM

To solve the nonlinear system of equations (2.1) we use Newton method. Indeed this method works well in the 1-dimensional case (1.4), (1.5), (1.6), (1.7) where the optimality system (2.2) can be rewritten

$$\begin{cases} -y'' + \sigma F(y) = o \\ -p'' + \sigma F'(y)\,(p - 1) = o \\ y(o)\, p'(o) = y(1)\, p'(1) = o \\ p(o) = p(1) = o \end{cases} \qquad (3.1)$$

Starting from $p = o$ and $y(x) = o$ for $o < x < 1$, $y(o) = y(1) = 1$, we find the absolute minimum $J \simeq -419$ which corresponds to $y(o) = y(1) \simeq 67$, and to y and p such that the 2 first equations of (3.1) are satisfied, together with the B.C.s

$$p(o) = p(1) = p'(o) = p'(1), \qquad y(o) = y(1) \simeq 67. \qquad (3.2)$$

Starting from $p = o$ and $y(x) = o$ for $o < x \leq 1$, $y(o) = 1$, we find y and p still satisfying the 2 first equations of (3.1), but this time with

$$p(o) = p(1) = p'(o) = o, \quad y(o) \simeq 166 \quad \text{and} \quad y(1) = o. \tag{3.3}$$

Starting from $y = p = o$, we find the absolute maximum of J, $J = o$, with $y = o$ and p such that

$$-p'' + \sigma (p - 1) = o, \quad p(o) = p(1) = o. \tag{3.4}$$

It took respectively 13, 20 and 2 iterations to achieve convergence in each case, with the cost function values $J \simeq -412$, $J \simeq -238$ and $J \simeq o$. However in the 2-dimensional case, when Ω is for example the unit square $\Omega =]o,1[\times]o,1[$, the method works well when the finite difference grid is coarse. But for 6 subintervals on each side]o,1[we have already $7 \times 7 \times 2 = 98$ variables and it takes 1 mn of VAX 780 CPU time to do 4 iterations.

4 - PENALTY METHOD I

We applied it to the problem

$$\begin{cases} -s''(x) + \sigma F(s(x)) = o \\ s(o) = v, \ \sigma s(1) = v \\ J(v,s) = +2s'(o) \end{cases} \tag{4.1}$$

Defining

$$\tilde{J}(v,z) = J(v,z) + \frac{1}{2} c_1 \int_\Omega |-s'' + \sigma F(s)|^2 dx + c_2 v^- \tag{4.2}$$

or rather a finite-difference approximation of it, the method consists in minimizing $\tilde{J}$ for greater and greater values of c_1, the starting point for each new value of c_1 being the optimal point previously found. In this way we find at optimality

$$v \simeq 70.116\ldots, \quad J \simeq -424.14\ldots, \quad s\left(\frac{1}{2}\right) \simeq 0.989\ldots \tag{4.3}$$

Although this example involved only one component for v, some useful observations were made :

i) the same optimal point was found whatever the starting point.

ii) the computer time was large. For example with 20 intervals on (o,1), $c_2 = 10$, starting from $s = v = 50$, taking sequentially $c_1 = 0.005$, 0.0125 and 0.125 and using the conjugate gradient method (VA 14 AD in Harwell Library) the optimal point (2.3) was attained after 1311 function calls and 19.57 seconds of VAX 780.

5 - PENALTY METHOD II (J.L. LIONS [3])

It consists in replacing the problem (1.4), (1.5) by the problem to minimize

$$J_\varepsilon(v,w,z) = -\sigma\,\theta\int_\Omega F(z)\,dx - \sigma\,(1-\theta)\int_\Omega w\,dx + \frac{1}{2\varepsilon}\int_\Omega (w - F(z))^2\,dx \qquad o \le \theta \le 1 \tag{5.1}$$

where z is related to v and w through

$$\begin{cases} -\Delta z + \sigma w = o \\ z/\Gamma = v. \end{cases} \tag{5.2}$$

Thus it consists in penalizing the constraint $w = F(z)$ and considering an optimal control problem for the linear system (5.2). The same observations as above can be made (KERNEVEZ [2]) : the method works well, but at the expense of a large number of function calls.

Remark 5.1 : For both penalty methods we only found solutions with $y(o) = y(1)$. ($\simeq$ 70 in the first, $\simeq$ 67 in the second). Each time we found $J \simeq -420$.

6 - PENALTY METHOD III (FORTIN et GLOWINSKI [1])

Define the augmented Lagrangian

$$\mathcal{L}_r(v,w,z,\mu) = -\theta\,\sigma\int_\Omega F(z)\,dx - (1-\theta)\,\sigma\int_\Omega w\,dx + \frac{r}{2}\int_\Omega (F(z) - w)^2\,dx + \int_\Omega \mu\,(F(z) - w)\,dx$$

where z is related to v and w through (5.2).
The method consisting in finding a saddle point to $\mathcal{L}_r$ has been used successfully by BOURGAT [1].

7 - CONTINUATION METHOD I

Consider the problem of minimizing

$$J(v,z), \qquad v \in \mathbb{R}^m, \qquad z \in \mathbb{R}^n \tag{7.1}$$

subject to

$$f(v,z) = o, \qquad f \in \mathbb{R}^n. \tag{7.2}$$

Freezing all the components of v except v_1, we consider the system of equations

$$\begin{cases} f(v_1,z) = o \\ \mu - J(v_1,z) = o \end{cases} \tag{7.3}$$

which can be rewritten

$$F(v_1,z,\mu) = o, \tag{7.4}$$

$$F : \mathbb{R}^{n+2} \longrightarrow \mathbb{R}^{n+1} \tag{7.5}$$

F is represented by a curve in $\mathbb{R}^{n+1}$. Suppose the plot of say $\|(v,z)\|$ as a function of μ looks like in figure 7.1.

Then $\mu = \mu_1$ is a minimum of $J(v_1,z)$, v_1 and z been constrained by $f(v,z) = o$. The condition for $((v_1,z),\mu)$ to be a turning point is "there exists a vector $v \in \mathbb{R}^n$ such that

$$F_{(v_1,z)} \begin{bmatrix} 1 \\ v \end{bmatrix} = o"$$

i.e.

$$\begin{bmatrix} f_{v_1} & f_z \\ -J_{v_1} & -J_z \end{bmatrix} \begin{bmatrix} 1 \\ v \end{bmatrix} = \begin{bmatrix} o \\ o \end{bmatrix}$$

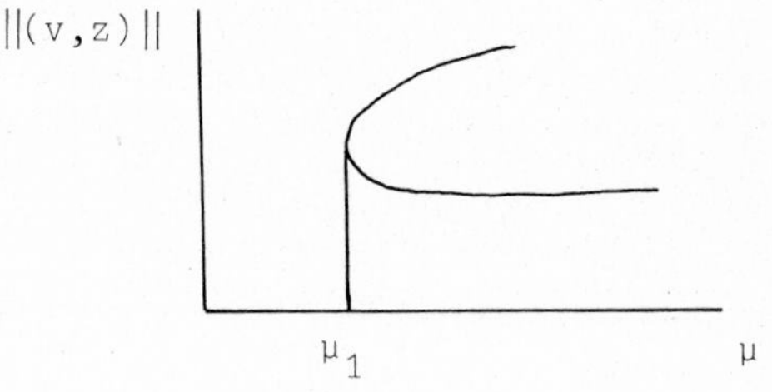

$\|(v,z)\|$ — μ_1 — μ

Figure 7.1

Equivalently the matrix

$$\begin{bmatrix} f^T_{v_1} & -J_{v_1} \\ f^T_z & -J^T_z \end{bmatrix}$$

is non singular, and its exists $p \in \mathbb{R}^n$ (the adjoint state) such that

$$\begin{cases} J_z + p^T f_z = 0 \\ J_{v_1} + p^T f_z = 0 \end{cases} \tag{7.6}$$

Now suppose there was a frozen variable v_2 which we free in the optimality system

$$\begin{cases} f(v_1,v_2,z) = 0 \\ J_z + p^T f_z = 0 \\ J_{v_1} + p^T f_{v_1} = 0 \\ \mu - J(v_1,v_2,z) \end{cases} \tag{7.7}$$

This system can be viewed as the equation

$$F((v_1,v_2,z,p),\mu) = 0 \tag{7.8}$$

where

$$F : \mathbb{R}^{2n+3} \longrightarrow \mathbb{R}^{2n+2} \tag{7.9}$$

is represented by a curve in $\mathbb{R}^{2n+2}$, and if this curve has a limit point with respect to μ, we have (7.7) with in plus

$$J_{v_2} + p^T F_{v_2} = 0. \tag{7.10}$$

This process of continuation of limit points can be repeated : if the last obtained limit point is

$$(v_1,\ldots,v_{i-1},z,p,\mu)$$

then its characterization is

$$\begin{cases} f(v_1,\ldots,v_{i-1},v_i,\ldots,v_m,z) = 0 \\ J_z + p^T f_z = 0 \\ J_{v_k} + p^T f_{v_k} = 0, \quad k = 1,\ldots,i-1 \\ \mu - J(v_1,\ldots,z) = 0. \end{cases} \tag{7.11}$$

When one frees v_i (7.11) can be written

$$F(v_1,\ldots,v_i,z,p,\ ) = o$$

where

$$F : \mathbb{R}^{2n+i+1} \to \mathbb{R}^{2n+i}, \quad \text{and so on.}$$

Application : This method has been applied to the one-dimensional case (2.1). The program for continuation was AUTO (DOEDEL [1,2]). AUTO has the capability of continuing limit points.
The method gave the already found minimum $J \simeq -419$ for $y(o) = y(1) \simeq 67$.

8 - CONTINUATION METHOD II

A code devised for continuation of solutions (KUBICEK [1]) can be modified in order to calculate optimal "state-control" pairs. The algorithm, already extensively described in KERNEVEZ, JOLY and SHARAN [1], consists in following a path in the subset of $\mathbb{R}^{m+n}$ defined by $f(v,z) = o$, at each step steering toward the steepest descent direction. In figure 8.1 $\|z\|$ is sketched as a function of v ($\in \mathcal{U}_{ad}$).

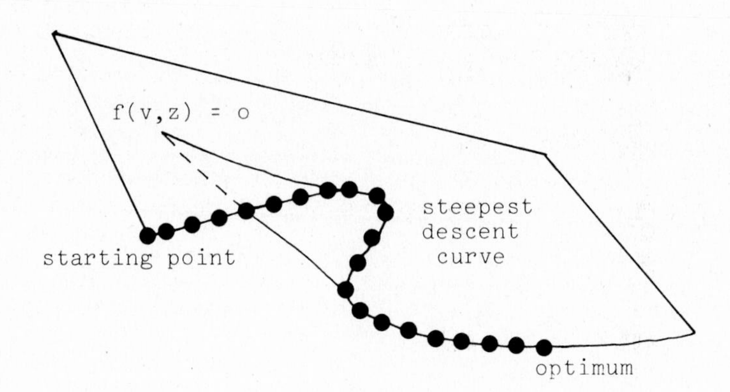

Figure 8.1

This direction is found as follows. Gauss elimination with complete pivoting applied to the n X (m + n) Jacobian matrix

$$\left[f_v \ \vdots \ f_z \right]$$

provides a reordering of the m + n variables (v,z) into vectors

$w \in \mathbb{R}^m$ and $y \in \mathbb{R}^n$ such that f_y is non singular. Also it yields a n X m matrix β satisfying

$$f_y \beta + f_w = o. \tag{8.1}$$

Thus, at least in theory, y can be expressed as a function of w and J becomes

$$\tilde{\tilde{J}}(w) = \tilde{J}(w,y(w)) \tag{8.2}$$

where

$$\tilde{J}(w,y) = J(v,z). \tag{8.3}$$

Then the gradient of $\tilde{\tilde{J}}$ is

$$g = \tilde{\tilde{J}}_w^T = \tilde{J}_w^T + y_w^T \tilde{J}_y^T = \tilde{J}_w^T + \beta^T \tilde{J}_y^T \tag{8.4}$$

since

$$\beta = y_w.$$

Let s denote arclength on this curve, and $\dot{w}(s) = dw/ds$, $\dot{y}(s) = dy/ds$. Then the steering direction is such that

$$\begin{cases} \dot{w}(s) = -\alpha_o \, g \\ \dot{y}(s) = -\alpha_o \, \beta g \\ \|\dot{w}\|^2 + \|\dot{y}\|^2 = 1. \end{cases} \tag{8.5}$$

Practically one <u>predicts</u> a point $(w(s) + h\,\dot{w}(s),\ y(s) + h\,\dot{y}(s))$ on the tangent to the curve, then <u>corrects</u> by Newton iterations to come back on $f(v,z) = o$.

Applied to the 1-dimensional case this method gives again the absolute minimum $J = -419$ for $y(o) = y(1) \simeq 67$. It works satisfactorily in the 2-D case too, giving several local minima, depending upon the starting point.

9 - CONCLUSION

We have described a system with multiple steady-states, with an associated control problem. We have discussed numerical algorithms for the solution of such systems. Among them, those based upon continuation methods seem to be promising. Indeed they can easily be extended to systems with multiple behaviors (time periodic or steady-state). Due to the existence of such multiple behaviors in non linear systems, and to the importance of being able to control them, there is no doubt that there will be an increasing need of such algorithms.

REFERENCES

BANKS H.T., DUBAN M.C. and KERNEVEZ J.P. [1], Optimal control of diffusion-reaction systems, p 47 - 59 in "Applied nonlinear analysis", edited dy V. Lakshmikantham, Academic Press, 1979.

BOURGAT [1] , Personal communication

DOEDEL E.J. [1], AUTO : a program for the automatic bifurcation analysis of autonomous systems, congressus numerantium, vol. 30 (1981) p 265 - 284.

DOEDEL E.J. [2], Continuation techniques in the study of chemical reaction schemes,to appear in : Proc. Special Year on Energy Math., Univ. of Wyoming, K.I. Gross, ed. SIAM Publ.

FORTIN M. et GLOWINSKI R. [1], Méthodes de Lagrangien augmenté. Collection MMI n° 9, Paris, Dunod (1982).

KERNEVEZ J.P. [1], Enzyme mathematics, North Holland, Amsterdam, 1980 X-262 pages.

KERNEVEZ J.P. [2], Optimal control of multistate systems,in Encyclopedia of systems and control, editor in chief : Prof. Madan Singh, Pergamon Press Ltd, Oxford.

KERNEVEZ J.P., JOLY G. and SHARAN M.[1], Control of systems with multiple steady-states, p 635 - 649 in Computing methods in applied sciences and engineering, North Holland, Amsterdam, 1982.

KERNEVEZ J.P. and LIONS J.L. [1], Book to appear.

KUBICEK M. [1], Dependence of solution of nonlinear systems on a parameter, ACM Transactions on mathematical software, vol. 2, n° 1, March 1976, p 98 - 107.

LIONS J.L. [1], Some methods in the mathematical analysis of systems and their control, Science Press, Beijing (1981).

LIONS J.L. [2], Contrôle des systèmes distribués singuliers, Gauthier Villars, Paris, 1983.

LIONS J.L. [3], Personal communication.

ERRATA

In section 7 the eigenvector $\begin{bmatrix} 1 \\ v \end{bmatrix}$ should be rewritten $\begin{bmatrix} -\hat{\alpha} \\ v \end{bmatrix}$ with

$v^t v + \hat{\alpha}^2 = 1$, the eigenvector $\begin{bmatrix} p \\ -1 \end{bmatrix}$ should be replaced by $\begin{bmatrix} p \\ -\alpha \end{bmatrix}$

with $p^t p + \alpha^2 = 1$, and the formula modified accordingly. Moreover the matrix $F^T_{(v_1, z)}$ <u>is singular</u>.

UNE EXTENSION DE LA PROGRAMMATION QUADRATIQUE SUCCESSIVE

Joseph Frédéric BONNANS* - Daniel GABAY**

RESUME : Les algorithmes de résolution de problèmes d'optimisation basés sur la programmation quadratique successive sont réputés pour leur efficacité. Ils peuvent cependant être mis en difficulté dans le cas où le sous problème quadratique est, à certaines itérations, mal posé. Nous donnons un moyen de calculer la direction de descente d'une fonctionnelle pénalisée (différentiable ou non), grâce à une approximation linéaire quadratique du critère et des contraintes. Le problème résolu à chaque itération est bien posé, même si les contraintes linéarisées sont incompatibles. Dans le cas de la pénalisation quadratique, on retrouve les formules de M.C. Bartholomew-Biggs [1]. Dans le cas de la pénalisation L^{∞} (resp. L^{1}) nous étendons le résultat de B. Pchenitchny (voir [9])(resp. S.P. Han [7]). Enfin, nous proposons une nouvelle fonction de pénalisation exacte basée sur la norme L^{2}.

ABSTRACT : Nonlinear programming codes based on successive quadratic programming enjoy a large popularity. They may encounter some difficulties, however, if the quadratic subproblem is non-well posed at some iterations. We give a mean to compute a descent direction of a penalized functional (differentiable or not), using a linear quadratic approximation of the criterion and constraints. At each iteration, the subproblem to be solved is well-posed, even if the linearized constraints are not consistent. If the penalization is quadratic, we get the formulae of M.C. Bartholomew-Biggs [1]. In the case of the L^{∞} (resp. L^{1}) penalization, we extend the result of B. Pchenitchny (see [9])(resp. S.P. Han [7]). We also propose a new exact penalty function based on the L^{2} norm.

---

* INRIA, Domaine de Voluceau, BP 105, Rocquencourt, 78153 LE CHESNAY Cédex (France)

** Laboratoire d'Analyse Numérique, Université Pierre et Marie Curie, Paris, et INRIA (France). Actuellement à l'Ambassade de France à Rome, palais Farnese, Piazzia Farnese, Roma (Italie).

I - DEFINITION DU PROBLEME ET PENALISATION

a) Pénalisation d'un problème d'optimisation

Soit $f : \mathbb{R}^n \to \mathbb{R}$ et $g : \mathbb{R}^n \to \mathbb{R}^m$ des applications de classe C^1. Nous avons en vue la résolution du problème

$$(P1) \qquad \begin{cases} \text{Min } f(x), \\ x \in \mathbb{R}^n, \; g_i(x) = 0, \; i = 1 \text{ à } m. \end{cases}$$

Nous supposons pour simplifier qu'il n'y a que des contraintes d'égalité, mais les méthodes étudiées ici s'étendent directement au cas où des contraintes d'inégalités sont présentes.

Soit $\bar{x}$ une solution de (P1). La condition nécessaire d'optimalité du premier ordre en $\bar{x}$ est : il existe $\bar{\lambda} \in \mathbb{R}^m$ tel que

$$(1.1) \qquad \begin{cases} \nabla f(\bar{x}) + \nabla g(\bar{x})\bar{\lambda} = 0, \\ g(\bar{x}) = 0. \end{cases}$$

Considérons maintenant le problème pénalisé

$$(P2) \qquad \begin{cases} \text{Min } \theta_r(x) = f(x) + r\,\phi(g(x)), \\ x \in \mathbb{R}^n, \end{cases}$$

où r est un paramètre de pénalisation positif, et où ϕ est une application de $\mathbb{R}^m$ dans $\mathbb{R}$ vérifiant :

$$(1.2) \qquad \begin{cases} \phi \text{ est convexe, continue, positive,} \\ \phi(y) \equiv 0 \iff y \equiv 0, \\ \phi(y) \to +\infty \text{ si } \|y\| \to +\infty. \end{cases}$$

Puisque ϕ est convexe et continue, elle est partout sous-différentiable. Les conditions nécessaires d'optimalité de (P2), écrites en $\hat{x}$, sont donc : il existe $\hat{\lambda} \in \mathbb{R}^m$ tel que

$$(1.3)\qquad \begin{cases} \nabla f(\hat{x}) + \nabla g(\hat{x})\,\hat{\lambda} = 0, \\ \hat{\lambda} \in r\ \partial\phi(g(\hat{x})). \end{cases}$$

b) Dualisation du problème pénalisé

Un problème équivalent à (P2) est

$$(P3)\qquad \begin{cases} \text{Min } f(x) + r\phi(y), \\ \text{où } x \in \mathbb{R}^n,\ y \in \mathbb{R}^m \text{ vérifient } g(x) = y. \end{cases}$$

Dualisons la contrainte d'égalité de (P3). Le lagrangien est

$$L(x,y,\lambda) = f(x) + r\phi(y) + \lambda^t(g(x) - y),$$

et le critère dual sera donc

$$\begin{aligned} -I^*(\lambda) &= \inf\{L(x,y,\lambda),\ x \in \mathbb{R}^n,\ y \in \mathbb{R}^m\}, \\ &= \inf_{x\in\mathbb{R}^n} (f(x) + g(x)^t\lambda) + \inf_{y\in\mathbb{R}^m} (r\phi(y) - y^t\lambda) \\ &= \inf_{x\in\mathbb{R}^n} (f(x) + g(x)^t\lambda) + \inf_{y\in\mathbb{R}^m} r(\phi(y) - y^t(\tfrac{1}{r}\lambda)). \end{aligned}$$

Soit

$$I_0(\lambda) = -\inf_{x\in\mathbb{R}^n} (f(x) + g(x)^t\lambda),$$

le critère dual de (P1). Il vient

$$(1.4)\qquad I^*(\lambda) = I_0(\lambda) + r\phi^*(\tfrac{1}{r}\lambda),$$

où ϕ^* est la fonction convexe conjuguée de ϕ(cf. [5]) définie par

$$\phi^*(\lambda) = \sup_{y\in\mathbb{R}^n} (y^t\lambda - \phi(y)).$$

II - UNE METHODE DE RESOLUTION DU PROBLEME PENALISE

a) Calcul d'une direction de descente

Donnons-nous $x \in \mathbb{R}^n$ et une matrice H $n\times n$, symétrique, définie positive. Considérons le problème convexe

$$(PC)\qquad \begin{cases} \text{Min } \nabla f(x)^t d + \frac{1}{2} d^t Hd + r\phi(g(x) + \nabla g(x)^t d), \\ d \in \mathbb{R}^n. \end{cases}$$

Les conditions nécessaires et suffisantes d'optimalité de (PC) sont : il existe $q \in \mathbb{R}^m$ tel que

$$(2.1)\qquad \begin{cases} \text{(i)} \quad \nabla f(x) + Hd + \nabla g(x)q = 0, \\ \text{(ii)} \quad q \in r\,\partial\phi(g(x) + \nabla g(x)^t d). \end{cases}$$

Théorème 2.1

La solution d de (PC) vérifie

$$\theta'_r(x,d) \leq - d^t Hd,$$

où $\theta'_r(x,d)$ est la dérivée de $\theta_r(x)$ dans la direction d. □

Remarque 2.1

Puisque θ_r est convexe et continu, $\theta'_r(x,d)$ existe et est finie pour tout d. □

Démonstration du Théorème 2.1

Il vient

$$\theta'_r(x,d) = \nabla f(x)^t d + p^t \nabla g(x)^t d,$$

où

$$p \in r\ \partial\phi(g(x)).$$

Additionnant le produit scalaire de (2.1.i) par -d il vient

(2.2) $$\theta'(x,d) = - d^t Hd + (p-q)^t \nabla g(x)^t d.$$

Par définition de p et q, il vient

$$r\phi(g(x) + \nabla g(x)^t d) \geq r\phi(g(x)) + p^t \nabla g(x)^t d$$

et

$$r\phi(g(x)) \geq r\phi(g(x) + \nabla g(x)^t d) - q^t \nabla g(x)^t d ,$$

d'où par addition :

$$0 \geq \frac{1}{r} (p-q)^t \nabla g(x)^t d,$$

d'où le résultat avec (2.2). □

Remarque 2.2

Posons $A = \nabla g(x)$. D'après (1.4), q est solution du problème dual

(PCD) $$\min_{p \in \mathbb{R}^m} Q(p) + r\phi^*(\frac{1}{r}p),$$

où Q(p) est le critère quadratique

$$Q(p) = \frac{1}{2} p^t A^t H^{-1} Ap + p^t (A^t H^{-1} \nabla f(x) - g(x)). \quad □$$

b) Une méthode globalement convergente de minimisation de la fonctionnelle pénalisée

Pour obtenir une méthode globalement convergente de résolution de (P2), il convient de choisir une règle de recherche linéaire convenable. Par exemple, définissons

$$\theta_{r,x}(x+d) = f(x) + \nabla f(x)^t d + r\phi(g(x) + \nabla g(x)^t d).$$

Règle 1 (Extension de la règle d'Armijo).

0) Fixer $\beta \in]0,1[$, $\sigma \in]0,1/2[$,

1) $\rho = \beta^\ell$, où ℓ est le plus petit entier tel que :

$$\theta_r(x + \beta^\ell d) - \theta_r(x) \leq \sigma(\theta_{r,x}(x + \beta^\ell d) - \theta_r(x)). \quad □$$

Remarque 2.3

La recherche linéaire s'arrête après un nombre fini de calculs de fonction. □

Théorème 2.2

On suppose que ϕ vérifie (1.2). Soit $\{H^k\}$ une suite bornée de matrices symétriques définies positives telle que $\{(H^k)^{-1}\}$ soit aussi bornée. Soit la suite x^k définie par

$$x^{k+1} = x^k + \rho^k d^k,$$

où d^k est solution de (PC) écrit en x^k, avec $H = H^k$, et où ρ^k est fixé par la règle de recherche linéaire ci-dessus.

Alors tout point d'adhérence $\bar{x}$ de $\{x^k\}$ vérifie les conditions nécessaires d'optimalité (1.3). □

Démonstration

(i) Vérifions que $\theta_r(x^k)$ est décroissante. Excluons le cas où il existe k_o tel que x^{k_o} vérifie (1.3). Alors $\forall k \geq 0$, $d^k \neq 0$ et par définition de (PC) :

$$f(x^k) + \nabla f(x^k)^t d^k + \frac{1}{2} d^{kt} H^k d^k + \phi(g(x^k) + \nabla g(x^k)^t d^k)$$

$$\leq f(x^k) + \phi(g(x^k)) = \theta_r(x^k),$$

d'où si $d^k \neq 0$:

$$f(x^k) + \nabla f(x^k)^t d^k + \phi(g(x^k + \nabla g(x^k)^t d^k) < \theta_r(x^k),$$

c'est-à-dire

$$\theta_{r,x^k}(x^k + d^k) - \theta_r(x^k) < 0.$$

Or $\theta_{r,x^k}(x^k + \rho d^k)$ est une fonction convexe de ρ donc puisque $\rho^k \in]0,1[$:

$$\theta_r(x^k + \rho^k d^k) - \theta_r(x^k) \leq \sigma(\theta_{r,x^k}(x^k + \rho^k d^k) - \theta_r(x^k)) < 0.$$

(ii) Si $\bar{x}$ est un point d'adhérence de $\{x^k\}$ on a d'après ce qui précède

$$\theta_r(\bar{x}) = \lim_{k\to\infty} \theta_r(x^k).$$

Vérifions que $\{p^k\}$ est bornée. Soit $A^k = \nabla g(x^k)$. D'après la remarque 2.2, p^k est solution de

$$\min_{p\in\mathbb{R}^k} \frac{1}{2} p^t A^{k^t} (H^k)^{-1} A^k p + p^t (A^{k^t} (H^k)^{-1} \nabla f(x^k) - g(x^k)) + r\phi^*(\frac{1}{r} p).$$

Le premier terme étant symétrique positif et $(A^k)^t (H^k)^{-1} \nabla f(x^k) - g(x^k)$ étant borné pour la sous-suite considérée, il suffit de prouver que

$$\frac{\phi^*(p)}{||p||} \to +\infty \text{ quand } ||p|| \to +\infty.$$

Si ceci n'est pas vérifié, il existe $a,b > 0$ et une suite q^k de $\mathbb{R}^m$ telle que

$$\phi^*(q^k) \leq a||q^k|| + b \text{ et } ||q^k|| \to +\infty. \tag{2.3}$$

Soit $a_1 > 0$ et la suite bornée

$$y^k = a_1 \frac{q^k}{||q^k||}.$$

Il vient

$$\begin{aligned}\phi(y^k) &= \sup_{q\in\mathbb{R}^m} (q,y^k) - \phi^*(q),\\ &\geq (q^k,y^k) - \phi^*(q^k) = a_1 \frac{q^{k^t} q^k}{||q^k||} - \phi^*(q^k).\end{aligned}$$

Les normes de $\mathbb{R}^n$ étant équivalentes, il existe $\gamma > 0$ tel que

$$\phi(y^k) \geq \gamma\, a_1 ||q^k|| - \phi^*(q^k).$$

Utilisant (2.3) on voit que si $a_1 > a/\gamma$, $\{y^k\}$ est bornée et $\phi(y^k) \to +\infty$, ce qui est en contradiction avec (1.2).

Donc, $\{p^k\}$ est borné, et aussi $\{d^k\}$ d'après les conditions d'optimalité de (PC), d'où par extraction de sous-suite

$$\begin{aligned}&x^k \to \bar{x},\\ &H^k \to \bar{H} \text{ inversible},\\ &d^k \to \bar{d},\\ &p^k \to \bar{p}.\end{aligned}$$

Passant à la limite dans les conditions d'optimalité de (PC) on déduit que $\bar{d}$ est solution de

$$\begin{cases} \text{Min } \nabla f(\bar{x})^t d + \frac{1}{2} d^t \bar{H} d + r\phi(g(\bar{x}) + \nabla g(\bar{x})^t d), \\ d \in \mathbb{R}^n. \end{cases}$$

Si $\bar{d} = 0$ on en déduit la conclusion. Si $\bar{d} \neq 0$, il existe $\bar{\ell}$ tel que $\bar{\rho} = \beta^{\bar{\ell}}$ vérifie strictement la condition de recherche linéaire. Soit

$$a = \sigma(\theta_{r,\bar{x}}(\bar{x} + \bar{\rho}\bar{d}) - \theta(\bar{x})).$$

Alors $a < 0$ et, par continuité de $\theta_{r,x}$ par rapport à x on en déduit que $\ell^k \leq \bar{k}$ pour la sous-suite considérée et que

$$\lim_{k\to\infty} \theta_r(x^k + \rho^k d^k) - \theta_r(x^k) < a < 0,$$

donc $\theta_r(x^k) \to -\infty$ ce qui est impossible puisque $\theta(\bar{x}) \leq \theta(x^k)$. □

Remarque 2.4

Dans le cas limite où $\{H^k\}$ est nulle, (PC) n'est en général pas bien posé. S'il l'est on peut encore obtenir la convergence globale (cf. E. Sachs [10]) mais on perd la possibilité d'une convergence superlinéaire. □

III - LE CAS DES FONCTIONNELLES PENALISEES NON EXACTES

a) Convergence globale vers une solution du problème de départ

La solution de (P2) n'est en général pas solution de (P1), en particulier si ϕ est différentiable. On rappelle que

$$\theta_r(x) = f(x) + r\,\phi(g(x)).$$

Soit $\bar{x}_r$ un minimum de $\theta_r(x)$. On sait que quand $r \to +\infty$, tout point d'adhérence de $\{\bar{x}_r\}_r$ est solution de (P1).

Le théorème suivant donne un exemple d'algorithme évitant le calcul exact d'une suite de minimums de $\theta_r(x)$.

Théorème 3.1

On suppose que $\{H^k\}$ et $\{(H^k)^{-1}\}$ sont bornés et

- f est bornée inférieurement,
- il existe $\alpha > 0$ tel que, $\forall x \in \mathbb{R}^n$:

(3.1) $$y^t \, \nabla g(x)^t \, \nabla g(x) y \geq \alpha ||y||^2, \; \forall y \quad \mathbb{R}^m.$$

Donnons nous $\sigma \in \,]0,1[$, $\alpha > 0$, $r^o > 0$ et reprenons l'algorithme du théorème 2.2 en remplaçant r par r^k vérifiant :

$$\begin{cases} r^k > r^{k-1} + \alpha \text{ si } ||d^k|| \leq \sigma \, \{\inf ||d^\ell||, \; \ell < k\}, \\ r^k = r^{k-1} \text{ sinon.} \end{cases}$$

Si $\{x^k\}$ est bornée, $\{r^k\} \to +\infty$ et un point d'adhérence de $\{x^k\}$ (au moins) vérifie les conditions nécessaires d'optimalité du premier ordre de (P1). □

Démonstration

Si $\{x^k\}$ est bornée, de (3.1), de la remarque 2.2 et du fait que $\phi^* \geq 0$ (car $\phi(0)=0$) on déduit que $\{p^k\}$ est bornée. De plus

$$\partial\phi(g(x^k) + \nabla g(x^k)^t d^k) \ni \frac{1}{r^k} p^k.$$

Si

$$\liminf_{k \to \infty} ||d^k|| > 0,$$

le paramètre r^k est constant pour $k > k_o$. On est alors dans le cadre du théorème 2.2, dont la démonstration implique la convergence d'une sous-suite de $\{d^k\}$ vers zéro. On a donc montré qu'une sous-suite de $\{d^k\}$ tend vers zéro. Soit $\{x^{k'}\}$ une sous-suite de $\{x^k\}$ telle que $\{d^{k'}\}$ tend vers zéro. Pour tout point d'adhérence $\bar{x}$ de $\{x^{k'}\}$, il vient (H. Brézis [4], p. 27) $\partial\phi(g(\bar{x})) \ni 0$, donc $g(\bar{x}) = 0$. Le passage à la limite dans

$$\nabla f(x^{k'}) + H^{k'} d^{k'} + \nabla g(x^{k'}) \, p^{k'} = 0$$

permet de retrouver la première condition d'optimalité de (1.1). □

b) Pénalisation quadratique

Nous allons expliciter les résultats précédents dans le cas où

$$\phi(y) = \frac{1}{2} \sum_{i=1}^{m} y_i^2 .$$

La fonctionnelle pénalisée est donc

$$\theta_r(x) = f(x) + \frac{r}{2} \sum_{i=1}^{m} (g_i(x))^2 . \tag{3.2}$$

Le problème (PC) s'écrit

$$\text{Min } \nabla f(x)^t d + \frac{1}{2} d^t H d + \frac{r}{2} \sum_{i=1}^{m} (g_i(x) + \nabla g_i(x)^t d)^2 , \tag{3.3}$$

et c'est un problème quadratique sans contraintes, mais mal conditionné si r est grand. Cependant, la solution s'exprime d'après la remarque 2.2 par

$$d = - H^{-1}(\nabla f(x) + \nabla g(x) q) , \tag{3.4}$$

où q est solution du problème dual

$$\text{Min } Q(q) + \frac{1}{2r} \sum_{i=1}^{m} (q_i)^2 . \tag{3.5}$$

La condition d'optimalité de (3.5) est

$$(\frac{1}{r} + \nabla g(x)^t H^{-1} \nabla g(x)) q = - \nabla g(x) H^{-1} \nabla f(x) + g(x) . \tag{3.6}$$

Les formules (3.4), (3.6) sont celles utilisées par M.C. Bartholomew-Biggs [1] pour obtenir une direction de descente de la fonctionnelle de pénalisation (3.2).

Remarque 3.2

Soit le problème quadratique

$$\text{(PQ)} \quad \begin{cases} \text{Min } \nabla f(x)^t d + \frac{1}{2} d^t H d, \\ d \in \mathbb{R}^n, \ g_i(x) + \nabla g_i(x)^t d = 0, \ i = 1 \text{ à } m. \end{cases}$$

Si l'hypothèse (3.1) est vérifiée, (PQ) a une solution unique $\bar{d}$ caractérisée par

$$\hat{d} = - H^{-1}(\nabla f(x) + \nabla g(x)\hat{\lambda}) ,$$

où λ est solution de

$$\nabla g(x)^t \, H^{-1} \nabla g(x) \lambda = - \nabla g(x) H^{-1} \, \nabla f(x) + g(x).$$

Utilisant (3.5) on vérifie alors qu'il existe $C_1 > 0$ indépendant de r tel que

$$||d - \hat{d}|| \leq \frac{C_1}{r} . \quad \square$$

IV - LE CAS DES FONCTIONNELLES PENALISEES EXACTES

a) Caractérisation des fonctionnelles pénalisées exactes

Nous nous intéressons ici aux fonctions telles que toute solution de (P2) soit aussi solution de (P1). Il est utile de caractériser ces fonctions (on étend des résultats de D.P. Bertsekas [2]).

Théorème 4.1

(i) Si $\bar{x}$ vérifie les conditions nécessaires d'optimalité de (P2) avec le multiplicateur $\bar{p}$ associé, alors

$$\bar{p} \in \overset{o}{\widehat{\partial\phi(0)}} \Longrightarrow \bar{x} \text{ vérifie les conditions nécessaires d'optimalité de (P1).}$$

(ii) Si $\bar{x}$ vérifie les conditions nécessaires d'optimalité de (P1) avec le multiplicateur $\bar{p}$ associé, alors

$$\bar{p} \in \partial\phi(0) \Longrightarrow \bar{x} \text{ vérifie les conditions nécessaires d'optimalité de (P2).} \quad \square$$

La démonstration du théorème 4.1 est directe. Pour le point (i), on utilise le fait que $\overset{o}{\widehat{\partial\phi(y_1)}} \cap \partial\phi(y_2) = \emptyset$ si $y_1 \neq y_2$.

Théorème 4.2

Soit $\bar{x}$ solution de (P1) telle que $\nabla g(\bar{x})$ soit de rang m. Alors il existe $\bar{p}$ unique tel que $(\bar{x},\bar{p})$ vérifient les conditions nécessaires d'optimalité de (P1) et

$\bar{p} \in \partial\phi(0) \iff \bar{x}$ vérifie les conditions d'optimalité de (P2). □

Remarque 4.1

Soit la fonction

$$\theta_r(x) = f(x) + r\phi(g(x)),$$

où $r > 0$. Si le convexe fermé $\partial\phi(0)$ (borné puisque ϕ est finie) vérifie

$$0 \in \overset{o}{\widehat{\partial\phi(0)}} \text{ dans } \mathbb{R}^m,$$

il existe r_o tel que $\forall r > r_o$, $\bar{x},\bar{p}$ vérifiant les conditions d'optimalité de (P1) vérifient aussi celles de (P2). □

Dans la suite, nous nous restreignons aux fonctions positivement homogènes de degré 1. Soit le convexe fermé

$$K^o = \{y \ ; \ \phi(y) \leq 1\}.$$

D'après l'hypothèse (1.2), l'intérieur de K^o contient 0. On vérifie que ϕ est la fonction jauge de K^o, définie par

$$\phi(y) = \inf \{r \geq 0 \ ; \ y \in r\, K^o\}.$$

La conjuguée de ϕ est

$$\phi^*(q) = \sup \{y^t q - \phi(y)\} = \begin{cases} +\infty & \text{si } \exists\, y \ ; \ \langle y,q\rangle > \phi(y) \ ; \\ 0 & \text{sinon.} \end{cases}$$

Soit $K = \{q \in \mathbb{R}^m \ ; \ \phi^*(q) = 0\}$. Alors $\phi^*(q) = I_K(q)$ donc ϕ, qui est la conjuguée de ϕ^* est la fonction support de K :

$$\phi(y) = S_K(y) = \sup \{q^t y \ , \ q \in K\}.$$

Si K^o est symétrique (c.a.d. $y \in K^o \Rightarrow -y \in K^o$), ϕ est une norme :

$$\phi(y) = ||y||.$$

Soit $||.||_D$ la norme duale, définie par

$$||q||_D = \sup \{q^t y, \ ||y|| \leq 1\} = \sup \{q^t y, \ y \in K^o\} = S_{K^o}(q) \ .$$

On vérifie alors que

$$K = \{q \in \mathbb{R}^m \ ; \ ||q||_D \leq 1\} \ ,$$

donc

$$\phi^*(q) = \begin{cases} 0 \text{ si } ||q||_D \leq 1, \\ +\infty \text{ sinon } . \end{cases}$$

b) Cas particuliers : normes L^∞, L^1, L^2

Nous avons vu que les normes de $\mathbb{R}^m$ représentent une classe importante, et assez générale, de fonctions de pénalisation. Le choix effectif d'une norme pour la mise en oeuvre de la méthode sera basé sur la facilité de résolution de (PC). Nous allons envisager le cas des normes L^∞ et L^1, ce qui permettre d'étendre au cas où (PQ) est inconsistant les théories de B. Pchénitchny (cf. [9]) et S.P. Han [7]. Puis nous donnerons quelques arguments en faveur de l'utilisation de la norme L^2.

Exemple 1 : Norme L^∞

B. Pchenitchny (cf. [9]) a proposé pour la résolution de (P1) de minimiser la fonction

$$\theta_{\infty,r}(x) = f(x) + r \max \{|g_i(x)|, \ i = 1 \text{ à } m\},$$

en observant que la solution d de (PQ) (si elle existe) donne une direction de descente de $\theta_{\infty,r}(x)$ (si r est assez grand). Le théorème 2.1 montre, plus généralement qu'une direction de descente de $\theta_{\infty,r}(x)$ est obtenue en résolvant le problème dual de (PC)

$$\text{Min}\{Q(\lambda), \ \lambda \in \mathbb{R}^m, \ \sum_{i=1}^{m} |\lambda_i| \leq r\},$$

où Q(.) est défini à la remarque 2.2, puis avec

$$(4.1) \qquad d = - H^{-1}(\nabla f(x) + \nabla g(x)\lambda). \qquad \square$$

Exemple 2 : Norme L^1

S.P. Han [7] a proposé une théorie similaire en utilisant la norme L^1 pour pénaliser les contraintes. La fonction de pénalisation est donc

$$\theta_{1,r}(x) = f(x) + r \sum_{i=1}^{m} |g_i(x)|.$$

Le problème dual de (PC) est

$$\text{Min } \{Q(\lambda),\ \lambda \in \mathbb{R}^m\ ,\ |\lambda_i| \le r, i = 1 \text{ à } m\},$$

puis d est obtenu grâce à (4.1).

Notons que R. Fletcher [6] a proposé une méthode de minimisation de $\theta_{1,r}(x)$, en minimisant la fonctionnelle pénalisée dans une région de confiance modifiée d'une façon adaptative. De fait, tous les résultats de cette communication peuvent être repris dans le cadre des algorithmes avec région de confiance de modèle. Par ailleurs, le problème dual de (PC) utilise aussi implicitement une notion de région de confiance. Ceci est mis à profit dans l'exemple suivant.

Exemple 3 : Norme L^2

Nous proposons d'utiliser la norme

$$||g(x)||_2 = (\sum_{i=1}^{m} g_i(x)^2)^{1/2}.$$

Le problème dual est alors

$$\text{Min}\{Q(\lambda),\ \lambda \in \mathbb{R}^m,\ ||\lambda||_2 \le r\}. \tag{4.2}$$

Soit α le multiplicateur associé à la contrainte ; la solution p de (4.2) vérifie donc

$$\text{Min}\{Q(\lambda) + \frac{\alpha}{2}||\lambda||_2^2,\ \lambda \in \mathbb{R}^m\}, \tag{4.3}$$

qui est un problème quadratique. Réciproquement, soit $\alpha \in \mathbb{R}^+$; la solution λ_α de (4.3) est solution du problème

$$\text{Min}\{Q(\lambda),\ \lambda \in \mathbb{R}^m\ ,\ ||\lambda||_2 \le ||\lambda_\alpha||_2\ \}.$$

Donc d_α solution de

$$d_\alpha = - H^{-1}(\nabla f(x) + \nabla g(x)\lambda_\alpha)$$

est une direction de descente de

$$f(x) + ||\lambda_\alpha||_2 \quad ||g(x)||_2 .$$

Comme la valeur de r n'est pas un paramètre critique du problème, on peut se contenter d'une approximation du paramètre α, quitte à faire varier r d'une itération sur l'autre. L'évaluation de α peut se faire par un algorithme de type Levenberg-Marquardt (cf. [8]). □

Discussion

Le problème dual de l'exemple 1 comporte une contrainte sur la norme L^1, difficile à prendre en compte. Celui de l'exemple 2 comporte des contraintes de borne. Notons que si des contraintes d'inégalité sont présentes, elles introduisent des contraintes de positivité sur le multiplicateur. Le seul inconvénient est que l'identification des contraintes actives peut être longue. Au contraire l'exemple 3 ne nécessite que l'évaluation grossière d'un paramètre. Notons aussi que le paramètre α améliore le conditionnement du problème dual et constitue donc un remède contre les instabilités numériques.

CONCLUSION

Ce papier analyse une classe de méthodes de résolution de problèmes sous contraintes, utilisant des fonctions de pénalisation (différentiables ou non), pour le calcul d'une direction de descente. Cette classe inclut en particulier les méthodes dites de programmation quadratique successive qui sont parmi les plus performantes. Nous montrons, par une technique de dualité, comment calculer une direction de descente de la fonction de pénalisation, même si les contraintes linéarisées sont incompatibles. Des méthodes globalement convergentes (même si la pénalisation n'est pas exacte) sont exposées. Les choix particuliers de la fonction de pénalisation permettent de retrouver les méthodes de M.C. Bartholomew-Biggs [1], B.P. Pchenitchny (cf. [9]) et S.P. Han [7]. ces deux dernières méthodes sont ainsi étendues d'une façon naturelle au cas où les contraintes sont incompatibles. Enfin, nous proposons une nouvelle fonction de pénalisation exacte, basée sur la norme L^2.

BIBLIOGRAPHIE

[1] M.C. BARTHOLOMEW-BIGGS. "On the convergence of some constrained minimization algorithms based on recursive quadratic programming". J. Inst. Maths. Applics. 21, pp.67-81, 1978.

[2] D.P. BERTSEKAS. "Necessary and sufficient conditions for a penalty method to be exact". Math. Prog. 9, pp. 87-99, 1975.

[3] J.F. BONNANS. "Asymptotic admissibility of the unity stepsize in exact penalty methods I : equality-constrained problems. Rapport INRIA, 1984.

[4] H. BREZIS. "Opérateurs maximaux monotones et semi-groupes de contraction dans les espaces de Hilbert". North-Holland, 1973.

[5] I. EKELAND, R. TEMAM. "Analyse convexe et problèmes variationnels". Dunod Gauthier-Villars, 1974.

[6] R. FLETCHER. "A model algorithm for composite non-differentiable optimization problems". Math. Prog. Study 17, pp. 67-76, 1982.

[7] S.P. HAN. "A globally convergent method for nonlinear programming". J. of Opt. Theory and Appl. 22, pp. 297-309, 1977.

[8] J.J. MORE. "The Levenberg-Marquardt algorithm : implementation and theory". Proc. of the Dundee Conference on Numerical Analysis, G.A. Watson ed., Springer-Verlag, 1978.

[9] B. PCHENITCHNY, Y. DANILINE. "Méthodes numériques dans les problèmes d'extrémum". Mir, Moscou, 1965 (édition française : 1977).

[10] E. SACHS. "Global convergence of quasi-Newton type algorithms for some nonsmooth optimization problems". J. of Opt. Theory and Appl. 40, pp. 201-219, 1983.

AN ADAPTIVE SINGULAR VALUE DECOMPOSITION ALGORITHM AND ITS APPLICATION TO ADAPTIVE REALIZATION.

J. Vandewalle, J Staar, B. De Moor, J. Lauwers.
ESAT laboratory Katholieke Universiteit Leuven
Kard. Mercierlaan 94
B-3030 Heverlee
Belgium
tel 016/220931

Abstract:

In this paper we present an algorithm ASVD for the computation of the singular value decomposition (SVD). The method presented is a power method for calculating the largest triplets of the SVD of a matrix A when multiplication is "cheap". The method used bears a lot of similitüde with the power method for finding the eigenvalues of a symmetric matrix M. The triplets are found one after another and also some deflation techniques (orthogonalization) are used. The algorithm can take profit of the SVD of slightly different matrices and it is based on the geometric properties of the SVD. Tests have shown that it is more efficient than Golub's algorithm if only the dominant part of the SVD of a long sequence of slowly varying matrices is needed. Also storage efficiency is obtained whenever the matrices are structured.
The paper describes the basic ASVD algorithm, its numerical properties using the shift mechanism, an acceleration method, a computer implementation and its use in adaptive state space realization of noisy impulse responses. It is expected that the new ASVD algorithm and its many strategies will be useful in the domain of signal processing, system theory and automatic control, where SVD is becoming more and more an important concept.

1. Introduction.

In signal processing, automatic control as well as in system theory, the singular value decomposition (SVD) is used increasingly (1-12) as well as for its conceptual as its numerical qualities. In order to compute the SVD, one always recommends the use of Golub's algorithm (2, 11, 13, 14), which is very efficient for the complete high precision SVD of a single full rank matrix. It is however the experience of the authors (3,4) that this algorithm is computationally too involved for most signal and system applications. In the context of adaptive state space realization for example (4), the SVD of a long sequence of slowly varying large block Hankel matrices of low numerical rank has to be computed. In general there exist many signal and system applications where only a small set of singular values are of interest (e.g. all those larger than a certain noise level), or where the required accuracy level of the singular values is limited (e.g. because the data are rather inaccurate), or where a good estimate for the SVD exists (e.g. from a slightly different matrix). It's the purpose of this paper to present a new algorithm for the SVD which, unlike Golub's algorithm, can benefit from these restrictions. The adaptive singular value decomposition algorithm (ASVD) converges one after another to the singular values and vectors in an iterative matrix multiplication process.

In section 2 we describe the basic algorithm and the convergence theorems. The numerical properties are analyzed in section 3 using the shift-mechanism, which is easy to visualize. This avoids many unnecessary computations in the iteration cycle. When the convergence is too slow, a special acceleration step can be applied (section 4). A strategy to exploit these speed ups is developped, implemented on a computer and evaluated in section 5. The use of this algorithm for adaptive realization is given in section 6. In the last section we present the conclusions and emphasize that the basic algorithm allows for many different strategies which should be selected according to the application at hand. The most general description (3) of this algorithm allows for a parallel data flow or analog computer implementation since each of the singular triplets of a singular value, a left and a right singular vector can be computed simultaneously. Here we will only present the recursive version where the triplets are computed one after another. This allows to obtain more useful convergence properties and is valid for all implementations on Von Neumann machines.

In computing the singular value decomposition of a matrix A, it is useful to keep in mind that the ASVD algorithm essentially follows certain trajectories of the discrete time system $x_{k+1} = A^t.A.x_k$. One should however be careful to distinguish the computation performed by ASVD from all the techniques which compute the SVD of a matrix A from the eigenvalue decomposition of $A^t.A$. It is known that the squaring performed in $A^t.A$ is numerically dangerous. It is stressed that the numerical qualities of ASVD are sound.

2. The basic ASVD algorithm.

Before presenting the algorithm, first the notion of singular value decomposition is recalled (1-3, 11, 13) and some notations are introduced.

Theorem 1. For any real mxn matrix A of rank r there exists a real factorization

$$A = U \cdot \Sigma \cdot V \tag{1.a}$$

where U and V are square orthonormal matrices and Σ is a pseudodiagonal mxn matrix

$$\Sigma = \begin{bmatrix} \mathrm{diag}(\sigma_1 \ \sigma_2 \ . \ . \ \sigma_r) & 0 \\ 0 & 0 \end{bmatrix} \tag{1.b}$$

with the singular values σ_i

$$\sigma_1 \geqslant \sigma_2 \geqslant . \ . \ . \ . \geqslant \sigma_r \tag{1.c}$$

The colums of u_i of U (resp. rows v_j of V) are called the left (resp. right) singular vectors. The set (u_i, σ_i, v_i) is called the i-th singular triplet. The spaces $S_U^i = \mathrm{Span}(u_1, . . . , u_i) \subset R^m$ (resp. $S_V^j = \mathrm{Span}(v_1, . . . , v_j) \subset R^n$) are called the i-th principal left (resp. j-th principal right) singular subspaces. For the uniqueness properties and the energy properties (14) of the SVD we refer to the literature. In order to understand the ASVD algorithm it is crucial to realize that any linear map $A{:}x \longrightarrow y = A \cdot x$ is the result of 3 linear operations $y = U \cdot (\Sigma \cdot (V \cdot x))$: an orthonormal transformation V, a scaling of the axes Σ and another orthonormal transformation U. Let the vector x be described with respect to the orthonormal basis $v_1, . . ., v_n$ as

$$x = \sum_{i=1}^{n} h_i . v_i \tag{2.a}$$

then from theorem 1 x is mapped by A into

$$y = A \cdot x = \sum_{i=1}^{r} (\sigma_i . h_i) . u_i \tag{2.b}$$

In other words, each of the coordinates in the V-bases is scaled with a singular value and reconstructed in the U-bases. Clearly, the components with the lowest index are more amplified than those with the largest. Analogously A^t maps

$$s = \sum_{i=1}^{m} \sigma_i . u_i \tag{3.a}$$

into

$$t = A^t . s = \sum_{i=1}^{r} (\sigma_i . \sigma_i) . v_i \quad (3.b)$$

Again the components σ_i with the lowest index are more amplified than those with the largest. The net result is that successive multiplications on the left with A and A^t forces a starting vector towards the singular vector with the greatest singular value. Based on these ideas it is quite natural to describe the basic ASVD algorithm as follows:

Computation of the R dominant triplets (u_i, σ_i, v_i) of matrix A:

For i = 1, 2, . . . , R perform the following process until it has converged:

Step 1: Set n=0, choose a vector $e_i^{(0)}$ as initial guess for u_i.

Step 2: Set n=n+1. Obtain the vector $f_i^{(n)}$ as a result of the multiplication $A^t.e_i^{(n-1)}$, orthogonalization with respect to S_V^{i-1} and normalization.

Step 3: Set n=n+1. Obtain the vector $e_i^{(n)}$ as a result of the multiplication $A.f_i^{(n-1)}$, orthogonalization with respect to S_U^{i-1} and normalization. $s_i^{(n)}$ is the lenght of the vector before normaliz.

Step 4: Verify whether the changes $e_i^{(n-1)} - e_i^{(n-3)}$, $s_i^{(n)} - s_i^{(n-1)}$, $f_i^{(n)} - f_i^{(n-2)}$ are sufficiently small. If not, return to Step 2.
If so the i-th singular triplet is obtained as

$$u_i = e_i^{(n-1)}, \sigma_i = s_i^{(n)}, v_i = f_i^{(n)} \quad (4)$$

The convergence of this process for sufficiently large n to the correct values (4) is guaranteed for almost all initial values by the following theorem and its corollary under the assumption of infinite precision arithmetic.

Theorem 2. For any mxn matrix with SVD (1) and for the vectors determined in the basic ASVD algorithm, one has:

$$\forall e \in S_U^p : (A^t.e) \in S_V^p \quad (5.a)$$

$$\forall f \in S_V^p : (A .f) \in S_U^p \quad (5.b)$$

$$e_p^{(0)} = \sum_{i=p}^{r} h_i.v_i : \quad e_p^{(n)} = \sum_{i=p}^{r} h_i \sigma_i^n v_i / \sqrt{\sum_{i=p}^{r} h_i^2 \sigma_i^{2n}} \quad (5.c)$$

$$s_p^{(n)} = \sqrt{\sum_{i=p}^{r} h_i^2 \sigma_i^{2n} / \sum_{i=p}^{r} h_i^2 \sigma_i^{2n-2}} \quad (5.d)$$

Proof: use the properties of the SVD and (3)&(4) (3,16,17)

Corollary 1: If $\sigma_p > \sigma_{p+1} > \sigma_{p+2}$ then the basic ASVD algorithm satisfies:

with $e_p^{(0)} = \sum_{i=p}^{r} h_i v_i$, $h_p \neq 0$, $h_{p+1} \neq 0$ and for sufficiently large n :

$$\|e_p^{(n)} - v_p\| = \|\Delta_p^{(n)}\| = \frac{h_{p+1}}{h_p} \left(\frac{\sigma_{p+1}}{\sigma_p}\right)^n \qquad (6.a)$$

$$s_p^{(n)} - \sigma_p = \frac{1}{2} \left(\frac{h_{p+1}}{h_p}\right)^2 . \left(1 - \frac{\sigma_{p+1}^2}{\sigma_p^2}\right) . \left(\frac{\sigma_{p+1}}{\sigma_p}\right)^{2n} \qquad (6.b)$$

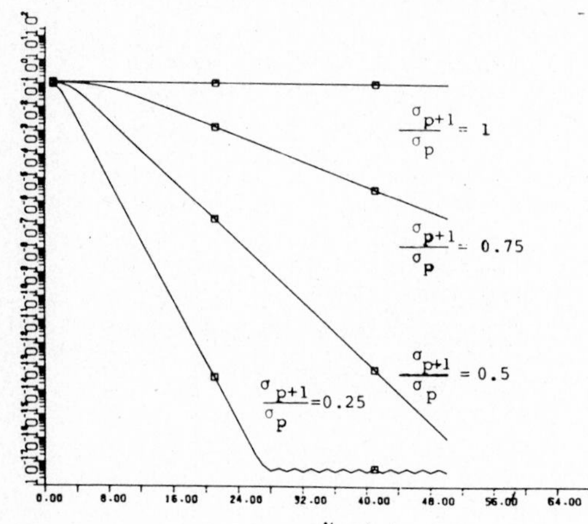

Fig.1. The error $\Delta_n = \| e_p^{(n)} - u_p \|$ decreases exponentially in terms of the number of iteration cycles. The rate however strongly depends on the ratio σ_{p+1}/σ_p.
For those cases where this ratio approximates 1, the acceleration step of section 4 is needed.

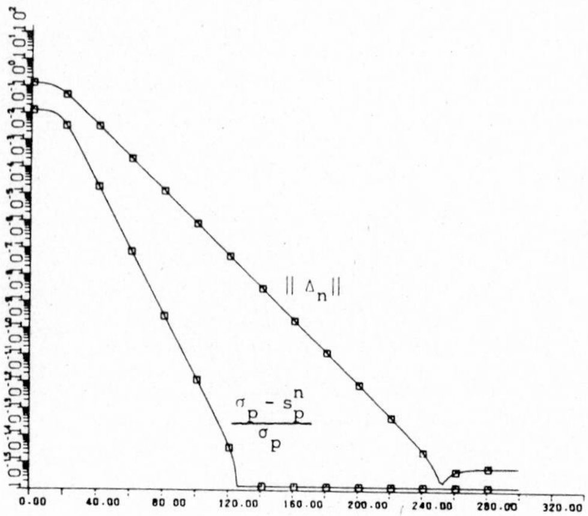

Fig.2. Both the error $\Delta_n = \| e_p^{(n)} - u_p \|$ on the singular vector u_p and the error on the singular value decrease exponentially but the singular values converge twice as fast as the vectors.

To conclude this section let us remind that the basic ASVD algorithm computes the triplets one after another. Unlike the Golub algorithm one can compute some singular triplets and one can use estimates for the singular values and vectors from nearby matrices and most of all one need not modify the matrix. Hence it is quite useful for sparse or structured (Hankel, Vandermonde, Toeplitz,. . .) matrices.

3. Analysis of the numerical properties using the shift mechanism.

Here we want to analyse the convergence of ASVD with floating point arithmetic (i.e. finite precision). From (2)&(3) and theorem 1, it is clear that the convergence can be easily analyzed if all vectors are represented in terms of the singular basis $u_1, \ldots, u_r$ or $v_1, \ldots, v_r$ and if one combines the two spaces such that $\Phi_i = u_i = v_i$. Though this representation of the vectors can only be given when the singular vectors are known, also for floating point arithmetic much insight in the mechanism and good strategies are based upon this representation. Any vector $e^{(n)}$ or $f^{(n)}$ can be described as

$$\sum_{i=1}^{r} \mu_i \Phi_i \qquad (7.a)$$

After a multiplication with A or A^t each component i is scaled by σ_i.

$$\sum_{i=1} \mu_i \sigma_i \Phi_i \qquad (7.b)$$

When the coordinates in the Φ_i-basis are represented on a logarithmic scale one sees that the multiplication causes for each coordinate i a shift by $\log(\sigma_i)$. The normalization then causes a shift of all components until the sum of the coordinates squared is 1. The floating point representation of the vectors (single precision 8 or double 16 decimals) then requires us to look at a window of 8 or 16 units (fig.3).

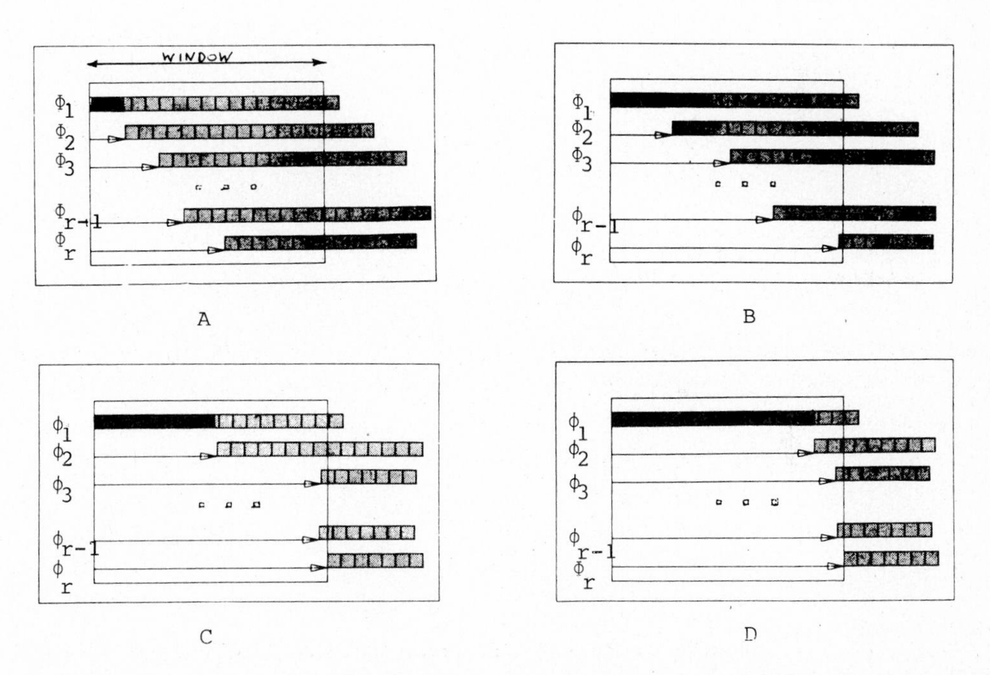

Fig.3. The shift mechanism of ASVD. Each figure describes in logarithmic scale the component of $\boldsymbol{\Phi}_i$ at a certain step in the convergence. As the algorithm proceeds the pattern moves from (A) to (B), (C) and (D). The black bar denotes the precision of the approximation of the singular vector. In each step the accuracy is increased with $\log_{10}(\sigma_1/\sigma_2)$.

Conclusion: The net effect of a multiplication followed by a normalization is that, up to second order effects, the coordinate i is in logarithmic scale shifted by $\log(\sigma_1/\sigma_i)$.

In order to have a practical and efficient algorithm, a number of issues have to be solved.

First, the computation of the p-th triplet can be greatly speeded up by making a good estimate for the p-th triplet during the convergence of the (p-1)-th triplet. The (p-1)-th convergence pattern may provide a good initial guess for the p-th triplet by taking the normalized difference

$$l_p^{(n)} = \frac{e_{p-1}^{(n+2)} - e_{p-1}^{(n)}}{\left\| e_{p-1}^{(n+2)} - e_{p-1}^{(n)} \right\|}$$

It is in general not optimal to take the value at the beginning of the iteration nor at the end (fig.4).

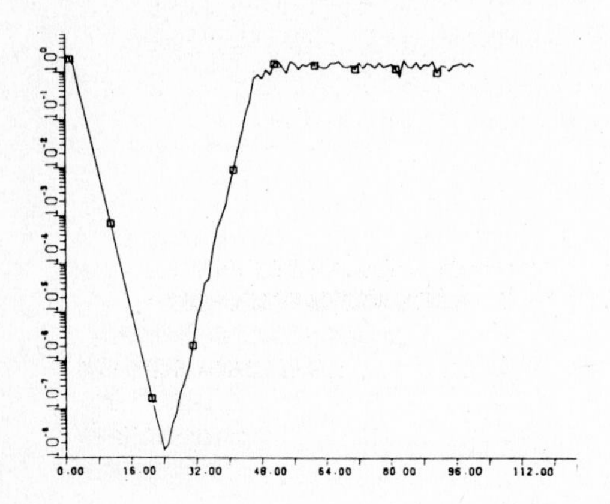

Fig.4. The error $\| l_p^{(n)} - u_p \|$ of the estimate $l_p^{(n)}$ of the p-th singular vector taken at the n-th step in the iteration of the (p-1)-th triplet. The optimal instant appears when the triplets p+1, p+2, . . have disappeared out of the window of finite precision.

Secondly, since the orthogonalization (Gramm-Schmidt or rather modified Gramm-Schmidt, (16-17)), is time consuming, one may wonder what happens to the convergence pattern of the p-th singular triplet if not during each step an orthogonalization is performed but every t steps (fig.5)

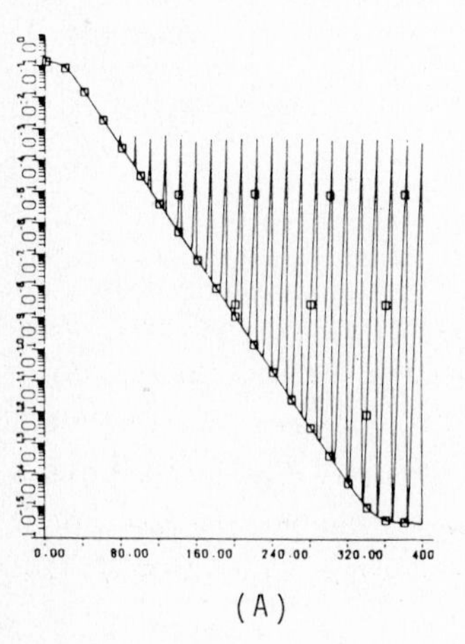

(A)

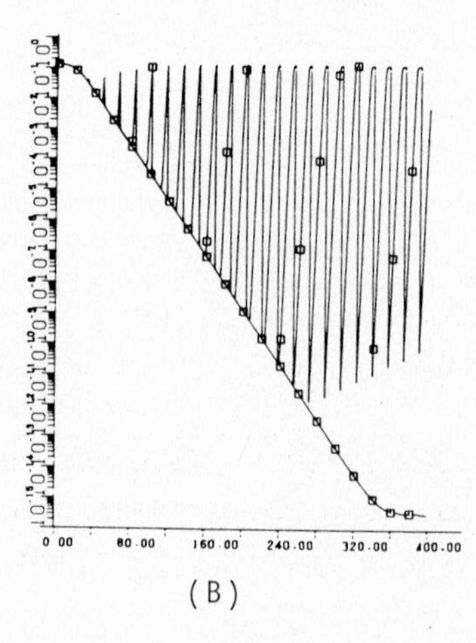

(B)

Fig.5. The error on the p-th singular vector by performing ASVD and orthogonalizing:
(A) every 16 steps.
(B) every 17 steps.

During the t steps between any two orthogonaliztions, a gradual deviation from the standard pattern (fig.1-2) occurs, which is completely eliminated by the orthogonalization. However, if too few orthogonalizations are performed, accuracy is lost (fig.5B). Again, this can be explained and analyzed by the shiftmechanism.

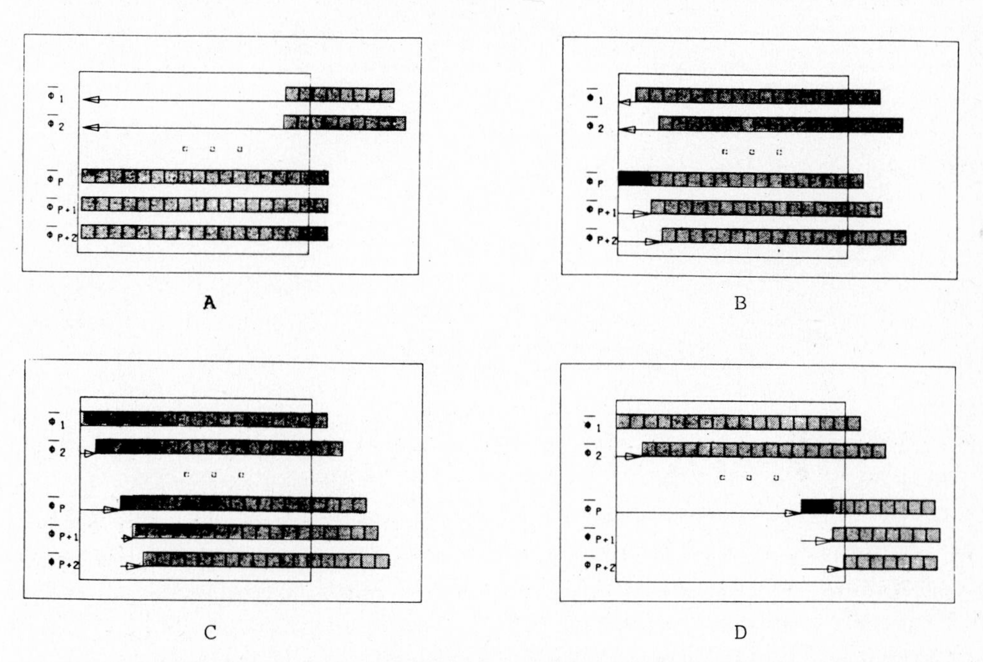

Fig.6. Analysis of the omission of some orthogonalizations in the iteration of the p-th triplet. After orthogonalization (A) the first p-1 contributions are at machine precision. The contributions of the p-1 first triplets grow faster (B&C). If one does not wait with the orthogonalization until the (p+1)-th contribution has disappeared (C) there is no los of accuracy. This condition has allowed to generate a period of t=16 for the example of fig.5. which in that case is the best possible.

4. An acceleration method.

From theorem 1 and fig.1 it is clear that the convergence for the p-th triplet is rather slow when σ_{p+1}/σ_p is close to 1. In this respect it is nice to remember that for $\sigma_{p+1} = \sigma_p$ only the left (resp. right) singular subspaces of the two singular vectors are unique. Then the left (resp. right) singular vectors can be freely chosen as orthogonal vectors in this subspace. Since for σ_{p+1}/σ_p close to 1 the basic ASVD algorithm has more difficulty in separating the p-th triplet from the (p+1)-th, it is more appropriate to iterate together on the p-th and the (p+1)-th triplet. This is the basic idea of the acceleration algorithm. Let Φ_p and Φ_{p+1} be the two singular vectors and let e_p and e_{p+1} be two estimates of these singular vectors that are in the same two-dimensional subspace as Φ_p and Φ_{p+1}. This subspace is drawn in fig.7. along with the vectors $A^t.e_p$, $A^t.e_{p+1}$, $f_p = A^t.e_p/\|A^t.e_p\|$ and $A.f_p$.

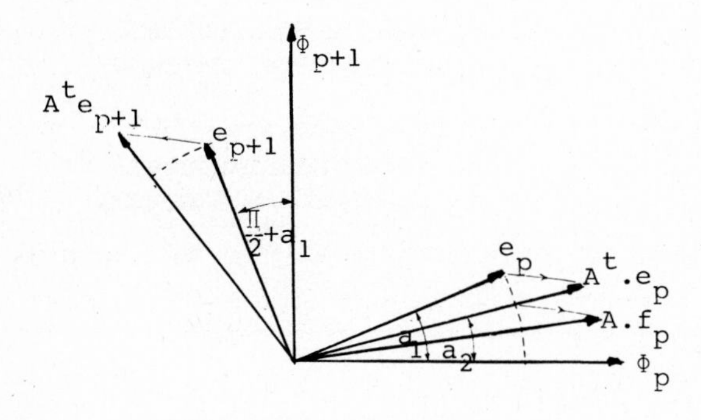

Fig.7. Geometric configuration used in the acceleration method.

Applying simple geometry and the basic properties (1)&(2) of multiplication with A or A^t one can set up (16,17) 4 nonlinear equations which relate the known parameters $s_1 = \|A^t.e_p\|$, $s_2 = \|A^t.e_{p+1}\|$, $t_1 = \|A.f_p\|$ to the unknown parameters σ_p, σ_{p+1} and the angles a_1 and a_2. These nonlinear equations can fortunately be solved analytically (16,17). Hence no further iteration is needed.

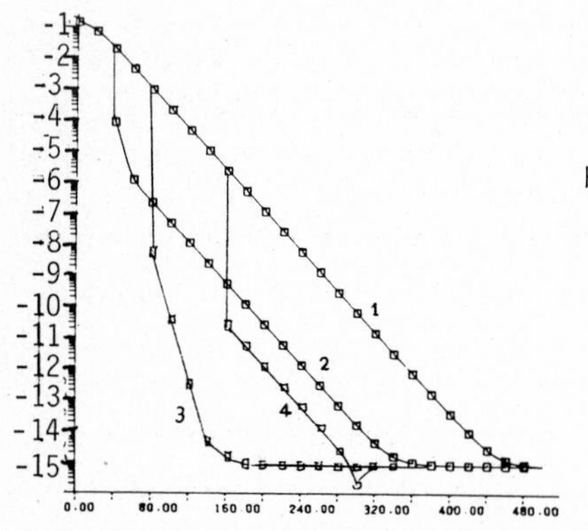

Fig.8. Curve 1 describes the exponential decrease of the error during the iteration of the basic ASVD algorithm. In curves 2, 3 & 4 one acceleration step is performed respectively at n=40, 80 & 160. Clearly, the acceleration has less effect when it is done too soon (curve 2) or too late (curve 4). Also it may happen that the convergence after an acceleration is faster (curve 3) than before. Based on the shift mechanism this implies that after an acceleration the contribution of the (p+2)-th triplet dominates that of the (p+1)-th.

A detailed analysis has shown that a single acceleration is best performed when the contribution in Φ_{p+1} after the acceleration with respect to that before is minimal(16). In some implementations it can be worth to do an acceleration in each step. Of course one acceleration step requires three matrix multiplications while a regular ASVD step only one.

5. Computer implementation of ASVD.

Based on the basic ASVD scheme many different versions of the algorithm with different trade offs between accuracy and efficiency can be generated. In order to dispose of a fully automatic program also techniques for computing σ_p, σ_{p+1}, σ_{p+2} en μ_{p+1}/μ_p in each triplet have been developped. A practical algorithm has been implemented on an IBM 3033 computer and the preliminary tests are promising. A first test example is a 10x10 matrix with singular values: σ_1 = 20.314622, σ_2 = 17.564908, σ_3 = 15.298623, σ_4 = 13.812922, σ_5 = 11.690173, σ_6 = 10.182562, σ_7 = 6.714472, σ_8 = 2.633963, σ_9 = 2.382086, σ_{10} = 2.176375.
The convergence pattern of some typical triplets are given in fig.9 (triplets 1, 2, 3, 7, 8). Typically the above described version of the ASVD algorithm requires for the computation of 3 singular triplets of a single 10x10 matrix as much time as Golub's algorithm.
Let us now evaluate what happens for a set of slightly different matrices.

6. Adaptive realization with ASVD.

The state space realization problem aims at finding a state space description for a system which is known by its impulse response. This problem often occurs in system theory and can also be a part of an identification. Algorithms to solve the state space realization problem have already been found in the sixties but have never been worked out in reliable software. Reasons for this are that the numerical qualities of these algorithms are not good (10) and that in practice the impulse responses are corrupted by noise which is also represented (4) in the state space realization. In order to tackle both problems Zeiger and Mc Ewen (8) proposed to find an approximate realization and to use the singular value decomposition. This has later on led to more general algorithms (3, 6, 7, 9) and the formulation of the problem as model reduction. Such algorithms basically perform the singular value decomposition of

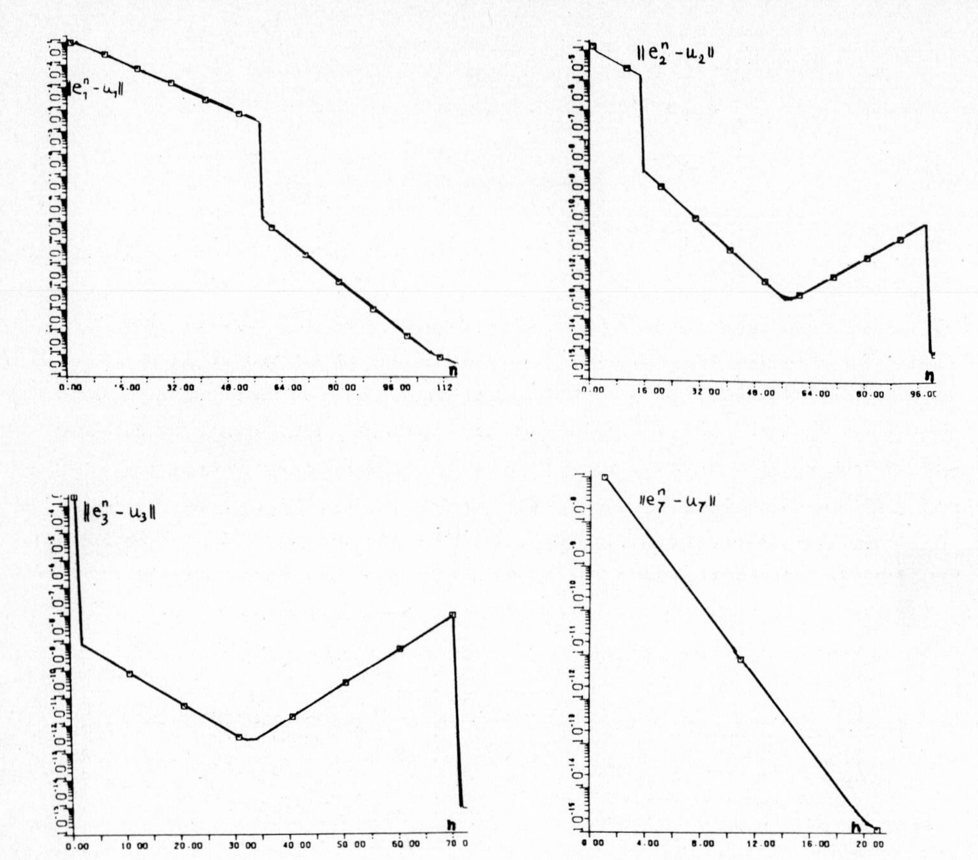

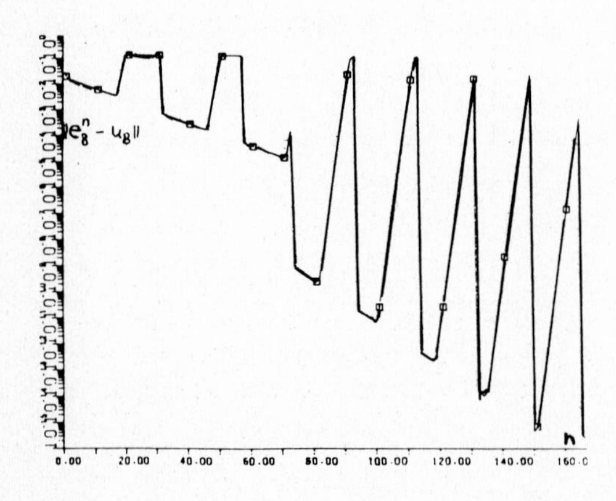

Fig.9 . The convergence $\|e_i^{(n)}-u_i\|$ *for the iteration of triplets* $i=1, 2, 3, 7, 8$*. Remark the effect of an acceleration in triplets 1,2 , 3,8 and of an orthogonalization in 2, 3 ,8.*

a block Hankel matrix:

$$\begin{bmatrix} H(1) & H(2) & . & . & . & . & . & H(M) \\ H(2) & H(3) & . & . & . & . & . & H(M+1) \\ . & . & . & . & . & . & . & . \\ H(N) & H(N+1) & & . & . & . & . & H(M+N-1) \end{bmatrix}$$

where H(1) . . H(M+N-1) are the lxm sample matrices of the impulse response of a system with m inputs and l outputs. Theoretically, for sufficiently large M and N, the rank of this matrix determines the degree n of the system. However, for noisy impulse responses the Hankelmatrix will be of full rank and hence the degree will be excessive. So, using the SVD (3, 6-9) only the singular triplets, which can be distinguished from the noise, should be considered. In (4) it has also been observed that the number of samples, which is theoretically irrelevant as long as $n < M,N$ dramatically affects the singular values and hence the amount of noise which can be tolerated (several orders of magnitude, fig.10)

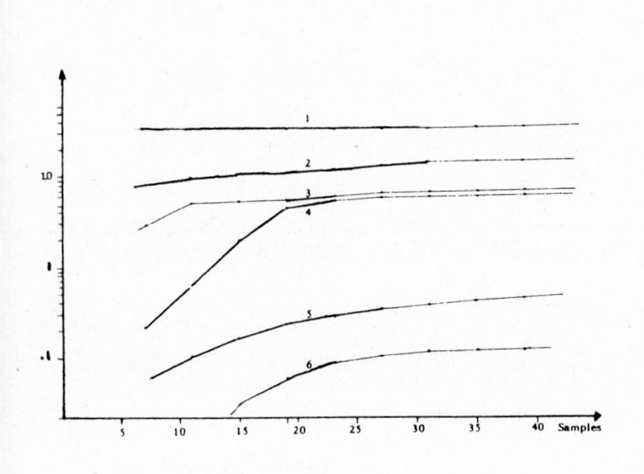

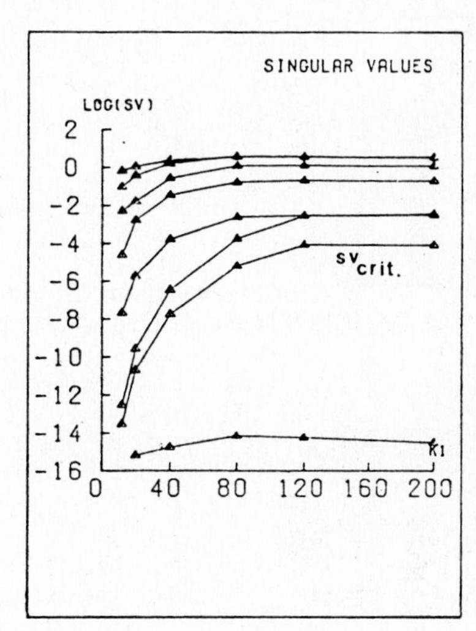

Fig.10. The singular values of two Hankelmatrices in terms of the number of samples of the impulse response. Remark that the last singular value emerges.

Although the experiments with this algorithm were quite promising (4) the on-line implementation with the Golub algorithm is quite time consuming and requires much storage. The ASVD algorithm was derived (3) with the aim of solving both problems. First of all, in an identification context, only few of the singular triplets of the Hankelmatrix have to be calculated. Using the link between energetical concepts and SVD (3) a so called noise criterion can be derived, which determines the number of triplets to be calculated:

$$\sigma_1^2 \geqslant \sigma_2^2 \geqslant \; . \; . \; . \; . \geqslant \sigma_n^2 > K.R_0.\sigma_\nu^2 \geqslant \sigma_{n+1}^2 \ldots$$

where K is a dimension dependent parameter, R_0 is the minimal energy ratio (3) and σ_r = (known noise %) . $(\max_k (\| H(k)\|_F, k=1,K))$.
In this way, the problem of determing the degree n of the system is reduced to a numerical rank problem. The degree is automatically derived from the partial SVD of the involved Hankelmatrix. An example is shown in fig.11.

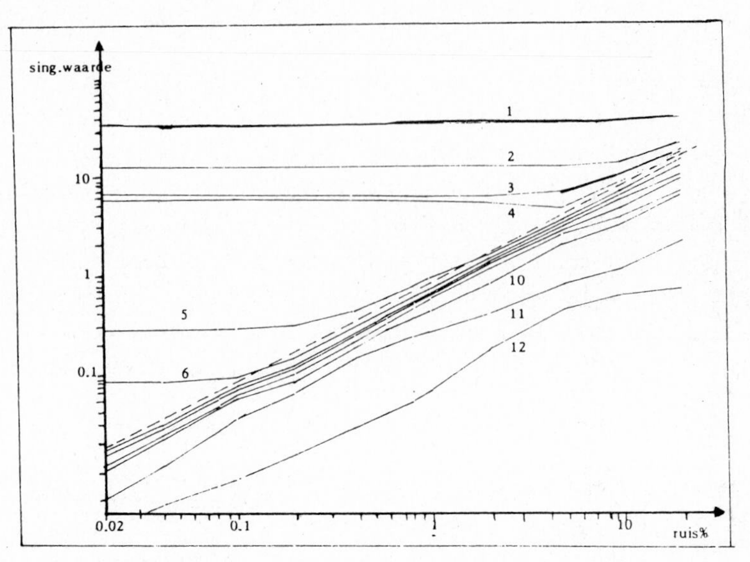

Fig.11. The impulse response of a monovariable 6th order system is stored in a 12x12 Hankelmatrix and poisened by several noise sequencies with increasing energy (standard deviation). The 12 singular values are plotted against the noise %. The noise criterion is plotted in dotted lines with R_0=2. Hence up to 0.1 % of noise the degree is determined correctly to be 6, between 0.1% and 1% the best model degree is 5 and up to 10% the degree is 4. In the neighbourhood of the noise criterion singular values cluster together. In those triplets the use of the acceleration algorithm is unavoidable.

Secondly, for on line identification the SVD of a sequence of slightly varying matrices has to be calculated. Making use of the recursivity of ASVD, one can take the SVD of a previous time step as a starting point for the calculation of the SVD at the present time step. (Fig.12).
Thirdly, the considerable gain in storage of ASVD with respect to the Golub algorithm is based upon the fact that Golub's algorithm operates on the Hankelmatrix entries while ASVD only requires the storage of H(1), ... , H(M+N-1). (Fig.13)
We may conclude that these partial results for adaptive realization with the ASVD algorithm are promising. Time saving properties are obtained by making use of the modular structure and the recursivity and by applying acceleration algorithms. Storage gain is obtained by exploiting the structure of the Hankelmatrices involved.

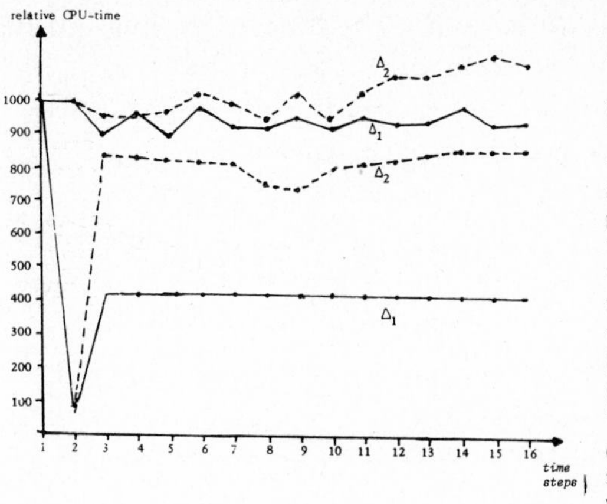

Fig.12. A set of 17 slightly varying 10x12 Hankelmatrices is considered of which the first and the second are equal. The upper curve is the time required for the calculation of 10 triplets of the SVD of each matrix with ASVD when no information of previous time steps is used. The other curves plot the time required for the SVD with ASVD when the starting vectors for timestep T are the singular vectors of timestep T-1. The curves markedwith Δ_1 are the results when the impulse response is changing rather slowly. The curves Δ_2 correspond to faster changes. Other speed improvements are still possible since none of the speed refinements of the previous sections are used here in order to show only the time saving by making use of the recursivity.

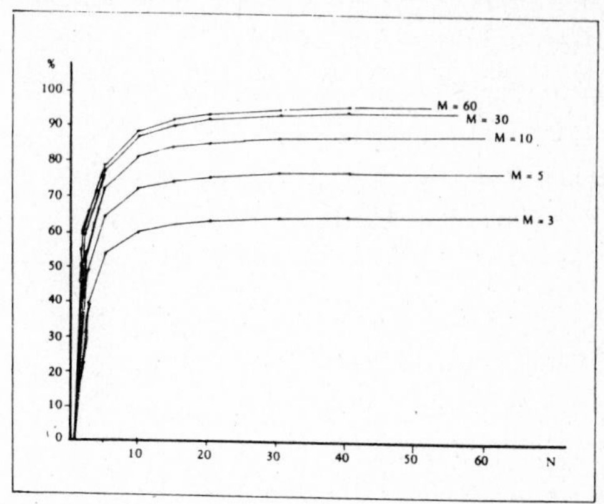

Fig.13. The storage gain obtained by ASVD with respect to the Golub algorithm for MxN matrices.

7. Conclusions.

ASVD is a recursive algorithm which gradually computes the dominant contributions in the SVD of a matrix A. The method bears a lot of similitude with the power method for the symmetric eigenvalue problem (21-22) Deflation techniques, similar to those described in Wilkinson (21) and Parlett (22),as well as acceleration methods are used.

The main advantages of the ASVD algorithm are: (i) Its natural and direct link with a basic property of SVD and hence its transparency for the potential user. (ii) Its simplicity, the only building blocks being the matrix-vector product, orthogonalization and normalization. (iii) Its complete recursivity. Any earlier approximation cuts down the remaining computation cost.(iv) Its modularity, and hence its flexibility for non-conventional implementations. (v) The fact that matrix A is not altered during the algorithm, and hence the storage efficiency in the case of highly structured or sparse matrices. (vi) Its numerical reliability (backward stable algorithm) and an easy understanding of the convergence by the shift mechanism.

Its main weak points are: (i) The algorithm is not competitive with Golub's algorithm for a single complete high precision SVD of a full rank matrix on a conventional machine. (ii) Convergence may be very slow for two triplets associated to very close singular values, if appropriate acceleration algorithms are not executed. Such full proof acceleration rules are not yet available for clusters of more than two singualar values.

Some tests have already shown that ASVD can be competitive with the Golub algorithm when only the dominant triplets of a sequence of large matrices are needed. Such situations arise in realization and identification, signal processing (12) and control theory: e.g. Quasi-Nyquist loci (20). For the Quasi-Nyquist loci the complex version of this algorithm is feasable and under consideration. Further work will be directed towards the evaluation of other strategies which may be less accurate but faster. In the case of the state space realization this may provide a unification of all recursive algorithms and allow a better trade off between speed and accuracy.

REFERENCES.

1. Special Issue on Multivariable Control Systems, IEEE Trans. Aut. Contr., Vol AC-26, No. 1, 1981.
2. V.C. Klema and A.J. Laub, "The singular Value Decomposition: Its computation and some applications." IEEE Trans. Aut. Contr., Vol AC-25, No 2, p.164-176, 1980.
3. J. Staar, "Concepts for reliable modelling of linear systems with application to on-line identification of multivariable state space descriptions", doct. diss., EAST laboratory, Kath. Univ. Leuven, June 1982.
4. J. Staar, J. Vandewalle & M. Wemans, "A Comparison of multivariable MBH-algorithms in the presence of multiple poles, and noise disturbing the Markov sequence." Proc. 4-th Int. Conf. on analysis and optimisation of systems (Lecture notes in Control and information Sciences), p.141-160, Springer Verlag, Versailles, December 1980.
5. J. Staar & J. Vandewalle, " On the numerical implication of the choice of a reference node in the nodal analysis of large circuits", Int. Jrnl. Circuit Theory and Applications, Vol 9, p.488-492, 1981.
6. J. Vandewalle and J. Staar, "On the recursive identification of multivariable state space models", Proc. Symposium on applications of multivariable system theory, Plymouth, October 1982.
7. J. Vandewalle & J. Staar, "Modelling of linear systems: critical examples and numerically reliable approaches", Proc. ISCAS, p.915-918, Rome May 1982.
8. H.P. Zeiger and A.J. Mc. Ewen, "Approximate linear realizations of given dimension via Ho's algorithm", IEEE Trans. Aut. Contr., Vol AC-19 p.153, 1974.
9. S.Kung, "A new identification and model reduction algorithm via singular value decomposition", Proc. 12-th Asilomar Conf. Circ. Syst. Nov 1978
10. L.S.De Jong. "Numerical aspects of the recursive realization algorithm", SIAM Journal on Control and optimization, Vol.16, p.646-659, 1978.
11. C.L. Lawson and R.J. Hanson, "Solving least squares problems", Prentice Hall Series in Automatic Computation, Englewood Cliffs 1974.
12. J. Vanderschoot, J. Vandewalle, J. Janssen, W. Sansen, H. Van Trappen: "Extraction of weak bioelectrical signals by means of singular value decomposition", sub. to this conference.
13. G.H. Golub and W.Kahan, "Calculating the Singular values and Pseudo-inverse of a matrix", SIAM J. Num. Anal. No.3, p.205-224, 1965.
14. G.H.Golub and C.Reinsch, "Singular Value Decomposition and Least squares solutions", Num. Math., Vol.14, p.403-420, 1970.
15. J.Staar and J. Vandewalle, "Oriented energy and signal to signal ratios concepts and use" in preparation for sub. to Systems and Control letters.
16. J.Lauwers, "Singuliere waardenontbinding in signaalverwerking; ontwerp en analyse van een efficient algoritme", Master's thesis, ESAT lab., Kath. Univ. Leuven, July 1983.
17. B.De Moor, "Een betrouwbaar adaptief realisatiealgoritme gebaseerd op recursieve singuliere waardenontbinding.", Master's thesis, ESAT Lab., Kath. Univ. Leuven, July 1983.
18. B.Garbow et al., "Matrix eigensystem routines -EISPACK guide extension", (Lecture Notes in Computing Sciences, Vol 51) New York: Springer 1977.
19. J.Dongerra et al., "Linpack User's guide", Philadelphia, SIAM 1979.
20. Y.S.Hung and A.G.J. Mac Farlane, "Multivariable feedback: A quasi-classical approach", Springer Lecture notes in Control and Information Sciences, Berlin 1982.
21. J.H.Wilkinson, "The algebraic eigenvalue problem", Clarendon Press, Oxford 1965.
22. Beresford N. Parlett, "The symmetric eigenvalue problem." Prentice Hall Series in computational mathemathics, 1980.

GENERAL SCHEMES FOR SOLUTION OF STATIC AND DYNAMIC EXTREMAL PROBLEMS

R.Gabasov
Byelorussian State University
Minsk, 220080 USSR

F.M.Kirillova, O.I.Kostyukova, A.V.Pokatayev
Institute of Mathematics
Surganov str., 11, Minsk 220604, USSR

Abstract. The history of optimal control processes has begun with the discovery of Pontryagin maximum principle [1]. This classical result has allowed to solve many applied problems and has become the basic means at constructing numerical methods of optimal control [2]. The given paper contains the recent results on solution of linear-quadratic extremal problems, optimization of linear dynamic systems with nonlinear input, linear-fractional problems, geometric programming problems.

1. THE ADAPTIVE METHOD. The analysis of the simplex-method [3] from the point of view of its possible use in optimal control theory displays the following drawbacks: 1) not arbitrary but special basic solutions are used on the iterations and this makes it difficult to use directly the information obtained from practice because this information can give basic solutions only in exceptional cases; 2) the heuristic principle of transformation of information adopted in the simplex-method consisting in obligatory movement along basic solutions is artificial and has no direct relation to the problem; 3) the simplex-method does not contain the rules of stopping the process of solution after attaining the given neighbourhood of the optimal solution which, to our point of view, essentially decreases the applied value of the method.

The wide practice of use of the simplex-method seems to give no ground to doubts about the merits of the simplex-method and from its point of view the given drawbacks seem to be artificial. To see the reverse consider a simple example of LP problem obtained from the terminal control problem

$$x(3) \to \min, \quad x^{(10)}(t) = u(t), \quad x^{(k)}(0) = x^{(k)}(5) = 0,$$
$$k = \overline{0,9}, \quad |u(t)| \le 1, \quad t \in T = [0,5],$$

in the class of impulse functions with the constant quantization period $h=1/100$. This problem is equivalent to LP problem with 500 variables and 10 constraints (n=500, m=10). From the point of view of modern software packages these sizes are dwarfish. Nevertheless the obtained problem was not solved by one of the most perfect and known systems LP ASU [4] intended, generally speaking, for the sizes $m \leq 16000, n=\infty$. This example shows that in spite of the existence of good and powerful systems additional investigations on constructing algorithms for solution of special problems for which general methods can turn to be inefficient are necessary. This situation takes place, as the given example shows, in the field of optimal control problems.

The adaptive method [5-7] for the solution of LP problems worked out by R.Gabasov, F.M.Kirillova and O.I.Kostyukova like the simplex-method is also primal, exact and relaxational. The details are given in [5-7]. Here we discuss the main principles of the method:

1) the concepts of support and support solution (the union of solution and support) form the basis of the method. The iteration of the method consists of two stages. On the first stage the transition from the current solution to a new solution with the smaller value of the cost function is realized. On the second stage the change of support takes place. The result of the iteration is receiving a new support solution. The support of the problem is similar to the basis from the simplex-method theory but it differs from the basis by the fact that it does not depend on the solution and is an independent element of the method. The support is mainly intended for the fulfilment of the main constraints of LP problem (the constraints of the form $Ax=b$ or $b_* \leq Ax \leq b^*$);

2) adaptive normalization of suitable directions of the change of the current solution makes the basis of the first part of the iteration. Unlike simplex normalization the adaptive normalization is not brought from outside. Adaptive normalization consists in weakening the constraints of LP problem by throwing away some constraints (constraints $d_{*j} \leq x_j \leq d_j^*$ on the support components of the solution x). For the obtained simplified LP problem it is easy to write the solution, i.e. pseudo-solution $\mathcal{X}$ and vector $l=\mathcal{X}-x$ is chosen as the admissible direction. If the support is successfully (optimally) chosen, i.e. vector $\mathcal{X}$ satisfies the thrown constraints then it means that $\mathcal{X}$ is the solution of LP problem;

3) as it is known the basis in the simplex-method being connected with the basic solution is transformed with the latter simultaneously. In the adaptive method the support does not depend on the solution

tion. Operations on the change of support are the second part of the iteration. Every support can be put in accordance with some dual solution accompanying it. So the operations on the change of support can be described in terms of improvement of accompanying dual solution. To do this various methods [7,8] using elementary and nonelementary directions, block change, methods with long and short dual step along suitable dual directions are applied;
4) in the course of iterations of the adaptive method there is a chance to follow the optimality measure of the current solution and stop computation at obtaining ε -optimal solution.

By now finite modifications of the adaptive method [9,18] without the usual assumptions on nondegeneracy have been constructed. In addition the exit procedures from the cycle in degenerate situations are based, like the adaptive method itself, on extremal principles and differ on principle from procedures worked out in the theory of simplex-method.

Among further ways of the rise of effectiveness of the adaptive method we must note multistep modifications [10].

Realizations of the adaptive method for the solution of special linear problems: transportation problems [7], piece-wise linear programming problems [11], dynamic models of linear programming problems [12] are worked out. The main ideas of the adaptive method are successfully applied for the solution of linear optimal control problems in the class of impulse functions [7,13] , piece-wise constant functions [14], problems of optimization of dynamic systems with inertial controls and phase constraints [15], quadratic programming problems [16].

Gradually after sufficiently complete test the algorithms are given to the Republican fund of algorithms and programmes of the BSSR in the form of scientific subprogrammes. The first collection of such "materialized" mathematical results appeared in 1983 [17].

2. <u>LINEAR-QUADRATIC EXTREMAL PROBLEMS</u>. A finite relaxational primal exact method for the solution of a general quadratic programming problem is described in terms of support solutions introduced by the authors earlier for linear problems and generalized on new problems. This method makes the nucleus of a new algorithm for the solution of a linear quadratic optimal control problem where the constraints on control and phase trajectory are taken into consideration. The algorithm consists of three procedures: the procedure of the change of admissible control, the procedure of the change of support and the finishing procedure. As a result the admissible control that satisfies maximum principle on the whole segment of definition is obtained.

Consider the problem

$$f(x^0) = \min f(x), \quad Ax = b, \quad d_* \le x \le d^*; \ \operatorname{rank} A = m, \ D \geqslant 0,$$

$$f(x) = c'x + x'Dx/2, \quad A = A(I, J), \ I = \{1,2,\dots,m\}, \ J = \{1,2,\dots,n\}.$$

Let x be the solution. The sets $J_{sup} \subset J$, $|J_{sup}| = m$; $J_f \subset \{j \in J: \Delta_j = 0\}$; $J_P = J_{sup} \cup J_f$ are said to be the supports of constraints and the cost function of the problem if $\det A_{sup} \ne 0$, $\det R \ne 0$. The support solution $\{x, J_P\}$ is nondegenerate if $d_{*j} < x_j < d^*_j$, $j \in J_P$. Here $\Delta = A'y - Dx - c$, $y' = (c'(J_{sup}) + x'D(J, J_{sup}))A^{-1}_{sup}$, $A_{sup} = A(I, J_{sup})$; R is the matrix with blocks $R_{11} = D(J_P, J_P)$, $R_{12} = A'(I, J_P) = R'_{21}$, $R_{22} = 0$.

The relations: $\Delta_j \geqslant 0$ at $x_j = d^*_j$; $\Delta_j \le 0$ at $x_j = d_{*j}$; $\Delta_j = 0$ at $d_{*j} < x_j < d^*_j$, $j \in J_n = J \setminus J_P$ are sufficient and in case of nondegeneracy also necessary for optimality of $\{x, J_P\}$. If

$$\beta(x, J_P) = \sum_{\Delta_j > 0} \Delta_j (d^*_j - x_j) + \sum_{\Delta_j < 0} \Delta_j (d_{*j} - x_j) \le \varepsilon$$

then x is ε-optimal solution.

Let $\varepsilon \geqslant 0$ be given but $\beta(x, J_P) > \varepsilon$. Assume $s_j = 1$ at $\Delta_j \geqslant 0$; $s_j = -1$ at $\Delta_j < 0$, $j \in J_n$. Realize the iteration $\{x, J_P\}, s(J_n) \longrightarrow \{\bar{x}, \bar{J}_P\}, \bar{s}(\bar{J}_n)$. For $x \to \bar{x}$ construct l, g: $l_j = d^*_j - x_j$ at $s_j = 1$; $l_j = d_{*j} - x_j$ at $s_j = -1$, $j \in J_n$; $(l(J_P), g) = -R^{-1}(D(J_P, J_n)\, l(J_n), A(I, J_n)\, l(J_n))$. Calculate $\theta = \min\{1, \theta_{j_0}, \sigma_{j_*}\}$; $\theta_{j_0} = \min \theta_j$, $j \in J_P$; $\theta_j = (d^*_j - x_j)/l_j$ at $l_j > 0$; $\theta_j = (d_{*j} - x_j)/l_j$ at $l_j < 0$; $\theta_j = \infty$ at $l_j = 0$; $\sigma_{j_*} = \min \sigma_j$, $j \in J_n$; $\sigma_j = -\Delta_j/\delta_j$ at $s_j \delta_j < 0$; $\sigma_j = \infty$ at $s_j \delta_j \geqslant 0$; $\delta = -Dl + A'g$. Assume $\bar{x} = x + \theta l$. Then $\beta(\bar{x}, J_P) = (1 - \theta)(\beta(x, J_P) - \theta l' D l)$.

Consider the case $\beta(\bar{x}, J_P) > \varepsilon$. At $J_P, s(J_n) \to \bar{J}_P, \bar{s}(\bar{J}_n)$ we distinguish 1) $\theta = \theta_{j_0}$, $j \in J_f$; 2) $\theta = \theta_{j_0}$, $j \in J_{sup}$; 3) $\theta = \sigma_{j_*}$, $\alpha = 0$ or $d^*_{j_*} - d_{*j_*} \le |\Delta_{j_*} + \delta_{j_*}|/\alpha$; 4) $\theta = \sigma_{j_*}$, $d^*_{j_*} - d_{*j_*} > |\Delta_{j_*} + \delta_{j_*}|/\alpha$; $\alpha = D(j_*, j_*) - (D(j_*, J_P), A'(I, j_*))\, R^{-1} (D(J_P, j_*), A(I, j_*))$.

1) Assume $\bar{J}_{sup} = J_{sup}$, $\bar{J}_f = J_f \setminus j_0$; $\bar{s}_{j_0} = \operatorname{sign} l_{j_0}$, $\bar{s}(J_n) = s(J_n)$.

2) Construct $v = e'_{j_0} A^{-1}_{sup} A(I, J_f)$. If $\exists v_k \ne 0$ then substituting J_{sup}, J_f for $(J_{sup} \setminus j_0) \cup k$, $(J_f \setminus k) \cup j_0$ we pass to 1). At $v = 0$ (or $J_f = \emptyset$) assume $\bar{\Delta} = \Delta + \theta \delta$, $p'(J_n) = e'_{j_0} A^{-1}_{sup} A(I, J_n) \operatorname{sign} l_{j_0}$, $\bar{\sigma}_j = -\bar{\Delta}_j / p_j$ at $s_j p_j < 0$, $\bar{\sigma}_j = \infty$ at $s_j p_j \geqslant 0$, $j \in J_n$. Let us put in order $\bar{\sigma}_j$, $j \in J_n$: $\bar{\sigma}_{j_1} \le \bar{\sigma}_{j_2} \le \dots \le \bar{\sigma}_{j_t}$ and find $\alpha_0 = (1 - \theta)|l_{j_0}|$, $\alpha_{i+1} = \alpha_i - |p_{j_{i+1}}(d^*_{j_{i+1}} - d_{*j_{i+1}})|$, $i = \overline{0, t-1}$. Let $\alpha_\nu \le 0$, $\alpha_{\nu-1} > 0$. Assume $\bar{J}_{sup} = (J_{sup} \setminus j_0) \cup j_\nu$, $\bar{J}_f = J_f$, $\bar{s}_{j_0} = \operatorname{sign} l_{j_0}$, $\bar{s}_j = -s_{j_i}$,

$i=\overline{1,\nu-1}$, $\bar{s}_{j_i}=s_{j_i}$, $i=\overline{\nu+1,t}$.

3) Assume $\bar{J}_{sup}=J_{sup}$, $\bar{J}_f=J_f$; $\bar{s}(J_n\setminus j_*)=s(J_n\setminus j_*)$, $\bar{s}_{j_*}=-s_{j_*}$.

4) Assume $\bar{J}_{sup}=J_{sup}$, $\bar{J}_f=J_f\cup j_*$, $\bar{s}(J_n\setminus j_*)=s(J_n\setminus j_*)$.

3. OPTIMIZATION OF LINEAR DYNAMIC SYSTEMS WITH NONLINEAR INPUT. Systems of control whose eigen dynamics is linear and whose nonlinearity is generated by the input device occupy a special position in optimal control theory [1]. For them many results of qualitative theory have almost completed form as can be attained for entirely linear systems [20]. However in the constructive theory these systems are often considered together with nonlinear systems [2]. In the given report the algorithm of optimization of the above systems is constructed. It is based on the same principles that were worked out earlier for linear problems [7,14]. Other algorithms belonging to the same class as the primal algorithm described below are not known to be published in literature.

Consider the problem

$$J(u)=c'x(t_*)\longrightarrow max,\quad \dot{x}=Ax+b(u),\ x(0)=x_0,$$

$$Hx(t_1)=g,\ u(t)\in U=[-1,1],\ t\in T=[0,t_*],\ x\in R^n,\ u\in R,$$

where $x=x(t)$ is the state of the control system, $u=u(t)$ is the value of controlling influence, g is m-vector of the given values of the output signals; A,H are n x n-, m x n-matrices of the parameters of the output device, c is n-vector of the parameters of the performance index, $rank H=m$.

With respect to the function $b(u)$, $u\in U$, describing the input device of the control system we shall suppose that it has the form

$$b(u)=f_k u+f^k,\ u\in\,]v_{k-1},v_k[\,,\ f_k,f^k\in R^n,$$

$$v_k\in U,\ v_0=-1,\ v_s=1,\ v_{k-1}<v_k,\ k=\overline{1,s}.$$

At the points v_k, $k=\overline{1,s}$ the function $b(u)$ is defined on continuity from the left and the right depending on the situation (see below).

Piece-wise continuous functions $u(t)$, $t\in T$, to which (admissible) trajectories $x(t)$, $t\in T$, of the system $\dot{x}=Ax+b(u)$, $x(0)=x_0$, satisfying the terminal constraint $Hx(t_*)=g$ correspond are said to be admissible controls.

The optimization problem consists in constructing optimal control $u^0(t)$, $t\in T$ that delivers maximal value among admissible controls to the performance index $J(u)$. For applications the construction of ε-optimal (suboptimal) control $u^\varepsilon(t)$, $t\in T$, that, being admissible, satisfies the inequality $J(u^0) - J(u^\varepsilon) \le \varepsilon$, is also of great value.

Denote: $S = \{0,1,\dots,s\}$, $S_* = \{k: 0<k<s, f_k v_k + f^k \ne f_{k+1} v_k + f^{k+1}\}$, $S_0 = S \setminus S_*$. Introduce the matrix $B = (b_j, j \in J)$, $|J| = |S_0| + 2|S_*|$, consisting of n-vector-columns

$$f_1 v_0 + f^1,\ f_k v_k + f^k,\ k \in S_0 \setminus \{0\};\ f_k v_k + f^k,\ f_{k+1} v_k + f^{k+1},\ k \in S_*.$$

Let us choose a finite set of points $T_{sup} \subset T$ and ascribe the set $J_{sup}(t) \subset J$, $|J_{sup}(t)| \ge 2$ to each point $t \in T_{sup}$.

The union $Q_{sup} = \{\{t, J_{sup}\}, t \in T_{sup}\}$ is said to be the support of problem [7] if $\det P_{sup} \ne 0$,

$$P_{sup} = (HF(t_*,t)(b_j - b_{j(t)}), j \in J_{sup}(t) \setminus j(t), t \in T_{sup}), \sum_{t \in T_{sup}} |J_{sup}(t)| = m + |T_{sup}|,$$

where $j(t)$ is any element from $J_{sup}(t)$, $\partial F(t,\tau)/\partial\tau = -F(t,\tau)A$, $F(t,t) = E$, $F \in R^{n \times n}$.

The pair $\{u, Q_{sup}\}$ of admissible control and support is support control.

According to the support Q_{sup} construct the vector of potentials $y' = (c' F(t_*,t)(b_j - b_{j(t)}), j \in J_{sup}(t) \setminus j(t), t \in T_{sup})' P_{sup}^{-1}$ and the solution $\psi(t)$, $t \in T$, of the conjugate system

$$\dot\psi = -A'\psi,\quad \psi(t_*) = H'y - c.$$

This will allow to calculate the increment of the performance index on admissible controls $u(t)$, $\bar u(t) = u(t) + \Delta u(t)$, $t \in T$:

$$\Delta J(u) = J(\bar u) - J(u) = \int_0^{t_*} \psi'(t)\, \Delta_{\bar u} b(u(t))\,dt,\quad \Delta_{\bar u} b(u) = b(\bar u) - b(u).$$

From here follows

ε -maximum principle. At any $\varepsilon > 0$ for ε -optimality of admissible control $u(t)$, $t \in T$, it is necessary and sufficient that such support Q_{sup} should exist that along the support control $\{u, Q_{sup}\}$ and corresponding trajectories $x(t), \psi(t)$, $t \in T$, of the original and conjugate systems hamiltonian $H(x,\psi,u) = \psi'(Ax + b(u))$ attains ε -minimal value

$$H(x(t),\psi(t),u(t)) = \min_{u \in U} H(x(t),\psi(t),u) + \varepsilon(t),\ t \in T;$$

$$\beta(u, Q_{sup}) = \int_0^{t_*} \varepsilon(t)\,dt \le \varepsilon.$$

The proof is analogous to [7].

The support Q_{sup} is said to be regular if

$$\gamma(t) = \psi'(t)b(u^*(t)) = \min_{u \in U} \psi'(t)b(u) = \psi'(t)b_{j(t)}, \quad t \in T_{sup}.$$

Optimal support is regular. That is why further we shall consider only regular supports.

Let for the initial support control $\{u, Q_{sup}\}$ ε-minimum principle be not fulfilled. Introduce the parameter $\alpha > 0$ and construct the sets $U_\alpha(t) = \{u \in U^0: u \neq u^*(t), \psi'(t)b(u) \leq \gamma(t) - \alpha\}$, $T_0 = \{t \in T: U_\alpha(t) \neq \emptyset\}$, $U^0 = \{v_k, k = \overline{0,s}\}$.
The set $T_* = T \setminus T_0$ is divided into subsets T_{*i}, $i = \overline{1,M}$; $T_{*i} \cap T_{*j} = \emptyset$, $i \neq j$; $\bigcup_{i=1}^{M} T_{*i} = T_*$. (The ways of subdivision can be different, for example: $T_{*i} = \{t \in T_*: \gamma_{i-1} \leq \varepsilon(t) \leq \gamma_i\}$ where $\gamma_0 \geq 0$, $\gamma_i > 0$, $i = \overline{1,M}$, are the parameters of the method).

Give the parameter h and divide T_0 into segments $[\tau_i, \tau^i]$, $i=\overline{1,K}$ in such a way that $\bigcup_{i=1}^{K} [\tau_i, \tau^i] = T_0$, $0 < \tau^i - \tau_i \leq h$, $]\tau_i, \tau^i[\cap]\tau_j, \tau^j[= \emptyset$, $i \neq j$, $u(t) = const = u_i$, $t \in]\tau_i, \tau^i[$, $i = \overline{1,K}$; $T_{sup} \subset \{\tau_i, i = \overline{1,K}\}$.

Assume $s = 1$ and consider the problem

$$\sum_{i=1}^{K} \bar{g}_i(\Delta u_i) + \bar{F}_s(\theta) \longrightarrow max, \tag{1}$$

$$\sum_{i=1}^{K} g_i(\Delta u_i) + F_s(\theta) = g - \sum_{j=1}^{s-1} F_j(1),$$

$$-1 - u_i \leq \Delta u_i \leq 1 - u_i, \quad i = \overline{1,K}; \quad \theta = 0 \vee 1.$$

Here $g_i(v) = q_{ik} v + \xi_{ik}$, $\bar{g}_i(v) = \bar{q}_{ik} v + \bar{\xi}_{ik}$,

$$v \in]v_{k-1} - u_i, v_k - u_i[, \quad q_{ik} = \int_{\tau_i}^{\tau^i} HF(t_*, t) b_k \, dt,$$

$$\bar{q}_{ik} = \int_{\tau_i}^{\tau^i} c'F(t_*, t) b_k \, dt, \quad \xi_{ik} = \int_{\tau_i}^{\tau^i} HF(t_*, t)(f_k u_i + f^k) dt,$$

$$\bar{\xi}_{ik} = \int_{\tau_i}^{\tau^i} c'F(t_*, t)(f_k u_i + f^k) dt,$$

$$F_j(\theta) = \int_{T_{*j}} H(t_*, t) b(u(t) + \theta(u^*(t) - u(t))) dt,$$

$$\bar{F}_j(\theta) = \int_{T_{*j}} c'F(t_*, t) b(u(t) + \theta(u^*(t) - u(t))) dt, \quad j = \overline{1,M}.$$

Let $(\Delta u_i^s, i = \overline{1,K}, \theta_s)$ be optimal solution of problem (1). If $\theta_s = 1$ then increase s: $s \longrightarrow s+1$ and once again solve problem (1).

If $\theta_s = 0$ or $s = M$ then construct a new control $\bar{u}(t),\ t \in T$:

$$\bar{u}(t) = u(t) + \Delta u_i^s,\ t \in [\tau_i, \tau^i],\ i = \overline{1,K};$$

$$\bar{u}(t) = u^*(t),\ t \in \bigcup_{j=1}^{s-1} T_{*j};\quad \bar{u}(t) = u(t),\ t \in \bigcup_{j=s}^{M} T_{*j}.$$

Pass to operations on the change of support Q_{sup} that coincide with the iteration of the dual method [7] .

4. LINEAR-FRACTIONAL PROBLEMS. (The results are obtained together with L.F.Dezhurko.) Consider the problem [7, 21]

$$f(x) \longrightarrow max,\quad x \in X, \tag{2}$$

$$f(x) = f_1(x)/f_2(x),\quad f_1(x) = p'x + \alpha_p,\ f_2(x) = q'x + \alpha_q,$$

$$X = \{x \in R^n: Ax = b,\ d_* \le x \le d^*\},\ A = A(I,J),\ I = \{1,2,\dots,m\}, J = \{1,2,\dots,n\}.$$

Let $\{x, J_{sup}\}$ be the initial support solution and let for the sake of definiteness $f_2(x) > 0$. Calculate $u_p, u_q, \Delta_p, \Delta_q, \Delta$:

$$A'_{sup} u_p = p_{sup};\ A'_{sup} u_q = q_{sup};\ \Delta_p = A'u_p - p,\ \Delta_q = A'u_q - q; \tag{3}$$

$$\Delta = \Delta_p - f(x)\Delta_q;\ A_{sup} = A(I, J_{sup}),\ p_{sup} = p(J_{sup}),\ q_{sup} = q(J_{sup}).$$

Optimality criterion. Relations

$$\Delta_j \ge 0 \text{ at } x_j = d_{*j};\ \Delta_j \le 0 \text{ at } x_j = d_j^*;\ \Delta_j = 0 \text{ at } d_{*j} < x_j < d_j^*,$$
$$j \in J_n = J \setminus J_{sup},$$

are sufficient and in case of nondegeneracy [7] also necessary for optimality of $\{x, J_{sup}\}$.

Together with problem (2) consider a derivative problem

$$f(x) \longrightarrow max,\ x \in X_{sup} = \{x \in R^n: Ax = b,\ d_{*j} \le x_j \le d_j^*,\ j \in J_n\}. \tag{4}$$

If $\sup f(x) < \infty,\ x \in X_{sup}$, then support J_{sup} is said to be admissible.

Suboptimality criterion. For ε -optimality of solution x [7] the existence of such support J_{sup} is necessary and sufficient that $\beta(x, J_{sup}) = f(x^0) - f(x) \le \varepsilon$.

The algorithm begins with the construction of admissible support. Then following [7] the iteration $\{x, J_{sup}\} \to \{\bar{x}, \bar{J}_{sup}\}$ is derived. At the first part of the iteration the change $x \to \bar{x}$ and at the second part the change $J_{sup} \to \bar{J}_{sup}$ are realized.

5. GEOMETRIC PROGRAMMING PROBLEMS. Nonlinear problems in the terms of posynomials are widely spread in applications. There is a lot of literature on the theory and methods of geometric programming (GP) [22]. The problems of this class are also interesting as convenient objects

for testing the principles of solution of general nonlinear problems.

Consider a GP problem in a separable form:

$$g_0(z) \to \min, \; g_K(z) \leq 1, \; K \in P, \; z \in R^m, \tag{6}$$

where $g_K(z) = \sum_{j \in J_K} e^{x_j}, \; K \in P \cup \{0\}, \; x = Az + b;$

$J_K = \{m_K, m_K+1, \ldots, n_K\}, \; K \in P \cup \{0\}; \; m_0 = 1, \; m_K = n_{K-1} + 1,$
$K \in P, \; n_p = n, \; n > m, \; \operatorname{rank} A = m, \quad A \; - \; n \times m$ -matrix, $x, b \in R^n; \; p = |P|$.

For the solution of problem (6) algorithms of linear and polynomial approximation are worked out.

The primal, dual and combined algorithms of linear approximation for the solution of an unconstrained GP problem ($P=\emptyset$) are based on a new principle of choice of suitable directions. To do this the cost function of the primal (dual) GP problem is approximated by minorant (majorant) for the construction of which the information about dual (primal) problem is attracted. The direction is chosen from the minimum (maximum) condition of this minorant (majorant).

Consider this procedure in the primal algorithm for the solution of an unconstrained GP problem in more detail:

$$g(z) \to \min_{z \in R^m}, \quad g(z) = \sum_{j \in J} e^{x_j}, \quad x = Az + b. \tag{7}$$

Let $J_0 \subset J = \{1, 2, \ldots, n\}, \; |J_0| = m, \; \det A_0 \neq 0, \; A_0 = A(J_0, I),$
$I = \{1, 2, \ldots, m\}, \; J_n = J \setminus J_0, \; A_n = A(J_n, I)$.
Assume $B(J_n, J_0) = A_n A_0^{-1}, \; d(J_n) = b(J_n) - B(J_n, J_0)\, b(J_0)$.
Then we have the problem equivalent to (7)

$$g(x) = \sum_{j \in J} e^{x_j} \to \min, \; x(J_n) = B(J_n, J_0)x(J_0) + d(J_n). \tag{8}$$

Let x be some solution of problem (8) and $\lambda \in R^n$ be dual solution which satisfies the relations: $\lambda > 0, \; \lambda(J_0) + B'(J_n, J_0)\lambda(J_n) = 0$.
The function $F(y) = \max\{\varphi_1(x+y, \lambda), \varphi_2(x, y)\}$ is minorant of the cost function $g(x+y)$,

$$\varphi_1(x, \lambda) = \sum_{j \in J_0} (e^{x_j} - \lambda_j x_j) + \sum_{j \in J_n} (\lambda_j + \lambda_j d_j - \lambda_j \ln \lambda_j),$$

$$\varphi_2(x, y) = \sum_{j \in J_0} (e^{x_j + y_j} + z_j y_j) + \sum_{j \in J_n} e^{x_j}, \; z_j = \sum_{i \in J_n} b_{ij} e^{x_i}, \; j \in J_0,$$

because $g(x) \geq \varphi_1(x, \lambda)$, $g(x+y) \geq \varphi_2(x, y)$ for all admissible solutions $x, x+y$ of the primal problem and λ of the dual one. The direction y of improvement of the solution x is found from a subproblem

$$F(y) \to \min. \tag{9}$$

Problem (9) has a simple structure, its solution presents no difficulties. Further operations on calculation of the step along the direction y are realized by one of the usual approaches.

The algorithm of linear approximation for the solution of a general GP problem is based on successive approximation of the original problem by interval LP problems for the solution of which the adaptive method is used. The procedure of approximation uses separable structure of constraint functions.

The use of curvilinear directions polynomials of s-th degree: $y(\theta) = z_1\theta + z_2\theta^2/2! + \dots + z_s\theta^s/s!$ is the characteristic feature of algorithms of polynomial approximation for the solution of GP problems. Degree s is the parameter of the algorithm, θ is the step along the direction, coefficients z_k, $k=\overline{1,s}$ are determined from the Kuhn-Tucker optimality conditions in a recurrent way.

The results of the numerical experiments [23,24] allow to conclude that the suggested algorithms are highly efficient.

6. CONCLUSION. Unfortunately, the text of the report does not include the results on methods of solution of general nonlinear programming and optimal control problems. The work in this field is approaching the final stage. The results will be presented at the conference. Here we only note that the schemes worked out essentially differ from the known ones and proceed (as well as all the results given above) from the idea of maximal adaptation to the situation (consideration of all the peculiarities of a concrete problem).

A wide numerical experiment has been conducted on all the worked out methods. Because of the size restrictions the results of these experiments are not included to the present paper. They will be presented at the conference.

REFERENCES

1. Pontryagin L.S., Boltyanski V.G., Gamkrelidze R.V., Mishtchenko E.F. (1961), Mathematical Theory of Optimal Processes, Fizmatgiz, M.
2. Fedorenko R.P. (1978), Approximate Solution of Optimal Control Problems, Nauka, M.
3. Dantzig G.B. (1963), Linear Programming, Princeton Univ. Press, N.Y.
4. Kuritski B.Y., Ed. (1980), Application of Software Packages on Mathematical and Economical Methods in ASU, Statistika, M.
5. Gabasov R., Kirillova F.M., Kostyukova O.I. (1979), Adaptive Method for Solving Large Problems of Linear Programming. Preprints of IFAC-IFOPS Symp. Optimiz. Methods, Applied Aspects, pp. 163-170, Bulgaria, Varna.

6. Gabasov R., Kirillova F.M. (1981), Constructive Methods of Parametric and Functional Optimization, in: Control Science and Technology for the Progress of Society. Preprints of IFAC 8th Triennial World Congress, v. 4, pp. 111-116, Japan, Kyoto.
7. Gabasov R., Kirillova F.M. (1977, 1978, 1980), Linear Programming Methods, v. 1-3, BGU Publ. House, Minsk.
8. Senko A.A. (1982), Algorithms of Solution of Interval Linear Programming Problems, Izvestiya AN BSSR, Ser. phys.-math. nauk, N 5, 21-26.
9. Gabasov R., Glushenkov V.S. (1982), Finite μ -adaptive Exact Method for the Solution of General Linear Programming Problem, Doklady AN BSSR. 26, N 7, 589-592.
10. Fam The Long (1981), To the Multistep Methods of Linear Programming, Doklady AN BSSR. 25, N 9, 784-787.
11. Shilkina E.I. (1981), Adaptive Method for Piece-Wise Linear Programming, in: R.Gabasov, F.M.Kirillova, Eds, Optimal Control Problems, pp. 349-356, Nauka i Technika, Minsk.
12. Kostyukova O.I., Tonev P.T. (1981), Primal and Dual Methods for Dynamic Transportation Problem, in: R.Gabasov, F.M.Kirillova, Eds, Optimal Control Problems, pp. 288-305, Nauka i Technika, Minsk.
13. Guminsky A.V. (1982), Finite Exact Algorithm for the Max-Min Optimal Control Problem, Preprint N 22(147), Institute of Mathematics of BSSR Acad. of Sci., Minsk.
14. Gabasov R., Kirillova F.M., Gnevko S.V. (1983), Primal Exact Algorithm of Construction of Optimal Control in Linear Problem, Avtomatika i Telemechanika, N 8, 30-38.
15. Gabasov R., Kirillova F.M., Kostyukova O.I. (1982), Optimization of Linear Systems with Inertial Controls, Preprint N 13(138), Institute of Mathematics of the BSSR Acad. of Sci., Minsk.
16. Gabasov R., Kirillova F.M., Raketski V.M. (1981), To the Methods of Solution of General Convex Quadratic Programming Problem, Doklady AN SSSR. 258, N 6, 1289-1292.
17. Gabasov R., Kirillova F.M., Senko A.A., Eds, (1983), Software, v. 43. Adaptive Optimization, Institute of Mathematics of BSSR Acad. of Sci.
18. Gabasov R., Kirillova F.M., Kostyukova O.I. (1981), Finite Primal Exact Linear Programming Algorithm, Doklady AN BSSR, 25, N 4, 301-304.
19. Tikhonov A.N., Arsenin V.Y. (1979), Methods of Solution of Noncorrect Problems, Nauka, M.
20. Krasovski N.N. (1971), Theory of Control of Movement, Nauka, M.
21. Charnes A., Cooper W.W. (1962), Programming with Linear Fractional Functionals, Naval Res. Logist. Quart. 9, 181-186.
22. Duffin R., Peterson E., Zener K. (1967), Geometric Programming - Theory and Applications, J.Wiley & Sons, N.Y.-London-Sydney.
23. Pokatayev A.V. (1982), Algorithm of Solution of Unconstrained Geometric Programming Problem, Izvest. AN SSSR, Techn. kibernetika, N 1, 39-46.
24. Pokatayev A.V. (1981), Linear Approximation Method for the Solution of a Geometric Programming Problem, in: Gabasov R., Kirillova F.M., Eds, Optimal Control Problems, pp. 335-349, Nauka i Technika, Minsk.

NUMERICALLY STABLE ALGORITHM AND PROGRAM FOR POLE ASSIGNMENT OF LINEAR SINGLE-INPUT SYSTEMS*

P.Hr. Petkov and N.D. Christov

Department of Automatics, Higher Institute of Mechanical and Electrical Engineering, 1156 Sofia, Bulgaria

M.M. Konstantinov

Institute of Technical Cybernetics and Robotics, Akad. G. Bonchev Str., Bl. 2, 1113 Sofia, Bulgaria

ABSTRACT

A numerically stable algorithm and a program for pole assignment of linear single-input systems are presented. The algorithm is based on orthogonal reduction of the closed-loop system matrix to upper quasi-triangular form whose diagonal blocks correspond to the desired poles. The algorithm performs equally well with real and complex, distinct and multiple desired poles. The number of the necessary computational operations is less than $6n^3$, the array storage being $2n^2 + 6n$ working precision words, where n is the order of the system. The program implementation of the algorithm consists of two subroutines written in FORTRAN, which are carefuly tested for various examples of order up to 50.

INTRODUCTION

Recently three computational methods for pole assignment have been proposed in [1], [2] and [3]. These methods exploit the quasitriangular (Schur) form of the closed-loop system matrix, which is preferable from numerical point of view since it may be obtained by orthogonal transformations. Unfortunately the numerical stability of these methods was not proved and thus the problem is still not completely solved.

In this paper a new efficient computational algorithm and a program for pole assignment of linear time-invariant systems are presented. The algorithm is based on orthogonal reduction of the closed-loop system matrix to upper Schur form whose 1×1 or 2×2 diagonal blocks correspond to the desired poles. The main feature of this algorithm is that it is numerically stable which makes it applicable to ill-conditioned and high order problems.

Consider the completely controllable time-invariant single-input linear system

$$\dot{x}(t) = Ax(t) + bu(t), \qquad (1)$$

where $x(t) \in \mathbb{R}^n$, $u(t) \in \mathbb{R}^1$, $A \in \mathbb{R}^{n.n}$ and $b \in \mathbb{R}^n$. Further on the system (1) is identified with the matrix pair (A,b).

It is necessary to find the matrix $k \in \mathbb{R}^{1.n}$ of the control law $u(t) = -kx(t)$ so that the closed-loop system matrix $A_c = A - bk$ to have desired eigenvalues $s_1,\dots,s_r$, $p_1+jq_1, p_1-jq_1, \dots, p_m+jq_m, p_m-jq_m$; $j^2 = -1$, $r+2m = n$.

* AMS Subject Classification 93 B 40

A preliminary step of the algorithm for computing k is a reduction of the pair (A,b) to the orthogonal form $(A^*,b^*) = (P^TAP, P^Tb)$, $P \in O(n)$ [4], where

$$A^* = \begin{bmatrix} & a_1 & & \\ a_{21} & & a_2 & \\ & \ddots & \ddots & \\ 0 & & a_{n,n-1} & a_n \end{bmatrix}, \quad b^* = \begin{bmatrix} b_{10} \\ \\ 0 \\ \\ \end{bmatrix},$$

where $a_i \in \mathbb{R}^{1 \cdot (n-i+1)}$ and $O(n)$ is the group of orthogonal $n \times n$ matrices. In view of the complete controllability of (A^*,b^*) one has $b_{10} \neq 0$, $a_{i,i-1} \neq 0$, $i = 2,\ldots,n$.
This form may be obtained efficiently using $n-1$ Householder reflections [4] similarly to the reduction of a general matrix to its Hessenberg form [5] , the corresponding algorithm being numerically stable.

THE ALGORITHM

The pole assignment algorithm presented in this section is based on the following idea. Since the open- and closed-loop system matrices A^* and $A^*_c = A^* - b^*k^*$, $k^* = kP$, are in Hessenberg form and differ only in their first rows, it is possible by seting a desired pole to find an eigenvector of A^*_c before computing the gain matrix k^*. Using sequences of plane rotations all elements of this eigenvector except the first one may be annihilated. Then by necessity the first column of the transformed matrix A^*_c will have zero elements below the (1,1)-element which must be equal to the desired eigenvalue. This gives an equation for the first element of the transformed gain matrix. An important fact here is that after the transformation the matrices A^* and A^*_c remain in Hessenberg form which permits to work in the same way at the next step. At each step the algorithm works with a subsystem of decreasing order thus reducing the computations.

The elements of the gain matrix k are computed in the following way.

Step 1. The eigenvector of A_c corresponding to s_1 is $v_1 = Pv^*_1$, where

$$A^*_c v^*_1 = v^*_1 s_1, \quad v^*_1 = (v_{11},\ldots,v_{n1})^T. \tag{2}$$

Since the matrices A^* and A^*_c have non-zero subdiagonal elements, the eigenvector v^*_1 may be determined from a triangular system of linear equations by back substitution (by necessity the element v_{n1} is non-zero and hence may be chosen equal to 1). However the elements of the eigenvector may be computed simultaneously with its transformation exploiting the fact that some of the previous elements are already annihilated. This reduces the number of the computational operations and improves the accuracy of the eigenvector. That is why the following technique is proposed.

1.1 Compute the eigenvector elements $v_{n-1,1}$ and (if $n > 2$) $v_{n-2,1}$ from

$$v_{n-1,1} = (s_1 - a_{nn})v_{n1}/a_{n,n-1},$$

$$v_{n-2,1} = ((s_1 - a_{n-1,n-1})v_{n-1,1} - a_{n-1,n}v_{n1})/a_{n-1,n-2},$$

where a_{ij} are the elements of A^*.
Construct a plane rotation [5] $R_1 \in O(n)$ in the $(n-1,n)$-plane such

that $R_1v_1^* = v_1^1 = (x,\dots,x,v_{n-2,1},v_{n-1,1}^*,0)^T$ ((x) denotes non-referenced elements).
Since $v_{n1} = 1$ and $\beta(v_1^1) = \beta(v_1^*)$ then $v_{n-1,1}^* \geq 1$, where $\beta(v)$ is the Euclidean norm of v. It follows from (2) that

$$R_1A_c^*R_1^Tv_1^1 = v_1^1s_1 \; . \tag{3}$$

This transformation introduces an additional non-zero element in the (n,n-2)-position of the matrices $R_1A^*R_1^T$ and $R_1A_c^*R_1^T$. If $n > 2$ the vector b^* is not affected.

1.2 If $n > 3$ compute the eigenvector element $v_{n-3,1}$ from

$$v_{n-3,1} = ((s_1-a_{n-2,n-2})v_{n-2,1}-a_{n-2,n-1}v_{n-1,1}^*)/a_{n-2,n-3} \; ,$$

where a_{ij} are now the corresponding elements of $R_1A_c^*R_1^T$ and hence of $R_1A^*R_1^T$.
Construct a plane rotation $R_2 \in O(n)$ in the (n-2,n-1)-plane which annihilates the (n-1)-element of v_1^1, i.e. $R_2v_1^1 = v_1^2$, $v_1^2 = (x,\dots,x,v_{n-3,1},v_{n-2,1}^*,0,0)^T$.
Since $v_{n-1,1}^* \geq 1$ and the norm of $(v_{n-2,1},v_{n-1,1}^*)^T$ is preserved, it follows that $v_{n-2,1}^* \geq 1$. Using (3) one obtains

$$R_2R_1A_c^*R_1^TR_2^Tv_1^2 = v_1^2s_1 \; . \tag{4}$$

Now it is easily verified that the (n,n-2)-elements of the matrices $R_2R_1A_c^*R_1^TR_2^T$ and $R_2R_1A^*R_1^TR_2^T$ must be zero. In fact the last equation of system (4) yields $a_{n,n-2}v_{n-2,1}^* = 0$, where $a_{n,n-2}$ is the (n,n-2)-element of $R_2R_1A_c^*R_1^TR_2^T$. Since $v_{n-2,1}^* \neq 0$ this implies $a_{n,n-2} = 0$. The (n,n-2)-element of $R_2R_1A^*R_1^TR_2^T$ also must be zero because it is not affected by the gain matrix.

1.(n-1) As a result of computations similar to those in substeps 1.1, ...,1.(n-2) one obtains the equation

$$Q_1^TA_c^*Q_1v_1^{n-1} = v_1^{n-1}s_1 \; , \tag{5}$$

where $Q_1 = R_1^TR_2^T\dots R_{n-1}^T$, $v_1^{n-1} = (v_{11}^*,0,\dots,0)^T$, $v_{11}^* \geq 1$. The matrix $Q_1^TA^*Q_1$ may be represented as

$$Q_1^TA^*Q_1 = \left[\begin{array}{c|ccc} a_{11} & x & \dots & x \\ \hline a_{21} & & & \\ 0 & & A^{*(2)} & \\ & & & \end{array}\right]$$

where $A^{*(2)} \in \mathbb{R}^{(n-1)\cdot(n-1)}$ is in Hessenberg form. At this step b^* is reduced to $Q_1^Tb^* = (b_1^*,b_2^*,0,\dots,0)^T$.
With regard to (5) the transformed closed-loop system matrix is to be in the form

$$Q_1^T A_c^* Q_1 = \left[\begin{array}{c|c} s_1 & x \; \dots \; x \\ \hline 0 & A_c^{*(2)} \end{array}\right], \tag{6}$$

where $A_c^{*(2)}$ is a Hessenberg matrix. Since the closed-loop system is also completely controllable the element b_2^* must be non-zero. Thus the matrices $A^{*(2)}$ and $A_c^{*(2)}$ differ in their first rows only. The relation (6) yields

$$b_1^* k_1 = a_{11} - s_1, \tag{7}$$

$$b_2^* k_1 = a_{21}, \tag{8}$$

where k_1 is the first element of the row vector $k^* Q_1$. The equations (7) and (8) are algebraically consistent but in some cases (7) may be zero identity. That is why k_1 is determined from

$$k_1 = (a_{11} - s_1)/b_1^* \quad \text{if} \quad ß(b_1^*) \geq ß(b_2^*),$$

$$k_1 = a_{21}/b_2^* \quad \text{if} \quad ß(b_1^*) < ß(b_2^*).$$

In this way as a result of step 1 one element of the transformed gain matrix is obtained and the problem is reduced to a problem of dimension n-1. Since the matrices $A^{*(2)}$ and $A_c^{*(2)}$ of the (n-1)th order subsystem are in Hessenberg form it is possible to proceed further by the same way. Note that the column transformations at the next steps are to be performed on the whole $n \times n$ matrix.

Using complex plane rotations the above technique may be applied also to determine the elements of the gain matrix in case of complex conjugate poles. However it is possible to solve the problem with slightly complicated technique using real arithmetic only. As a result the transformed closed-loop system matrix will have 2×2 blocks on its diagonal. This technique is described in the following double step.

Steps (r+1),(r+2) The computation of the real x_1 and imaginery y_1 parts of the complex eigenvectors $x_1 + jy_1$, $x_1 - jy_1$ of the matrix $A_c^{*(r+1)}$, corresponding to the poles $p_1 \pm jq_1$, may be performed by the equation

$$A_c^{*(r+1)} z_1 = z_1 S_1; \; z_1 = [x_1 \;\; y_1] \in \mathbb{R}^{n.2}, \; S_1 = \begin{bmatrix} p_1 & q_1 \\ -q_1 & p_1 \end{bmatrix}. \tag{9}$$

The examination of equation (9) shows that in the case of small imaginery part of the poles the vectors x_1 and y_1 will tend to be linearly dependent which will deteriorate the solution. For this reason equation (9) is modified taking

$$S_1 = \begin{bmatrix} p_1 & 1 \\ -q_1^2 & p_1 \end{bmatrix}. \tag{10}$$

The matrix S_1 in (10) has the same eigenvalues as this in (9) but using (10) the vectors x_1 and y_1 will be linearly independent even when $q_1 = 0$. In the latter case y_1 will be determined as a principal eigenvector.

Now two elements of the transformed gain matrix may be determined simultaneously applying plane rotations to annihilate appropriate elements of x_1 and y_1. Similarly to the real case the elements of the complex eigenvector may be computed simultaneously with the annihilation of the previous elements thus reducing the number of computations.

<u>(r+1).1</u> Compute the elements $x_{n-1,1}$, $y_{n-1,1}$ and (if $n>2$) $x_{n-2,1}$, $y_{n-2,1}$ from

$$x_{n-1,1} = ((p_1-a_{nn})x_{n1}-q_1^2y_{n1})/a_{n,n-1} ,$$

$$y_{n-1,1} = (x_{n1}+(p_1-a_{nn})y_{n1})/a_{n,n-1} ,$$

$$x_{n-2,1} = ((p_1-a_{n-1,n-1})x_{n-1,1}-a_{n-1,n}x_{n1}-q_1^2y_{n-1,1})/a_{n-1,n-2} ,$$

$$y_{n-2,1} = (x_{n-1,1}+(p_1-a_{n-1,n-1})y_{n-1,1}-a_{n-1,n}y_{n1})/a_{n-1,n-2} .$$

Construct a plane rotation $U_1\in O(n)$ in the $(n-1,n)$-plane such that $U_1x_1 = x_1^1$, $x_1^1 = (x,\ldots,x,x_{n-2,1},x^*_{n-1,1},0)^T$. This transformation must be applied also on the vector y_1 and it introduces a non-zero element in the $(n,n-2)$-position of the matrices $U_1A^{*(r+1)}U_1^T$ and $U_1A_c^{*(r+1)}U_1^T$.

<u>(r+1).2</u> If $n>3$ compute the elements $x_{n-3,1}$, $y_{n-3,1}$. Construct a plane rotation $U_2\in O(n-r)$ such that

$$U_2x_1^1 = x_1^2, \quad x_1^2 = (x,\ldots,x,x_{n-3,1},x^*_{n-2,1},0,0)^T .$$

This rotation is applied also on y_1 and it introduces a non-zero element in the $(n-1,n-3)$-position of $U_2U_1A_c^{*(r+1)}U_1^TU_2^T$.

<u>(r+2).1</u> Construct a plane rotation $V_1\in O(n)$ in the $(n-1,n)$-plane to annihilate the element y_{n1}, i.e. $V_1y_1 = y_1^1$, $y_1^1 = (x,\ldots x, y^*_{n-1,1},0)^T$. This transformation does not affect the transformed vector x_1. The matrix $V_1U_2U_1A_c^{*(r+1)}U_1^TU_2^TV_1^T$ acquires a non-zero element in the $(n,n-3)$-position.

<u>(r+1).3</u> Construct a plane rotation $U_3\in O(n-r)$ to annihilate the element $x^*_{n-2,1}$ and apply this transformation to y_1. It will not destroy the form of y_1. The $(n,n-3)$-element of the matrix $U_3V_1U_2U_1A_c^{*(r+1)}U_1^TU_2^TV_1^TU_3^T$ becomes non-zero.

<u>(r+2).2</u> Compute a plane rotation $V_2\in O(n-r)$ to annihilate the element $y^*_{n-1,1}$. Then the $(n,n-2)$- and $(n-1,n-3)$-elements of the matrix $V_2U_3V_1U_2U_1A_c^{*(r+1)}U_1^TU_2^TV_1^TU_3^TV_2^T$ become zero.

This process may be continued untill the elements $x^*_{r+2,1}$ and $y^*_{r+3,1}$ are annihilated, and clearly may be considered as a QR-decomposition of the vectors x_1, y_1. Since these vectors are linearly independent ($x^*_{r+1,1},y^*_{r+2,1}\geq 1$) one obtains

$$W^T_{r+1,r+2}A_c^{*(r+1)}W_{r+1,r+2}\begin{bmatrix}1 & 0\\ 0 & 1\\ \vdots & \vdots\\ 0 & 0\end{bmatrix} = \begin{bmatrix}1 & 0\\ 0 & 1\\ \vdots & \vdots\\ 0 & 0\end{bmatrix}\overline{S}_1, \tag{11}$$

where $W_{r+1,r+2}\in O(n)$ are the transformations accumulated at the double step $(r+1),(r+2)$, and $\overline{S}_1$ is a matrix, similar to S_1. Denoting $Q_{r+1,r+2} = \mathrm{diag}(I_r,W_{r+1,r+2})\in O(n)$ it follows from (11) that the transformed closed-loop system matrix is to be in the form $Q^T_{r+1,r+2} \cdots Q^T_1 A^*_c Q_1 \cdots Q_{r+1,r+2} =$

$$
\begin{bmatrix}
s_1 & x & \cdot & \cdot & \cdot & \cdot & \cdot & \cdot & \cdot & x \\
 & \cdot & & & & & & & & \\
 & & \cdot & & & & & & & \\
 & & & \cdot & & & & & & \\
 & & & & s_r & x. & \cdot & \cdot & \cdot & x \\
 & 0 & & & & \bar{s}_1 & x. & \cdot & \cdot & x \\
 & & & & & & & A^{*(r+3)}_c & &
\end{bmatrix}
$$

where $A^{*(r+3)}_c$ is a Hessenberg matrix. The vector $(b^*_{r+1},0,\ldots,0)^T \in \mathbb{R}^{n-r}$ is reduced to $(x,b^*_{r+2},b^*_{r+3},0,\ldots,0)^T$ and the complete controllability ensures that $b^*_{r+3} \neq 0$.

Now equation (11) may be used to determine the elements k_{r+1}, k_{r+2} of the transformed gain matrix $k^*Q_1 \cdots Q_{r+1,r+2}$. As a result one obtains

$$
\begin{aligned}
b^*_{r+2}k_{r+1} &= a_{r+2,r+1} + q_1^2 y^*_{r+2,1}/x^*_{r+1,1}\,, \\
b^*_{r+2}k_{r+2} &= a_{r+2,r+2} - p_1 - q_1^2 y^*_{r+1,1}/x^*_{r+1,1}
\end{aligned}
\tag{13}
$$

and

$$
\begin{aligned}
b^*_{r+3}k_{r+1} &= a_{r+3,r+1}\,, \\
b^*_{r+3}k_{r+2} &= a_{r+3,r+2}\,,
\end{aligned}
\tag{14}
$$

where $a_{r+i,r+j}$; $i = 2,3$; $j = 1,2$, are elements of the transformed open-loop system matrix $Q^T_{r+1,r+2} \cdots Q^T_1 A^* Q_1 \cdots Q_{r+1,r+2}$.

Equations (13) and (14) are algebraically consistent and may be solved as equations (7) and (8) in the real case.

It may be observed that at this step the real and the imaginery parts of the eigenvectors are obtained as a solution of a 4-diagonal system of linear equations.

In this way the complex poles are treated in a similar manner as the real poles at the cost of a small increase of the number of computational operations.

The next steps are performed in the same way. At steps $(n-1),n$ the vector $x_m \in \mathbb{R}^2$ is transformed only once, No element of y_m is to be annihilated. The elements k_{n-1}, k_n are obtained from equations of type (13) which cannot be zero identities since the closed-loop system must be completely controllable.

Finally one obtains $k^* = (k_1,\ldots,k_n)Q^T$ and $k = k^*P^T$, where $Q = Q_1 \cdots Q_{r+1,r+2} \cdots Q_{n-1,n}$.

The algorithm presented has many common with the deflation technique [5] used to eliminate a known eigenvalue from an eigenvalue problem. This technique is very stable although the approximate eigenvector may be far from the accurate one.

The algorithm proposed also has very good numerical properties due to the fact that the computation of an eigenvector, its transformation and the determination of the gain matrix elements correspond to a small residual in the equation for this eigenvector. In this way the subdiagonal elements of the triangular form obtained are negligible and since it is exact for a matrix close to the closed-loop system matrix, this ensures the numerical stability of the algorithm (the full proof is available from the authors).

The number of necessary computational operations (including the reduction to orthogonal canonical form) is $17n^3/3$, the array storage being $2n^2 + 6n$ words.

The program implementation of the algorithm consists of the FORTRAN subroutines TRSCF and POLSC. Their single precision versions are given in the Appendix. TRSCF reduces the pair (A,b) to (A*,b*) using Householder reflections. The subroutines ORTHES and ORTRAN from EISPACK are envoked for orthogonal reduction of a matrix to upper Hessenberg form and for accumulation of the transformations. Once (A*,b*) is obtained and a desired spectrum is chosen, the subroutine POLSC determines the gain matrix k for the original system.

EXTENSION TO THE MULTI-INPUT CASE

The above algorithm may be extended to multi-input systems (A,B) using the orthogonal canonical form (A*,B*) obtained by QR-decomposition also [4]. Once (A*,B*) is found, an eigenvector of the closed-loop system is obtained by the same way as in the single-input case. This time more elements of the eigenvector are free, while the rest ones are determined from systems of linear equations, which may be underdetermined. These systems are again solved by QR-decomposition. The annihilation of the eigenvector elements is performed exactly as in the single-input case. After the transformations the resulting subsystems are again in orthogonal canonical form, and thus the process continues in the same way. However, due to the non-uniqueness of the solution for the gain matrix K, different choices of the free eigenvector elements will lead to different matrices K.

REFERENCES

1. A. Varga. A Schur method for pole assignment. IEEE Trans. Automat. Contr., AC-26 (1981), 517-519.
2. M.M. Konstantinov, P.Hr. Petkov, N.D. Christov. A Schur approach to pole assignment problem. Proc. 8th IFAC Congress, Kyoto, August 1981, 3, 1587-1592.
3. G.S. Miminis, C.C. Paige. An algorithm for pole assgnment of time invariant linear systems. Int. J. Contr., 35 (1982), 341-354.
4. M.M. Konstantinov, P.Hr. Petkov, N.D. Christov. Orthogonal invariants and canonical forms for linear controllable systems. Proc. 8th IFAC Congress, Kyoto, August 1981, 1, 49-54.
5. J.H. Wilkinson. The Algebraic Eigenvalue Problem. Clarendon Press, Oxford, 1982.

APPENDIX. PROGRAM LISTINGS

```
      SUBROUTINE TRSCF(NM,N,A,B,MATZ,Z,ORT)
C
      INTEGER I,J,K,N,II,JJ,NM,NP1
      REAL A(NM,N),B(N),Z(NM,N),ORT(N)
      REAL F,G,H,SCALE
      REAL SQRT,ABS,SIGN
      LOGICAL MATZ
C
C     THIS SUBROUTINE REDUCES THE LINEAR TIME-INVARIANT SINGLE-INPUT
C     SYSTEM
C
C        DX / DT = A * X + B * U,
C
C     WHERE  A  IS A N X N AND  B  IS A N X 1 MATRIX, INTO ORTHOGONAL
C     CANONICAL FORM USING AND OPTIONALY ACCUMULATING ORTHOGONAL
C     SIMILARITY TRANSFORMATIONS.
C
C     ON INPUT-
C
C        NM   IS AN INTEGER VARIABLE SET EQUAL TO THE ROW DIMENSION
C             OF THE TWO-DIMENSIONAL ARRAY  A  (AND  Z , IF MATZ IS
C             TRUE) AS SPECIFIED IN THE DIMENSION STATEMENT FOR  A
C             (AND  Z) IN THE CALLING PROGRAM,
C
C        N    IS AN INTEGER VARIABLE SET EQUAL TO THE ORDER OF THE
C             MATRIX  A . N MUST BE NOT GREATER THAN NM,
C
C        A    IS A WORKING PRECISION REAL TWO-DIMENSIONAL ARRAY WITH
C             ROW DIMENSION NM AND COLUMN DIMENSION AT LEAST N
C             CONTAINING THE MATRIX  A ,
C
C        B    IS A WORKING PRECISION REAL ONE-DIMENSIONAL ARRAY OF
C             DIMENSION AT LEAST N CONTAINING THE MATRIX  B ,
C
C        MATZ IS A LOGICAL VARIABLE SET EQUAL TO .TRUE. IF THE
C             ACCUMULATION OF THE ORTHOGONAL TRANSFORMATIONS IS
C             DESIRED AND SET EQUAL TO .FALSE. OTHERWISE.
C
C     ON OUTPUT-
C
C        A    CONTAINS THE CANONICAL FORM OF THE MATRIX  A . THE
C             ELEMENTS BELOW THE SUBDIAGONAL ARE SET EQUAL TO ZERO,
C
C        B    CONTAINS THE CANONICAL FORM OF THE MATRIX  B . THE
C             ELEMENTS BELOW THE FIRST ONE ARE SET EQUAL TO ZERO,
C
C        Z    IS, IF MATZ IS TRUE, A WORKING PRECISION REAL TWO-
C             DIMENSIONAL ARRAY WITH ROW DIMENSION NM AND COLUMN
C             DIMENSION AT LEAST N CONTAINING THE ORTHOGONAL
C             TRANSFORMATION MATRIX PRODUCED IN THE REDUCTION TO THE
C             CANONICAL FORM. IF MATZ IS FALSE  Z  IS NOT REFERENCED
C             AND CAN BE A DUMMY (WORKING PRECISION) VARIABLE,
C
C        ORT  IS A WORKING PRECISION REAL TEMPORARY ONE-DIMENSIONAL
C             ARRAY OF DIMENSION AT LEAST N USED TO HOLD AN INFORMATION
C             ABOUT THE ORTHOGONAL TRANSFORMATIONS DURING THE REDUCTION.
C
C     SUBROUTINES CALLED-
C
C        ORTHES, ORTRAN [EISPACK]
```

```
C
C     P.HR.PETKOV, HIGHER INSTITUTE OF MECHANICAL AND ELECTRICAL
C     ENGINEERING, SOFIA, BULGARIA.
C     THIS VERSION DATED  MAY, 1981.
C
      IF (N .EQ. 1) GO TO 190
C
C     REDUCE B
C
      H = 0.0E0
      SCALE = 0.0E0
C
      DO 10 I = 1, N
   10 SCALE = SCALE + ABS(B(I))
C
      IF (SCALE .EQ. 0.0E0) GO TO 90
      NP1 = N + 1
C
      DO 20 II = 1, N
         I = NP1 - II
         ORT(I) = B(I) / SCALE
         H = H + ORT(I) * ORT(I)
   20 CONTINUE
C
      G = -SIGN(SQRT(H),ORT(1))
      H = H - ORT(1) * G
      ORT(1) = ORT(1) - G
C
C     TRANSFORM  A
C
      DO 50 J = 1, N
         F = 0.0E0
C
         DO 30 II = 1, N
            I = NP1 - II
            F = F + ORT(I) * A(I,J)
   30    CONTINUE
C
         F = F / H
C
         DO 40 I = 1, N
   40    A(I,J) = A(I,J) - F * ORT(I)
C
   50 CONTINUE
C
      DO 80 I = 1, N
         F = 0.0E0
C
         DO 60 JJ = 1, N
            J = NP1 - JJ
            F = F + ORT(J) * A(I,J)
   60    CONTINUE
C
         F = F / H
C
         DO 70 J = 1, N
   70    A(I,J) = A(I,J) - F * ORT(J)
C
   80 CONTINUE
C
      ORT(1) = SCALE * ORT(1)
      B(1) = SCALE * G
```

```
C
C     REDUCE  A
C
   90 CALL ORTHES(NM,N,1,N,A,ORT)
C
C     ACCUMULATE THE TRANSFORMATIONS OF  A
C
      IF (MATZ) CALL ORTRAN(NM,N,1,N,A,ORT,Z)
      IF (N .LT. 3) GO TO 120
C
      DO 110 I = 3, N
         K = I - 2
C
         DO 100 J = 1, K
  100    A(I,J) = 0.0E0
C
  110 CONTINUE
C
  120 IF (B(1) .EQ. 0.0E0) GO TO 200
      IF (.NOT. MATZ) GO TO 170
C
C     ACCUMULATE THE TRANSFORMATION OF  B
C
      DO 130 I = 2, N
  130 ORT(I) = B(I)
C
      DO 160 J = 1, N
         G = 0.0E0
C
         DO 140 I = 1, N
  140    G = G + ORT(I) * Z(I,J)
C
         G = (G / ORT(1)) / B(1)
C
         DO 150 I = 1, N
  150    Z(I,J) = Z(I,J) + G * ORT(I)
C
  160 CONTINUE
C
  170 DO 180 I = 2, N
  180 B(I) = 0.0E0
C
      GO TO 200
C
C     1 X 1 CASE
C
  190 IF (MATZ) Z(1,1) = 1.0E0
  200 RETURN
      END
```

```
      SUBROUTINE POLSC(NM,N,A,B,G,WR,WI,Z,IERR,RV1,RV2)
C
      INTEGER I,J,K,L,N,KK,LL,NI,NJ,NL,NM,LM1,LP1,IERR
      REAL A(NM,N),B(N),G(N),WR(N),WI(N),Z(NM,N),RV1(N),RV2(N)
      REAL P,Q,R,S,T,B1,ZZ
      REAL SQRT,ABS
      INTEGER MAXO,MINO
      LOGICAL COMPL
C
C     THIS SUBROUTINE DETERMINES THE STATE FEEDBACK MATRIX  G  OF THE
C     LINEAR TIME-INVARIANT SINGLE-INPUT SYSTEM
C
C        DX / DT = A * X + B * U,
C
C     WHERE  A  IS A N X N AND  B  IS A N X 1 MATRIX, SUCH THAT THE
C     CLOSED-LOOP SYSTEM
C
C        DX / DT = (A - B * G) * X
C
C     HAS DESIRED POLES. THE SYSTEM MUST BE PRELIMINARY REDUCED TO
C     ORTHOGONAL CANONICAL FORM USING THE SUBROUTINE TRSCF.
C
C     ON INPUT-
C
C        NM     IS AN INTEGER VARIABLE SET EQUAL TO THE ROW DIMENSION
C               OF THE TWO-DIMENSIONAL ARRAYS  A  AND  Z  AS SPECIFIED
C               IN THE DIMENSION STATEMENTS FOR  A  AND  Z  IN THE
C               CALLING PROGRAM,
C
C        N      IS AN INTEGER VARIABLE SET EQUAL TO THE ORDER OF THE
C               MATRICES  A  AND  Z . N MUST BE NOT GREATER THAN NM,
C
C        A      IS A WORKING PRECISION REAL TWO-DIMENSIONAL ARRAY WITH
C               ROW DIMENSION NM AND COLUMN DIMENSION AT LEAST N
C               CONTAINING THE CANONICAL FORM OF THE MATRIX  A ,
C
C        B      IS A WORKING PRECISION REAL ONE-DIMENSIONAL ARRAY OF
C               DIMENSION AT LEAST N CONTAINING THE CANONICAL FORM OF
C               THE MATRIX  B ,
C
C        WR,WI ARE WORKING PRECISION REAL ONE-DIMENSIONAL ARRAYS OF
C               DIMENSION AT LEAST N CONTAINING THE REAL AND IMAGINERY
C               PARTS, RESPECTIVELY, OF THE DESIRED POLES. THE POLES
C               CAN BE UNORDERED EXCEPT THAT THE COMPLEX CONJUGATE
C               PAIRS OF POLES MUST APPEAR CONSECUTIVELY. NOTE THAT
C               ON OUTPUT THE IMAGINERY PARTS OF THE POLES MAY BE
C               MODIFIED,
C
C        Z      IS A WORKING PRECISION REAL TWO-DIMENSIONAL ARRAY WITH
C               ROW DIMENSION NM AND COLUMN DIMENSION AT LEAST N
C               CONTAINING THE ORTHOGONAL TRANSFORMATION MATRIX PRODUCED
C               IN TRSCF WHICH REDUCES THE SYSTEM TO CANONICAL FORM.
C
C     ON OUTPUT-
C
C        A      CONTAINS THE UPPER (QUASI) TRIANGULAR FORM OF THE CLOSED-
C               LOOP SYSTEM MATRIX  A - B * G , THAT IS TRIANGULAR EXCEPT
C               OF POSSIBLE 2 X 2 BLOCKS ON THE DIAGONAL,
```

```
C
C        B      CONTAINS THE TRANSFORMED MATRIX  B,
C
C        G      IS A WORKING PRECISION REAL ONE-DIMENSIONAL ARRAY OF
C               DIMENSION AT LEAST N CONTAINING THE STATE FEEDBACK
C               MATRIX G OF THE ORIGINAL SYSTEM,
C
C        Z      CONTAINS THE ORTHOGONAL MATRIX WHICH REDUCES THE
C               CLOSED-LOOP SYSTEM MATRIX  A - B * G  TO THE UPPER
C               (QUASI) TRIANGULAR FORM,
C
C        IERR   IS AN INTEGER VARIABLE SET EQUAL TO
C               ZERO  AT NORMAL RETURNING,
C               1     IF THE SYSTEM IS NOT COMPLETELY CONTROLLABLE,
C
C        RV1    IS A WORKING PRECISION REAL TEMPORARY ONE-DIMENSIONAL
C               ARRAY OF DIMENSION AT LEAST N. IT HOLDS THE REAL PARTS
C               OF THE EIGENVECTORS DURING THE REDUCTION,
C
C        RV2    IS A WORKING PRECISION REAL TEMPORARY ONE-DIMENSIONAL
C               ARRAY OF DIMENSION AT LEAST N. IT HOLDS THE IMAGINERY
C               PARTS OF THE EIGENVECTORS DURING THE REDUCTION.
C
C     SUBROUTINES CALLED-
C
C        NONE
C
C     P.HR.PETKOV, HIGHER INSTITUTE OF MECHANICAL AND ELECTRICAL
C     ENGINEERING, SOFIA, BULGARIA.
C     THIS VERSION DATED  MAY, 1981.
C
      IERR = 0
C
C     CHECK FOR COMPLETE CONTROLLABILITY
C
      IF (B(1) .EQ. 0.0E0) GO TO 250
      IF (N .EQ. 1) GO TO 240
C
      DO 10 I = 2, N
         IF (A(I,I-1) .EQ. 0.0E0) GO TO 250
   10 CONTINUE
C
      B1 = B(1)
      B(1) = 1.0E0
      L = 0
   50    L = L + 1
         COMPL = WI(L) .NE. 0.0E0
         IF (L .EQ. N) GO TO 140
         LM1 = L - 1
         LP1 = L + 1
         NL = N - L
         IF (NL .GT. 1) NL = NL + 1
         IF (COMPL) GO TO 60
         RV1(N) = 1.0E0
         GO TO 70
   60    RV1(N) = 1.0E0
         RV2(N) = 1.0E0
         T = WI(L)
         WI(L) = 1.0E0
         WI(L+1) = T * WI(L+1)
```

```
C
C        COMPUTE AND TRANSFORM EIGENVECTOR
C
   70    DO 130 KK = 1, NL
            LL = 1
            K = N - KK
            IF (K .EQ. LM1) GO TO 80
            RV1(K) = (WR(L) - A(K+1,K+1)) * RV1(K+1)
            IF (KK .GT. 1) RV1(K) = RV1(K) - A(K+1,K+2) * RV1(K+2)
            IF (COMPL) RV1(K) = RV1(K) + WI(L+1) * RV2(K+1)
            RV1(K) = RV1(K) / A(K+1,K)
            IF (.NOT. COMPL) GO TO 80
            RV2(K) = (WR(L+1) - A(K+1,K+1)) * RV2(K+1)
     X                                         + WI(L) * RV1(K+1)
            IF (KK .GT. 1) RV2(K) = RV2(K) - A(K+1,K+2) * RV2(K+2)
            IF (KK .GT. 2) RV2(K) = RV2(K) - A(K+1,K+3) * RV2(K+3)
            RV2(K) = RV2(K) / A(K+1,K)
   80       IF (KK .EQ. 1 .AND. NL .GT. 1) GO TO 130
            IF (KK .GT. 1 .AND. LL .EQ. 1) K = K + 1
            P = RV1(K)
            IF (LL .EQ. 2) P = RV2(K)
            Q = RV1(K+1)
            IF (LL .EQ. 2) Q = RV2(K+1)
            S = ABS(P) + ABS(Q)
            P = P / S
            Q = Q / S
            R = SQRT(P*P+Q*Q)
            T = S * R
            IF (K .GT. L .AND. LL .EQ. 1) RV1(K-1) = RV1(K-1) / T
            RV1(K) = 1.0E0
            IF (LL .EQ. 2) RV2(K) = T
            P = P / R
            Q = Q / R
            NJ = MAX0(K-1,L)
            IF (LL .EQ. 2) NJ = MAX0(K-2,L)
C
C           TRANSFORM  A
C
            DO 90 J = NJ, N
               ZZ = A(K,J)
               A(K,J) = P * ZZ + Q * A(K+1,J)
               A(K+1,J) = P * A(K+1,J) - Q * ZZ
   90       CONTINUE
C
            NI = MIN0(K+2,N)
            IF (COMPL .AND. LL .EQ. 1) NI = MIN0(K+3,N)
C
            DO 100 I = 1, NI
               ZZ = A(I,K)
               A(I,K) = P * ZZ + Q * A(I,K+1)
               A(I,K+1) = P * A(I,K+1) - Q * ZZ
  100       CONTINUE
C
            IF (K .GT. LP1) GO TO 110
            IF (K .EQ. LP1 .AND. LL. EQ. 1) GO TO 110
C
C           TRANSFORM  B
C
            ZZ = B(K)
            B(K) = P * ZZ
            B(K+1) = -Q * ZZ
```

```
C
C           ACCUMULATE TRANSFORMATIONS
C
  110       DO 120 I = 1, N
               ZZ = Z(I,K)
               Z(I,K) = P * ZZ + Q * Z(I,K+1)
               Z(I,K+1) = P * Z(I,K+1) - Q * ZZ
  120       CONTINUE
C
            IF (.NOT. COMPL .OR. LL .EQ. 2) GO TO 130
            ZZ = RV2(K)
            RV2(K) = P * ZZ + Q * RV2(K+1)
            RV2(K+1) = P * RV2(K+1) - Q * ZZ
            IF (KK .GT. 1 .AND. K .GT. L) RV2(K-1) = RV2(K-1) / T
            RV2(K) = RV2(K) / T
            RV2(K+1) = RV2(K+1) / T
            IF (KK .LT. 3) GO TO 130
            RV2(K+2) = RV2(K+2) / T
            K = K + 1
            LL = LL + 1
            GO TO 80
  130    CONTINUE
C
  140    IF (COMPL) GO TO 170
C
C        FIND ONE ELEMENT OF  G
C
         K = L
         R = B(L)
         IF (L .EQ. N) GO TO 150
         IF (ABS(B(L+1)) .LE. ABS(B(L))) GO TO 150
         K = L + 1
         R = B(L+1)
  150    P = A(K,L)
         IF (K .EQ. L) P = P - WR(L)
         P = P / R
C
         DO 160 I = 1, LP1
  160    A(I,L) = A(I,L) - P * B(I)
C
         G(L) = P / B1
         IF (L .EQ. N) GO TO 200
         GO TO 50
C
C        FIND TWO ELEMENTS OF  G
C
  170    L = L + 1
         IF (L .LT. N) LP1 = L + 1
         K = L
         R = B(L)
         IF (L .EQ. N) GO TO 180
         IF (ABS(B(L+1)) .LE. ABS(B(L))) GO TO 180
         K = L + 1
         R = B(L+1)
  180    P = A(K,L-1)
         IF (K .EQ. L) P = P - (RV2(L) / RV1(L-1)) * WI(L)
         Q = A(K,L)
         IF (K .EQ. L) Q = Q - WR(L) + (RV2(L-1) / RV1(L-1)) * WI(L)
         P = P / R
         Q = Q / R
```

```
C
         DO 190 I = 1, LP1
            A(I,L-1) = A(I,L-1) - P * B(I)
            A(I,L) = A(I,L) - Q * B(I)
  190    CONTINUE
C
         G(L-1) = P / B1
         G(L) = Q / B1
         IF (L .EQ. N) GO TO 200
      GO TO 50
C
C     TRANSFORM  G
C
  200 DO 220 I = 1, N
         S = 0.0E0
C
         DO 210 J = 1, N
  210    S = S + G(J) * Z(I,J)
C
         RV1(I) = S
  220 CONTINUE
C
      DO 230 I = 1, N
         B(I) = B1 * B(I)
         G(I) = RV1(I)
  230 CONTINUE
C
      GO TO 260
C
C     1 X 1 CASE
C
  240 P = A(1,1) - WR(1)
      A(1,1) = A(1,1) - P
      G(1) = P / B(1)
      GO TO 260
C
C     SET ERROR -- THE SYSTEM IS NOT COMPLETELY CONTROLLABLE
C
  250 IERR = 1
  260 RETURN
      END
```

Session 12

STOCHASTIC CONTROL

CONTRÔLE STOCHASTIQUE

STOCHASTIC CONTROL WITH STATE CONSTRAINTS AND NON LINEAR ELLIPTIC EQUATIONS WITH INFINITE BOUNDARY CONDITIONS.

by Jean-Michel LASRY and Pierre-Louis LIONS,
CEREMADE, Université Paris-Dauphine.

§ I . Introduction.

Let us start with a standard problem of stochastic control in a bounded regular open set O in $\mathbb{R}^N$:

(1) $$\text{Minimize } J(x,\xi) \quad , \quad \xi \in C$$

where the cost function J is defined by

(2) $$J(x,\xi) = E \int_0^{T} e^{-\lambda t} \left[f(X_t)+g(\xi(X_t)) \right] dt + e^{-\lambda T} \varphi(X_T)$$

where $x \in \mathbb{R}^p$, $f \in C(O,\mathbb{R})$, $g \in C(\mathbb{R}^N,\mathbb{R}^N)$, $\lambda > 0$ are given data , and where the state ξ is driven by the stochastic differential equation

(3) $$X_o = x \quad , \quad dX_t = \xi(X_t)dt + \sqrt{2}\, dB_t$$

Here T is the exit time (first t such that $X_t \in \partial O$) .

Under reasonnable regularity and growth condition on f,g,φ and provide the class C of admissible feedback is large enough, the Bellman function V_λ defined by

(4) $$V_\lambda(x) = \inf J(x,\xi) \quad , \quad \xi \in C$$

is the unique solution of (5), (6) :

(5) $\quad - \Delta u + \lambda u + g(\nabla u) = f \quad$ in 0

(6) $\quad u = \varphi \quad$ on the boundary $\partial 0$

(7) $\quad$ with $g^{\star}(y) = \text{Sup}\,\{xy - g(x)\ ;\ x \in \mathbb{R}^{p}\}$

We will be interested here by different form of the following requirement : keep the state X_t in 0 . This constraint can be explicit in the definition of C if we restrict the class of admissible controls to those for which $T \equiv +\infty$ (with probability 1). The requirement can be implicit : put $\varphi \equiv +\infty$ then by the definition of (2) the minimization (1) will lead to consider only the controls such that $T \equiv +\infty$. "More implicit" will be the case where the function f blow up at the boundary $\partial 0$ so fast that the state X cannot reach the boundary with finite cost $J(x,\xi)$. These general considerations will be detailed in [. 1] . They motivate the study of equation (5) with the unusual boundary condition : $u = +\infty$ on $\partial 0$, or with blowing up function f (see § II).

§ II . Infinite boundary conditions.

For the sake of simplicity we shall reduce here to the case where

(8) $\quad g(a) = \frac{1}{q}\,|a|^{q} \qquad \forall\, a \in \mathbb{R}^{N} \qquad (1 < q < +\infty)$

so that the dual function $g^{\star}$ (see (7)) is :

(9) $\quad g^{\star}(b) = \frac{1}{p}\,|b|^{p} \qquad \forall\, b \in \mathbb{R}^{N} \qquad \frac{1}{p} + \frac{1}{q} = 1$

Theorem 1. We suppose $1 < p \leqslant 2$, $f \in C^{\gamma}(0)$ with $0 < \gamma \leqslant 1$, f bounded below and :

(10) $\quad f(x)\, d(x)^{-q} \to 0 \qquad$ when $\quad d(x) \to 0$

(11) $\quad$ where $d(x)$ is the distance to $\partial 0$,

(regularity of O will mean : d is C^2 in a neighborhood of ∂O). Then there exists a unique solution u of (12), (13) :

$$(12) \qquad - \Delta u + \lambda u + \frac{1}{p} |\nabla u|^p = f \qquad \text{in } O \qquad (u \in C^2(O))$$

$$(13) \qquad u(x) \to +\infty \qquad \text{when} \qquad d(x) \to 0$$

Actually this solution will verify the more precise boundary condition

$$(14) \qquad u(x)\, d(x)^{-\alpha} \to C_o \qquad \text{when} \qquad d(x) \to 0$$

with $\alpha = (2-p)/(p-1)$ and $C_o = (p-1)^{-\alpha} (2-p)^{-1/(p-1)}$.

The proof use mainly comparison arguments : one of them being that $(C_o+\varepsilon)d^{-\alpha} + C_\varepsilon$ (resp. $(C_o-\varepsilon)d^{-\alpha} - C_\varepsilon$) is a supersolution (resp. subsolution) of (12).

Hence this case (see § III) corresponds to the idea that $\varphi \equiv +\infty$ (i.e. : (6) replaced by (13)) traduces well the state constraint "$T \equiv +\infty$" . But this is not always the case ; for example if $p > 2$ the results are completely different :

<u>Theorem 2</u>. Let $p > 2$ and f be lipschitz continuous. Then any solution u of (12) in O which is bounded below can be extended to $\overline{O}$ with $u \in C^\alpha(\overline{O})$, $\alpha = (2-p)/(p-1)$. Moreover there exists a maximum solution $\bar{u} \in C^\alpha(\overline{O})$ (i.e. : $\bar{u} \geq u$ for all other solution u of (12)).

When the open set O and the function f are convex we can prove that

$$(15) \qquad \frac{\partial \bar{u}}{\partial n} \to +\infty \qquad \text{when} \qquad d(x) \to 0$$

(n is the exterior unit normal). We conjecture that (15) always hold.

The proof of Theorem 2 relies heavily on estimates of $|\nabla u|$ due to P.L. Lions (see [3]).

Let us turn now to the case of blowing up functions f :

Theorem 3. If $p > 1$, f locally lipschitz and

(16) $$f(x)\, d(x)^{-\beta} \to c_1 \quad \text{when} \quad d(x) \to 0$$

with $c_1 > 0$ and $\beta > \max(p,q)$ then there exists a unique solution u of (12) such that

(17) u is bounded below .

In this case (17) stands as a very weak boundary condition (without (17) unicity does not hold). In fact the solution u of (12)-(17) verifies

(18) $$u(x)\, d(x)^{-\alpha} \to c_2 \quad \text{when} \quad d(x) \to 0$$

with $\alpha = (\beta-p)/p$ and $c_2 = c_1^{1/p}\, p^{1/p}\, \alpha^{-1}$.

The first step of the proof is to show that any solution of (12)-(17) verifies: $u(x) \to +\infty$ when $d(x) \to 0$ (this is done by comparison with explicit subsolutions). Then the proof is very similar to that of Theorem 1.

§ III . Stochastic control.

Let us now come back to the stochastic control problem. Let C_o be the class of all $\xi \in W^{1,\infty}_{loc}(0)$ such that the solution X_t of the differential equation (3)

(3) $$X_o = x \quad , \quad dX_t = \xi(x)dt + \sqrt{2}\, dB_t$$

does not reach the boundary, i.e. : the exit time T is infinite with probability 1. Hence for $\xi \in C_o$ the cost J writes

(2) $$J(x,\xi) = \int_o^{+\infty} e^{-\lambda t} \left[f(X_t)+g(\xi(X_t)) \right] dt$$

Define the Bellman function V_λ by

(19) $$V_\lambda(x) = \inf \{J(x,\xi) , \xi \in C_o\}$$

Then in the cases investigated previously (§ II) the Bellman function V_λ is characterized by the various boundary conditions which appeared. More precisely :

Theorem 4.

a) Under the hypothesis of theorem 1, the Bellman function V_λ is the unique solution of

(12) $$- \Delta V_\lambda + \lambda V_\lambda + \frac{1}{p} |\nabla V_\lambda|^p = f \qquad \text{in } 0$$

such that : $V_\lambda \to +\infty$ when $d(x) \to 0$.

b) Under the hypothesis of Theorem 2, the Bellman function V_λ is the maximum solution of (12) (i.e. : $V_\lambda = \bar{u}$ - see theorem 2).

c) Under the hypothesis of theorem 3, the Bellman function V_λ is the unique solution of (12) which is bounded below.

Of course in all cases there is a unique optimal feedback $\xi_o \in C_o$. This optimal feedback is given as usual by $\xi_o = -Dg^\star \circ DV_\lambda$, here :

$$\xi_o(x) = - |DV_\lambda(x)|^{p-2} DV_\lambda(x) \qquad \forall x \in 0$$

(as a side effect this confirm that $C_o \neq \emptyset$) .

Finally let us mention that this type of results hold for the "ergodic" problem, i.e. : $\lambda \to 0_+$ (see [1]).

[1] J.M. LASRY et P.L. Lions, note au C.R.A.S., 1984, and detailed work in preparation.

[2] P.L. LIONS, Arch. Rat. Mech. Anal., 74 (1980), p. 335-353.

[3] P.L. LIONS, On quasilinear elliptic equations, in preparation.

OPTIMAL STOPPING WITH CONSTRAINT

Monique PONTIER

Département de Mathématiques
Université d'Orléans
45046 - ORLEANS CEDEX

Jacques SZPIRGLAS

Centre National d'Etudes des
Télécommunications, PAA/TIM/MTI
38-40, rue du Général Leclerc
92131 - ISSY-LES-MOULINEAUX

Abstract. In this paper, we give a solution to the following constrained optimal stopping problem. Let Y and Y' be two bounded non negative right-continuous-left-limited processes, and let a be a non negative real number. The average reward $E(Y_T)$ is maximized within the class of stopping times T satisfying the constraint

$$E(Y'_T) \geqq \sup_S E(Y'_S) - a.$$

Under some regularity conditions, it is shown that there exist solutions in the set of randomized stopping times, including the set of stopping times, by Lagrangian saddle-point methods.
This can be applied to cases where Y' is a sub-martingale, an upper-martingale, or a characteristic function as for example $\mathbb{1}(t < D)$ where D is a totaly inaccessible stopping time.

I. INTRODUCTION.

Let $(\Omega, \underline{\underline{A}}, \mathbb{P})$ be a probability space endowed with a filtration $\underline{\underline{F}} = (\underline{\underline{F}}_t \; ; \; t \geq 0)$ satisfying the usual conditions (5), Y and Y' be two reward processes defined on $(\Omega, \underline{\underline{A}}, \underline{\underline{F}}, \mathbb{P})$ that we assume bounded and non negative adapted right-continuous-left-limited (cadlag) up to infinity.

Let $\underline{\underline{T}}$ denote the set of $\underline{\underline{F}}$-stopping times (s.t.) and $\underline{\underline{T}}^a$ the set of s.t. T such that :

$$E(Y'_T) \geq \sup_{S \in \underline{\underline{T}}} E(Y'_S) - a$$

for a strictly positive real number a .

The problem is to find a.s.t. T^* which maximizes within class $\underline{\underline{T}}^a$ the average reward $E(Y_T)$, i.e. :

$$E(Y_{T^*}) = \sup (E(Y_T) \; ; \; T \in \underline{\underline{T}}^a)$$

We call such a s.t. T^* an optimal stopping time for the a-constrained stopping problem.

The cases where Y' is a characteristic function provide many applications: the goal is to find stopping rules so that the expected cost is minimized while (because of limiting values depending on the state of the system or the environment) a target set has to be reached with a sufficiently large probability. For instance, for a failure detection problem (cf (18)), the average cost is to be minimized while the probability of false alarm has to be smaller than a threshold; or, for an economic system, the average product is to be maximized and the probability that max inflation rate exceeds some fixed rate has to be smaller than a threshold.

As in the general optimal stopping problems (see (2), (7), (18)) no solution can be found in $\underline{\underline{T}}$ but there is one in the convex hull of $\underline{\underline{T}}$ where Convex Analysis methods (see (6), (17)) can be applied.

Let V denote the set of cadlag adapted bounded functions up to infinity and V' the set of continuous positive linear functional μ on V such that $\mu(1) = 1$. Let us note that Y and Y' belong to V ; then it is possible to define subset V^a of V' :

$$V^a = \{\mu \in V' \; ; \; \mu(Y') \geq \sup (\nu(Y') \; ; \; \nu \in V') - a\}$$

It can be easily seen that to each T in $\underline{\underline{T}}$ is associated a functional of V' :

$$Z \longrightarrow E(Z_T)$$

Therefore $\underline{\underline{T}}^a$ may be embedded in V^a and we define a new optimization problem:

(*) To find a functional μ^* in V^a which maximizes within set V^a $\mu(Y)$, i.e. :

$$\mu^*(Y) = \sup\ (\mu(Y)\ ;\ \mu \in V^a\ ).$$

Actually we will find a solution in the set of randomized stopping times, i.e. :

Definition I.1. A linear functional μ of V' is called a randomized stopping time (r.s.t.) if it is associated to triplet (b, T_1, T_2) where b is a real number in $[0,1]$, T_1 and T_2 are two stopping times, such that:

$$Z \in V\ ,\quad \mu(Z) = b\, E(Z_{T_1}) + (1-b)E(Z_{T_2})$$

The constrained optimal stopping problems have not been studied very much. Let us quote ROBIN (16) who was concerned with particular case $Y'_t = e^{-\alpha t}$ and KENNEDY (10) who studied by a similar method the problem with discrete time :

$$\sup\ \{\ E(Y_T)\ ;\ \ T\ \text{s.t.}\ /\ E(T) < n\ \}$$

Our paper replaces in a general framework the examples of (18) concerned with sequential testing and disorder problem with a threshold for the false alarm probability. Let us quote some papers concerned with constrained stochastic control : average or sample constraints on the final state of the controlled process (see (4), (8), (9), (11), (13), (15)), or constraints on the control (see (2), (9)).

At the end of this part we recall some general definitions and results on optimal stopping. In the second part, by means of Convex Analysis, we give some existence results for optimal and ε-optimal linear functionals. In the last part three examples are given.

We write down first some definitions and results in the theory of optimal stopping that we need.

Definition I.2. Process Z is a regular cadlag process if and only if for all monotonous sequence of s.t. T_n converging a.s. to T :

$$E(Z_T) = \lim_{n \to \infty} E(Z_{T_n})$$

We assume by now that processes Y and Y' are bounded adapted non negative regular cadlag processes. More general hypotheses can be found in (14).

The main tool in optimal stopping theory is the Snell envelope (see (7), (12)) :

Definition I.3. We call Snell envelope J of process Z the smallest strong uppermartingale bigger than Z such that :

$$\forall\, T \in \underline{\underline{T}} \qquad J_T = \operatorname*{ess\,sup}_{S \geq T} E(Z_S \,/\, \underline{\underline{F}}_T).$$

We mean by strong uppermartingale a process J such that :

$$\forall\, S \geq T\,,\quad S \text{ and } T \in \underline{\underline{T}}\,,\quad E(J_S \,/\, \underline{\underline{F}}_T) \leq J_T$$

Then an optimal s.t. for the classical optimal stopping problem associated to a reward process Z exists :

Proposition I.4. Let Z be a cadlag regular bounded non negative adapted process and J its Snell envelope.
Then the entry time D in the set $\{ (t,\omega) \,/\, J_t(\omega) = Z_t(\omega)\}$:

$$D = \inf\{t \geq 0 \,/\, J_t = Z_t\}$$

is an optimal stopping time for the stopping problem associated to Z, i.e. :

$$E(Z_D) = \sup\, (E(Z_T) \;;\; T \in \underline{\underline{T}})$$

II. OPTIMAL STOPPING WITH CONSTRAINTS ; LAGRANGIAN FORMULATION.

We first notice that the new optimization problem (*) in V' has a solution. First, the set V^a is not void because a is strictly positive. Then it is easy to see that V^a is a weakly compact set of V' where the continuous application $\mu \longrightarrow \mu(Y)$ realizes its supremum. By a Lagrangian approach we obtain a more precise expression of an optimal solution in the set of randomized stopping times.

Let us define the Lagrangian L on $V' \times \mathbb{R}^+$:

$L(\mu,p) = \mu(Y) + p(\mu(Y') - \sup\,(\nu(Y') \;;\; \nu \in V') + a)$

Let Y^p denote in the sequel:

$Y^p = Y + pY'$,

J, J' and J^p the Snell envelopes of Y, Y' and Y^p respectively. Let us notice that (see (3)) :

$$E(J'_o) = \sup\,(\nu(Y') \;;\; \nu \in V').$$

By convexity considerations it is easy to see that L is endowed with a saddle point (μ^*, p^*) :

$$\forall\, p \in \mathbb{R}^+,\quad \mu \in V'\,,\quad L(\mu,p^*) \leq L(\mu^*, p^*) \leq L(\mu^*, p).$$

Functional μ^* is then optimal for problem (*). The next proposition will provide a characterization of such optimal functionals.

Proposition II.1. A linear functional μ^* of V' is optimal for problem (*) if and only if one of the two conditions is satisfied :

(i) $\mu^*(Y) = \sup (\mu(Y) \; ; \; \mu \in V')$ and $\mu^*(Y') \geqq E(J'_o) - a$.

(ii) There exists a strictly positive real number p^* such that :

$$\mu^*(Y^P) = \sup (\mu \; (Y^P) \; ; \; \mu \in V') \text{ and } \mu^*(Y') = E(J'_o) - a.$$

This is a direct consequence of the saddle point inequalities. Conditions (i) and (ii) correspond to a saddle point (μ^*, p^*) with p^* equal to zero or p^* strictly positive, respectively.

We use then the constructive method of (8) and (16) to find an ε-optimal μ_a^ε of V' , i.e. such that :

$$\mu_a^\varepsilon \in V^a \text{ and } \mu_a^\varepsilon(Y) \geqq \sup \; (\mu(Y) \; ; \; \mu \in V^a)-\varepsilon.$$

By means of hypotheses on Y and Y' and proposition I.4, we know that the stopping problem with respect to reward process Y^P has a solution D_p :

$$D_p = \inf \{t \geqq 0 \; / \; J_t^p = Y_t^p \}$$

The following lemma is essential in the sequel:

Lemme II.2. Mapping $p \to E(Y'_{D_p})$ is increasing to $E(J'_o)$.

The increasingness is easy to prove using the optimality of D_p. The limit $E(J'_o)$ is deduced from:

$$E(Y'_{D_p}) \geqq E(Y'_T) + \frac{1}{p} \; (E(Y_T) - E(Y_{D_p}))$$

for any T in $\underline{\underline{T}}$.

Remark. From proposition I.4, when D_o satisfies the constraint, D_o is optimal for the constrained problem. The same is valid for s.t. D_p, when p is strictly positive and :

$$E(Y'_{D_p}) = E(J'_o) - a.$$

We may deduce from that the following corollary :

Corollary II.3. Assume that mapping $p \to E(Y'_{D_p})$ is continuous. Then there exists an optimal stopping time for the constrained stopping problem.

This is the case in the disorder problem for a Wiener process (18) when we do not have this continuity, we generally get :

Proposition II.4. <u>For all strictly positive</u> ε <u>and</u> a, <u>an</u> ε<u>-optimal randomized stopping time</u> μ_a^ε <u>can be constructed such that</u> :

$$\mu_a^\varepsilon(Y) \geqq \sup(\mu(Y) ; \mu\varepsilon V^a) - \varepsilon \text{ and } \mu_a^\varepsilon(Y') = E(J'_o) - a.$$

Proof. Assume D_o is not a solution. Then from lemma II.2 there exist two real numbers :

$$p = n \ \varepsilon/a \quad \text{and} \quad q = (n+1)\varepsilon/a$$

such that:

$$E(Y'_{D_p}) < E(J'_o) - a \quad \text{and} \quad E(Y'_{D_q}) \geqq E(J'_o) - a.$$

Therefore there exists a real number b in $[0,1]$ such that:

$$b\ E(Y'_{D_p}) + (1-b)\ E(Y'_{D_q}) = E(J'_o) - a.$$

It is then easy to check that r.s.t. $\mu_a^\varepsilon = (b, D_p, D_q)$ is ε-optimal.

Remark. Integer n entering the definition of p and q is necessarily bounded by sup Y/ε. Then for example in a Markovian situation only a finite number of Snell "reduites" and Dirichlet problems are needed to compute an ε-optimal r.s.t.

To go on we need some more assumptions on the regularity of the mapping $p \longrightarrow D_p$. In part III we will give two examples of the situation.

Proposition II.5. <u>Let us assume that the mapping</u> $p \longrightarrow D_p$ <u>is left and right limited. Then we can construct an optimal randomized stopping time for the constrained stopping problem associated to a.</u>

Proof. If s.t. D_o is not optimal, from lemma II.2, there exists a real number p_a such that :

$$\forall\ \varepsilon > 0,\ E(Y'_{D_{p_a-\varepsilon}}) < E(J'_o)-a \text{ and } E(Y'_{D_{p_a+\varepsilon}}) \geqq E(J'_o)-a$$

Let ε tend to zero ; then from the regularity of Y' and the existence of right and left limits for D_p we get :

$$E(Y'_{D_{p_a-}}) \leqq E(J'_o)-a \text{ and } E(Y'_{D_{p_a+}}) \geqq E(J'_o)-a.$$

Therefore there exists a real number b in $[0,1]$ such that :

$$bE(Y'_{D_{p_a-}}) + (1-b)\ E(Y'_{D_{p_a+}}) = E(J'_o)-a.$$

It remains to prove that s.t. D_{p_a-} and D_{p_a+} are optimal for the unconstrained problem associated to process Y^{p_a}. For example, from definition of $D_{p_a-\varepsilon}$, we get :

$$\forall T \in \underline{\underline{T}},\quad E(Y_{D_{p_a-\varepsilon}}) + (p_a-\varepsilon)E(Y'_{D_{p_a-\varepsilon}}) \geqq E(Y_T) + (p_a-\varepsilon)\, E(Y'_T).$$

It is enough now to let ε tend to zero in order to conclude that D_{p_a-} is optimal. So r.s.t. (b, D_{p_a-}, D_{p_a+}) is a solution of problem (✱) as it satisfies condition (ii) of proposition II.1.

III. EXAMPLES.

Two examples are given where the assumption of proposition II.5 is satisfied; we recall that processes Y and Y' are cadlag regular bounded non negative processes defined up to infinity.

Proposition III.1. Let us assume that Y' is an uppermartingale (resp. submartingale). Then mapping $p \to D_p$ is right continuous decreasing (resp. left continuous increasing). So $p \to D_p$ is left and right limited and there exists an optimal randomized stopping time for the constrained problem associated to a.

Proof. Let us assume that Y' is an uppermartingale. Then we have up to an evanescent set, for all positive numbers p,q :

$$J^{p+q} \leqq J^p + qY'$$

So we have the inclusion :

$$\{ (t,\omega) \;/\; J^p = Y+pY' \} \subset \{ (t,\omega) \;/ J^{p+q} = Y + (p+q)Y' \}$$

So mapping $p \to D_p$ is decreasing.

Let us show the right continuity. Let p_n be a decreasing sequence to p ; s.t. D_{p_n} are increasing and bounded by D_p. Then it tends to s.t.D. From the regularity of Y and Y' we get for any T of $\underline{\underline{T}}$:

$$\lim_{n \to \infty} E(Y_{D_{p_n}} + p_n Y'_{D_{p_n}}) = E(Y_D + pY'_D) \geqq E(Y_T + pY'_T)$$

Therefore D is an optimal s.t. for process Y^p, so bigger than D_p which is the smallest one. This implies that $D = D_p$.

The second example is devoted to the case where Y' is a characteristic function, for example of a stochastic interval $[\![S,T]\!]$ with S,T two totally inaccessible s.t. .

Proposition III.2. Let us assume that Y' has only for values 0 or 1. Let us define s.t. D_p, D_p^o, D_p^1 :

$$D_p = \inf\{t \geqq 0 \;/\; J^P = Y^P\}$$

$$D_p^o = \inf\{t \geqq 0 \;/\; J^P = Y \text{ and } Y' = 0\}$$

$$D_p^1 = \inf\{t \geqq 0 \;/\; J^P = Y+p \text{ and } Y' = 1\}$$

Then mapping $p \longrightarrow D_p^o$ and $p \longrightarrow D_p^1$ are increasing and decreasing respectively and D_p is right and left limited as the infinimum of D_p^o and D_p^1. So there exists an optimal randomized stopping time for the constrained problem associated to a.

Proof. First it is obvious that $D_p = D_p^o \wedge D_p^1$. Notice that $p \to J^P$ is increasing up to an evanescent set. Therefore we have:

$$\forall \quad p,q \geqq 0 \;, \quad \{Y' = 0,\; J^{p+q} = Y\} \subset \{Y' = 0,\; J^P = Y\}$$

So $p \to D_p^o$, is increasing. Furthermore, we can show that up to an evanescent set :

$$\forall \; p,q \geqq 0 \;, \; \{ Y' = 1,\; J^P = Y + p\} \subset \{Y'=1,\; J^{p+q} = Y + p + q\}$$

So $p \to D_p^1$ is decreasing and we can conclude the proof.

The last example can be illustrated by the following : let us consider a Poisson process $(P_t,\; t \geqq 0)$ on $(\Omega, \underline{\underline{A}}, \underline{\underline{F}}, \mathbb{P})$; Y_t is a bounded function $f(P_t)$; f is defined on $\mathbb{N}$ such that ;

$$f(2) = \sup_n f(n) \text{ and } f(4) = \sup_{n>2} f(n) = f(2) - 1.$$

The constraint process Y' is supposed to be :

$$Y' = \mathbb{1}_{[3,5]}(P_t).$$

Then the Snell envelope of Y^p can be easily expressed by :

$$J_t^P = q^P(X_t)$$

where

$$q^P(x) = \sup_{y \geqq x} (f(y) + p\, \mathbb{1}_{[3,5]}(y)).$$

The sequence of s.t. D_p is given by :

$$D_p = \inf \{ t \geqq 0 ; P_t = 2 \} \text{ with } P_{D_p} = 2 \text{ if } p \leqq 1 ,$$

$$D_p = \inf \{ t \geqq 0 ; P_t = 4 \} \text{ with } P_{D_p} = 4 \text{ if } p > 1 .$$

Therefore the r.s.t. $(a, D_1 ; D_{1+})$ is optimal for the constrained problem.

Remark. Another example is next to the last one. Let Y^2 be a cadlag regular bounded non negative uppermartingale and D a totally inaccessible s.t. Process Y' is defined by :

$$Y'_t = Y^2_t \; 1_{[\![D + \infty[\![}(t)$$

Such a process is too a cadlag regular bounded non negative process. As in the last example it is easy to prove that s.t. D_p is there the infimum for two s.t. D^o_p and D^1_p such that $p \to D^i_p$ are monotonous.

$$D^o_p = \inf \{ t \geqq 0 ; t < D \text{ and } J^p_t = Y_t \}$$

$$D^1_p = \inf \{ t \geqq 0 ; t > D \text{ and } J^p_t = Y_t + p Y^2_t \}$$

R E F E R E N C E S

(1) A. BENSOUSSAN, J.L. LIONS : "Applications des inéquations variationnelles en contrôle stochastique", Dunod 1978.

(2) J.M. BISMUT : "An Example of optimal control with constraints", SIAM J. Control, vol. 12 n° 3 (1974), 401-418.

(3) J.M. BISMUT : "Temps d'arrêt optimal, quasi-temps d'arrêt et retournement du temps", Annals of Proba. vol. 7 n° 6 (1979)

(4) N. CHRISTOPEIT : "A stochastic control model with chance constraints", SIAM J. Control optim. Vol. 16 n° 5 (1978), 702-714.

(5) C. DELLACHERIE, P.A. MEYER : "Probabilités et potentiels", tome 1 (1975), tome 2(1980), Hermann.

(6) I. EKELAND, R. TEMAN : "Convex analysis and variational problems", North-Holland Publ. 1976.

(7) N. EL KAROUI : "Les aspects probabilistes du contrôle stochastique", Ecole d'été de St Flour IX-1979, Lect. Notes in Math. n° 876, Springer-Verlag, 1981.

(8) E.B. FRID : "On optimal strategies in control problems with constraints", Theory Prob. Appl. Vol. XVII n° 1 (1972), 188-192.

(9) U.G. HAUSSMANN : "Some example of optimal stochastic controls or : the stochastic maximum principle at work", SIAM Review vol. 23 n° 3 (1981), 292-307.

(10) D.P. KENNEDY : "On a constrained optimal stopping problem", J. Appl. Prob. 19 (1982), 631-642.

(11) H.J. KUSHNER : "On the stochastic maximum principle with average constraints", J. of Math. Anal. Applic. 12 (1965), 13-26.

(12) J.F. MERTENS : "Théorie des processus stochastiques généraux. Applications aux surmartingales", Z. f. Wahr. V. Geg. 22 (1972), 45-68.

(13) N.K. OZGOREN, R.W. LONGMAN; C.A. COOPER : "Probabilistic inequality constraints in stochastic optimal control theory", J. of Math. Anal. Applic. 66 (1978), 237-259.

(14) M. PONTIER, J. SZPIRGLAS : "Arrêt optimal avec contrainte", J. Applied Probability 15 (1983), 798-812.

(15) J.P. QUADRAT : "Existence de solution et algorithme de résolution numérique de problème de contrôle optimal de diffusion stochastique dégénérée ou non", SIAM Control Optimization vol. 18 n° 2 (1980), 199-266.

(16) M. ROBIN : "On optimal stochastic control problems with constraints", Game theory and related topics, North-Holland, (1979), 187-202.

(17) K.T. ROCKAFELLAR : "Convex analysis", Princeton university Press, 1970.

(18) A.N. SHIRYAYEV : Optimal stopping rules", Appl. of Math. n° 8, Springer-Verlag, 1977.

Etude de la stabilité de la solution d'une E D S bilinéaire à coefficients périodiques. Application au mouvement des pales d'hélicoptère .

E.PARDOUX* et M. PIGNOL**

* Université de Provence et INRIA

** Université de Provence,3,Place Victor Hugo,13331 MARSEILLE Cedex 3

Résumé. L'étude de la stabilité du mouvement des pales d'un hélicoptère en vol d'avancée, en atmosphère turbulente, conduit à l'étude de la stabilité de la solution d'une équation différentielle stochastique bilinéaire, à coefficients fonctions périodiques du temps. On indique comment des résultats récents de ARNOLD et KLIEMAN [2] se généralisent à cette situation, puis on étudie un algorithme qui permet de décider si la solution est stable ou non .

Abstract. The study of the stability of the movement of helicopter blades during a flight, taking into account the turbulence of the wind, leads to the study of the stability of the solution of a bilinear stochastic differential equation, whose coefficients are periodic functions of time. We show that, under a hypoellipticity condition, such an equation possess exactly one Lyapounov exponent λ ,thus adapting recent results of ARNOLD and KLIEMAN [2] for the constant coefficients case. We restrict the present analysis to the " white noise model ", which permits us to use the result of FURSTENBERG [4] on products of i.i.d. random matrices. Finally, we show that all Lyapounov exponents of a suitable time-discretized version of the initial stochastic differential equation, tend to λ ,as the discretization step tends to zero. This suggests a numerical algorithm for the computation of an approximate value of λ .

1 - Introduction

Si l'on ne retient comme degrés de liberté pour la pale d'un hélicoptère que les mouvements de battement et de torsion, les équations linéarisées du mouvement se mettent sous la forme :

$$\frac{d^2Z(t)}{dt^2} = C(t) \begin{pmatrix} \frac{d}{dt} & Z(t) \\ & Z(t) \end{pmatrix} + H(t) \tag{1.1}$$

où $Z(t)$ prend ses valeurs dans $\mathbb{R}^2$, $C(t)$, $D(t)$ et $H(t)$ sont des fonctions périodiques de t, à valeurs respectivement matrices 2 x 4 et vecteurs de dimension 2 . La période commune de C(t) et H(t) est la période T de rotation des pales de l'hélicoptère, laquelle est constante. Si l'on modélise les modifications turbulentes de l'angle d'incidence du vent par rapport aux pales sous la forme d'un processus stochastique ξ_t , on est amené à remplacer l'E.D.O.(1,1) par l'E.D.S :

$$\frac{d^2 Z_t}{dt^2} = C(t) \begin{pmatrix} \frac{dZ_t}{dt} \\ Z_t \end{pmatrix} + H(t) + \xi_t \left[D(t) \begin{pmatrix} \frac{dZ_t}{dt} \\ Z_t \end{pmatrix} + J(t) \right] \tag{1.2}$$

Si l'on pose $Y_t = \begin{bmatrix} \frac{dZ_t}{dt} \\ Z_t \end{bmatrix}$ on obtient pour Y_t l'E.D.S. :

$$\frac{dY_t}{dt} = A(t)\, Y_t + F(t) + \xi_t \left[B(t) Y_t + G(t) \right] \tag{1.3}$$

où en particulier $A(t) = \begin{pmatrix} C(t) \\ 1\;0\;0\;0 \\ 0\;1\;0\;0 \end{pmatrix}$, $B(t) = \begin{pmatrix} D(t) \\ 0\;0\;0\;0 \\ 0\;0\;0\;0 \end{pmatrix}$

L'étude de la stabilité de la solution de (1.3) peut se faire dans les deux cas suivants :

(a) - ξ_t est un processus de diffusion.

(b) - ξ_t est un bruit blanc, et on interprète l'équation (1.3) au sens de Stratonovitch .

Bien que la seconde modélisation soit beaucoup moins réaliste dans notre problème, c'est celle-ci que nous allons adopter dans le cadre de cet article. L'étude du cas(a) conduit à des résultats tout à fait similaires, mais les démonstrations sont plus longues . Elle sera présentée dans [11] .

Rappel sur les intégrales stochastiques . Nous utiliserons trois types d'intégrale stochastique : l'intégrale de Stratonovitch, l'intégrale de Ito, et l'intégrale de Stratonovitch " rétrograde ". Soit $[a,b] \subset \mathbb{R}$, et $\{W_t,\ t \in [a,b]\}$ un processus de Wiener réel standard (i.e. tel que $E[(W_t - W_s)^2] = |t-s|$). Pour tout $t \in [a,b]$, on définit les deux tribus :

$$\mathcal{F}_t^a = \sigma\{ W_s - W_a,\quad a \leq s \leq t \}$$

$$\mathcal{F}_b^t = \sigma\{ W_s - W_b,\quad t \leq s \leq b \}$$

On pourrait grossir ces tribus en leur "rajoutant" une tribu indépendante de $\{W_t,\ t \in [a,b]\}$ .Soit maintenant $\{\varphi_t, t \in [a,b]\}$ une semi-martingale p.s. continue . Si φ_t est $\mathcal{F}_t^{\circ}$ -mesurable,$\forall\, t \in [a,b]$, on peut définir l'intégrale de Stratonovitch par :

$$(1.4)\quad \int_a^b \varphi_t \circ dW_t = P\text{-}\lim_{n\to\infty} \sum_{k=0}^{n-1} \frac{\varphi_{t_k^n} + \varphi_{t_{k+1}^n}}{2} \left(W_{t_{k+1}^n} - W_{t_k^n} \right)$$

et l'intégrale de Ito par :

$$(1.5)\quad \int_a^b \varphi_t \, dW_t = P\text{-}\lim_{n\to\infty} \sum_{k=0}^{n-1} \varphi_{t_k^n} \left(W_{t_{k+1}^n} - W_{t_k^n} \right)$$

où $t_k^n = a + (b-a)\frac{k}{n}$.

Si φ_t est $\mathfrak{F}_b^t$-mesurable, $\forall\, t \in [a,b]$, alors la convergence dans (1.4) a encore lieu, et nous appellerons l'intégrale correspondante intégrale de Stratonovitch rétrograde, que nous ne distinguerons pas, du point de vue notation, de l'intégrale de Stratonovitch usuelle.

2 - Etude théorique de la stabilité

Considérons donc le système différentiel stochastique :

$$(2.1)\quad d\,Y_t = (A(t)Y_t + F(t))\,dt + \sum_{i=1}^{k} (B_i(t)Y_t + G_i(t)) \circ d\,W_t^i$$

où les W_t^i sont des processus de Wiener scalaires standards indépendants, $A(t)$, $B_i(t)$, $F(t)$ et $G_i(t)\,(i = 1,\dots,k)$ sont respectivement des matrices $d \times d$ et des vecteurs de dimension d, fonctions périodiques du temps de période T, et le signe $\circ$ signifie que l'intégrale stochastique est au sens de Stratonovitch .

On associe à ce système le système différentiel" sans second membre " :

$$(2.2)\quad d\,X_t = A(t)X_t\,dt + \sum_{i=1}^{k} B_i(t)Y_t \circ d\,W_t^i$$

et la solution fondamentale du système sans second membre $\Phi(t,s)$, défini pour $0 \leqslant s < t < \infty$ par :

$$(2.3)\quad \begin{cases} d\,\Phi(t,s) = A(t)\,\Phi(t,s)\,dt + \sum_{i=1}^{k} B_i(t)\,\Phi(t,s) \circ d\,W_t^i, & t \geqslant s \\ \Phi(s,s) = I \end{cases}$$

où $\Phi(t,s)$ est un processus à valeurs dans l'espace des matrices $d \times d$, et I est la matrice identité. On notera $\Phi(t)$ pour $\Phi(t,o)$. Définissons $\Phi\,(t,s)$ pour $t \leqslant s$ par la même équation(2.3),où cette fois"o" signifie l'intégrale de Stratonovitch rétrograde .

On peut vérifier directement l'identité suivante(qui est aussi une conséquence de la théorie des flots stochastiques):

$$\Phi(t,s) = \Phi(s,t)^{-1} \quad \text{p.s.,} \quad \forall\, s,t \in \mathbb{R}_+$$

Théorème 2.1. Supposons qu'il existe $\lambda < 0$ tel que

$$\limsup_{t\to+\infty} \frac{1}{t} \operatorname{Log} \|\Phi(t)\| \leqslant \lambda \text{ p.s.}$$

Alors l'équation (2.1) admet une solution périodique de période T $\{\overline{Y}_t\}$, et $\forall \{Y_t\}$ solution de (2.1),

$$Y_t - \overline{Y}_t \to 0 \quad \text{p.s., quand } t \to \infty$$

Remarque : Par solution périodique, nous entendons une solution $\{\overline{Y}_t, t \geq 0\}$ telle que $\forall n \in \mathbb{N}$, $\forall 0 \leq t_1 < \ldots < t_n$, la loi du vecteur aléatoire$(\overline{Y}_h, \overline{Y}_{t_1+h}, \ldots, \overline{Y}_{t_n+h})'$ soit une fonction périodique de h, de période T. La dernière affirmation du théorème entraîne que cette solution périodique est unique en loi .

Preuve : Toute solution de l'équation (2.1) s'écrit :

$$Y_t = \Phi(t) \left[Y_0 + \int_0^t \Phi(s)^{-1} \left(F(s)ds + \sum_{i=1}^{k} G_i(s) \circ d W_s^i\right) \right]$$

Il est donc immédiat que sous l'hypothèse du théorème, la différence de deux solutions tend vers zéro p.s., quand $t \to +\infty$.
Etant donnés $\{V_t^i, t \geq 0; i = 1 \ldots k\}$ des processus de Wiener réels standards indépendants entre eux et des $\{W_t^i\}$, on définit W_t^i pour tout $t \in \mathbb{R}$ en posant $W_t^i = V_{-t}^i$, $t \leq 0$; $i = 1 \ldots k$.

On définit alors $\Phi(t)$ pour tout $t \in \mathbb{R}$, et il résulte de l'hypothèse du théorème :

$$\limsup_{t \to -\infty} \frac{1}{t} \operatorname{Log} \| \Phi(t)^{-1} \| \leq \lambda \quad \text{p.s.}$$

Alors la formule :

$$\overline{Y}_t = \Phi(t) \int_{-\infty}^{t} \Phi(s)^{-1} \left(F(s)ds + \sum_{i=1}^{k} G_i(s) \circ d W_s^i\right)$$

(où l'intégrale stochastique est rétrograde de 0 à $-\infty$, et progressive de 0 à t) définit un processus stochastique solution de (2.1), pour $t \in \mathbb{R}$. Pour montrer que $\{\overline{Y}_t\}$ est périodique, il suffit, compte tenu de la propriété de Markov de ce processus, de montrer que la loi marginale de $\overline{Y}_t$ est une fonction périodique de t . Ceci résulte des propriétés suivantes : la loi de $\{\Phi(t), t \in \mathbb{R}\}$ coïncide avec celle de $\{\Phi(t+nT, nT), t \in \mathbb{R}\}$ celle de $\{W_t, t \in \mathbb{R}\}$ avec celle de $\{W_{t+nT} - W_{nT}, t \in \mathbb{R}\}$et F et les G_i sont périodiques de période T .

□

Le théorème 2.1 nous permet de ramener l'étude de la stabilité de la solution de (2.1)à celle de (2.2)(i.e. à la vérification de l'hypothèse du théorème). L'étude de la stabilité p.s.de la solution d'une équation du type (2.2) ,dans le cas de coefficients constants, est menée dans le cas elliptique dans KHASMINSKII [7], et dans le cas hypoelliptique dans ARNOLD-KLIEMAN [2] . Nous allons indiquer ci-dessous comment ces derniers résultats s'adaptent au cas des coefficients périodiques. Le choix de la modélisation "bruit blanc" va nous permettre de simplifier les démonstrations ,en faisant appel à des résultats sur les produits de matrices aléatoires .

On peut remarquer que l'étude de la limite quand $t \to \infty$ de :

$$\frac{1}{t} \operatorname{Log} \| X_t \|$$

est équivalente à celle quand $n \to \infty$ de :

$$\frac{1}{nT} \operatorname{Log} \| X_{nT} \|$$

Or X_{nT} s'écrit :

$$X_{nT} = \Phi(nT,(n-1)T) \ \Phi((n-1)T,(n-2)T) \dots \ \Phi(T,0)X_0$$

où les matrices aléatoires $\Phi((k+1)T,kT), k=0,1\dots.$ sont indépendantes et identiquement distribuées .

Pour pouvoir appliquer le célèbre théorème de Furstenberg, il nous faut normaliser ces matrices .

On pose : $\overline{\Phi}((k+1)T,kT) = \dfrac{\Phi((k+1)T,kT)}{(\text{dét } \Phi((k+1)T,kT))^{1/d}}$, et on a :

$$X_{nT} = \overline{X}_{nT} \exp\left(\frac{1}{d} \int_0^{nT} \operatorname{Tr} A(s)ds + \frac{1}{d} \sum_{i=1}^{k} \int_0^{nT} \operatorname{Tr} B_i(s)d W_s^i \right)$$

$$\overline{X}_{nT} = \overline{\Phi}(nT,(n-1)T) \dots \overline{\Phi}(T,0)X_0$$

On a le :

<u>Théorème 2.2.</u> (FURSTENBERG [4])

Si la loi de $\overline{\Phi}(T)$ est irréductible, alors $\exists \ \overline{\lambda} \geqslant 0$ tel que :

$$\frac{1}{nT} \operatorname{Log} \| \overline{X}_{nt} \| \to \overline{\lambda} \qquad \text{p.s.,} \quad \forall X_0$$

<u>Corollaire 2.3</u> : Sous l'hypothèse du théorème 2.2.,

$$\frac{1}{t} \operatorname{Log} \| X_t \| \to \overline{\lambda} + \frac{1}{dT} \int_0^T \operatorname{Tr} A(s)ds \quad \text{p.s.,} \quad \forall X_0$$

et

$$\frac{1}{t} \operatorname{Log} \| \Phi(t) \| \to \overline{\lambda} + \frac{1}{dT} \int_0^T \operatorname{Tr} A(s)ds \quad \text{p.s.}$$

□

L'hypothèse d'irréductibilité est que le sous-groupe de $SL(d;\mathbb{R})$ engendré par le support de la loi de $\overline{\Phi}(T)$ ne doit laisser invariant aucun sous espace propre de $\mathbb{R}^d$. Une condition suffisante est que $\forall\, x \in \mathbb{R}^d$, la loi de $\overline{\Phi}(T)x$ admette une densité . Ceci est assuré dès que l'opérateur $\frac{\partial}{\partial t} + \overline{L}(t)$ est hypoelliptique, où $\overline{L}(t)$ est le générateur infinitésimal du processus de Markov $\overline{X}_t$ solution de :

$$d\,\overline{X}_t = \overline{A}(t)\overline{X}_t\,dt + \sum_{i=1}^{k} \overline{B}_i(t)\overline{X}_t \circ d\,W_t^i$$

avec la notation $\overline{C} = C - \frac{1}{d}\,\mathrm{Tr}\ C$.

Remarque : On aboutirait à des conditions un peu plus faible en cherchant à vérifier une condition similaire pour le processus

$U_t = \dfrac{\overline{X}_t}{\|\overline{X}_t\|} = \dfrac{X_t}{\|X_t\|}$ (cf.ci-dessous pour l'équation du processus U_t, qui évolue sur la sphère S^{d-1}). Cependant, ces conditions sont très difficiles à vérifier en pratique, et on est amené à vérifier des conditions plus fortes, qui sont précisément celles que nous allons indiquer ci-dessous. Remarquons enfin que dans le cas des coefficients constants, il suffit de vérifier que l'opérateur $\overline{L}$ est hypoelliptique, ce qui conduit à des conditions plus faibles que les nôtres (cf. ARNOLD-KLIEMAN [2]).

□

On définit récursivement les fonctions matricielles de t suivantes :

$$A\,d^{o}_{\overline{A}}\,(\overline{B}_i)(t) = \overline{B}_i(t)\ ,\ i = 1\ldots k$$

$$A\,d^{j+1}_{\overline{A}}\,(\overline{B}_i)(t) = [\overline{A}(t),\, A\,d^{j}_{\overline{A}}\,(\overline{B}_i)(t)] + \frac{d}{dt}\,A\,d^{j}_{\overline{A}}\,(\overline{B}_i)(t),\ i = 1\ldots k$$

où le crochet $[C,D]$ des matrices C et D est défini par :

$$[C,D] = DC - CD$$

On considère alors pour tout $t \in [0,T]$ l'algèbre de Lie de matrices :

$$\overline{Q}(t) = A.L.\{A\,d^{j}_{\overline{A}}\,(\overline{B}_i)(t)\ ;\ i = 1\ldots k\ ;\ j = 0,1,2,\ldots\}$$

La condition d'hypoellipticité de l'opérateur $\frac{\partial}{\partial t} + \overline{L}$ est alors :

(H) $\dim\ \overline{Q}(t)x = d,\quad \forall\, x \in \mathbb{R}^d - \{0\}\ ,\forall\, t \in [0,T]$

En fait, il suffit de vérifier la condition ci-dessus pour un $t \in [0,T]$, et avec $\overline{Q}(t)$ remplacée par $Q(t)$, qui est définie de façon analogue en remplaçant partout $\overline{A}$ et les $\overline{B}_i$ par A et les B_i. Mais la condition (H) est trop difficile à vérifier en pratique, et nous lui préférons la condition plus forte :

(H') $\exists\, t \in [0,T]$ t.q. $\dim \overline{Q}(t) = d^2-1$

La condition analogue à (H') en terme de $Q(t)$ serait encore plus forte . Sous la condition (H') le support de la loi de $\overline{\Phi}(T)$ est d'intérieur non vide dans $SL(d;\mathbb{R})$, d'où le groupe engendré est $SL(d;\mathbb{R})$, et on en déduit (cf.GUIVARC'H [6]) :

Proposition 2.4 : Sous l'hypothèse (H'), $\exists \lambda \in \mathbb{R}$ t.q.

$$\frac{1}{t} \operatorname{Log} \| X_t \| \to \lambda \qquad \text{p.s.}, \quad \forall X_o$$

et de plus $\quad \lambda > \frac{1}{dT} \int_o^T \operatorname{Tr} A(t) dt$

□

On peut argumenter que la condition(H') est satisfaite génériquement . Ceci n'est pas convainquant pour notre exemple , où les matrices A et B contiennent beaucoup de zéros.Cependant, la vérification de cette hypothèse peut se faire sur chaque exemple particulier, à l'aide d'un langage de calcul formel, du type MACSYMA ou REDUCE.

Remarque : Nous étudions la stabilité p.s., qui est celle qui importe en pratique. On pourrait aussi déduire la stabilité p.s. de la stabilité du moment d'ordre 2. Pour une étude des liens entre stabilité p.s. et stabilité des moments, nous renvoyons à ARNOLD [1] , et à [9].

□

Il reste à calculer l'exposant de Lyapounov λ - au moins de façon approchée . L'algorithme que nous proposons consiste à simuler- pendant un temps suffisamment long - le processus $\{X_t\}$. Cependant , nous ne pouvons simuler qu'une approximation de $\{X_t\}$, et il reste à vérifier que l'exposant de Lyapounov correspondant est bien une approximation de λ .

Nous allons tout d'abord donner une autre expression de λ . On pose $U_t = \frac{X_t}{\| X_t \|}$. Tenant en compte le fait que X_t évolue dans l'ouvert $\mathbb{R}^d - \{0\}$, on déduit du calcul différentiel stochastique :

$$\| X_t \| = \| X_o \| \exp\left[\int_o^t q(s,U_s) ds + \sum_{i=1}^k \int_o^t p_i(s,U_s) d W_s^i \right]$$

avec

$$q(s,u) = (A(s)u,u) + \frac{1}{2} \sum_{i=1}^k \left[(B_i^2(s)u,u) + | B_i(s)u |^2 - 2(B_i(s)u,u)^2 \right]$$

$p_i(s,u) = (B_i(s)u,u)$,l'intégrale stochastique ci-dessus étant une intégrale de Ito. Par ailleurs,

$$d U_t = \left[A - (A U_t, U_t) \right] U_t \, dt + \sum_{i=1}^k \left[B_i - (B_i U_t, U_t) \right] \circ d W_t^i$$

Le résultat suivant est alors immédiat :

Lemme 2.5. Sous l'hypothèse (H'), $\frac{1}{t} \int_o^t q(s,U_s) ds \xrightarrow{\text{p.s.}} \lambda$, quand $t \to \infty$, pour toute condition initiale U_o .

□

Soit ψ un difféomorphisme de S^1 sur $[0,T[$, et soit $x_o \in S^1$ tel que $\psi(x_o) = 0$. Pour $x \in S^1$, $u \in S^{d-1}$, on pose : $\tilde{q}(x,u) = q(\psi(x),u)$.

Soit $\{R_t\}$ le processus à valeurs dans S^1 défini par :

$$\frac{d R_t}{dt} = 1, \quad R_o = x_o$$

Alors $\frac{1}{t}\int_o^t q(s,U_s)ds = \frac{1}{t}\int_o^t \tilde{q}(R_s,U_s)ds$, où le processus

$\left\{\binom{R_t}{U_t}\right\}$,$t \geq 0$ est un processus de diffusion homogène à valeurs dans la variété compacte $S^1 \times S^{d-1}$. Ce processus admet donc au moins une probabilité invariante . On a le :

<u>Théorème 2.6 :</u> $\forall\, \mu$ probabilité invariante du processus $\left\{\binom{R_t}{U_t}\right\}$, on a :

$$\lambda = \int_{S^1 \times S^{d-1}} \tilde{q}(r,u)d\mu(r,u)$$

<u>Preuve :</u> Il suffit de montrer le résultat avec μ probabilité invariante extrémale. Soit donc μ une telle probabilité . Alors il existe une condition initiale $\binom{R_o}{U_o}$ qui entraîne :

$$\frac{1}{t}\int_o^t \tilde{q}(R_s,U_s)ds \xrightarrow{p.s} \int_{S^1 x\, S^{d-1}} \tilde{q}(r,u)d\mu(r,u)$$

Le résultat découle alors aisément du lemme 2.5.

□

Remarquons que toute probabilité invariante μ se désintègre sous la forme :

$\mu(dr, du) = \rho(dr) \quad \nu(r,du)$ où ρ est la probabilité uniforme sur S^1 .

<u>§ 3 - Calcul approché de l'exposant de Lyapounov</u>

Pour alléger les notations , nous allons nous limiter au cas $k = 1$.

Considérons une approximation de l'équation (2.2), associée à un pas de discrétisation en temps de longueur $\frac{T}{n}$:

$$\bar{X}^n_{k+1} = C^n_k \quad \bar{X}^n_k \quad , \quad \bar{X}^n_o = X_o$$

où $C_k^n = \exp\left[\frac{T}{n} A(\frac{kT}{n}) + (W(\frac{k+1}{n}) - W(\frac{kT}{n}))\, B(\frac{kT}{n})\right]$

On définit alors le processus $\{X_t^n, t \in \mathbb{R}_+\}$ par :

$$X_t^n = \bar{X}_k^n \text{ , pour } t \in [\frac{k}{n} T, \frac{k+1}{n} T[\text{ , } k = 0,1,\ldots$$

On pose de plus :

$$\Phi^n((j+1)T, jT) = C_{(n+1)j} \cdots\cdots C_{nj+1} C_{nj}$$

Une variante des méthodes de [10] permet d'établir la :

Proposition 3.1 : On a les deux convergences suivantes en moyenne quadratique :

$$X_t^n \to X_t \text{ , } \forall\, t \in \mathbb{R}_+$$

$$\Phi^n(T,0) \to \Phi(T,0)$$

□

Définissons :

$$\lambda^n(X_0) = \lim_{j \to \infty} \frac{1}{jT} \operatorname{Log} \| X_{jT}^n \|$$

$$= \lim_{j \to \infty} \frac{1}{jT} \operatorname{Log} \| \Phi^n(jT,(j-1)T) \ldots \Phi^n(T,0) X_0 \|$$

Les matrices $\{\Phi^n(jT,(j-1)T); j = 1,2,\ldots\}$ étant indépendantes et identiquement distribuées, et vérifiant :

$$E\,[\operatorname{Log}^+ \| \Phi^n(T,0) \| + \operatorname{Log}^+ \| (\Phi^n(T,0))^{-1} \|] < \infty$$

la limite ci-dessus existe p.s., et prend ses valeurs dans un ensemble fini $\{\lambda_0^n, \lambda_1^n, \ldots, \lambda_r^n\}$, avec $\lambda_0^n > \lambda_1^n > \ldots\ldots > \lambda_r^n$, et ,

$$\lambda_0^n = \lim_{j \to \infty} \frac{1}{jT} \operatorname{Log} \| \Phi^n(jT,(j-1)T) \ldots \Phi^n(T,0) \| \quad \text{p.s.} \tag{3.1}$$

cf. FURSTENBERG-KIFER [5] . Si la loi de $\Phi^n(T,0)$ vérifie la condition d'irréductibilité du § 2 ,alors $r = 0$. De plus, on peut montrer que si la loi de X_0 est absolument continue par rapport à la mesure de Lebesgue, alors :

$$\lambda^n(X_0) = \lambda_0^n \quad \text{p.s.} \tag{3.2}$$

Mais KIFER [8] et FURSTENBERG-KIFER [5] ont montré que, sous la condition d'irréductibilité de la loi de la limite $\Phi(T,0)$, et d'autres propriétés que l'on vérifie aisément dans notre cas(notamment grâce à la proposition 3.1), on a le résultat :

Théorème 3.2 : $\lambda_0^n \to \lambda$, quand $n \to \infty$

□

Les relations (3.1) et (3.2) nous donnent donc deux méthodes pour simuler une valeur approchée de λ . En fait, même si la loi de $\Phi^n(T,0)$ n'est pas irréductible, on peut simuler une valeur approchée de λ en simulant, pour $x \in \mathbb{R}^d - \{0\}$ arbitrairement choisi :

$$\lim_{j \to \infty} \frac{1}{jT} \operatorname{Log} \| \Phi^n(jT,(j-1)T) \ldots \Phi^n(T,0)x \|$$

En effet , on a le résultat suivant :

<u>Théorème 3.3.</u> : Supposons que la loi de $\Phi(T)$ est irréductible. Alors , $\forall\, k, 0 \leqslant k \leqslant r$,

$$\lambda_k^n \to \lambda \quad , \text{ quand } n \to \infty$$

<u>Preuve</u> : (esquisse) ; On considère la chaîne de Markov en temps discret $\{(R_j^n , U_j^n), j = 0,1,\ldots\}$ à valeurs dans $S^1 \times S^{d-1}$ définie par :

$$U_{j+1}^n = \frac{C_j^n \; U_j^n}{\| C_j^n \; U_j^n \|} \quad , \quad U_o^n = \frac{x}{\|x\|} \; ; \; R_j^n = R_{\frac{jT}{n}}$$

Il est clair que l'on a :

$$\| \overline{X}_{k+1}^n \| = \| X_o \| \times \prod_{j=1}^{k} \| C_j^n \; U_j^n \|$$

On montre, par le même argument que dans CRAUEL [3] , qu'il existe une probabilité invariante μ_k^n de la chaîne $\{(R_j^n, U_j^n)\}$ telle que :

$$\lambda_k^n = \int_{S^1 \times S^{d-1}} \varphi_n(r,u) \; d\, \mu_k^n(r,u)$$

$$\text{où : } \varphi_n(r,u) = \frac{n}{T} \int_{\mathbb{R}} \operatorname{Log} \| e^{\frac{T}{n} A(r) + v \sqrt{\frac{T}{n}} B(r)} u \| \; d\, \theta(v)$$

avec θ mesure de Gauss de moyenne nulle et de variance 1 .

Au vu du théorème 2.6, il reste à montrer que quand n

(i) $\varphi_n(r,u) \to \tilde{q}(r,u)$ uniformément sur $S^1 \times S^{d-1}$

(ii) $\{\mu_k^n, n \geqslant 1\}$ est tendue,et toute limite étroite d'une sous-suite convergente est une probabilité invariante du processus (R_t, U_t).

(i) se vérifie en remarquant que

$$\frac{1}{2} \operatorname{Log} \| e^{\frac{T}{n} A(r) + v \sqrt{\frac{T}{n}} B(r)} u \|^2 = \frac{1}{2} \operatorname{Log} (\| (I + \frac{T}{n} A(r) + v \sqrt{\frac{T}{n}} B(r) +$$

$$+ \frac{v^2}{2} \frac{T}{n} B^2(r)) u \|^2 + O(\frac{1}{n^{3/2}}))$$

$$=\sqrt{\frac{T}{n}}\, v(B(r)u,u) + \frac{T}{n}\left[(A(r)u,u) + \frac{v^2}{2}\left(B^2(r)u,u) + |B(r)u|^2 - v^2(B(r)u,u)^2\right)\right] + O\left(\frac{1}{n^{3/2}}\right)$$

D'où l'on tire :

$$\varphi_n(r,u) = \tilde{q}(r,u) + O\left(\frac{1}{n^{1/2}}\right)$$

(ii) résulte de ce que si l'on définit le processus $\{(\bar{R}^n_t , \bar{U}^n_t), t \geq 0 \}$ par :

$$(\bar{R}^n_t , \bar{U}^n_t) = (R^n_j , U^n_j) \text{ , pour } t \in [\tfrac{j}{n} T, \tfrac{j+1}{n} T[$$

avec la condition initiale $(\bar{R}^n_o , \bar{U}^n_o) = (R^n_o , U^n_o) = (R_o, U_o)$.

Alors $\{(\bar{R}^n_t, \bar{U}^n_t), t \geq 0 \} \to \{(R_t,U_t), t \geq 0 \}$ en loi .

Or la suite $\{ \mu^n_k ,n \geq 1 \}$ de probabilités sur le compact $S^1 x S^{d-1}$ est tendue , et d'après la convergence ci-dessus toute limite étroite d'une sous-suite est une probabilité invariante du processus (R_t,U_t) .

□

Nous renvoyons à [11] pour des résultats numériques .

BIBLIOGRAPHIE

[1] L. ARNOLD *A formula connecting sample and moment stability of linear stochastic systems.Bremen Univ. Report(1983).*

[2] L.ARNOLD- W.KLIEMAN *Lyapounov exponents of linear stochastic systems . Bremen Univ. Report(1983).*

[3] H. CRAUEL *Lyapounov numbers of markov solutions of linear stochastic systems. Bremen Univ.Report (1983).*

[4] H.FURSTENBERG *Non commuting random products. Trans.Amer.Math.Soc. 108,377-428(1963).*

[5] H.FURSTENBERG-Y.KIFER *Random matrix products and measures on projective spaces. Preprint .*

[6] Y.GUIVARC'H *Quelques propriétés asymptotiques des produits de matrices aléatoires,in : Lecture Notes in mathematics 774, Springer (1980)*

[7] R.Z. KHASMINSKII *Stochastic stability of differential equations*, Sijthoff & Noordhoff (1980)

[8] Y. KIFER *Perturbations of random matrix products*. Z. Wahrschein. 61, 83-95 (1982).

[9] E. PARDOUX *Relation between a.s. and moment stability of linear stochastic differential equations*, en préparation.

[10] E. PARDOUX-D.TALAY *Discretization and simulation of stochastic differential equations*. Acta Applicandae Mathematicae (1984), à paraître.

[11] M. PIGNOL *Thèse de 3° cycle*, Université de Provence, en préparation.

Markov Decision Processes with Constraints

Keith W. Ross
University of Michigan
and
INRIA, France

Abstract

This article addresses the Markov decision problem with long-run average reward V_u when there is a global constraint to be satisfied: $I_u \leq \alpha$, where I_u is also a long-run average. Using Lagrange multiplier techniques, existence of an optimal stationary policy is proven. Unlike the unconstrained theory, optimal stationary policies are in general randomized. Structural properties of an optimal policy are determined and the corresponding dynamic programming equations are derived. Finally, conditions are given for the existence of an optimal pure policy and an optimal "almost" bang-bang policy.

1. Introduction

Consider the Markov decision problem of maximizing the long-run average cost

$$V_u(x) = \underline{\lim_t} \frac{1}{t} E_u [\sum_{s=0}^{t-1} C(x_s, a_s) \mid x_0 = x] \tag{1.1}$$

over all policies u, including those which are randomized and/or past-dependent. If the state and action spaces are finite, we know [1, chap.7] from the theory of Markov Decision Processes (MDP's) that there exists an optimal pure policy (non-randomized and stationary) that optimizes (1.1) for all x. There are several available algorithms which obtain an optimal pure policy in a finite number of iterations. Furthermore, Markov decision theory furnishes the dynamic programming (DP) equations, from which the structure (monotone, bang-bang) of an optimal policy can often be obtained.

In this article we consider maximizing the average reward (1.1) over a smaller class of policies, namely, the policies u that satisfy

$$I_u(x) = \overline{\lim_t} \frac{1}{t} E_u [\sum_{s=0}^{t-1} D(x_s, a_s) \mid x_0 = x] \leq \alpha \tag{1.2}$$

for all x. Optimal stochastic control problems of this type arise naturally in applications; for example, in queueing systems it may be of interest to maximize the throughput subject to a constraint on the average time-delay [2].

Derman [3, chap.7] has also studied Markov decision processes with constraints for the long-run average cost. He assumed finite action and state spaces, and established existence of optimal randomized stationary policies using the concept of expected state-action frequencies.

We study MDP's with the optimizatization criterion (1.1)-(1.2) using Lagrangian techniques. Assuming finite state and compact action spaces, we prove the existence of an optimal stationary policy with a very simple structure, i.e., at each decision epoch one of two pure policies is applied according to a biased coin. Because of this special structure, we are able to give sufficient conditions for the existence of optimal pure policies. Furthermore, the Lagrangian techniques enable us to derive the DP equations from which conditions are obtained for the existence of "almost" bang-bang optimal policies.

In the next section of this article we define the problem and collect several results from the theory of Markov chains. In the last section we indicate the method of approach and then give the existence results for the constrained problem.

2. The Constrained Problem

Definitions of MDP's are given in [1], [6] and our setup is basically the same. The state space $S = \{0,..,N\}$ is finite and the action space A is a compact subset of a metric space (X,ρ). Let C and D be two mappings from $S \times A$ into the nonnegative reals that are continuous on A for each fixed $x \in S$. Denote $p(y \mid x,a)$ with $x,y \in S$ and $a \in A$ for the law of motion, assumed to be continuous on A for each fixed $x,y \in S$.

Define the history sets

$$H_t = \{x_0 a_0 x_1 .... a_{t-1} x_t : x_s \in S,\ a_s \in A\}.$$

Then a policy u is a sequence $\{u_0, u_1, u_2, ...\}$ such that $u_m = u_m(\cdot \mid h_m)$ is a probability measure on $(A, B(A))$ depending on $h_m \in H_m$. We require that $u_m(B \mid \cdot)$ be measurable on $(H_m, B(H_m))$ for all $B \in B(A)$, for all $m \geq 0$. Each policy induces an unique probability measure on the probability space $(H_\infty, B(H_\infty))$ such that the expectations (1.1) and (1.2) are well-defined [2, chap.2]. A policy is said to be stationary if for all $m \geq 0$, all $h_m \in H_m$ and all $B \in B(A)$ we have $u_m(B \mid h_m) = u(B \mid x_m)$, where x_m is the last component of h_m. Finally, a policy is said to be pure if it is stationary and $u(\cdot \mid x)$ is concentrated on an element a_x of A for each x. Any pure policy can be uniquely represented by a mapping from S to A. Denote U,F,G for the class of policies, stationary policies and pure policies, respectively.

Consider $V_u(x)$ and $I_u(x)$ defined by (1.1) and (1.2). A policy $u \in U$ is said to be feasible if (1.2) is satisfied for all $x \in S$. The policy u is said to be optimal for the constrained problem if u is feasible and maximizes $V_u(x)$, for all x, over the class of feasible policies. We will soon address the existence and structure of optimal policies, but first we need to collect some results pertaining to Markov chains.

2.1. Controlled Markov Chains

Under any stationary policy f, $\{x_t\}_{t=0}^{\infty}$ is a homogeneous Markov chain; its transition matrix is

$$P_f(x,y) = \int_A p(y \mid x,a) u_f(da \mid x)$$

where $u_f(\cdot \mid \cdot)$ is any component of $f = (u_f, u_f, u_f, ...)$. Put

$$P_f^* = \lim_m \frac{1}{m} \sum_{k=0}^{m} P_f^m. \tag{2.1}$$

P_f^* is a stochastic transition matrix with the following properties [7]:

$$P_f^* = P_f P_f^* = P_f^* P_f = P_f^* P_f^* \tag{2.2}$$

$$(I - P_f + P_f^*) \text{ is invertible.} \tag{2.3}$$

The following assumption will be in force throughout this article:

Accessibility Assumption: the state $0 \in S$ is accessible from every other state regardless of which $g \in G$ is being used.

The next theorem shows that without loss of generality we can replace $g \in G$ in the accessibility assumption by the stronger condition $f \in F$. In the proof, and what follows, we will use

$$||P|| = \max_{x \in S} |\sum_{y \in S} P(x,y)|$$

for the norm of $N+1 \times N+1$ matrices.

Theorem 2.1: The state $0 \in S$ is accessible from every other state regardless of which $f \in F$ is being used.

Proof: Fix on $f \in F$ and put

$$c = \min \{ P_f(x,y): x,y \in S, P_f(x,y)>0 \}.$$

Take a ε-net $\{a_1, a_2, \ldots, a_n\}$ in the totally bounded set A and using the equicontinuity of the functions $p(y|x,\cdot)$, choose numbers $q_{xk} \geq 0$, such that

$$||\tilde{P}-P|| < \delta(\varepsilon)$$

where

$$\tilde{P}(x,y) \triangleq \sum_{k=1}^{n} q_{xk} p(y|x,a_k)$$

and

$$\sum_{k=1}^{n} q_{xk} = 1 \qquad \forall x \in S.$$

We may choose ε such that $\delta(\varepsilon) < \frac{c}{4}$, which means that $\tilde{P}_f(x,y)>0$ whenever $P_f(x,y)>0$. Next, modify $\tilde{P}_f$ to $\hat{P}_f$ such that

$$0 < \hat{P}_f(x,y) \leq \tilde{P}_f(x,y) - \frac{c}{2} < P_f(x,y) \tag{2.4}$$

whenever $P_f(x,y)>0$; this is done by eliminating for each x all but one of the q_{xk}, and replacing that q_{xk} by a smaller multiple (if necessary) called q. Thus

$$\hat{P}_f(x,y) = qp(y|x,a_{k_x}), \quad q>0. \tag{2.5}$$

where k_x is the index that corresponds to the q_{xk} that has not been eliminated. Consider now $g \in G$ described by $g(x) = a_{k_x}$ for each $x \in S$. From (2.5) we get $\hat{P}_f = qP_g$. By our hypothesis, 0 is accessible from x under P_g. Thus, for fixed $x \in S$, there exists m and states $i_1, i_2, \ldots, i_m$ such that

$$P_g(x,i_1)P_g(i_1,i_2) \cdots P_g(i_m,0) > 0 \tag{2.6}$$

Combining (2.4), (2.5) and (2.6) we arrive at

$$P_f(x,i_1)P_f(i_1,i_2) \cdots P_f(i_m,0) > 0$$

which proves the theorem.

Because of the accessibility assumption, and its consequence theorem 2.1, the Markov chain $\{X_n\}_{n=0}^{\infty}$ under fixed $f \in F$ will have several important properties. The matrix P_f^* has identical rows each of which is given by the unique probability row vector π_f that is a solution to

$$\pi_f P_f = \pi_f \tag{2.7}$$

The following two theorems are also consequences of the accessibility assumption and will be needed to obtain the main results of the next section.

Theorem 2.2: Suppose $\{f_n\}_{n=0}^{\infty}$ is a sequence in F such that

$$\lim_n P_{f_n} = P_{f_0}. \tag{2.8}$$

Then $\lim_n \pi_{f_n} = \pi_{f_0}$ ∎

Proof: For convenience, write π_n and P_n for π_{f_n} and P_{f_n}, respectively. Let $\{\pi_{n'}\}$ be a subsequence of $\{\pi_n\}$. Since the set of probability vectors forms a compact set in R^{N+1}, there is a subsequence $\{\pi_{n''}\}$ of $\{\pi_{n'}\}$ that converges to a probability vector π. But

$$\pi = \lim_{n''} \pi_{n''} = \lim_{n''} \pi_{n''} P_{n''} = \pi P_0$$

which, by uniqueness of solutions to (2.7), implies $\pi = \pi_0$. Hence, every subsequence of $\{\pi_n\}$ in turn has a subsequence converging to π_0, which proves the theorem ∎

In the proof of the next theorem, and what follows, we will refer to (G,ρ_G) for the metric space G with metric

$$\rho_G(g_1,g_2) = \sum_{x \in S} |g_1(x)-g_2(x)|.$$

(G,ρ_G) is a compact metric space since A is a compact subset.

Theorem 2.3: $\sup_{x \in S, g \in G} E_g[T \mid x_0=x] < \infty$ where $T = \inf\ \{ t \geq 0: x_t=0 \}$ ∎

Proof: Since G is compact, it suffices to prove that

$$E_g[T \mid x_0=x] \tag{2.9}$$

is a continuous function of g on G for each $x \in S$. Fix $x \in S$, and let $\{g_n\}_{n=0}^{\infty}$ be such that $g_n \to g_0$. Since $P_g(x,y) = p(y \mid x, g(x))$ is continuous on A, it follows that $P_{g_n} \to P_{g_0}$. Express P_{g_n} as

$$P_{g_n} = \begin{bmatrix} P_n(0,0) & P_n(0,1) \ \ldots\ldots & P_n(0,N) \\ R_n & & Q_n \end{bmatrix}$$

which gives $Q_n \to Q_0$. Because of the accessibility assumption [7, p51] we have

$$E_{g_n}[T \mid x_0=x] = [(I-Q_n)^{-1} e]_x$$

and

$$E_{g_0}[T \mid x_0=x] = [(I-Q_0)^{-1} e]_x$$

where e is the column vector of all ones. But $(I-Q_n)^{-1} \to (I-Q_0)^{-1}$ by a standard result in operator theory [8, p31], which shows that (2.9) is continuous in g on G ∎

3. Main Results

In this section we use the theory developed to prove the existence theorems. But first we indicate the method of approach by introducing the Lagrangian for the Markov decision problem with constraints.

3.1. Langrange Multiplier Techniques for MDP's with Constraints

The main tool employed to resolve the constrained problem (1.1)-(1.2) is the Lagrangian

$$J_u^\lambda(x) = \lim_t \frac{1}{t} E_u\left[\sum_{s=0}^{t-1} B^\lambda(x_s, a_s) \mid x_0 = x \right] \tag{3.1}$$

where $B^\lambda(x,a) = C(x,a) - \lambda D(x,a)$. We will study the unconstrained problem (3.1), to which we may apply the classical theory of MDP's, as the multiplier $\lambda \geq 0$ varies. In fact, if we put

$$J^\lambda(x) = \sup_{u \in U} J_u^\lambda(x) \tag{3.2}$$

we have the classical result

Theorem 3.1: For each $\lambda \geq 0$, there exists a unique w and a mapping h on S, unique up to an additive constant, such that the following DP equations are satisfied:

$$w + h(x) = \max_{a \in A} \{ B^\lambda(x,a) + \sum_y p(y \mid x,a) h(y) \} \tag{3.3}$$

Furthermore, $w = J^\lambda(x) = J_g^\lambda(x)$ for all x, where $g \in G$ is any pure policy such that

$$B^\lambda(x, g(x)) + \sum_y p(y \mid x, g(x)) h(y) = \max_{a \in A} \{ B^\lambda(x,a) + \sum_y p(y \mid x,a) h(y) \} \tag{3.4}$$

for any h satisfying (3.3).

Proof: Except for the unicity of h satisfying (2.3), all of the statements of the above theorem follow from theorem 2.3 and [6, theorem 6.19]. For the unicity of h, let h_1 and h_2 be two solutions to (3.3), and let g_1 and g_2 be solutions to (3.4) with h replaced by h_1 and h_2, respectively. Thus, using vector notation,

$$J^\lambda e + h_1 = B_{g_1}^\lambda + P_{g_1} h_1 \tag{3.5}$$

and

$$J^\lambda e + h_2 = B_{g_2}^\lambda + P_{g_2} h_2 \tag{3.6}$$

where $B_{g_i}^\lambda(x) \triangleq B^\lambda(x, g_i(x))$. Taking $v = h_1 - h_2$, we obtain

$$v \geq P_{g_1} v \tag{3.7}$$

which implies

$$v \geq P_{g_1}^* v. \tag{3.8}$$

By a similar argument we have

$$v \leq P_{g_2}^* v. \tag{3.9}$$

If we recall that π_g is any row in P_g^*, we get from (3.8) and (3.9)

$$\pi_{g_1} v \leq \min_{x \in S} [v(x)], \quad \max_{x \in S} [v(x)] \leq \pi_{g_2} v. \tag{3.10}$$

The first inequality is maintained only if $v(0) \leq \min_{x \in S}[v(x)]$, since $\pi_{g_1} > 0$ by the accessibility assumption. Similarly, $\max_{x \in S} [v(x)] \leq v(0)$; hence $v(x) = v(0)$ for all $x \in S$, and h is unique up to an additive constant.

We need to introduce the notation

$$C_f(x) = \int_A C_f(x,a) u_f(da \mid x) \tag{3.11}$$

for $f \in F$.The expressions $D_f(x)$ and $B_f^\lambda(x)$ are defined analogously. Because of the accessibility assumption, $V_f(x)$, $I_f(x)$ and $J_f^\lambda(x)$ do not depend on x (write V_f, I_f and J_f^λ) and are given by

$$V_f e = P_f^* C_f \quad I_f e = P_f^* D_f \quad J_f^\lambda e = P_f^* B_f^\lambda . \tag{3.12}$$

It is useful to note that V_g and I_g are continuous functions of g on G. This follows from the continuity of $C(x,\cdot)$, $D(x,\cdot)$, $p(y \mid x,\cdot)$, theorem 2.2 and equations (3.12).

The following theorem gives sufficient conditions of optimality for the constrained problem and is the underlying idea behind this article: we should search for a feasible policy $f \in F$ that optimizes the unconstrained problem, for some $\lambda \geq 0$, and that achieves $I_f = \alpha$.

Theorem 3.2: Suppose for some $\lambda \geq 0$ and $f \in F$ we have $I_f = \alpha$ and $J_f^\lambda = J^\lambda$. Then the constrained problem is optimized by f.

Proof: $J_f^\lambda \geq J_u^\lambda(x) \quad \forall u \in U, \quad \forall x \in S.$

Thus $$V_f \geq V_u(x) + \lambda(\alpha - I_u(x)) \quad \forall u \in U, \quad \forall x \in S \tag{3.13}$$

Now, if u is a feasible policy, then $\alpha - I_u(x)$ is nonnegative, which implies

$$V_f \geq V_u(x) \quad \forall x \in S$$

for all feasible $u \in V$.

The following series of lemmas will lead to some geometrical insight for the solution of the constrained problem. First denote G^λ for the set of all pure policies $g \in G$ such that $J_g^\lambda = J^\lambda$. By theorem (3.1), G^λ is not empty; but it may not be a singleton. We will say a mapping g from $[0,\infty)$ to G is a version of $\lambda \to G^\lambda$ if $g^\lambda \in G^\lambda$ for all $\lambda \geq 0$.

Lemma 3.1: $\lambda \to J^\lambda$ and $\lambda \to I_{g^\lambda}$ are decreasing no matter which version $\lambda \to g^\lambda$ is considered.

Proof : $$0 > J_{g^{\lambda+\eta}}^{\lambda+\eta} - J_{g^{\lambda+\eta}}^{\lambda} \geq J_{g^{\lambda+\eta}}^{\lambda+\eta} - J_{g^\lambda}^{\lambda} \geq J_{g^\lambda}^{\lambda+\eta} - J_{g^\lambda}^{\lambda} \tag{3.13}$$

From (3.13)

$$0 > -\eta I_{g^{\lambda+\eta}} \geq J^{\lambda+\eta} - J^\lambda \geq -\eta I_{g^\lambda} \tag{3.14}$$

which proves the lemma.

Lemma 3.2: $\lambda \to J^\lambda$ is uniformly continuous.

Proof: From (3.14) $|J^{\lambda+\eta} - J^\lambda| \leq \eta I_{g^\lambda} \leq \eta I_{g^0}$

which implies the lemma since $I_{g^0} \leq \sup_{x,a} D(x,a) < \infty$.

Lemma 3.3: Let $E = \{ \lambda \geq 0 : I_g \neq I_{\tilde{g}} \text{ for any } g, \tilde{g} \in G^\lambda \}$. Then E has measure zero.

Proof: Fix a version $\lambda \to g^\lambda$. Since monotonicity extends to any version

$$I_{g^{\lambda+\varepsilon}} \leq I_{\tilde{g}^{\lambda}} \leq I_{g^{\lambda-\varepsilon}}$$

where $\tilde{g}$ is any element of G^λ. If λ is a continuity point of $\lambda \to I_{g^\lambda}$, then $I_{g^\lambda} = I_{\tilde{g}^\lambda}$. But since $\lambda \to I_{g^\lambda}$ is monotone, it is continuous almost everywhere. Therefore E has measure zero ▪

If we were able to show that for some version of $\lambda \to g^\lambda$, that $\lambda \to I_{g^\lambda}$ is continuous, and that $I_{g^{\lambda_0}} \leq \alpha \leq I_{g^{\lambda_1}}$ for some λ_0, λ_1, then, by theorem 3.2, there would exist an optimal pure policy for the constrained problem. However, we do not in general have the continuity. For example, if A is a finite set, then I_{g^λ} takes on a finite number of values and the mapping $\lambda \to I_{g^\lambda}$ is only piecewise continuous. Therefore, we need a more detailed analysis. Denote

$$\gamma = \inf \ \{ \lambda \geq 0 : I_{g^\lambda} \leq \alpha \}$$

By lemma 3.3, γ is independent of the version $\lambda \to g^\lambda$ taken.

3.2. Existence and Structure of Optimal Policies

Suppose there exists a feasible policy $g \in G^0$; clearly g would be an optimal pure policy for the constrained problem. We henceforth assume that each $g \in G^0$ is infeasible.

Now suppose that $I_g > \alpha$ for all $g \in G$, and consider, for the moment, the Markov decision problem of minimizing $I_u(x)$ over all $u \in U$. Using the techniques in the proof of theorem 3.1, we can show that there exists a $g \in G$ with $I_g = \inf_{u \in U} I_u(x)$ for all x. Thus $I_u(x) > \alpha$ for all x, and all $u \in U$, i.e., all policies in U are infeasible. So if $I_g > \alpha$ for all $g \in G$, the problem is not well-posed. We henceforth make the natural assumption that there exists a $\hat{g} \in G$ with $I_{\hat{g}} < \alpha$. This condition can often be directly verified from $D(x,a)$ and $p(y|x,a)$.

In order to prove the existence theorem, and its corollaries we need to carefully chose a version $\lambda \to g^\lambda$. Henceforth, let g^λ be any one g that satisfies (3.4); thus $\lambda \to g^\lambda$ is a well-defined mapping. By the above assumption $I_{g^0} > \alpha$; we also have

Lemma 3.4: There exists $\lambda \geq 0$ such that $I_{g^\lambda} \leq \alpha$.

Proof: Suppose for all $\lambda \geq 0$, $I_{g^\lambda} > \alpha$. Then

$$J^\lambda = V_{g^\lambda} - \lambda I_{g^\lambda} < V_{g^\lambda} - \lambda\alpha.$$

Also $\quad J^\lambda \geq V_u(x) - \lambda I_u(x) \quad \forall u \in U, \quad \forall x \in S.$

So, $\quad V_{g^\lambda} - V_u(x) > \lambda(\alpha - I_u(x)) \quad \forall u, \quad \forall x.$

In particular, $\quad V_{g^\lambda} - V_{\hat{g}} > \lambda(\alpha - I_{\hat{g}}) \quad \forall u.$ (3.15)

But $\alpha - I_{\hat{g}} > 0$ and $V_{g^\lambda} - V_{\hat{g}} \leq \max_{a,x} 2C(x,a) < \infty$, which contradicts (3.15) since λ can be made arbitrarily large ▪

One of the consequences of the previous lemma is that γ is finite. Now let $\{\lambda_n\}$ and $\{\lambda'_n\}$ be two sequences such that $\lambda_n \uparrow \delta$, $\lambda'_n \downarrow \delta$, $\lim_n g^{\lambda_n} = \bar{g}$ exists in G and $\lim_n g^{\lambda'_n} = \underline{g}$ exists in G. Such sequences can be found since G is compact. Furthermore,

$$I_{\underline{g}} \leq \alpha \leq I_{\bar{g}}. \tag{3.16}$$

Now denote $[g_1, g_2, q]$ for the policy in F which chooses, at each decision epoch, g_1 or g_2 with probability q and $1-q$, respectively. Such policies will be referred to as random mixes. For convience, denote $f_q = [\bar{g}, \underline{g}, q]$.

Theorem 3.5: There exists a $q \in [0,1]$ such that $I_{f_q} = \alpha$.

Proof: By (3.16), it suffices to show that $\lambda \to I_{f_q}$ is continuous on $[0,1]$. Since $P_{f_q} = qP_{\bar{g}} + (1-q)P_{\underline{g}}$, we have $q \to P_{f_q}$ is continuous on $[0,1]$, which, by theorem 2.2, implies that $q \to \pi_{f_q}$ is continuous on $[0,1]$. But

$$I_{f_q} = \pi_{f_q}[qD_{\bar{g}} + (1-q)D_{\underline{g}}]$$

which shows that $q \to I_{f_q}$ is continuous on $[0,1]$.

Lemma 3.5: $J_{\bar{g}}^{\gamma} = J_{\underline{g}}^{\gamma} = J^{\gamma}$.

Proof: $J_{\bar{g}}^{\gamma} = V_{\bar{g}} - \gamma I_{\bar{g}} = \lim_n V_{g^{\lambda_n}} - \lambda_n I_{g^{\lambda_n}} = \lim_n J^{\lambda_n} = J^{\gamma}$

where we have used the continuity of $g \to I_g$, $g \to V_g$ and $\lambda \to J^{\lambda}$. Similary for $\underline{g}$.

Now, by theorem 3.2 and theorem 3.5, if we can show that $J_{f_q}^{\gamma} = J^{\gamma}$ for all $q \in [0,1]$, then there would exist a q such that f_q is optimal for the constrained problem. Lemma 3.5, however, does not imply this directly. In fact, the following counterexample shows that if g optimizes the unconstrained problem, it does not necessarily satisfy the DP equations; and that the random mix of two optimal policies is not necessarily optimal.

Counterexample: Consider the unconstrained problem $V = \sup_u V_u(x)$ with $S = \{0,1\}$, $A = \{a,b\}$, $p(0|0,a) = p(0|1,a) = 1$, $C(0,a) = 2$, $C(1,a) = 1$, $p(y|x,b) = \frac{1}{2}$ and $C(x,b) = 2$ for all x,y. It is clear that $V = 2$ and that g_1, g_2 are both optimal, where

$$g_1(0) = g_1(1) = a \qquad g_2(0) = g_2(1) = b.$$

The corresponding DP equations are

$$h(0)+2 = \max\ [\ 2+h(0);\ 2+\frac{1}{2}h(0)+\frac{1}{2}h(1)\ ]$$

$$h(1)+2 = \max\ [\ 1+h(0);\ 2+\frac{1}{2}h(0)+\frac{1}{2}h(1)\ ]$$

for which $h(0) = h(1) = 0$ is a solution. It is clear that g_1 does not satisfy the DP equations, although it is optimal. Now consider the random mix $f = [g_1, g_2, q]$. It is easy to show that $\pi_f = (\frac{3}{4}, \frac{1}{4})$ and $C_f = [2, \frac{3}{2}]^T$, which gives $V_{f_q} = \frac{15}{8}$. Hence, f_q is suboptimal, although g_1 and g_2 are optimal.

We are now in position to give the main result:

Theorem 3.6: There exists a random mix $[g_1, g_2, q]$ which is optimal for the constrained problem. Furthermore, there are real numbers $\gamma \geq 0$, w and a function h on S such that the following DP equations hold

$$\begin{aligned} w+h(x) &= \max_{a \in A}\ \{\ B^{\gamma}(x,a)+\sum_y p(y|x,a)h(y)\ \} \\ &= B^{\gamma}(x,g_1(x))+\sum_y p(y|x,g_1(x))h(y) \\ &= B^{\gamma}(x,g_2(x)+\sum_y p(y|x,g_2(x))h(y). \end{aligned} \tag{3.17}$$

Proof: We first want to show

$$J^{\gamma}e+h = B_{\bar{g}}^{\gamma}+P_{\bar{g}}h \tag{3.18}$$

and

$$J^{\gamma}e+h = B_{\underline{g}}^{\gamma}+P_{\underline{g}}h. \tag{3.19}$$

for some h. To do this, for each $\lambda \geq 0$, let h^{λ} be the unique solution to (3.3) with

$$P_{g^{\lambda}}^{*}h^{\lambda} = 0 \tag{3.20}$$

Thus

$$h^{\lambda}+J^{\lambda}e = B_{g^{\lambda}}^{\lambda}+P_{g^{\lambda}}h^{\lambda}. \tag{3.21}$$

Equations (2.3), (3.12), (3.20) and (3.21) give

$$h^{\lambda} = (I-P_{g^{\lambda}}+P_{g^{\lambda}}^{*})^{-1}(I-P_{g^{\lambda}}^{*})B_{g^{\lambda}}^{\lambda} \tag{3.22}$$

after some matrix algebra. Now $B_{g^{\lambda_n}}^{\lambda_n}$, $P_{g^{\lambda_n}}$ and $P_{g^{\lambda_n}}^{*}$ converge to $B_{\bar{g}}^{\gamma}$, $P_{\bar{g}}$ and $P_{\bar{g}}^{*}$, respectively. Since $(I-P_{g^{\lambda_n}}+P_{g^{\lambda_n}}^{*})$ converges to the invertible matrix $(I-P_{\bar{g}}+P_{\bar{g}}^{*})$, the inverse converges to $(I-P_{\bar{g}}+P_{\bar{g}}^{*})^{-1}$, by a standard result in operator theory [8, p31]. Thus

$$\lim_{n} h^{\lambda_n} = (I-P_{\bar{g}}+P_{\bar{g}}^{*})^{-1}(I-P_{\bar{g}}^{\gamma})B_{\bar{g}}^{\gamma} \underline{\Delta} \bar{h} \tag{3.23}$$

from which

$$(I-P_{\bar{g}}+P_{\bar{g}}^{*})\bar{h} = (I-P_{\bar{g}}^{*})B_{\bar{g}}^{\gamma}. \tag{3.24}$$

Premultiplying both sides of (3.24) by $P_{\bar{g}}^{*}$, and using (2.2) and (3.12) gives

$$P_{\bar{g}}^{*}\bar{h} = 0 \tag{3.25}$$

and

$$J^{\gamma}e+\bar{h} = B_{\bar{g}}^{\gamma}\bar{h}. \tag{3.26}$$

Now, because $h^{\lambda_n} \to \bar{h}$, we have

$$J^{\gamma}+h(x) = \lim_{n} \sup_{a} \{ B^{\lambda^n}(x,a)+\sum_{y}p(y|x,a)h^{\lambda_n}(y) \}. \tag{3.27}$$

It is easy to show

$$\lim_{n} B^{\lambda_n}(x,a)+\sum_{y}p(y|x,a)h^{\lambda_n}(y) = B^{\gamma}(x,a)+\sum_{y}p(y|x,a)\bar{h}(y)$$

where the convergence is uniform in a on A; thus we can interchange lim and sup in (3.25), which gives

$$J^{\gamma}+\bar{h}(x) = \sup_{a} \{ B^{\gamma}(x,a)+\sum_{y}p(y|x,a)\bar{h}(y) \}. \tag{3.28}$$

From (3.28) and theorem 3.1, $\bar{h}$ and h^{γ} are equal up to an additive constant, and taking into acount 3.26, we obtain (3.18) with h replaced by h^{γ} . A parallel argument leads to (3.19) with h replaced by h^{γ} .

To complete the proof, we multiply (3.18) and (3.19) by q and $1-q$, respectively, and then adding gives

$$J^{\gamma}e+h^{\gamma} = B_{f_q}^{\gamma}+P_{f_q}h^{\gamma}. \tag{3.29}$$

Finally, premultiplying both sides of (3.29) by $P_{f_q}^{*}$, we get

$$J^\gamma = J^\gamma_{f_q}. \tag{3.20}$$

Equation (3.30) combined with theorems 3.2 and 3.5 finish the proof if we take $g_1 = \underline{g}$ and $g_2 = \bar{g}$.

As a result of theorem 3.6, we can present two corollaries which give conditions for the existence of optimal pure policies and of optimal "almost" bang-bang policies. But first we need to consider the

Linearity Assumption:

a) $X = R^k$ (with which we equip an inner product $<\cdot, \cdot>$).

b) $p(y|x,a)$, $C(x,a)$ and $D(x,a)$ can be expressed as

$$p(y|x,a) = <r_1(x,y),a> + r_2(x,y)$$

$$D(x,a) = <d_1(x),a> + d_2(x)$$

$$C(x,a) = <c_1(x),a> + c_2(x)$$

for some mappings $r_1: S\times S \to X$, $r_2: S\times S \to R$, $d_1: S \to X$, $d_2: S \to R$, $c_1: S \to X$ and $c_2: S \to R$ ▪

Corollary 3.1: Suppose A is convex and the linearity assumptions are satisfied. Then there exists an optimal pure policy for the constrained problem.

Proof: For each $x \in S$, let $g(x) = qg_1(x)+(1-q)g_2(x)$ where $[g_1,g_2,q]$ is the optimal mixture of the previous theorem. Then

$$P_g(x,y) = q<r_1(x,y), g_1(x)> + (1-q)<r_1(x,y), g_2(x)> + r_2(x,y)$$
$$= qP_{g_1}+(1-q)P_{g_2} = P_{f_q}$$

with $f_q = [g_1,g_2,q]$. Similarly, we have $C_g = C_{f_q}$ and $D_g = D_{f_q}$. Thus $V_g = V_{f_q}$ and $I_g = V_{f_q}$, showing the optimality of g for the constrained problem ▪

Existence of optimal pure policies, as in the above corollary, for stochastic control problems with constraints is unusual. Such conditions implying optimality of pure policies were not obtained for the discounted [5] nor the optimal stopping [4] analogs of (1.1)-(1.2). In what follows, we will assume that $X = R$. Then a pure policy is said to be bang-bang if $g(x)$ takes on one of the two extreme values of A.

Corollary 3.2: Suppose the linearity assumptions is satisfied. Then g_1 and g_2 may be taken to be bang-bang.

Proof: Let $\bar{a} = \sup A$ and $\underline{a} = \inf A$. Then, by the hypotheses, equation (3.3) becomes

$$\max_{a\in A} \{ B^\lambda(x,a)+\sum_y p(y|x,a)h(y) \} = Z_2^\lambda(x)+\max [\bar{a}Z_1^\lambda(x), \underline{a}Z_1^\lambda(x)] \tag{3.31}$$

where $Z_i^\lambda(x) = c_i(x)-\lambda d_i(x)+\sum_y h(y)r_i(x,y) \quad i=1,2.$

It follows from (3.31) that for each $\lambda \geq 0$, there exists a bang-gang policy g^λ that satisfies (3.4). Let $\lambda \to g^\lambda$ be a version such that each g^λ is bang-bang and satisfies (3.4). It then follows that the limits $\bar{g}$ and $\underline{g}$ are both bang-bang ▪

Thus, we have found an optimal policy which is "almost" bang-bang in the sense that it randomly chooses between two bang-bang policies. Therefore, if the conditions of corollary 3.2 are satisfied, we may reduce A to the two element set $\{\underline{a}, \bar{a}\}$, which could be nice for computational purposes.

Acknowledgements:
The author would like to thank F. Beutler, M. Robin and F. Delebecque for their useful discussions.

References

[1] Dynkin, E.B. and Yushkevich, A.A., "Controlled Markov Processes," Springer-Verlag, Berlin, 1979.

[2] Lazar, A., "Optimal Flow Control of a Class of Queueing Networks in Equilibrium," IEEE AC-28, November 1983.

[3] Derman, C., "Finite State Markovian Decision Processes," Academic Press, New York, 1970.

[4] Robin, M. "On Optimal Stochastic Control with Constraints," Game Theory and Related Topics, North-Holland, 1979.

[5] Frid, E.B., "On Optimal Strategies in Control Problems with Constraints," Theory of Prob. Appl., Vol. XVIII, No. 1, 1972.

[6] Ross, Sheldon, "Applied Probability Models with Optimization Applications," Holden-Day, San Francisco, 1970.

[7] Kemeny, J. and Snell, J., "Finite Markov Chains," D. Van Nostrand Company, New York, 1960.

[8] Kato, T., "A Short Introduction to Perturbation Theory for Linear Operators," Springer-Verlag, New York, 1982.

Session 13

LINEAR SYSTEMS II

SYSTÈMES LINÉAIRES II

SOME CONNECTIONS BETWEEN ALGEBRAIC PROPERTIES OF PAIRS OF MATRICES AND 2D SYSTEMS REALIZATION

E. Fornasini, G. Marchesini
Istituto di Elettrotecnica e di Elettronica
6/A Via Gradenigo, Padova, Italy

ABSTRACT

This paper is concerned with some properties of transfer functions in two variables which can be realized by classes of 2D systems characterized by pairs of state updating matrices which generate algebras with special structures. Two situations are mainly considered. The first deals with pairs of matrices which generate a solvable Lie algebra (i.e. are simultaneously triangularizable). The second refers to pairs of matrices which generate abelian Lie algebras (i.e. the matrices commute).

The analysis of the connections between the properties of 2D realizations and transfer functions is based on the representation algorithms of non-commutative rational power series.

1. INTRODUCTION

It is well known [1,2,3] that any proper rational transfer function in two variables can be realized by a finite dimensional 2D system (A_1, A_2, B, C) described by the following state updating and read-out equations:

$$x(h+1,k+1) = A_1x(h+1,k)+A_2x(h,k+1)+B_1u(h+1,k)+B_2u(h,k+1)$$
$$y(h,k) = Cx(h,k) \tag{1}$$

In general it should be expected that any constraint we assume on the structure of the pairs (A_1,A_2) translates into a restriction of the class of transfer functions which can be realized by (1).

In this communication we shall concentrate our attention on pairs of matrices which can be simultaneously reduced by similarity to upper (lower) triangular form and, in particular, on pairs of commutative matrices.

Commutative matrices have been first considered by Attasi [4], with reference to the special class of systems given by the following equations

$$x(h+1,k+1) = A_1\, x(h+1,k) + A_2\, x(h,k+1) - A_1A_2\, x(h,k) + B\, u(h,k)$$
$$y(h,k) = C\, x(h,k) \tag{2}$$

with $A_1A_2 = A_2A_1$. The transfer functions realizable by this model are (causal) separable functions, that is they can be written in the form $n(z_1,z_2)/p(z_1)q(z_2)$, where n is in $(z_1,z_2)\, K[z_1,z_2]$, p in $K[z_1]$ and q in $K[z_2]$. The converse is also true, in the sense that any (causal) separable transfer function is realizable in the class of Attasi's models.

As we shall see, the main feature of the transfer functions we obtain from (1) when A_1 and A_2 commute, is that their denominators factor completely in the complex field into linear factors [5].

The same is true when the commutativity assumption is weakened and we assume that A_1 and A_2 are simultaneously triangularizable. The difference between the two cases is that the commutativity of A_1 and A_2 imposes some constraints on the numerator of the transfer function while triangularizability does not.

In order to make our analysis simpler, we shall assume that either B_1 or B_2 is the zero vector. So doing the analysis developed in the sequel, applies also to the following models [6,7]:

$$x(h+1,k+1) = A_1x(h+1,k) + A_2x(h,k+1) + B\, u(h,k) \tag{3}$$
$$y(h,k) = C\, x(h,k)$$

and:

$$x(h+1,k+1) = A_1x(h+1,h) + A_2x(h,k+1) + B\, u(h+1,k+1) \tag{4}$$
$$y(h,k) = C\, x(h,k)$$

If we don't take into account the multiplicative factors z_1, z_2 or z_1z_2, which are unessential to our discussion, the structure of the transfer functions of systems (2) (with B_1 or $B_2 = 0$), (3) and (4) reduces to the following form

$$s = C(I-A_1z_1-A_2z_2)^{-1}B \tag{5}$$

The possibility of representing a proper rational function in the form (5), allows us to associate its realization (A_1,A_2,B,C) with the series

$$\sigma = C(I-A_1\xi_1-A_2\xi_2)^{-1}B \tag{6}$$

in the non-commutative variables ξ_1 and ξ_2 and to exploit known results from the theory of non-commutative power series [8].

2. REALIZABILITY AND SIMULTANEOUS TRIANGULARIZATION

Two matrices A_1 and A_2 are simultaneously triangularizable if they can be reduced by similarity transformation to upper (lower) triangular form.

Simultaneous triangularizability - also referred in the literature as property P - has been related to other algebraic properties of pairs of matrices. We summarize the principal results in the following theorem [9,10]:

Theorem 1. *Let* A_1 *and* A_2 *belong to* $\mathbb{C}^{n\times n}$. *Then the following statements are equivalent:*

(i) *there is an invertible matrix* T *such that* $P^{-1}A_1P$ *and* $P^{-1}A_2P$ *are upper (lower) triangular;*

(ii) *the Lie algebra* $\mathscr{L}$ *defined by matrices* A_1 *and* A_2 *is solvable;*

(iii) *for every scalar polynomial* $\pi(\xi_1,\xi_2)$ *in the non-commutative variables* ξ_1,ξ_2, *each of the matrices* $\pi(A_1,A_2)[A_1,A_2]$ *is nilpotent;*

(iv) *there is an ordering of the eigenvalues* λ_i *of* A_1 *and* μ_i *of* A_2 *such that the eigenvalues of any scalar polynomial* $\pi(A_1,A_2)$ *are* $\pi(\lambda_i,\mu_i)$, $i=1,2,\dots,n$.

As an obvious consequence of property P, we have that the polynomial $\det(I-A_1z_1-A_2z_2)$ factors completely in the complex field into linear factors:

$$\det(I-A_1z_1-A_2z_2) = \prod_i(1-\lambda_iz_1-\mu_iz_2) \tag{7}$$

The factorization property (7) - also called property L [10] - is weaker than property P, if $n > 2$.

The role played by pairs of matrices with property P in the realization of 2D systems is defined by the following theorem.

Theorem 2. *Let* $W(z_1,z_2) = p(z_1,z_2)/d(z_1,z_2)$, $d(0,0) = 1$ *and* p *and* q *coprime polynomials. Then* $W(z_1,z_2)$ *is realizable by a 2D system with* A_1 *and* A_2 *having property* P *if and only if* $d(z_1,z_2)$ *factors completely in the complex field into linear factors.*

Proof. Assume A_1 and A_2 have property P. By (6), since $d(z_1,z_2)$ divides $\det(I-A_1z_1-A_2z_2)$, it factors into linear elements. Conversely, note that starting from 2D systems with A_1 and A_2 having property P, and connecting them in series and parallel, the A_1 and A_2 matrices of the resulting systems still have property P. So, we need only to take into account transfer functions $W_{ij}(z_1,z_2) = z_1^i z_2^j/1-\alpha z_1-\beta z_2$. The following 2D system, with A_1 and A_2 in triangular form,

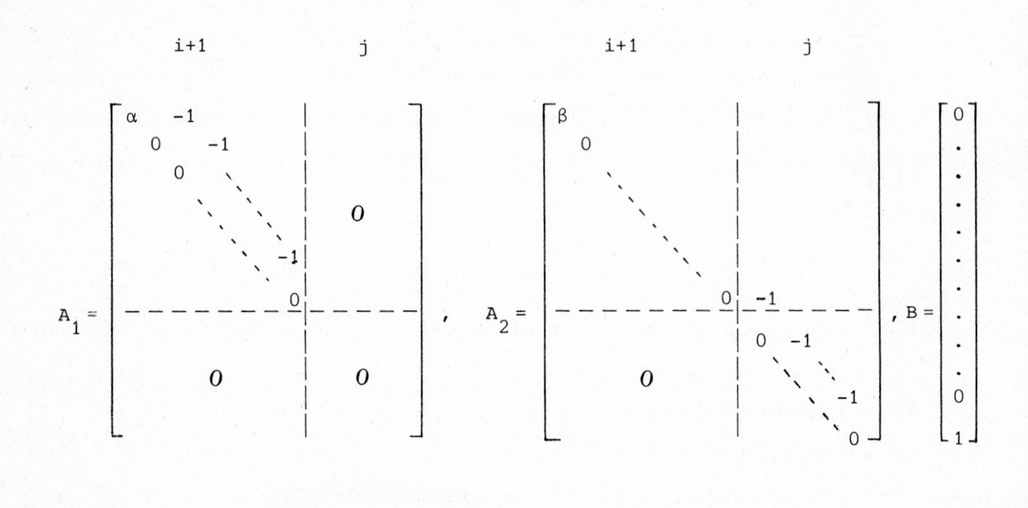

$C = [1\ 0\ \dots\dots\dots\ 0]$

provides a 2D realization of the elementary transfer function W_{ij}.

A classical result due to Frobenius [11] states that any pair of commutative matrices satisfies property P. This fact can be viewed as a corollary of Theorem 1, since the commutativity hypothesis $[A_1,A_2] = 0$ implies (iii).

Consequently, as 2D systems with commutative matrices A_1 and A_2 are a subclass of 2D systems with triangular matrices, the denominator of their transfer function factors completely into linear elements.

Nevertheless, as we shall see, it is not true that any transfer function with denominator factorizable into linear elements, can be realized by a 2D system with commutative matrices A_1 and A_2. This is due to the fact that when we look for 2D realizations of this type, the numerator of the transfer function cannot be arbitrarily assigned.

The analysis of the constraints to be imposed on the transfer functions for obtaining 2D system realizations with A_1 and A_2 commuting, will be developed in the next section by resorting to non-commutative power series. This will allow us also to give a first insight into the problem of understanding how property P and commutativity affect the structure of minimal realizations.

3. COMMUTATIVITY AND PROPERTY P IN THE REPRESENTATION OF NON-COMMUTATIVE POWER SERIES

Simultaneous triangularization and commutativity assumptions on A_1 and A_2 impose structural constraints on the coefficients of the non-commutative power series (6). The nature of these constraints is relevant for the analysis of 2D systems having A_1 and A_2 matrices with the same properties. In fact we can associate any non-commutative power series σ with its commutative image induced by the algebra morphism $\phi : K\langle\langle \xi_1,\xi_2\rangle\rangle \longrightarrow K[[z_1,z_2]]$, assigned by $\phi(k) = k$, $\forall k \in K$, $\phi(\xi_1) = z_1$, $\phi(\xi_2) = z_2$. Then, assuming σ to be represented as in (6), the map ϕ associates the non-commutative series σ with the 2D system (A_1,A_2,B,C) whose transfer function is $\phi(\sigma) = C(I-A_1z_1-A_2z_2)^{-1}B$.

In order to analyze in detail these facts, we need some properties of non-commutative power series that we shall briefly recall in the sequel.

Let K be the ground field. A generic element σ of the algebra $K\langle\langle \xi_1,\xi_2\rangle\rangle$ of formal power series in the noncommuting variables ξ_1 and ξ_2 with coefficients in K is written as

$$\sigma = \sum_{w \in \{\xi_1,\xi_2\}^*} (\sigma,w)w$$

where $\{\xi_1,\xi_2\}^*$ is the free monoid generated by ξ_1 and ξ_2 and (σ,w) in K is the <u>coefficient</u> of w in the series σ. The series $\phi(\sigma)$ in $K[[z_1,z_2]]$ is called the <u>commutative image</u> of σ.

A series σ in $K\langle\langle \xi_1,\xi_2\rangle\rangle$ is <u>exchangeable</u> if the words which have the same commutative image have the same coefficient in σ.

A series σ in $K\langle\langle \xi_1,\xi_2\rangle\rangle$ is <u>rational</u> if there exist a positive integer n and matrices A_1, A_2 in $K^{n\times n}$, B in $K^{n\times 1}$, C in $K^{1\times n}$ such that

$$\sigma = C \sum_{k=0}^{\infty} (A_1\xi_1+A_2\xi_2)^k B = C(I-A_1\xi_1-A_2\xi_2)^{-1} B \tag{8}$$

A 4-tuple (A_1,A_2,B_1,C) is called a <u>representation</u> of σ if (8) holds.

The following Theorem [8] shows how the commutativity assumption $[A_1,A_2] = 0$ made on the representation (6) of a non-commutative series σ, can be expressed as a condition on the coefficients of σ itself.

<u>Theorem 3</u>. *Let σ be in $K\langle\langle \xi_1,\xi_2 \rangle\rangle$. Then the following facts are equivalent:*

i) *σ is rational and exchangeable*

ii) *σ is a linear combination of shuffle products*(*) *of the following form*

$$p(\xi_1)q(\xi_1)^{-1} \sqcup\!\sqcup \; r(\xi_2)t(\xi_2)^{-1} \tag{9}$$

where p, q, r, t are polynomials

iii) *there exists a representation (A_1,A_2,B,C) of σ with $A_1A_2 = A_2A_1$, that is*

$$(\sigma,w) = C A_1^{|w|_1} A_2^{|w|_2} B, \; \forall\, w \in \{\xi_1,\xi_2\}^*$$

where $|w|_i$ denotes the number of ξ_i in w, i = 1,2.

A further characterization of exchangeable rational series is given in terms of separable rational functions.

<u>Theorem 4</u>. *Let $\sigma \in K\langle\langle \xi_1,\xi_2 \rangle\rangle$ be exchangeable and define the map $\bar{\phi}$: by the assignment*

$$\bar{\phi} : \sum_w (\sigma,w)w \mapsto \sum_{i,j=0}^{\infty} (\sigma,\xi_1^i\xi_2^j) z_1^i z_2^j$$

Then σ is rational if and only if $\bar{\phi}(\sigma)$ is (the power series expansion of) a separable rational function.

(*) For any f and g in $\{\xi_1,\xi_2\}^*$ the shuffle product of f and g is defined as $f \sqcup\!\sqcup g \triangleq \sum \{f_1g_1\ldots f_Kg_K \mid f = f_1\ldots f_K,\ g = g_1\ldots g_K\}$. By linearity, the definition extends to $K\langle\langle \xi_1,\xi_2 \rangle\rangle$.

Assume now that the series σ admits a representation with A_1 and A_2 having property P. The following Theorem shows how this assumption reduces to a condition on the coefficients of σ.

Theorem 5. *Let σ be a rational series in $K\langle\langle \xi_1,\xi_2 \rangle\rangle$ and admit a representation of dimension n. Then σ admits a representation with A_1 and A_2 having property P if and only if for any (n+1)-tuplet $w_1,\ldots,w_{n+1}$ in $\{\xi_1,\xi_2\}^*$ we have*

$$\sum_{i_1,\ldots,i_n \in \{1,2\}} (-1)^{i_1+i_2+\ldots+i_n} (\sigma, w_1\gamma_{i_1} \ldots\ldots w_n\gamma_{i_n} w_{n+1}) = 0 \tag{10}$$

where $\gamma_1 \stackrel{\Delta}{=} \xi_1\xi_2$ *and* $\gamma_2 \stackrel{\Delta}{=} \xi_2\xi_1$.

Proof. Let (A_1,A_2,B,C), with A_1 and A_2 having property P, be a representation of σ. It is not restrictive to assume that the dimension of this representation is less than or equal to n. In fact any minimal representation $(\bar{A}_1,\bar{A}_2,\bar{B},\bar{C})$ of σ can be obtained (modulo a similarity transformation) from (A_1,A_2,B,C) by standard reducing procedures without destroying property (iii) of Theorem 1 and hence property P. Then, for any (n+1)-tuplet $w_1,\ldots,w_{n+1}$, in $\{\xi_1,\xi_2\}^*$, we have

$$w_1(A_1,A_2)\,[A_1,A_2]\,\ldots\ldots\, w_n(A_1,A_2)\,[A_1,A_2]\, w_{n+1}(A_1,A_2) = 0 \tag{11}$$

as we can check directly by assuming A_1 and A_2 in triangular form.

Let now multiply (11) by C on the left and by B on the right to get (10). Conversely, let (A_1,A_2,B,C) be a minimal representation of σ of dimension $m \leq n$. It is known from [8] that there exist m^2 matrices $M_{ij} \in K^{m\times m}$ and two sets of m words, each word with length less than m, $\{d_1,\ldots,d_m\}$ and $\{g_1,\ldots,g_m\}$ such that for any $w \in \{\xi_1,\xi_2\}^*$, it results

$$w(A_1,A_2) = \Sigma_{h,k}\, M_{h,k}(\sigma, g_h w\, d_k)\ .$$

Then, for any n-tuplet $w_1,\ldots,w_n$ in $\{\xi_1,\xi_2\}^*$ we have

$$\begin{aligned}\sum_{i_1,\ldots,i_n} (-1)^{i_1+\ldots+i_n} w_1(A_1,A_2)\gamma_{i_1}(A_1,A_2)\ldots\ldots w_n(A_1,A_2)\gamma_{i_n}(A_1,A_2) &= \\ = \sum_{h,k} M_{h,k} \sum_{i_1,\ldots i_n} (-1)^{i_1+\ldots+i_n} (\sigma, g_h w_1\gamma_{i_1}\ldots w_n\gamma_{i_n} d_k) &= 0\end{aligned} \tag{12}$$

Now take any polynomial π in $K<\xi_1,\xi_2>$ and consider the matrix

$$(\pi(A_1,A_2)\ [A_1,A_2])^n \quad .$$

This turns out to be zero since it is a linear combination of terms of the same type as those in the summation on the left side of (12).

By applying criterion (iii) in Theorem 1 we conclude that A_1 and A_2 satisfy property P.

In view of the applications we shall made, it is worthwhile to state by a separate Theorem the following fact we already used in the proof of Theorem 5.

Theorem 6. *Assume that the rational series* σ *in* $K<<\xi_1,\xi_2>>$ *admits a representation* (A_1,A_2,B,C) *with* A_1 *and* A_2 *simultaneously triangularizable (commutative). Then the matrices* A_1 *and* A_2 *appearing in any minimal representation of* σ *are simultaneously triangularizable (commutative).*

4. COMMUTATIVE REALIZATIONS

Let's now go back to the problem of the existence of commutative realizations. Consider a 2D rational transfer function s and denote by $\mathcal{M}$ the set of the 2D systems $\Sigma = (A_1,A_2,B,C)$ which realize s. Denote by $\mathcal{N}$ the set of noncommutative rational power series whose commutative image is s.

Then any system $\Sigma = (A_1,A_2,B,C)$ in $\mathcal{M}$ is associated with a representation of a noncommutative series σ in $\mathcal{N}$, i.e. the series $\sigma = C(I-A_1\xi_1-A_2\xi_2)^{-1}B$.

Viceversa, any series σ in $\mathcal{N}$ admits representations (A_1,A_2,B,C) and, since $\phi(\sigma) = s$, the corresponding 2D systems $\Sigma = (A_1,A_2,B,C)$ are realizations of s, that is elements of $\mathcal{M}$ [3].

It is now clear that there exists a commutative realization of s if and only if $\mathcal{N}$ contains an exchangeable series, or, in other terms, if and only if the (unique) exchangeable series σ^* having s as commutative image is rational. Moreover the full class of the commutative realizations of s is identified with the class of the commutative representations (8) of σ^*.

Theorem 4 provides another condition for the existence of a commutative realization of s in terms of separability of a commutative power series.

Given $s = \sum_{i,j} s_{ij} z_1^i z_2^j$, introduce the series

$$\bar{s} = \bar{\phi}(\sigma^*) = \sum_{i,j} \bar{s}_{ij} z_1^i z_2^j \quad , \qquad \bar{s}_{ij} = \binom{i+j}{j}^{-1} s_{ij} \tag{13}$$

Assume s have a commutative realization $\Sigma = (A_1, A_2, B, C)$. Then, from

$$s = C(I - A_1 z_1 - A_2 z_2)^{-1} B = \sum_{i,j=0}^{\infty} \binom{i+1}{j} C A_1^i A_2^j B z_1^i z_2^j \tag{14}$$

we have

$$\bar{s} = \sum_{i,j=0}^{\infty} C A_1^i A_2^j B z_1^i z_2^j = C(I - A_1 z_1)^{-1} (I - A_2 z_2)^{-1} B \tag{15}$$

which shows that $\bar{s}$ is separable.

For the converse, assume $\bar{s}$ be separable. Then $\bar{s}$ can be represented as in (15), with $A_1 A_2 = A_2 A_1$ (see, for instance, |8|), and we go back to (14) following the previous steps in the reverse order.

Remark. If s admits a commutative realization, the commutative representations (8) of the associated exchangeable series σ^* are in one to one correspondence with the commutative representations (15) of the separable series $\bar{s}$. This shows that the series σ^* and $\bar{s}$ play essentially the same role in the solution of the commutative realization problem.

The existence of commutative realizations of a transfer function s and their construction are essentially based on the properties of Hankel matrices.

The Hankel matrix [8] of a non-commutative series σ (a commutative series r) is an infinite matrix whose rows and columns are indexed by the words of the free monoid $\{\xi_1, \xi_2\}^*$ (by the monomials $z_1^i z_2^j$). The matrix element indexed by the pair (u,v) (by the pair $(z_1^i z_2^j, z_1^h z_2^k)$) is the coefficient (σ, uv) of the word uv (the coefficient $r_{i+h,j+k}$ of the monomial $z_1^{i+h} z_2^{j+k}$).

Denoting by H(r) the Hankel matrix of r, we have that:

i) r is separable if and only if rank H(r) is finite

ii) rank H(r) gives the dimension of minimal, commutative representations (15) of r

iii) minimal, commutative representations (15) are algebraically equivalent. They can be computed from H(r) via Ho's algorithm [4].

Analogously, let $H(\sigma)$ be the Hankel matrix of σ. Then

i) σ is rational if and only if rank $H(\sigma)$ is finite

ii) rank $H(\sigma)$ gives the dimension of minimal representations (8) of σ

iii) minimal representations (8) are algebraically equivalent and can be derived from $H(\sigma)$ via Ho's algorithm [3].

By Theorem 4, minimal representations of the exchangeable series σ^* are necessarily commutative and coincide with minimal representations (15) of $\bar{s}$. So we have rank $H(\sigma)$ = rank $H(\bar{s})$.

The rank finiteness of $H(\bar{s})$ is equivalent to the existence of commutative realizations of s, and the 4-tuples (A_1,A_2,B,C) which provide minimal, commutative representations (15) of $\bar{s}$ constitute the minimal commutative realizations of s. Since minimal representations (15) are algebraically equivalent, minimal commutative realizations are essentially unique, modulo a change of basis in the local state space. This make a strong difference between commutative and non-commutative realizations, since non-commutative realizations are not necessarily algebraically equivalent [6].

The realizability condition based on the rank of $H(\bar{s})$ allows us to give a negative answer to the question whether structure conditions on the denominator of the transfer functions s are sufficient to guarantee the existence of commutative realizations.

This is done by considering the following rational function

$$s = \frac{1}{(1-z_1)(1-z_1-z_2)} = \sum_{i,j=0}^{\infty} \binom{i+j+1}{j+1} z_1^i z_2^j \tag{16}$$

So, by (13), we have

$$\bar{s} = \sum_{i,j=0}^{\infty} \frac{i+j+1}{j+1} z_1^i z_2^j$$

In the Hankel matrix

$$H(\bar{s}) = \begin{bmatrix} H_{00} & H_{01} & H_{02} & \cdots \\ H_{10} & H_{11} & H_{12} & \cdots \\ H_{20} & H_{21} & \cdots & \cdots \\ \cdots & \cdots & \cdots & \cdots \end{bmatrix}$$

the diagonal block matrices are given by:

$$H_{00} = [1], \quad H_{11} = 3\begin{bmatrix} 1 & 1/2 \\ 1/2 & 1/3 \end{bmatrix}, \quad \ldots \quad H_{nn} = (2n+1)\begin{bmatrix} 1 & 1/2 & \ldots & 1/n+1 \\ 1/2 & 1/3 & \ldots & 1/n+2 \\ \ldots & \ldots & \ldots & \ldots \\ 1/n+1 & 1/n+2 & \ldots & 1/2n+1 \end{bmatrix} \ldots$$

Now notice that $H_{nn}/(2n+1)$, $n = 0,1,2\ldots$ are the $(n+1)x(n+1)$ submatrices appearing in the upper left hand corner of the Hankel matrix associated with the nonrational power series $-\log(1-x) = \sum_{n=1}^{\infty} x^n/n$.

Letting n go to infinity in rank $H(\bar{s}) > \text{rank } H_{nn}$, we obtain rank $H(\bar{s}) = \infty$. This implies that (16) cannot be realized using commutative matrices A_1 and A_2, despite the denominator of s factorizes as a product of linear factors.

An existence condition for commutative realizations may be obtained by using jointly the following facts:

i) a rational function s admits a commutative realization if and only if it is the commutative image of a <u>rational</u> exchangeable noncommutative series σ

ii) a noncommutative series can be represented as a linear combination of series with structure (9).

By exploiting partial fraction expansion of rational functions in one variable, the series having structure (9) reduce to linear combinations of the noncommutative series $\xi_1^m \sqcup\!\sqcup\, \xi_2^n$, $\xi_1^m \sqcup\!\sqcup\, (1-b\xi_2)^{-n}$, $(1-a\xi_1)^{-m} \sqcup\!\sqcup\, \xi_2^n$, $(1-a\xi_1)^{-m} \sqcup\!\sqcup\, (1-b\xi_2)^{-n}$, m, n $\in$ N.

Thus the commutative image of a rational exchangeable series is the power series expansion of a linear combination of the following functions:

$$z_1^m z_2^n, \quad \frac{\partial^m}{\partial z_2^m} \frac{(z_1 z_2)^m}{(1-bz_2)^n}, \quad \frac{\partial^n}{\partial z_1^n} \frac{(z_1 z_2)^n}{(1-az_1)^m}, \quad \frac{\partial^{m+n}}{\partial z_1^m \partial z_2^n} \frac{z_1^m z_2^n}{1-a z_1 - b z_2}. \tag{17}$$

Viceversa, any linear combination of rational functions (17) is the commutative image of an exchangeable rational series, hence it admits a commutative realization.

5. FURTHER REMARKS

In general, given a rational transfer function, the class of its realizations with matrices A_1 and A_2 having property P, does not share all properties with the class of commutative realizations.

For instance, minimal realizations with A_1 and A_2 having property P, need not be algebraically equivalent.

Example. The following 2D systems

$$\Sigma_1 : \left(\begin{bmatrix} 1 & 1 \\ 1 & 0 \end{bmatrix}, \begin{bmatrix} 1 & 0 \\ 0 & 0 \end{bmatrix}, \begin{bmatrix} 1 \\ 1 \end{bmatrix}, \begin{bmatrix} 1 & 0 \end{bmatrix}\right)$$

$$\Sigma_2 : \left(\begin{bmatrix} 1 & 1 \\ 0 & 1 \end{bmatrix}, \begin{bmatrix} 0 & -1 \\ 0 & 1 \end{bmatrix}, \begin{bmatrix} -1 \\ 1 \end{bmatrix}, \begin{bmatrix} 1, 2 \end{bmatrix}\right)$$

are minimal realizations of (16) with A_1 and A_2 triangular matrices. Yet, Σ_1 and Σ_2 are not algebraically equivalent. This follows checking that the non-commutative power series associated with Σ_1 and Σ_2 are different.

Moreover, Σ_1 and Σ_2 represent (modulo similarity transformations) the whole class of minimal realizations of (16) which is then wholly constituted by 2D systems with A_1 and A_2 triangularizable.

This is not surprising. In fact, minimal realizations of any rational transfer function whose denominator factors into linear elements, have matrices A_1 and A_2 with property P, if their dimension is 2. If the dimension is greater than 2 the following example shows that matrices A_1 and A_2 of minimal realizations need not simultaneously triangularize.

Example. The following 2D systems

$$\Sigma_1 : \left(\begin{bmatrix} 0 & 1 & 0 \\ 0 & 0 & -1 \\ 0 & 0 & 0 \end{bmatrix}, \begin{bmatrix} 0 & 0 & 0 \\ 1 & 0 & 0 \\ 0 & 1 & 0 \end{bmatrix}, \begin{bmatrix} 1 \\ 0 \\ 0 \end{bmatrix}, \begin{bmatrix} -1 & 0 & 0 \end{bmatrix}\right)$$

$$\Sigma_2 : \left(\begin{bmatrix} 0 & -1 & 0 \\ 0 & 0 & 0 \\ 0 & 0 & 0 \end{bmatrix}, \begin{bmatrix} 0 & 0 & 0 \\ 0 & 0 & -1 \\ 0 & 0 & 0 \end{bmatrix}, \begin{bmatrix} 1 \\ 0 \\ 1 \end{bmatrix}, \begin{bmatrix} 1 & 0 & 0 \end{bmatrix}\right)$$

are minimal realizations of the polynomial $1-z_1z_2$. It is easy to check that A_1 and A_2 from Σ_1 do not have property P. Actually $[A_1, A_2]$ is not nilpotent.

Finally, we observe that minimal commutative realizations of a transfer function have higher dimension than minimal realizations with property P and, a fortiori, than minimal unconstrained realizations of the same transfer function. As an example, $z_1^m z_2^m$ has minimal commutative realizations of dimension $(m+1)^2$, while the dimension of minimal realizations with property P is 2m+1 [5].

REFERENCES

|1| Fornasini E., and Marchesini G. (1978) "Doubly Indexed Dynamical Systems: State Space Models and Structural Properties", Mathematical Systems Theory, vol. 12, n. 1.

|2| Fornasini E., and Marchesini G. (1976) "State Space Realization Theory of Two--Dimensional Filters", IEEE Trans on Automat. Contr., vol. AC-21, pp. 484-492.

|3| Fornasini E., and Marchesini G. (1980) "On the Problem of Constructing Minimal Realizations for Two-Dimensional Filters", IEEE Trans PAMI, vol. 2, n. 2, pp. 172-76.

|4| Attasi S. (1973) "Systèmes linéaries homogènes à deux indices", Rapport LABORIA, 31.

|5| Bisiacco M., Fornasini E., and Marchesini G. (1983) "On Commutative Realizations of 2D Transfer Functions", Proc. of 1983 EES-MECO Congress, Athens.

|6| Fornasini E., and Marchesini G. (1976) "Two Dimensional Filters: New Aspects of the Realization Theory", Third Int. Joint Conf. on Pattern Recognition, Coronado, California, Nov. 8-11.

|7| Sontag E.D. (1978) "On First-Order Equations for Multi-Dimensional Filters", IEEE Trans ASSP, vol. 26, pp. 480-82.

|8| Fliess M. (1974) "Matrices de Hankel", J. Math. Pures et Appl. 53, pp. 197-224.

|9| McCoy N.H. (1936) "On the Characteristic Roots of Matric Polynomials", Bull. Amer. Math. Soc., pp. 592-600.

|10| Motzkin T.S., Taussky O. (1952) "Pavis of Matrices with Property L", Trans Amer Math. Soc, vol. 73, pp. 108-114.

|11| Suprunenko D.A., Tyshkevich R.I. (1968) "Commutative Matrices", Acad. Press..

CONSERVATION DE LA MINIMALITE

par

ECHANTILLONNAGE ALEATOIRE

C. DENIAU
Université Aix-Marseille III
G R E Q F
41,rue des dominicaines
13001 MARSEILLE FRANCE.

G. OPPENMEIM Cl. VIANO
Université Paris V
Equipe statistique Appliquée Université Paris Sud . Bât 425 91405 ORSAY-Cédex France

Abstract :

The controllability preservation of controlled non stochastic continuous time linear system after discretization has been studied in [6] then in [10],[11],[12].

We study a similar problem : the controllability preservation of a non controlled stochastic linear system when the discretization process is run by renewal process.

This study is concerned with discrete time and set necessary and sufficient conditions for the preservation of minimality of linear system representation :

$$X_{(t+1)\Delta} = F_\Delta X_{t\Delta} + \varepsilon_{t\Delta} \quad ; \; Y_{t\Delta} = H X_{t\Delta} \quad , \; t \in \mathbb{Z}$$

by a renewal process $(T_t)_{t \in \mathbb{N}}$ (when Δ is yhe basis sampling step).

Conditions are on the injectivity property of the generating function of $(T_{t+1} - T_t)$.

We have similar results with continuous time and with unstable systems. They are not included here.

0. INTRODUCTION.

0.1 La première observation d'où découle ce travail concerne la stabilité des systèmes linéaires (au moins ceux dont le spectre de la matrice de transition est inclus dans le disque unité ouvert) par un échantillonnage aléatoire. La procédure d'échantillonnage étudiée est un processus de renouvellement T sur le temps ; T est indépendant des processus d'état et d'observation du système linéaire.

Dans tout système linéaire, la minimalité du système (qui est la possibilité de commander et d'observer tous les états en un temps fini) est une propriété essentielle. Les dangers causés par l'incommandabilité sont bien connus par exemple en résistance de structures qui risquent la rupture lorsqu'elles vibrent. L'inobservabilité entraîne la méconnaissance de l'état quelle que soit la durée de l'observation. Une question se pose donc immédiatement : un système linéaire minimal aléatoirement échantillonné conserve-t-il sa propriété de minimalité ?

Nous disposons d'une condition nécessaire et suffisante portant sur une fonction ϕ (extension au disque unité de la fonction caractéristique de la loi d'échantillonnage $\phi(Z) = \sum_{1}^{\infty} L_j Z^j$) pour que la propriété de minimalité soit héréditaire par échantillonnage aléatoire. L'établissement de cette propriété fait l'objet de la *seconde partie*.

La *troisième partie* concerne l'étude des conséquences sur la perte de minimalité d'un système (F,G,H) donné, de la non injectivité ou de la séparation (Splitting en anglais) de ϕ sur le spectre de F.

Notation : Si U $p \times q$ est une matrice, [U] désigne le sous-espace vectoriel de $\mathbb{R}^p$ engendré par les vecteurs colonnes de U, $\rho[U]$ est le rang de U, ${}^tA$ est la transposée de A, $\lambda(F)$ désigne une valeur propre de F et Spect(F) le spectre de F.

On utilise les abréviations :

vp : valeur propre, VP : vecteur propre, VPG : vecteur propre généralisé.

Enfin on note ϕ' (resp. $\phi^{(k)}$) la dérivée première (resp. $k^{\text{ième}}$) de la fonction ϕ et [a] la partie entière de $a \in \mathbb{R}$.

02. *Ensemble des systèmes linéaires étudiés* .

Soit S un système linéaire homogène à temps discret $t \in \mathbb{Z}$

$$(1) \qquad (S) \begin{cases} X_{t+1} = FX_t + G\varepsilon_t \\ Y_t = HX_t \end{cases}$$

où F $n \times n$, G $n \times r$, H $s \times n$ sont des matrices fixes, quel que soit $t \in \mathbb{Z}$ Y_t et X_t sont des vecteurs aléatoires du second ordre et $(\varepsilon_t)_{t \in \mathbb{Z}}$ un processus de bruit blanc centré innovation du processus X tel que :

$$E\, \varepsilon_s \varepsilon'_r = \Sigma_\varepsilon \delta_r^s \quad , \Sigma_\varepsilon > 0.$$

On suppose dans ce qui suit que, avec $D = D(0,1) = \{Z \mid Z \in \mathbb{C}, |Z| < 1\}$,

$$(2) \qquad \text{Spect}(F) \subset D.$$

On dit que le système S est minimal si le couple (F,G) est commandable et le couple (F,H) observable c'est à dire. Si l'on a : (3) et (4) :

$$(3) \qquad \rho[G, FG, F^2G, \ldots, F^{n-1}G] = n$$

$$(4) \qquad \rho[{}^tH, {}^tF\,{}^tH, {}^tF^2\,{}^tH, \ldots, {}^tF^{n-1}\,{}^tH] = n$$

$\mathcal{S}$ est l'ensemble des systèmes linéaires S satisfaisant (2),(3) et (4).

03. *Processus de renouvellement.*

Soit $T = (T_t)_{t \in \mathbb{N}}$, $T_0 = 0$ un processus à valeur dans $\mathbb{N}$ tel que la suite $(T_{t+1} - T_t)_{t \in \mathbb{N}}$ est une suite de variables aléatoires mutuellement indépendantes équidistribuées dont on note L la loi :

$\forall (j,t) \in \mathbb{N}^2$: $L_j = P(T_{t+1} - T_t = j)$.

$L(0) = 0,\ L(\mathbb{N}^*) = 1$.

On suppose enfin que les processus X et T sont mutuellement indépendants.

On notera $\tilde{X}_t = X_{T_t}$, $\tilde{Y}_t = Y_{T_t}$ quel que soit t élément de $\mathbb{N}$.

1. ETUDE DU COUPLE $(\tilde{X},\tilde{Y})$

Théorème 1 :

Le couple $(\tilde{X}_t,\tilde{Y}_t)$ constitue un système linéaire homogène que l'on peut toujours prendre commandable.

C'est à dire qu'il existe : $\tilde{F}$ $n \times n$, $\tilde{G}$ $n \times u$, $\tilde{H}$ $s \times n$ des matrices fixes, un bruit blanc $(\eta_t)_{t \in \mathbb{Z}}$ avec $\Sigma_\eta = I_u$ tel que

$$(5) \quad \begin{cases} \tilde{X}_{t+1} = \tilde{F}\,\tilde{X}_t + \tilde{G}\eta_t \quad , t \in \mathbb{N} \\ \tilde{Y}_t = \tilde{H}\,\tilde{X}_t \\ \tilde{F} = \sum_{j=1}^{\infty} L_j\, F^j \quad , \tilde{H} = H \end{cases}$$

et le couple $\tilde{F},\tilde{G}$ vérifiant (3)

démonstration :

Elle se déduit immédiatement des propriétés 1 et 2 suivantes.

Propriété 1 :

Le processus $\tilde{X}$ est un A.R.(1) centré stationnaire physiquement réalisable de même matrice de covariance instantannée que X.

Démonstration :

Le processus X étant lui même un A.R.(1) centré stationnaire et physiquement réalisable on a :

$$\forall j > 0 \ : \ \Sigma_X(j) = E\, X_{t+j}\, {}^tX_t = F\, \Sigma_X(j-1)$$

$$(6) \quad \Sigma_X(j) = F^j\, \Sigma_X(0)$$

où $\Sigma_X(0)$ est solution de l'équation : $\Sigma_X(0) = F\,\Sigma_X(0)\,{}^tF + G\,\Sigma_\varepsilon\,{}^tG$ qui existe d'après (2)

De plus $\tilde{X}$ est centré et sa fonction de covariance est donnée par :

$$\Sigma_{\tilde{X}}(j) = E\, \tilde{X}_{t+j}\, {}^t\tilde{X}_t = E\,[\,E^{T_t,T_{t+j}}(X_{T_{t+j}}\ {}^tX_{T_t})\,]$$

$$= E(\Sigma_X(T_{t+j} - T_t))$$

et d'après (6) :

$$\Sigma_{\tilde{X}}(j) = E(F^{T_{t+j} - T_t})\, \Sigma_X(0)$$

Or $[F^{(T_{t+k+1} - T_{t+k})}]_{0 \leq k \leq j-1}$ constitue une suite de v.a.i.i.d.

Vérifions que leur espérance existe. En effet

$$(7) \quad \phi(z) = \sum_{k=1}^{\infty} L(k) z^k$$

est définie et analytique sur D. On peut donc en déduire l'existence de

$$(8) \quad E(F^{T_{t+k+1} - T_{t+k}}) = \sum_{k=1}^{\infty} L(k) F^k = \phi(F)$$

d'où :

(9) $E(\tilde{X}_{t+j}\ {}^t\tilde{X}_t) = (\phi(F))^j\ \Sigma_X(0)$.

En particulier pour $j=0$ on obtient $\Sigma_{\tilde{X}}(0) = \Sigma_X(0)$.

Le calcul de la fonction de covariance de $\eta'_t = \tilde{X}_{t+1} - \tilde{F}\tilde{X}_t$ et des covariances croisées $E\tilde{X}_t\ {}^t\eta'_{t+j}$, $j \geq 0$, montrent que η'_t est un bruit blanc, innovation de $\tilde{X}_t$.

(10) La relation $\Sigma_{\tilde{X}}(0) = \tilde{F}\ \Sigma_{\tilde{X}}(0)\ {}^t\tilde{F} + \Sigma_{\eta'_t}$, définit $\Sigma_{\eta'_t}$

Notons que :

(11) $\operatorname{Spect} \phi(F) = \phi(\operatorname{Spect}(F))$.

Comme de plus $|\phi(Z)| \leq |Z|\ \forall Z \in D$ on conclut que

(12) $\forall i \in \{1,\dots,n\} : |\lambda_i(\phi(F))| \leq |\lambda_i(F)| \leq 1$.

On notera par la suite :

(13) $\tilde{F} = \phi(F)$.

Propriété 2 :

Le processus $\tilde{X}$ est représentable sous la forme $\tilde{X}_{t+1} = \tilde{F}\tilde{X}_t + \tilde{G}\eta_t$ où (η_t) est un bruit blanc $(\Sigma_{\eta_t} = I_u)$, le couple $(\tilde{F},\tilde{G})$ étant commandable.

Démonstration :

(14) Soit $\tilde{G}$ $n \times u$ l'une quelconque des matrices définies par $\Sigma_{\eta'_t} = \tilde{G}\ {}^t\tilde{G}$ où $u = \operatorname{rang} \Sigma_{\eta'_t}$.

(15) Montrons que $\rho[G, FG,\dots,F^{n-1}G] = \rho[\tilde{G}, \tilde{F}\tilde{G},\dots,\tilde{F}^{n-1}\tilde{G}]$.

Il en résulte que puisque (F,G) est commandable $(\tilde{F},\tilde{G})$ l'est.

En effet :

$$\forall m \geq n : \rho\Big(\sum_{j=o}^{m} F^j G\ {}^tG\ {}^tF^j\Big) = \rho[G,FG,\dots,F^mG] = \rho[G,FG,\dots,F^{n-1}G]$$

d'après le théorème de Cayley Hamilton.

Donc : $\rho[G,FG,\dots,F^{n-1}G] = \rho[\sum_{j=o}^{\infty} F^jG\ {}^tG\ {}^tF^j]$, la convergence de la série étant assurée par (2).

De même en utilisant (2) et (12) on obtient :

$$\rho[\tilde{G},\tilde{F}\tilde{G},\dots,\tilde{F}^{n-1}\tilde{G}] = \rho[\sum_{j=o}^{\infty}\tilde{F}^j\tilde{G}\ {}^t\tilde{G}\ {}^t\tilde{F}^j]$$

Considérons alors l'équation $X = F\ X\ {}^tF + G\ {}^tG$. D'après l'hypothèse (2) elle admet $\sum_0^{\infty} F^j\ G\ {}^tG\ {}^tF^j$, comme solution unique. De même $\tilde{X} = \tilde{F}\ \tilde{X}\ {}^t\tilde{F} + \tilde{G}\ {}^t\tilde{G}$ admet $\sum_0^{\infty} \tilde{F}^j\ \tilde{G}\ {}^t\tilde{G}\ {}^t\tilde{F}^j$ comme solution unique. Comme à l'évidence ces deux équations sont satisfaisantes par $\Sigma_X(0) = \Sigma_{\tilde{X}}(0)$ il en résulte l'égalité des deux séries et donc des rangs dans (15).

La démonstration de la propriété est achevée en prenant $(\eta_t)_{t\in\mathbb{Z}}$ blanc, $\Sigma_{\eta_t} = I_u, \eta'_t = \tilde{G}\ _t$. Remarquons que si l'on avait choisi une autre décomposition $\Sigma_{\eta'_t} = \tilde{G}_1\ {}^t\tilde{G}_1$, puisque $\tilde{G}_1 = \tilde{G}\ R$ où R est une matrice orthogonale, le couple $(\tilde{F},\tilde{G}_1)$

serait aussi commandable.

2. HEREDITE DE LA MINIMALITE PAR ECHANTILLONNAGE ALEATOIRE.

Le théorème 1 montre que le système $\tilde{S} = (\tilde{F}_{nn}, \tilde{G}_{nr}, \tilde{H}_{sn})$ associé par échantillonnage aléatoire à $S = (F_{nn}, G_{nr}, H_{sn})$ reste commandable. Nous étudions dans ce paragraphe l'application τ_ϕ définie sur l'ensemble des systèmes linéaires minimaux $\mathcal{S}$ qui à S associe $\tilde{S}$, (plus exactement qui à S associe l'un des représentants.
La classe des systèmes $\tilde{S}$ tels que $\tilde{G}^t\tilde{G} = \Gamma_X(0) - \tilde{F}\,\Gamma_X(0)^t\tilde{F}$).

Théorème 2 :

L'application τ_ϕ préserve la minimalité dans $\mathcal{S}$ si et seulement si l'application ϕ est injective dans D.

Démonstration :

i) Condition nécessaire :

Montrons que si ϕ est non injective, il existe un S de $\mathcal{S}$ tel que $\tau_\phi(S) = \tilde{S}$ n'est pas minimal. Puisque ϕ est non injective, il existe deux éléments distincts de D z_1 et z_2 tels que

$$\phi(z_1) = \phi(z_2).$$

Considérons une matrice F 4 x 4, réelle, diagonale de valeurs propres $z_1, \bar{z}_1, z_2, \bar{z}_2$. Choisissons G de sorte que (F,G) soit commandable, et H une matrice 2 x 4 dont les colonnes h_1 h_2 h_3 et h_4 (qui correspondent aux v.p z_1, $\bar{z}_1$, z_2, $\bar{z}_2$) sont telles que $h_1 = h_3$, $h_2 = h_4$ et $\rho[h_1, h_2] = 2$.
(F,H) est observable, même si z_1 est réel ($z_1 = \bar{z}_1 \neq z_2$ et $\neq \bar{z}_2$) ou si $z_1 = \bar{z}_2 \in \mathbb{C}$.
La matrice $\tilde{F}$ est aussi diagonale, ses termes diagonaux étant $\phi(z_1), \phi(\bar{z}_1), \phi(z_2), \phi(\bar{z}_2)$.
Or $\phi(z_1) = \phi(z_2)$ tandis que les colonnes h_1 et h_3 correspondantes de $\tilde{H} = H$ sont égales donc dépendantes. $(\tilde{F}, H)$ n'est pas observable.

ii) La réciproque est une conséquence de la :

Propriété 3 :

Si ϕ est injective dans D,$(\tilde{F},\tilde{G},\tilde{H})$ est observable si et seulement si (F,G,H) l'est.

Démonstration :

Elle utilise ([5],tome 1, ch 6) et de lourdes notations.

Commençons par noter $[(\sigma_{i,j})_{1 \leq j \leq \theta_i}]_{1 \leq i \leq I}$ les *indices d'observabilité* associés aux diviseurs élémentaires de F où :

i) $(\lambda_i)_{1 \leq i \leq I}$ sont les valeurs propres de F, chacune d'ordre θ_i.

ii) $(z - \lambda_i)^{\alpha_{i,1}}, \ldots, (z - \lambda_i)^{\alpha_{i,s_i}}$ les s_i diviseurs élémentaires associes à λ_i ,

avec $\sum_{j=1}^{s_i} \alpha_{i,j} = \theta_i$ et les α_{ij} rangés par ordre décroissant.

iii) $\sigma_{i,1} \triangleq 1$, $\sigma_{i,j} \triangleq (\sum_{k=1}^{j-1} \alpha_{i,k}) + 1$ $\quad (2 \leqslant j \leqslant s_i)$ définit la suite des indices d'observabilité associée à λ_i.

iv) $\{U_{\sigma_{i,j}}, 1 \leqslant j \leqslant s_i\}$ une famille des s_i vecteurs propres associés à λ_i.

$U_{i,j} \triangleq \{ U_{\sigma_{i,j}+k} \; ; \; o \leqslant k \leqslant \alpha_{i,j} - 1\}$ une famille libre de $\alpha_{i,j}$ vecteurs propres généralisés [3] associés au diviseur élémentaire $(z - \lambda_i)^{\alpha_{i,j}}$.

v) $U_i \triangleq \bigoplus_{1 \leqslant j \leqslant s_j} U_{ij}$, $\mathbb{R}^n = \bigoplus_{i=1}^{I} U_i$

vi) . F_i $\theta_1 \times \theta_i$ matrice de la restriction de F à U_i

.. H_i $s \times \theta_i$ matrice de la restriction de H à U_i exprimée dans la base U

... $h_k^{(i)}$ $1 \leqslant k \leqslant \theta_i$ les colonnes de H_i

(16) $Sq_i \triangleq (h^{(i)}_{\sigma_{ij}})_{1 \leqslant j \leqslant s_i}$ est le *squelette d'observabilité* de H associé à λ_i.

On note que : $\rho[Sq_i] = s_i$ pour tout i tel que $1 \leqslant i \leqslant I$ est la condition d'observabilité de(F,H)

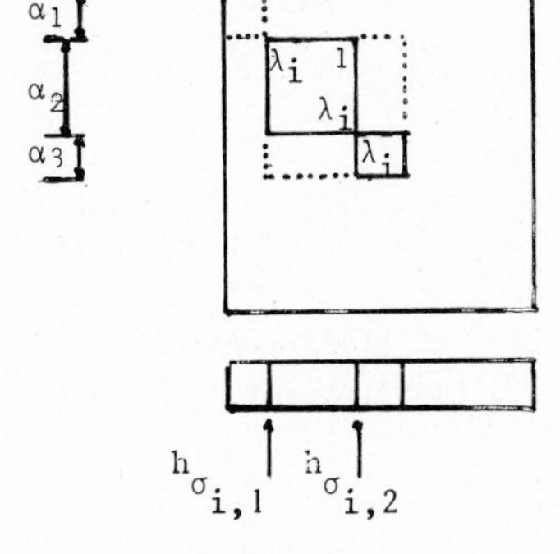

: F_i matrice de Jordan associée à λ_i et au sous-espace U_i ; elle est partitionnée en s_i blocs Jordan de taille $\alpha_{i,j} \times \alpha_{i,j}$ $(1 \leqslant j \leqslant s_i)$.

Fig 1

L'observabilité s'étudie séparément pour des valeurs propres distinctes, ϕ étant injective, on étudie séparément les diverses valeurs propres $\phi(\lambda_i)$ de $\tilde{F}$.
On suppose donc que $I = 1$, on omet l'indice i : on raisonne comme si F n'avait qu'une seule valeur propre λ.
Par ailleurs, puisque ϕ est injective dans D, sa dérivée ne s'y annule pas (Titchmar-(Titchmarsh p. 198 [8]). Puisque $0 \notin \phi'(\mathrm{Spect}(F))$, les diviseurs élémentaires de $\tilde{F}$ se déduisent de ceux de F, [5]:

$\phi(\lambda)$	: Valeur propre de $\tilde{F}$
θ	: Ordre de $\phi(\lambda)$. Il est égal à celui de λ.
$(z - \phi(\lambda))^{\alpha_1}, \ldots (z-\phi(\lambda))^{\alpha_s}$	: Diviseurs élémentaires de $\tilde{F}$ associés à $\phi(\lambda)$
$U_{\sigma_1}, U_{\sigma_2}, \ldots U_{\sigma_s}$	: Sont encore (les) s vecteurs propres de $\tilde{F}$.

Remarquons que les indices d'observabilité des formes de Jordan de $(\tilde{F},\tilde{H})$ sont ceux de (F,H). Soit $=(T_1,T_2,\dots,T_\sigma)$ une base de V.P.G dans laquelle la matrice de $\tilde{F}$ est sous forme de Jordan. Les $T_{\sigma_j} = U_{\sigma_j}$ $1 \leq j \leq s$ sont les seuls V.P. de $\tilde{F}$.

A T_{σ_j} est associé un bloc de Jordan $\alpha_j \times \alpha_j$ $(1 \leq j \leq s)$.

La matrice T $(\theta \times \theta)$ de changement de base (des U_k au T_k) admet pour colonnes d'indice $\sigma_1,\dots\sigma_s$, les vecteurs d'indice $\sigma_1,\dots\sigma_s$ de la base canonique de $\mathbb{R}^\theta$.

Par ce changement de base : $(\tilde{F},H)$ devient $(T^{-1}\tilde{F}T, HT)$. En notant $K = HT$ et k_i les colonnes de K :

$$k_{\sigma_j} = h_{\sigma_j}, \forall j : 1 \leq j \leq s$$

D'où $Sq(HT) = Sq(H)$, où Sq est défini en (16).
Puisque (F,H) est observable, $(h_{\sigma_j})_{1 \leq j \leq s}$ donc $(k_{\sigma_j})_{1 \leq j \leq s}$ est une famille libre ; alors $(T^{-1}\tilde{F}T, HT)$ est observable, et il en est de même de $(\tilde{F},H)$.

Pour la même raison, si (F,H) n'est pas observable, il est exclu que $(\tilde{F},H)$ le soit car alors le système $(h_{\sigma_j})_{1 \leq j \leq s}$ serait libre. Ceci achève la démonstration.

3. EFFET D'UN ECHANTILLONNAGE ALEATOIRE SUR UN SYSTEME S.

3.1. Dans la partie précédente, nous avons étudié globalement l'application qui, à un système linéaire S de $\mathcal{S}$ associe le système $\tilde{S}$. Dans la présente partie, nous considérons un système S particulier et étudions en détail, l'effet de l'échantillonnage aléatoire sur S.

Remarquons que cette étude est bien distincte de la précédente. En effet le système $\tilde{S}$ peut-être minimal lorsque S l'est, même si ϕ n'est pas injective dans D. Nous étudions dans ce contexte quelques conditions nécessaires et suffisantes de conservation de la minimalité dans différents cas associés aux propriétés suivantes de ϕ :

<u>Cas 0</u> : ϕ est injective dans D

<u>Cas 1</u> : ϕ est injective sur Spect(F) et $o \notin \phi'(\mathrm{Spect}(F))$

<u>Cas 2</u> : ϕ n'est pas injective sur Spect(F) et $o \notin \phi'(\mathrm{Spect}(F))$

<u>Cas 3</u> : ϕ est injective sur Spect(F) et il y a séparation d'au moins une valeur propre.

<u>Cas 4</u> : ϕ n'est pas injective et il y a séparation d'au moins une valeur propre

Rappelons qu'il y a *séparation d'ordre k* de la valeur propre λ (d'ordre $\theta > 1$) de F s'il existe $k \in \mathbb{N}$, $1 \leq k <$ tel que :

$$\phi'(\lambda) = \dots \phi^{(k-1)}(\lambda) = o \quad , \phi^{(k)}(\lambda) \neq o$$

ou si

$$\phi'(\lambda) = \dots = \phi^{(\theta-1)}(\lambda) = o$$

Remarquons que dans le cas 0 il n'y a pas de séparation.

3.2. Etude du cas 0.

Théorème 3 :

Si ϕ est injective dans D, une condition nécessaire et suffisante pour qu
que $\tilde{S} \in \mathcal{S}$ est que $S \in \mathcal{S}$.

Démonstration : La condition suffisante a été établie dans la propriété 3, ainsi que la partie de la condition nécessaire portant sur l'observabilité. Pour la commandabilité on a :

$$n = \rho[\tilde{G}\ \tilde{F}\tilde{G}\ \dots\ \tilde{F}^{n-1}\ \tilde{G}] = \rho[\sum_{0}^{n-1} \tilde{F}^k\ \tilde{G}\ {}^t\tilde{G}\ {}^t\tilde{F}^k] = \rho(\mathcal{X}_{\tilde{X}}) = \rho(\mathcal{X}_X) =$$

$$= \rho(\sum_{0}^{\infty} F^k\ G\ {}^tG\ {}^tF^k) = \rho[G\ FG\ \dots\ F^{n-1}\ G]$$

3.3. Etude des cas sans séparation. Cas 1 et 2.

Les cas 1 et 2 peuvent être regroupés.

Notons

$$\mathrm{Spect}(\tilde{F}) = \{\psi_1, \dots \psi_A\} = \phi(\mathrm{Spect}(F))\ ;$$

$$I_a = \{i \mid \phi(\lambda_i) = \psi_a\}\ ,\ 1 \leqslant a \leqslant A.$$

$$H_a = \Sigma^*\{\text{ Matrices } H_i \mid i \in I_a\}$$

où Σ^* désigne la concatenation par juxtaposition des matrices dans l'ordre des indices i ou la somme des sous-espaces dans l'expression :

$$\Sigma^*\{[(h^{(i)}_{\sigma_{i,j}})_{1 \leqslant j \leqslant s_i}] \mid i \in I_a\}$$

Théorème 4 :

Soit $S \in \mathcal{S}$. Une condition nécessaire et suffisante pour que $\tilde{S} \in \mathcal{S}$ est : pour tout $a \in \{1, \dots A\}$, les $\Sigma(s_i \mid i \in I_a)$ vecteurs $h^{(i)}_{\sigma_{i,j}}$ $(1 \leqslant j \leqslant s_i,\ i \in I_a)$ soient libres.

Démonstration :

Si $\phi'(\lambda) \neq 0$ au diviseur $(z-\lambda)^\nu$ de F est associé $(z-\phi(\lambda))^\nu$ de $\tilde{F}$; tout vecteur propre de F reste vecteur propre de $\tilde{F}$ et aucun nouveau n'apparaît ; la commandabilité est assurée; une condition nécessaire et suffisante d'observabilité exprime que la dimension du sous-espace engendré par le squelette d'observabilité de $(\tilde{F},\tilde{H})$ est égale au nombre de vecteurs du squelette.

Les conséquences suivantes sont sans doute plus intéressantes. Pour des observations scalaires on obtient une condition nécessaire et suffisante particulièrement sinple.

Corollaire 1 : (cas 1 et 2).

1. *Soit $S \in \mathcal{S}$. Une condition nécessaire pour que $\tilde{S} \in \mathcal{S}$ est que*

$$\forall a,\ 1 \leqslant a \leqslant A,\ \ \Sigma(s_i \mid i \in I_a) = s = \dim Y.$$

2. *Si $s = 1$, soit $S \in \mathcal{S}$. Une condition nécessaire et suffisante pour que $\tilde{S} \in \mathcal{S}$ est que ϕ soit injective sur Spect(F).*

Démonstration :

Le point 1 est évident; le point 2 se déduit de 1 pour sa partie condition nécessaire (si ϕ est non injective, il existe a tel que Card $I_a > 1$)et de la démonstration du théorème 4 pour la condition suffisante.

Remarque : Si ϕ est injective sur Spect(F), sans séparation, la minimalité est conservée.

3.4. Etude du cas 3.

L'application ϕ étant injective $\phi(\lambda_i)$ prend p valeurs distinctes. Il suffit de considérer ces valeurs séparément. On omet l'indice i. Notons $k(>1)$ l'ordre de la séparation relative à λ, et $\alpha_1,\ldots\alpha_M$ les degrés des diviseurs élémentaires de F associés à λ.

Pour tout α $(\alpha_m)_{1\leq m\leq M}$ définissons les indices $q,\ell,(\sigma_j)_{1\leq j\leq \alpha\wedge k}; \beta$.

(17) Pour q et ℓ: $\alpha = qk+\ell$ où $0\leq q = [\frac{\alpha}{k}]$; $0\leq\ell\leq k$.

. Pour les σ_j:	$1<k\leq\alpha$	
(18)	. si $\ell\neq 0$	$1\leq j\leq k-\ell$ $\quad\sigma_j=(j-1)q+1$
(19)		$k-\ell<j\leq k$ $\quad\sigma_j=(k-\ell)q+[(j-1)-(k-\ell)](q+1)+1$
(20)	. si $\ell=0$	(19) est inutile
(21)	$\alpha+1\leq k$	$1\leq j\leq\alpha$ $\quad\sigma_j=j$

(22). Pour β : $\beta=\alpha_k\wedge k$

Théorème 5 :

i) Supposons que F ait une seule valeur propre λ. Soient $m\in\mathbb{N}$, $1\leq m\leq M$, et les indices α_m, q_m, ℓ_m définis comme en (17),$(\sigma_{m,j})_{1\leq j\leq\beta_m}$ définis comme en (18),(19),(20),(21) et (22).

Soient enfin le sous espace vectoriel $\mathcal{H}_m$ défini par :

$$_m = [h_{\sigma_{m,1}},\ldots,h_{\sigma_{m,\beta_m}}]$$

Si $S\in\mathcal{S}$ une condition nécessaire et suffisante pour que $\tilde{S}\in\mathcal{S}$ est que :

(23) $$\dim\Big[\bigcup_{1\leq m\leq M}\mathcal{H}_m\Big]=\sum_{m=1}^{M}\beta_m .$$

ii) Si F possède plusieurs valeurs propres distinctes, la propriété (23) doit être vérifiée pour toutes les valeurs propres qui se séparent. Pour les autres valeurs propres la propriété se réduit à celle du théorème 4.

Démonstration :

Pour la clarté de l'exposé, une remarque s'avère utile avant la démonstration proprement dite qui ne portera que sur le point 1 du théorème.

a) Remarque :

Supposons $k\leq n-1$.

Montrons d'abord que si à la v.p. λ est associé le diviseur élémentaire $(z-\lambda)^n$ et si $U=\{U_1,\ldots U_n\}$ est une base de V.P.G de F alors $U_1,\ldots U_k$ sont V.P de $\tilde{F}$ associés à $\phi(\lambda)$, si k est l'ordre de la séparation de λ.

En effet $\tilde{F}$ est, dans la base $\{U_1,\ldots,U_n\}$ une matrice de Toeplitz triangulaire supérieure déterminée par sa première ligne $(a_j)_{1\leqslant j\leqslant n}$ où $a_j = \phi^{(j-1)}(\lambda)/(j-1)!$ avec $a_1=\phi(\lambda)$, $a_j = o$ $2\leqslant j\leqslant k$, $a_{k+1} \neq o$ un vecteur propre ${}^tX = (x_1,\ldots,x_n)$ associé à $\phi(\lambda)$ vérifie

$$o = (\tilde{F} - \phi(\lambda)I_n)X$$

Puisque $a_{k+1} \neq o$ on en déduit que : $x_{k+1}=\ldots\ldots = x_n = o$. Donc $U_1,\ldots,U_k$ sont des vecteurs propres de $\tilde{F}$ associés à $\phi(\lambda)$.

. Si $k\geqslant n$, $\tilde{F}-\phi(\lambda)I_n = o$ et tout l'espace est propre associé à la valeur propre multiple $\phi(\lambda)$.

b) Déterminons le squelette associé à l'une des valeurs propres de $\tilde{F}$.

En utilisant les diviseurs élémentaires de F :$(z-\lambda)^{\alpha_m}$ $1\leqslant m\leqslant M$, ceux de $\tilde{F}$ sont : $(z-\phi(\lambda))^{q_m+1}$,ℓ_m fois et $(z-\phi(\lambda))^{q_m}$ $(k-\ell_m)$ fois.

Les indices σ_j étant définis par (18) à (21) une base $\mathcal{W}$ du sous-espace invariant associé au diviseur élémentaire de degré α_m dans laquelle $\tilde{F}$ prend une forme de Jordan est, d'après la remarque, constituée de

$$W_1 = U_1,\ W_{\sigma_{m,j}} = U_j,\ 2\leqslant j\leqslant k,$$

et, éventuellement, d'autres vecteurs qu'il est inutile de déterminer.

Soit W la matrice, de passage à $\mathcal{W}$:

$\tilde{F}$ devenant $W^{-1}\tilde{F}W$ et H devenant HW, le squelette associé à $\phi(\lambda)$ (étant diviseur élémentaire de degré α_m) est constitué par les colonnes de H lui-même, d'indice :

$$\sigma_{m,1},\ldots,\sigma_{m,\beta_m} .$$

Comme on le voit en explicitant W $\alpha_m \times \alpha_m$ puis en calculant HW. Le résultat énoncé se déduit alors directement.

$$W = \left[\begin{array}{ccccccccc}
1 & X & X & 0 & X & X & 0 & & 0\\
0 & X & X & 1 & X & X & 0 & & .\\
0 & & & 0 & X & X & 1 & & .\\
0 & X & X & . & & & 0 & \ldots & 0\\
. & & & & & & . & & 1\\
. & & & & & & . & & .\\
. & & & & & & . & & .\\
. & & & & & & . & & .\\
. & & & & & & . & & .\\
0 & X & X & 0 & X & X & 0 & & 0
\end{array}\right]$$

$\underbrace{\quad}_{q_m+1}$ $\underbrace{\quad}_{q_{m+1}}$ $\underbrace{\quad}_{q_m}$

ℓ_m fois $\quad k-\ell_m$ fois

Fig 2

Le corollaire suivant est sans doute plus intéressant que le théorème 5.

Corollaire 2 :

1. *Soit* $S\in\mathcal{S}$. *Une condition nécessaire pour que* $\tilde{S}\in\mathcal{S}$ *est que :* $\forall i\ (1\leqslant i\leqslant I)$: $\Sigma\{\beta_{i,m}\ |1\leqslant m\leqslant s_i\} \leqslant s$.

2. *Lorsque* $s=1$; *si* $S\in\mathcal{S}$, *une condition nécessaire et suffisante*

pour que $\tilde{S} \in \mathcal{S}$ est que les v.p. λ_i tels que $\phi'(\lambda_i) = 0$ soient simples.

Démonstration :

Démontrons le point 2 seul, le point 1 est immédiatement déduit du théorème. Puisque $s = 1$ et que S est minimal, il y a un seul diviseur élémentaire par v.p :

$s_i = 1 \quad \forall i \quad 1 \leqslant i \leqslant I.$

Pour que $\tilde{S}$ soit minimal il faut, d'après 1, que : $\beta_i = 1$ pour tout i, $1 \leqslant i \leqslant I$.
Or si $\phi'(\lambda_i) = 0$: $k_i > 1$, puisque $\beta_i = k_i \wedge \alpha_i = 1$ on a nécessairement $\alpha_i = 1$, la valeur propre est donc simple, la condition nécessaire est donc prouvée.
La condition suffisante est triviale puisque ϕ étant injective, les v.p. s'étudient une par une, la perte d'observabilité pourrait provenir de la séparation d'une valeur propre multiple ; ce cas est exclu.

3.5 Etude du cas 4 :

C'est le cas où ϕ est non injective sur Spect(F) et ou une valeur propre de F, au moins, se sépare. Alors on peut sans difficulté regrouper les études des cas 2 et 3 pour énoncer les résultats souhaités. Compte tenu de la longueur des écritures nous laissons au lecteur le soin d'effectuer cette synthèse.

4. BIBLIOGRAPHIE .

[1] AKAIKE H.: Markovian Representation of stochastic processes and its application to the analysis of Autoregressive moving average processes. Annals of Inst. Math. Stat. 1974, 26, p. 363-387.

[2] ASH R.B., GARDNER M.G.: Topics in stochastic processes. Academic Press 1975, 323 p.

[3] CHEN C.T.: Introduction to linear system theory. Holt-Rinehart-Winston 1970, 431 p.

[4] FOSSART A. (GUEGEN C): Commande des systèmes multidimensionnels. Dunod 1972, 350 p.

[5] GANTMACHER F.R.: Théorie des matrices. Tome 1 - 2. Traduction française. (370 p. et 268 p.) Dunod 1966.

[6] KALMAN R.E.,HO Y.C., NARENDRA K.S.: Controllability of linear Dynamical systems in contributions to differential equations, vol. 1 n° 2 p. 189-213, Wiley 1963.

[7] SHAPIRO H., SILVERMAN R.A.: Alias-free sampling of random noise. Journal Ind. Publ. Math. 1960, 8-2 p. 225-248.

[8] TITCHMARSH E.C.: The theory of functions. Oxford University Press. 2e Ed. 1939, 454 p.

[9] WOLOWICH W.A.: Linear Multivariable systems. Springer Verlag. 1974, 374 p.

[10] Y.BAR-NESS et G.LANGHOLZ (1975): Preservation of Controlability under sampling. Int. J. Control. Vol. 22 n° 1 p. 39-47.

[11] J.A.GIBSON et T.T.HA (1980): Further to the preservation of controllability under sampling.
Int. J. Control. Vol 31 n° 6 p. 1013-1026.

[12] M.L.J.HAUTUS (1970): Stabilization Controllability and Observability of linear Autonomous systems.
Koninkl. Nederl. Akademie Van Wetensshappen Série A, 73 n° 5.

APPLICATION DE LA THEORIE DES SYSTEMES IMPLICITES

A L'INVERSION DES SYSTEMES

José GRIMM

Institut National de Recherche en
Informatique et Automatique
Centre de Sophia-Antipolis
06560 Valbonne

Résumé:

Dans cet article nous commencerons par rappeler la définition d'un système implicite et certaines de leur propriétés. Nous définirons ensuite la notion de système inverse, en prouvant l'existence et l'unicité du point de vue de la représentation externe. Dans un deuxième temps, nous montrerons comment la théorie des systèmes implicites permet d'écrire facilement le système inverse, et d'en déduire certaines des ses propriétés. Par exemple, nous montrerons qu'un système classique est inversible à droite ou à gauche, si et seulement si, dans la forme de Kronecker ne figurent pas certains types de blocs, ce qui permettra de retrouver des conditions déja obtenues dans [8].

Abstract:

In this paper we first recall the definition of an implicit system (sometimes also called "generalized state-space system") and its most important properties. We then define the notion of inverse system, proving existence and uniqueness from an external point of view. Next we show how the theory of implicit systems yields inverse systems and some of their properties in an easy way. For instance we examine the conditions under which the inverse system is minimal, and show also that a state-space system is right or left invertible if some blocks are missing in its Kronecker form. This allows us to find conditions already established in [8].

1. INTRODUCTION.

Cet article a pour objectif d'étudier certaines propriétés des systèmes inverses à l'aide de la théorie des systèmes implicites.

Pour les notions concernant les systèmes implicites nous renvoyons le lecteur à la bibliographie ([1], [3], [13], [14]) et plus particulierement à [2] pour les résultats et notations utilisés ici. Dans le second paragraphe nous rappellerons les principaux résultats de [2].

Considérons un système Kalmanien Σ : $\dot{x} = Fx + Gu$, $y = Hx + Ju$ et essayons de définir son inverse. Ce sera un système Σ^{-1} tel que, si y est une sortie de Σ associée à la commande u, la sortie de Σ^{-1} associée à la commande y sera u.

Il est bien connu qu'en général, pour Σ il existe plusieurs commandes donnant la même sortie y. D'autre part, toujours en général, toutes les fonctions y ne sont pas des sorties de Σ. D'où Σ^{-1} n'accepte pas toutes les commandes et associe plusieurs sorties à une commande admissible.

Le système inverse cherché, s'il existe, est donc un système implicite. Il est d'ailleurs facile à voir qu'il n'existe de système inverse Kalmanien que si J est inversible.

Nous donnerons la définition précise d'un système inverse dans le cas général dans la suite du texte. Il est facile de construire le système inverse d'un système Kamanien:

En effet considérons $\dot{x} = Fx + Gu$, $y = Hx + Ju$.

Posons $x_2 = u$. Le système s'écrit :

$$\begin{aligned} \dot{x} &= Fx + Gx_2 \\ 0 &= Hx + Jx_2 - y \\ u &= x_2 \end{aligned}$$

Considérant ce système comme un système implicite de commande y, de sortie u, d'état $\binom{x}{x_2}$, on voit que l'on a obtenu un système inverse du système initial. On voit par conséquent que la théorie des systèmes implicites est l'outil naturel pour l'étude des systèmes inverses.

Dans une première partie de cet article, nous définirons en toute rigueur la notion de système inverse (du point de vue représentation externe), nous montrerons l'existence et l'unicité du système inverse (à équivalence près bien entendu) et nous donnerons les formules permettant de calculer (au moins théoriquement) la fonction de transfert du système inverse. Il est à remarquer qu'au point de vue représentation externe, l'existence d'un système inverse n'est pas évidente, alors qu'au point de vue représentation interne c'est une trivialité, (nous venons en effet de donner le système inverse dans un cas particulier, celui des systèmes Kalmaniens, mais le cas général est identique).

Nous verrons ensuite, dans quelle mesure ce système inverse particulier décrit précédemment est minimal.

Certains auteurs, en particulier Silverman et Hautus [8] ont introduit les notions d'inversibilité à gauche et à droite. Nous verrons dans la suite ce que cela signifie pour le système inverse et retrouverons par une méthode différente les conditions pour que le système soit inversible à gauche ou à droite.

2. RAPPELS SUR LES SYSTEMES IMPLICITES

1) Au niveau de la représentation interne, un système implicite s'écrit sous la forme :

$$E\dot{x}(t) = Fx(t) + Gu(t) \qquad y(t) = Hx(t) + Ju(t)$$

ou en discret : $Ex(t+1) = Fx(t) + Gu(t) \qquad y(t) = Hx(t) + Ju(t)$

2) Au niveau de la représentation externe, un système implicite est défini par la donnée d'une correspondance entre les entrées et les sorties. De façon plus précise, en transformée de Laplace (ou en transformée en z dans le cas des systèmes discrets) l'ensemble de définition de cette correspondance est un sous-espace rationnel $\mathrm{Ker}\mathcal{M}$ de l'espace des commandes. A chaque commande u de cet ensemble (appelée commande admissible) la correspondance associe l'ensemble $\mathcal{K}u + \mathrm{Im}\mathcal{L}$, où $\mathrm{Im}\mathcal{L}$ est un sous-espace rationnel, et $\mathcal{K}$ une application linéaire rationnelle (non nécessairement propre). On peut supposer $\mathcal{L}$ et $\mathcal{M}$ rationnelles propres. Un tel triplet $(\mathcal{K}, \mathcal{L}, \mathcal{M})$ sera appelé un <u>triplet de tranfert</u>. On peut supposer et on le fera dans la suite que $\mathcal{L}$ est injective et $\mathcal{M}$ surjective).

3) Une façon de définir $(\mathcal{K}, \mathcal{L}, \mathcal{M})$ à partir de la représentation interne est la suivante: En transformée de Laplace le système est le suivant: $(zE-F)x=Gu$, $y=Hx+Ju$. Pour que u soit commande admissible, donc pour qu'il existe x tel que $(zE-Fx)=Gu$ il faut que Gu Im $(zE-F)$. Soit $\mathcal{M}_1$ tel que $\mathrm{Im}(zE-F)=\mathrm{Ker}\,\mathcal{M}_1$ et posons $\mathcal{M} = \mathcal{M}_1 G$. Alors u est commande admissible si et seulement si $\mathcal{M}u=0$. On peut alors choisir une solution x de $(zE-F)x=Gu$ linéairement en u, disons $x=\mathcal{K}_1 u$. Toutes les autres solutions x se déduisent de celle-ci en lui ajoutant une solution quelconque de l'équation linéaire homogène $(zE-F)x=0$. Soit $\mathcal{L}_1$ une application linéaire telle que $\mathrm{Im}\,\mathcal{L}_1=\mathrm{Ker}(zE-F)$. Les solutions de $(zE-F)x=Gu$ sont alors $x=\mathcal{K}_1 u + \mathcal{L}_1 v$ où v est quelconque. Posons $\mathcal{K}=H\mathcal{K}_1+J$ et $\mathcal{L}=H\mathcal{L}_1$. on obtient alors le résultat suivant:
u est commande admissible $\Longleftrightarrow$ $\mathcal{M}u=0$
Les sorties associées à u sont $y=\mathcal{K}u+\mathcal{L}v$ où v est arbitraire.
Par conséquent $(\mathcal{K}, \mathcal{L}, \mathcal{M})$ est un triplet de transfert.

4) Une autre façon de procéder consiste à mettre le système sous la forme qui suit, qui est proche de la forme de Kronecker. On pourra se rapporter à [2] pour l'obtention de cette forme et ses propriétés.

$$\dot{x}_1 = Fx_1 + G_1u$$

$$N\dot{x}_2 = x_2 + G_2u$$

$$\dot{x}_3 = Ax_3 + Bv + G_3u$$

$$\dot{x}_4 = Cx_4 + G_4u$$

$$0 = Dx_4 + G_5u$$

$$y = H_1x_1 + H_2x_2 + H_3x_3 + H_4x_4 + H_5v + Ju$$

où la paire (A,B) est complètement accessible, la paire (D,C) est complètement observable et la matrice N est nilpotente.
Dans ce cas on obtient un triplet de transfert en posant:

$$\mathcal{K} = H_1(zI-F)^{-1}G_1 + H_2(zN-I)^{-1}G_2 + H_3(zI-A)^{-1}G_3 + H_4(zI-C)^{-1}G_4 + J$$

$$\mathcal{L} = H_3(zI-A)^{-1}B + H_5$$

$$\mathcal{M} = D(zI-C)^{-1}G_4 + G_5$$

5) Nous appellerons <u>fonction de transfert</u> d'un système l'ensemble de ses triplets de transfert. Ceux-ci peuvent être obtenus à partir de l'un d'entre eux $(\mathcal{K}, \mathcal{L}, \mathcal{M})$ en prenant $(\mathcal{K} + \mathcal{L}L + M\mathcal{M}, \mathcal{L}P, Q\mathcal{M})$ où L, M, P et Q sont des matrices rationnelles, $\mathrm{Im}\,\mathcal{L}P = \mathrm{Im}\,\mathcal{L}$, $\mathrm{Ker}\,Q\mathcal{M} = \mathrm{Ker}\,\mathcal{M}$, L et M quelconques.
Remarquons que si le système est Kalmanien, de fonction de transfert $\mathcal{K}$, il n'y a qu'un seul triplet de transfert, à savoir $(\mathcal{K}, 0, 0)$.

6) Nous définissons deux notions d'équivalence des systèmes implicites donnés sous forme de représentation interne. La première, appelée équivalence faible (ou plus simplement équivalence), consiste en fait à dire que deux systèmes sont équivalents s'ils ont même fonction de transfert. On dit que deux systèmes sont <u>fortement équivalents</u> si le deuxième système peut se déduire du premier en remplaçant E, F, G, H et J respectivement par UEV, UFV, U(G + FN), (H + MF)V, J + HN + MG + MFN où U et V sont inversibles, EN = 0 et ME = 0.

Cette notion d'équivalence forte est une généralisation naturelle de la notion de changement de base. Elle correspond à la notion de "strict system equivalence" de Rosenbrock dans le cas où les matrices sont scalaires.

7) On dit qu'un système est minimal si ce système minimise à la fois le nombre de lignes et de colonnes de la matrice E, dans l'ensemble de tous les systèmes qui lui sont équivalents.

8) On dit qu'un système est canonique si les trois conditions suivantes sont vérifiées

(CA1) $\forall x \quad Ex = 0 \Rightarrow Fx \in \mathrm{Im}E$

(CA2) $\forall(\lambda,\mu) \in \mathbb{C} \times \mathbb{C}, \quad (\lambda,\mu) \neq (0,0), \quad \begin{pmatrix} \lambda E - \mu F \\ H \end{pmatrix}$ est injective

(CA3) $\forall(\lambda,\mu) \in \mathbb{C} \times \mathbb{C} \quad (\lambda,\mu) \neq (0,0), \quad (\lambda E - \mu F \ G)$ est surjective

Remarque

Supposons (CA1) vérifiée. Les deux propriétés (CA2) et (CA3) sont alors respectivement équivalentes à l'observabilité et la commandabilité du système. Dans le cas où (CA1) n'est pas vérifiée cela devient faux.
La propriété (CA1) signifie qu'il n'y a pas de variables non dynamiques dans le système. Si l'on veut, comme par exemple [13], minimiser le rang de la matrice E plutôt que sa taille, il faut supprimer cette condition et remplacer les deux autres par des conditions légèrement plus compliquées.

9) Le résultat principal de la théorie des systèmes implicites est le suivant:
Tout système est équivalent à un système canonique. Celui-ci est unique à une équivalence forte près, et les notions de minimalité et de canonicité coïncident.

3. DEFINITION ET ETUDE AU NIVEAU FONCTION DE TRANSFERT.

Définition

Deux systèmes Σ_1 et Σ_2 sont dits inverses l'un de l'autre si :

a) Les sorties de Σ_1 sont les commandes admissibles de Σ_2 et réciproquement les sorties de Σ_2 sont les commandes admissibles de Σ_1.

b) Si u est commande admissible de Σ_1 et si y est une sortie associée à cette commande, le système Σ_2 commandé par y admet u comme sortie (i.e. u fait partie de l'ensemble des sorties).
De plus la même condition est demandée en échangeant les indices 1 et 2.

Donnons-nous un triplet de transfert $(\mathcal{K}_i, \mathcal{Y}_i, \mathcal{M}_i)$ du système Σ_i pour i=1,2. Pour un triplet de transfert $(\mathcal{K}, \mathcal{L}, \mathcal{M})$ on introduit les matrices $\bar{\mathcal{L}}$ et $\bar{\mathcal{M}}$ définies par $\mathrm{Im}\, \mathcal{L} = \mathrm{Ker}\, \bar{\mathcal{L}}$ et $\mathrm{Ker}\, \mathcal{M} = \mathrm{Im}\, \bar{\mathcal{M}}$. Ces matrices ne sont bien entendu pas uniques.

Ecrivons la condition a). Dire que y est une sortie de Σ_1 équivaut à dire qu'il existe u et v tels que $y = \mathcal{K}_1 u + \mathcal{L}_1 v$ avec $\mathcal{M}_1 u = 0$. Mais le noyau de $\mathcal{M}_1$ est l'image de $\bar{\mathcal{M}}_1$, d'où dire que y est sortie de Σ_1 équivaut à dire que y est dans l'image de $(\mathcal{K}_1 \bar{\mathcal{M}}_1 \quad \mathcal{L}_1)$. Par conséquent la condition a) s'écrit :

$$\mathrm{Ker}\, \mathcal{M}_2 = \mathrm{Im}\, (\mathcal{K}_1 \bar{\mathcal{M}}_1 \quad \mathcal{L}_1) \qquad (1)$$

$$\mathrm{Ker}\, \mathcal{M}_1 = \mathrm{Im}\, (\mathcal{K}_2 \bar{\mathcal{M}}_2 \quad \mathcal{L}_2) \qquad (2)$$

Ecrivons la condition b). Supposons que y soit une sortie de Σ_1 associée à la commande u. Il existe alors v tel que $y = \mathcal{K}_1 u + \mathcal{L}_1 v$ avec $\mathcal{M}_1 u = 0$. Dire que u fait partie de l'ensemble des sorties de Σ_2 associées à la commande y équivaut à l'existence de z vérifiant : $u = \mathcal{K}_2 y + \mathcal{L}_2 z$.

Portons l'expression de y dans cette formule. On obtient :

$$\exists z \qquad u = \mathcal{K}_2 \mathcal{K}_1 u + \mathcal{K}_2 \mathcal{L}_1 v + \mathcal{L}_2 z$$

On peut choisir z linéairement en fonction de u et v ; écrivons donc $z = Au + Bv$ où A et B sont des matrices (qu'on peut choisir rationnelles). On obtient par conséquent :

$$\mathcal{M}_1 u = 0 \qquad (\mathcal{K}_2 \mathcal{K}_1 + \mathcal{L}_2 A - I)u + (\mathcal{K}_2 \mathcal{L}_1 + \mathcal{L}_2 B)v = 0.$$

Ce qui peut encore s'écrire :

$$\exists A\ \exists C \quad \mathcal{K}_2 \mathcal{K}_1 + \mathcal{L}_2 A + C\mathcal{M}_1 = I \quad (3) \quad \exists B \qquad \mathcal{K}_2 \mathcal{L}_1 + \mathcal{L}_2 B = 0 \qquad (4)$$

En échangeant les indices 1 et 2 on obtient les formules (5) et (6) suivantes

$$\exists D \qquad \mathcal{K}_1 \mathcal{L}_2 + \mathcal{L}_1 D = 0 \quad (5) \quad \exists E\ \exists F \qquad \mathcal{K}_1 \mathcal{K}_2 + \mathcal{L}_1 E + F\mathcal{M}_2 = I \qquad (6)$$

Par conséquent dire que les deux systèmes Σ_1 et Σ_2 sont inverses l'un de l'autre équivaut à dire que les conditions (1) à (6) sont vérifiées.
Remarquons que ces conditions entraînent les suivantes :

$$\mathrm{Im}\, \mathcal{L}_2 = \mathrm{Ker} \begin{pmatrix} \bar{\mathcal{L}}_1 \mathcal{K}_1 \\ \mathcal{M}_1 \end{pmatrix} \quad (7) \qquad \mathrm{Im}\, \mathcal{L}_1 = \mathrm{Ker} \begin{pmatrix} \bar{\mathcal{L}}_2 \mathcal{K}_2 \\ \mathcal{M}_2 \end{pmatrix} \qquad (8)$$

Comme manifestement les équations (7) et (8) impliquent les équations (5) et (4) on a prouvé:

Théorème: Les deux systèmes sont inverses l'un de l'autre si et seulement si

$$\text{(I)}\quad \begin{aligned}
&\mathrm{Ker}\,\mathcal{M}_2 = \mathrm{Im}\,(\mathcal{K}_1\bar{\mathcal{M}}_1 \;\; \mathcal{L}_1) && (1)\\
&\mathrm{Ker}\,\mathcal{M}_1 = \mathrm{Im}\,(\mathcal{K}_2\bar{\mathcal{M}}_2 \;\; \mathcal{L}_2) && (2)\\
&\mathrm{Im}\,\mathcal{L}_2 = \mathrm{Ker}\begin{pmatrix}\bar{\mathcal{L}}_1\mathcal{K}_1\\ \mathcal{M}_1\end{pmatrix} && (7)\\
&\mathrm{Im}\,\mathcal{L}_1 = \mathrm{Ker}\begin{pmatrix}\bar{\mathcal{L}}_2\mathcal{K}_2\\ \mathcal{M}_2\end{pmatrix} && (8)\\
&\exists E\ \exists F\quad \mathcal{K}_1\mathcal{K}_2 + \mathcal{L}_1E + F\mathcal{M}_2 = I && (6)\\
&\exists A\ \exists B\quad \mathcal{K}_2\mathcal{K}_1 + \mathcal{L}_2A + B\mathcal{M}_1 = I && (3)
\end{aligned}$$

Les équations sont symétriques par rapport aux deux systèmes (elles sont invariantes si l'on échange les indices 1 et 2). Nous allons donner un système d'équations non symétrique permettant de calculer la fonction de transfert du système Σ_2 en fonction de celle du système Σ_1.

Théorème :

Les deux systèmes sont inverses l'un de l'autre si et seulement si:

$$\text{(II)}\quad \begin{aligned}
&\mathrm{Ker}\,\mathcal{M}_2 = \mathrm{Im}\,(\mathcal{K}_1\bar{\mathcal{M}}_1 \;\; \mathcal{L}_1) && (1)\\
&\mathrm{Im}\,\mathcal{L}_2 = \mathrm{Ker}\begin{pmatrix}\bar{\mathcal{L}}_1\mathcal{K}_1\\ \mathcal{M}_1\end{pmatrix} && (7)\\
&\exists A\ \exists B\quad \mathcal{K}_1\mathcal{K}_2\mathcal{K}_1 - \mathcal{K}_1 = \mathcal{L}_1A + B\mathcal{M}_1 && (9)\\
&\exists C\quad \mathcal{K}_1\mathcal{K}_2\mathcal{L}_1 = \mathcal{L}_1C && (10)\\
&\exists D\quad \mathcal{M}_1\mathcal{K}_2\mathcal{K}_1 = D\mathcal{M}_1 && (11)\\
&\mathcal{M}_1\mathcal{K}_2\mathcal{L}_1 = 0 && (12)
\end{aligned}$$

Prouvons maintenant que le système (II) admet une solution unique, c'est-à-dire qu'il admet au moins une solution, et que quel que soit le triplet de transfert $(\mathcal{K}_1, \mathcal{L}_1, \mathcal{M}_1)$ du système Σ_1 choisi, quel que soit le triplet $(\mathcal{K}_2, \mathcal{L}_2, \mathcal{M}_2)$ solution des équations (II), le système Σ_2 ayant $(\mathcal{K}_2, \mathcal{L}_2, \mathcal{M}_2)$ comme triplet de transfert est unique.

Il est clair que (1) et (7) définissent de façon unique les sous-espaces Im $\mathcal{L}_2$ et Ker $\mathcal{M}_2$ à partir de la fonction de transfert de Σ_1. Etudions maintenant $\mathcal{K}_2$. Comme $\mathcal{L}_1$ est injective et $\mathcal{M}_1$ surjective par changement de base on peut supposer $\mathcal{L}_1 = \binom{I}{0}$ et $\mathcal{M}_1 = (I\ 0)$

Ecrivons dans ces bases $\mathcal{K}_1 = \begin{pmatrix} E & F \\ G & H \end{pmatrix}$.

Remarquons que remplacer $\mathcal{K}_1$ par $\mathcal{K}_1 + \mathcal{L}_1 L + M\mathcal{M}_1$, consiste à prendre n'importe quelles matrices pour E, F et G, sans changer H.
Considérons maintenant deux matrices inversibles T et S telles que $THS = \begin{pmatrix} I & 0 \\ 0 & 0 \end{pmatrix}$.

Ecrivons $T = \begin{pmatrix} \bar{T}_1 \\ \bar{T}_2 \end{pmatrix}$ $\quad S = (\bar{S}_1, \bar{S}_2)$ $\quad T^{-1} = (T_1\ T_2)$ $\quad S^{-1} = \begin{pmatrix} S_1 \\ S_2 \end{pmatrix}$

Alors $H = T_1 S_1$ Ker H = Ker S_1 = Im $\bar{S}_2$ et Im H = Im T_1 = Ker $\bar{T}_2$

Dans ces nouvelles bases on peut prendre $\bar{\mathcal{M}}_1 = \begin{pmatrix} 0 \\ I \end{pmatrix}$ et $\bar{\mathcal{L}}_1 = (0\ I)$.

On peut par conséquent choisir $\mathcal{M}_2 = (0\ \bar{T}_2)$. De même on peut choisir $\mathcal{L}_2 = \begin{pmatrix} 0 \\ \bar{S}_2 \end{pmatrix}$.

Cherchons $\mathcal{K}_2$ sous la forme $\mathcal{K}_2 = \begin{pmatrix} X & Y \\ Z & U \end{pmatrix}$. Les relations (9) à (12) sont équivalentes à l'existence de matrices A B C D telles que $X=0$ $Y=A\bar{T}_2$ $Z=\bar{S}_2 B$ $U=\bar{S}_1\bar{T}_1+\bar{S}_2 C+D\bar{T}_2$.

Par conséquent

$$\mathcal{K}_2 = \begin{pmatrix} 0 & 0 \\ 0 & \bar{S}_1\bar{T}_1 \end{pmatrix} + \begin{pmatrix} A \\ D \end{pmatrix} (0\ \bar{T}_2) + \begin{pmatrix} 0 \\ \bar{S}_2 \end{pmatrix} (B\ C)$$

$$= \begin{pmatrix} 0 & 0 \\ 0 & \bar{S}_1\bar{T}_1 \end{pmatrix} + \begin{pmatrix} A \\ D \end{pmatrix}\mathcal{M}_2 + \mathcal{L}_2\,(B\ C)$$

Par conséquent les équations (9) à (12) admettent une solution, et les triplets $(\mathcal{K}_2, \mathcal{L}_2, \mathcal{M}_2)$ solutions du système (II) sont des représentants d'une même fonction de transfert.

4. ETUDE AU NIVEAU REPRESENTATION INTERNE.

Comme dit dans l'introduction l'existence d'un système inverse est triviale. En effet supposons que le système Σ soit

$$E\dot{x} = Fx + Gu \qquad (13)$$
$$y = Hx + Ju$$

Rajoutons l'équation $u = x_2$. Le système s'écrit donc:

$$E\dot{x} = Fx + Gx_2$$
$$0 = Hx + Jx_2 - y$$
$$u = x_2$$

Posons $E_1 = \begin{pmatrix} E & 0 \\ 0 & 0 \end{pmatrix}$ $F_1 = \begin{pmatrix} F & G \\ H & J \end{pmatrix}$ $G_1 = \begin{pmatrix} 0 \\ -I \end{pmatrix}$ $H_1 = (0\ I)$ et $x_1 = \begin{pmatrix} x \\ x_2 \end{pmatrix}$

Ce système s'écrit donc $E_1\dot{x}_1 = F_1x_1 + G_1y \quad u = H_1x_1$ (14)

Il est donc manifeste que le système Σ_1

$$E_1\dot{x}_1 = F_1x_1 + G_1u \qquad y = H_1x_1 \tag{15}$$

est le système inverse de Σ.

Théorème: Σ_1 est canonique si et seulement si Σ l'est, E est inversible et J nul.

Il est clair que pour que Σ_1 soit canonique il faut déja que Σ le soit. Supposons donc Σ canonique et regardons à quelle condition Σ_1 est canonique.

La première condition de canonicité (CA1) s'écrit:

$$E_1x_1 = 0 \Rightarrow F_1x_1 \in \operatorname{Im} E_1$$

donc $$Ex = 0 \Rightarrow \exists y \quad \begin{matrix} Fx + Gx_2 = Ey \\ Hx + Jx_2 = 0 \end{matrix}$$

On en déduit que J doit être nul. La condition (CA1) est alors manifestement vérifiée si E est inversible, ce que nous prouverons dans la suite.

La condition (CA2) s'écrit

si $(\lambda,\mu) \neq (0,0)$ $\begin{pmatrix} \lambda E_1 - \mu F_1 \\ H_1 \end{pmatrix}$ injective

donc si $(\lambda,\mu) \neq (0,0)$ $\begin{pmatrix} \lambda E - \mu F & -\mu G \\ -\mu H & 0 \\ 0 & I \end{pmatrix}$ injective

donc pour $\mu = 0$: E injective

pour $\mu \neq 0$ $\begin{pmatrix} \mu E - F \\ H \end{pmatrix}$ injective.

La condition (CA3) s'écrit de même: E surjective et $(\lambda E - F \quad G)$ surjective.

Par conséquent E doit être inversible. On peut donc supposer que E est l'identité. Les conditions $\begin{pmatrix}\lambda E - F\\ H\end{pmatrix}$ injective, $(\lambda E - F \quad G)$ surjective sont alors équivalentes à la canonicité du système Σ (qui est dans ce cas un système Kalmanien).

5. INVERSIBILITE A DROITE ET A GAUCHE : ETUDE GEOMETRIQUE.

Définition:

Un système Kalmanien Σ $\quad \dot{x} = Fx + Gu$ (16)

$$y = Hx + Ju$$

de fonction de transfert $\mathcal{H} = H(zI - F)^{-1}G + J$ est appelé inversible à droite ou à gauche si sa fonction de transfert $\mathcal{H}$ est inversible à droite ou à gauche respectivement (donc surjective ou injective).

Dire que $\mathcal{H}$ est surjective équivaut à dire que tout y est sortie du système, donc que pour le système inverse Σ^{-1} toute commande est commande admissible. Dire que est injective équivaut à dire que toute sortie ne peut être obtenue que par une seule entrée, donc que pour Σ^{-1}, pour toute commande admissible il n'y a qu'une seule sortie.

En fait soit $(\mathcal{K}, \mathcal{L}, \mathcal{M})$ un triplet de transfert de Σ^{-1}. D'après la partie 3., ce triplet vérifie les équations :

$$\begin{aligned} \operatorname{Ker}\mathcal{M} &= \operatorname{Im}\mathcal{H} \\ \operatorname{Im}\mathcal{L} &= \operatorname{Ker}\mathcal{H} \\ \mathcal{H}\mathcal{K}\mathcal{H} &= \mathcal{H} \end{aligned} \tag{17}$$

Donc Σ est inversible à gauche (resp. à droite) équivaut à $\mathcal{L} = 0$ (resp. $\mathcal{M} = 0$). Ceci équivaut également à dire que pour le système Σ^{-1} écrit sous forme (15) il n'y a pas de blocs ε (resp. η).

Remarquons bien que s'il n'y a pas de blocs ε la matrice $\mathcal{L}$ est nulle, mais en général la réciproque est fausse. Ce n'est pas le cas ici car la forme (15) n'est pas la forme générale d'un système implicite. Ecrivons (15) sous la forme :

$$\begin{aligned} \dot{x}_1 &= Fx_1 + Gx_2 \\ 0 &= Hx_1 + Jx_2 - u \\ y &= x_2 \end{aligned} \tag{18}$$

Supposons $\mathcal{L}$ nul. Ceci signifie que lorsque la commande u est nulle la sortie y est nulle (il y a unicité de la sortie). Par conséquent x_2 est nul, et donc la première équation $x_1 = Fx_1$ montre que x_1 est nul (on suppose ici que la condition initiale est nulle).

Or pour un système implicite $\begin{matrix} E\dot{x} = Fx + Gu \\ y = Hx + Ju \end{matrix}$ dire qu'il n'a pas de blocs ε signifie que l'équation $E\dot{x} = Fx$ admet comme seule solution la solution nulle si la condition initiale est nulle. Par conséquent pour le système (18) les conditions "$\mathcal{L}$ nul" et "il n'y a pas de blocs ε" sont équivalentes.

On prouve de même que "$\mathcal{M}$ nul" et "il n'y a pas de blocs η" sont équivalentes.

L'étude géométrique faite dans [2] montre à quelle condition il n'y a pas de blocs ε ou η dans le système (18). Cette condition est $V_* \cap W^* = 0$ (resp. $V_* + W^*$ est l'espace tout entier et la matrice $(E_1\ F_1)$ est surjective), où

W^* est le plus grand sous-espace W tel que $F_1W \subset E_1W$

V_* est le plus petit sous-espaxe V tel que $E_1^{-1}F_1V \subset V$.

Déterminons ces deux espaces.

Ecrivons $W^* = \mathrm{Im}\begin{pmatrix} A \\ B \end{pmatrix}$. Alors $F_1W^* = \mathrm{Im}\begin{pmatrix} FA + GB \\ HA + JB \end{pmatrix}$ et $E_1W^* = \mathrm{Im}\begin{pmatrix} A \\ 0 \end{pmatrix}$.

Dire que $F_1W^* \subset E_1W^*$ équivaut à l'existence d'une matrice C telle que

$$\begin{aligned} FA + GB &= EAC \\ HA + JB &= 0 \end{aligned} \qquad (19)$$

Soit T une matrice inversible telle que $AT = (A_1\ 0)$ avec A_1 injectif.
Si nous écrivons $BT = (B_1\ B_2)$ nous aurons

$$W^* = \mathrm{Im}\begin{pmatrix} A_1 & 0 \\ B_1 & B_2 \end{pmatrix} \quad \text{et} \quad \begin{matrix} FA_1 + GB_1 = A_1C_1 \\ HA_1 + JB_1 = 0 \end{matrix} \ (20.a) \quad \begin{matrix} GB_2 = A_1C_2 \\ JB_2 = 0 \end{matrix} \qquad (20.b)$$

Comme A_1 est injectif l'équation (20.a) équivaut à l'existence de K telle que :

$$\begin{aligned} (F - GK)A_1 &= A_1C_1 \\ (H - JK)A_1 &= 0 \end{aligned} \qquad (21.a) \qquad \text{où } B_1 = KA_1.$$

Soit $\mathcal{W}$ l'espace introduit par exemple par Silverman et Hautus [8], qui est le plus grand sous-espace de l'espace d'état (du système initial Σ) tel qu'il existe un feedback K tel que $(F - GK)$ (21.b)

$$(H - JK) = 0$$

D'où par (21.a) $\mathcal{W} = \operatorname{Im} A_1$. Introduisons $\mathcal{W}_1$ le plus grand sous-espace de l'espace des commandes tel que

$$G\mathcal{W}_1 \subset \mathcal{W} \quad J\mathcal{W}_1 = 0 \tag{21.c}$$

Par (20.b) on déduit $\operatorname{Im} B_2 = \mathcal{W}_1$.

D'où:

Un vecteur $\begin{pmatrix} x \\ u \end{pmatrix}$ est dans W^* si et seulement si x est dans l'image de A_1 et $Kx - u$ est dans l'image de B_2. Par conséquent W est l'ensemble des vecteurs de la forme $\begin{pmatrix} x \\ Kx + v \end{pmatrix}$ où $x \in \mathcal{W}$ et $v \in \mathcal{W}_1$.

Déterminons maintenant V_*. Ecrivons à nouveau $V_* = \operatorname{Im}\begin{pmatrix} A \\ B \end{pmatrix}$.

On a $E_1^{-1} F_1 V_* \subset V_* \iff$ si $E_1 y \in F_1 V_*$ alors $y \in V_*$

Donc en prenant $y = \begin{pmatrix} x \\ u \end{pmatrix}$: si $\begin{pmatrix} x \\ 0 \end{pmatrix} \in F_1 V_*$ alors $\begin{pmatrix} x \\ u \end{pmatrix} \in V_*$.

Donc si $\begin{array}{l} x = Fy + Gv \\ 0 = Hy + Jv \end{array}$ avec $\begin{pmatrix} y \\ v \end{pmatrix} \in V_*$ alors $\begin{pmatrix} x \\ u \end{pmatrix} \in V_*$.

Ce qui s'écrit encore $\begin{pmatrix} y \\ v \end{pmatrix} \in V_*$ $Hy + Jv = 0 \Rightarrow \begin{pmatrix} Fy + Gv \\ u \end{pmatrix} \in V_*$.

Donc $HAx + JBx = 0 \Rightarrow \exists y \quad FAx + GBx = Ay \quad u = By$

On constate que B doit être surjective. Il existe donc T inversible telle que $BT = (0 \; I)$. Posons $AT = (A_1 \; A_2)$.

Alors $V = \operatorname{Im}\begin{pmatrix} A_1 & A_2 \\ 0 & I \end{pmatrix}$ et

$$HA_1x_1 + HA_2x_2 + Jx_2 = 0 \Rightarrow \exists y_1 \; \exists y_2 \quad \begin{array}{l} FA_1x_1 + FA_2x_2 + Gx_2 = A_1y_1 + A_2y_2 \\ \text{et} \quad u = y_2 \end{array}$$

Regardant cette condition pour $x_1 = x_2 = 0$ on constate que $\operatorname{Im} A_2$ $\operatorname{Im} A_1$ et donc qu'il existe K_1 tel que $A_2 = A_1K_1$. Par conséquent:

$$HA_1x_1 + Jx_2 = 0 \qquad y_1 \quad FA_1x_1 + Gx_2 = A_1y_1$$

Ceci implique l'existence de trois matrices L M_1 M_2 telles que:

$$(F - LH)A_1 = A_1 M_1 \qquad (22)$$
$$G - LJ = A_1 M_2$$

Soit $\mathcal{V}$ l'espace introduit par Silverman et Hautus qui est le plus petit sous-espace de l'espace d'état (de Σ) tel que $(F - LH)\mathcal{V} \subset \mathcal{V}$

$\mathrm{Im}\,(G - LJ) \subset \mathcal{V}$

Comme $V_* = \mathrm{Im}\begin{pmatrix} A_1 & A_1K_1 \\ 0 & I \end{pmatrix} = \mathrm{Im}\begin{pmatrix} A_1 & 0 \\ 0 & I \end{pmatrix}$

et que c'est le plus petit sous-espace de cette forme où A_1 vérifie les équations (22) on en déduit $\mathrm{Im}\,A_1 = \mathcal{V}$.

Par conséquent V_* est l'ensemble des vecteurs $\binom{x}{u}$ où x est dans $\mathcal{V}$ et u quelconque. Ayant maintenant déterminé les espaces W^* et V_* on peut écrire les conditions $V_* \cap W^* = 0$ et $V_* + W^* =$ l'espace en entier.

Ecrivons d'abord $V_* \cap W^* = 0$. Soit $\binom{x}{u} \in W^* \cap V_*$. Ceci équivaut à $x \in \mathcal{V} \cap \mathcal{W}$ et $Kx - u \in \mathcal{W}_1$. Pour que x soit nul il faut que $\mathcal{V} \cap \mathcal{W} = 0$. Pour que u soit nul il faut en plus que $\mathcal{W}_1 = 0$.

Comme $\mathcal{W}_1$ est le plus grand sous-espace tel que $\begin{matrix} G\mathcal{W}_1 \subset \mathcal{W} \\ J\mathcal{W}_1 = 0 \end{matrix}$, pour que $\mathcal{W}_1$ soit nul il faut que $\binom{G}{J}$ soit injectif, car son noyau vérifie cette relation. Cette condition est en fait suffisante.

En effet par (22) $\mathrm{Im}\,(G - LJ) \subset \mathcal{V}$ d'où $(G - LJ)\mathcal{W}_1 \subset \mathcal{V}$.
Par (21.c) $(G - LJ)\mathcal{W}_1 \subset \mathcal{W}$. Comme $\mathcal{V} \cap \mathcal{W} = 0$ on en déduit $(G - LJ)\mathcal{W}_1 = 0$; associé au fait que $J\mathcal{W}_1 = 0$ et que $\binom{G}{J}$ est injectif on en déduit $\mathcal{W}_1 = 0$.

Par conséquent $V_* \cap W^* = 0 \iff \mathcal{V} \cap \mathcal{W} = 0$ et $\binom{G}{J}$ injectif.
Ecrivons maintenant que $V_* + W^*$ est l'espace en entier. Un élément de cet espace est de la forme

$\begin{pmatrix} x_1 + x_2 \\ u_1 + u_2 \end{pmatrix}$ où $x_1 \in \mathcal{W}$, $u_1 - Kx_1 \in \mathcal{W}_1$, $x_2 \in \mathcal{V}$ et u_2 est quelconque. Une condition nécessaire et suffisante pour que $V_* + W^*$ soit l'espace en entier est donc que $\mathcal{V} + \mathcal{W}$ soit l'espace d'état (de Σ) en entier.

Il s'agit encore de voir sous quelle condition la matrice $(E_1 \; F_1)$ est surjective.

Or $(E_1\ F_1) = \begin{pmatrix} I & 0 & F & G \\ 0 & 0 & H & J \end{pmatrix}$. Elle est donc surjective si (H J) l'est.

En résumé

Le système Σ est inversible à gauche $\Leftrightarrow$ $\mathcal{V} \cap \mathcal{W} = 0$ et $\begin{pmatrix} G \\ J \end{pmatrix}$ injective

Le système Σ est inversible à droite $\Leftrightarrow$ $\mathcal{V} + \mathcal{W} = \mathcal{X}$ et (H J) surjective

où $\mathcal{X}$ est l'espace d'état du système Σ.

Ces conditions sont celles de [8] (théorèmes (3.24) et (3.26)).

6. CONCLUSION

Le problème sous-jacent à l'inversion des systèmes est le suivant: Etant donné un système Σ et un comportement y, on aimerait savoir si y est une sortie de Σ et calculer une commande u qui donne y en sortie. Ayant introduit le système inverse, le problème est devenu celui de savoir si une commande u est admissible et de calculer au moins une sortie associée.

Pour résoudre ce problème on peut mettre le système sous forme de Kronecker, sous la forme citée dans la partie 2.2) ou sous forme normale (cf [2]). Pour obtenir des deux dernières formes, le plus simple est de commencer par mettre le système sous forme de Kronecker. C'est pourquoi nous citons dans la bibliographie un certain nombre d'articles concernant la mise sous forme de Kronecker.

D'un autre côté, comme nous l'avons montré dans cet article, de nombreux problèmes de la théorie des systèmes peuvent se résoudre au moyen de la théorie des systèmes implicites.

Bibliographie

1 D. COBB : "Descriptor variable and generalized singularly perturbed systems : a geometric approach" PhD Dissertation, Department of Electrical Enginearing, University of Illinois, Urbana-Champaign, 1980

2 J. GRIMM : "Sur les systèmes dynamiques linéaires implicites singuliers" thèse de troisième cycle, Université de Paris IX Dauphine, Paris, 1983.

3 LUENBERGER, STENGEL, LARSON, CLINE : "A Descriptor Variable Approach to modeling and optimisation of Large Scale Systems" Proceedings of the Engineering Foundation Conference on System Engineering for power Organizational Forms for Large Scale Systems, vol 7, Davos ed. Switzerland, 1979

4 MOLINARI : "Structural Invariants of Linear Multivariable Systems" Int. J. Cont. vol 28, pp 493-510, 1978

5 MORSE : "Structural Invariants of Linear Multivariable Systems" SIAM Journal on Control, vol 11, pp 446-465, 1973

6 MOYLAN : " Stable Inversion of Linear Systems " IEEE Tr. Aut. Contr., AC-22, pp 74-78, 1977

7 H.H. ROSENBROCK "State-Space and Multivariable Theory" Nelson, 1970

8 L.M. SILVERMAN & M.L.J. HAUTUS : " System structure and singular control", Linear Algebra and its Applications, vol 50, avril 1983.

9 L.M. SILVERMAN "Discrete Riccati Equations: alternative algoritms, asymptotic properties, and system theory interpretations" Control & Dynamic Systems, New-York, Ac.Press, 1976

10 THORP : "the Singular Pencil of a Linear Dynamical System " Int. J. Contr. p 577-596, 12973

11 VAN DOOREN, VERGHESE, KAILATH "Properties of the System Matrix of a Generalized State-Space System" Int. J. Contr., vol 30, pp 235-243, 1979

12 VAN DOOREN : "The generelized Eigenstructure Problem in a Linear System Theory" IEEE, AC-26, pp 111-129, 1981

13 G.C. VERGHESE, B. LEVY, T. KAILATH : "A generalized state-space for singular systems", IEEE Transactions on Automatic Control, AC 26, pp 811-831, 1981.

14 VERGHESE : "Infinite-frequency Behavior in Generalized Dynamical Systems" Ph.D. Dissertation,Dep. Electrical Engineering ,Stanford Univ., Dec 1978

15 J.H. WILKINSON "Linear Differential Equations and Kronecker's Canonical Form" Recent advances in numerical analysis, De Boor and Golub ed., Ac. Press 1979, Poceedings of a conference held in Madison, 1978.

SUR L'IDENTIFICATION DES SYSTEMES CYCLIQUES

L. BARATCHART
INRIA. Route des lucioles
Sophia-Antipolis
06560 VALBONNE
FRANCE

S. STEER
INRIA. Domaine de Voluceau
Rocquencourt
78153 LE CHESNAY
FRANCE

SUMMARY.

In order to take into account the fact that the data generaly given for a linear system identification are truncated and subject to numerical errors, the paper considers the identification problem as an approximation one.

We use a L_2 norm, and suppose that the system is stable so that the truncation error may be considered small (at least theoretically).

Using results of [2] we examine the (generic) case where the system to be identified is cyclic. If the errors on the data are sufficently small, we prove the uniqueness of a best approximant and his cyclicity. We also prove that it is the unique solution of an equation, for which we propose a linear heuristic.

Some numericals aspects of this procedure are then examines and numerical examples are presented.

One of the features of this approach is that it identifies the transfer function of the system under a form which allows one to find easily a state space representation.

I. INTRODUCTION

Le but de cet article est d'examiner comment le problème de l'identification d'un système linéaire stable et cyclique peut être considéré comme un problème d'approximation, pour lequel on propose une heuristique de résolution, avant de présenter des résultats numériques . Le lecteur est supposé familier avec les notions de base de la théorie des systèmes linéaires.

2. IDENTIFICATION ET APPROXIMATION

Le problème de l'identification tel qu'on se le pose usuellement, est le suivant: étant donné une suite $(A_k) = (A_1, A_2, \dots)$ de matrices dans $R^{p\times m}$, on cherche $H \in R^{p\times m}$, $F \in R^{n\times n}$, $G \in R^{n\times m}$ telles que :

a) $A_k = H F^{k-1} G$, $k=1,2,\dots$ b) n est minimal

Si $L = \begin{pmatrix} A_1 & A_2 & \cdots \\ A_2 & A_3 & \cdots \\ \vdots & \vdots & \end{pmatrix}$ est la matrice de Hankel associée à (A_k), il est bien connu que ce problème admet des solutions si et seulement si le rang de L est fini. Ce rang est alors le nombre n cherché. Dans ce cas, la suite (A_k) est dite rationelle, (H,F,G) est une réalisation minimale de (A_k), dont l'ordre est n. La matrice rationelle $\mathcal{H} = H(zI-F)^{-1}G$ est dite fonction de transfert de (A_k).

Dans la pratique, les (A_k) représentent la réponse impulsionelle d'un système discret, ou la réponse fréquentielle d'un système continu, et on n'en connaît qu'un nombre fini. Des conditions sont connues [1] pour que cette suite finie se prolonge de façon que toute l'information nécessaire soit contenue dans les données, autrement dit que le rang maximal soit celui d'une sous-matrice construite avec des A_k connus. Même dans ce cas cependant, les erreurs numériques entachant les (A_k) font que le rang est à peu près égal à la taille de la plus grande matrice carrée que l'on peut écrire avec les données. Les méthodes classiques de résolution s'attachent alors à déterminer un rang "vraisemblable" pour L, qui diffère de son rang numérique, puis à factoriser une sous matrice convenable de L pour déterminer algébriquement (H,F,G). Les aspects numériques des procédures de ce type ont également été étudiés [9], [10].

Dans cette optique, on considère que les données sont exactement les premiers coefficients d'un système linéaire, et qu'on en possède suffisamment pour résoudre le problème, qui est alors purement algébrique. Une optique différente est la suivante : on suppose que l'on identifie un système rationel (A_k), mais qu'on ne connaît en réalité que $(A_k+\Delta A_k)$, (ΔA_k) représentant l'erreur de troncature et les imprécisions numériques. Supposons aussi que l'on connaisse un majorant n de l'ordre de (A_k), et que (A_k) et (ΔA_k) appartiennent à un même espace vectoriel donné $(E, \|\ \|)$. Si on sait trouver (A'_k) rationelle d'ordre au plus n dans E, telle que $\|(A_k+\Delta A_k)-(A'_k)\|$ soit minimal on a :

$$\|(A_k)-(A'_k)\| \le \|(A_k)-(A_k+\Delta A_k)\| + \|(A_k+\Delta A_k)-(A'_k)\| \le \|(\Delta A_k)\| + \|(A_k+\Delta A_k)-(A_k)\|$$

soit : $$\|(A_k)-(A'_k)\| \le 2\|(\Delta A_k)\|. \qquad (1)$$

Ainsi, en identifiant (A_k) comme étant (A'_k), l'erreur commise au sens de la norme est petite si $\|(\Delta A_k)\|$ l'est.

Un problème voisin est de trouver un système d'ordre minimal approchant $(A_k+\Delta A_k)$ à ε arbitraire fixé près. Une solution explicite est developpée dans [11] pour la norme de Hankel. Cependant le calcul ne peut être effectif que si $(A_k+\Delta A_k)$ est rationelle et qu'on en connaît une réalisation (éventuellement non minimale).

Le cas envisagé ici est celui des suite de carré sommable, c'est à dire que si $A_k = (a_{ij}^k)$, on a $\|(A_k)\| = \sum_{i,j} \sum_{k=1}^{\infty} (a_{ij}^k)^2 < \infty$. Les (A_k) représentent alors soit la réponse impulsionelle d'un système discret stable, soit le résultat de la substitution de $\frac{z+1}{z-1}$ à z dans la réponse fréquentielle d'un système continu stable.

Il est vrai dans ce cas qu'en prenant suffisamment de termes A_k, et si les erreurs numériques sont assez faibles, $\|(\Delta A_k)\|$ peut théoriquement devenir arbitrairement petit. Cependant, on a remplacé le problème du calcul algébrique de (H,F,G) par un problème d'approximation nettement plus difficile, et, à la connaissance des auteurs, non résolu. Nous utiliserons dans la suite des résultats de [2] pour aborder la question dans le cas (générique) où le système (A_k) est cyclique, et où l'ordre exact est supposé connu (et en pratique calculé comme dans les méthodes classiques). Une heuristique sera présentée, qui ne constitue certes pas un algorithme de réalisation. Toutefois, des exemples numériques montrent que lorsque les hypothèses "génériques" sont satisfaites de façon suffisamment nette, la méthode fournit des résultats plus rapides et plus précis (au sens L^2) que l'algorithme de HO[3]. Un intérêt de la procédure est qu'elle fournit aussi la fonction de transfert du système à identifier, alors que cette information est numériquement délicate à obtenir à partir d'un triplet (H,F,G).

3. PROPRIETES DE L'APPROXIMANT

Introduisons quelques notations : T désigne le cercle unité de C, U le disque unité ouvert. H_2^- sera l'espace vectoriel des fonctions holomorphes dans le complémentaire de $\bar{U}$, et qui s'annulent à l'infini, et s'écrivent $f(z) = \sum_{k=1}^{\infty} a_k z^{-k}$, avec $a_k \in R$, $\sum a_k^2 < \infty$. Toute fraction rationelle $\frac{p}{q}$ dont les pôles sont dans U, et telle que $d^o p < d^o q$, est dans H_2^-. L'application $f \to f^*$ définie par $f^*(e^{i\sigma}) = \sum_{k=1}^{\infty} a_k e^{-ki\sigma}$ (égalité au sens L^2) est une isométrie [4] de H_2^- sur le sous espace de $L^2(T)$ formé des fonctions dont les coefficients de Fourier d'indice positif sont nuls. En outre, on a $\lim_{r\to 1} f(re^{i\sigma}) = f^*(e^{i\sigma})$ presque partout. On identifiera H_2^- à son image dans $L^2(T)$, dont il est un sous espace de Hilbert. L'espace E du paragraphe précédent sera $(H_2^-)^{p\times m}$, sous espace de Hilbert de $L^2(T)^{p\times m}$, muni de la norme associée au produit scalaire :

$$<(f_{k,\ell}),(g_{k,\ell})> = \sum_{k,\ell} < f_{k,\ell}, g_{k,\ell} >_2$$

où " $<, >_2$" désigne le produit scalaire usuel de $L^2(T)$. Une suite (A_k) de carré sommable s'identifie à un élément de E par son developpement en série convergent hors de $\bar{U}$: $\sum_{k=1}^{\infty} A_k z^{-k}$. Si (A_k) est rationelle, ceci est aussi le développement en série de sa fonction de transfert, ce qui permet de les identifier. On notera $\mathcal{S}_n$ l'ensemble des suites (on dira aussi "systèmes") rationelles d'ordre au plus n, tandis que Σ_n sera l'ensemble des suites d'ordre exactement n. Le premier résultat tiré de [2] que nous énoncerons est le suivant :

<u>Théorème 1</u> : $\forall n \in N$, $\forall f \in E$, $\exists \mathcal{H}_o$ (non unique en général) $\in \mathcal{S}_n \cap E$ tel que $\|f-\mathcal{H}_o\| = \min_{\mathcal{H} \in \mathcal{S}_n \cap E} \|f-\mathcal{H}\|$.

Si $\mathcal{H}_o \in \mathcal{S}_{n-1}$, alors $f \in \mathcal{S}_{n-1}$.

Nous voyons donc que notre problème, qui est de trouver un meilleur approximant dans $\mathcal{S}_n \cap E$ pour $(A_k+\Delta A_k)$ admet une solution. Par ailleurs nous avons supposé que nous connaissons l'ordre exact de (A_k). Cette hypothèse permet d'obtenir un renseignement de plus sur l'approximant :

Proposition 1 : Si $(A_k) \in \Sigma_n$, et $\|(\Delta A_k)\|$ assez petit, tout meilleur approximant de $(A_k+\Delta A_k)$ est dans Σ_n.

preuve : Soit (A_k') un meilleur appproximant ; Si (A_k') est dans $\mathcal{S}_{n-1}$, on a :

$$\|(A_k)-(A_k')\| \geq \operatorname*{Inf}_{\in \mathcal{H}_{n-1} \cap E} \|(A_k)-\mathcal{H}\|$$

et le Inf, qui est atteint d'après le théorème, est nécessairement strictement positif. L'inégalité (1) du paragraphe précédent conclut.

D'après la proposition, nous sommes donc ramenés, lorsque $\|(\Delta A_k)\|$ est suffisamment petit à chercher notre approximant sur Σ_n qui est de dimension n(m+p) en tant que variété C^∞ [5]. Pour énoncer le prochain résultat, introduisons une notion de plus : Si (A_k) est rationelle, il est connu que toutes les matrices F intervenant dans une réalisation minimale (H,F,G) sont conjuguées. Elles ont donc mêmes polynômes invariants [6], de sorte que cela a un sens de parler de la structure cyclique de (A_k). On dira que (A_k) est cyclique s'il n'y a qu'un polynôme invariant.

Proposition 2 : Si (A_k) est cyclique, et $\|(\Delta A_k)\|$ assez petit, tout meilleur approximant de $(A_k+\Delta A_k)$ dans Σ_n est cyclique.

Preuve : l'ensemble des systèmes cycliques étant ouvert dans Σ_n [2], la proposition résulte de l'inégalité (1) et du lemme suivant :

lemme 1 : Sur $E \cap \Sigma_n$, la métrique induite par la norme est compatible avec la topologie de Σ_n.

Preuve : Soit $(A_k) \in \Sigma_n \cap E$ (qui est un ouvert de Σ_n car il correspond aux systèmes dont les pôles sont dans U), et L sa matrice de Hankel. Soit $(A_k') \in \Sigma_n \cap E$ tel que $\|(A_k')-(A_k)\| < \varepsilon$. Si L' est la matrice de Hankel de (A_k'), L'-L a tous ses coefficients inférieurs à ε. Pour ε assez petit, une sous matrice de rang maximal n sera au même endroit dans L et L'. Comme on peut obtenir rationellement une réalisation minimale à partir d'une telle sous matrice et de certains autres éléments de la matrice de Hankel (par exemple par l'algorithme de Silverman), (A_k) et (A_k') admettent des réalisations minimales arbitrairement voisines pour ε assez petit, ce qui montre que la topologie de $\Sigma_n \cap E$ est moins fine que celle induite par E. Réciproquement si (H,F,G) et (H',F',G') sont des réalisations minimales de (A_k) et (A_k'), $\mathcal{H}$ et $\mathcal{H}'$ les fonctions de transfert, on a $\mathcal{H} = \frac{N}{d}$, $\mathcal{H}' = \frac{N'}{d'}$ avec $d = \det(zI-F)$, $d' = \det(zI-F')$, N et N' matrices polynomiales de degré (n-1). Si les coefficients de nos deux réalisations sont arbitrairement voisins, ceux de ces polynômes le sont aussi. En particulier les zéros de d' sont voisins de ceux de d, qui sont dans U. On peut donc supposer :

$$\operatorname*{Inf}_{z\in T} |d'(z)| \geq \operatorname*{Inf}_{z\in T} |d(z)| - \varepsilon \geq r > 0$$

Si les coefficients des polynômes de N-N' sont inférieurs à η, on a :

$$\|(A'_k)-(A_k)\|^2 = \frac{1}{2\pi} \text{ trace } \left(\int_{-\pi}^{\pi} {}^t(\overline{\mathcal{H}'-\mathcal{H}})(\mathcal{H}'-\mathcal{H})\,d\sigma \right) \le \frac{1}{r^2}\, mp(n\eta)^2 .$$

Ceci montre que les deux topologies coïncident.

Nous voyons donc que si $\|(\Delta A_k)\|$ est assez petit, nous pouvons supposer notre meilleur approximant cyclique. Nous tirons parti de ce fait dans le paragraphe suivant.

4. EQUATIONS D'OPTIMUM

Rappelons brièvement quelques faits de théorie des systèmes pour lesquel on peut consulter [12] :

Si est la fonction de transfert d'un système d'ordre n, toute factorisation : $\mathcal{H} = N'D^{-1}N$, où N',D,N sont des matrices polynomiales telles que (N',D) soient premières entre elles à droite et (D,N) premières entre elles à gauche, est telle que :

i) deg (dét D) = n

ii) les polynômes invariants différents de 1 de D sont ceux du système. Introduisons à présent quelques notations.

Si a et b sont des polynômes dans R[z], b ≠ 0, on a par division a = bq+r ou encore $\frac{a}{b} = q + \frac{r}{b}$. On pose $E(\frac{a}{b}) = q$, $PP(\frac{a}{b}) = \frac{r}{b}$, qu'on appelle respectivement partie entière et partie propre de $\frac{a}{b}$. On va s'appuyer sur les résultats suivants, tirés de [2] :

Si $\mathcal{H}$ est la fonction de transfert d'un système cyclique de Σ_n, il existe des matrices régulières $R_1 \in R^{p\times p}$, $R_2 \in R^{m\times m}$, telles que l'on ait :

$$\mathcal{H} = R_1\, D^{-1}\, N\, R_2 \tag{2}$$

avec D et N matrices polynômiales du type suivant :

$$D = \begin{pmatrix} 1 & & & a_1 \\ & 1 & 0 & \vdots \\ & 0 & \ddots\; 1 & a_{p-1} \\ & & & a_p \end{pmatrix} \qquad N = \begin{pmatrix} E(a_1 b_1/a_p) & \cdots & E(a_1 b_m/a_p) \\ \vdots & & \vdots \\ E(a_{p-1} b_1/a_p) & & E(a_{p-1} b_m/a_p) \\ b_1 & & b_m \end{pmatrix}$$

avec $d^\circ a_p = n$, $d^\circ a_i < n$ si $1 \le i \le p-1$, $d^\circ b_j < n$ pour $1 \le i \le m$.

En outre, il existe des matrices D_1 et N_1 telles que $D^{-1}N = N_1 D_1^{-1}$, où D_1 et N_1 ont des formes transposées de celles de D et N respectivement. Une condition suffisante pour que R_1 et R_2 permettent l'écriture de (2) et que le système dont la fonction de transfert est $R_1^{-1}\mathcal{H}R_2^{-1}$ soit commandable par sa dernière entrée et observable par sa dernière sortie. Ces conditions sont génériques, et, lorsqu'elles sont satisfaites, tout système dans Σ_n suffisamment voisin de $\mathcal{H}$ est factorisable selon (2) avec les mêmes R_1 et R_2. Pour $\mathcal{H}$, R_1 et R_2 fixés, les matrices D et N (resp. D_1 et N_1) sont uniques et les coefficients des polynômes qui y figurent forment un système de coordonnées sur Σ_n , qui sera celui utilisé dans ce qui suit. En outre, on peut imposer à R_1 et R_2 d'être orthogonales. Si $\mathcal{H} \in \Sigma_n \cap E$, on définit enfin :

$$\hat{V}_{D,D_1} = \{M \in R[z]^{p\times m} \; ; \; D^{-1}MD_1^{-1} \in E\}$$

$D^{-1}\hat{V}_{D,D_1} D_1^{-1}$ est un sous espace de E de dimension n(m+p) qu'on notera W_{D,D_1}. On peut montrer que les dérivés de $D^{-1}N$ par rapport aux coordonées en sont une base [2].

Soit à présent $F \in E$. Supposons qu'un meilleur approximant $\bar{\mathcal{H}}$ de F dans $\mathcal{S}_n$ soit en fait dans Σ_n et soit cyclique. $\bar{\mathcal{H}}$ s'écrit $R_1\bar{D}^{-1}\bar{N}$, avec R_1 et R_2 orthogonales, de sorte que l'application $x \to R_1 \, x \, R_2$ est une isométrie de E (cela se voit par exemple sur l'expression intégrale du produit scalaire mentionnée à la fin de 3). Quitte à changer F en $R_1^{-1} F R_2^{-1}$ on peut supposer $R_1 = R_2 = I$ (matrice identité). $(\bar{D},\bar{N})$ est point critique de la fonction :

$$(D,N) \xrightarrow{\psi_F} \langle F-D^{-1}N, \; F-D^{-1}N \rangle$$

Ce qui signifie que pour toute coordonée λ, on a :

$$\langle F-\bar{D}^{-1}\bar{N}, \; \frac{\partial}{\partial\lambda}(D^{-1}N)_{(\bar{D},\bar{N})} \rangle = 0 \qquad (3)$$

ou encore, d'après ce qui précède, et en notation abrégée :

$$\langle F-\bar{D}^{-1}N, \; W_{\bar{D},\bar{D}_1} \rangle = 0 \qquad (4)$$

En remarquant que $\bar{D}^{-1}\bar{N} \in W_{\bar{D},\bar{D}_1}$, on voit que $\bar{\mathcal{H}}$ est la projection de F sur cet espace, et qu'en particulier $\|\bar{\mathcal{H}}\| \le \|F\|$. Par ailleurs, en appelant A_{D,D_1} l'orthogonal, dans W_{D,D_1}, de l'ensemble des éléments qui s'écrivent $D^{-1}N$, on voit que (4) entraîne :

$$\langle F, \; A_{\bar{D},\bar{D}_1} \rangle = 0 \qquad (5)$$

Réciproquement, si pour deux matrices $\bar{D}$ et $\bar{D}_1$ (5) est vraie, la projection de F sur $W_{\bar{D},\bar{D}_1}$ est du type $\bar{D}^{-1}\bar{N}$. Si ce système, qui est dans $\mathcal{S}_n$, est en fait dans Σ_n, on peut, quitte à multiplier à droite F et $\bar{N}$ par une matrice réelle du type $\binom{I\,x}{0\,I}$ et $\bar{D}_1$ à gauche par ${}^tM$ (ce qui ne modifie ni (4) ni (5)), écrire $\bar{D}^{-1}N = N_1D_1^{-1}$ [2]. Si $D_1 = \bar{D}_1$, (3) est vérifiée, et on simplifie M pour conclure que le point original $(\bar{D},\bar{N})$ est point critique de ψ_F. L'équation (5) généralise celle trouvée par Rosencher [7] dans le cas $m = p = 1$.

L'équation (3) n'est qu'une condition nécessaire d'optimum. Le théorème suivant montre que dans le cas étudié ici, elle donne cependant localement la solution de notre problème.

Théorème 2 : Soit (A_k) cyclique dans $\Sigma_n \cap E$, s'écrivant $\bar{D}^{-1}\bar{N}$. Il existe un voisinnage $\mathcal{V}$ de (A_k) dans Σ_n, et un nombre $\varepsilon > 0$, tels que pour $F \in E$ vérifiant $\|F-\bar{D}^{-1}\bar{N}\| < \varepsilon$, la fonction : $\phi_F : \mathcal{V} \to R^{n(m+p)}$ définie par $\phi_F(D,N) = \text{Gradient}(\psi_F)_{(D,N)}$ admet un unique zéro dans $\mathcal{V}$ qui correspond à l'unique meilleur approximant de F dans $\mathcal{S}_n$.

Preuve : Posons $g(F,D,N) = \phi_F(D,N)$; g est définie sur $E \times W$, où W est un voisinnage de $(\bar{D},\bar{N})$. Appelons $\lambda_1,\lambda_2,\ldots,\lambda_{n(m+p)}$ nos coordonées. Il est facile de voir que l'application $\frac{\partial g}{\partial(D,N)}(\bar{D}^{-1}\bar{N},(\bar{D},\bar{N}))$ est représentée par une matrice dont l'élément (i,j) s'écrit : $< \frac{\partial}{\partial\lambda_i}(D^{-1}N)_{(\bar{D},\bar{N})}, \frac{\partial}{\partial\lambda_j}(D^{-1}N)_{(\bar{D},\bar{N})} >$. Cette matrice est donc la matrice de Gram d'un système de vecteurs dont on a dit qu'ils étaient indépendants, et elle est régulière. Le théorème de redressement local [8] assure l'existence de $\mathcal{V}$ et ε comme dans l'énoncé, tels que ψ_F ait un unique point critique dans $\mathcal{V}$ si $\|F-\bar{D}^{-1}\bar{N}\| < \varepsilon$. On sait par ailleurs d'après (1) et le lemme 1, que pour ε assez petit, tous les meilleurs approximants de F seront dans $\mathcal{V}$, ce qui achève la preuve. (Notons que la dépendance en F de notre approximant est C^∞).

5. UNE HEURISTIQUE DE RESOLUTION

Dans la pratique, on ne connaît pas les matrices R_1 et R_2 qui permettent de factoriser l'optimum. On sait que génériquement toute matrice convient, donc en particulier $R_1 = R_2 = I$. Cependant on verra un exemple où ce choix n'est pas acceptable numériquement. Nous ne discuterons pas ici d'une procédure de choix systématique. Indiquons simplement qu'une approche (lourde) consiste à chercher des coefficients pour combiner les lignes et les colonnes de la matrice de Hankel associée à (A_k) de sorte que la nouvelle matrice de Hankel ainsi obtenue soit encore de rang n. Dans les exemples présentés on s'est bornés à essayer plusieurs matrices lorsque $R_1 = R_2 = Id$ ne convenait pas.

Nous proposons à présent une heuristique linéaire pour tenter de résoudre (5) tout en vérifiant les conditions additionelles qui la rendent équivalentes à (3).

1) déterminer n, R_1 et R_2.
2) choisir D et D_1
3) calculer une base $\mathcal{B}$ de A_{D,D_1}
4) résoudre l'équation linéaire en les coefficients de la matrice $\hat{D}$:

$$< F, {}^t\hat{D}(\tfrac{1}{z})({}^tD^{-1}(\tfrac{1}{z})\mathcal{B}) > = 0 \tag{6}$$

5) remplacer D par $\hat{D}$ et calculer $\hat{N}$ tel que $\hat{D}^{-1}\hat{N}$ soit la projection de F sur l'espace vectoriel des $\hat{D}^{-1}N$.
6) calculer D_1 tel que $\hat{D}^{-1}\hat{N} = N_1D_1^{-1}$
7) retourner en (3)

Si les matrices D et D_1 convergent, la limite vérifie (5) et les conditions additionelles, de sorte qu'on a résolu (3).

Nous ne voulons nullement prétendre que ceci la meilleure façon de traiter l'équation (3). Cette méthode cependant fournit certains résultats, et est basée sur la remarque suivante.

A_{D,D_1} est précisément constitué des matrices de $\hat{V}_{D,D_1}$ qui, multipliés à gauche par ${}^tD^{-1}(\frac{1}{z})$ donnent un élément de $L_2^2(T)^{p\times m}$ qui est en fait dans E [2]. Puisque N_o qui est un polynôme est orthogonal à E, on en déduit que si $F = D_o^{-1}N_o$, l'équation (6) admet D_o pour solution indépendamment de D et D_1. En d'autres termes, on retrouve dans ce cas la fraction initiale en résolvant un système linéaire.

6. ASPECTS NUMERIQUES.

Dans cette partie, nous proposons des méthodes numériques pour effectuer les étapes (3,4,5) le l'heuristique de résolution de l'équation d'optimum. Les autres étapes si ce n'est le choix de R_1 et R_2, ne posent aucun problème de méthode numérique.

a) Détermination d'une base de A_{DD_1}

Une fois D et D_1 choisis, il faut déterminer une base de A_{DD_1} orthogonal de $D^{-1}V_D = \{D^{-1}N,\ N/D^{-1}N \in E\}$ dans $W_{DD_1} = D^{-1}\hat{V}_{DD_1}D_1^{-1} = \{D^{-1}M\,D_1^{-1},\ M/D^{-1}M\,D_1^{-1} \in E\}$

Dans le cas m=n=1 Rosencher [7] a mis en évidence une base explicite de cet orthogonal. N'ayant pu trouver un tel résultat dans le cas multivariable nous proposons dans la suite une détermination numérique.

Notons : $V_{D_1D_1^{-1}} = \{N_1D_1^{-1},\ N_1/N_1D_1^{-1} \in E\}$.

et considérons

$\mathcal{W} = \{\mathcal{V}_r,\ r \in \Omega_w\}$ une base de $D^{-1}V_D$

$\mathcal{V} = \{\mathcal{V}_s,\ s \in \Omega_v\}$ une base de $V_{D_1D_1^{-1}}$

par exemple les bases "naturelles" de $D^{-1}V_D$ et $V_{D_1}D_1^{-1}$ définies par :

$$\Omega_v = \{(i,\ell)\ ;\ i=1,\dots m\ ;\ \ell=o,\dots n-1\}$$

$$V_{\substack{i\ell \\ i\neq m}} = \left(\begin{array}{c|c} 0 & 0 \\ -pp\left(\frac{z^\ell c_j}{a_m}\right) & \frac{z^\ell}{a_m} \\ 0 & 0 \end{array}\right) \longleftarrow i^{\text{ième}} \text{ ligne}$$

($j^{\text{ième}}$ colonne ; dernière colonne)

$$V_{m\ell} = \left(\begin{array}{c|c} \frac{a_i}{a_m}pp\left(\frac{z^\ell c_j}{a_m}\right) & -\frac{z^\ell a_i}{a_m^2} \\ \hline -\frac{1}{a_m}pp\left(\frac{z\,c_j}{a_m}\right) & \frac{z}{a_m^2} \end{array}\right) \begin{array}{l} \longleftarrow \text{ligne } i \neq m \\ \\ \longleftarrow \text{ligne } m \end{array}$$

(colonne $j \neq p$)

$$\Omega_w = \{(j,k),\ j=1,\dots p\ ,\ k=0,\dots n-1\}$$

$$W_{jk} = \left(\begin{array}{ccc} 0 & -pp\left(\frac{a_i z^k}{a_m}\right) & 0 \\ \hline 0 & \frac{z^k}{a_m} & 0 \end{array}\right) \begin{array}{l} \longleftarrow \text{ligne } i\neq m \\ \\ \longleftarrow \text{ligne } m \end{array}$$

(colonne j)

En utilisant la relation $D^{-1}V_{DD_1}D_1^{-1} = D^{-1}V_D \oplus V_{D_1}D_1^{-1}$, [2] il vient alors qu'une base de $(D^{-1}V_D)^{\perp}$ peut être engendrée par les vecteurs $A_s = V_s - \sum_{\Omega_w} W_r \lambda_{rs}$, $s \in \Omega_v$, les λ_{rp} étant déterminés de sorte que $\forall s \in \Omega_v \quad \forall r \in \Omega_w$, $A_s \perp w_r$ □

La première méthode numérique qui vient à l'esprit pour déterminer les λ_r $r \in \Omega_w$ est de résoudre en λ le système d'équation linéaire :

$$\{ < w_i, A_s > = 0\} \; r \in \Omega_w, \; s \in \Omega_v\} \qquad \text{(orthohonalisation de Gram-Schmidt)}$$

Cependant, cette méthode d'orthogonalisation est reconnue pour être numériquement sensible aux choix des bases. En particulier à celui de $\mathcal{W}$, qui peut entrainer, si elle "n'engendre pas bien numériquement" le sous espace, la singularité numérique de la matrice de gram. □

Pour essayer de s'affranchir de cette difficulté, nous avons utilisé la caractérisation de l'orthogonal de $D^{-1}V_D$ m ntionné à la fin du § précédent qui s'est avéré plus précise numériquement.

Avec les notations de ce paragraphe une formulation possible est : $A_s \in (D^{-1}V_D)^{\perp}$ si et seulement si ${}^tD^{-1}(\frac{1}{z})A_s$ est une matrice propre. (7)

Nous déterminons alors les coefficients d'orthogonalisation λ_{rs} (qui sont uniques) en satisfaisant les conditions de propreté des matrices ${}^tD^{-1}(\frac{1}{z})A_s$.

Si l'on choisit pour base et les "bases naturelles", on peut vérifier aisément que :

$$\forall i \in \{0,\ldots m\} \quad \forall j \in \{0,\ldots p\} \quad , \forall k \in \{0,\ldots n-1\} \quad ,\forall \ell \in \{0,n-1\}$$

les coefficients des m-1 premières lignes des matrices ${}^tD^{-1}(\frac{1}{z})\, V_{i\ell}$ et ${}^tD^{-1}(\frac{1}{z})\, W_{jk}$ sont propres.

Donc les m-1 premières lignes des matrices ${}^tD^{-1}(\frac{1}{z})A_{i\ell}$ sont propres quelque soient les coefficients $\lambda_{ijk\ell}$.

En notant r(x), le reste de la division euclidienne de x par a_m, la dernière ligne de la matrice ${}^tD^{-1}(\frac{1}{z})A_{i\ell}$ s'écrit :

$$\alpha_{i\ell} = \frac{1}{a_m \tilde{a}_m} \{\tilde{a}_i r(c_j z^{\ell}) + \sum_k \lambda_{ijk\ell}(\sum_u \tilde{a}_u r(a_u z^k) + z^{n+k}); -\tilde{a}_i z^{\ell} + \sum_k \lambda_{ipk\ell}(\sum \tilde{a}_u r(a_u z^k) + z^{n+k})\}$$

si $i \neq m$

$$\alpha_{m\ell} = \frac{1}{a_m^2 \tilde{a}_m}\{-(\sum_u a_u \tilde{a}_u + z^n) r(c_j z^{\ell}) + a_m \sum_k \lambda_{mjk\ell}(\sum_u \tilde{a}_u r(a_u z^k) + z^{n+k});$$

$$(\sum_u a_u \tilde{a}_u + z^n) z^{\ell} + a_m \sum_k \lambda_{mpk\ell}(\sum \tilde{a}_u r a_u z^k + z^{n+k})\} \qquad \text{si } i = m$$

On détermine alors les λ_{ijk} en résolvant les équations linéaires déduites des conditions :

i) le numérateur de $\alpha_{i\ell}$ doit être divisible par $\tilde{a}_m$, et le quotient de la division est de degré strictement inférieur au degré de a_m

ii) le numérateur de $\alpha_{m\ell}$ doit être divisible par $\tilde{a}_m$, et le quotient de la division doit être de degré strictement inférieur au degré de a_m^2

Puisque les coefficients d'orthogonalisation $\lambda_{ijk\ell}$ sont uniques, ce système d'équation est régulier. On peut d'autre part remarquer, pour simplifier la résolution que les équations sont découplées par rapport aux indices i, j et ℓ

b) détermination de l'équation lineaire (6)

D'après (7), il vient :

$\hat{D}$ est solution du système d'équations linéaires :

$$\{<f, {}^t\hat{D}(\tfrac{1}{z})({}^tD^{-1}(\tfrac{1}{z})A_s)> = 0 \quad , s \in \Omega_v\} \tag{8}$$

Une fois les matrices A_p calculées comme indiquées précédemment, il suffit de savoir évaluer des produits scalaires de la forme :

$$P_{ij\alpha} = < f_{ij}, \frac{z^\alpha}{q} >$$

ou q est égal à a_m ou a_m^2

et $\alpha <$ degré (q).

(résultat immédiat d'après l'expression des matrices ${}^tD^{-1}(\frac{1}{z})A_p$)

En remarquant que $P_{ij\alpha}$ est le terme constant du produit des séries $f_{ij}(\frac{1}{z})$ et s, ou s est le développement en série de $\frac{z^\alpha}{q}$, alors ces produits scalaires peuvent être définis, et calculés, comme étant la réponse à l'instant $-\alpha$ du filtre discret, dont la transformation en Z est $\frac{1}{q}$, excité par le signal $f_{ij}(-k)$. □

Remarque : Si l'on prend comme valeur initiale de D la matrice :

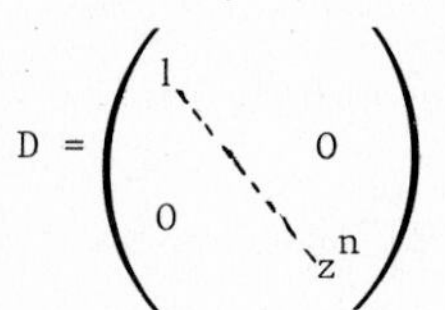

$$D = \begin{pmatrix} 1 & & 0 \\ & \ddots & \\ 0 & & z^n \end{pmatrix}$$

alors le système d'équations :(8) se simplifie considérablement, en particulier ce système se découple par rapport à chacun des polynômes a_i de $\hat{D}$ et les équations d'orthogonalité ne font plus intervenir que les composantes f_{ip} de f ! □

c) Déterminations des paramètres de la matrice $\hat{N}$.

Si l'on suppose connue la matrice $\hat{D}$ la condition d'optimalité locale s'écrit :

$$<f-\hat{D}^{-1}\hat{N}, \hat{D}^{-1}\frac{\partial\hat{N}}{\partial b_{j\ell}}> = 0 \qquad \forall j \in \{1,\ldots p\} \quad ; \forall \ell \in \{0,\ldots n-1\}$$

soit encore sous forme matricielle scalaire :

$$M_a.B = - F_a$$

où :

- . M_a est la matrice de gram du système $\{\hat{D}^{-1}\frac{\partial\hat{N}}{\partial b_{j\ell}}, \ell \in \{0,\ldots n-1\}\}$.$\forall j \in 1\ldots p$ on peut vérifier que les $\hat{D}^{-1}\frac{\partial\hat{N}}{\partial b_{j\ell}}$ sont indépendants donc que M_a est régulière).
- . B la matrice $B = [b_{j\ell}]_{\ell j}$

. F_a la matrice $[< f, \hat{D}^{-1} \frac{\partial \hat{N}}{\partial b_{j\ell}} >]_{\ell j}$

l'évaluation de ce système linéaire passe par le calcul des produits scalaires : $< f, \hat{D}^{-1} \frac{\partial \hat{N}}{\partial b_{j\ell}} >$ pour lesquels nous avons déjà proposé une méthode et

$$< \hat{D}^{-1} \frac{\partial \hat{N}}{\partial b_{j\ell}}, \hat{D}^{-1} \frac{\partial \hat{N}}{\partial b_{j\ell'}} > = \sum_{i=m}^{m-1} < pp(\frac{a_i z^\ell}{a_m}), pp(\frac{a_i z^{\ell'}}{a_m})> + < \frac{z^\ell}{a_m}, \frac{z^{\ell'}}{a_m} >$$

que l'on saura évaluer si l'on connait la matrice de Gram du système

$$\{\frac{z^\ell}{a_m}, \ \ell \in 0,\ldots n-1\}$$

Nous proposons ici une méthode pour effectuer ce calcul.

Il est facile de voir par la formule de Cauchy que :

$$<f,g> = \frac{1}{2i\pi} \text{ trace} \oint_T {}^t f(\frac{1}{z}) \, g(z) \frac{dz}{z}$$

Par conséquent :

$$< \frac{z^\ell}{a_m}, \frac{z^{\ell'}}{a_m} > = \oint_T \frac{z^{n-\ell+\ell'}}{a_m \tilde{a}_m} \frac{dz}{2i\pi z} .$$

Le polynôme a_m ayant tous ses zéros dans le disque unité Γ, $\tilde{a}_m$ a ses zéros hors du disque Γ. Donc a_m et $\tilde{a}_m$ sont premiers et il existe deux polynômes λ et μ uniques tels que : (th de Bezout)

$$\lambda a_m + \mu \tilde{a}_m = 1$$

$$\text{et } d°(\lambda) < d°(\tilde{a}_m) \quad ; \ d\ (\mu) < d°(a_m)$$

d'où :

$$< \frac{z^\ell}{a_m}, \frac{z^{\ell'}}{a_m} > = \oint_T \frac{\lambda z^{n-\ell+\ell'-1}}{\tilde{a}_m} \frac{dz}{2i\pi} + \oint_T \frac{\mu z^{n-\ell+\ell'-1}}{a_m} \frac{dz}{2i\pi}$$

soit encore puisque $\tilde{a}_m$ a tous ses zéros à l'extérieur de

$$< \frac{z^\ell}{a_m}, \frac{z^{\ell'}}{a_m} > = \oint_T \frac{\mu z^{n+\ell-\ell'-1}}{a_m} \frac{dz}{2i\pi}$$

ce qui est encore égal d'après le théorème des résidus à :

$$\sum_{j=o}^{n-2} \mu_j \alpha_{n+\ell-\ell'+j} \quad \text{avec } \mu = \sum_{j=o}^{n-2} z^j \mu_j \quad \text{et } \frac{1}{a_m} = \sum_{i=o}^{\infty} \alpha_i \, z^{-i}$$

7. RESULTATS NUMERIQUES.

Pour étudier la validité numérique de cet algorithme de réalisation (appelons le ARL2), nous avons comparé ses résultats avec ceux fournis par l'algorithme de B.L. HO [3], (La factorisation de la matrice de Hankel étant obtenue par sa décomposition en valeurs singulières), qui est réputé pour fournir les meilleurs résultats numériques si l'on ne prend pas en compte le volume mémoire et le temps de calcul nécessaire.

Les résultats des deux algorithmes sont comparés pour des ordres de la réalisation identiques.

Pour mesurer la qualité des résultats nous avons dans les deux cas calculé les erreurs quadratiques entre la réponse impulsionnelle donnée et la réponse impulsionelle simulée à partir de la réalisation.

Rappelons que l'algorithme de HO nécessite une place mémoire de l'ordre de $2mp(N/2)^2$ si N est le nombre de points de la réponse impulsionnelle. Alors que ARL2 utilise $(n\times\max(m,p))^2 + mpN$ mots mémoire. On peut ainsi traiter des réponses plus longues (amortissement faible). Cependant l'algorithme de HO, en particulier l'utilisation de la matrice de HANKEL, permet une détermination de l'ordre de la réalisation. Pour l'algorithme ARL2 cette grandeur doit être préalablement connue, ou déterminée par essais successifs validés par la mesure de l'erreur quadratique.

Nous avons étudié le comportement numérique de ces deux algorithmes pour deux types de données.

i) Des réponses impulsionnelles de fonction de transfert de la forme $D^{-1}N$ (ce qui garantit la validité des hypothèses).

Pour ce type de données, pour un ordre de réalisation égal à l'ordre du système initial, les résultats sont constants : l'algorithme ARL2 conduit à une erreur quadratique toujours sensiblement inférieure à celle obtenue par l'algorithme de HO.

Nous ne donnerons ici que le résultat obtenu pour un système à 4 entrées et 3 sorties de dimension d'état 9 dont les pôles sont : 0.638 ; 0.2 ± 0.8i ; 0.45 ± 0.7i ; 0.65 ± 0.4i ; 0.91 ; 0.5

et la matrice de transfert :

$$F(z) = \begin{pmatrix} -pp(\frac{a_1 b_1}{a_z}) & -pp(\frac{a_1 b_2}{a_z}) & -pp(\frac{a_1 b_3}{a_z}) \\ -pp(\frac{a_2 b_1}{a_z}) & -pp(\frac{a_2 b_2}{a_z}) & -pp(\frac{a_2 b_3}{a_z}) \\ \frac{b_1}{a_z} & \frac{b_2}{a_z} & \frac{b_3}{a_z} \end{pmatrix}$$

avec :

$$a_1 = 2.34 - 6.79z + 0.45z^2 - 23.6z^3 + 12z^4 + 7.92z^5 - 0.0345z^6 + 1.32z^7$$

$$a_2 = 2.45z^2 - 45.9z^3 - 0.023z^4 + 0.0012z^5 + 0.34z^6 + 27z^7 + 2z^8$$

$$a_3 = 0.0796-0.45z+1.06z^2-1.16z^3-0.269z^4+3.11z^5-5.554z^6+5.57z^7-3.37z^8+z^9$$

$$b_1 = 0.568 + 0.34z^3 + 62.4z^4 + 0.35z^7 - 0.5z^8$$

$$b_2 = 3z$$

$$b_3 = 1$$

$$b_4 = 0.5z^7 + z^8$$

Si l'on tronque la réponse après ses 50 premiers points, maximum possible pour HO dans ce cas, les erreurs quadratiques moyennes valent respectivement :

0.1 10^{-4} pour ARL2

0.1 pour HO

Si l'on tronque après les 120 premiers points l'erreur pour ARL2 devient :

0.2 10^{-10}; □

ii) Des réponses impulsionnelles, relatives aux systèmes d'états :

$$\{U^{-1}F\ U : U^{-}GR_e,\ R_s\ H\ U\} .$$

où F est la matrice compagnon de dimension 7

$$F = \begin{bmatrix} 0 & I \\ 0 & X \end{bmatrix}$$

dont les valeurs propres sont :

$0.815 \pm 0.33i$; $0.309 \pm 0.737i$; $0.488 \pm 0.815i$; 0

où,

$$H = \begin{bmatrix} e_1^T \\ \\ e_7^T \end{bmatrix} ; \ G = [e_7,\ e_1]$$

si e_i est le vecteur colonne dont tous les éléments sont nuls sauf le $i^{ième}$ et où :

U est une matrice régulière quelconque,

R_e et R_s des matrices orthogonales

on remarque que le triplet ($U^{-1}FU$, $U^{-1}G$,HU) n'est pas observable, (commandable) par sa dernière sortie, (entrée), et de fait l'algorithme ARL2 n'aboutit pas (le système d'équations linéaires qui devrait fournir les coefficients de D est singulier).

La matrice F étant cyclique il est possible, par un choix très large de matrices orthogonales R_e et R_s, de rendre le triplet commandable et observable par rapport à ses dernières entrée et sortie.

Nous avons choisi au hasard les matrices R_e et R_s :

$$R_e = \begin{bmatrix} -0.9414 & 0.3372 \\ -0.3372 & -0.9414 \end{bmatrix}$$

$$R_s = \begin{bmatrix} -0.9887 & -0.1500 \\ -0.1500 & 0.9887 \end{bmatrix}$$

Alors pour une matrice U de conditionnement $K(u) = 10^{-2}$ (le conditionnement des matrices d'observabilité et de commandabilité du triplet ($U^{-1}FU$, $U^{-1}GRe$, R_s HU- est sensiblement proportionnel à K(u)).

Les erreurs pour une réponse impulsionnelle tronquée après ses 85 premiers points sont :

$\begin{cases} \text{ARL2} : 0.2 \ 10^{-3} \\ \text{HO} \ \ \ : 0.4 \ 10^{-1} \end{cases}$

pour les 250 premiers points ARL2 : 0/1 10^{-10}. □

L'algorithme est aussi résistant aux erreurs de troncature ; pour ce même exemple et une réponse tronquée après ses 40 premiers moints (ce qui correspond à un amortissement de seulement 80%) l'algorithme permet d'obtenir une erreur moyenne égale à 0.24 10^{-1} contre 0.36 pour l'algorithme de HO.

Par contre les résultats sont plus divers dans le cas où la réponse impulsionnelle donnée n'est pas exactement dans la classe de l'approximant recherché -par exemple si l'ordre n'est pas correct-

Les difficultés provenant d'un très mauvais conditionnement du système d'équations (8). (le conditionnement de ce système est étroitement lié au degré d'indépendance des vecteurs de base de $\{ \frac{\partial D^{-1}N}{\partial \lambda}$, λ paramètre de $D^{-1}N\}$ donc au choix de la paramétrisation.).

Dans tous les cas où ce système est bien conditionné, l'algorithme permet de trouver une bonne réalisation, en particulier si l'on utilise la possibilité d'itérer.

Pour un système de degré d'état 7, de trois entrées et deux sorties et une perturbation aléatoire de variance $0.1*f_{ij}(k)$ sur le coefficient $f_{ij}(k)$ de la réponse

ARL2 donne à la première iteration une erreur de 0.26 10^{-1}
à la seconde 0.90 10^{-2} et pour les suivantes 0.88 10^{-2}.

et HO donne une erreur de 0.89 ;

BIBLIOGRAPHIE

[1] KALMAN, R.E ; "On minimal partial realizations of a Linear Input/Output map", Aspects of Network and System theory, R.E. Kalman and N. Declaris eds., New-York, Holt, Rinehart and Winston, pp. 385-407, 1971.

[2] BAKATCHART, L ; "Une structure différentielle pour certaines classes de systèmes ; application à l'approximation L^2". Thèse de docteur ingénieur. ENSHP. 1982.

[3] HO,B.L.et KALMAN, R.E. ; Effective construction of linear state variable model from Input/Output Data. Rroc 3e Allertion Conférence (1965), pp. 449-459/

[4] RUDIN. W ; "Real and complex Analysis". Mc Graw-Hill series in higher Mathematics 1966. ch.17.

[5] HAZEWINKEL, M. and Kalman, R.E. ; "Moduli and canonical forms for linear systems, Report 7504, Econometric Institute. Erasmus Univ. Rotterdam (1974).

[6] JACOBSON, N. : "Lectures in Abstract Algebra". T.II. "Linear Algebra". Sringer-Verlag 1953. ch.3.

[7] ROSENCHER. E. ; "Approximation Rationelle des filtres à un ou deux indices : Une Approche Hilbertienne". Thèse de docteur-ingénieur. Paris IX. 1978.

[8] LANG, S. : "Differential Manifolds". Addison Wesley series in Mathematics n° 4166, 1972, 17.

[9] DE JONG, L.S. ; "Numerical aspects of realization algorithms in linear system theory". Thesis Eindhoven Univ. of technologie the Netherlands 1975.

[10] DE JONG, K.S. ; "Numerical aspects of recursive realization algorithms". SIAM J. on Control and optimisation vol. 16 n°4 July 1978.

[11] KUNG, S.Y. et LIN. D.W. ; Optimal Hankel norm model reduction : multivariable systems : IEEE Trans on automatic control August 1981- Vol. 26, n°4.

[12] FUHRMANN, P.A. ; Linear systems and operators in hilbert space, Mc Graw-Hill, New-York, 1981.

Session 14

COMPUTER AIDED CONTROL SYSTEM DESIGN I

CAO EN AUTOMATIQUE I

THEORETICAL AND SOFTWARE ASPECTS OF OPTIMIZATION-BASED CONTROL SYSTEM DESIGN

by

E. POLAK
Department of Electrical Engineering and Computer Sciences
and the Electronics Research Laboratory
University of California, Berkeley, California 94720

D. Q. MAYNE
Department of Electrical Engineering
Imperial College, London SW 2BT, England

1. *INTRODUCTION.*

Parametric optimization is a powerful tool for the selection of favorable values for design variables. In linear control system design, parametric optimization has been used for at least twenty years: since the introduction of the linear-quadratic regulator problem. Although linear quadratic regulator theory was derived in the context of optimal control theory rather than mathematical programming theory, it is nevertheless true that the optimal gain matrix for the linear-quadratic regulator problem is a solution to an unconstrained parametric optimization problem of the form

$$\min_{K \in \mathbb{R}^n \times \mathbb{R}^n} f(K) \tag{1.1}$$

where $f(K)$ is the largest eigenvalue of a symmetric matrix of the form

$$\int_0^\infty exp[t(A+BK]^T(Q+K^TRK)exp[t(A+BK]dt \tag{1.2}$$

with Q symmetric and positive semi-definite and R symmetric and positive definite.

In keeping with the state of the art in constrained optimization of the sixties (see e.g., [Ath.1]), the cost function $f(K)$, in the linear quadratic regulator problem, expresses a penalty function approach to the satisfaction of performance requirements. Over the last eight years a much more powerful approach has become possible with the development of semi-infinite optimization algorithms for engineering design (see e.g., [Gon.1, Pol.1, Pol.2, Pol.3]). The major advantage of semi-infinite optimization over the penalty function approach implicit in LQR theory, is that it accepts as constraints exact bounds on time and frequency responses over time and frequency intervals. These bounds take on the form of infinite systems of inequalities which are parametrized by time or frequency, that the finite number of designable compensator parameters must satisfy (see e.g. [Pol.1]). Such inequalities are often refered to as *semi-infinite* inequalities. Clearly, semi-infinite optimization opens up totally new possibilities in control system design. This is particularly so since the introduction of nondifferentiable optimization theory has led to algorithms that can solve not only constraints on time and frequency responses which are (pointwise) differentiable, but also on eigenvalues of system matrices and singular values of transfer function matrices, which are not differentiable everywhere (see [Gon.1, Pol.2, Pol.3]).

2. CONTROL SYSTEM DESIGN VIEWED AS AN OPTIMIZATION PROBLEM

Whenever design specifications include envelope constraints on time and frequency responses, control system design problems transcribe into semi-infinite optimization problems with, possibly, a dummy cost. These problems have the form

$$\min_{x \in \mathbb{R}^n} \{ f(x) \mid g^j(x) \leq 0, j = 1,2,\ldots,m;\ \varphi^k(x,\alpha_k) \leq 0,\ k = 1,2,\ldots,l,\ \forall\, \alpha_k \in A_k \} \qquad (2.1)$$

where $x \in \mathbb{R}^n$ is the design vector (designable compensator parameters); $f: \mathbb{R}^n \to \mathbb{R}, g^j: \mathbb{R}^n \to \mathbb{R}, \varphi^k: \mathbb{R}^n \times \mathbb{R}^{p_k} \to \mathbb{R}$ are locally Lipschitz continuous and the sets of $A_k \subset \mathbb{R}^{p_k}$ are compact. Most often, $p_k = 1$ so that the A_k are intervals (of time or frequency). Only in the case of parametric dynamic model uncertainty does one need to introduce sets A_k of higher dimension.

A simple example will illustrate the genesis of forms such as (2.1) in control system design. Consider the control system in Fig. 1, for which it is necessary to design a compensator C. The dynamics of this system are given by

$$\begin{aligned} \dot{z}_P &= A_P z_P + B_P u_P \\ y_P &= C_P z_P + D_P u_P \end{aligned} \qquad (2.2a)$$

$$\begin{aligned} \dot{z}_C &= A_C z_C + B_C u_C \\ u_C &= C_C z_C + D_C u_C \end{aligned} \qquad (2.2b)$$

$$\begin{aligned} u_p &= y_C \\ u_C &= -y_P - d + r \end{aligned} \qquad (2.2c)$$

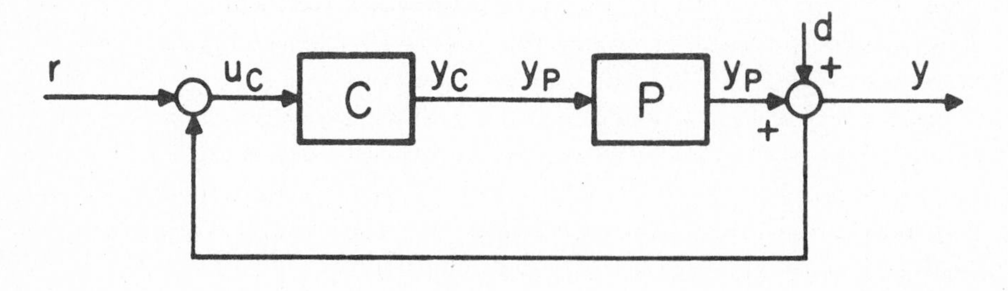

Fig. 1

Where (2.2a) specifies the plant dynamics, (2.2b) specifies the compensator dynamics and (2.2c) specifies the system interconnection. The designer chooses the dimension of the compensator state vector z_C and has to compute the compensator matrices A_C, B_C, C_C, D_C, whose elements eventually form the components of the design vector x. In order to reduce the dimension of the design vector x, the designer may specify the system matrix A_C in block diagonal form

$$A_C = \mathrm{diag}(A_{1C}, A_{2C}, \ldots, A_{kC}, \lambda_{2k+1}, \ldots, \lambda_N) \tag{2.3a}$$

where the λ_j are real (some may be frozen at zero for integral action), while

$$A_{jC} = \begin{bmatrix} 0 & 1 \\ a_{ij} & a_{oj} \end{bmatrix} \tag{2.3b}$$

Some structural simplification of the B matrix is also possible.

Now consider typical constraints.

(i) **Time Domain**: Given step inputs $r_i(t) = (0,0,\ldots,1,0\ldots0)$ (with the 1 in the ith place), we require that the corresponding step responses $y^i(t,x,r_j)$ remain within the envelopes shown in Fig. 2. This leads to the two semi-infinite constraints

$$y^i(x,t,r_j) - b_u^{ij}(\psi) \leq 0 \quad \forall t \in [0,T] \tag{2.4a}$$

$$-y^i(x,t,r_j) + b_l^{ij}(\psi) \leq 0 \quad \forall t \in [0,T] \tag{2.4b}$$

We note that when $i \neq j$ (2.4a,b) express limitations on the permissible interaction.

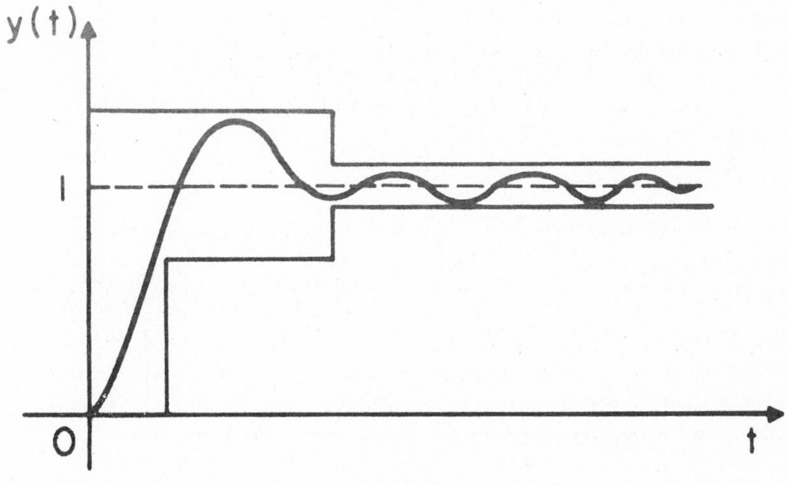

Fig. 2

(ii) **S-plane:** The system matrix of the closed loop system has the form

$$A(x) = \begin{bmatrix} A_P - B_p(I+D_C D_P)^{-1} D_C C_P & -B_P(I+D_C D_P)^{-1} C_C \\ B_C(I+D_P D_C)^{-1} C_P & A_C - B_C(I+D_P D_C)^{-1} D_P C_C \end{bmatrix} \tag{2.5}$$

We may require that all the eigenvalues of this matrix lie in a cone in $\dot{C}_-$. Denoting the eigenvalues of $A(x)$ by $\lambda^j[A(x)]$, we are lead to the system of inequalities

$$\mathrm{Im}[\lambda^j[A(x)] + \xi[\mathrm{Re}\lambda^j[A(x)] + \zeta \le 0 \quad \text{for } j = 1,2,\ldots,N \tag{2.6}$$

where $\xi, \zeta > 0$. Note that in (2.6) we are exploiting the fact that complex eigenvalues must come in conjugate pairs. Alternatively, the same result could have been specified in semi-infinite inequality form via the modified Nyquist criterion described in [Pol. 4].

(iii) **Frequency Domain:** Assuming that there is some unstructured uncertainty in the plant model, so that (2.2a) represents only the structures part P_0, while the actual plant has a transfer function matrix of the form $P(s) = P_0(s)(I + L(s))$, with $P_0(s)$ the transfer function matrix of (2.2a) and $L(s)$ a perturbation known only to the extent that

$$\bar{\sigma}_M[L(j\omega)] \le b(\omega) \ \forall \omega \ge 0 \tag{2.7}$$

where $\bar{\sigma}[\cdot]$ denotes the largest singular value. In that case (see [Che. 1]), for the close loop system to be stable, we require, in addition to (2.6), that

$$\bar{\sigma}[H^0_{yr}(x,j\omega) - \frac{1}{b(\omega)} \le 0 \ \forall \omega \ge 0 \tag{2.8}$$

where $H^0_{yr} = (I + P_0 C) P_0 C$. Clearly, (2.8) is a semi-infinite inequality.

Finally, we may elect to minimize the influence of the disturbance over a critical frequency range, which leads to the definition of the cost function (see [Doy. 1])

$$f(x) = \max_{\omega \in [\omega', \omega'']} \bar{\sigma}[H^0_{yd(x,j\omega)}] \tag{2.9}$$

where $H^0_{yd} = (I + P_0 C)^{-1}$.

We see that this simple design example has the form of the problem (2.1). Note that in view of this example, one may not change the hypothesis that the functions in (2.1) are only locally Lipschitz continuous to the assumption that they are continuously differentiable. It should be pointed out that a number of the functions in the constraints are locally Lipschitz continuous only in the subset of the parameter space where the closed loop system is stable. Because of this, the algorithms discussed in the next section have to be slightly modified so as to preserve stability throughout the entire optimization.

3. *SEMI-INFINITE OPTIMIZATION ALGORITHMS FOR CONTROL SYSTEM DESIGN*

In the preceeding section, we saw that control system design problems reduce to optimization problems with constraints of the form $\max_{y_j \in Y_j} \varphi^i(x, y_j) \le 0$, where the y_j represent either time or frequency. To simplify notation, we concentrate on the simplest problem in this class:

$$\min\{f(x) \mid \varphi(x,y) \le 0 \ \ \forall y \in Y\} \tag{3.1}$$

where $f: \mathbb{R}^n \to \mathbb{R}$ is continuously differentiable and $\varphi: \mathbb{R}^n \times \mathbb{R} \to \mathbb{R}$ is locally Lipschitz continuous and $Y \subset \mathbb{R}$ is a compact interval.

Our favorite algorithms for solving (3.1) are semi-infinite phase I-phase II methods [Pol.5, Gon.1, Pol.2, Pol.3] obtained by extension of simple methods of feasible directions (see [Pol.7]). These methods compute a feasible point (i.e., a design satisfying specifications) very rapidly. After that, they reduce the cost without violation of specifications. Since to a large measure, design consists of simply satisfying specifications, the advantage of the phase I-phase II methods is clear.

It is easiest to understand the principles governing these phase I-phase II algorithms by considering at first simpler problems such as

$$\min\{f(x) \mid g^j(x) \le 0, j \in \underline{m}\} \tag{3.2}$$

where $f: \mathbb{R}^n \to \mathbb{R}$, $g^j: \mathbb{R}^n \to \mathbb{R}$, $j \in \underline{m}$ are continuously differentiable and $\underline{m} \triangleq \{1,2,\ldots,m\}$. For any $x \in \mathbb{R}^n$ let

$$\psi(x) \triangleq \max_{j \in \underline{m}} g^j(x) \tag{3.3a}$$

$$\psi_+(x) \triangleq \max\{0, \psi(x)\} \tag{3.3b}$$

and for any $\varepsilon \ge 0$, and $x \in \mathbb{R}^n$ let

$$I_\varepsilon(x) \triangleq \{j \in \underline{m} \mid \psi(x)_+ - f^j(x) \ge \varepsilon\} \tag{3.3c}$$

Next we recall that the directional derivatives of $f(\cdot)$ and $\psi_+(\cdot)$ at x in the direction $h \ne 0$, are given by $df(x;h) = \langle f(x), h\rangle$ while $d\psi_+(x;h) = \max_{j \in I_0(x)} \{\langle \nabla g^j(x), h\rangle, 0\}$ when $\psi(x) = 0$, $d\psi_+(x;h) = \max_{j \in I_0(x)} \langle \nabla g^j(x), h\rangle$ when $\psi(x) > 0$, and $d\psi_+(x;h) = 0$ when $\psi(x) < 0$. In order to avoid zigzagging, it is common to introduce an ε-directional derivative of $\psi_+(x)$, defined, for $\varepsilon \ge 0$ by

$$d_\varepsilon \psi_+(x;h) \triangleq \begin{cases} \max_{j \in I_\varepsilon(x)} \langle \nabla g^j(x), h\rangle & \text{if } \psi(x) \ge -\varepsilon \\ 0 \text{ if } \psi(x) < -\varepsilon \end{cases} \tag{3.4}$$

Note that for any x such that $\psi(x) \ge 0$, $d\psi(x;h) \le d_\varepsilon \psi_+(x;h)$ for all $h \in \mathbb{R}^n$ and every $\varepsilon \ge 0$. A phase I-phase II *search direction* can now be defined by

$$h_\varepsilon(x) \triangleq \operatorname{argmin} \{½\|h\|^2 + \max\{df(x;h) - \gamma\psi(x)_+, d_\varepsilon\psi(x;h)\} \quad \text{where } \gamma > 0 \tag{3.5}$$

Let $\vartheta_\varepsilon(x)$ denote the *value* of the quadratic program (3.5). Clearly, if $\vartheta_\varepsilon(x) < 0$, then a) if $\psi(x)_+ = 0$, then $h_\varepsilon(x)$ is a feasible usable direction (can decrease cost without constraint violation); b) if $\psi(x)_+ > 0$, we get a direction of reduction of constraint violation, mitigated by the need to reduce the cost $f(x)$ as $\psi(x)$ approaches zero. Finally, it is necessary to reduce the anti-zigzagging precautions as a solution point is approached. This can be done to defining, with $\nu \in (0,1)$,

$$E \triangleq \{0,1,\nu,\nu^2,\nu^3, \cdots \} \tag{3.6a}$$

and

$$\varepsilon(x) = \max\{\varepsilon \in E \mid \vartheta_\varepsilon(x) \le -\varepsilon\} \tag{3.6b}$$

Putting together the elements we have, we obtain

Algorithm 3.1:

Parameters: $\alpha,\beta,\gamma \in (0,1), \gamma > 0$.

Data: $x_0 \in \mathbb{R}^n$.

Step 0: Set i = 0.

Step 1: Compute $\varepsilon(x_i)$ and the search direction $h_i \triangleq h_{\varepsilon(x_i)}(x_i)$.

Step 2: Compute the step size λ_i,

$$\lambda_i = \arg\max_{k \in \mathbb{N}_+}\{\beta^k \mid \psi(x_i + \beta^k h_i) - \psi(x_i) \le -\beta^k \alpha\varepsilon(x_i)\} \text{ if } \psi(x_i) > 0,$$

$$\lambda_i = \arg\max_{k \in \mathbb{N}_+}\{\beta^k \mid f(x_i + \beta^k h_i) - f(x_i) \le -\beta^k \alpha\varepsilon(x_i); \psi(x_i + \beta^k h_i) \le 0\} \text{ if } \psi(x_i) \le 0 \tag{3.7}$$

Step 3: Update: set $x_{i+1} = x_i + \lambda_i h_i$, replace i by $i + 1$ and go to step 1. ▪

Referring to [Pol.5] we find the following result.

Theorem 3.1: Suppose that $0 \notin \mathrm{co}\{\nabla g^j(x)\}_{j \in I_0(x)}$ for all $x \in \mathbb{R}^n$ such that $\psi(x) \ge 0$. If $\hat{x}$ is an accumulation point of $\{x_i\}_{i=0}^\infty$ constructed by Algorithm 3.1, then $\psi(\hat{x}) \le 0$ and $\vartheta_0(\hat{x}) = 0$ (i.e., $\hat{x}$ satisfies the F. John condition of optimality (see [Clar. 1]). ▪

In order to extend Algorithm 3.1 to nondifferentiable problems, we begin by refining the ε-search direction finding problem (3.5) by introducing an additional real variable h^0 so that we move to $\mathbb{R}^{n+1}$ and introduce the vector $\bar{h} = (h^0,h)$, with $h \in \mathbb{R}^n$, and we replace $\gamma\psi(x)_+$ by the computationally more desirable factor $\sqrt{\gamma\psi(x)_+}$. The new ε-search direction finding problem becomes

$$\bar{h}_\varepsilon(x) = (h_\varepsilon^0(x),h_\varepsilon(x)) \triangleq \arg\min\{\tfrac{1}{2}\|\bar{h}\|^2 + \max\{df(x;h) - \sqrt{\gamma\psi_+(x)}h^0, d_\varepsilon\psi(x;h)\}\} \tag{3.8a}$$

making use of the Von Neumann minmax theorem, we obtain from (3.8a) that

$$\bar{h}_\varepsilon(x) = \arg\min\{\tfrac{1}{2}\|\bar{\xi}\|^2 \mid \bar{\xi} \in G_\varepsilon^{f,\psi}(x)\} \tag{3.8b}$$

where

$$G_\varepsilon^{f,\psi}(x) \triangleq \text{co}\left\{\begin{bmatrix}\sqrt{\gamma\psi_+(x)}\\ \nabla f(x)\end{bmatrix}, \begin{bmatrix}0\\ \nabla g^j(x)\end{bmatrix}\right\}_{j\in I_\varepsilon(x)} \tag{3.9}$$

It is easy to show that replacing $h_\varepsilon(x)$, as computed by (3.5), by $h_\varepsilon(x)$, as computed by (3.8b) and $\vartheta_\varepsilon(x)$, the value of (3.5), by $\vartheta_\varepsilon(x)$, the value of (3.8b), in Algorithm 3.1, does not affect the truth of Theorem 3.1. We therefore assume from now on that $h_\varepsilon(x)$ and $\vartheta_\varepsilon(x)$ in Algorithm 3.1 are computed using (3.8b). First we extend Algorithm 3.1 to the case of problem (3.1) where $\nabla_x\varphi(x,y)$ exists and is continuous (cf. [Gon. 1]). For this case, for any $x \in I\!R^n$ and $\varepsilon \geq 0$, we define

$$\tilde{Y}_\varepsilon(x) \triangleq \{y \in Y \mid \psi_+(x) - \varphi(x,y) \leq \varepsilon\} \tag{3.10a}$$

where, now, $\psi(x) \triangleq \max_{y\in Y} \varphi(x,y)$ and $\psi_+(x) = \max\{0,\psi(x)\}$, as before, and

$$\hat{Y}_\varepsilon(x) \triangleq \{y \in Y_\varepsilon(x) \mid y \text{ is a local maximizer of } \varphi(x,\cdot) \text{ in } Y\} \tag{3.10b}$$

Engineering considerations allow us to assume that $Y_\varepsilon(x)$ is a finite set (see [Pol.1]). To extend Algorithm 3.1 to the case of Problem (3.1) in question, we define

$$G_\varepsilon^{f,\psi}(x) \triangleq \text{co}\left\{\begin{bmatrix}\sqrt{\gamma\psi_+(x)}\\ \nabla f(x)\end{bmatrix}, \begin{bmatrix}0\\ \nabla_x\varphi(x,y)\end{bmatrix}\right\}_{y\in Y_\varepsilon(x)} \tag{3.11}$$

substituting into (3.8b), we get a finite quadratic program in baricentric co-ordinates, to solve in computing $\bar{h}_\varepsilon(x)$. Referring to [Gon. 1], we see that the conclusions of theorem 3.1 remain valid for this case, with the modification that the optimality condition now satisfied is a generalization of the F. John condition (see [Pol.7, Cla.1]).

Next, we turn to the special case where $\varphi(x,y) = (\bar{\sigma}[H(x,jy)])^2$ with $\bar{\sigma}$ the maximum singular value of a transfer function matrix whose coefficients are differentiable in x (see [Pol.3]). We define the symmetric positive semi-definite matrix $Q(x,y) \triangleq H(x,jy)^* H(x;jy)$, with * denoting the complex conjugate transpose. Noting that in (3.11) $\text{co}\{\nabla_x\varphi(x,y)\}_{y\in Y_\varepsilon(x)}$ is a superset of the Clarke generalized gradient of $\partial_x\varphi(x,y) = \text{co}\,\{\nabla_x\varphi(x,y)\}_{y\in Y_\varepsilon(x)}$, [Cla. 1], we proceed by analogy again.

Let $\varphi(x,y) \triangleq \lambda^1(x,y) \geq \lambda^2(x,y) \geq \cdots \geq \lambda^n(x,y)$ be the eigenvalues of $Q(x,y)$, let $\psi(x) \triangleq \max_{y\in Y}\varphi(x,y)$, as before, and let $Y_\varepsilon(x)$ be defined as in (3.10b). Next, we define

$$F_\varepsilon(x,y) \triangleq \begin{cases} \{\xi\in I\!R^n \mid \xi^i = \langle U_\varepsilon(x,y)z, \dfrac{\partial Q(x,y)}{\partial x^i} U_\varepsilon(x,y)x\rangle,\ \|z\| = 1\} \text{ if } \psi(x)\geq-\varepsilon \\ = \emptyset \text{ otherwise} \end{cases} \tag{3.12}$$

where $U_\varepsilon(x,y)$ is a unitary matrix whose columns span the eigen space corresponding to the eigenvalues $\lambda^k(x,y)$, $k = 1,2,\ldots,k_\varepsilon(x,y)$ such that $\lambda^1(x,y) - \lambda^k(x,y) \leq \varepsilon$. Then we define

$$G_\varepsilon^{f,\psi}(x) \triangleq \mathrm{co}\left\{\begin{bmatrix}\sqrt{\gamma\psi_+(x)} \\ \nabla f(x)\end{bmatrix}, \begin{bmatrix}0 \\ \xi\end{bmatrix}\right\}_{\substack{\xi \in F_\varepsilon(x,y) \\ y \in Y_\varepsilon(x)}} \tag{3.13}$$

Inserting into (3.8b) we obtain a formula for computing a search direction. This defines Algorithm 3.1 for constraints on $(\overline{\sigma}[H(x,jy)])^2$. The formulas for constraints on $\overline{\sigma}[H(x,jy)]$ are similar, but no longer symmetrical (i.e., are uses matrices of singular vectors $V_\varepsilon(x,y)$ and $U_\varepsilon(x,y)$ in (3.12)).

Again, referring to [Pol.3, Pol.8], we see that the conclusions of Theorem 3.1 remain valid with the modification that x satisfies an appropriate optimality condition in terms of generalized gradients (see [Cla. 1]). However, the computation of the search direction $\overline{h}_\varepsilon(x)$ is no longer a finite quadratic program. Hence $\overline{h}_\varepsilon(x)$ has to be computed by means of some nearest point algorithm such as the ones described in [Pol.3, Pol.7].

4. *SOFTWARE FOR OPTIMIZATION-BASED CONTROL SYSTEM DESIGN*

A software system, DELIGHT.MIMO (see [Pol.6]), implementing an optimization-based control system design methodology is currently being developed jointly by research teams at the University of California, Berkeley, and Imperial College, London. Some specific contributions to DELIGHT.MIMO are also being made at other institutions as well (see acknowledgement at the end of this paper).

DELIGHT.MIMO is a member of a family of optimization-based CAD packages currently being implemented in the DELIGHT system [Nye.1, Nye.2]. Hence a description of DELIGHT.MIMO must begin with a brief description of DELIGHT. DELIGHT can be thought of as a highly portable operating system for a FORTRAN or C machine. As can be expected from an operating system, it provides a certain number of commonly found features such as a text editor, a read and write files command, an ability to install and execute FORTRAN and C programs, a *help* command, a *history* command, a *repeat* command, hard interrupts, etc.

In addition, DELIGHT provides a number of rather special features. The most important of these are the following.

1) **Color graphics** for interaction with data and programs. The graphics provide a number of low level commands, such as *viewport*, *window*, *vector*, *move*, *cursor* and *text*. These are used when display improvisation is necessary. In addition there are also a number of high level commands of the form *plot* data according to the options specified, which can be used to produce various orthodox type plots.

2) **High level language**, RATTLE, for the programming of optimization algorithms as well as information display options. RATTLE requires about 1/10 of the number of program lines compared to FORTRAN, but it executes considerably slower than FORTRAN. Because of this, in design packages a mixture of RATTLE and FORTRAN code is always used.

RATTLE compiles incrementally, it has binary matrix operation capability, it uses defines for command simplification and macros for producing simple RATTLE calls to complex FOR-

TRAN programs (e.g. linprog $x = argmin\{<c,z>|\,Az = 0, Bz \geq 0, z \geq 0\}$). It is easy to use RATTLE to construct code for conversational data entry.

3) **Soft interrupts** for program debugging and temporary algorithm modification. Unlike hard interrupts which suspend a program the instant the break key is depressed, soft interrupts suspend a program only at designated break points in the program. When either a hard or a soft interrupt is executed, it is possible to *enter* suspended subprocedures and display and modify both local and global variables. After an interrupt the user may start up a totally unrelated computation or resume execution of the suspended program. To return to a suspended program after an unrelated side computation, the user executes the *reset* (a given number of interrupt levels) command.

4) **A modular, RATTLE code, optimization algorithm library** is being assembled. To use this library, the user assembles an algorithm from optional blocks, such as step size and direction finding procedures, via a menu. The problem to be solved must be described by means of several files containing either dimensional information or RATTLE code for: the cost function, ordinary inequality constraints, functional inequality constraints, and gradients of the appropriate functions. The optimization problem and algorithm are linked by means of the *solve* command, e.g., *solve pid using polak_wardi*, when neither the problem pid nor algorithm polak_wardi has been compiled, or *solve pid* (or *solve using polak_wardi)* when the algorithm (problem) have been compiled earlier. Algorithms can be executed a desired number of iterations by means of the *run k* command, or they can be executed atomically, step by step, by means of the *stepk* command. When execution of an optimization program is interrupted by means of a soft or hard interrupt, the user may adjust algorithm parameters, completely replace the algorithm, modify the problem description files, display variable values or plot response graphs.

DELIGHT.MIMO adds to the basic DELIGHT system a data base for control system interconnection description, programs for control system time and frequency response simulation, a symbolic differentiator for obtaining derivatives of these responses with respect to design parameters, interactive programs for initial design generation, an interactive program which assists the user in forming the RATTLE problem description files from design specifications, as required by the optimization algorithm library format, and both alpha-numeric and graphical means for entering the control system configuration. The optimization algorithm currently used for control system design is the Polak-Wardi method described in [Pol.3]; it has the form of the last algorithm described in Section 3.

5. THE DATA-BASE

The DELIGHT.MIMO data-base allows a system to be represented as an interconnection of subsystems. The subsystems may be either symbolic or state space representations. When the subsystems are represented symbolically, their names and interconnection data are stored in a *link table*. For the system in Fig. 3, where the block R generates the external system input, the link table consists of two blocks, as shown below:

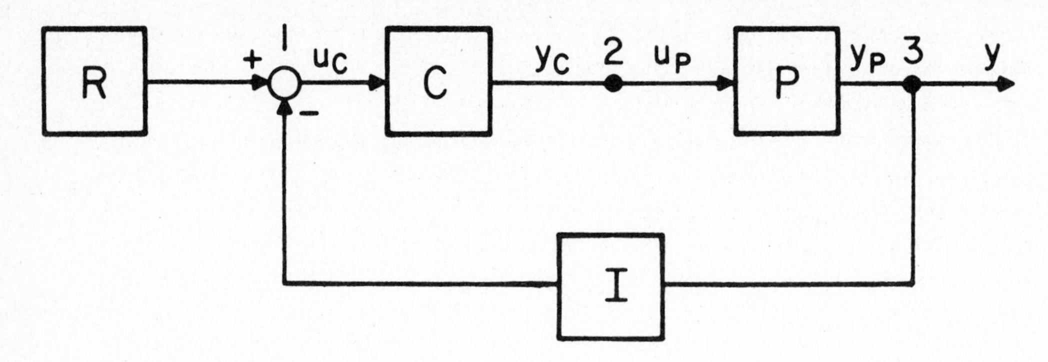

Fig. 3

Subsystem	"From" Node	"To" Node	Sign
P	2	3	+
C	1	2	+
I	3	1	-

Signal Generator	"To" Node
R	1

The link table can be constructed either alpha-numerically via the terminal keyboard or by means of the graphical block diagram editor.

When the subsystems are represented in state space form, as

$$S_i \begin{cases} \dot{z}_i = A_i z_i + B_i u_i \\ y_i = C_i z_i + D_i u_i \end{cases} \tag{5.1}$$

they define (assuming there are N subsystems) an *assembly of subsystems* S of the form

$$S \begin{cases} \dot{z} = Az + Bu \\ y = Cz + Du \end{cases} \tag{5.2}$$

where
$A = diag\,(A_1,A_2,\ldots A_N)$, $B = diag\,(B_1,B_2,\ldots B_N)$, $C = diag\,(C_1,C_2,\ldots,C_N)$, $D = diag\,(D_1,D_2,\ldots,D_N)$. In this case, the interconnections between the subsystems are expressed algebraically:

$$u = Ey + Jr \tag{5.3}$$

where r is a vector of external inputs and E and J are matrices whose elements are zeros and ones. It should be clear that once the dimensions of the inputs and outputs are defined, the matrices E and J can be constructed from the data in the link table.

The matrices A_i, B_i, C_i, D_i, specifying the subsystems may be given either in numerical form or in parametric form. When given in parametric form, their elements must be multinomials in the elements of the design parameter vector x. A symbolic differentiator is available for computing their derivatives with respect to the parameters.

The interconnection equation (5.3) can be eliminated by means of the *link* command which produces a reduced description of the form

$$S \begin{cases} \dot{z} = A_c z + B_c r \\ y = C_c z + D_c r \end{cases} \tag{5.4}$$

where $A_c = A+B[I-ED]^{-1}EC$, $B_c = B[I-ED]^{-1}J$, $C_c = C+D[I-ED]^{-1}EC$, and $D_c = D[I-ED]^{-1}J$, in terms of the matrices in (5.2). The link command can only be executed when specific values have been assigned to the design parameters.

In addition to the *link* command, three other commands are used in conjunction with the data-base. The first is the command which enables the user to *load* into the data-base numerical or parametrized descriptions of subsystems. The second is the *replace* command which associates subsystems in the data base for symbolically defined subsystems in the link table. The third is the *transfer* command, which can be used to transfer parametrized compensator descriptions and their initial values from a design initialization program.

6. COMPUTATION OF SYSTEM RESPONSES AND THEIR DERIVATIVES

Since the closed loop system (5.4) always has distinct eigenvalues (at least with probability 1), the computation of responses can be considerably simplified by diagonalization (more robust techniques, based on Schur decomposition, are also being contemplated). Thus, rewriting (5.2) with the parameters made explicit, we get

$$\dot{z}(t,x) = A_x(x)x(t,x) + B_x(x)r(t) \tag{6.1a}$$

$$y(t,x) = C_x(x)x(t,x) + D_x(p)r(t) \tag{6.1b}$$

We begin with the time responses to inputs $r(t)$ which are polynomials in t. With $W(x)$ a matrix of eigenvectors of $A(x)$, we obtain,

$$z(t,x) = W(x)e^{\Lambda(x)t}W(x)^{-1}x(0) + \int_0^t W(x)e^{\Lambda(x)(t-s)}W(x)^{-1}r(s)ds \tag{6.2}$$

The output $y(t,x)$ is then computed according to (6.1b). Because the input $r(t)$ is a polynomial, the integral in (6.2) can be and is evaluated analytically (not numerically).

Next, the symbolic differentiator produces formulas for the components of the matrices A_c, B_c, C_c, D_c with respect to the components of the design parameter vector x. Numerical values for the derivatives are obtained by substituting current parameter values. We note that the derivatives with respect to x of $z(t,x)$ and $y(t,x)$ in (6.1) satisfy

$$(d/dt)(\partial z(t,x)/\partial x) = A_c(x)(\partial z(t,x)/\partial x) + (\partial A_c(x)/\partial x)z(t,x) + (\partial B_c(x)/\partial x)r(t) \quad (6.3a)$$

$$\partial y(t,x)/\partial x = C_c(x)(\partial z(t,x)/\partial x) + (\partial C_c(x)/\partial x)z(t,x) + (\partial D_c(x)/\partial x)r(t) \quad (6.3b)$$

The diagonalization matrix $W(x)$ can be used again to produce fairly simple formulas for the derivatives $(\partial z(t,x)/\partial x)and(\partial y(t,x)/\partial x)$. Numerical substitution into these formulas yields efficient derivative evaluations.

Next we turn to the frequency response of the interconnected system. The input output transfer function of the interconnected system is given by

$$G(jw,x) = C_c(x)[jwI - A_C(x)]^{-1}B_C(x) + D_C(x) \quad (6.4a)$$

Since the derivative of G with respect to x is not a matrix, it is easiest to obtain component-wise expressions for it, viz.,

$$\begin{aligned}\partial G(jw,x)/\partial x &= \partial C_c(x)/\partial x)[Ijw - A_c(x)]^{-1}, B_c(x) + D_C(x) \quad (6.4b)\\ &+ C_c(x)[jwI - A_c(x)]^{-1}(\partial A_c(x)/\partial x)[jwI - A_c(x)]^{-1}\\ &+ C_c(x)[jwI - A_c(x)]^{-1}(\partial B_c(x)/\partial x) + (\partial D_c(x)/\partial x)\end{aligned}$$

Assuming that the time response derivatives are computed first, the only major computation left in the evaluation of the frequency responses and their derivatives as specified by (6.4a) (6.4b) is the evaluation of the matrix $[jwI - A_c(x)]^{-1}$. Since a diagonalization for $A_c(x)$ is already available, this computation can be considerably simplified by making use of the formula

$$[jwI - A_c(x)]^{-1} = W(x)[jwI - \Lambda(x)^{-1}]W(x) \quad (6.5)$$

7. *DESIGN INITIALIZATION TECHNIQUES*

Design via optimization in not a totally automatic process. The designer is not only required to transcribe design specifications into semi-infinite inequalities, he or she is also required to decide on an initial compensator configuration as well as to produce a set of initial values for the compensator. This is a creative process which is very designer dependent. To facilitate the design initialization task, the DELIGHT.MIMO system will incorporate software implementing some of the more popular techniques, for example, such as those described in [Des.1, Doy.1, Mac.1, Moo.1, Ros.1, Saf.1, Saf.2, Ste. 1]. At the present time, there is software in DELIGHT.MIMO enabling design of compensators via LQG techniques as well as some model reduction algorithms. In the simplest case, these replace the observer dynamics with its DC gain matrix. It should be noted that in order to introduce into the design integrators for the elimination of steady state errors, a certain amount of ingenuity must be exercised in using LQG techniques. For example, consider the case in Fig.4. For the purpose of designing a state

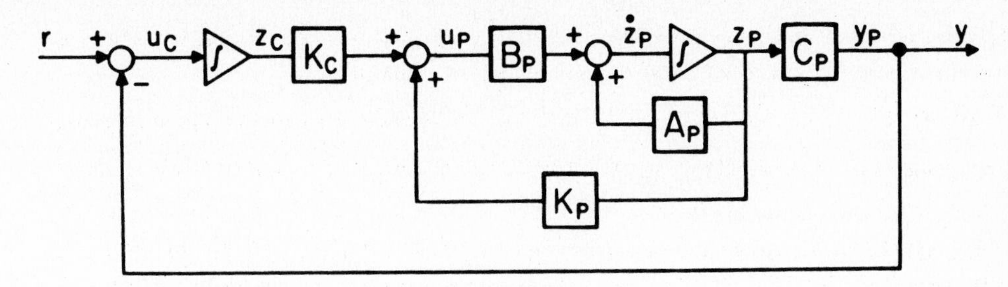

Fig. 4

feedback matrix K, the external input r must be neglected, while the plant input is used as the feedback channel. Thus, suppose that the plant has dynamics given by

$$\dot{z}_P = A_P z_P + B_P u_P \tag{7.1}$$

$$y_P = C_P z_P$$

Next, the integrator of the compensating block has dynamics

$$\dot{z}_C = u_C \tag{7.2}$$

$$y_C = z_C$$

and the interconnection is specified by

$$u_C = -u_P \tag{7.3}$$

Thus the assembly satisfies the state equation

$$\frac{d}{dt}\begin{bmatrix} z_P \\ z_C \end{bmatrix} = \begin{bmatrix} A_P & 0 \\ -C_P & 0 \end{bmatrix} + \begin{bmatrix} B_P \\ 0 \end{bmatrix} U_P \tag{7.4}$$

LQR techniques can now be used to compute a state feedback matrix $K = [K_P | K_C]$ and the feedback law then becomes

$$u_P = K_P z_P + K_C z_C \tag{7.5}$$

Since the state of the integrator block is available, an observer is needed only for estimating the plant state in this scheme.

ACKNOWLEDGEMENT:

1. The DELIGHT.MIMO software was constructed by three teams:

At the University of California, Berkeley:

E. Polak (overall system specification and project coordination),

T.S. Wuu, (optimization library, alphanumeric interaction tools),

P. Siegel (Graphical block diagram editor specification),

T. Baker (LQG design initialization tools),

W. T. Nye (DELIGHT consultant).

At Imperial College, London:

D. Q. Mayne (overall system specification, initialization tools, symbolic differentiator and system response evaluator),

A. J. Heunis (Symbolic differentiator and system response evaluator),

At the Lawrence Livermore National Laboratory:

C. J. Herget, D. Gavel, D. Tilly, S. Bly (graphical block diagram editor implementation).

Control system design subroutines were contributed by:

M. J. Denham, Kingston Polytechnic, and

A. J. Laub, University of California, Santa Barbara.

Over the years, the DELIGHT.MIMO project was supported by The National Science Foundation (ECS-7913148, ECS-8121149), The Joint Services Electronics Program (F49620-79-C-0178), The Office of Naval Research (N00014-83-K-0602), The Air Force Office of Scientific Research (AFOSR-83-0361), The Kirtland Air Force Weapons Laboratory, The Lawrence Livermore National Laboratory (I03403805, I03247705), The Semiconductor Research Consortium (SCR-82-11-008), and The Science and Engineering Research Council of Great Britain.

9. REFERENCES

[Ath.1] Athans, M., "The role and use of stochastic linear-quadratic-gaussian problem in control system design", *IEEE Trans.* vol. AC-16, no. 6, 1971.

[Bec.1] Becker, R. G., Heunis, A. J., and Mayne, D. Q., "Computer-Aided Design of Control Systems via Optimization, *Proc. IEE*, vol. 126, no. 6, 1979.

[Che.1] Chen, M. J. and Desoer, C. A., "Necessary and Sufficient Conditions for Robust Stability of Linear Distributed Feedback Systems", *Int. Journal on Control*, Vol. 35, No. 2, pp 255-267, 1982.

[Cla.1] Clarke, F. H., *Optimization and Nonsmooth Analysis*, Wiley-Interscience, New York, N.Y., 1983.

[Den.1] Denham, M. J., and Benson, C. J., "SLICE: a subroutine library for control system design," Internal Report 01/82, School of Electronic Engineering and Computer Science, Kingston Polytechnic, Kingston upon Thames KT1 2EE, 1982.

[Des.1] Desoer, C. A., and Gustafson C. L.,,"Algebraic Design of Two-Input Controllers for Linear Multivariable Feedback Systems", to appear *IEEE Transaction on Automatic Control.*

[Doy.1] [Doy.1]. Doyle, J. C., and Stein, G. "Multivariable Feedback Design: Concepts for a Classical/Modern Synthesis", *IEEE Trans. on Control*, Vol. AC-26, No. 1, pp. 4-16,

1981. Academic Press, NY, 1963.

[Gon.1] Gonzaga, C., Polak, E., and Trahan, R., "An Improved Algorithm for Optimization Problems with Functional Inequality Constraints", *IEEE Trans.*, Vol. AC-25, No. 1, 1980.

[Mac.1] MacFarlane, A. G. J., and Postlethwaite, I., "Generalizes Nyquist Stability Criterion and Multivariable Root Loci," *International Journal of Control*, vol. 25(1), 1977.

[Moo.1] Moore, B. C., "Principal component analysis in linear systems: controllability, observability and model reduction", *IEEE Trans., Vol. AC-26, No.1 pp. 17-32, 1981.*

[Nye.1] Nye, W.T., Polak, E., Sangiovanni-Vincentelli, A., and Tits, A., "DELIGHT: an Optimization-Based Computer-Aided-Design System" Proc. IEEE Int. Symp. on Circuits and Systems, Chicago, Ill, April 24-27, 1981.

[Nye.2] Nye, W. T., "DELIGHT: An interactive system for optimization-based engineering design", Electronics Research Laboratory, University of California, Berkeley, Memo No. UCB/ERL M83/33, May 31, 1983.

[Pol.1] Polak, E., " Semi-infinite optimization in engineering design", in Lecture Notes in Economics and Mathematical Systems, Vol. 215: Semi-Infinite Programming and Applications, Edited by A. V. Fiacco and K. O. Kortanek, Springer-Verlag, Berlin, New York, Tokyo, 1983.

[Pol.2] Polak, E., and Mayne, D. Q., "An Algorithm for Optimization Problems with Functional Inequality Constraints," *IEEE Trans.*, vol. AC-21, no. 2, 1976.

[Pol.3] Polak, E., and Wardi, Y. Y., "A nondifferentiable optimization algorithm for the design of control systems subject to singular value inequalities over a frequency range", *Automatica*, Vol. 18, NO. 3, pp. 267-283, 1982.

[Pol.4] Polak, E., "A Modified Nyquist Stability Criterion for Use in Computer-Aided Design", ERL Memo No. M83/11, IEEE Trans. on Automatic Control, Vol. AC-28, No. 11, March 1984.

[Pol.5] Polak, E., Trahan, R., and Mayne, D. Q., "Combined phase I - phase II methods of feasible directions", *Math. Programming*, Vol. 17, No. 1, 1979, pp 32-61.

[Pol.6] Polak, E., Siegel P., Wuu, T., Nye, W. T., and Mayne, D. Q., "DELIGHT-MIMO an interactive, optimization based multivariable control system design package", *IEEE Control Systems Magazine*, Vol.2, No.4, Dec. 1982, pp 9-14.

[Pol.7] Polak, E., *Computational Methods in optimization: A Unified Approach*, Academic Press, N.Y., 1971.

[Pol.8] Polak, E., and Mayne, D. Q., "Algorithm Models for Nondifferentiable Optimization," University of Calfornia, Electronics Research Laboratory Memo No. UCB/ERL M82/34, 10 May 1982.

[Ros.1] Rosenbrock, H. H., *Computer-Aided Control System Design*, Academic Press, London, 1974.

[Saf.1] Safonov, M. G., Laub, A. J., and Hartman, G. L., "Feedback Properties of Multivariable Systems: The Role and Use of the Return Difference Matrix," *IEEE Trans. on Control*, vol. AC-26, 1981.

[Saf.2] Safonov, M. G., "Choice of quadratic cost and noise matrices and the feedback properties of multiloop LQG regulators", *Proc. Asilomar Conf. on Circuits, Systems, and Computers*, Pacific Grove, California, 1979. Vol. 15, 1979.

[Ste.1] Stein, G., "Generalized quadratic weights for asymptotic regulator properties", *IEEE Trans.*, Vol. AC-24, 1979.

CTRL-C AND MATRIX ENVIRONMENTS FOR THE COMPUTER-AIDED DESIGN OF CONTROL SYSTEMS

J.N. Little, A. Emami-Naeini, and S.N. Bangert
Systems Control Technology, Inc., 1801 Page Mill Rd., Palo Alto, CA 94303, USA

Abstract

A computer-aided control system design package, called CTRL-C, provides a matrix workbench for the analysis and design of multivariable systems. CTRL-C is an interactive environment with a comprehensive set of tools for analysis, identification, design, and evaluation.

A unified software system is possible for matrix analysis, engineering graphics, control system design, and digital signal processing. A common thread in these disciplines is the role of a single data object: the complex matrix. CTRL-C demonstrates that a matrix environment can lead to a powerful, natural, and extensible software system.

1.0 INTRODUCTION

A Workbench is a collection of tools (a "toolbox") and a suitable environment in which to perform a job. Several computer workbenches have been available for some time under operating systems like Unix. Professional writers have a writer's workbench. The writer's workbench is a collection of tools that include editors, spelling checkers, grammar critiquing and document preparation. Professional programmers have a programmer's workbench. The programmer's workbench provides editors, beautifiers, verifiers, timing analyzers, and source code control systems. Both of these workbenches exist in an environment (Unix) that provides excellent file handling and text manipulation capabilities. Inspired by these two examples, CTRL-C is intended to be a control designer's workbench.

A useful workbench for a control designer should not be limited to control design. His workbench should encompass other important, related fields. The fields considered by CTRL-C include: Matrix Analysis; Engineering Graphics; Control System Design and Analysis; and Digital Signal Processing.

Historically, separate, stand-alone programs have been employed within each of these disciplines. A unified approach to these disciplines is possible, however, based upon a simple observation: matrices are important objects in all four fields. That is, a single data type, a rectangular matrix with complex elements, can be used to represent the important objects in each of these fields. Scalars, when needed, are simply represented as 1-by-1 matrices, while 1-by-n and m-by-1 matrices represent row and column vectors.

The principle goals of this paper are: (1) to demonstrate the CTRL-C system and (2) to show how the use of a matrix environment can lead naturally to a unified, versatile software system. Section 2 of this paper describes the fundamental principles and concepts that were used in the development of the CTRL-C interactive environment. Sections 3-6 show the use of CTRL-C for each of: (1) matrix analysis, (2) engineering graphics, (3) control design and analysis, and (4) digital signal processing. Section 7 describes extensibility concepts in CTRL-C, and the use of CTRL-C as a programming language. Section 8 concludes with a description of some of the important numerical algorithms.

2.0 PRINCIPLES

In the 1970s, the first computer-aided control system design environments emerged. These early packages were most often menu-driven or of the question-answer dialog variety. Recognizing the limitations of these primitive environments, more advanced command-driven environments have been developed. Unfortunately, most of these are specialized and limited, often utilizing complex and arbitrary data structures. The data structures are not generally transportable between programs, nor are they usually understood by the casual user. The result is that most environments are not extensible, they do what they are designed to do, and very little more.

To try and overcome these difficulties, related computer science fields have been examined. The result is a set of four principles upon which the CTRL-C interactive environment is based: easy matrix manipulation; uniform file handling; direct manipulation; and extensibility.

Master the Matrix

Since the matrix is an important data object, then an appropriate interactive environment should be one where matrices are treated naturally. After learning a few simple concepts for matrix manipulation, the user becomes able to work throughout four disciplines. It is not necessary to learn four separate environments. The idea that matrix manipulation environments are powerful is not new; the small but dedicated group of APL users have been saying so since the 1960s. For numerous reasons, however, APL has not been widely popular within the four disciplines. One reason is that APL does not use the standard ASCII character set. Another, more serious reason is that APL code is often too concise and subtle. For these reasons, APL has been referred to as a "write-only" language.

A matrix program called MATLAB offers an alternative to APL. CTRL-C is based on MATLAB, a program which was originally developed by Cleve Moler of the University of New Mexico [1]. MATLAB was written as a convenient "laboratory" for compu- tations involving matrices. Applying some of the concepts of SPEAKEASY to APL resulted in an environment where the only primitive data object is the complex array. It is command driven; that is single line commands are accepted from the user, processed immediately, and the result displayed.

Uniform File Handling

In the CTRL-C environment, all variables are stored in a large stack. This stack resides in semiconductor memory (or managed virtual memory). It is necessary, however, to allow data communication between this stack and disk files. It is important for the utility of the system that file manipulation commands be powerful yet simple. To provide a uniform user interface, all commands that read or write disk files use the Unix-like notation [2] of left and right angle bracket symbols < and >, and the hyphen "-" for switches. Roughly translated, the brackets mean "get input from" and "send output to", respectively. Thus, file operations, which are cumbersome in a pure matrix environment, can be accomplished using commands that excel at file operation.

Direct Manipulation

A principle described as direct manipulation [3] has been used to characterize traits often associated with popular software. It has been observed that some systems evoke "glowing enthusiasm" from their users, while others result in "grudging acceptance or outright hostility". The good systems usually are easy to learn, inspire confidence in their use, install an eagerness to teach others, and develop a desire to explore. Examples of these types of systems include display editors (e.g. EMACS, EDT, VI, FSE, WORDSTAR), spreadsheet programs (VISCALC, 1-2-3), and certain operating systems or languages (UNIX, LOGO). Many of these systems are aptly described with the expression "what you see is what you get". For all of these systems, the user is able to apply intellect directly to the task; the tool seems to disappear.

A feeling of direct manipulation is found in CTRL-C. For systems involving matrices, "what you see is what you get", that is, matrix algebra is performed naturally.

Extensibility

Certain computer languages inspire a unique view of programming. In the traditional FORTRAN sense, programming consists of writing a main program, and then writing subroutines. In other languages, including LISP, LOGO, and FORTH, there is a subtle, but important difference in the approach to programming. The user thinks of programming as consisting of creating new "words" in the language. Once a new word is created, it is used the same way as a permanent word. This principle can be described as extensibility of the environment. In CTRL-C this is achieved through the Define Function capability.

With these concepts in mind, the next three sections demonstrate the CTRL-C system for (1) matrix analysis, (2) engineering graphics, 3) control system design and analysis, and (4) digital signal processing. The examples are intended to show how the use of a matrix environment can lead naturally to a useful and simple interaction with the computer.

3.0 MATRIX ANALYSIS

CTRL-C has a natural matrix environment. To enter a matrix, a simple list is used. The list is surrounded by brackets, '[' and ']', and uses the semicolon ';' to indicate the ends of the rows. For example, the input line a = [1 5 9 13; 2 6 11 14; 3 7 11 16; 4 8 12 18] results in the output

```
A =
       1.    5.    9.   13.
       2.    6.   11.   14.
       3.    7.   11.   16.
       4.    8.   12.   18.
```

The matrix A will be saved for later use.

In CTRL-C, matrix algebra is easy -- it is accomplished the way it is normally written on the back of an envelope. For example, the matrix transpose is obtained as b = a' which results in

```
B =
       1.    2.    3.    4.
       5.    6.    7.    8.
       9.   11.   11.   12.
      13.   14.   16.   18.
```

Matrix multiplication is obtained by typing c = a * b which produces

```
C =
     276.  313.  345.  386.
     313.  357.  393.  440.
     345.  393.  435.  488.
     386.  440.  488.  548.
```

Simple matrix functions are easily obtained, for example, the determinant is found by typing det(a) which results in ANS = 4.0000.

A complete set of common matrix functions is available in CTRL-C. Largely inherited from MATLAB, they represent the basic tools for matrix analysis. Typical functions include:

```
eig(x)     - eigenvalues and eigenvectors
geig(a,b)  - generalized eigenvalues
exp(x)     - matrix exponential
inv(x)     - inverse
svd(x)     - singular value decomposition
schur(x)   - schur decomposition
```

Polynomials can be represented in a matrix environment as row vectors containing the coefficients ordered by descending powers.

Polynomial multiplication may be accomplished using convolution. If A and B are polynomials, then Y = CONV(A,B) calculates the polynomial product. For example, typing

```
[> a = [1 2 1];   b = [1 2];
[> c = conv(a,b)                    yields the polynomial product

C =   1.    4.    5.    2.
```

Polynomial division, root finding, and other polynomial operations are similarly accomplished.

In summary, a matrix environment allows matrix algebra operations to be written directly, with no cumbersome syntax. Dimensioning of variables is accomplished automatically by the software. Polynomials can be represented in a matrix environment and polynomial arithmetic is performed readily.

4.0 ENGINEERING GRAPHICS

Graphics abilities are a requirement for a useful computer-aided engineering package. Rather than being an afterthought, as with many CAD packages, the graphics facilities in CTRL-C are useful in their own right as a stand-alone system. Data are graphed using the same natural syntax with which matrices are manipulated.

Engineering X-Y plots are created with separate commands for data plotting, titling and labeling. For example, a sine curve might be generated, plotted, and titled with

```
[>  t = 0:.05:4*pi;
[>  y = sin(t);
[>  plot(t,y)
[>  title('sine(t)')
```

The first statement generates a vector consisting of elements running from 0.0 to 4pi in increments of 0.05. The second statement creates a vector y containing the sine of each of the elements of t. The third statement plots y versus t and, together with the fourth statement, results in Figure 1.

Three-dimensional surface plots can be useful to "look at" large matrices. An intuitive understanding of the structure of a matrix can often be found that is not clear from just looking at numbers. For example, the state dynamics matrix of a 59th order aircraft model is too large to display conveniently on a CRT screen. The command p3d(a) produces the 3-dimensional plot of Figure 2, where the value of each element represents the height Z above the X-Y plane. This yields a perspective on the matrix structure not evident from looking at 3600 numbers.

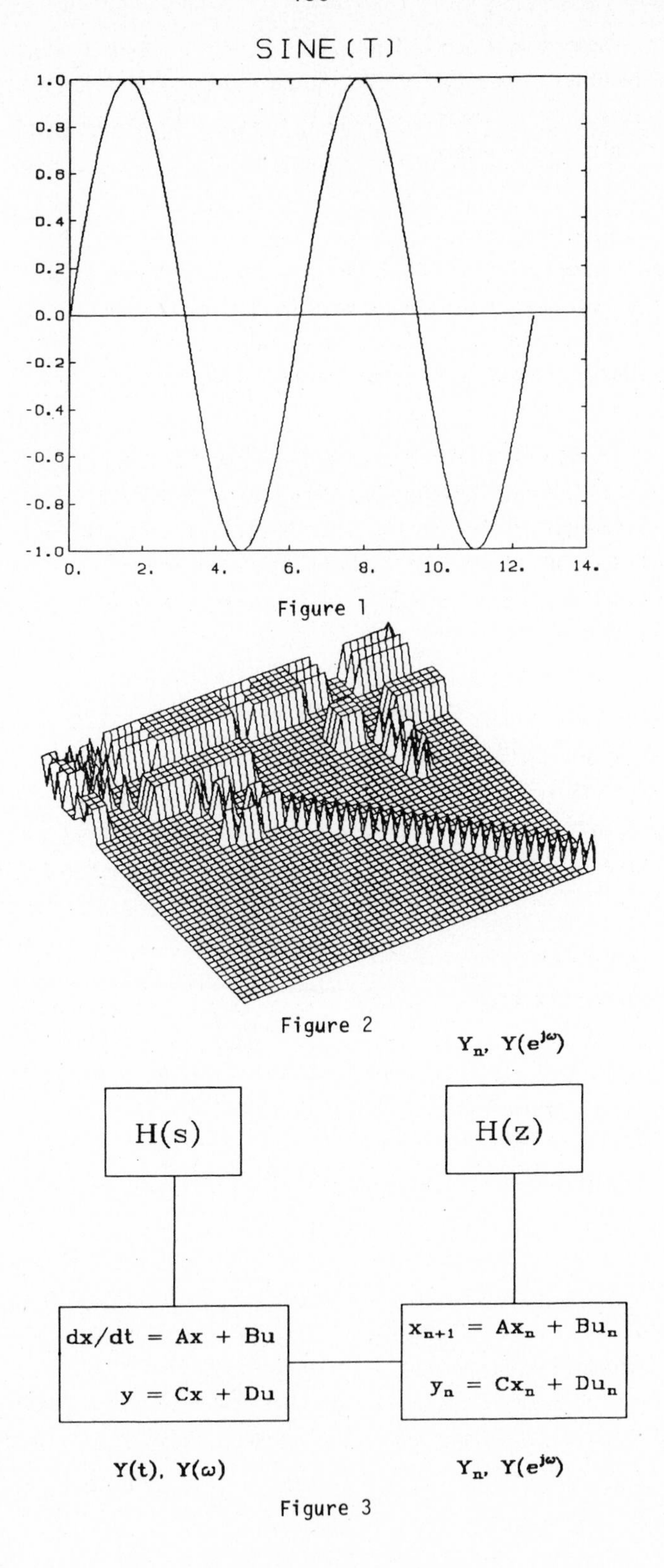

Figure 1

Figure 2

Figure 3

In summary, some simple plots have been created. Other commands and options are available in CTRL-C for log-log plots, overplots, axis labeling and other basic graphics functions. It is demonstrated that engineering graphics can be a natural extension to a matrix environment.

5.0 CONTROL SYSTEMS

Matrix environments are particularly convenient for working with linear systems that can be represented in state-space form. Systems may be described in discrete-time or in continuous-time. Systems may also be described in polynomial notation as a Laplace transfer function for continuous time, or as a Z-transform transfer function for discrete time.

In CTRL-C, transformations between these representations are provided, as well as tools for the calculation of time and frequency domain measures. Other primitives implement various control design algorithms. Pictorially, the system representations in CTRL-C are shown in Figure 3. The remainder of this section consists of two simple examples, each selected to illustrate basic concepts of the use of matrix environments and CTRL-C for control design and analysis.

Example 1

The first example demonstrates the input of a system described by a Laplace transfer function, the conversion to state-space, the calculation of time and frequency responses, and finally the design of a simple controller. Consider the system described by a simple Laplace transfer function in Figure 4. To describe this system in CTRL-C, the numerator and denominator coefficients for the first block are entered:

```
[> num = [1 2]; [> den = [1 .4 1]
```

If it is desired to find the pole locations, ROOT is used. Typing:

```
[> dr = root(den)     results in
DR =

   -0.2000 + 0.9798i
   -0.2000 - 0.9798i
```

The natural frequency and damping factor are easily found:

```
[> Wn = abs(dr)
WN    =

   1.0000
   1.0000
[> Zeta = cos(imag(log(dr)))
ZETA  =

  -0.2000
  -0.2000
```

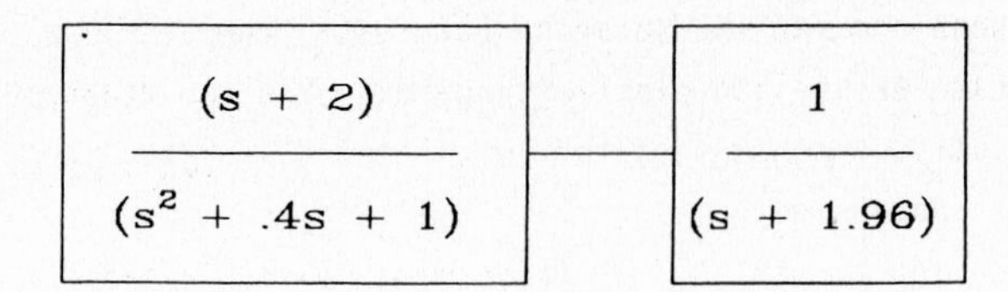

Figure 4

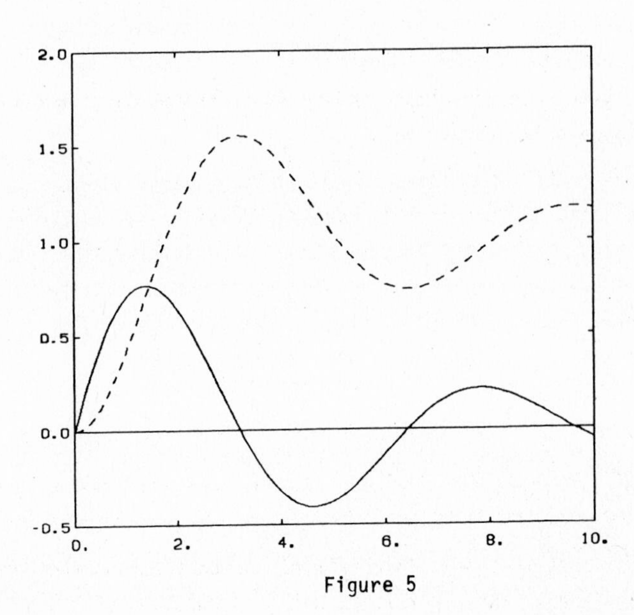

Figure 5

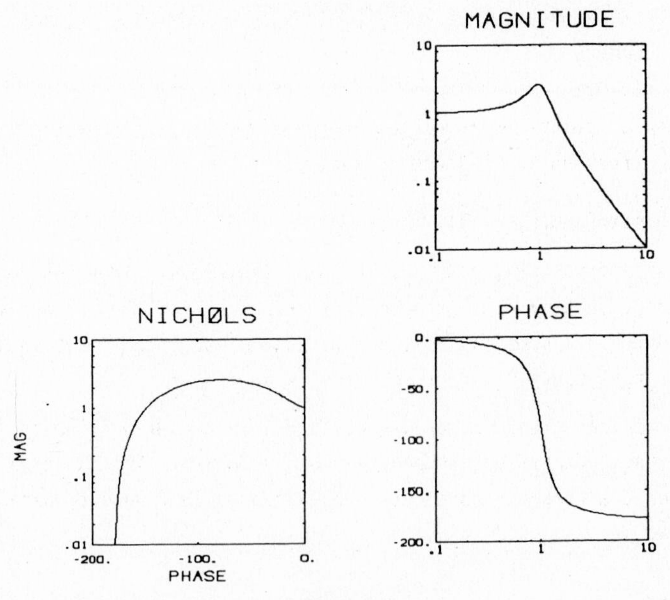

Figure 6

These commands show some of the power of complex arithmetic in a matrix environment. To cascade the second block of Figure 4, the denominator term is formed for the new block:

```
[> den2 = [1 1.96]
```

The series connection is achieved by polynomial multiplication (convolution) of the two denominators. Typing:

```
[> den  = conv(den,den2)      results in
DEN =
    1.0000    2.3600    1.7840    1.9600
```

This combined system can be transformed to state-space using the transfer-function to state-space primitive. Typing

```
[> [a,b,c,d] = tf2ss(num,den)
```

results in the controller canonical form description:

```
D =     0

C =     0.    1.    2.

B =     1.
        0.
        0.

A =
        -2.3600   -1.7840   -1.9600
        1.0000    0.0000    0.0000
        0.0000    1.0000    0.0000
```

With the system in state-space, a variety of common time and frequency domain measures can be calculated. The first step in the calculation of a time response is to define the time base. In CTRL-C, this is done using the colon ":" operator. The command

```
[> t=0:.1:10;
```

creates a vector of points from 0.0 to 10.0 seconds in increments of 100 ms. Impulse and step responses are found by typing:

```
[> yi = impulse(a,b,c,1,t);
[> ys = step(a,b,c,d,1,t);
```

Matrices yi and ys now contain output time histories. The rows correspond the rows of the output vector, while the columns correspond to the successive time points from 0.0 to 10.0 seconds. The two responses are graphed by typing

```
[> plot(t,yi,t,ys)        which results in Figure 5.
```

Frequency response measures are calculated in a manner similar to time response functions. First, a frequency vector is formed. The function LOGSPACE is provided to create a vector with points evenly spaced in frequency between two decades. Typing

```
[> w=logspace(-1,1);
[> [mag,phas]=bode(a,b,c,d,1,w);
```

creates matrices mag and phas containing the magnitude and phase responses at the frequencies in vector w. This magnitude response is plotted on log -log scales and titled in the upper right corner of the screen with the commands

```
[> window('222')
[> plot(w,mag,'loglog')
[> title('magnitude')
```

Similar commands plot the phase and Nichols responses, resulting in Figure 6.

The pole placement formula of Ackerman allows arbitrary pole placement for single input systems. In CTRL-C the primitive PLACE is used on a vector P containing the desired pole locations, to calculate the gain vector K:

```
[> p = 3 * [-1; (-1 + i); (-1 - i)];
[> k = place(a,b,p)

K =
    6.6400    34.2160    52.0400
```

We can check the closed loop eigenvalues:

```
[> e = eig(a - b*k)

E =
    -3.0000 + 3.0000i
    -3.0000 - 3.0000i
    -3.0000 + 0.0000i
```

and indeed they are at the prescribed locations.

The reference feedforward matrix N is calculated to provide unity DC gain

```
[> n = 1/(d-(c-d*k)/(a-b*k)*b)

N =
    27.0000
```

The closed loop system matrices are built within CTRL-C:

```
[> Ap = a-b*k;   Bp = b*n;   Cp = c-d*k;   Dp = d*n;
```

and the closed loop impulse and step responses are found

```
[> yi = impulse(Ap,Bp,Cp,1,t);
[> ys = step(Ap,Bp,Cp,Dp,1,t);
[> plot(t,yi,t,ys)
```

which results in Figure 7.

Optimal Control Solutions

The standard linear quadratic regulator problem is solved with the CTRL-C function LQR. For the linear system described by:

$$\dot{x} = Ax + Bu$$

Typing [> k = lqr(a,b,q,r)
finds the gain matrix K such that the control law u = -Kx minimizes the quadratic cost function:

$$J = 1/2 \int [x'u'] \begin{bmatrix} Q & N \\ N' & R \end{bmatrix} \begin{bmatrix} x \\ u \end{bmatrix} dt$$

The LQR function is a good example of how optional arguments and calling sequences provide flexibility within CTRL-C. For example, the cross weighting term N often is not needed, so it is an optional input argument. It can be included as: K = LQR(A,B,Q,R,N). The Riccati solution matrix S is an optional output argument. It is obtained with [K,S] = LQR(A,B,Q,R).

A method is provided to switch between different algorithms. Typing LQR('qz') switches to the QZ algorithm. Typing LQR('qr') changes back to the default QR algorithm.

Primitives to solve the optimal estimator problem, the discrete time problem, and the implicit model following and output weighting formulations are also available.

In summary, two examples have been shown that introduce some of the CTRL-C control design and analysis primitives. The state-space representation of systems lends itself naturally to a matrix environment. CTRL-C provides primitives to convert to and from other representations, and to perform common analysis and design tasks. Many control design methodologies are possible using a combination of the matrix primitives and the control primitives. The matrix environment results in a very simple dialogue with the computer.

6.0 DIGITAL SIGNAL PROCESSING

Digital signal processing (DSP) is concerned with the representation and processing of signals that are represented by sequences of numbers. The purpose of processing a signal, in general, can be to identify some model or model parameters that characterize the signal. It can also be to enhance a signal or to remove undesirable components of the signal.

A matrix environment is ideal for the development and use of signal processing techniques. Vectors are used to represent arbitrary sampled-data signals. The natural mathematical interaction with vectors provided in a matrix environment makes it very convenient to process and manipulate sampled data sequences. Primitives for filtering, FFT analysis, identification, and other digital signal processing calculations become very conversational using the complex (Real + Imaginary) vector manipulation concepts.

Consider the implementation of a simple digital filter. The difference equation for a general causal linear time-invariant (LTI) digital filter is given by

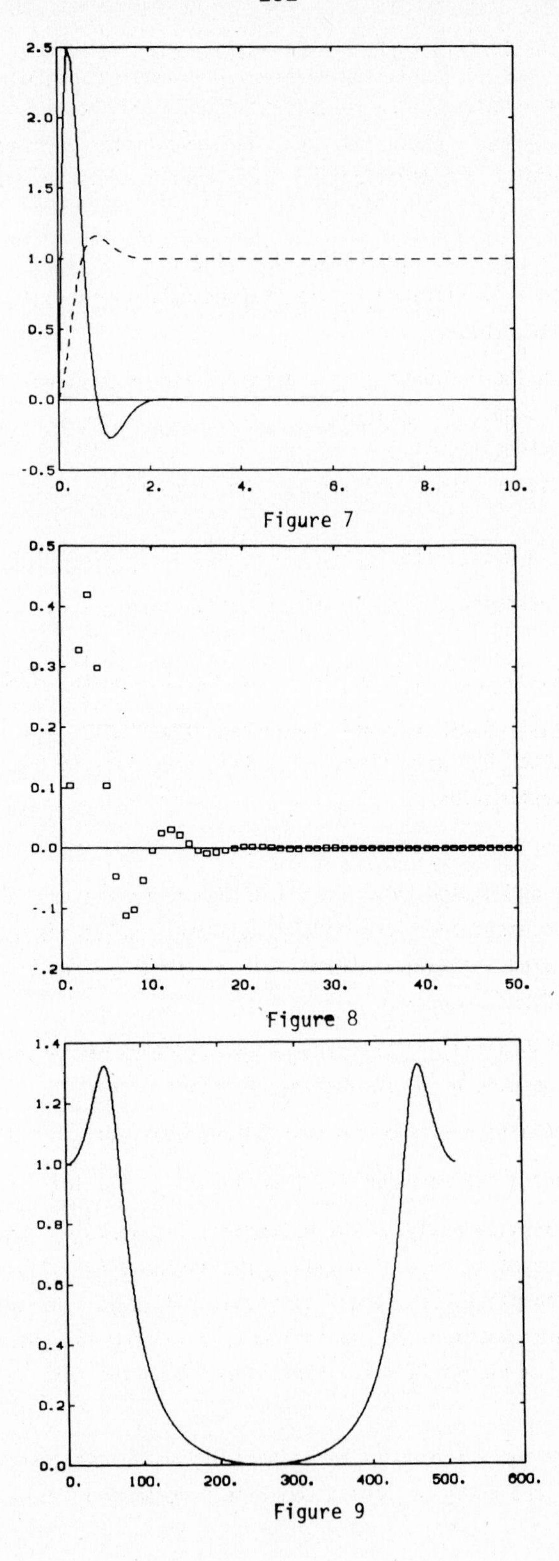

Figure 7

Figure 8

Figure 9

$$y(n) = b(1)*x(n) + b(2)*x(n-1) + \ldots + b(nb)*x(n-nb+1) - a(2)*y(n-1) - \ldots - a(na)*y(n-na+1)$$

where x is the input signal, y is the output signal, and the constants $b(i)$, $i=1,2,3,\ldots,nb$, $a(i)$, $i=1,2,3,\ldots,na$ are the filter coefficients. In CTRL-C, if the numerator and denominator filter coefficients are contained in vectors B and A,

```
B = [0.1042   0.2083   0.1042];  A = [1.0000   -1.1430   0.5596];
```

then a data sequence x is filtered with a "tapped delay-line" filter by typing

```
[> y = tdlf(x,a,b);
```

Suppose the impulse response of a digital filter is desired. An input vector representing a unit sample is created:

```
[> u = [1 0.*ones(1,511)];
```

which in this case is of length 512 points. The impulse response is found and graphed with

```
[> y = tdlf(u,a,b);  plot(y(1:40),'point=3')
```

producing Figure 8.

The frequency response of the filter is easily found using a fast Fourier transform (FFT):

```
[> yy = fft(y);   plot(abs(yy))
```

which results in Figure 9.

This simple filter example demonstrates the versatility of a matrix environment. In CTRL-C, other DSP primitives allow filter design, system identification, and power spectrum estimation.

7.0 EXTENSIBILITY

CTRL-C is most often used in a command driven mode; the user types single-line commands, CTRL-C processes them immediately, and the results are displayed. CTRL-C is also capable of executing sequences of commands that are grouped together to form a short "procedure".

In some other CAD packages, the word "Macro" is used to describe what is referred to as a Procedure or a User-Defined Function in CTRL-C.

There are three different types of procedures available in CTRL-C:

(1) DO Procedures; (2) User-Defined Functions; and (3) Text Macros

The first type operates globally on the workspace. It works by simply redirecting the input from the keyboard to a disk file. The second type allows the user define his own functions, complete with local and global variables, and argument passing. Once defined, the new functions are indistinguishable from the native CTRL-C primitives. The third is a simple facility for interpreting the text contained in a CTRL-C variable.

Together, these three procedure types form a powerful interpretive environment. Other syntax and commands form a complete programming language, similar in spirit to other popular interpretive languages. This notion of extensibility is one of the most powerful features of CTRL-C. Many applications can be developed directly in the CTRL-C language, without resorting to time-consuming "low-level languages" like Fortran.

8.0 ALGORITHMS

Careful attention has been paid to the selection of reliable and algorithms. Subroutines from EISPACK and LINPACK provide state-of-the-art algorithms for matrix analysis, decompositions, and eigenvalue problems.

The numerically stable staircase algorithm [4] is used to compute controllable (observable) and uncontrollable (unobservable) modes as well as the Kronecker (controllability) indices for a system. The method also provides an orthogonal matrix for the associated subspaces. The staircase algorithm can be used together with SVD to provide a reliable algorithm for minimal realization as well as complete canonical decomposition of a linear system.

The matrix pencil reduction algorithm [6] is used to compute the transmission and decoupling zeros of multivariable systems. This algorithm treats the most general case of the problem as it handles non-square and degenerate problems as well.

The Lyapunov and Riccati equations arise in many control and estimation problems. The most reliable and efficient algorithm to solve the Lyapunov equation is the method of Bartlet and Stewart [7]. A modification of this method solves the unsymmetric Lyapunov equations in an efficient manner. For the Riccati equation, the extended generalized eigenvalue approach [4], [10] is used. The Schur vector approach [5] is also available. These algorithms are the latest techniques for solving Riccati equations; they provide a good balance between numerical reliability and efficiency.

Various pole-placement algorithms exist in the literature. CTRL-C contains an algorithm for robust eigenstructure assignment [11]. The poles of a multivariable system can be assigned while the eigenvectors are selected using various strategies. One possibility is to find a set of eigenvectors which are as close to orthogonal as possible.

For frequency response and various singular value measures, CTRL-C uses the efficient algorithm based on the Hessenberg form [9]. This algorithm is numerically stable and avoids problems encountered when methods based on the Jordan structure are used.

The "squaring-down" algorithm [5] is used to compute discrete equivalents of continuous systems. This is a reliable algorithm and compares very well to the numerous other techniques of discretizing continuous systems.

The identification algorithms in CTRL-C include the Levenberg-Marquardt modification of Gauss-Newton method for maximum likelihood identification. Trankle [13] has suggested a further modification of the algorithm which makes it extremely efficient.

9.0 CONCLUSIONS

A unified software system is possible for matrix analysis, control system design, digital signal processing, and engineering graphics. A common thread in these disciplines is the role of a single data object: the complex matrix. The use of a matrix environment produces a powerful, natural, and extensible software system.

The concept of direct manipulation should be a design goal for the development of user interfaces for computer-aided control system design (CACSD) packages. The user must be able to apply intellect directly to the task; the tool should seem to disappear.

The identification of the basic tools required is a crucial step in the design of a CACSD package. A well-designed CACSD system has a minimum set of reliable baseline primitives, plus a mechanism for extensibility. Extensibility is a method whereby new primitives are constructed out of existing ones. It must be possible to use the new primitives as if they were baseline primitives.

REFERENCES

1. Moler, C. and C. Van Loan, Nineteen Dubious Ways to Compute the Exponential of a Matrix, SIAM REVIEW, 20, 4, 1978.
2. Thomas, R. and Yates, J., "A User Guide to the UNIX System", Osborne/McGraw-Hill, Berkeley, CA, 1982.
3. Shneiderman, B. "Direct Manipulation: A Step Beyond Programming Languages", IEEE Computer Magazine, August 1983.
4. Emami-Naeini, A., "Application of the Generalized Eigenstructure Problem to Multivariable Systems and the Robust Servomechanism for a Plant which Contains an Implicit Internal Model," Ph.D. Dissertation, Dept. of Electrical Engineering, Stanford University, April 1981.
5. Franklin, G.F. and J.D. Powell, Digital Control of Dynamic Systems, Addison-Wesley, 1980.
6. Emami-Naeini, A. and P. Van Dooren, "Computation of Zeros of Linear Multivariable Systems," Automatica, Vol. 18, No. 4, pp. 415-430, July 1982.
7. Bartels, R.H. and G.W. Stewart, "A Solution of the Equation AX+XB=C", Commun. ACM, Vol. 15, pp. 820-826, 1972.
8. Laub, A.,J., "A Shur Method for Solving Algebraic Riccati Equations", Laboratory for Information and Decision Systems Report 859, MIT, October 1978.
9. Laub, A.J., "Efficient Multivariable Frequency Response Computations", IEEE Transactions on Automatic Control, Vol. AC-26, No. 2, April 1981.
10. Van Dooren, P., "A Generalized Eigenvalue Approach for Solving Riccati Equations," SIAM J. Sci. Stat. Comput., Vol. 2, pp. 121-135, 1981.
11. Kautsky, J. et al., "Numerical Methods for Roust Eigenstructure Assignment in Control System Design," in Proc. Workshop on Numerical Treatment of Inverse Problem..., Heidelberg, 1982.
12. C.B. Moler, MATLAB User's Guide, University of New Mexico, Computer Science Department, 1981.
13. Trankle, T.L., Vincent, J.H., Franklin, S.N., "Systems Identification of Nonlinear Aerodynamic Models", AGARDOgraph No. 256, Advances in the Techniques and Technology of the Application of Nonlinear Filters and Kalman Filters, 1982

SIRENA : UN OUTIL DE CAO POUR L'AUTOMATIQUE

Y.YEM, K.CHOUMLIVONG, A.BARRAUD
Société RSI, chemin du Pré Carré
ZIRST,38240 MEYLAN - FRANCE

ABSTRACT

The first commercial system of CAD in this field, SIRENA gives the user access to advanced techniques in simulation, signal analysis and model synthesis. SIRENA has been built to help all those who are involved in the conception, development and realisation of systems where the control of the dynamic evolution of the process is of prime importance. SIRENA belongs to a new generation of simulators used by design and development departments, which requires no previous knowledge of programming nor any computer language. It is operated by an auto-guided dialogue which allows the operator to access calculating methods and sophisticated treatments.

RESUME

Premier système à caractère industriel de CAO dans son domaine, SIRENA rend accessible dès aujourd'hui, les techniques de demain en matière de simulation, d'analyse de signaux et de synthèse de modèles. SIRENA s'adresse à tous ceux qui ont à concevoir, développer et réaliser des systèmes ou procédés dont la maitrise et la prévision de l'évolution dynamique sont primordiales. SIRENA constitue une nouvelle génération de simulateurs de conception à l'usage des bureaux d'études, manipulable sans aucune connaissance informatique, ne nécessitant l'apprentissage d'aucun langage, tout en donnant l'accès, à travers un dialogue (auto) guidé, à des outils de calculs et traitements graphiques extrêmement sophistiqués.

INTRODUCTION

Le système SIRENA dont on présente la version Industrielle est issu d'une collaboration universitaire faisant intervenir l'INSA de Rennes, l'Ecole Supérieure d'Electricité (Rennes) et l'Ecole Nationale Supérieure de Mécanique de Nantes [8].

La version Industrielle de SIRENA découle plus directement du prototype développé à l'INSA de Rennes [6]. Elle reprend dans son esprit les fonctionnalités répondant aux besoins immédiats du CNET de Grenoble-Meylan qui en a été le premier demandeur.

Dans cette communication nous développerons sucessivement les fonctionnalités du système SIRENA vues de l'utilisateur , nous aborderons la description technique de ce logiciel avant d'en aborder les caractéristiques informatiques. Nous présenterons ensuite quelques exemples illustrant certaines des fonctionnalités de SIRENA. Dans la conclusion nous dégagerons les perspectives et les développements envisagés pour ce système.

I. FONCTIONNALITES DU SYSTEME SIRENA

SIRENA constitue une nouvelle génération de simulateurs de conception à l'usage des bureaux d'études tant en milieu industriel qu'en milieu universitaire et d'enseignement.

L'usage de SIRENA se caractérise par une manipulation sans aucune connaissance informatique et ne nécessite l'apprentissage d'aucun langage. Il donne ainsi accès, à travers un dialogue (auto) guidé, à des outils de calculs et de traitements graphiques extrêmement sophistiqués [9].

SIRENA permet actuellement la description, l'analyse, la simulation de systèmes dynamiques continus, discrets et/ou échantillonnés. La description fonctionnelle des systèmes s'effectue sous forme de blocs diagrammes. Chacune des entitées définies (entrée, sortie, transfert, bloc non linéaire, paramètre formel...) est reconnue par son nom. Le dialogue utilisateur-calculateur reste aussi proche que possible de la démarche naturelle, que ce soit du point de vue vocabulaire que de l'enchainement des directives qui ont été regroupées en sept grandes classes dans le tableau 1.

Une des possibilités du système est de pouvoir construire et gérer des systèmes de complexité plus ou moins arbitraires à concurrence de saturation de la base de données. Les figures qui suivent illustrent les possibilités élémentaires du système:

Exemples de description de systèmes sous SIRENA :

⇒DECLARATION-SYSTEME
TYPE: CONTINU
NOM: S1

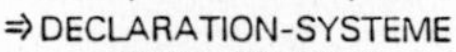

ENTREE (S) : U
SORTIE (S) : Y

Y = G * U

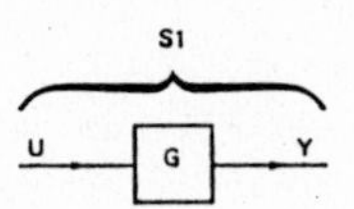

⇒DECLARATION-SYSTEME
TYPE : CONTINU
NOM : S2

ENTREE (S) : EC
SORTIE (S) : CC

CC = INT * VIT
VIT = FNL (XC) FNL est un bloc non linéaire
XC = HC * UC
UC = KC * EC

⇒ASSOCIATION-SYSTEME
TYPE: CONTINU
NOM: S
SYST. ASSOCIE (S) : S1, S2
ENTREE (S) : RC
SORTIE (S) : CC

U = CC
EC = RC - Y

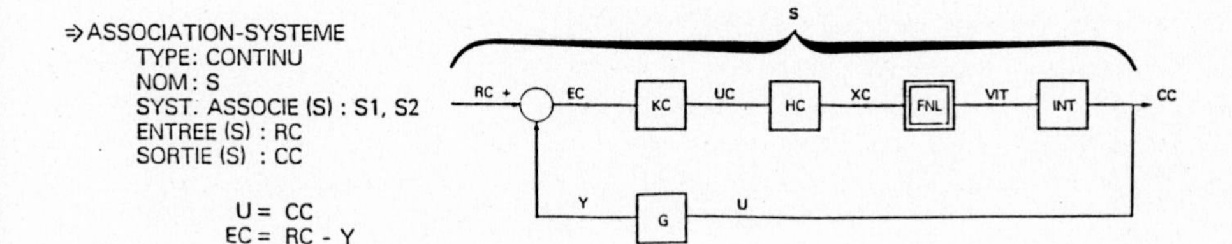

Au niveau du dialogue il est à remarquer que l'utilisateur peut abréger arbitrairement les directives tant qu'elles restent non ambigues et anticiper sur les questions du système quand il en connait à l'avance le déroulement. Par exemple, l'utilisateur déclare la commande suivante :

=> DEC C S2 EC CC

Dans ces conditions le système lui redonne en clair l'ensemble des questions et réponses qui ont été enregistrées à savoir ici :

TYPE : CONTINU
NOM : S2
ENTREE(S) : EC
SORTIE(S) : CC
(etc...)

Une fois un ou plusieurs systèmes déclarés, l'utilisateur est à même de réaliser un ensemble de simulations. En standard, chaque simulation donne lieu à une visualisation des signaux simulés. Cette visualisation peut être soit numérique (tableaux de valeurs) soit graphique. Dans la seconde hypothèse SIRENA adopte une présentation standard des résultats qui peut par la suite être modifiée à la requête de l'utilisateur au moyen d'un éditeur graphique intégré au système. Cet éditeur se présente également comme un ensemble de directives que l'on a regroupé en quatre classes dans le tableau 2.

Après toute simulation l'utilisateur a le choix de conserver ou non sous forme de fichier les résultats numériques correspondants. Ces fichiers sont reconnus dans ou à l'extérieur de SIRENA par le nom qui leur a été affecté automatiquement par le système. Pour réaliser une simulation, l'utilisateur dispose d'un ensemble type de signaux pré-définis (impulsion, échelon, rampe, créneau, sinus, bruit uniforme, bruit gaussien, PRBS, PRTS) ainsi qu'une possibilité d'accès à des entrées quelconques définies par points dans un fichier. Les caractéristiques dynamiques d'un système à simuler peuvent être décrites par des fonctions de transfert, des modèles d'états, des non linéarités statiques (relais, saturation, retard, seuil, courbe points par points) ainsi que des fonctions du type FORTRAN.

Il est important de souligner la possibilité de paramètrer formellement tous ces éléments. Par exemple :

$$H(s) = \frac{G}{(1+10s)\ \{[(R+jI) - s]\ [(R-jI) - s]\}^{K}}$$

où G, R, I et K sont des paramètres formels dont on demandera la ou l'ensemble de valeurs avant de réaliser une simulation numérique. Exemple :

G = 10
K = 1;2;3
etc...

II. DESCRIPTION TECHNIQUE

SIRENA est organisée autour d'une base de donnée, d'un dialogue et d'un ensemble de programmes numériques permettant de réaliser simulations, conceptions et tracés graphiques.

II.1. LA BASE DE DONNEES

La STRUCTURE DE DONNEES est constituée de 2 types de blocs :
- des blocs dits standards de taille fixe
- des blocs de dimension variable.

a) Les blocs standards comprennent :
- bloc SYSTEME
- bloc VARIABLE dont on distingue :
  - . variable explicite
  - . variable implicite
- bloc OPERATEUR : RELATION FONCTION DE TRANSFERT, on distingue :
  - . FONCTION DE TRANSFERT explicite
  - . FONCTION DE TRANSFERT implicite
- bloc OPERATEUR : MATRICE, dont on distingue :
  - . MATRICE explicite
  - . MATRICE implicite
- bloc FONCTION
- bloc PARAMETRE
- blocs constituant :
  - . la table SYSTEME
  - . la table SOUS-SYSTEME
  - . la table des VARIABLES fondamentales telles Table :
    ENTREE, SORTIE, ETAT.

b) Les blocs de taille variable sont exclusivement constitués par les tables DONNEES NUMERIQUES.

II.1.1. DEFINITION DES DIFFERENTS BLOCS STANDARDS

Chaque bloc élémentaire (standard) sera constitué de :
- 6 caractères : sémantique du bloc (NOM, TYPE, ...).

  Remarque :

  NOM est une chaine de caractères alphanumériques majuscules (de longueur < 4) dont le premier caractère est une lettre alphanumérique.

- 4 entiers : liens entre les blocs.

Le caractère étant une entité FORTRAN à part entière, un bloc "élémentaire" pour mini 16 et 32 bits aura la représentation suivante:

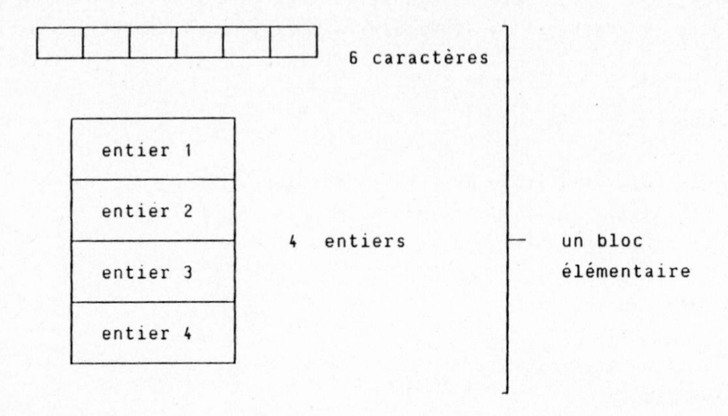

II.1.2. DEFINITION DES TABLES

a) Table SYSTEME

C'est une suite de blocs standards chaînes ; chaque fois que l'on déclare un nouveau système, on crée un nouvel élément (bloc standard) que l'on insère dans la table. Chaque élément de la table a les caractéristiques suivantes :

PTSYST
NB
PTRST
PTSV

PTSYST : Pointeur vers le bloc SYSTEME
NB : Nombre de blocs alloués pour ce système
PTRST : Pointeur vers le bloc PARAMETRE
PTSV : Pointeur vers l'élément suivant de la Table SYSTEME.

b) Table DONNEES NUMERIQUES

Cette table sera créée soit :

- lors de la définition des valeurs numériques des transmittances, des matrices,
- lors de la définition des fonctions non linéaires déclarées,
- et plus généralement, lors de l'affectation de valeurs numériques aux paramètres formels.

Elle est obtenue à partir d'une allocation dynamique de la zone mémoire réservée aux données numériques.

Table DONNEES NUMERIQUES

Pointe vers Table valeur numérique suivante
Taille de la table
. . . .

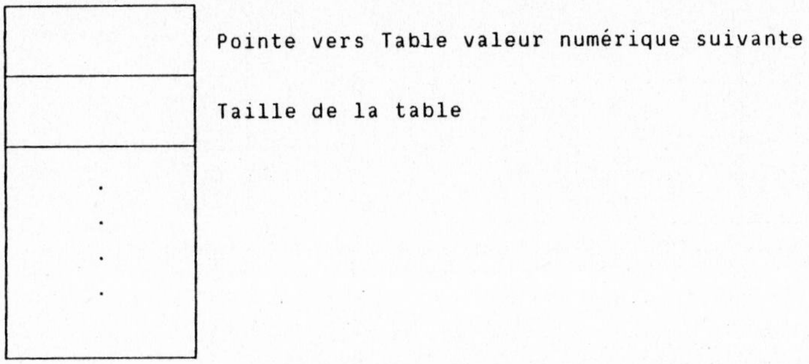

II.1.3. ORGANISATION DE LA STRUCTURE DE DONNEES

La STRUCTURE DE DONNEES est constituée de deux zones dont la frontière peut varier dynamiquement :

- zone réservée pour les blocs de taille fixe :
  . bloc SYSTEME, bloc VARIABLE
  . bloc OPERATEUR : Fonction de transfert
  . bloc OPERATEUR : Matrice
  . et blocs constituant la table SYSTEME, SOUS-SYSTEME, ENTREE, SORTIE, ETAT...

- zone réservée pour les blocs de taille variable (Table DONNEES NUMERIQUES).

a) Représentation de la STRUCTURE DE DONNEES

C'est :

. un tableau entier ISTR (TBLOC, NSTR) correspondant au tableau des pointeurs avec :

TBLOC = 4
NSTR = taille maximum de la Structure de Données

. un tableau de chaines de caractères CHSTR*6 (NBLOC) correspondant au tableau des NOMS, TYPES, OPERATEURS avec :

NBLOC = nombre maximum de blocs standards.

En somme, le Iième bloc sera donc représenté par :

ISTR (TBLOC,I)
+
CHSTR (I)

I S T R

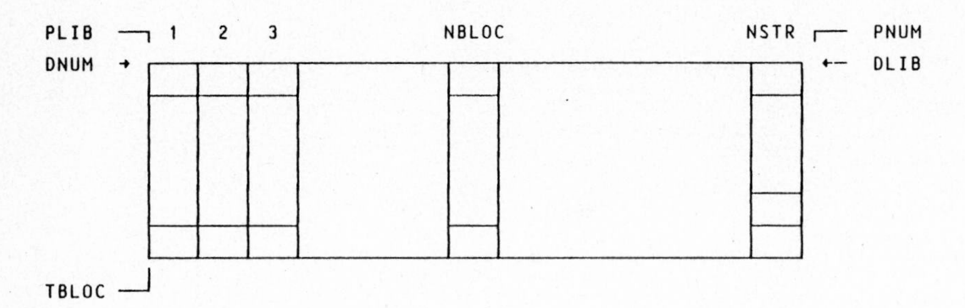

Etat initial de la Structure de Données

PLIB : Pointe vers le premier bloc libre de la zone des blocs standards

PNUM : Pointe vers le premier bloc libre de la zone DONNEES NUMERIQUES

DLIB, DNUM : Pointe vers les frontières entre la zone des blocs standards et la zone de Données Numériques.

DLIB et DNUM (2 pointeurs) varient dynamiquement au cours de l'éxécution tout en respectant DNUM < DLIB.

II.2. LE DIALOGUE

Le dialogue utilisateur calculateur s'effectue sous une forme de questions réponses, respectant la démarche naturelle de l'utilisateur, ne demandant du point de vue vocabulaire aucune notion d'Informatique. SIRENA a, pour cette phase fondamentale d'utilisation, privilégié l'aspect "outil-intéractif", en donnant accès à tout instant, au cours du dialogue, à un guide d'utilisation.

L'utilisateur a ainsi a sa disposition la liste des réponses possibles lui permettant de progresser dans le travail en cours. Nous avons exposé dans un paragraphe précédant les facilités accordées pour anticiper les commandes lorsque l'enchainement d'une phase de dialogue est parfaitement connu de l'utilisateur.

Exemple de dialogue guidé :

```
=> DEF-PARAM
   TYPE : C
   NOM  : S
   TYPE : CONTINU
   NOM  : S

   PARAMETRE-SYSTEME : U
   NATURE : ?
```

*** LISTE DES NATURES REPERTORIEES ***

. CRENEAU
. ECHELON
. IMPULSION
. POINT-PAR-POINT
. PRBS
. PRTS
. RAMPE
. SINUSOIDE
. BRUIT-UNIFORME

NATURE : ...

II.3. LES ALGORITHMES DE SIMULATION

On distinguera successivement les systèmes discrets, continus, échantillonnés.

Le cas des systèmes discrets ne présente aucune difficulté numérique spécifique, qu'il soit décrit sous forme de fonction de transfert ou d'états. En effet leurs calculs se résument à des produits scalaires pour lesquels ils existent des routines parfaitement rodées [5].

Quand aux non linéarités, elles se résument à des expressions FORTRAN élémentaires. Les systèmes continus, en l'absence de non linéarités sont intégrés à pas constant en calculant un système discret d'états équivalent à travers la formulation suivante:
Si le système est décrit par une fonction de transfert H (s), celle-ci est convertie en une forme d'état continue :

$$\underline{x} = A\underline{x} + B\underline{u}$$
$$\underline{y} = C\underline{x}$$

Le système d'état continu étant connu, on calcule le système discret équivalent :

$$\underline{x}_{t+1} = \Phi\, \underline{x}_t + \Gamma\, \underline{u}_t$$
$$\underline{y}_t = C\underline{x}_t$$

avec :

$$\begin{bmatrix} \Phi & \Gamma \\ 0 & 1 \end{bmatrix} = \exp \begin{bmatrix} A & B \\ 0 & 0 \end{bmatrix} h$$

où, h est le pas d'échantillonnage imposé par l'utilisateur ou calculé automatiquement par le système suivant des critères de stabilité numérique et de la dynamique du système connu à travers le spectre A. L'exponentielle est elle-même calculé par un approximant de Padé d'ordre approprié après mise à l'echelle (en base 2) de A. Dans l'hypothèse où $\|B\| > \|A\|$, B est également mise à l'echelle (en base 2), Γ étant corrigé en conséquence [1]. En présence de non linéarités, le système est intégré par la méthode de Runge Kutta Feldberg (RKF45) qui sélectionne automatiquement le pas en combinant une évaluation d'ordre 4 et 5 et en tenant compte de la précision requise et géré automatiquement par SIRENA en fonction de la précision de la machine hôte [2]. Finalement les systèmes échantillonnés exploitent ces deux techniques de simulation en dissoçiant le pas d'intégration numérique des parties continues de la période (ou des périodes) des échantillonneurs - bloqueurs présents dans la configuration du système à simuler. SIRENA étant à même de simuler des systèmes de grande complexité (bouclages multiples, etc...) les simulations numériques opèrent dans les faits sur une structure simplifiée obtenue par une ou plusieurs réduction formelle des parties linéaires. En l'absence de non linéarités la structure globale décrite, quelle qu'en soit la complexité, est réduite à un seul bloc linéaire.

II.4. LES PROGRAMMES D'APPLICATIONS

Le système SIRENA peut intégrer n'importe quel type de programmes d'applications moyennant une interface appropriée. Ceci confère au système tout son caractère de souplesse et d'ouverture aux évolutions futures. Actuellement SIRENA intègre un certain nombre de modules de conception qui sont notamment :

- un logiciel de synthèse de filtres numériques [3]
- un lieu des pôles
- une procédure d'identification dans le domaine fréquentiel decrit par gain et phase
- une procédure d'identification dans le domaine temporel basée sur la méthode du modèle.

Les procédures d'identifications opèrent par minimisation d'un critère réalisée par programmation non linéaire. Leur mise en oeuvre dans SIRENA est du type quasi-Newton BFGS [7] avec mise à jour sous forme factorisée d'une approximation du Hessien. L'initialisation de la procédure itérative peut-être laissée aux soins de l'utilisateur ou réalisée automatiquement par SIRENA. Dans ce dernier cas le jeu initial de paramètre est obtenu par minimisation d'une erreur d'équation conduisant à un problème de moindres carrés classiques. Celui-ci est obtenu par factorisation orthogonale [4], dans une version séquentielle du type "Information Square Root Filter". Son application à une approche moindres carrés étendus testée hors SIRENA est en cours d'intégration. C'est, semble-t-il, une démarche originale, numériquement stable, conduisant <u>directement</u>, au niveau identification, à la forme globalement stable de l'algorithme [10]. La formulation classique:

$$\underline{z}_t = [u_t,\dots,u_{t-m},-y_{t-1},\dots,-y_{t-n},\tilde{\varepsilon}_{t-1},\dots,\tilde{\varepsilon}_{t-p}]$$

$$\varepsilon^{*}_{t} = y_t - \underline{\theta}^{T}_{t-1}.\underline{z}_t \quad ; \quad \tilde{\varepsilon}_t = y_t - \underline{\theta}^{T}_{t}.\underline{z}_t$$

$$\underline{\theta}_t = \underline{\theta}_{t-1} + M_{t-1}\underline{z}_t(\underline{z}_t M_{t-1}\underline{z}_t + 1)^{-1}\varepsilon^{*}_{t} \quad ; \quad M_{t+1} = M_t - M_t\underline{z}_{t+1}(\underline{z}_{t+1}M_t\underline{z}_{t+1} + 1)^{-1}\underline{z}_{t+1}M_t$$

a été remplacée par l'algorithme:

$$W_t = \left[\begin{array}{c|c} U_t & \underline{v}_t \\ \hline 0\ldots0 & e_t \\ \hline \underline{z}^{T}_{t} & y_t \end{array}\right] \quad ; \quad S_t = Q.W_{t+1} = \left[\begin{array}{c|c} U_{t+1} & \underline{v}_{t+1} \\ \hline 0\ldots0 & e_{t+1} \\ \hline 0\ldots0 & 0 \end{array}\right]$$

$$\tilde{\varepsilon}_t = y_t - \underline{\theta}^{T}_{t}.\underline{z}_t \quad ; \quad \underline{\theta}_t = U^{-1}_{t}\underline{v}_t$$

où Q est une matrice orthogonale implicite construite par rotaions de Givens de façon à annuler la dernière ligne de W_t. A l'instant suivant W_{t+1} est obtenu à partir de S_t en remplaçant la dernière ligne par $\underline{z}^{T}_{t+1}$ et y_{t+1}. Les conditions initiales sont simplement W = 0. Le tout occupe (n+1).(n+4)/2 réels la matrice U étant triangulaire supèrieure. On a noté n le nombre de paramètres à estimer.

II.5. LE GRAPHIQUE

En standard, toute simulation ou tout programme d'application active l'éditeur graphique intégré du système SIRENA. Dans une phase initiale, l'évolution de toutes les variables accessibles dans le système simulé, est visualisée d'une manière automatique. L'utilisateur a ensuite accès au logiciel graphique proprement dit, lui permettant une mise en page intéractive des tracés de courbes. Les directives graphiques dont les fonctionnalités sont exposées dans le tableau 2 permettent notamment, de choisir le cadre pour son tracé, de plusieurs courbes ou de tracer l'évolution de variables en fonction d'autres variables. D'autres directives permettent de spécifier les graduations sur les axes (Linéaire, Logarithmique, Degrés), de choisir son mode de tracé (en ligne brisée, en escalier), de repérer les coordonnées d'un point, de choisir une fenêtre de tracé (fonction loupe), etc... La syntaxe des directives graphiques est homogène à celle adaptée pour les directives du système SIRENA notamment en ce qui concerne les facilités de dialogue (abréviation, anticipation) ainsi qu'en ce qui concerne le guide d'utilisation intégré.

II.6. CARACTERISTIQUES INFORMATIQUES

SIRENA est un progiciel écrit en FORTRAN norme ANSI X3.9-1978 (Fortran 77). La conformité au standard garantit sa portabilité sur tout calculateur supportant cette norme. SIRENA est destiné a priori aux machines à mots de longueur au moins égale à 32 bits mais est implantable sur tout mini-calculateur 16 bits ayant des possibilités suffisantes d'overlay ou d'adressage étendu. La version actuelle de SIRENA représente approximativement 50 000 lignes de code FORTRAN décomposé environ en 700 modules. L'implémentation de SIRENA sur une nouvelle machine a été grandement facilitée par un regroupement dans quelques modules fonctionnels caractéristiques des machines hôtes qui se résument essentiellement à trois points :

- dépendances du système d'exploitation (nom de fichiers, etc...)
- spécification des variables de type REAL et INTEGER (overflow, underflow, précision machine, etc...)
- paramètres spécifiques au support graphique ainsi que les routines de bases (adressage, tracé élémentaire, etc...).

Le système SIRENA est actuellement opérationnel sur les calculateurs suivants:

NORSK DATA	ND 100	(16 bits)
HEWLETT PACKARD	HP 1000	(16 bits)
NORSK DATA	ND 500	(32 bits)
DEC	Série VAX	(32 bits)
CDC	Série CYBER 170	(60 bits)
UNIVAC	Série 1100	(36 bits)

et sera prochainement disponible sur:

APOLLO
MICROMEGA
BFM

CONCLUSIONS

Nous avons, dans cette communication, présenté le système SIRENA qui est un outil logiciel destiné à l'analyse et à la simulation de systèmes dynamiques. L'analyse des besoins des utilisateurs, les orientations choisies pour la réalisation de cet outil en font un système ouvert aux évolutions futures, aussi bien sur le plan des fonctionnalités que sur le plan des techniques numériques. Le développement de SIRENA se poursuit dans les secteurs suivants :

- description et validation de systèmes logiques et d'automates
- interconnexion de systèmes logiques et de systèmes dynamiques
- l'exemple significatif étant la description d'un procédé et de ses automates de commande (Robotique, Productique,...)
- extension des programmes d'applications en Traitement du signal (par exemple algorithmes de calculs à nombre de digits fixés pour la simulation d'implantation sur microprocesseurs,...)
- intégration de logiciels d'identification notamment identification multivariable
- intégration de logiciels de synthèse de commande...

Les domaines d'applications de SIRENA sont aussi nombreux que diversifiés, tels l'électronique, les télécommunications, les procédés énergétiques de toute nature, l'avionique, les automatismes complexes, etc..., et plus généralement tout système pouvant être décrit sous forme de blocs linéaires et non linéaires incluant à terme les systèmes combinatoires et les automates.

EXEMPLE D'UTILISATION

REPONSE TEMPORELLE D'UN SYSTEME CONTINU NON LINEAIRE

```
=> DECL-SYSTEME CONT S1 E POS

   TYPE : CONTINU
   NOM  : S1
   ENTREE(S) : E
   SORTIE(S) : POS

         POS = INT*VIT
         VIT = MOT*X
         X   = REL(U)
         U   = E - POS - TAC*VIT
```

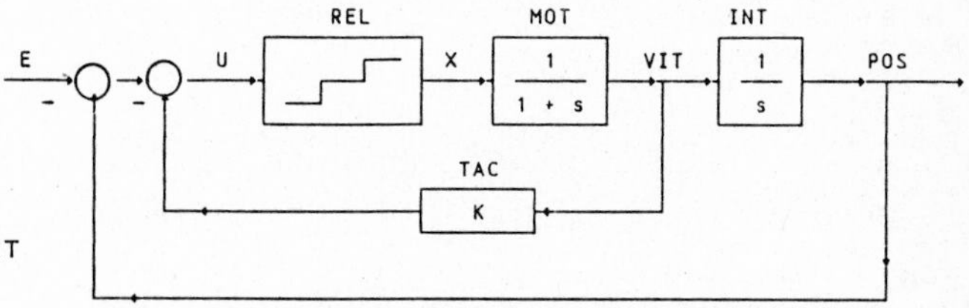

```
=> DEF-PAR
   TYPE : CONTINU
   NOM  : S1

   PARAMETRE-SYSTEME : E
   NATURE     : ECH
   RETARD     : 0
   AMPLITUDE  : 1

   PARAMETRE-SYSTEME : REL
   NATURE     : RELAIS
   ABSCISSE   : 0.1
   ORDONNEE   : 1

   PARAMETRE-SYSTEME : INT
   NATURE     : TR
   COEF-GAIN  : 1
   FORME-NUMERATEUR    (POLYNOME/RACINE) : POL
   COEFFICIENT(S) PAR ORDRE DE PUISSANCE  DECROISSANTE  :
             : 1
   FORME-DENOMINATEUR  (POLYNOME/RACINE) : POL
   COEFFICIENT(S) PAR ORDRE DE PUISSANCE  DECROISSANTE  :
             : 1;0

   PARAMETRE-SYSTEME : MOT
   NATURE     : TR
   COEF-GAIN  : 1
   FORME-NUMERATEUR    (POLYNOME/RACINE) : POL
   COEFFICIENT(S) PAR ORDRE DE PUISSANCE  DECROISSANTE  :
             : 1
   FORME-DENOMINATEUR  (POLYNOME/RACINE) : POL
   COEFFICIENT(S) PAR ORDRE DE PUISSANCE  DECROISSANTE  :
             : 1;1

   PARAMETRE-SYSTEME : TAC
   NATURE     : GAIN
   COEF-GAIN  : K

   PARAMETRE-SYSTEME : £
```

```
=) REPONSE-TEMP

   TYPE : CONTINU
   NOM  : S1
   ENTREE : E
   SORTIE : POS

)) DEFINITION PARAMETRES-FORMELS
   PARAMETRE-SYSTEME : TAC
            K = 0;0.1

)) REPONSE TEMPORELLE
   DOMAINE TEMPOREL (T0;TF) : 0;10
   NOMBRE DE POINTS A VISUALISER : 1001
   VARIABLE(S) SUPPLEMENTAIRE(S) A VISUALISER : VIT

)) CALCUL DE LA REPONSE POUR
                  . K  = 0.0
                  . T0 = 0.0 ; TF = 10.0
                  . PAS D'ECHANTILLONAGE = 0.01
                  . NOMBRE DE POINTS =1001
)) CALCUL DE LA REPONSE POUR
                  . K  = 0.1
                  . T0 = 0.0 ; TF = 10.0
                  . PAS D'ECHANTILLONAGE = 0.01
                  . NOMBRE DE POINTS =1001
)) VISUALISATION ( OUI/NON ) ? O
   NUMERIQUE OU GRAPHIQUE ( NUM/GRA ) ? GRA
```

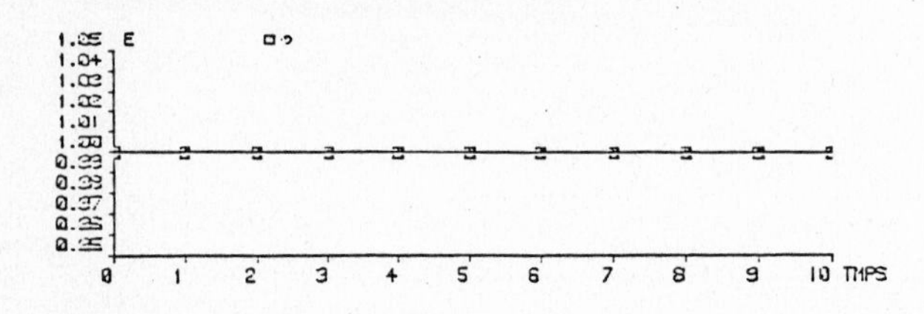

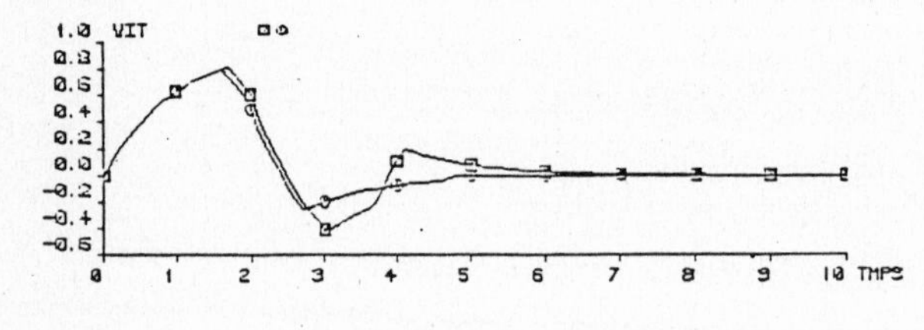

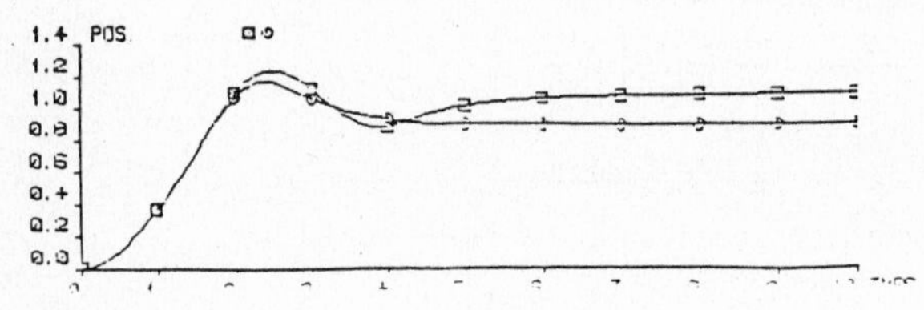

TYPE DE DIRECTIVE	FONCTION
Directives de déclaration	* Déclaration d'un système. * Association de systèmes. * Modification d'un système. * Modification d'un nom. * Modification de paramètres. * Destruction d'un système. * Destruction de tous les systèmes de l'utilisateur.
Directive de définition	* Définition des paramètres "système" (transmittance, matrice, fonction non-linéaire, entrée, sortie).
Directive de calcul formel	* Relation formelle entre deux variables d'un système.
Directives de calcul numérique	* Réponse fréquentielle. * Réponse temporelle. * Modélisation. * Synthèse de filtres. * Lieu d'Evans.
Directives de sauvegarde	* Liste des sauvegardes. * Sauvegarde de la structure de données. * Rappel d'une sauvegarde.
Directives d'aide	* Liste des directives. * Liste des systèmes déclarés. * Liste des équations d'un sytème. * Liste des valeurs numériques des paramètres systèmes.
Directive d'arrêt	* Arrêt du système SIRENA.

<u>Tableau 1 : Classification des directives</u>

TYPE	FONCTIONS
Directives d'information	* Noms des variables visualisables. * Noms et valeurs des paramètres formels. * Choix des itérations de paramètres formels.
Directives de configuration	* Choix du nombre de cadres par écran. * Choix de représentation. * Choix du tracé.
Directives de visualisation	* Visualisation graphique. * Choix d'une fenêtre (loupe). * Demande de coordonnées d'un point. * Positionnement d'un symbole. * Ecriture de commentaire. * Tracé d'horizontales. * Tracé de verticales.
Directives de gestion d'écran	* Effacement d'écran. * Copie.

Tableau 2 : Directives graphiques

BIBLIOGRAPHIE

[1] A. BARRAUD : "More on the conversion problem of discrete - continuous models" Appl. Math. Modelling, 1981, vol 5, December 1981. pp 414- 416.
[2] L. F. SHAMPINE H.A. WATTS S. DAVENPORT : "Solving non-stiff ordinary differential equation - The state of the art " Sandia Laboratories report SAND75-0182, 1975.
[3] G.F. DEHNER : "Program for the design of recursive digital filters" In-Programs for digital signal processing, IEEE PRESS, 1979.
[4] C.L. LAWSON, R.J. HANSON : "Solving least squares problems". Prentice Hall 1974.
[5] C.L. LAWSON, R.J. HANSON, D.R. KINCAID, F.T. KROGH : "Basic Linear Algebra Subprograms for FORTRAN usage". SAND 77 - 0898, Sandia Laboratories U.S.A.
[6] J.P. LE BARON : "Conception assistée en Automatique" Thèse Doct. Ing. - Rennes, Juin 1979.
[7] M.J.D POWELL : "Some global convergence properties of a variable metric algorithm for minimization without exact line searches". C.S.S. 15 Report AERE Harwell 1975.
[8] R.S.I. : "Réalisation d'une version Industrielle de l'outil logiciel SIRENA" Rapport de fin de contrat DAII/CNET, Juin 1983.
[9] R.S.I. : "SIRENA : manuel d'utilisation" Septembre 1983.
[10] C. DONCARLI Ph. de LARMINAT : "Analyse de la stabilité globale d'un algorithme d'identification récursive des systèmes linéaires stochastiques discrets". RAIRO Automatique, vol. 12, n° 3, 1978,pp. 269-276.

CASAD - AN INTERACTIVE PACKAGE FOR COMPUTER AIDED SYSTEM ANALYSIS AND DESIGN

A. Davidoviciu, A. Varga

Central Institute for Management
and Informatics
Bd. Miciurin, No.8-10, 71316 Bucharest
Romania, Telex: 11891

ABSTRACT

This paper presents an interactive package - CASAD, for the Computer Aided Systems Analysis and Design. CASAD implements a complete methodology for CAD of linear multivariable control systems by state-space methods. The main steps of this methodology are: (1) plant modeling; (2)analysis of system properties; (3) design of robust compensators; and (4) simulation of control configurations. CASAD is based on two powerful, portable Fortran subroutines packages BIMAS and BIMASC, which implement the latest advances in numerical algorithms using the highest quality available numerical software. CASAD offers many new facilities, compared with the existing similar packages. All functions are performed by the means of a command language. A flexible data organization allows an easy communication among the various CASAD programs. The CASAD package is implemented on the romanian family of minicomputers I-100, 102F, CORAL-4011, 4030, compatible with the DEC PDP-11 systems.

1. INTRODUCTION

In the last few years,major developments have been achieved in developing efficient and reliable algorithms for most computational problems of modern multivariable control theory [1] - [8]. The existence of high performance linear algebra packages LINPACK [9] and EISPACK [10], [11] contributed decisively to reliable computer implementations of these algorithms. Recently, two powerful packages of portable Fortran subprograms, BIMAS [12] and BIMASC [13], have been developed for the computer aided control system design (CACSD). BIMAS is destined to solve the basic mathematical problems in CACSD. BIMASC extends BIMAS with subprograms which solve specific problems of modern multivariable control theory. BIMAS and BIMASC implement the latest advances in numerical algorithms using the highest quality available numerical software. These packages represent powerful software tools which cover most of the computational problems in CACSD. For a thorough discussion of this theme see [14].

CASAD is an interactive Computer Aided System Analysis and Design package based on BIMAS and BIMASC packages. CASAD implements a complete methodology for CAD of linear multivariable systems using state-space methods. The main steps of this methodology are: (1) plant modeling; (2) system analysis; (3) compensator design; and (4) performance evaluation.

The purpose of modeling is to determine the appropriate model to be used by the analysis and design methods. Some of the model transformations implemented in CASAD are: computation of minimal and non-minimal state-space realizations of transfer matrices, discretization of continuous models, similarity transformations, balancing, model reduction, transfer matrix evaluation.

The analysis of model properties permits to determine the existence of the solution of certain design problems. This step involves the analysis of open-loop system stability, its controllability-stabilizability and observability-detectability properties.

The computation of poles is used to determine the system stability margin, while zeros computation is useful in determining the existence of solution of regulation problem. The simulation of open-loop systems may serve in many cases as a powerful tool for analysis. It can reveal for example which input influences which output, the nature of step response, the rise time and other quantitative or qualitative measures of system dynamics.

The design methods implemented in CASAD are: design of robust state-feedback controllers by pole-assignment and linear-quadratic optimization techniques, design of robust output feedback controllers by parameter optimization techniques, design of minimal and full order state estimators, design of feed-forward controllers. The structure of the compensator can also include a reference model.

The evaluation of performances of different control configurations is performed by simulation. All design and simulation procedures are available for both continuous and discrete systems. The simulation of hybrid or non-linear systems can also be performed.

Graphical facilities can be used to display simulation results.

Some of the functions implemented in CASAD are similar with the functions of other interactive packages (SYNPAC [15], CLADP [16], DAREK [17]), but CASAD offers many useful facilities which are not included in any of the existing packages. CASAD is based on a reliable numerical software implemented in the powerful Fortran subroutine packages BIMAS and BIMASC, which offer many of the standard functions required by the CACSD applications.

CASAD consists of 20-30 programs selected according to the user options. All implemented functions are performed by the means of a command language. Each program fetches its input data from the disk and puts the results on disk. The outputs of most of programs are used as inputs to other programs of the package. In its present version, CASAD can solve problems with maximum 35-40 state variables, 5 command inputs, 5 measured outputs and 3 measurable disturbances. All computations are performed in double precision.

CASAD is implemented on the romanian family of minicomputers I-100, 102F, CORAL 4011, 4030 under AMS or MINOS operating systems. CASAD can be installed also on the DEC PDP-11 family under the RSX-11M V3.2 or V4.0 operating systems.

2. DESCRIPTION OF CASAD

2.1. PURPOSE AND FEATURES OF CASAD

The main purpose for developing the CASAD package was to have a powerful suite of interactive CAD programs which use extensively the subprograms from the BIMAS and BIMASC packages. The main topics covered by BIMAS are: computation and ordering of Schur forms, computation of block diagonal form, solution of matrix equations (Lyapunov, Sylvester, and Riccati), stabilization and eigenvalue assignment, computation of matrix exponentials. BIMASC extends BIMAS with specific functions for analysis, modeling, design and simulation of multivariable systems, as for example: computation of system poles and zeros, minimal realization, balancing of state space models, evaluation of transfer matrices, discretization of continuous systems, design of state-feedback and feed-forward controllers, design of minimal order state estimators, simulation of linear and nonlinear systems.

The algorithms implemented in BIMAS and BIMASC have been rigorously selected in order to accomplish the attributes of generality, reliability, numerical stability, accuracy, efficiency. Many subprograms make calls to subroutines from EISPACK and LINPACK packages. BIMASC also includes two powerful ordinary differential equation solvers RKF [18] and LSODE [19] for non-stiff and stiff problems, respectively.

The structural approach to solve complex problems, used in BIMAS and BIMASC, is reflected also in CASAD organization, both at the level of the package as well as at the level of the individual programs. Due to the high modularity of BIMAS and BIMASC packages, the segmentation of the CASAD programs was a very easy task. In this way the CASAD programs can be used to solve on minicomputers problems with 35-40 state variables, all computations being done in double precision. All functions in CASAD are performed by means of a command language. The command line contains all informations needed for a program to be executed. All program parameters and options have default values, thus, the usual operations can be very easily performed. HELP facilities are available for all implemented commands.

The CASAD programs are written in Fortran, excepting several routines used to define the syntax of commands, to get and to parse command lines. These routines are written in MACRO-11 language, implemented on the PDP-11 family of minicomputers and could be replaced in principle by analogue routines written in Fortran-77 to make the package more portable. Other system facilities used in conjunction with CASAD, contribute to a very flexible use of the package.

2.2 SYSTEM MODELS

Most of computations in CASAD are performed on continuous or discrete, linear time-invariant systems described by state-space models of the form

$$\begin{aligned} \lambda x(t) &= Ax(t) + Bu(t) + Ew(t) \\ y(t) &= Cx(t) \\ y_r(t) &= C_r x(t) \end{aligned} \qquad (1)$$

where x is the n-dimensional state vector, u is the m-dimensional control vector, w is the q-dimensional disturbance vector, y is the p-dimensional vector of measurable outputs, y_r is the l-dimensional vector of regulated outputs, and where λ is the differential operator d/dt for continuous systems or the forward shift operator $\lambda x(t) = x(t+1)$ for discrete systems. The system (1) will be reffered to as the quadruple (A,B,C,E) or as the triple (A,B,C) if E = 0. We shall assume that the regulated outputs are also measurable, that is the matrix C_r has the form $C_r = [I_1 \quad 0]C$.

An alternative system description accepted by several CASAD programs is the input-output description given by

$$Y(\lambda) = G(\lambda)U(\lambda) + G_d(\lambda)W(\lambda) \qquad (2)$$

where Y, U and W are the transform output, input and disturbance vectors, respectively. In continuous case the Laplace-transform is used, while in discrete case the Z-transform is used. The input-output transfer matrix $G(\lambda)$ and the disturbance-output transfer matrix $G_d(\lambda)$ are related to the model (1) by the formulas

$$G(\lambda) = C(\lambda I-A)^{-1}B, \; G_d(\lambda) = C(\lambda I-A)^{-1}E \qquad (3)$$

The use of CASAD programs assumes the availability of the linearized model of the process to be controlled. This model could be obtained either from model building based on basic physical principles or through system identification.

2.3. DATA STRUCTURES

CASAD deals with data of many different types. The basic data structure used in CASAD is the matrix, having as elements real or complex numbers, or transfer functions. Vectors are usually considered as matrices having one column or one row. More complex data structures can be constructed from two or more matrices. For example, one of the most widely used data structure is formed from the matrices of the quadruple (A,B,C,E) which defines the system (1). Twelve different data structures are used presently in CASAD. Some of them are: state model, input-output model, estimator model, controller matrices, time series, real or complex matrices etc.

Each data structure is stored as a source file on disk. Each file contains besides the matrix data, information about the format used for storage, structure identification number, matrix dimensions, system type (continuous or discrete). A predefined yet flexible file format is constructed for each data structure which allows an easy change of data between programs. The basic data management operations (creating, deleting, updating, listing, copying) are performed through the standard editing and file operation utilities offered by the minicomputer operating system.

The files are identified by a file description of the form

filename . filetype; version

controllers. This design involves the stabilization of an extended system formed by the open-loop system (1) coupled with an internal model or servocompensator [21], [22]. The internal model is described by the equation

$$\lambda x_i(t) = A_i\, x_i(t) + B_i\, e(t) \tag{4}$$

where x_i is the internal model state vector, e(t) is the regulation error vector. e(t) has the following expression

$$e(t) = r(t) - y_r(t) \tag{5}$$

where r is the reference vector for the regulated system outputs y_r. PAL and LQN determines a stabilizing gain matrix $H' = [H_I, H]$ for the extended system

$$\lambda x'(t) = A'\, x'(t) + B'\, u(t)$$

where

$$x' = \begin{bmatrix} x_i \\ x \end{bmatrix}, \quad A' = \begin{bmatrix} A_i & -B_i C_r \\ 0 & A \end{bmatrix}, \quad B' = \begin{bmatrix} 0 \\ B \end{bmatrix}$$

The robust controller is given by the control law

$$u(t) = H\, x(t) + H_I\, x_i(t) \tag{6}$$

Internal models corresponding to first or second order, continuous or discrete-time, integrators can be automatically generated. More general internal models can be also used.

5.2 NME - Full order (non-minimal) state estimator design

NME is used for the design of full order state estimators by pole assignment. The estimator is described by the equation

$$\lambda z(t) = Fz(t) + Gy(t) + Lu(t) + Kw(t) \tag{7}$$
$$x_e(t) = z(t)$$

where $x_e(t)$ is the estimate of the state vector. An important feature of NME is its ability to determine F in an upper quasitriangular form (real Schur form) or in a block diagonal form. This facility is implemented by performing orthogonal or well-conditioned similarity transformations on the system formed from the state equations (7) and output equations (3) or (6). By this procedure, an important saving of operations requirement is obtained in on-line implementation of the state estimator.

The file type provides a convenient means for distinguishing different forms of the same system, or different elements of the same control configuration. For example, the standard state-space model of a paper machine might be named PM.SSM, the transfer matrix description of the same system might be named PM.TRM, and the different elements of the control configuration such as feedback gain matrices, state estimator, feed-forward gain matrices, reference model, might be named as PM.GAM, PM.ESM, PM.FFM, PM.REM, respectively. The use of file type offers a very convenient way to define default file types for input and output data specified in CASAD commands. We must note that the file identification described above is the same with the file identification convention used by the minicomputer operating system.

2.4 CASAD COMMAND LANGUAGE

The command language of CASAD reflects the nature of the interaction between the user and the computer for the considered application type. The user makes a choice of operation to be performed, specifies the input and output data descriptions and makes several options in order to accomplish the specified operation in the desired way.

Most of CASAD command have the following format

    opr [output][/sw] = input[/sw][,inp2][,inp3] ...

where opr specifies the operation command code, input, inp2, inp3,... are input data file identifiers, output is the output file identifier and /sw are input or output switches. The switches can be attached to both output and input file identifiers and are used to specify some problem parameters or user options. The entities contained between right parenthesis are optional. All optional entities have default values. For example, the default file name for inp2, inp3,... is the same as specified in input. Each file description has default file type. The extensive use of defaults, makes the common operation in CASAD very easy.

As an example consider the following command

    MNE PM/PR = PM/LI/BD/SM:0.7/TOL:1.E-6,POLES

This command is used for the design of a minimal order state estimator by pole assignment for a paper machine. The state-space model of the paper machine is contained in the file PM.SSM and the desired estimator poles are contained in the file POLES.COM (SSM and COM are the default input file types). The resulted estimator matrices will be stored in the file PM.ESM. The switches used in the command line specify several user options. Thus, the input data and input parameters will be displayed on the terminal (/LI), the results will be printed on the line printer (/PR), the resulted estimator state matrix will be in a block-diagonal form (/BD), the stability margin to be used for pole assignment is 0.7 (/SM:0.7), and the tolerance to be used for observability/detectability tests is 10^{-6} (/TOL:1.E-6). All switches have default values, and logical swiches, as /LI,/PR, /BD can be used in negated forms /-LI, /-PR, /-BD.

If the desired estimator poles are contained in the file PM.COM, then the estimator design can be performed using the much simpler command

```
MNE  PM = PM
```

In this case, the input and output data are not displayed or printed, the resulted estimator state matrix will be in a real Schur form, the default stability margin used is -0.5 for a continuous system and 0.5 for a discrete system, and the tollerance for observability test is 10^{-5}. The simple form of all commands ensures an easy operation for standard computations.

An important feature of CASAD environment is the possibility to use predefined command sequences. This macro-like facility offered by the minicomputer operating system can be efficiently used to solve complex design problems or to avoid the repeated entering of long command lines with many parameters. This facility can also be used to develop conversational mode of work for all implemented commands.

3. MODEL TRANSFORMATIONS

3.1 TMCD - Discretization of continuous input-output models

TMCD is used for the computation of the sampled-data (discrete) input-output model corresponding to a description of the form (2). Given the Laplace transformed transfer matrices G(s) and $G_d(s)$, and a sampling period T, TMCD computes the corresponding Z-transformed transfer-matrices by using a zero-order hold as a discrete to continuous-time converter for the input signals of the continuous system. If G(s) or $G_d(s)$ have elements with time-delays, then by an appropriate choice of T, the irrational factors can be eliminated. The matrix exponential method [8] is used for the discretization of the state space realization of each transfer matrix element.

3.2 TSCD - Discretization of continuous state-space models

TSCD is used for the computation of the sampled-data state-space model corresponding to a description of the form (1). The discrete to continuous-time converter for the inputs signals of the continuous system is a zero-order hold. The discretization is performed using the matrix exponential method with Padé approximations [8], combined with a very efficient block diagonalization procedure [27].

3.3 NMR - Non-minimal realizations of transfer matrices

NMR is used to compute non-minimal state-space realizations for a continuous or discrete input-output model of the form (2). The resulted state-space model can be both uncontrollable and unobservable, but also controllable or observable state-space realizations can be determined. The computed state-space model has the form (1) in which the state matrix A has a block-diagonal structure. Each subsystem formed from a diagonal block of A, the corresponding rows of the matrix [B,E] and the corresponding columns of C, is controllable and observable. Optionally, a balancing procedure based on the Moore's method [20] can be performed at the level of each subsystem.

3.4 NMR - Minimal realization of state-space models

MNR is used to compute a minimal order (controllable and observable) state-space realization from a non-minimal one. MNR usually follows NMR and removes succesively the uncontrollable and unobservable parts of a non-minimal system using the algorithm described in [2]. In the implemented version of this algorithm, the rank determinations are based on the QR decompositions with column pivoting. MNR uses only orthogonal similarity transformations and the overall minimal realization procedure (NMR followed by MNR) is numerically stable.

3.5 Other modeling facilities

Besides the above programs, other modeling facilities are implemented in CASAD. TSIM can be used to perform non-orthogonal similarity transformations on a state-space model in order to put the state matrix A in various simpler forms (balanced, Hessenberg, Schur, block-diagonal). TSO can be used to put A in a Hessenberg or an ordered Schur form by orthogonal similarity transformations. Another program TCF uses orthogonal transformations to compute the standard controllability or observability forms of a state-space model [2]. TBAL can be used to perform balancing transformations on a state-space model and TRED can compute reduced order models using the internally balanced system state-space model [20],[29].TSM can be used to evaluate the transfer matrix corresponding to a given state-space model by a non-orthogonal implementation of the method described in [28].

4. SYSTEM ANALYSIS

4.1 SMA - State-space model analysis

SMA is used for the analysis of stability, controllability/ stabilizability and observability/detectability properties of a state-space model of the form (1). SMA computes also the system poles (the eigenvalues of A) and for uncontrollable or unobservable systems, it determines also the uncontrollable or unobservables poles, respectively.

4.2 MZE - Computation of multivariable system zeros

MZE is used to compute the invariant zeros of a state space-model. The zeros of a multivariable system play an important role in several control problems as for example the regulation problem and robust control. MZE can be also used to compute the uncontrollable poles (input decoupling zeros) or unobservable poles (output decoupling zeros) of the system. MZE is based on the algorithm proposed in [30].

5. MULTIVARIABLE SYSTEM DESIGN

5.1 PAL and LQN - Design of state-feedback controllers

PAL is used to compute the gain matrix H in the control law

$$u(t) = Hx(t) \tag{3}$$

which assigns a set of desired eigenvalues for the spectrum of the closed-loop system matrix A + BH. A very useful feature of the implemented algorithm in PAL [7] is its ability to modify only the "bad" eigenvalues of the matrix A. This algorithm is based on the use of orthogonal similarity transformations on system matrices and in certain conditions it is numerically stable. In the implemented algorithm, uncontrollable, but stable eigenvalues are automatically deflated. Information about the degree of controllability of each modified "bad" eigenvalue are computed.

LQN is used to determine H in (3) using linear-quadratic optimization techniques [25]. The quadratic performance function can be defined either in terms of state or output variables. LQN uses the Newton method to solve the Riccati equations.

PAL and LQN can be also used for the design of robust state-feedback

5.3 MNE - Minimal order state estimator design

MNE is used for the design of minimal order state estimators by pole assignment. The estimator is described by the equations

$$\lambda z(t) = Fz(t) + Gy(t) + Lu(t) + Kw(t) \quad (8)$$
$$x_e(t) = My(t) + Nz(t)$$

The estimator state-matrix F is determined either in real Schur or block-diagonal form. The implemented algorithm is described in [23].

5.4. SFF and MFF - Feed-forward controllers design

SFF and MFF are used for the design of feed-forward controllers for systems described by either the equations (1) or (2), respectively. The feed-forward controller which ensures zero steady-state regulation errors for step disturbances and references is given by

$$u(t) = H_r\, r(t) + H_d\, w(t) \quad (9)$$

5.5 Other available design facilities

Besides the above programs, other design facilities are implemented in CASAD. STA can be used alternatively to determine the state-feedback gain matrix H from (3) which ensures a prescribed stability margin using stabilization techniques [24]. KBE can be used to design Kalman-Bucy filters or predictors as full order state-estimators in continuous or discrete case, respectively. OPI can be used to determine output proportional-integral centralized or descentralized multivariable compensators and OIF determines output integral feedback matrices for stable plants. Both OPI and OIF use tuning procedures based on direct search optimization algorithms.

6. SIMULATION OF LINEAR CONTROL SYSTEMS

Consider the linear time-invariant system described by (1). The control u can be a step or ramp signal in open loop simulation or can be computed as

$$u(t) = u_1(t) + u_2(t) \quad (10)$$

where u_1 is the feedback term and u_2 is the feed-forward term. Both terms in (10) are optional.

The feedback term can be generated in one of the following forms

$$
\begin{aligned}
u_1(t) &= H\, x_e(t) \\
u_1(t) &= H\, y_r(t) \\
u_1(t) &= H_I\, x_i(t) \\
u_1(t) &= H\, x_e(t) + H_I x_i(t) \\
u_1(t) &= H\, y_r(t) + H_I x_i(t)
\end{aligned}
\tag{11}
$$

In (11) x_e is either the system state vector (if estimator is not used) or the estimate of the state vector generated either by the full order estimator (7) or the minimal order estimator (8). x_i is the state vector of the internal model (4), where the regulation error is computed either in the form

$$e(t) = r(t) - y_r(t) \tag{12}$$

or

$$e(t) = y_m(t) - y_r(t) \tag{13}$$

In (13) y_m is the output of a reference model described by

$$
\begin{aligned}
\lambda x_m(t) &= A_m\, x_m(t) + B_m\, r(t) \\
y_m(t) &= C_m\, x_m(t) + D_m\, r(t)
\end{aligned}
\tag{14}
$$

The feed-forward term in (10) can have one of the following forms :

$$
\begin{aligned}
u_2(t) &= H_r\, r(t) \\
u_2(t) &= H_d\, w(t) \\
u_2(t) &= H_r\, r(t) + H_d\, w(t)
\end{aligned}
\tag{15}
$$

The equations (10) - (15) completed with (7) or (8), and (4), form the extended system

$$
\begin{aligned}
\lambda v(t) &= A'v(t) + E'w(t) + E''r(t) \\
y_r(t) &= C'v(t)
\end{aligned}
\tag{16}
$$

where v is the extended state vector formed from the state vectors x, z, x_i and x_m. All simulation programs evaluate the right hand side vector of (16) without forming explicitly the matrices A', E', E", C'. In order to reduce the operation count for evaluating this vector, the state matrices A, F, A_m and A_i are reduced to upper Hessenberg form by similarity transformations, as the first step in performing simulation.

SDS is a simulation program for discrete control systems. SCS and SSCS are used to simulate continuous control configurations. SCS is based on the RKF package which uses the Runge-Kutte-Fehlberg-45 method [18] and is appropriate for non-stiff or mildly stiff systems. For stiff systems or for high accuracy requirements, SSCS is recommended,which is based on linear multistep methods implemented in the LSODE package [19]. A very effective block diagonal approximation for the Jacobian matrix A' has been proposed in [26] and is implemented in SSCS. To simulate hybrid control configurations formed from a continuous plant driven by a discrete controller, the SHS program can be used.

Future developments of CASAD will include the simulation of non-linear systems described by ordinary differential equations in explicit or implicit forms, linearization of non-linear models, linear (continuous or discrete) control of non- linear systems.

The results computed by the simulation programs can be plotted using the program PLOT. Only input and output variable are displayed, the maximum number of variable which can be ploted simultaneously is ten. Several useful options are implemented in PLOT such as the individual plotting of variables, use of given scales for variables, use of non-standard, user specified character set etc.

REFERENCES

1. Van Dooren, P., The generalized eigenstructure problem in linear system theory, IEEE Trans. Autom. Control, vol. AC-26, p.111-129, 1981.
2. Varga, A.,Numerically stable algorithm for standard controllability form determination, Electronics Letters, vol.17, p.74-75, 1981.
3. Laub, A.J., A Schur method for solving the algebraic matrix Riccati equations, IEEE Trans. Autom. Control, vol. AC-24, p.913-921, 1979.
4. Pappas, T., Laub. A.J., and Sandell, N.R., On the numerical solution of the discrete-time algebraic Riccati equation, IEEE Trans. Autom. Control, vol.AC-25, p.631-641, 1980.
5. Van Dooren, P.,A generalized eigenvalue approach for solving the Riccati equations, Rep. NA- 80-02, Comp. Scie. Dept., Stanford Univ., 1980.
6. Miminis, G.S. and Paige, C.C., An algorithm for pole assignment of time-invariant multi-input linear systems, 21-st IEEE Conf. on Decision and Control, San Diego, 1982.
7. Varga, A.,A Schur method for pole assignment, IEEE Trans. Autom. Contr., vol.AC-26, p.517-519, 1981.
8. Van Loan, C.F.,Computing integrals involving matrix exponentials, IEEE Trans. Autom. Control, vol.AC-23, p.395-404, 1978.
9. Dongarra, J.J., Bunch, J.R., Moler, C.B. and Stewart, G.W., LINPACK User's Guide, SIAM, Philadelphia, 1979.
10. Smith, B.T.,Boyle, J.M., Dongarra, J.J., Garbow, B.S., Ikebe, Y., Klema, V.C. and Moler, C.B., Matrix eigensystem routines - EISPACK Guide, Lect. Notes in Comp. Scie., vol.6, Springer Verlag, Berlin, 1974.

11. Garbow, B.S., Boyle, J.M., Dongarra, J.J., and Moler, C.B., Matrix eigensystem routines - EISPACK Guide Extension, Lect. Notes in Comp. Scie., vol.51, Springer Verlag, Berlin, 1977.

12. Varga, A. and Sima, V., BIMAS - A basic mathematical package for computer-aided systems analysis and design, Preprints of the IFAC 9th World Congress, Budapest, 2-6 June, 1984.

13. Varga, A., BIMASC - A package of Fortran subprograms for analysis, design and simulation of control systems, Report ICI, TR-10.83, 1983.

14. Åström, K.J., Computer aided modeling, identification and control system design - a perspective, IEEE Control Systems Magazine, Nov., 1983.

15. Wieslander, J., Interaction in computer aided analysis and design of control systems, Ph. D. Thesis, Dept. of Automatic Control, Lund, 1979.

16. Maciejowski, J.M. and MacFarlene, A.G.J., CLADP - The Cambridge linear analysis and design programs, IEEE Control Systems Magazine, Dec., 1982.

17. Pedersen, J.O., Pohner, F. and Solheim, O.A., Computer aided design of multivariable control systems, Preprints of IFAC 5th World Congress, Paris, 1972.

18. Forsythe, G.E., Malcolm, M.A. and Moler, C.B., Computer methods for mathematical computations, Prentice Hall, Englewood Cliffs, 1977.

19. Hindmarsh, A.C., LSODE and LSODI, two new initial value ordinary differential equation solvers, ACM Signum Newsletter, vol. 15, p.10-11, 1980.

20. Moore, B.C. Principal component analysis in linear systems: controllability, observability and model reduction, IEEE Trans. Autom. Contr., vol.AC-26, p.17-32, 1981.

21. Wonham, W.M., Linear multivariable control. A geometric approach, Springer Verlag, Berlin, 1979.

22. Davison, E.J. and Goldenberg, A., The robust control of a general servomechanism problem: the servocompensator, Automatica, vol.11, p.461-471, 1975.

23. Varga, A.,Computer aided design of robust compensators by pole assignment, Preprints of SOCOCO'82 Symp., Sept. 1982, Madrid.

24. Varga, A.,On stabilization algorithms for linear time-invariant systems, Rev. Roum. Scie. Tech.-Electrotech. and Energ., vol.26, p.115-124, 1981.

25. Sima, V.,On the real Schur form in linear control system design, Rev. Roum. Scie. Tech. - Electrotech. et Energ., vol.25, p. 625-632, 1980.

26. Varga, A., Sima, V. and Varga, C.V.,On numerical simulation of linear continuous control systems, Preprints of SIMULATION'83 Symposium, Prague, June, 1983.

27. Bavely, C.A. and Stewart, G.E., An algorithm for computing reducing subspaces by block diagonalization, SIAM J.Numer. Anal., vol.10, p.359-367, 1979.

28. Varga, A. and Sima V., A numerically stable algorithm for transfer matrix evaluation, Int. J. Control, vol.33, p.1123-1133, 1981.

29. Laub, A.J., On computing "balancing" transformations, Preprints of JACC Symp., San Francisco, Aug., 1980.

30. Emami-Naeini, A. and Van Dooren P., Computation of zeros of linear multivariable systems, Automatica, vol.18, p.415-430, 1982.

Session 15

SIGNAL PROCESSING

TRAITEMENT DU SIGNAL

ON THE SELECTION OF MEMORYLESS ADAPTIVE LAWS FOR BLIND EQUALIZATION IN BINARY COMMUNICATIONS

Sergio Verdú
Coordinated Science Laboratory
University of Illinois at Urbana-Champaign
Urbana, IL 61801 USA

ABSTRACT

We consider the adaptive equalization of an unknown linear time-invariant channel without observations of the input sequence, by updating the impulse response coefficients of the equalizer with the output of the channel times a memoryless nonlinear function of the equalizer output. To date, no such function is known to result in global convergence to the inverse of the channel when the input consists of binary data. The effect of the selection of the memoryless nonlinearity in the convergence properties of the adaptive scheme is studied, and it is shown that for a large class of laws (including the continuous functions), unequalized impulse responses with few nonzero coefficients are points of convergence, and that there exist undesired local minima for a subset of functions that includes those previously proposed.

I. PRELIMINARIES

Noiseless observations, $\underline{x} = \{x_t, t \in \mathbf{Z}\}$, of the output of an unknown discrete-time linear time-invariant system driven by a Bernoulli sequence $\underline{u} = \{u_t, t \in \mathbf{Z}\}$ are used in order to adjust the impulse response of an (IIR) linear equalizer $\theta = \{\theta_t, t \in \mathbf{Z}\}$ such that when driven by $\underline{x}$ its output is the original Bernoulli sequence. A possible strategy is to use a memoryless adaptive law of the type

$$\underline{\theta}^{n+1} = \underline{\theta}^{n} - \tau_n \psi(c_n) \underline{x}^{n} \tag{1}$$

where τ_n is a sequence of small positive scalars, c_n is the current output of the equalizer, $\underline{x}^n$ is an n-delayed version of $\underline{x}$ and $\psi(\cdot)$ is a real-valued function to be specified.

Use of the scheme (1) for the blind equalization problem has been made in the past in [1]-[3], and it is based on the fact that (1) is a stochastic approximation [8] for the minimization of the risk $R(\underline{\theta}) = E[\Omega(c)]$, where $\Omega(x) = \Omega(0) + \int_0^x \psi(t)dt$.

As popularized by Ljung [9], the analysis of the convergence of (1) can be carried out, if ψ is smooth enough, by first studying the steepest descent lines of $R(\underline{t}) = E[\Omega(\underline{u}\,\underline{t})] = \sum_{i=-\infty}^{\infty} t_i u_{-i}$, and then analyzing the behavior of (1) with respect to that of the integral curves of

$$\frac{d\underline{t}}{dx} = -\text{ grad } R(\underline{t}). \tag{2}$$

Ideally, the designer's goal would be to select an adaptive law $\psi(\cdot)$ such that all the steepest descent lines of $R(\underline{t})$ converge to $(\ldots,0,\pm 1,0,\ldots)$ - a sign uncertainty is inevitable since the input distribution is symmetric. However, to date, no function $\psi(\cdot)$ has been found to satisfy such a requirement. In fact, since the existence of steepest descent lines that converge to unstable equilibria (crest lines) may be a source of loss of efficiency of the algorithm but may not destroy the qualitative convergence properties of (1), we could require an adaptive law with less stringent properties:

Definition 1

ψ is *admissible* if the steepest descent lines of $R(\underline{t})$ converge and its local minima belong to the set $\{\underline{t} \in \mathbb{R}^{\mathbb{Z}}, \exists\, k \neq 0,\ \underline{t} = (\ldots,0,K,0,\ldots)\} \neq \phi$.

In their seminal work [1], Benveniste, Goursat and Ruget showed that if the distribution (λ) of the i.i.d. input sequence is sub-Gaussian rather than Bernoulli and ψ satisfies:

(a) $\psi(x) = -\gamma \operatorname{sgn}(x) + \tilde{\psi}(x)$

(b) $\int x\tilde{\psi}(x)\lambda(dx) = \gamma \int |x|\lambda(dx)$

(c) $\tilde{\psi}$ is odd, twice differentiable and convex on $(0,\infty)$.

Then ψ is an admissible adaptive law, and the only local minima are $(\ldots,0,\pm 1,0,\ldots)$. However, the existence of an admissible adaptive law when the input is Bernoulli - the most important case in practice - remains an open problem. As a step in that direction, the goal of this work is to impose conditions on $\psi(\cdot)$ such that an adaptive scheme with the above desirable properties - admissibility (and inexistence of crest lines) - can be obtained. The nature of the input distribution allows us to work in a simple algebraic framework, and to rule out large classes of functions by studying the behavior of the risk around systems with few nonzero coefficients. On the other hand, only restricted types of infinite impulse responses can be dealt with since a general characterization of the distribution of $\underline{u}\,\underline{t}$ in the Bernoulli case is not known [6].

II. DEVELOPMENT

Without significant loss of generality we restrict our attention to laws that satisfy:

(a) Integrable in any finite interval

(b) Left-hand and right-hand limits

$$\psi^{+}(x) = \lim_{h \downarrow 0} \psi(x+h)$$

$$\psi^{-}(x) = \lim_{h \downarrow 0} \psi(x-h)$$

exist at every point (i.e., only discontinuities of the first kind are allowed).

(c) ψ is odd. (Note that since u_t has symmetric distribution, any function Ω results in the same risk as the even function $\Omega(x) + \Omega(-x)$.)

Conditions (a)-(b) ensure the existence of the left-hand and right-hand derivatives of $\Omega(z)$, which is necessary for the analysis of the steepest descent minimization of $R(\underline{t})$. Let the (unnormalized) Gateaux directional derivative of $R(\cdot)$ at point $\underline{t}$ in the direction $\underline{\delta}$ be defined by

$$\rho(\underline{t},\underline{\delta}) = \lim_{h\downarrow 0} \frac{1}{h} [R(\underline{t} + h\underline{\delta}) - R(\underline{t})] \quad . \tag{3}$$

The steepest descent algorithm [7] for the minimization of $R(\underline{t})$ selects at each point the direction at which $\rho(\underline{t},\underline{\delta})/\|\delta\|$ - for some norm $\| \ \|$ - is minimum. If such a minimum exists and is negative then a new approximation to the sought-after global minimum of $R(\underline{t})$ is obtained. Hence the sinks of the steepest descent lines have the following properties:

Definition 2

$\underline{t}$ is a *point of convergence* if

(a) for all $\underline{\delta}$, $\rho(\underline{t},\underline{\delta}) \geqslant 0$

(b) for every neighborhood $N(\underline{t})$ there exists $\underline{t}' \in N(\underline{t})$ such that $R(\underline{t}) < R(\underline{t}')$.

Our first result will be invoked again and again in the sequel and reduces the fulfillment of condition (a) in the last definition to a purely finite dimensional problem when - as will be the case later on - there is only a finite number of non-zero terms in $\underline{t}$.

Proposition 1

If for every $\underline{\delta}$ such that $t_i = 0 \Rightarrow \delta_i = 0$ we have

$$\rho(\underline{t},\underline{\delta}) \geqslant 0 \quad , \tag{4}$$

then (4) holds for every direction $\underline{\delta}$ such that the directional derivative $\rho(\underline{t},\underline{\delta})$ exists.

Proof

For an arbitrary $\underline{\delta}$, we have

$$\begin{aligned}\rho(\underline{t},\underline{\delta}) &= \lim_{h\downarrow 0} \frac{1}{h} E[\Omega(\underline{u}(\underline{t}+h\underline{\delta})) - \Omega(\underline{u}\,\underline{t})] \\ &= \lim_{h\downarrow 0} \frac{1}{h} E[E[\Omega(\sum_{t_i=0} h\delta_i u_i + \sum_{t_i\neq 0} (t_i+h\delta_i)u_i) - \Omega(\underline{u}\,\underline{t}) \,|\{u_i, \text{s.t. } t_i=0\}]] \\ &= \lim_{h\downarrow 0} \frac{1}{h} E[E[\Omega(-\sum_{t_i=0} h\delta_i u_i + \sum_{t_i\neq 0} (t_i+h\delta_i)u_i) - \Omega(\underline{u}\,\underline{t}) \,|\{u_i, \text{s.t. } t_i=0\}]],\end{aligned}$$

where the last equation follows because Ω is even. Therefore it is enough to show that for any positive scalar ε,

$$\lim_{h\downarrow 0} \frac{1}{h} E[\Omega(h\varepsilon + \sum_{t_i\neq 0} (t_i+h\delta_i)u_i) - \Omega(\underline{u}\,\underline{t})] \geqslant 0 \tag{5}$$

if $\rho(\underline{t},\underline{\delta}) \geqslant 0$ for every $\underline{\delta}$ such that $\delta_i = 0$ unless $t_i \neq 0$. Let $\tilde{\delta}$ be such that $\tilde{\delta}_i = \delta_i/\varepsilon$ if $t_i \neq 0$, and $\tilde{\delta}_i = 0$ if $t_i = 0$.

$$\begin{aligned}\lim_{h\downarrow 0} \frac{1}{\varepsilon h} E[\Omega(h\varepsilon + \sum_{t_i=0} (t_i+h\delta_i)u_i) - \Omega(\underline{u}\,\underline{t})] &= E[\psi^+(\underline{u}\,\underline{t})(1+\underline{u}\,\tilde{\underline{\delta}})\,|-1 \leqslant \underline{u}\,\tilde{\underline{\delta}}]P[-1 \leqslant \underline{u}\,\tilde{\underline{\delta}}] \\ &+ E[\psi^-(\underline{u}\,\underline{t})(1 + \underline{u}\,\tilde{\underline{\delta}})\,|\underline{u}\,\tilde{\underline{\delta}} < -1]P[\underline{u}\,\tilde{\underline{\delta}} < -1] =\end{aligned}$$

$$= E[\psi^+(\underline{u}\,\underline{t})\,(1+\underline{u}\,\tilde{\underline{\delta}})\,|-1 \leq \underline{u}\,\tilde{\underline{\delta}} \leq 1]P[-1\leq \underline{u}\,\tilde{\underline{\delta}} \leq 1]$$

$$+ E[\psi^+(\underline{u}\,\underline{t})\,(2\underline{u}\,\tilde{\underline{\delta}})\,|\,1 < \underline{u}\,\tilde{\underline{\delta}}]\,P[1< \underline{u}\,\tilde{\underline{\delta}}]$$

where we used $\psi^+(x) = -\psi^-(-x)$. Now we show that the right hand side of the last equation is $\frac{1}{2}\,[\rho(\underline{t},\tilde{\underline{\delta}}^+) + (\underline{t},\tilde{\underline{\delta}}^-)]$ where $\tilde{\delta}_i^{\pm} = \tilde{\delta}_i$ for $i \neq 0$ and $\tilde{\delta}_0^{\pm} = \tilde{\delta}_0 \pm 1$ (without loss of generality we may assume $\tilde{\delta}_0 \neq 0$).

$$\begin{aligned}
\rho(\underline{t},\tilde{\underline{\delta}}^+) &= E[\psi^+(\underline{u}\,\underline{t})\,(\underline{u}\,\tilde{\underline{\delta}}+ u_0)\,|\underline{u}\,\tilde{\underline{\delta}} >1]P[\underline{u}\,\tilde{\underline{\delta}}> 1] \\
&+ P[-1\leq\underline{u}\,\tilde{\underline{\delta}} \leq 1,u_0 = 1]E[\psi^+(\underline{u}\,\underline{t})(\underline{u}\,\tilde{\underline{\delta}} + 1)\,|-1\leq\underline{u}\,\tilde{\underline{\delta}} \leq 1,u_0 = 1] \\
&+ P[-1\leq\underline{u}\,\tilde{\underline{\delta}} \leq 1,u_0 = -1]E[\psi^-(\underline{u}\,\underline{t})(\underline{u}\,\tilde{\underline{\delta}} - 1)\,|-1\leq\underline{u}\,\tilde{\underline{\delta}} \leq 1,u_0 = -1] \\
&+ P[\underline{u}\,\tilde{\underline{\delta}} <1]E[\psi^-(\underline{u}\,\underline{t})(\underline{u}\,\tilde{\underline{\delta}} +u_0)\,|\,\underline{u}\,\tilde{\underline{\delta}}< 1] \\
&= 2\,E[\psi^+(\underline{u}\,\underline{t})(\underline{u}\,\tilde{\underline{\delta}} + u_0)\,|\underline{u}\,\tilde{\underline{\delta}} >1]P[\underline{u}\,\tilde{\underline{\delta}}< 1] \\
&+2\,E[\psi^+(\underline{u}\,\underline{t})(\underline{u}\,\tilde{\underline{\delta}} +1)\,|\,-1\leq\underline{u}\,\tilde{\underline{\delta}} \leq 1,u_0 = 1]P[-1\leq\underline{u}\,\tilde{\underline{\delta}} \leq 1,u_0 = 1] \quad .
\end{aligned}$$

Analogously, we obtain

$$\begin{aligned}
\rho(\underline{t},\tilde{\underline{\delta}}^-) &=2\,E[\psi^+(\underline{u}\,\underline{t})(\underline{u}\,\tilde{\underline{\delta}} - u_0)\,|\underline{u}\,\tilde{\underline{\delta}} >1]P[\underline{u}\,\tilde{\underline{\delta}}< 1] \\
&+ 2\,E[\psi^+(\underline{u}\,\underline{t})(\underline{u}\,\tilde{\underline{\delta}} + 1)\,|-1\leq\underline{u}\,\tilde{\underline{\delta}} \leq 1,u_0 = -1]P[-1\leq\underline{u}\,\tilde{\underline{\delta}} \leq 1,u_0 = -1].
\end{aligned}$$

Therefore,

$$\begin{aligned}
&E[\psi^+(\underline{u}\,\underline{t})\,(1 + \underline{u}\,\tilde{\underline{\delta}})\,|\,-1\leq\underline{u}\,\tilde{\underline{\delta}} \leq 1]P[-1\leq\underline{u}\,\tilde{\underline{\delta}} \leq 1] \\
&\quad + E[\psi^+(\underline{u}\,\underline{t})(2\underline{u}\,\tilde{\underline{\delta}})\,|1<\underline{u}\,\tilde{\underline{\delta}}\,]P[1<\underline{u}\,\tilde{\underline{\delta}}\,] \\
&\quad = \tfrac{1}{2}\,[\rho(\underline{t},\tilde{\underline{\delta}}^+) + \rho(\underline{t},\tilde{\underline{\delta}}^-)]\geq 0
\end{aligned}$$

where the inequality follows from the assumption of the proposition since $\tilde{\delta}_i^+ = \tilde{\delta}_i^- = 0$ unless $t_i \neq 0$.

Next, our emphasis is on the search for necessary conditions on the adaptive law such that points of convergence that do not correspond to equalized systems are avoided. To that end, we focus attention on the behavior of the risk around systems with two or three nonzero coefficients with equal magnitude. (Recall that the risk is invariant under permutation or change of sign of the impulse response coefficients.)

Proposition 2

If ψ is admissible and continuous at the origin then there exists $c \neq 0$ such that $(\ldots,0,c,c,0,\ldots)$ is a point of convergence.

Proof

Let d be any nonzero scalar. Proposition 1 implies that in order to show

$$\rho((\ldots,0,d,d,0,\ldots),\underline{\delta}) \geq 0$$

for every $\underline{\delta}$ it is enough to restrict attention to $\underline{\delta} = (\ldots,0,\delta_1,\delta_2,0,\ldots)$. Hence

$$\rho((\ldots,0,d,d,0,\ldots),(\ldots,0,\delta_1,\delta_2,0,\ldots)) = \begin{cases} 2\psi^+(2d)(\delta_1 + \delta_2) & \text{if } \delta_1 + \delta_2 \geq 0 \\ 2\psi^-(2d)(\delta_1 + \delta_2) & \text{if } \delta_1 + \delta_2 \leq 0 \end{cases}$$

where the continuity at the origin has been used. Since ψ is admissible, there exists $c \neq 0$, such that $(\ldots,0,2c,0,\ldots)$ is a local minimum of $R(\underline{t})$ and therefore $\psi^-(2c) \leq 0 \leq \psi^+(2c)$. In order to show that $(\ldots,0,c,c,0,\ldots)$ is not a local maximum, it is enough to consider

$R((\ldots,0,c+\varepsilon,c+\varepsilon,0,\ldots)) - R((\ldots,0,c,c,0,\ldots)) = 2[\Omega(2c+2\varepsilon) - \Omega(2c)]$,

which is strictly positive for some ε in every neighborhood of 0, because 2c is a local minimum of $\Omega(\cdot)$. □

The requirement of having a discontinuity at the origin (in order to avoid the existence of points of convergence at unequalized systems) is fulfilled by all the adaptive laws proposed in [1]-[3]. However, as the following result shows, other discontinuities are necessary when the location of the stable points of convergence is restricted to $(\ldots,0,\pm 1,0,\ldots)$.

Proposition 3

Suppose that ψ is admissible and that there exists $K > 0$ such that $(\ldots,0,c,0,\ldots)$ is not a local minimum of $R(\underline{t})$ if $|c| \neq K$. Then, there exists no $t \in [K/3,K]$ such that $(\ldots,0,t,t,t,0,\ldots)$ is a point of convergence if and only if $\psi^-(3K) + \psi^-(K) > 0$ and at least one of the following is satisfied.

(a) $\psi^+(K) + \psi^+(K/3) < 0$ and there exists $z \in (K/3,K)$ such that

(a1) $\psi^+(3z) > \psi^-(3z)$

(a2) $\psi^+(z) < \psi^-(z)$

(a3) z is a strict local minimum of $\Omega(3\cdot) + 3\Omega(\cdot)$

(b) $\psi^+(K/3) < \psi^-(K/3)$ and for every $t \in (K/3,K)$ at least one of the following is true:

(b1) $\psi^+(3t) + \psi^+(t) < 0$

(b2) $\psi^-(3t) + \psi^-(t) > 0$

(b3) $\psi^+(t) < \psi^-(t)$

(b4) $(\ldots,0,t,t,t,0,\ldots)$ is a local maximum.

Proof

First we obtain necessary and sufficient conditions for

$\rho((\ldots,0,t,t,t,0,\ldots),\underline{\delta}) \geq 0$ for all $\underline{\delta}$.

Invoking Proposition 1, we only need to take into account $\underline{\delta} = (\ldots,0,\delta_1,\delta_2,\delta_3,0,\ldots)$. Using the definition of ρ and taking expectation with respect to the input sequence, we obtain

$$4\rho((\ldots,0,t,t,t,0,\ldots),(\ldots,0,\delta_1,\delta_2,\delta_3,0,\ldots)) =$$

$$= \begin{cases} [\psi^+(3t)+\psi^+(t)](\delta_1+\delta_2+\delta_3), & \text{if } 0 \leq \delta_1,\delta_2,\delta_3 \\ [\psi^-(3t)+\psi^-(t)](\delta_1+\delta_2+\delta_3), & \text{if } 0 \geq \delta_1,\delta_2,\delta_3 \\ [\psi^+(3t)+\psi^+(t)](\delta_1+\delta_2+\delta_3)+[\psi^+(t)-\psi^-(t)](\delta_i-\delta_j-\delta_k), & \text{if } \delta_i \geq 0, \delta_i \geq |\delta_j+\delta_k| \\ [\psi^-(3t)+\psi^-(t)](\delta_1+\delta_2+\delta_3)+2\delta_i[\psi^+(t)-\psi^-(t)] & \text{if } 0 \leq \delta_i \leq -\delta_j-\delta_k \\ [\psi^+(3t)+\psi^+(t)](\delta_1+\delta_2+\delta_3)-2\delta_i[\psi^+(t)-\psi^-(t)] & \text{if } -\delta_j-\delta_k \leq \delta_i \leq 0 \\ [\psi^-(3t)+\psi^-(t)](\delta_1+\delta_2+\delta_3)+[\psi^+(t)-\psi^-(t)](-\delta_i+\delta_j+\delta_k), & \text{if } \delta_i \leq 0, \delta_i \leq |\delta_j+\delta_k| \end{cases}$$

From the last expression it is straightforward to check that for the directional derivative to be nonnegative in all directions it is necessary and sufficient that

$$\text{(i)} \quad \psi^+(3t) + \psi^+(t) \geq 0 \tag{6}$$

$$\text{(ii)} \quad \psi^-(3t) + \psi^-(t) \leq 0 \tag{7}$$

(iii) $\psi^+(t) - \psi^-(t) \geqslant 0$ (8)

Since all the steepest descent lines converge and there are no local minima of $\Omega(\cdot)$ other that $\pm K$, it is necessary that

(i) $\psi^+(x) \geqslant 0 \quad x \in [K,\infty)$ (9)

(ii) $\psi^-(x) \leqslant 0 \quad x \in [0,K]$ (10)

If $\psi^-(3K) + \psi^-(K) \leqslant 0$ then particularizing (9) at K and 3K and (10) at K, we obtain that conditions (6)-(8) are satisfied and that $(\ldots,0,K,K,K,0,\ldots)$ is not a local maximum. Hence, we obtain that $\psi^-(3K) + \psi^-(K) > 0$ is necessary and sufficient for the inexistence of points of convergence at $(\ldots,0,K,K,K,0,\ldots)$. Analogously, using (10) we have that $(\ldots,0,K/3,K/3,K/3,0,\ldots)$ is a point of convergence if and only if $\psi^+(K) + \psi^+(K/3) \geqslant 0$ and $\psi^+(K/3) \geqslant \psi^-(K/3)$. (It is easy to check that $(\ldots,0,K/3,K/3,K/3,0,\ldots)$ is not a local maximum, e.g., $R((\ldots,0,K/3,K/3,K/3,0,\ldots)) < R((\ldots,0,K/3-\varepsilon,K/3-\varepsilon,K/3-\varepsilon,0,\ldots))$ for sufficiently small $\varepsilon > 0$.)

In order to find out the conditions for existence of points of convergence at $(\ldots,0,t,t,t,0,\ldots)$ for $t \in (K/3,K)$ when they do not exist at $t = K/3,K$, we first consider the case:

(a) $\psi^-(3K) + \psi^-(K) > 0$ and $\psi^+(K) + \psi^+(K/3) < 0$.

These imply that there exists a strict local minimum (t) of $\Omega(3\cdot) + 3\Omega(\cdot)$ in the interval (K/3,K)-t is an upcrossing of $\psi(3\cdot) + \psi(\cdot)-$. On the one hand this implies that $\Omega(3t + 3\varepsilon) + 3\Omega(t+\varepsilon) > \Omega(3t) + 3\Omega(t)$ for either $\varepsilon > 0$ or $\varepsilon < 0$ sufficiently small. Therefore $(\ldots,0,t+\varepsilon,t+\varepsilon,t+\varepsilon,0,\ldots)$ has strictly higher risk than $(\ldots,0,t,t,t,0,\ldots)$. On the other hand, we have that $\psi^-(3t) + \psi^-(t) \leqslant 0 \leqslant \psi^+(3t) + \psi^+(t)$; so $(\ldots,t,t,t,0,\ldots)$ is a point of convergence if and only if $\psi^+(t) \geqslant \psi^-(t)$. (Note that condition (a1) in the statement of Proposition 3 follows from (a2) + (a3).) The alternative condition for avoiding points of convergence at $(\ldots,0,K/3,K/3,K/3,0,\ldots)$ is

(b) $\psi^-(3K) + \psi^-(K) > 0$ and $\psi^+(K/3) < \psi^-(K/3)$

Conditions (b1)-(b4) are necessary and sufficient for the inexistence of a point of convergence at $(\ldots,0,t,t,t,0,\ldots)$. It can be shown that for $(\ldots,0,t,t,t,0,\ldots)$ to be a local maximum such that (b1)-(b3) are not satisfied it is necessary that $\psi(\cdot)$ be continuous at t and 3t and that t be a downcrossing of $\psi(\cdot) + \psi(3\cdot)$. If, furthermore, $\psi(\cdot)$ is differentiable at 3t and t, then it is necessary and sufficient that $\psi'(t) = 0$ and $\psi'(3t) < 0$. □

Proposition 3 shows that the discontinuity of the adaptive law is necessary for the inexistence of points of convergence at systems $(\ldots,0,t,t,t,0,\ldots)$. This implies that even in the event that there exists an adaptive law without points of convergence at unequalized systems, in practice, the adaptive scheme will lack the necessary robustness for assuring a given speed of convergence. Note also that the above points of convergence occur at systems that are not far from the memoryless ones, and therefore central limit arguments for guaranteeing the behavior of the risk function cannot be used (cf. [1,Remark 4]).

A more crucial point than the inexistence of unequalized points of convergence is the admissibility of the adaptive law. Analogously to the last result, we find a large class of inadmissible functions by examining the systems with three equal-magnitude nonzero coefficients. Here, in order to simplify matters we restrict our attantion to adaptive laws whose derivative $\psi'(x)$ exists for $x \in (0,\infty)$.

Proposition 4

If there exists $t \in (0,\infty)$ such that

(i) $\psi(t) + \psi(3t) = 0$

(ii) $\psi'(t) + \psi'(3t) \geqslant 0$

(iii) $\psi'(t) \geqslant 0$

and at least one inequality is strict, then ψ is not admissible.

Proof

Under the above conditions there exists a local minimum at $(\ldots,0,t,t,t,0,\ldots)$. In order to show this, we prove that

$$\frac{\partial}{\partial t_i} R(\underline{t}) \Bigg|_{\underline{t} = (\ldots,0,t,t,t,0,\ldots)} = 0 \qquad i = 1,2,3$$

if and only if (i) holds (Equations (6)-(8) reduce to (i) if ψ is continuous), and that

$$H = \left[\left(\frac{\partial^2 R(\underline{t})}{\partial t_i \partial t_j}\right)_{ij}\right] \geqslant 0, \qquad \underline{t} = (\ldots,t,t,t,0,\ldots)$$

and $H \neq \underline{0}$ if and only if (ii) and (iii) are true. Notice that this is enough because of Proposition 1 and $\frac{\partial^2 R(\underline{t})}{\partial t_i \partial t_j} = 0$ at $\underline{t} = (\ldots,0,t,t,t,0,\ldots)$ if $t_i = 0$. In order to show the nonnegativity of the Hessian, we have that

$$4\,\frac{\partial^2 R(\underline{t})}{\partial t_i t_j}\Bigg|_{\underline{t} = (\ldots,0,t,t,t,0,\ldots)} = \begin{cases} \psi'(3t) + 3\psi'(t) & i = j \\ \psi'(3t) - \psi'(t) & i \neq j \end{cases}$$

and therefore ($|H| \geqslant 0$ if the first two principal minors are nonnegative) $H \geqslant 0$ if and only if

$$\psi'(3t) + 3\psi'(t) \geqslant 0$$

and

$$(\psi'(3t) + 3\psi'(t))^2 \geqslant (\psi'(3t) - \psi'(t))^2$$

which is equivalent to

$$\psi'(3t) + \psi'(t) \geqslant 0$$

and

$$\psi'(t) \geqslant 0.$$

□

Corollary

If $\psi'(t) > 0$ for every $t \in (0,\infty)$ then ψ is not admissible.

Proof

Suppose that ψ is admissible. Then since it is strictly increasing there exists a point K such that $\psi(x) < 0 < \psi(y)$ for $0 < x < K < y$. Now, using the fact that ψ is continuous, there exists a point $t \in (K/3,K)$ such that $\psi(3t) + \psi(t) = 0$. □

This corollary shows that no continuously differentiable convex - in $(0,\infty)$ - cost function $\Omega(\cdot)$ is admissible. This is particularly interesting since the adaptive laws considered by Godard [3] and Sato [2] (see also [1, Sect. VI]) are ruled out.

III. EXTENSIONS AND CONCLUDING REMARKS

A class of functions (including those adaptive laws proposed in the communications literature) has been shown to be inadmissible from the viewpoint of global convergence. Furthermore, a much larger class of laws (including the continuous functions) has shown to result in points of convergence at unequalized systems. All this has been accomplished by restricting attention to the behavior of the cost around systems whose impulse response have no more than three nonzero coefficients and hence avoiding the need to specify the underlying space of systems. It is plausible that by studying more complex systems much larger classes of functions can be excluded, and hopefully more light can be shed into the problem of the existence of an admissible memoryless adaptive law. Following the approach taken in Section II, the sufficient conditions for the existence of local minima (assuming differentiability of ψ, for example) could be investigated. While it is straightforward to show that if n equal-magnitude nonzero coefficients are allowed the gradient of the risk is zero for all directions at $(\ldots,0,t,\ldots,t,0,\ldots)$ if and only if

$$\psi(nt) + (n-2)\psi((n-2)t) + \sum_{i=2}^{\lceil \frac{n}{2}-1 \rceil} \left[\binom{n-2}{i}-\binom{n-2}{i-2}\right]\psi((n-2i)t) = 0 \quad ,$$

it is tedious to find conditions for the nonnegativity of the Hessian for generic n. On the other hand, when $\underline{t}$ has more than three or four nonzero coefficients it is equally tedious to obtain necessary and sufficient conditions for $\rho(\underline{t},\underline{\delta}) \geq 0$ for all $\underline{\delta}$ (cf. (6)-(8) for n = 3). However, it is possible to systematize the derivations of such conditions by using the following result.

Proposition 4

Let V be the set of all vertices of the simplices defined by subsets of the following hyperplanes in $\mathbb{R}^n$:

$$\begin{array}{ll} x_i \geq -1 & i = 1,\ldots,n \\ x_i \leq 1 & i = 1,\ldots,n \\ \underline{u}\,\underline{x} \geq 0 & \underline{u} \in \{-1,1\}^n \end{array} .$$

If $\rho(\underline{t},\underline{v}) \geq 0$ for all $v \in V$, then $\rho(\underline{t},\underline{\delta}) \geq 0$ for all $\underline{\delta} \in \mathbb{R}^n$.

Proof

Select any $\underline{\delta} \in \mathbb{R}^n - \{\underline{0}\}$ and let $U_\delta = \underline{u} \in \{-1,1\}^n$, s.t. $\underline{u}\,\underline{\delta} \geq 0\}$, then $\underline{\tilde{\delta}} = \underline{\delta}/\|\underline{\delta}\|_\infty$ belongs to the simplex (with vertices $V_\delta \subset V$) defined by

$$S_\delta = \{\underline{x} \in \mathbb{R}^n,\ \|\underline{x}\|_\infty \leq 1\} \cap \bigcap_{\underline{u} \in U_\delta} \{\underline{x}, \underline{u}\,\underline{x} \geq 0\} ,$$

and therefore $\underline{\tilde{\delta}}$ can be put as a convex combination of the vertices in $V_\delta = \{\underline{v}^1,\ldots,\underline{v}^k\}$:

$$\underline{\tilde{\delta}} = \sum_{i=1}^{K} a_i \underline{v}^i, a_i \geq 0, \sum_{i=1}^{K} a_i = 1.$$

The proof is completed by showing that

$$\rho(\underline{t},\underline{\delta}) = \|\underline{\delta}\|_\infty \sum_{i=1}^{K} a_i \rho(\underline{t},\underline{v}^i).$$

To that end we have

$$(\underline{t},\underline{\delta})/\|\delta\|_\infty = \rho(\underline{t},\underline{\tilde{\delta}}) = 2\, E[\psi^+(\underline{u}\,\underline{t})\, \underline{u}\, \underline{\tilde{\delta}} \mid \underline{u}\, \underline{\delta} \geq 0] P[\underline{u}\, \underline{\delta} \geq 0]$$

$$= 2 \sum_{i=1}^{K} a_i E[\psi^+(\underline{u}\,\underline{t})\, \underline{u}\, \underline{v}^i | \underline{u}\, \underline{\delta} \geq 0] P[\underline{u}\, \underline{\delta} \geq 0]$$

$$= 2 \sum_{i=1}^{K} a_i E[\psi^+(\underline{u}\,\underline{t})\, \underline{u}\, \underline{v}^i | \underline{u}\, \underline{v}^i \geq 0] P[\underline{u}\, \underline{v}^i \geq 0]$$

$$= \sum_{i=1}^{K} a_i \rho(\underline{t},\underline{v}^i)$$

where we have used $\psi^+(x) = -\psi^-(-x)$, and the fact that both $\underline{v}^i$ and $\underline{\tilde{\delta}}$ belong to the simplex S_δ. □

The import of Proposition 4 is that when $\underline{t}$ has a finite number of nonzero components (recall Prop. 1) it is enough to restrict attention to a finite set of directions given by vertices of polytopes in the unit cube (note that any other norm defined by hyperplanes e.g. $\| \ \|_1$ would result in analogous conclusions) which can be generated systematically. Once the conditions corresponding to every vertex have been generated they can be reduced to a smaller linearly independent set.

Figure 1 illustrates the case n = 3 for which we derived conditions (6), (7), and (8). It turns out that these equations are generated by the vertices $\underline{v}^1$, $-\underline{v}^1$ and $\underline{v}^2$ respectively. In particular, note that not every vertex must be investigated because if $\underline{v} \in V$, then $\underline{w} \in V$ where $w_i = v_{p(i)}$ and $p(i)$ is any bijective function on $\{1,\ldots,n\}$, and $\underline{v}$ and $\underline{w}$ result in the same condition. The geometrical insight of this approach and its connections with linear programming could be exploited for obtaining an algorithm that generates efficiently necessary and sufficient conditions for the nonnegativity of the directional derivative of the risk.

Overall, it appears that the approach taken in this work is effective for showing the existence of unequalized points of convergence and the inadmissibility of classes of adaptive laws. Nevertheless, it seems that a substantially different approach is needed to prove or disprove the existence of admissible laws if a reasonably general space of impulse responses is allowed (note that requiring that the system and its inverse have finite energy - cf. [1] - may be too restrictive). Another point is that although the concept of admissibility used here is perhaps more realistic than the more restrictive one used in [1] (that requires convergence to an equalized system with a priori known gain), it introduces a further degree of freedom which makes it difficult to obtain similar results to those obtained with

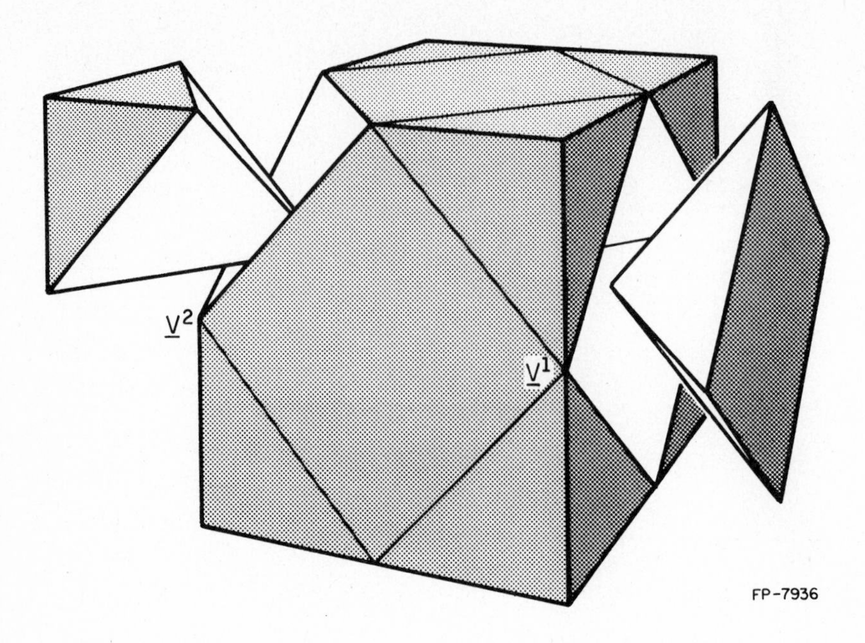

Figure 1. Simplices in unit cube for generating necessary and sufficient conditions for nonnegativity of the directional derivative of the risk.

the narrower sense definition (e.g. Proposition 3). Note finally that the class of memoryless adaptive laws considered here only takes advantage of the one-dimensional distribution of the output of the unequalized linear system; it is plausible that the consideration of dynamic laws which take into account the stochastic dependence of the equalizer outputs, can result in notably improved convergence properties.

ACKNOWLEDGMENT

This work was supported in part by an IBM Pre-doctoral Fellowship and by the U. S. Office of Naval Research under Contract N0014-81-K-0014.

REFERENCES

[1] A. Benveniste, M. Goursat and G. Ruget, "Robust identification of a nonminimum phase system: Blind adjustment of a linear equalizer in data communications," *IEEE Trans. Automatic Control,* vol. AC-25, no. 3, pp. 385-399, June 1980.

[2] Y. Sato, "A method of self-recovering equalization for multilevel amplitude modulation," *IEEE Trans. Communications,* pp. 679-682, June 1975.

[3] D. N. Godard, "Self-recovering equalization and carrier tracking in two-dimensional data communication systems," *IEEE Trans. Communications*, vol. COM-28, no. 11, pp. 1867-1875, Nov. 1980.

[4] A. Benveniste, M. Goursat and G. Ruget, "Analysis of stochastic approximation schemes with discontinuous and dependent forcing terms with applications to data communication algorithms," *IEEE Trans. Automatic Control*, vol. AC-25, no. 6. pp. 1042-1058, Dec. 1980.

[5] G. Ungerboeck, "Comments on 'Self-recovering equalization and carrier tracking in two-dimensional data communication systems'," *IEEE Trans. Communications*, vol. COM-30, no. 3, pp. 557, Mar. 1982.

[6] F. S. Hill, Jr. and M. A. Blanco, "Random geometric series and intersymbol interference," *IEEE Trans. Information Theory*, vol. IT-19, no. 3, pp. 326-335, May 1973.

[7] L. V. Kantorovich and G. P. Akilov, *Functional Analysis*, Second Edition, Oxford: Pergamon Press, 1982.

[8] L. Ljung, "Analysis of a general recursive prediction error identification algorithm," *Automatica*, vol. 17, no. 1, Jan. 1981.

[9] L. Ljung, "Analysis of recursive stochastic algorithms," *IEEE Trans. Automatic Control*, vol. AC-22, no. 4, Aug. 1977.

An efficient implementation of square root filtering : Error analysis, complexity and simulation on flight-path reconstruction

M. H. Verhaegen*
ESAT Laboratory
Katholieke Universiteit Leuven
Kardinaal Mercierlaan 94
B-3030 Heverlee
Belgium

P. Van Dooren
Philips Research Laboratory
Brussels
Av. Van Becelaere, 2
B-1170 Brussels
Belgium

Abstract : In this paper we first develop a new kind of numerical implementation of the so-called "square root filters" used for Kalman filtering. The method is based on the choice of an initial state transformation which "condensed" the model. This preliminary transformation can be done without loss of numerical precision and leads to substantial savings in computing time for the subsequent implementation of the filter. The new filter implementation is also compared with the conventional Kalman filter (KF). For this purpose a detailed analysis is done of the propagation of computational errors in the recursive algorithms. In a second part we use this method for implementing an efficient on-line flight-path reconstruction technique based on a two-step method. The above method is shown to be particularly suited for this problem, even though the model here is inherently time varying.

1. Introduction

The incentive of this paper is the on-line reconstruction of flight paths, a problem which is generally solved via non-linear estimation techniques [1-2]. Since these classical methods require much computational effort and experience, attempts have been made to develop more efficient methods. Recent investigations were based on KF via a reformulation of the problem in a two-step procedure [3],[9]. In order to be suited for on-line use, special care has to be put into the implementation of these filters.

In the first section of this paper a new method to implement time invariant square root filters (SFR), is presented. Although the SRF are known to be superior to the conventional KF in several aspects (preservation of symmetry, semi-positivity, etc [6],[7]) they turn out somewhat slower [6]. U-D covariance filters have been developed that retain numerical stability but are more efficient than the SRF.

By an appropriate choice of initial state-space transformation one can considerably "condense" the model. Three types of such forms are considered : the Schur form, the observer- and controller- Hessenberg forms. These "condensed forms" are the basis of considerable savings in computations during the filter recursion. Since unitary transformations can be used to obtain the initial condensed forms, no loss of accuracy is induced for the subsequent filter implementation. A detailed analysis of the error propagation is performed in order to compare the conventional KF with these

* supported by the IWONL

"cost efficient" SRF's.

In a third section we apply these ideas to the two-step procedure for aircraft model identification [3],[9] . Here the system model is linear (but time varying) due to an appropriate system model reformulation. The model happens to be in Schur form and can also be transformed into observer Hessenberg form despite its time varying nature. The condensed forms prove to be advantageous here and yield significantly faster results than the conventional KF (about 30% savings in computing time). These simulations also confirm the results obtained from the error analysis.

In the interests of brevity, proofs are left out here.

2. Efficient square root filters

In this section we first quickly review the recursions of the CKF (a) and its square root implementation (b,c) and then discuss the computational aspects of their new implementations from the point of view of complexity (d) and numerical stability (e).

a. Linear state reconstruction by Kalman filtering

Let us consider the following discrete time system :

$$\text{process} \quad : \quad x_{k+1} = A_k x_k + F_k u_k + B_k w_k \tag{1}$$

$$\text{observation} \quad : \quad y_k = C_k x_k + v_k \tag{2}$$

where $x_k \in \mathbb{R}^n$, $w_k \in \mathbb{R}^m$ and $v_k \in \mathbb{R}^p$. The deterministic input signal is given by u_k. The sequences w_k and v_k are Gaussian white noises with zero mean and covariances Q_k and R_k respectively.

The conventional filter equations, given for the system model (1-2) can be summarized in the following "covariance form" [8] (measurement and time update are separated here) :

MEASUREMENT UPDATE - filtered estimate

$$R_k^e = R_k + C_k P_{k|k-1} C_k^T \tag{3}$$

$$K_k = P_{k|k-1} C_k^T [R_k^e]^{-1} \tag{4}$$

$$P_{k|k} = [I-K_k C_k] P_{k|k-1} \tag{5}$$

$$\hat{x}_{k|k} = \hat{x}_{k|k-1} + K_k [y_k - C_k \hat{x}_{k|k-1}] \tag{6}$$

TIME UPDATE - one-step predicted estimate

$$\hat{x}_{k+1|k} = A_k \hat{x}_{k|k} + F_k u_k \tag{7}$$

$$P_{k+1|k} = A_k P_{k|k} A_k^T + B_k Q_k B_k^T \tag{8}$$

with $P_{0|-1} = P_0$, the initial state covariance matrix and $\hat{x}_{0|-1} = \hat{x}_0$, the initial state estimate.

This leads to the following algorithm for updating $P_{k|k-1}$ and K_k. The number of "flops" (1 flop = 1 multiplication + 1 addition) for each step is given between brackets (only the leading terms of these expressions are given :

PC	$<= P_{k\|k-1} * C_k^T$	n^2p
REPS	$<= C_k * PC + R_k$	$p^2\ n/2$
K	$<= PC * (REPS)^{-1}$ solving a set of linear equations	$p^3/6 + n\ p^2$
$P_{k+1\|k}$	$<= A_k * (P_{k\|k-1} - K_k * PC^T) * A_k^T + B_k * Q_k * B_k^T$	$3/2\ n^3 + 1/2\ n^2(p+m) + m^2 n$
$\hat{x}_{k+1\|k}$	$<= A_k(I - K_k C_k)\ \hat{x}_{k\|k-1} + A_k\ K_k\ y_k + F_k u_k$	$n^2 + 2np + nm$

b. The Square Root Covariance Filter

Square root covariance algorithms propagate the square root of the error covariance matrix P. The square roots S, $R^{1/2}$ and $Q^{1/2}$ are chosen to be lower triangular :

$$P_{k|k-1} = S_k\ S_k^T \tag{9}$$

$$R_k = R_k^{1/2}\ R_k^{T/2} \tag{10}$$

$$Q = Q^{1/2}\ Q^{T/2} \tag{11}$$

A combined time and measurement update is then summarized for the square root covariance form [8] as follows :

$$T \cdot \begin{bmatrix} R^{T/2} & 0 \\ 0 & Q_k^{T/2}\ B_k^T \\ S^T C_k^T & S_k^T\ A_k^T \end{bmatrix} = \begin{bmatrix} R_k^{eT/2} & K_k'^T\ A_k^T \\ 0 & S_{k+1}^T \\ 0 & 0 \end{bmatrix} \tag{12}$$

(pre-array) (post-array)

$$\hat{x}_{k+1|k} = A_k[\hat{x}_{k|k-1} + K_k'\ R_k^{e-1/2}(y_k - C_k\ \hat{x}_{k|k-1})] + F_k u_k \tag{13}$$

with $S_0\ S_0^T = P_0$. Here T is an arbitrary orthogonal transformation that triangularizes the pre-array.

c. The Square Root Information Filter

The information filter accentuates the recursive least squares nature of filtering. The SRIF propagates the SR of $P_{k|k}^{-1}$, i.e. S_k^{-1}. When choosing upper triangular square roots of the following inverses :

$$Q_k^{-1} = Q_k^{-1/2} \; Q_k^{-T/2} \tag{14}$$

$$R_k^{-1} = R_k^{-1/2} \; R_k^{-T/2} \tag{15}$$

$$P_{k|k}^{-1} = S_k^{-1} \; S_k^{-T} \tag{16}$$

then the combined measurement and time update is given by [8]:

$$\bar{T} \cdot \begin{bmatrix} Q_k^{-1/2} & 0 & Q_k^{-1/2} \, E\{w\} \\ S_k^{-1} \, A_k^{-1} \, B_k & S_k^{-1} \, A_k^{-1} & S_k^{-1} \, \hat{x}_{k|k} \\ 0 & R_{K+1}^{-1/2} \, C_{K+1} & R_{K+1}^{-1/2} \, y_{K+1} \end{bmatrix} = \begin{bmatrix} Q_k^{e1/2} & * & * \\ 0 & S_{k+1}^{-1} & S_{k+1}^{-1} \, \hat{x}_{k+1|k+1} \\ 0 & 0 & r_{k+1} \end{bmatrix} \tag{17}$$

where $E\{.\}$ denotes the mathematical expectation and with the filtered state estimate given by,

$$\hat{x}_{k+1|k+1} = S_{k+1} \, [S_{k+1}^{-1} \, \hat{x}_{k+1|k+1}] \, .$$

d. The Condensed Forms

Three so-called "condensed forms" of an arbitrary dense state-space model (subscripts here denote dimensions of the matrices) :

$$\{A_{nn} \, , \, B_{nm} \, , \, C_{pn}\} \tag{18}$$

can be obtained using a unitary state-space transformation :

1. the Schur form, where A is upper triangular
2. the controller-Hessenberg form and observer-Hessenberg form, where (B,A) and $(A^T,C^T)^T$, respectively, are in upper trapezoidal form.

These three forms are illustrated below for n=2, n=6, p=2 :

$$\left[\begin{array}{c|c} B_s & A_s \\ \hline & C_s \end{array}\right] = \left[\begin{array}{cc|cccccc} x & x & x & x & x & x & x & x \\ x & x & 0 & x & x & x & x & x \\ x & x & 0 & 0 & x & x & x & x \\ x & x & 0 & 0 & 0 & x & x & x \\ x & x & 0 & 0 & 0 & 0 & x & x \\ x & x & 0 & 0 & 0 & 0 & 0 & x \\ \hline & & x & x & x & x & x & x \\ & & x & x & x & x & x & x \end{array}\right] \tag{19}$$

$$\left[\begin{array}{c|c} B_c & A_c \\ \hline & C_c \end{array}\right] = \left[\begin{array}{cc|cccccc} x & x & x & x & x & x & x & x \\ 0 & x & x & x & x & x & x & x \\ 0 & 0 & x & x & x & x & x & x \\ 0 & 0 & 0 & x & x & x & x & x \\ 0 & 0 & 0 & 0 & x & x & x & x \\ 0 & 0 & 0 & 0 & 0 & x & x & x \\ \hline & & x & x & x & x & x & x \\ & & x & x & x & x & x & x \end{array}\right] \qquad (20)$$

$$\left[\begin{array}{c|c} B_0 & A_0 \\ \hline & C_0 \end{array}\right] = \left[\begin{array}{cc|cccccc} x & x & x & x & x & x & x & x \\ x & x & x & x & x & x & x & x \\ x & x & x & x & x & x & x & x \\ x & x & 0 & x & x & x & x & x \\ x & x & 0 & 0 & x & x & x & x \\ x & x & 0 & 0 & 0 & x & x & x \\ \hline & & 0 & 0 & 0 & 0 & x & x \\ & & 0 & 0 & 0 & 0 & 0 & x \end{array}\right] \qquad (21)$$

The number of "flops" needed to obtain these forms is roughly $n^2(5kn+m+p)$, $n^2(3n+p+m)$ and $n^2(3n+p+m)$ respectively. Here k is the average number of "QR steps" used per eigenvalue, which is usually between 1 and 2 [14]. These operation counts include the construction of the corresponding state-space transformation yielding the condensed forms. Notice that the three forms have as many zeros as each other, namely $n(n-1)/2$.

These forms can now be used to reduce the complexity of the SRCF and the SRIF algorithms. Let the system $\{A_k^T, B_k^T, C_k^T\}$ be in the Schur form (19), then this structure can be exploited to yield the following algorithm, discussed in [15]. The complexity is also included.

$M_{11} \Leftarrow R_k^{T/2}$	
$M_{22} \Leftarrow Q_k^{T/2} B_k^T$	$m^2 n/2$
$M_{31} \Leftarrow S_k^T C_k^T$; $M_{33} \Leftarrow S_k^T A_k^T$ (together)	$n^3/3+pn^2/2$
step 1 (anihilating M_{31} with Givens)	$2np^2+2n^2p$
step 2 (further triangularization with Householder)	pn^2
$\mathrm{diag}\,\{P_{k+1\|k}\} = \mathrm{diag}\,\{S_k S_k^T\}$	$n^2/2$
$\hat{x}_{k+1\|k} = A_k \hat{x}_{k\|k-1} + A_k K_k (y_k - C_k \hat{x}_{k\|k-1}) + F_k u_k$	$n^2+2np+nm$

When using an observer Hessenberg form $\{A_k^T, B_k^T, C_k^T\}$ the algorithm and the complexity becomes :

$M_{11} \Leftarrow R_k^{1/2}$	
$M_{22} \Leftarrow B_k^T Q_k^{T/2}$	$m^2n/2$
$M_{31} \Leftarrow S_k^T C_k^T$, $M_{32} \Leftarrow S_k^T A_k^T$	$n^3/3+pn^2/2$
step 1 (anihilation of M_{31} with Householder)	$p^2n+p^3/3$
step 2 (further triangularization with Householder)	$n^2(p+m)$
diag $p_{k\|k+1}$	$n^2/2$
$\hat{x}_{k\|k+1}$	$n^2+2np+nm$

The use of condensed forms speeds up all three methods considerably but also makes the SRCF algorithms more effective than the other two. This follow from the operation counts summarized in Table 1. (see also simulation results)

e. Error propagation

All the methods discussed in the previous section are algebraically equivalent. We will now concentrate on the convergence properties of the recursive Ricatti equation (8), under the presence of roundoff errors.
The following theorems prove the inherent numerical stability of these equations. the first theorem considers the propagation of an error δP_k on $P_{k|k-1}$ to an error δP_{k+1}, when no additional rounding errors are performed during this step.
Let $\bar{P}_k$ and $\bar{R}_k^e$ be defined as follows :

$$\bar{P}_k = P_{k|k-1} + \delta P_k \tag{23a}$$

$$\bar{R}_k^e = R_k + C_k [\bar{P}_k] C_k^T = R_k^e + C_k\, \delta P_k\, C_k^T \tag{23b}$$

then we have the following Lemma, where $|.|$ is any consistent norm.

Lemma 1 : The following relations between $\bar{R}_k^{e-1}$ and R_k^{e-1} hold :

$$\bar{R}_k^{e-1} = R_k^{e-1} + \delta R_1 \tag{24}$$

$$\bar{R}_k^{e-1} = R_k^{e-1} - R_k^{e-1}(\bar{R}_k^{e-1} - R_k^{e-1})R_k^{e-1} + \delta R_2 \tag{25}$$

where $|\delta R_1| = O(\delta)$
$|\delta R_2| = O(\delta^2)$

□

Using this, we obtain the following theorem :

Theorem 1 : If the error δP_k and δx_k are sufficiently small (namely $\delta = \max\{|\delta P_k||C_k^T R_k^{e-1} C_k|, |\delta x_k|/|x_k|\} << 1$)

then the error propagation on $P_{k+1|k}$ and $\hat{x}_{k+1|k}$ are given by ,

$$\delta P_{k+1} = A_k^f \delta P_k A_k^{fT} + O(\delta^2) = \bar{A}_k^f \delta P_k \bar{A}_k^{fT} + O(\delta^2) \qquad (27)$$

$$\delta x_{k+1} = \bar{A}_k^f (\delta x_k + \delta P_k C_k^T \bar{R}_k^{e-1} [y_k - C_k \hat{x}_{k|k-1}]) + O(\delta^2) \qquad (28)$$

□

Corollary 1. The theorem is very useful when convergence is almost achieved, i.e. when $\bar{P}_k - P_\infty$ is small. When the pair $[A,C]$ is detectable and the pair $[A, BQ^{T/2}]$ is stabilizable, the resulting filter $\bar{A}_k^f$ will be close to A_∞^f and hence exponentially stable. For this case one has,

$$\delta P_{k+1} \simeq \bar{A}_\infty^f \delta P_k \bar{A}_\infty^{fT} \qquad (29)$$

$$\delta x_{k+1} \simeq \bar{A}_\infty^f (\delta x_k + \delta P_c C_k^T \bar{R}_k^{e-1} [y_k - C_k \hat{x}_{k|k-1}]) \qquad (30)$$

This means that the KF has a forgetting factor built in for errors resulting from the previous steps.

□

In order to be more rigorous, one should also take the errors of step k+1, and more importantly their interaction with errors of the previous step, into account. The total errors $\delta_t P_{k+1}$ and $\delta_t x_{k+1}$ in step k+1, can be modelled as,

$$\delta_t P_{k+1} = \bar{A}_k^f \delta_t P_k \bar{A}_k^{fT} + \Delta_p\{A_k, B_k, C_k, R_k, \Delta t, \delta_t P_k\} \qquad (31)$$

$$\delta_t x_{k+1} = \bar{A}_k^f (\delta_t x_k + \delta P_k C_k^T \bar{R}_k^{e-1} [y_k - C_k \hat{x}_{k|k-1}]) + \Delta_x \{A_k, B_k, C_k, Q_k, R_k, \Delta t, \delta_t P_k\} \qquad (32)$$

where $\Delta\{.\}$ is a nonlinear perturbation due to the interaction of roundoff errors in step k+1. A similar formula for the product $\bar{A}_k \bar{K}_k$ can also be made. Notice that this equation is general enough to include model errors and the errors made by discretization as well, see [13]. For these roundoff errors one can now formulate upper bounds that are independent of $\delta_t P_k$ and $\delta_t x_k$ as is shown in the next theorem, based on [16][17] .

Theorem 2. Denoting the errors due to roundoff in the construction of $P_{k+1|k}$, $A_k K_k$ and $\hat{x}_{k+1|k}$ by Δ_p, Δ_{ak} and Δ_x respectively, we obtain the following upper bounds (for some constants $\cos\phi$, $\cos\psi$ and $\cos\chi$ between 0 and 1) :

(i) CKF

$$|\Delta_p| \leqslant \varepsilon c_1 |P_{k+1}| \kappa(R_k^e) \qquad (33)$$

$$|\Delta_k| \leqslant \varepsilon c_2 |K_k| \kappa(R_k^e) \qquad (34)$$

$$|\Delta_x| \leqslant |\Delta_k||A_k|\{|C_k||x_k|+|y_k|\}+\varepsilon\{|A_k^f||x_k|+|A_k||K_k||y_k|+|F_k||u_k|\} \qquad (35)$$

(ii) SRCF

$$|\Delta_p| \leqslant \varepsilon c_3 (1 + \frac{\bar{\sigma}_1}{\bar{\sigma}_p}) |P_{k+1}|/\cos\phi \qquad (36)$$

$$|\Delta_{ak}| \leqslant \varepsilon \frac{c_4}{\bar{\sigma}_p} (\frac{\bar{\sigma}_1}{\bar{\sigma}_p} |S_{k+1}| + \bar{\sigma}_1 |A_k K_k| + |S_{k+1}|/\cos\phi) \qquad (37)$$

$$|\Delta_x| \leqslant |\Delta_{ak}|\{|C_k||x_k|+|y_k|\}+\varepsilon\{c_5|A_k^f||x_k|+|A_k K_k||y_k|+|F_k||x_k|\} \qquad (38)$$

(iii) SRIF (here $A_k K_k$ is not computed explicitely)

$$|\Delta_s| \leqslant \varepsilon c_6\{\kappa(A_k)+\kappa(R_k^{1/2}) + \frac{\bar{\sigma}_1}{\bar{\sigma}_p}[\kappa(Q_k^{1/2}) + \kappa(A_k)]\}|S_{k+1}^{-1}| \cos\psi \qquad (39)$$

$$|\Delta_x| \leqslant |\Delta_s|\{\kappa(P_{k+1}^{-1})|r_{k+1}|+\kappa(S_{k+1}^{-1})|x_{k+1}|+|r_{k+1}|/\cos\chi\} \qquad (40)$$

with

$$|\Delta_p| \leqslant 2|\Delta_s||S_{k+1}^{-1}|$$

where $\bar{\sigma}_i$ is the ith singular value of $\bar{R}_k^{e1/2}$ and $\kappa(.)$ stands for the condition number of a matrix □

In the literature SRF are generally claimed (e.g. [6,7])to behave better because their conditioning is the "square root" of that of the CKF. However, this is only valid for the error on the covariance matrix as is easily seen from (33-40). (See also [16] for an explanation of this phenomenon).

Corollary 2. From this theorem the behaviour of the errors in the computer is modelled by a linear time varying system with bounded external input. In this case estimates for $\delta_t P_{k+1}$ and for $\delta_t K_k$ can be given [18] :

$$|\delta P_{k+1}| \simeq |\Delta_p|/(1-\bar{\gamma}^2) \qquad (41)$$

$$|\delta K_k| \simeq |\Delta_p|/(1-\bar{\gamma}) + |\Delta_k| \qquad (42)$$

if and only if the spectral norm $\bar{\gamma}$ of $\bar{A}_k^f$ is smaller than 1. □

Corollary 3. The usefulness of this theorem depends again on the convergence of $\bar{A}_k^f$ to A_∞^f. In this context the assumption of corollary 2 is rather severe and we can only assume the spectral radius $\bar{\rho}$ of $\bar{A}_k^f$ to be smaller than 1. In this case though, one still obtaines a similar result to (40-41) (with $\bar{\gamma}$ replaced by $\bar{\rho}$). □

The above corollaries indicate the extreme robustness of Kalman filters. Even in the presence of rounding errors, the computational solution does not diverge from the exact solution when the conditions of theorem 2 and corollary 2 (or 3) are satisfied. As indicated by the error analysis (33)-(40) and also by the experiments of the next subsection, divergence is more likely to occur with the CKF than with the SR versions. In fact, it never occurred with the SR filter in the performed experiments.

f. Computational tests

We now demonstrate a number of effects mentioned in previous subsections by simulations. The relevant variables that influence the error propagation to be considered, are :

1. $\kappa(Q_k)$
2. $\kappa(A_k)$. This can be affected by a state space transformation T with a large condition number.
3. $\kappa(R_k^e)$. In the simulations, it turns out that the initial choice of $\kappa(R_k)$ approximately determines $\kappa(R_k^e)$ during the whole run.
4. $\kappa(A_k^f)$. This can be affected by "weighting" of the system state matrix A_k by a factor EWF.
5. $\kappa(S_k)$. Is hard to control.

The simulations are carried out for time invariant system matrices and only the CKF and the Schur SRCF are considered here. The system model is the mathematical aircraft model used for flight-path reconstruction in a two-step indentification procedure, discussed in the next section. This model is ideally suited for demonstrating the result obtained by the error analysis. The eigenvalues of the state matrix A_k^f are indeed very close to the unit circle, whence this example is sensitive to numerical roundoff.

To study roundoff errors, mixed precision computations were carried out. Double precision results are considered to be exact. In the experiments the following quantities were compared :

1. the infinite solution of the Ricatti equation P_∞ was calculated with the eigenvector-eigenvalue method described in [19].
2. the recursive solution of (8) both in double and single precision for CKF and the SRCF. P_k and $\bar{P}_k$ denote the double, respectively the single precision results. $|\bar{P}_k - P_k|/|\bar{P}_k|$ therefore reflects the computational errors accumulated at step k.

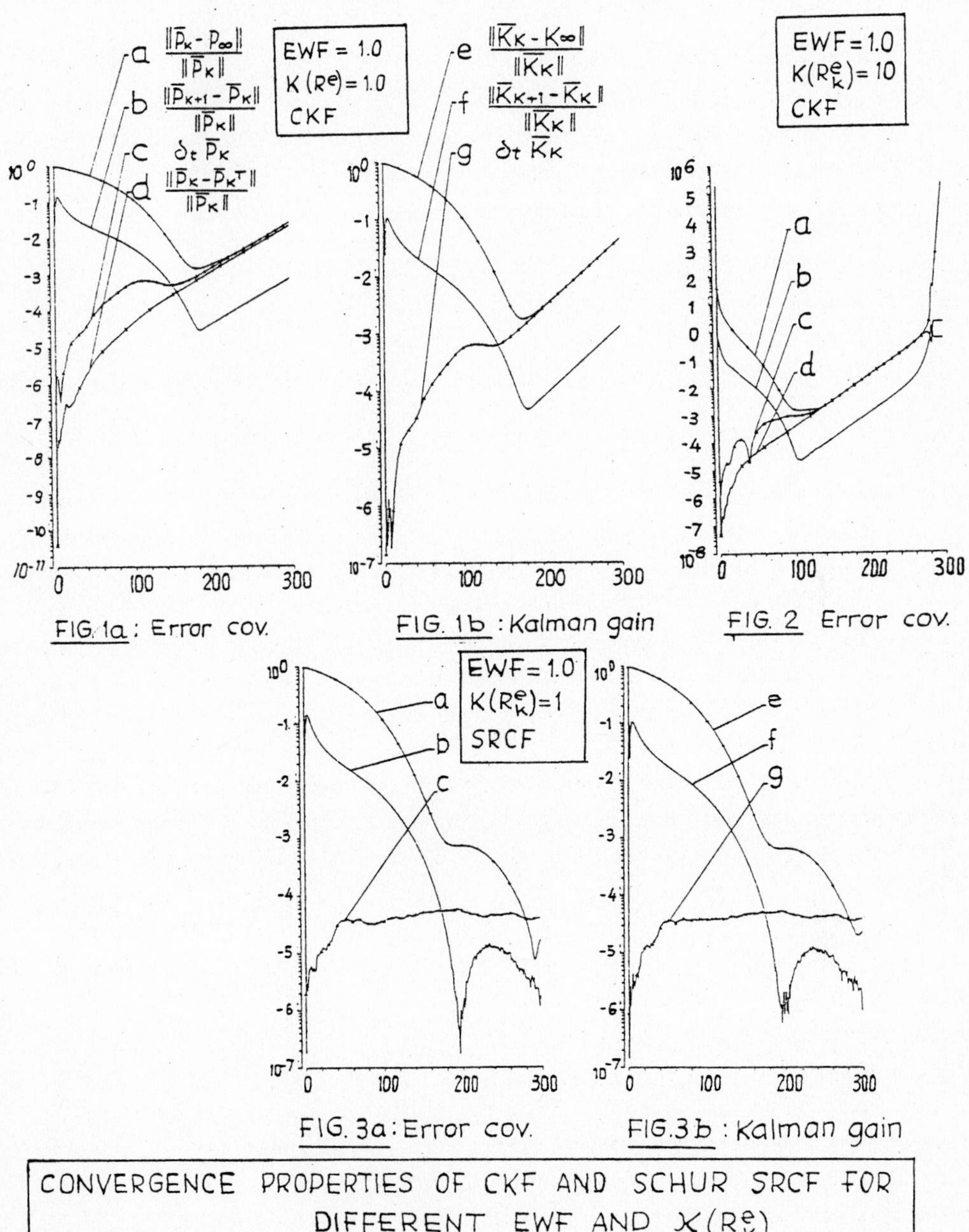

CONVERGENCE PROPERTIES OF CKF AND SCHUR SRCF FOR DIFFERENT EWF AND $\mathcal{K}(R^e_k)$

$|\bar{P}_k - P_\infty| / |\bar{P}_k|$ therefore indicates the exact convergence.

We also plotted $|\bar{P}_k - \bar{P}_{k+1}| / |\bar{P}_k|$ to show that it also indicates convergence reasonably well.

Similar quantities are plotted for $\bar{K}_k$ as well.

Test 1 : (EWF=1, $\kappa(R^e)$=1)

Figures 1a,b give the result for the CKF and fig. 3a,b for the SRCF. When $\kappa(R_k^e)$ is changed to 10 the result is graphed in fig.2.

From these figures the following observations are made :

1. (fig. 1a -CKF) when convergence is achieved, i.e. when the result $\bar{P}_k$ becomes very close to P_∞, A_k^f is not sufficiently stable and the solution start to diverge. (the spectral radius of $\bar{A}_k^f$ comes in the vicinity of 0.995. Notice that the loss of accuracy goes along with the loss of symmetry.
2. (fig. 2 - CKF) with the same spectral radius of $\bar{A}_k^f$, the effect of increasing $\kappa(R_k)$ clearly makes $\delta_t P_k$ higher and divergence starts earlier.
3. (fig. 3a - SRCF) $\delta_t\ \bar{P}_k$ does not accumulate at all even when $\bar{P}_k$ approaches P_∞ and $\bar{A}_k^f$ has eigenvalues very close to 1. In this case the error level $\delta_t\ P_k$ remains small enough to guarantee numerical convergence. Changing $\kappa(R_k)$ to 10 doesn's cause divergence but the residual error is larger as expected.

The convergence of the Kalman gain is given in fig. 1b and fig. 3b. Since $\kappa(R_k)$ is 1, $|\delta_t\ \bar{K}_k|$ is comparable to $|\delta_t\ \bar{P}_k|$ as predicted by theorem 2.

Test 2 : (EWF = 0.9, $\kappa(R^e)$=100)

In this experiment the spectral radius ρ of A_k^f tends to become smaller than 0.9. The simulation results are given in fig. 4a,b and fig. 5.a,b. From these figures we make the following observations :

1. (fig. 4a,b - CKF) from eq. 33 the error $\delta_t\ \bar{P}_k$ will be of a higher level than in the previous experiment, because of the higher $\kappa(R_k^e)$.
2. (fig. 5a,b - SRCF) from eq. 36 the constant level of $\delta_t\ \bar{P}_k$ for large k is indeed smaller. The ratio between the constant level for the CKF and the SRCF is approximately $\kappa(R_k^e)^{1/2}$.

Also the similar behaviour of $\delta_t\ \bar{K}_k$ with $\delta_t\ \bar{P}_k$ is apparent now and the level for large k is higher for both the CKF and the SRCF.

Test 3 : (EWF=1.1, $\kappa(R^e)$=1)

The choice of an EWF larger than 1, results in overweighting more recent data [8] . In this case when the system is detectable and stabilizable, we theoretically know that a stable filter will result even when the system under consideration is unstable.

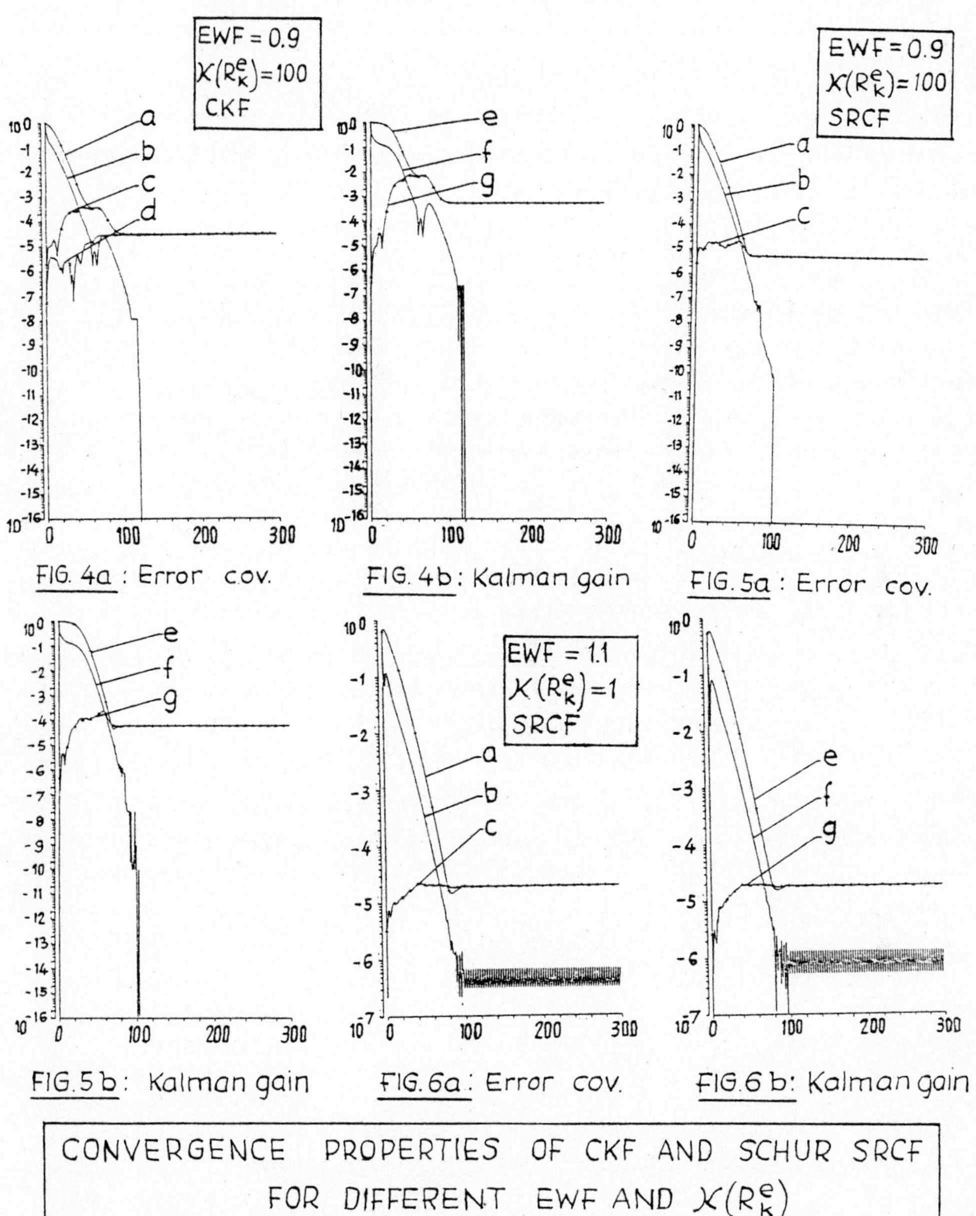

FIG. 4a : Error cov.

FIG. 4b : Kalman gain

FIG. 5a : Error cov.

FIG. 5b : Kalman gain

FIG. 6a : Error cov.

FIG. 6b : Kalman gain

CONVERGENCE PROPERTIES OF CKF AND SCHUR SRCF FOR DIFFERENT EWF AND $\mathcal{K}(R_k^e)$

However, some modes of the system under consideration will remain unstable over some time interval during the recursions. This may depend on how controllable these modes are. During this time interval the error $\delta_t \bar{P}_k$ in the CKF now grows so fast that complete divergence results (overflow).

For the SRCF, the numerical computations resulted in a stable filter, i.e. the spectral radius became smaller than 0.9. From fig. 6a,b the convergence behaviour is similar to the previous experiment for both $\delta_t \bar{P}_k$ and $\delta_t \bar{K}_k$. We can however observe the rather wild transient behaviour.

3. Flight path reconstruction

On-line algorithms have to satisfy very severe requirements of efficiency and realiability. These seemingly contradicting features are both obtained in the problem of flight path reconstruction when using a two-step procedure.

First, this reformulation allows to reconstruct flight path using linear filtering of a stochastic process generated by a time varying system. Secondly, this time varying system can be transformed via a constant transformation to a Schur or observer Hessenberg form. The latter then enables us to use the cost efficient SR filters described in the previous section.

a. Condensed system models of the mathematical aircraft description

In this paper we restrict ourselves to longitudinal aircraft motions. The modelling of a rigid aircraft is based on Newton's second law. Only small perturbations from some stationary flight condition are considered.

A linear state space model with the same input and output signals as the physical system is based on a linearization of the areodynamical forces and moments. This state space description (input, state and output variable are listed in table 2) is the basis of the ONE-STEP identification method [9]. The system matrices now contain unknown aerodynamic derivatives. In [9] it is shown that when measuring the inertial signals very accurately, a new state space model can be defined, using new state, input and output variables (table 3) :

$$\begin{cases} \dot{x}_{fp} = A_{fp}\, x_{fp} + B_{fp}\, u_{fp} \\ y_{fp} = C_{fp}\, x_{fp} \end{cases} \tag{43}$$

Here f_p indicates that the model is only of mathematical importance for reconstructing the flight-path. This model has interesting properties [9]. The system matrix A_{fp} is nilpotent (with index of nilpotency 3) and the matrices,

$$A_k = \exp(A_{fp}\,\Delta t)\ ;\ \ B_k = [\,{}_0\!\int^t \exp(A_{fp}\,\tau)d\tau\,]B_{fp}\ ;\ \ C_k = C_{fp}$$

of the discretized version of (43) :

Filter	Implementation	complexity
CKF	full	$3/2n^3+n^2(3/2p+1/2m)+n(3/2p^2+m^2)+1/6p^3$
	Schur	$3/4n^3+n^2(\ \ 2p+\ \ \ m)+n(\ \ 2p^2+m^2)+1/6p^3$
	Hess	$3/4n^3+n^2(\ \ 3p+\ \ \ m)+n(3/2p^2+m^2)+2/3p^3$
SRCF	full	$7/6n^3+n^2(5/2p+\ \ \ m)+n(2p^2+1/2m^2)+2/3p^3$
	Schur	$1/3n^3+n^2(7/2p\ \ \ \ \ \ )+n(2p^2+1/2m^2)$
	Hess	$1/3n^3+n^2(3/2p+\ \ \ m)+n(\ p^2+1/2m^2)+1/2p^3$
SRIF	full	$5/3n^3+n^2(7/2m+\ \ \ p)+n(2m^2+1/2p^2)+5/6m^3$
	Schur	$1/3n^3+n^2(7/2m\ \ \ \ \ \ )+n(2m^2+1/2p^2)+1/6m^3$
	Hess	no reduction

Table 1: Operation counts for different filter implemtation

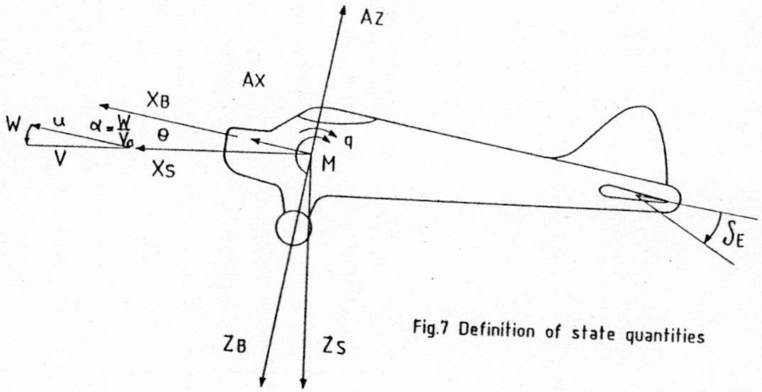

Fig.7 Definition of state quantities

	quantity	symbol	unit
input	elevator angle	δ_e	rad
state	speed along X-axis	u	m/s
	angle of attack	α	rad
	angle of pitch	Θ	rad
	rate of pitch (dimensionless)	$\frac{q\bar{c}}{V_0}$	/
	altitude deviation	Δh	m
output	speed along X-axis	u	m/s
	altitude deviation	Δh	m
	acceleration along x-axis	A_x	m/s^2
	acceleration along	A_z	m/s^2
	rate of pitch	q	rad/s

Table 2: System state variables of the physical model representation (ONE-STEP method)

	quantity	symbol	unit
input	acceleration along x-axis	A_x	m/s^2
	acceleration along	A_z	m/s^2
	rate of pitch	q	rad/s
state	speed along X-axis	u	m/s
	angle of attack	α	rad
	angle of pitch	Θ	rad
	altitude deviation	Δh	m
	bias on A_x	λ_x	/
	bias on A_z	λ_z	/
output	speed along X-axis	u	m/s
	altitude deviation	Δh	m

Table 3: System state variables of the mathematical model for flightpath reconstruction (TWO-STEP method)

	quantity	1-σ	unit	
output	speed along X-axis	2.06e-1	m/s	diagonal elements
	altitude deviation	4.10e-1	m	R-matrix
	acceleration along x-axis	2.78e-3	m/s^2	diagonal elements
	acceleration along	6.98e-3	m/s^2	Q-matrix
	rate of pitch	6.97e-5	rad/s	

Table 4: Measurement error statistics

$$\begin{cases} x_{k+1} = A_k\, x_k + B_k\, u_k \\ y_k = C_k\, x_k \end{cases} \tag{44}$$

can be computed <u>analytically</u>. Furthermore, condensed forms are easily obtained for this model. A reordering of the state vector of table 3, according to

$$x^T = [\lambda_z, \lambda_q, \theta, \alpha, u, \Delta h]$$

results in a lower <u>Schur</u> form for the matrix A_k, and an additional

constant state-space transformation :

$$z_k = T\, x_k = \begin{bmatrix} 1 & 0 & 0 & 0 & 0 & 0 \\ 0 & 1 & 0 & 0 & 0 & 0 \\ 0 & 0 & 1 & 0 & \Delta t/2 & 0 \\ 0 & 0 & -1 & 1 & 0 & \Delta t/2 \\ 0 & 0 & 0 & 0 & 1 & 0 \\ 0 & 0 & 0 & 0 & 0 & 1 \end{bmatrix} x_k \tag{45}$$

yields a (lower) observer Hessenberg form.

b. <u>Performance analysis</u>

Three different types of SRF, i.e. the <u>Schur SRCF</u>, <u>observer Hessenberg SRCF</u> and the <u>Schur SRIF</u> are evaluated. In the simulations a linear system model of the aircraft was used. The unknown aerodynamic derivatives in the system matrices were taken from previous flight test experiments on the DHC-2 Beaver Laboratory aircraft [5] of the Delft University of Technology. The main advantage is now that we could use the <u>real</u> flight path in the performance analysis. In the preliminary experiments we choose some optimal input signal [9] for the elevator angle δ_e and no system model errors were considered. The output signals of table 2 are corrupted by zero mean white noise with variances given in table 4, taken from sensor calibrations. The experiments were performed in double precision only, on an IBM 360 computer of the Catholic University Leuven.

The constant <u>bias errors</u> λ_z and λ_q on the inertial signals, included in the state vector, are not controllable. Whence the filter problem is not solvable. From [4] a solution is forced by introducing a "random walk" process. In this case we model the bias terms as,

$$\lambda_{k+1} = \lambda_k + w_k \tag{46}$$

where the zero mean white noise term w_k allows "small" time variations.

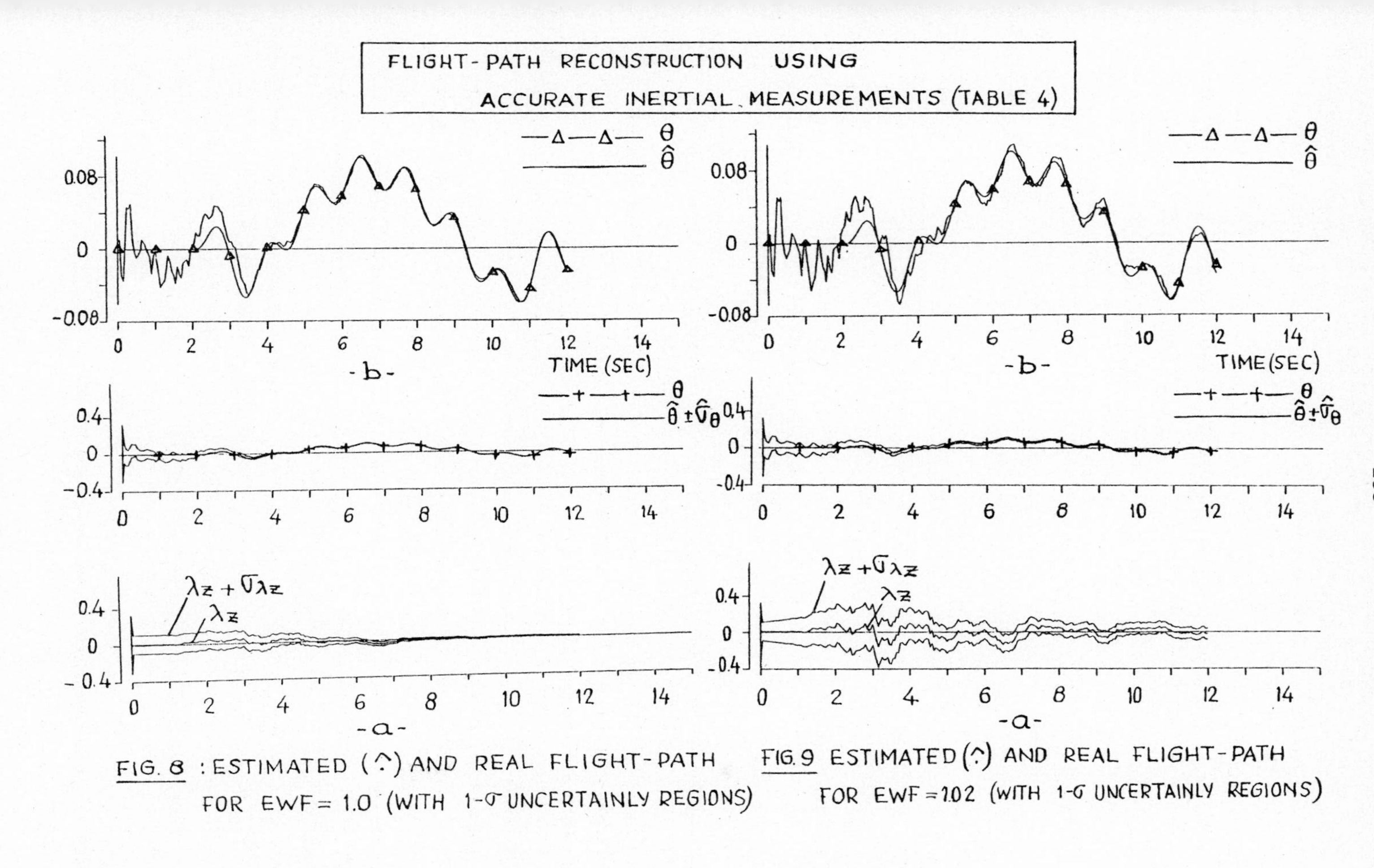

FIG. 8 : ESTIMATED (^) AND REAL FLIGHT-PATH FOR EWF = 1.0 (WITH 1-σ UNCERTAINLY REGIONS)

FIG. 9 ESTIMATED (^) AND REAL FLIGHT-PATH FOR EWF = 1.02 (WITH 1-σ UNCERTAINLY REGIONS)

In the first experiment the Schur SRCF, SRIF and observer Hessenberg SRCF were tested. They all use the same noise statistics for Q, R as in the simulation program. The results of the three different filters were the same and that of the Schur SRCF is pictured in fig. 8 (here only θ and λ_z are shown).
In figure 8a we plotted the 1-σ uncertainty region around the estimated state. For the considered circumstances (without model errors) the real flight path lies completely in this region. And through the use of very accurate measurements the estimated flight path in fig. 8b coïncides with the real flight path with almost zero uncertainty. The main difference for the SR state reconstruction algorithms was their execution time :

Schur SRCF (m=5) : 1765 ms CPU for 241 iterations
O-Hes SRCF (m=5) : 1592 ms CPU for 241 iterations
Schur SRIF (m=5) : 2645 ms CPU for 240 iterations
Schur SRCF (m=3) : 1848 ms CPU for 300 iterations
CKF (m=3) : 2345 ms CPU for 300 iterations

In the second experiment we demonstrated the effect of exponential weighting only for the Schur SRCF with the same Q and R covariance matrices. From fig. 9 it becomes apparent that the use of exponential weighting introduces higher input noise, as described in [8]. However, through the weak controllability of the bias term λ_z the 1-σ uncertainty is much greater than in the previous experiment, although a very small EWF (1.02) was used. Through the bad controllability of this parameter this increase in uncertainty is necessary to stabilize the system (to place the pole within the unit circle). The use of a much higher EWF has a disastrous effect on the estimation results, and requires further investigation.

4. Conclusions

The numerical implementation of SRF, based on condensed forms, was shown to have a much lower complexity than the normal ones. Due to this complexity reduction, they become faster than the CKF, while still preserving their better numerical performance. The latter is illustrated by a detailed error analysis and by simulations as well. This method was then used for recursive flight path reconstruction where its advantages become apparent.
Even more considerable savings can be expected for sparse systems since the most time consuming step in the new SRCF implementations is the multiplication $S_k^T A_k^T$.
Finally, in order to make this method useful for on-line applications one should be able to check and control convergence of the algorithm in a robust manner. For this, a more detailed analysis has to be performed of the effect of exponential weighting [12], measures of the controllability of the system, estimating $\rho(A_k^f)$, estimating $\delta_t(.)$, the effect of diagonal balancing, sampling rate, etc. These topics are still being studied.

References

1. A.E. Bryson, Mini-issue on NASA's advanced control law program for the F-8c DFBW aircraft. IEEE Trans. AC-22, N° 5, Oct. 1977, pp. 752-805.

2. G. L. Hartmann et al. , F-8C adaptive flight control laws. NASA report CR-2880, 1977.

3. O. H. Gerlach, Analysis of a method for determination of performances and stability derivatives in non-stationary symmetric flights. Delft University of Technology. VTH-Report 117 (in Dutch)

4. P. C. Young, A second generation adaptive autostabilization system for airborn vehicles. Automatica, vol. 17, N° 3, pp. 459-490, 1981.

5. H. L. Jonckers, Application of Kalman filter to flight path reconstruction from flight test data including estimation of instrumental bias error corrections. Delft University of Technology.

6. P. G. Kaminski et al., Discrete square root filtering, A survey of current techniques. IEEE Trans. AC-16, pp. 727-735, Dec. 1971.

7. G. J. Bierman et al., Numerical comparison of Kalman filter algorithms : Orbit determination case study. Automatica, Vol. 13, pp. 23-35, 1977.

8. B. D. O. Anderson and J. B. Moore, Optimal filtering, Prentice-Hall, Information and System Sciences Series, 1979.

9. M. H. Verhaegen, An analysis of two methods for analysis of dynamic flight test manoeuvres. ACI'83 IASTED symposium Copenhagen, 1983.

10. P. Van Dooren and M. H. Verhaegen, The use of condensed forms in linear system theory. IEE Worshop on Robust Numerical Software in Control Systems Design, London, Nov. 1983.

11. H. Kwakernaak and R. Sivan, Linear optimal control systems, New York, Wiley Interscience, 1972.

12. M. Verhaegen and P. Van Dooren, A robust and efficient technique for dealing with time-varying instrumental bias in linear filtering. MIC'84 IASTED Symposium Innsbrück, 1984.

13. A. H. Jazwinski, Stochastic processes and filtering theory. Ac. Press, New York, London, 1970.

14. J. Wilkinson and C. Reinsch, Handbook for automatic computation II Linear algebra. Springer Berlin, 1971.

15. M. H. Verhaegen, Fortran implementations of efficient square root algorithms. Catholic University Leuven.

16. A. Van der Sluis, Stability of the solutions of Linear Least squares problems. Numer. Math. 23, pp. 241-254, 1975.

17. G. Stewart, On the perturbation of pseudo-inverses, projections and linear least squares problems. SIAM Review, Vol. 19, N° 4, Oct. 1977.

18. P. Henrici, Discrete variable methods for ODE. Wiley New York, 1962.

19. P. Van Dooren, A generalized eigenvalue approach for solving Riccati equations, SIAM Sciec., Vol. 2, N° 2, June 1981.

ON THE USE OF VARIOGRAMS IN LEVINSON PREDICTORS

Michel Gevers
Department of Systems Engineering
Australian National University
Canberra, Australia

Abstract

We consider the prediction of stationary stochastic processes with non-zero mean. When the covariance of the process is known, but the mean is not, the classical approach is to first estimate the mean from the past data, and then apply an optimal predictor to the zero-mean residuals. Bastin and Henriet [1] showed that an alternative was to use a predictor based on "variograms" rather than covariance information, thus avoiding the estimation of the mean. We show here that the two predictors are identical, when the unknown mean is replaced by its minimum variance estimate. We also argue that in a practical situation, where the covariance is unknown, it is wiser to use the variogram approach.

1 Introduction

We consider a discrete scalar stochastic process $\{y_0, y_1, \ldots, y_t, \ldots\}$ with constant but not necessarily zero mean:

$$E\{y_t\} = m \qquad (1)$$

We define the "variogram" of the process as the semi-variance of the increments

$$\gamma(t,t-\tau) = \tfrac{1}{2} E\{(y_t - y_{t-\tau})^2\} \qquad (2)$$

We shall assume throughout that the variogram is stationary, i.e.

$$\gamma(t,t-\tau) = \gamma(\tau) = \tfrac{1}{2} E\{(y_t - y_{t-\tau})^2\} \tag{3}$$

A process with the properties (1) and (3) is called intrinsic (see e.g. [2]). If the process is wide-sense stationary, then the (auto)covariance can be defined as

$$R(\tau) = E\{(y_t - m)(y_{t-\tau} - m)\} \tag{4}$$

For a wide-sense stationary process, the covariance function and the variogram are related by

$$\gamma(\tau) = \sigma^2 - R(\tau) \text{ , with } \sigma^2 = R(0) \tag{5}$$

Note that an intrinsic random process is always wide-sense stationary, but the converse is not true. The class of intrinsic processes is larger and includes Wiener processes as a special case. Except when specifically stated, we shall from now on assume that the process $\{y_t\}$ is wide-sense stationary.

In this paper we shall derive different minimum variance unbiased (MVU) expressions for (d+1)-step ahead predictors of the $\{y_t\}$ process under a variety of assumptions. We shall not make any assumption about the existence of an underlying finite-dimensional model. Hence we shall consider Levinson predictors with growing memory: the predicted value at time t=N uses all available past data. In Section 2 we shall briefly recall the expressions of the classical Levinson predictor (CLP) for y_{N+d}, given $\{y_0, y_1, \ldots, y_{N-1}\}$, under the assumptions that the constant mean m and the covariance (or the variogram) are known. In Section 3 we consider the case where the mean is unknown. Two different MVU predictors can be used in this case:

1) one can compute an unbiased estimate $\hat{m}$, and then replace the mean by its estimate $\hat{m}$ in the expressions of the CLP of Section 2. This predictor will be called Approximate Levinson Predictor (ALP);

2) alternatively, one can use a MVU predictor of the form $\sum_{i=1}^{N} b_i y_{N-i}$, in which the b_i are computed from the variogram function. This predictor, derived by Bastin and Henriet [1], who called it the modified Levinson predictor, does not involve the mean or its estimate in any way.

In Section 4 we show that the two predictors are exactly identical if the MVU estimate is used for $\hat{m}$ in the ALP. This may appear as a surprising result, considering that the first involves an estimate of the mean, while the other does not. Our result corrects a result of Bastin and Henriet, who claimed that the MLP leads to a strictly smaller prediction error variance than the ALP. We shall further derive a number of interesting expressions for the MLP and its prediction error variance, and show that all these quantities can be expressed in terms of either the covariance function or the variogram. In Section 5, we shall give a number of reasons why the MLP, written in terms of the variogram, is to be preferred in practical situations, where both the mean and the covariance function are unknown, and have to be estimated from the data.

2 Processes with known mean and covariance function

When the mean is known, the classical (d+1)-step ahead Levinson predictor (CLP) for y_{N+d} given $y_0,\dots,y_{N-1}$ has the form (see e.g. [3]):

$$\hat{y}_{N+d} = m + \sum_{i=1}^{N} \mathbf{a}_i \, (y_{N-i}-m) = m + \mathbf{a}^T(Y-mU) \qquad (6)$$

where : $\mathbf{a}^T = (a_1,\dots,a_N)$, $U^T = (1,1,\dots 1)$, $Y^T = (y_{N-1},\dots,y_0)$. The prediction error $\tilde{y}_{N+d} \triangleq \hat{y}_{N+d} - y_{N+d}$ is unbiased. The minimization of the prediction error variance $E\{(\tilde{y}_{N+d})^2\}$ with respect to the a_i leads to the following system of N linear equations:

$$R\mathbf{a} = R_0 \qquad (7)$$

where

$$R = E\{(Y-mU)(Y-mU)^T\} \qquad (8a)$$

$$R_0 = E\{(Y-mU)(y_{N+d}-m)\} \qquad (8b)$$

Using (5), we can also write:

$$R = \sigma^2 UU^T - \Gamma \;,\; R_0 = \sigma^2 U - \Gamma_0 \qquad (9)$$

where

$$\Gamma = \begin{vmatrix} \gamma(0) & \gamma(1) & \dots & \gamma(N-1) \\ \gamma(1) & \gamma(0) & & \gamma(N-2) \\ & \cdot & & \\ & & \cdot & \\ & & & \cdot \\ \gamma(N-1) & \gamma(N-2) & \dots & \gamma(0) \end{vmatrix} \;,\; \Gamma_0 = \begin{vmatrix} \gamma(d+1) \\ \gamma(d+2) \\ \cdot \\ \cdot \\ \cdot \\ \gamma(d+N) \end{vmatrix} \qquad (10)$$

Note that $\gamma(0) = 0$. Substituting (9) in (7) yields an alternative system of equations for $\mathbf{a}$ in terms of the variogram:

$$\begin{vmatrix} \Gamma & U \\ U^T & \frac{1}{\sigma^2} \end{vmatrix} \begin{vmatrix} \mathbf{a} \\ \alpha\sigma^2 \end{vmatrix} = \begin{vmatrix} \Gamma_0 \\ 1 \end{vmatrix} \tag{11}$$

where $\alpha \triangleq 1 - \mathbf{a}^T U = 1 - \sum_{i=1}^{N} a_i$. (12)

Using matrix and vector notations, (11) can be rewritten as $A_1 \lambda_1 = \mathbf{c}$, with obvious definitions for A_1, λ_1 and $\mathbf{c}$. The optimal prediction error variance, $V_c \triangleq E\{(\tilde{y}_{N+d})^2\}$, can be written in a number of ways:

$$\begin{aligned} V_c &= \sigma^2 - \mathbf{a}^T R_0 = \alpha\sigma^2 + \mathbf{a}^T \Gamma_0 \\ &= \sigma^2 - R_0^T R^{-1} R_0 \\ &= \mathbf{c}^T A_1^{-1} \mathbf{c} \end{aligned} \tag{13}$$

3 Processes with unknown mean and known second order statistics

We now consider the case where the constant mean m is unknown, but where the covariance function $R(\tau)$ or the variogram $\gamma(\tau)$ are assumed known. Recall that they are related by (5).

3.1 The approximate Levinson predictor

The most obvious strategy is to replace the mean in (6) by a linear unbiased estimate based on the past data:

$$\hat{m} = \sum_{i=1}^{N} \zeta_i y_{N-i} = \boldsymbol{\zeta}^T Y \tag{14}$$

with the unbiasedness condition:

$$\boldsymbol{\zeta}^T U = 1 \tag{15}$$

The CLP of (6) is then replaced by the approximate Levinson predictor (ALP):

$$y^*_{N+d} = \hat{m} + \mathbf{a}^T (Y - \hat{m}U) \tag{16}$$

where **a** is the solution of (7), as before. This estimator was called approximate classical Levinson predictor (ACLP) in [1]. The prediction error can be written

$$\tilde{y}_{N+d} \triangleq y^*_{N+d} - y_{N+d} = m + a^T(Y-mU) - y_{N+d} + \alpha(\hat{m}-m) \qquad (17)$$

The sum of the first two terms in (17) is the optimal CLP, so that the sum of the first three terms is the CLP prediction error. Therefore, the prediction error variance $V_a \triangleq E\{(y^*_{N+d} - y_{N+d})^2\}$ is:

$$V_a = V_c + \alpha^2\sigma^2_{\hat{m}} + 2\alpha E\{[m+\mathbf{a}^T(Y-mU)-y_{N+d}][\hat{m}-m]\}$$

$$= V_c + \alpha^2\sigma^2_{\hat{m}} + 2\alpha E\{[\mathbf{a}^T(Y-mU)-y_{N+d}][(Y-mU)^T\boldsymbol{\zeta}]\}$$

since $\hat{m}$ is unbiased. $\sigma^2_{\hat{m}}$ denotes the variance of $\hat{m}$. The third term in the above expression is zero by (7). Therefore

$$V_a = V_c + \alpha^2\sigma^2_{\hat{m}} \qquad (18)$$

It follows from (18) that V_a will be minimum if $\sigma^2_{\hat{m}}$ is minimum, i.e. if $\hat{m}$ is chosen as the MVU estimate of m. The corresponding estimator will be called MVALP, for Minimum Variance Approximate Levinson Predictor. The MVU estimate for $\hat{m}$ is obtained as follows:

$$\sigma^2_{\hat{m}} = E\{\hat{m} - m)^2\} = E\{[\boldsymbol{\zeta}^T(Y-mU)]^2\}$$

$$= \boldsymbol{\zeta}^T R\, \boldsymbol{\zeta} = \sigma^2 - \boldsymbol{\zeta}^T \Gamma\, \boldsymbol{\zeta} \qquad (19)$$

Minimizing $\boldsymbol{\zeta}^T R \boldsymbol{\zeta}$ w.r.t. $\boldsymbol{\zeta}$ subject to (15) yields

$$\boldsymbol{\zeta} = (U^T R^{-1} U)^{-1} R^{-1} U \qquad (20)$$

After lengthy manipulations, using (5), $\boldsymbol{\zeta}$ can also be expressed in terms of $\gamma(\tau)$:

$$\boldsymbol{\zeta} = (U^T \Gamma^{-1} U)^{-1} \Gamma^{-1} U \qquad (21)$$

The corresponding minimum error variance for the mean estimate will be denoted $\sigma^2_{\hat{\underline{m}}}$:

$$\sigma^2_{\hat{\underline{m}}} = (U^T R^{-1} U)^{-1} = \sigma^2 - (U^T \Gamma^{-1} U)^{-1} \qquad (22)$$

The expression (18) shows that the ALP yields a larger error variance than the

CLP. This is to be expected, since the CLP is the optimal estimator. The two error variances coincide when $\sigma^2_{\hat{m}} \to 0$, as should be expected. We shall denote by $\underline{V}_a$ the prediction error variance of the MVALP. Using

$$\alpha = 1 - \mathbf{a}^T U = 1 - R_0^T R^{-1} U \tag{23}$$

together with (22) and (13), we can write

$$\underline{V}_a = V_c + \alpha^2 (U^T R^{-1} U)^{-1} \tag{24a}$$

$$= V_c + (1 - R_0^T R^{-1} U)^2 (U^T R^{-1} U)^{-1} \tag{24b}$$

$$= \sigma^2 - R_0^T R^{-1} R_0 + (1 - R_0^T R^{-1} U)^2 (U^T R^{-1} U)^{-1} \tag{24c}$$

Note that this last expression is entirely in terms of the covariance function of $\{y_t\}$.

3.2 The modified Levinson predictor (MLP)

When the mean m is unknown, an alternative strategy is to seek the MVU predictor of the following form:

$$\hat{y}_{N+d} = \sum_{i=1}^{N} b_i y_{N-i} = \mathbf{b}^T Y \tag{25}$$

together with the constraint, imposed by unbiasedness, that

$$\sum_{i=1}^{N} b_i = 1 \quad , \text{ i.e. } \quad \mathbf{b}^T U = 1 \tag{26}$$

Minimization of $E\{(\tilde{y}_{N+d})^2\}$ w.r.t. $\mathbf{b}$ subject to (26) yields the following linear system of (N+1) equations:

$$\begin{vmatrix} R & -U \\ U^T & 0 \end{vmatrix} \begin{vmatrix} \mathbf{b} \\ \mu \end{vmatrix} = \begin{vmatrix} R_0 \\ 1 \end{vmatrix} \tag{27}$$

where μ is a Lagrange coefficient. Alternatively, using (5), we get

$$\begin{vmatrix} \Gamma & U \\ U^T & 0 \end{vmatrix} \begin{vmatrix} \mathbf{b} \\ \mu \end{vmatrix} = \begin{vmatrix} \Gamma_0 \\ 1 \end{vmatrix} \tag{28a}$$

or, with obvious matrix and vector notations,

$$A_2 \lambda_2 = \mathbf{c} \tag{28b}$$

The prediction error variance can be written in a number of ways

$$V_m = \sigma^2 + \mathbf{b}^T R \mathbf{b} - 2 \mathbf{b}^T R_0 = 2 \mathbf{b}^T \Gamma_0 - \mathbf{b}^T \Gamma \mathbf{b} \tag{29a}$$

$$= \mu + \sigma^2 - \mathbf{b}^T R_0 = \mu + \mathbf{b}^T \Gamma_0 \tag{29b}$$

$$= \mathbf{c}^T A_2^{-1} \mathbf{c}$$

Note that the last expression is the only one that does not involve **b** or μ. An expression similar to (29c), involving only the covariance function, can of course be obtained from (27).

The predictor MLP was proposed by Bastin and Henriet [1], who showed that

$$V_m = V_c + \alpha^2 (U^T R^{-1} U)^{-1} \quad (30)$$

with α defined by (12). This, together with (24a), shows that

$$V_m = \underline{V}_a \quad (31)$$

In fact, we show in the next section that the MVALP and the MLP are identical, which of course explains (31). In the process, we shall derive a number of interesting expressions for **b** and μ.

4 The MVALP and the MLP are identical

<u>Proof:</u> It follows from (16) and (14) that the MVALP can be written:

$$y^*_{N+d} = (1 - \mathbf{a}^T U)\hat{m} + \mathbf{a}^T Y$$

$$= (\alpha\, \zeta^T + \mathbf{a}^T) Y \quad (32)$$

Therefore, we need to prove that

$$\mathbf{a} + \alpha\, \zeta = \mathbf{b} \quad (33)$$

where **a**, $\boldsymbol{\zeta}$, **b** are solutions of (7), (20) and (27) respectively, and with α given by (12).

The solution of (27) is unique, because it is the vector **b** that minimizes $E\{[y_{N+d}]^2\} = b^T R b + \sigma^2 - 2 b^T R_0$ subject to (26). It is also trivial to see that $\mathbf{a}+\alpha\zeta$ satisfies the last equation of (27). Therefore, to prove (33), it remains to be shown that there exists a μ such that

$$R(\mathbf{a} + \alpha\, \zeta) - \mu U = R_0 \quad (34)$$

It follows immediately from (7), (20) and (22) that this is so for

$$\mu = \alpha\, \underline{\sigma}^2_{\hat{m}} \quad (35)$$

This concludes the proof.

In the process of proving our main result, we have shown that the solution μ of (27) and (28) is $\alpha\, \underline{\sigma}^2_{\hat{m}}$. This yields some closed form expressions for **b** and some new expressions for μ in terms of either $R(\tau)$ or $\gamma(\tau)$. From (35), (27) and (28) we get

$$\mathbf{b} = R^{-1}(R_0 + \alpha\sigma^2_{\hat{m}} U) = \Gamma^{-1}(\Gamma_0 - \alpha\sigma^2_{\hat{m}} U) \tag{36}$$

Using (35), (12) and (22), we have

$$\mu = \frac{1-U^T R^{-1} R_0}{U^T R^{-1} U} = \frac{U^T \Gamma^{-1} \Gamma_0 - 1}{U^T \Gamma^{-1} U} \tag{37}$$

The second equality in (37) is obtained by multiplying (36) to the left by U^T, and using (26), (12) and (22). This then allows us to write two expressions for **b** in terms of $R(\tau)$ or $\gamma(\tau)$ only:

$$\mathbf{b} = R^{-1} R_0 + \frac{1-U^T R^{-1} R_0}{U^T R^{-1} U} R^{-1} U \tag{38a}$$

$$= \Gamma^{-1} \Gamma_0 + \frac{1-U^T \Gamma^{-1} \Gamma_0}{U^T \Gamma^{-1} U} \Gamma^{-1} U \tag{38b}$$

The formulas (38a) and (38b) are remarkable because the expressions in terms of (R, R_0) and (Γ, Γ_0) are identical. Note that the first term of (38a) is **a** while the second term is $\alpha\,\zeta$. (Recall (33)). This is not so, however, for (38b), because $R^{-1}R_0 \neq \Gamma^{-1}\Gamma_0$. Finally, note that the expressions (37) and (38) can also be obtained from (27) using a matrix inversion lemma.

In Bastin and Henriet [1], it was claimed that $\underline{V}_a - V_m = \alpha^2\sigma^2 \geqslant 0$, and that therefore the MLP was better than the MVALP. This was based on an erroneous expression for V_a: the correct expression is (18) above. Note, however, that the equality $V_m = \underline{V}_a$ holds only if the unbiased estimator $\hat{m}$ is chosen to be the minimum variance estimator. If the sample mean estimator

$$\hat{m} = \frac{1}{N} \sum_{i=1}^{N} y_{N-i} \tag{39}$$

is used in (16), then $V_m \leqslant V_a$.

5 Motivation for the use of the variogram and the MLP

We have shown that when the mean is unknown, but the covariance or variogram is known, the MVALP and the MLP are identical. We have also shown that the predictor coefficients of all three predictors that we discussed can be computed either from $R(\tau)$ or from $\gamma(\tau)$.

In practice, neither the mean nor the covariance or the variogram are known, and they have to be estimated from the data. Two issues must then be raised:

a) what predictor should one use?

b) should one estimate its coefficients using sample estimates of $R(\tau)$ or $\gamma(\tau)$?

Since we have shown that the MVALP and the MLP are identical, the practical choice is between:

- the ALP (16) with $\hat{m}$ replaced by its sample mean (39), or
- the MLP (25).

In both cases, the parameter vectors (**a** for (16), **b** for (25)) can be computed either by covariance equations ((7) and (38.a)) or by variogram equations ((11) and (38.b)), and in practice covariances and variograms would be replaced by their sample estimates. Of course, other predictors could be used in practice, such as least-squares predictors derived directly from the data rather than from sample covariance or variogram estimates, but our aim in this discussion is only to compare the two predictors just mentioned.

We give a number of reasons why we believe that the MLP should be used with its coefficients b_i computed from the variogram formulas (38.b).

1) The MLP does not require an estimate of the mean. The ALP would use an estimate of the sample mean which, in almost all cases, would not have minimum variance. We have shown that, with known covariances, this would lead to an ALP that has larger error variance than the MLP. It is likely that this result will still hold when exact covariances are replaced by estimated ones, although this has not been proved.

2) So far, we have assumed that the process is known to be stationary. In practice, however, this is not always easy to validate. Failure to detect that a process is nonstationary may lead to completely erroneous results if covariances are used, as the following example, due to Matheron [5], shows. Consider a Wiener process with variogram $\gamma(\tau)=|\tau|$, and suppose we have observed $\{y_0,\ldots,y_{N-1}\}$. Now suppose that the user believes thet the process he or she observes is wide-sense stationary, and that he or she estimates $\hat{m}$ by (39) and $\hat{R}(\tau)$ by:

$$\hat{R}(\tau) = \frac{1}{N-\tau} \sum_{t-\tau}^{N-1} (y_t - \hat{m})(y_{t-\tau} - \hat{m}) \tag{40}$$

Then it can be shown that, for $\tau \geqslant 0$,

$$E\{\hat{R}(\tau)\} = \frac{N^2-1}{3N} - \frac{4}{3}\tau + \frac{2}{3}\frac{\tau^2}{N} \tag{41}$$

It is a parabola; an apparent variance of $\hat{R}(0) = \hat{\sigma}^2 = \frac{N^2-1}{3N}$ will be found, whereas the true variance is infinite. The experimental variogram, on the other hand, is unbiased:

$$\hat{\gamma}(\tau) = \frac{1}{2(N-\tau)} \sum_{t-\tau}^{N-1} (y_t - y_{t-\tau})^2 \tag{42}$$

3) For intrinsic (but not necessarily wide-sense stationary) processes, the equations (28) and the estimator (38b) can be derived directly without going through the relation (5). Note that in such case, the covariance $R(\tau)$ and the variance σ^2 may not exist. Therefore, the MLP (38b) covers a wider class of random processes.

6 Concluding remarks

All these elements seem to indicate that the MLP should be preferred in situations where the mean and the covariance function of a process are unknown. Of course, a definite answer can only be given by comparing the expected predictors when estimated means, covariances and variograms are used. A theoretical comparison appears to be very difficult.

Acknowledgements

The author wishes to thank Georges Bastin for interesting comments on an earlier draft of this paper.

References

1. G. Bastin and E. Henriet, "A minimum variance property of Levinson predictors using variograms". Systems and Control Letters, Vol.2, No.1, July 1982, pp. 25-29.
2. G. Matheron, "Les variables regionalisees et leur estimation", Masson Ed., 1965.
3. B.D.O. Anderson and J.B. Moore, "Optimal Filtering", Prentice Hall, 1979.
4. T.W. Anderson, "The Statistical Analysis of Time Series", Wiley, 1958.
5. G. Matheron, "The Theory of Regionalized Variables and its Applications", Cahiers du Centre de Morphologie Mathematique de Fontainebleau, 1971.

UNOBSERVED RATIONAL EXPECTATIONS AND THE GERMAN HYPERINFLATION

WITH ENDOGENOUS MONEY SUPPLY: A PRELIMINARY REPORT*

Edwin Burmeister
Department of Economics
University of Virginia
and
Kent D. Wall
Department of Systems Engineering
University of Virginia

1. Introduction and Summary

Two important economic issues are addressed in this paper:

(1) Was the money supply process during the German hyperinflation exogenous, or did the expected rate of inflation influence the rate of growth of the nominal money supply?

(2) Assuming that price expectations were formed rationally, were these rational expectations dynamically stable during the German hyperinflation?

While the first question is of interest for its own sake, our primary concern is with (2). Our interest in (1) stems from the fact that most previous work on (2), e.g., Burmeister and Wall (1982) and Flood and Garber (1980), has embodied the postulate that the money supply process was exogenous. Therefore if this postulate is wrong, the conclusions regarding question (2) are suspect. Moreover, there is substantial evidence to indicate that the German money supply process was not exogenous; see, for example, Sargent and Wallace (1973) for one technical discussion and Guttman and Meehan (1976) for a historical perspective.

The approach taken here follows Burmeister and Wall (1982) and uses a Kalman filtering prediction error technique which enables us to retain the explicit reference to the unobserved variables. That is, a state space form representation is employed wherein the unobserved variables become state variables. Drawing upon a well-developed theory from control engineering, it is then possible to obtain simultaneously estimates of both the model parameters and the unobserved expectations. This permits a much easier test of certain hypotheses than would otherwise be the case. It also enables us to deal easily with the case where the expected inflation rate influences the rate of growth of the money supply.

Our primary empirical results may be briefly summarized; details are provided in Sections 3 and 4 below.

There is very strong evidence that the money supply process is not exogenous. The null hypothesis that $\alpha_2 = 0$, where α_2 is the coefficient on the expected rate of inflation in the money supply equation, was rejected at a 0.05 level in all but one

*The authors thank the National Science Foundation (SES-82-18229) for financial support and James C. Spall for research assistance. We have benefited greatly from comments received on an earlier version presented to the macroeconomics workshop at the University of Virginia.

of ten instances tested, and it was rejected at a 0.02 level in six instances. The evidence to reject the hypothesis that rationally formed expectations are always convergent is even stronger; in all ten instances tested the null hypothesis of always convergent expectations was rejected at the 0.02 level, and in eight of ten instances it was rejected at the 0.01 level.

We conclude that the common assumption that rationally formed expectations are always convergent cannot be maintained in this model of the German hyperinflation. As in Burmeister and Wall (1982), our primary conclusion is that estimation techniques which impose restrictions implied by the assumption of convergent expectations, and which therefore are conditional upon this convergent expectations assumption, are suspect without additional verification of the underlying stability hypothesis. We have demonstrated the feasibility of an alternative estimation methodology which does not preclude the possibility that rationally formed expectations are unstable. We have shown how to test an important hypothesis and to obtain parameter estimates which are not conditional upon a perhaps invalid assumption.

2. Estimation Technique and Results

The model, using notation from Burmeister and Wall (1982), consists of

$$g(t) - \pi(t) = b - a[\pi^*(t+1,t+1) - \pi^*(t,t)] + \mu(t), \tag{2.1a}$$

$$g(t) = \alpha_0 + \alpha_1 g(t-1) + \alpha_2 \pi^*(t,t) + \varepsilon(t), \tag{2.1b}$$

where

$$\pi^*(t,t) = [m_{21} - \frac{m_{11}m_{22}}{m_{12}}] c_{1,t} + [g(t-1)-\bar{z}_1]\frac{m_{22}}{m_{12}} + \bar{z}_2, \tag{2.2}$$

and

$$c_{1,t+1} = \lambda_1 c_{1,t} + \eta(t). \tag{2.3}$$

This model presents us with a combination of inference problems. An endogenously determined money supply manifests itself through the presence of the parameter α_2, and testing for endogenous money reduces to a traditional econometric problem in determining the statistical significance of a model coefficient. The possibility of nonconvergent inflationary expectations is embodied in the presence of the parameter $c_{1,t}$ which, when indexed by time, may be viewed as an unobserved variable. Testing for the significance of this second parameter involves the development of estimates of an unobserved, dynamically evolving parameter which may embody stochastic variations. While econometric methods have been established for a random parameter in certain special cases, its simultaneous incorporation with the endogenous money supply leads to a non-standard econometric problem.

2.1 State Space Model

In Burmeister and Wall (1982) a state space formulation was employed to circumvent the computational complications induced by the presence of an unobserved, possibly stochastic parameter. The resulting representation then could be estimated using a recursive prediction error algorithm based on the Kalman filter. For similar reasons the same approach is taken here. A state space representation of the form

$$\underline{x}(t+1) = F\underline{x}(t) + G\underline{u}(t) + \underline{w}(t) \tag{2.4}$$

$$\underline{y}(t) = H\underline{x}(t) + D\underline{u}(t) + \underline{v}(t) \tag{2.5}$$

is obtained for the model (2.1) by incorporating the rational expectations restrictions (2.2) - (2.3) and employing the following definitions:

$$\begin{aligned}
\underline{x}(t) &= [c_{1,t}] \\
\underline{w}(t) &= [\eta(t)] \\
\underline{y}(t) &= [g(t),\pi(t)]' \\
\underline{u}(t) &= [g(t),\pi(t),g(t-1),1]' \\
\underline{v}(t) &= [\varepsilon(t),d\varepsilon(t) + \beta\eta(t) - \mu(t)]'
\end{aligned} \tag{2.6}$$

where $d = 1 + a\, m_{22}/m_{12}$ and $\beta = a[m_{12}m_{21}-m_{11}m_{22}]/m_{12}$. Thus the state vector, $\underline{x}(t)$, and state disturbance, $\underline{w}(t)$, are scalars. The output, $\underline{y}(t)$ is a 2 x 1 vector as is the measurement noise, $\underline{v}(t)$, while the input, $\underline{u}(t)$, is a 4 x 1 vector. The coefficient matrices are therefore defined by:

$$F = [\lambda_1]$$

$$G = [\ 0\ 0\ 0\ 0\ ]$$

$$H = \frac{m_{12}m_{21}-m_{11}m_{22}}{m_{12}} \begin{bmatrix} \alpha_2 \\ d\alpha_2 + a(\lambda_1-1) \end{bmatrix} \tag{2.7}$$

$$D = \begin{bmatrix} 0 & 0 & \alpha_1 + \alpha_2 m_{22}/m_{12} & \bar{\alpha}_0 \\ 0 & 0 & d(\alpha_1+\alpha_2 m_{22}/m_{12}) - a\, m_{22}/m_{12} & d\bar{\alpha}_0-b \end{bmatrix} \tag{2.8}$$

$$Q = E\{\underline{w}(t)\underline{w}'(t)\} = \sigma_\eta^2$$

$$R = E\{\underline{v}(t)\underline{v}'(t)\} = \begin{bmatrix} \sigma_\varepsilon^2 & d\sigma_\varepsilon^2 \\ d\sigma_\varepsilon^2 & d^2\sigma_\varepsilon^2+\beta^2\sigma_\eta^2+\sigma_\mu^2 \end{bmatrix} \tag{2.9}$$

$$S = E\{\underline{w}(t)\underline{v}'(t)\} = [\ 0 \quad \beta\sigma_\eta^2\ ]$$

where $\bar{\alpha}_0 = \alpha_0 - \alpha_2\bar{z}_1 m_{22}/m_{12} + \alpha_2\bar{z}_2$.

The nonzero 1 x 2 matrix S gives the contemporaneous correlation between the state disturbances, $\underline{w}(t)$, and the measurement noise, $\underline{v}(t)$, and requires a minor modification of the Kalman filter recursive state estimator algorithm (see Jazwinski (1970, pg. 210)) or a slight redefinition of the state equation (see Anderson and Moore (1979, pg. 114)) or Gelb (1974, pg. 124)). We take the latter approach. The result is a new state equation with the previous F, G, and Q replaced by

$$F = [\lambda_1-\gamma(dh_{11}-h_{21})] \tag{2.10}$$

$$G = [-\gamma d,\gamma,\gamma d d_{13}-\gamma d_{23},\gamma d d_{14}-\gamma d_{24}] \tag{2.11}$$

$$Q = \sigma_\eta^2\alpha_\mu^2/(\alpha_\mu^2+\beta^2\sigma_\eta^2) \quad (2.12)$$

where $h_{11}, h_{21}, d_{13}, d_{14}, d_{23}$, and d_{24} denote the respective elements of the matrices H and D as defined previously, and $\gamma = \beta\sigma_\eta^2/(\sigma_\mu^2+\beta^2\sigma_\eta^2)$. The resulting modified state equation has a redefined disturbance $\underline{w}(t)$ that is now uncorrelated with the measurement noise $\underline{v}(t)$.

2.2 Estimation Technique

With the model transformed to (2.6) - (2.12) the estimation algorithm described in Burmeister and Wall (1982) can be applied directly. The model contains eight unknown parameters: $\alpha_0, \alpha_1, \alpha_2, a, b, \sigma_\varepsilon^2, \sigma_\mu^2$, and σ_η^2. The initial state estimate and its variance constitute another two unknowns but are not estimated. They comprise the prior on the state, $c_{1,t}$, and should not affect the estimation results for sufficiently long data sets.[1]

Upon convergence of the estimation algorithm for the unknown parameters, a Kalman filter-smoother combination is called upon to produce efficient estimates of the state variable conditioned on the values estimated for the model parameters. These smoothed estimates, denoted $\hat{x}(t|T)$ are useful in our tests of convergent expectations in Section 4.

Finally, we note that $\sigma_\varepsilon^2, \sigma_\mu^2$, and σ_η^2 were not estimated directly. In order to guarantee that the estimates for Q and R be positive definite, the lower triangular Cholesky factor of each matrix was employed during estimation. Thus Q was represented as

$$Q = [\sigma_\eta\sigma_\mu/\sqrt{\sigma_\xi^2 + \beta^2\sigma_\eta^2}]\,[\sigma_\eta\sigma_\mu/\sqrt{\sigma_\mu^2 + \beta^2\sigma_\eta^2}\,]$$

and R as

$$R = \begin{bmatrix} \sigma_\varepsilon & 0 \\ d\sigma_\varepsilon & \sqrt{\alpha_\mu^2+\beta^2\sigma_\eta^2} \end{bmatrix} \begin{bmatrix} \sigma_\varepsilon & d\sigma_\varepsilon \\ 0 & \sqrt{\sigma_\mu^2+\beta^2\sigma_\eta^2} \end{bmatrix}$$

The unknown variances were thus replaced by their respective unknown standard deviations and the unknown parameter vector, $\underline{\theta}$, of the estimation problem described in Burmeister and Wall (1982) was thus defined by

$$\underline{\theta} = [\alpha_0, \alpha_1, \alpha_2, a, b, \sigma_\varepsilon, \sigma_\mu, \sigma_\eta]'.$$

2.3 Estimation Results

Estimation of (2.4) - (2.5) was carried out using price data taken from Zahlen zur Geldentwertung in Deutschland 1914 bis 1923, Reimar Hobbing, Berlin, 1925. Money supply figures were taken from table VII of Flood and Garber (1980). The same period for both price and money covered January 1919 through June 1923. Five different bubble phenomena were considered, all lasting until June 1923 but beginning in different months. Beginning dates for the bubbles were: January 1919, October 1919, July 1920, June 1921, and June 1922. The third and fifth correspond exactly with those treated by Flood and Garber (1980).

Initial attempts at estimating all eight parameters over all five price bubbles failed to yield convergence in the last two cases. Convergence in all five cases was obtained by dropping α_0 and b from $\underline{\theta}$. Both did not appear significant according to the results obtained in the first three cases: each attained a relatively small value during estimation and was never more than 1.9 times its estimated standard error. Estimation results for the reduced parameter vector are given in Table 2.1, 2.2 and Figure 1.[2] Convergence of the parameter estimation algorithm was obtained in every case with σ_η taking on significant nonzero values.

The trajectories pictured in Figure 1 appear significant when compared with their standard errors which are computed by the smoothing algorithm. We note, however, that such an inference is valid only under a normal distribution assumption which may not be well founded in this situation. Yet, the fact that $c_{1,T}$ is at least ten times its asymptotic estimated standard error in each case points to significance. Finally, it should be noted that the negative values for $c_{1,t}$ are a consequence of the estimated values for the elements of the H matrix in (2.7); these are negative so that their values multiplied by the values estimated for $c_{1,t}$ give an expectations effect of opposite sign to that in the figure. Hence, as intuition would suggest, $c_{1,t}$ tends to add a positive impetus to $\pi(t)$ and $g(t)$ (except, perhaps, for the second, third, and fourth bubbles during the middle of 1921).

The eigenvalues corresponding to the expectation formation mechanisms for π^* and g^* given the estimates in Table 2.1 are presented in Table 2.2. We cannot give any standard errors, asymptotic of otherwise, for these values, but in all cases we find real distinct eigenvalues. Instability is also indicated in all but the last three bubbles, and even in these cases the largest eigenvalue is uncomfortably close to unity.

Tables 2.3, 2.4, and Figure 2 depict the estimation results obtained by imposing a deterministic evolution on $c_{1,t}$. Here σ_η is restricted to zero. Significant parameters are obtained in all cases with the values remaining relatively stable across all bubbles. Table 2.2 may be referred to for the eigenvalues associated with the parameter values reported in Table 2.3. Again, all cases give distinct eigenvalues with one inside the unit circle and one outside the unit circle. Table 2.4 reports the smoothed values for $c_{1,t}$, along with their variances. Here, again, we find very significant values for c_{1,t_0} and $c_{1,T}$.

3. Empirical Results I: Tests on Endogenous Money Supply

We wish to test whether or not the expected current rate of inflation, $\pi^*(t,t)$, influenced the current rate of growth of the money supply, $g(t)$. In view of equation (2.1b), we formulate the null hypothesis

$$H_0: \alpha_2 = 0$$

versus the alternative

$$H_1: \alpha_2 \neq 0.$$

Three types of empirical evidence all support a rejection of H_0, namely (i) evidence based on standard errors for α_2, (ii) tests based on the Chebyshev inequality, and (iii) two sample Kolmogorov-Smirnov (hereafter K-S) tests.

(i) Referring to Tables 2.1 and 2.3 we find that the estimates for α_2 are at least 5.8 times their standard errors and, while these t-statistics are only of consequence under a normal distribution assumption for the estimation errors, they are indicative of the importance of $\pi^*(t,t)$ in explaining $g(t)$. Indeed, this observation is borne out by the nonparametric test results which follow.

(ii) Using the parameter estimates obtained when α_2 is not restricted, we then constrain $\alpha_2 = 0$ and calculate the normalized innovations[3] $\underline{\nu}_t^* = (\nu_{1t}^*, \nu_{2t}^*)$ which are the realizations obtained under the restriction on α_2. We then ask what is the probability that

$$\max[|\sum_t \nu_{1t}|, |\sum_t \nu_{2t}|] \geqq \max[|\sum_t \nu_{1t}^*|, |\sum_t \nu_{2t}^*|] \tag{3.1}$$

where $\{\underline{\nu}_{t_0}, \cdots, \underline{\nu}_T\}$ is the normalized innovations process, $\underline{\nu}_t = (\nu_{1t}, \nu_{2t})$, and where $\sum_t$ denotes summation from $t = t_0$ to $t = T$. We thus obtain a measure of the loss of explanatory power under the $\alpha_2 = 0$ restriction because we can calculate a bound on the probability that (3.1) holds. If this probability is sufficiently small, then H_0 is rejected. To this end we employ Chebyshev's inequality, i.e.

$$\Pr(|x|>\varepsilon) \leqq E(x^2/\varepsilon^2).$$

In particular, the p-value corresponding to the Chebyshev inequality test is p_1 in Table 3.1 and is given by

$$\Pr[\max_i |\sum_t \nu_{it}| \geqq \max_i |\sum_t \nu_{it}^*|]$$

$$\leqq 2 \min_j \Pr[|\sum_t \nu_{jt}| \geqq |\sum_t \nu_{jt}^*|]$$

$$\leqq 2 \min_j [E((\sum_t \nu_{jt})^2/(\sum_t \nu_{jt}^*)^2)]$$

$$= 2\,(T-t_0+1)/\max_j(\sum_t \nu_{jt}^*)^2$$

$$\equiv p_1.$$

(iii) The Kolmogorov-Smirnov (K-S) test compares the sample distribution functions of two sets of observations, and, most importantly, no distributional assumptions are required. Specifically, consider two sets of observations $X_1, \cdots X_n$ and $Y_1, \cdots, Y_n$ with distribution functions $F(\cdot)$ and $G(\cdot)$, respectively. We wish to test H_0: $F(x) = G(x)$ for all x versus H_1: not H_0. The empirical distribution function of, say, the $\{X_i\}$ sample is defined as

$$F_n(x) = \frac{\text{no. of } X_i\text{'s} \leqq x}{n}$$

The measure of difference between F_n and G_n,

$$D_n = \max_x |F_n(x) - G_n(x)|,$$

has a known distribution for both finite n and for $n\to\infty$. If $\Pr(D_n \geq D_n^* | H_0 \text{ true})$ is small, H_0 is rejected. The K-S test is used here because it is distribution free and because it has reasonable power against a wide range of alternative hypotheses.

Application of the K-S test relies upon an assumption that each set of observations is drawn from a corresponding common distribution function. This will be only approximately true in the case considered here. We have, by definition, forced the observations (the $\{\underline{\nu}_t\}$) to have common first and second moments. But this, of course, does not mean that they necessarily have the same distribution (i.e., the underlying distribution of $\underline{\nu}_{t_0+1}$ may be different from that of $\underline{\nu}_{t_0+2}$ even though both have identical first and second moments and were generated from the same model). However, the assumption seems reasonably weak, and we make it.

Specifically the K-S test is implemented here as follows: Let $\underline{\nu}_t = (\nu_{1,t}, \nu_{2,t})$ and construct EDF's for $\nu_{1,t_0}, \ldots, \nu_{1,T}$ and $\nu^*_{1,t_0}, \nu^*_{1,T}$, where $\underline{\nu}_t$ denotes the innovation from the $\alpha_2 \neq 0$ model and $\underline{\nu}^*_t$ is that of the $\alpha_2 = 0$ model. Denote the EDF's of two samples as $F^{(1)}_{T-t_0+1}(x)$ and $G^{(1)}_{T-t_0+1}(x)$, respectively, and compute

$$D^{(1)}_{T-t_0+1} = \max_x |F^{(1)}_{T-t_0+1}(x) - G^{(1)}_{T-t_0+1}(x)|$$

Similarly, compute $D^{(2)}_{T-t_0+1}$. The p-value reported in Table 4.1 is then p_2 in the relationship

$$\Pr[\max_i D^{(i)}_{T-t_0+1} \geq \max_i D^{(i)*}_{T-t_0+1}] \leq 2 \min_j \Pr[D^{(j)}_{T-t_0+1} \geq D^{(j)*}_{T-t_0+1}]$$

$$\equiv p_2.$$

This value of p_2 would correspond to $\max_j D^{(j)*}_{T-t_0+1}$.

The results presented in Table 3.1 provide strong evidence of the significance of α_2. Even if one is not willing to make the assumption of a common distribution for each of $\{\nu^*_t\}$ and $\{\nu_t\}$ necessary for the K-S test (as discussed above), the Chebyshev inequality p-values are all less than the usual significance level of .05.

We conclude, therefore, that the expected rate of inflation did influence the money growth during the German hyperinflation. Accordingly, previous empirical results obtained by Burmeister and Wall (1982), Flood and Garber (1980), and others are based upon a misspecified model, and consequently conclusions concerning the convergence or nonconvergence of rationally formed expectations may be suspect. We now turn to this question for a model that is free of the $\alpha_2 = 0$ restriction.

4. Empirical Results II: Tests on Convergent Rational Expectations

It is commonly assumed that rationally formed expectations always converge. This assumption is crucial for at least three reasons: First, in many models convergent expectations are needed to determine a unique momentary equilibrium at each instant; without this convergence assumption, the future path of the economy might be indeterminate. Second, the assumption of convergent expectations is commonly used to provide cross-equation restrictions which facilitate identification and econometric estimation; thus, most econometric estimates of rational expectations models are conditional upon this assumption. Third, if markets are not always in equilibrium, then there exist many cases in which the assumption of convergent expectations is untenable because it implies a contradiction.[4]

Since the postulate of convergent rational expectations carries with it such important economic implications, and because the postulate is testable, a careful empirical investigation is essential. Moreover, as noted at the end of the previous section, tests of convergent expectations using data from the German hyperinflation should be made for a model which is specified with an endogenous money supply. Our empirical evidence as to whether or not $c_{1,t} = \underline{0}$ for all t -- which is equivalent to the assumption that expectations are always convergent in this model -- is based on two types of tests. Hereafter we denote $c_{1,t}$ by c_t for simplicity.

(i) Our first test relates directly to the statistical significance of the estimates for σ_η that are recorded in Table 2.1. In each case this parameter exhibits a value roughly ten times its standard error suggesting that c_t obeys a stochastic process. As such, it is impossible that c_t can be identically zero over any time interval of finite length. Thus the very significance of σ_η provides strong evidence for rejecting the hypothesis that c_t is identically zero.

(ii) If expectations are always convergent, it is necessary that $c_t = 0$ for all $t = t_0, \cdots, T$. Hence it is sufficient to test the null hypothesis

$$H_0: c_T = 0$$

versus the alternative

$$H_1: c_T \neq 0.$$

As in Table 3.1, the value of p_1 is the Chebyshev p-value corresponding to the upper bound of the inequality

$$\Pr[|\hat{c}_T - c_T| \geqq \hat{c}_T^*|\ c_T=0] \leqq P_T/(\hat{c}_T^*)^2, \text{ where } P_t = E[(\hat{c}_t - c_t)(\hat{c}_t - c_t)'].$$

The p-values under Gaussian distribution assumptions, namely the probability p_2 that

$$\Pr[|c_T - c_T| \geqq c_T|\ c_T=0] \text{ and } [c_T - c_T] \sim N(0, P_T),$$

are also reported in Table 4.1 to illustrate the conservative nature of the p_1 values. Given the generally conservative nature of the Chebyshev inequality, the p_1-values reported in Table 4.1 provide very strong evidence for rejecting H_0.

References

Anderson, B. D. O. and J. B. Moore, 1979, Optimal Filtering (Prentice-Hall, Englewood Cliffs, NJ).

Black, Fischer, 1974, Uniqueness of the price level in monetary growth models with rational expectations, Journal of Economic Theory, Jan.

Box, George E. P. and Gwilym M. Jenkins, 1973, Time Series Analysis: Forecasting and Control, rev. ed. (Holden Day, San Francisco, CA).

Bryson, A. E. and Y.-C. Ho, 1969, Applied Optimal Control (Blaisdell, Waltham, MA).

Bucy, R. S. and P. D. Joseph, 1968, Filtering for Stochastic Processes With Applications (Wiley, New York).

Burmeister, Edwin, 1980, On some conceptual issues in rational expectations modelling, Journal of Money, Credit and Banking, Vol. 12, no. 4.

Burmeister, Edwin, 1984, Indeterminacy and stability in both rational expectations and perfect foresight models, Paper presented at Summer Econometric Society Meetings, San Diego, June (revised 1982). Forthcoming in a volume edited by G. Feiwell, MacMillian.

Burmeister, Edwin, Robert P. Flood and Stephen J. Turnovsky, 1979, Rational expectations and stability in a stochastic monetary model of inflation. Working paper (University of Virginia, Charlottesville, VA).

Burmeister, Edwin, Robert P. Flood and Peter M. Garber, 1983, On the equivalence of solutions in rational expectations mode, Journal of Economic Dynamics and Control, forthcoming.

Burmeister, Edwin and K. D. Wall, 1982, Kalman filtering estimation of unobserved rational expectations with an application to the German hyperinflation, Journal of Econometrics, vol. 20, no. 1, 255-284.

Cagan, P., 1956, "The Monetary Dynamics of Hyperinflation," in Studies in the Quantity Theory of Money. M. Friedman (ed.), Chicago: University of Chicago Press.

Caines, P. E., 1976, Prediction error methods for stationary stochastic processes, IEEE Transactions on Automatic Control AC-21, no. 4, Aug., 500-505.

Cooley, Thomas F., Barr Rosenberg and Kent D. Wall, 1977, A note on optimal smoothing for time varying coefficient problems, Annals of Eco. Soc. Meas. 6, no. 4.

Feller, W., 1971, An Introduction to Probability Theory and Its Applications, vol. 2, Wiley, New York.

Flood, Robert P. and Peter M. Garber, 1980, Market fundamentals vs. price level bubbles: The first tests, Journal of Political Economy 88, no. 4, Aug., 745-770.

Gelb, A. (ed.), 1974, Applied Optimal Estimation, Cambridge, Mass.: MIT Press.

Godfrey, L. G., 1978, Testing against general autoregressive and moving average error models when the regressors include lagged dependent variables, Econometrica 46, no. 6, Nov.

Goodrich, R. L. and Peter F. Caines, 1979, Linear system identification from non-stationary cross-sectional data, IEEE Transactions on Automatic Control AC-24, no. 3, June, 403-410.

Guttmann, William, and Patricia Meehan, 1976, The Great Inflation, Gordon & Cremones; London.

Hansen, Lars P. and Thomas J. Sargent, 1980, Formulating and estimating dynamic linear rational expectations models, Journal of Economic Dynamics and Control 2, no. 1, 7-46.

Jazwinski, A. H., 1970, Stochastic Processes and Filtering Theory (Academic Press, New York).

Lehmann, E. L., 1975, Nonparametrics: Statistical Methods Based on Ranks, Holden Day, San Francisco.

Ljung, Lennart, 1978, Convergence analysis of parametric identification methods, IEEE Transactions on Automatic Control AC-23, 770-783.

Ljung, Lennart and Peter E. Caines, 1979, Asymptotic normality of prediction error estimators for approximate systems models, Stochastics 3, 29-46.

McCallum, Bennett T., 1980, Rational expectations and macro-economics stabilization policy: An overview, Journal of Money, Credit and Banking, Nov.

Muth, J. F., 1967, Rational expectations and the theory of price movements. Econometrica, July.

Samuelson, Paul A., 1957, Intertemporal price equilibrium: A prologue to the theory of speculation, Weltwirtschaftliches Archiv 79, no. 2. Reprinted in Joseph E. Stiglitz, ed., 1966, The collected scientific papers of Paul A. Samuelson. Vol. 2 (M.I.T. Press, Cambridge, MA).

Samuelson, Paul A., 1967, Indeterminacy of developments in a heterogeneous-capital model with constant saving propensity, in: Karl Shell, ed., Essays on the theory optimal economic growth (M.I.T. Press, Cambridge, MA). Reprinted in: Robert C. Merton, ed., 1972, The collected scientific papers of Paul A. Samuelson, Vol. 3 (M.I.T. Press, Cambridge, MA).

Sargent, Thomas J. and Neil Wallace, 1973, Rational expectations and the dynamics of hyperinflation, International Economic Review 4, June.

Wall, Kent D., 1980, Generalized expectations modeling in econometrics, Journal of Economic Dynamics and Control 2, no. 2, May.

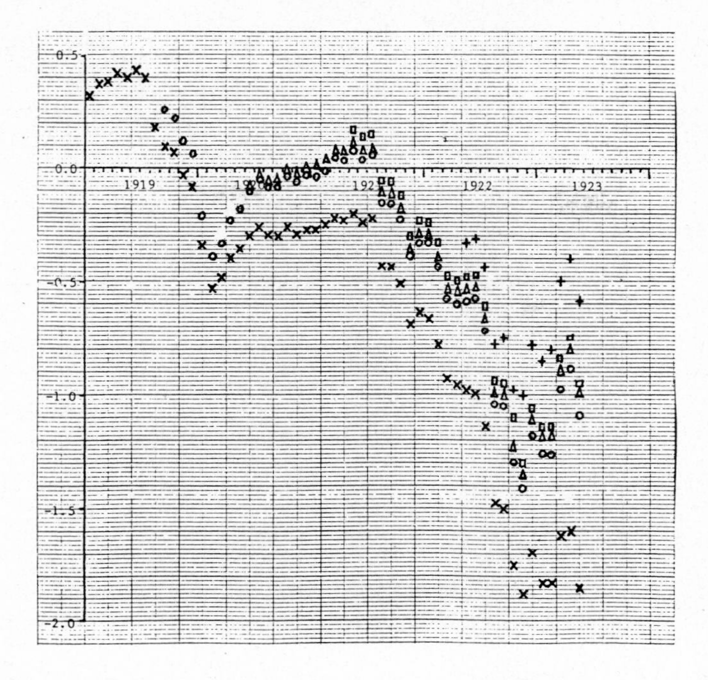

Figure 1: Smoothed estimates for $c_{1,t}$ (stochastic case)

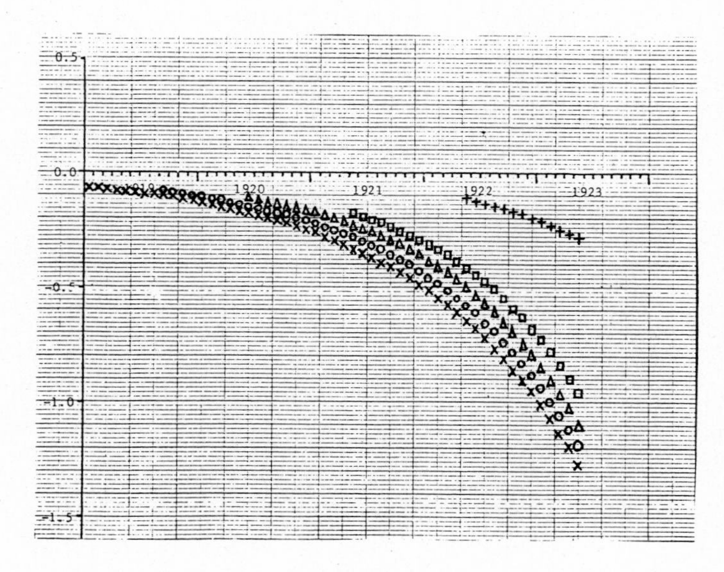

Figure 2: Smoothed estimates for $c_{1,t}$ (deterministic case)

TABLE 2.1: ESTIMATION RESULTS FOR THE STOCHASTIC CASE[a]

Date of Bubble	α_1	α_2	a	σ_ε	σ_μ	σ_η	D.W.[c]	Q[b]
2/19 - 6/23	0.5743 ($\pm$.0566)	0.3935 ($\pm$.0577)	7.3907 ($\pm$1.867)	0.0595 ($\pm$.0058)	1.14×10^{-6} ($\pm$.1292)	0.1117 ($\pm$.0110)	1.967 1.470	8.19 19.89
10/19 - 6/23	0.5145 ($\pm$.0845)	0.4826 ($\pm$.0808)	5.0768 ($\pm$.9706)	0.0553 ($\pm$.0055)	2.30×10^{-5} ($\pm$.0392)	0.1184 ($\pm$.0121)	2.161 1.500	9.88 21.71
7/20 - 6/23	0.5136 ($\pm$.0821)	0.4872 ($\pm$.0794)	4.7517 ($\pm$.8809)	0.0547 ($\pm$.0057)	1.89×10^{-7} ($\pm$.0426)	0.1191 ($\pm$.0118)	2.222 1.473	12.61 21.06
6/21 - 6/23	0.5051 ($\pm$.0878)	0.4959 ($\pm$.0855)	4.5003 ($\pm$.9322)	0.0543 ($\pm$.0054)	1.12×10^{-5} ($\pm$.0511)	0.1202 ($\pm$.0123	2.230 1.486	11.83 19.37
6/22 - 6/23	0.5164 ($\pm$.0788)	0.4992 ($\pm$.0779)	3.5356 ($\pm$1.125)	0.0538 ($\pm$.00563)	1.01×10^{-5} ($\pm$.0684)	0.1226 ($\pm$.0129)	2.343 1.366	16.64 22.49

[a]Asymptotic estimates of standard errors are in parentheses.

[b]Box-Pierce Q-statistic with 24 degrees of freedom [see Box and Jenkins (1973), p. 291]. The first line gives the Q value for the $q(t)$ equation while the second line indicates the Q value associated with the $\pi(t)$ equation.

[c]Durbin-Watson test statistic.

TABLE 2.2: EIGENVALUES FOR THE A MATRIX

Date of Bubble	Deterministic λ_1	Deterministic λ_2	Stochastic λ_1	Stochastic λ_2
2/19 - 6/23	1.058	0.707	1.012	0.644
10/19 - 6/23	1.062	0.711	1.002	0.615
7/20 - 6/23	1.068	0.721	0.999	0.622
6/21 - 6/23	1.072	0.726	0.999	0.618
6/22 - 6/23	1.079	0.843	0.987	0.672

TABLE 2.3: ESTIMATION RESULTS FOR THE DETERMINISTIC CASE[a]

Date of Bubble	α_1	α_2	a	σ_ε	σ_μ	D.W.[c]	Q[b]
2/19 - 6/23	0.6091 (±.0530)	0.3169 (±.0508)	4.3818 (±1.4253)	0.06394 (±.0065)	0.1610 (±.0157)	2.052 1.373	8.19 20.15
10/19 - 6/23	0.6059 (±.0548)	0.3217 (±.0522)	4.0577 (±1.2757)	0.06371 (±.0064)	0.1616 (±.0160)	2.074 1.352	8.19 22.36
7/20 - 6/23	0.6058 (±.0556)	0.3246 (±.0522)	3.6961 (+1.092)	0.06373 (±.0064)	0.1619 (±.0162)	2.086 1.344	9.62 22.10
6/21 - 6/23	0.5935 (±.0595)	0.3430 (±.0587)	3.2216 (±.8757)	0.0631 (±.0063)	0.1617 (±.0160)	2.100 1.344	9.88 22.23
6/22 - 6/23	0.5673 (±.1288)	0.4121 (+.1420)	1.6593 (±.5587)	0.0650 (±.0069)	0.1650 (±.0162)	2.080 1.299	8.45 20.28

[a]Asymptotic estimates of standard errors are in parentheses.

[b]Box-Pierce Q-statistic with 24 degrees of freedom [see Box and Jenkins (1973, p. 291)]. The first line gives the Q value for the g(t) equation while the second line indicates the Q value associated with the $\pi(t)$ equation.

[c]Durbin-Watson test statistic.

TABLE 2.4: SMOOTHED ESTIMATES FOR $c_{1,t}$ AND ITS ERROR VARIANCE IN THE DETERMINISTIC CASE

Date of Bubble	c_{1,t_0}	$var(c_{1,t_0})$	$c_{1,T}$	$var(c_{1,T})$
2/19 - 6/23	-.0687	.000113	-1.268	.0385
10/19 - 6/23	-.0855	.000178	-1.196	.0348
7/20 - 6/23	-.1125	.000327	-1.107	.0317
6/21 - 6/23	-.1818	.000849	-0.961	.0237
6/22 - 6/23	-.1172	.001050	-0.292	.0065

TABLE 3.1

P-VALUES FOR TEST OF $\alpha_2 = 0$
IN STOCHASTIC CASE

t_0	$\max_j \lvert \sum_t \nu_i^{(j)*} \rvert$	$\max_j D^{(j)*}_{T-t_0+1}$	Chebyshev (p_1)	K-S[1,2] (p_2)
2	50.9	.481	.0401	<.0002
10	52.9	.568	.0315	<.0002
19	49.0	.400	.0291	.0148
30	49.1	.708	.0199	<.0002
42	37.2	.583	.0173	.0628

[1]The p-values shown here come from Lehmann [1975], Table E for t_0=30,42 and Table F for T_0=2,10,19.

[2]The fact that $p_1<p_2$ for t_0=42 reflects the different means of detecting "unusual" behavior in $\{\nu_t\}$ being considered.

TABLE 4.1

P-VALUES FOR TEST OF $c_T = 0$

t_0	$\hat{c}_T$	P_T	Chebyshev p_1	Gaussian p_2
2	-1.457	.0317	.0149	$<2\cdot10^{-7}$
10	-1.172	.0142	.0103	$<2\cdot10^{-7}$
19	- .974	.0020	.0022	$<2\cdot10^{-7}$
35	-1.306	.0050	.0029	$<2\cdot10^{-7}$
42	- .283	.00075	.0094	$<2\cdot10^{-7}$

Footnotes

[1]The effects of the prior on the state are only transient and die out sufficiently rapidly for a completely observable and completely controllable state model. See footnote 9 of Burmeister and Wall (1982).

[2]All standard errors reported in Table 3.1 require a special interpretation similar to that given in Flood and Garber (1980, fn. 18) because the state space representations associated with (2.6)-(2.12) are unstable (i.e., each has an F matrix with one eigenvalue which lies outside the unit circle), and result in a stituation analogous to the 'exploding regressor' case in econometrics. Thus, we must view our data sample as one drawing from a cross-section of repeated hyperinflations, all with the same pre-1920 events and behavioral parameters. In this sense the length of the data sample is fixed at 54 observations (Jan. 1919 through June 1923), while the number of repetitions, N, of this sample tends to infinity. The estimates obtained here then are both asymptotically consistent and normal [see Goodrich and Caines (1970].

[3]The following notation is needed to define the normalized innovations process: Let $\hat{\underline{x}}_{t|t}$ and $\hat{\underline{x}}_{t|t-1}$ denote the Kalman filter estimate of $\underline{x}_t$ based on observations $\{\underline{y}_1,\cdots,\underline{y}_t\}$ and $\{\underline{y}_1,\cdots,\underline{y}_{t-1}\}$ respectively; likewise $P_{t|t}=E[\hat{\underline{x}}_t-\underline{x}_t)\quad(\hat{\underline{x}}_t-\underline{x}_t)']$ and and $P_{t|t-1}=E[(\hat{\underline{x}}_{t|t-1}-\underline{x}_t)\ (\hat{\underline{x}}_{t|t-1}-\underline{x}_t)']$. The normalized innovations process is $\underline{\nu}_t \equiv (H_t P_{t|t-1} H_t' + R_t)^{-\frac{1}{2}} x\ (\underline{y}_t - H\hat{\underline{x}}_{t|t-1} - D\underline{u}_t)$; note that $E\underline{\nu}_t = 0$ and $E(\underline{\nu}_t \underline{\nu}_s') = \delta_{ts} I$ where δ_{ts} is the Dirac delta function. Henceforth all quantities with an "*" denote sample values, i.e., those quantities evaluated using the given data under the given restrictions.

[4]See Burmeister, Flood and Turnovsky (1979). Burmeister (1980, 1982) provides a survey of some conceptual issues in rational expectations modelling, including the question of convergent expecations. Burmeister, Flood and Garber (1983) have shown that there is only one type of indeterminacy in rational expectation models; in the model considered here, there is indeterminacy whenever expectations are not convergent because there exist an infinity of equilibrium rational expecations paths with divergent expectations.

STOCHASTIC IDENTIFICATION OF CRSD MODELS FROM ARMA REPRESENTATIONS OR COVARIANCES [1]

Fabrice J. Clara [2] and Leonard M. Silverman

Department of Electrical Engineering-Systems
University of Southern California
Los Angeles, California 90089-0272

ABSTRACT

In this paper, we propose two methods to identify the parameters of a Causal, Recursive, Separable in Denominator (CRSD) model from either an Auto-Regressive, Moving-Average (ARMA) representation or the covariance function of a 2-D stationary stochastic process.

1. INTRODUCTION

In the last decade, there has been a growing interest in recursive modeling and estimation of images, and a large number of techniques has been proposed [1,2,3,4,...]. In most of these approaches, some a-priori information on the statistics of the 2-D signal is assumed available, and used to derive the parameters of a state-space representation of the image. Typically, one will characterize the image by the coefficients of some partial difference equation, describing the image formation process [1,2], or by its covariance function [4,5].
Because no unique concept of causality can be introduced in the discrete plane, several state-space models can be proposed, which can be roughly classified in two groups :

1. Half-Plane, or Non-Symetric Half-Plane models,
2. Quarter-Plane (QP) models.

---

1 This research was supported by the National Science Foundation Grant ECS-8011 911.

2 F.J. Clara is now with Institut National de la Recherche en Informatique et Automatique, Rocquencourt, BP 105, 78153 Le Chesnay Cédex, France.

These classes refer to the chosen causality, for the 2-D models.
In this presentation, we are concerned with a class of Quarter-Plane models, called Causal, Recursive, Separable in Denominator (CRSD) models, originally proposed by Roesser [6] and extensively studied by Hinamoto [7], then Lashgari et al [8,9]. This model is defined as :

$$\begin{pmatrix} x^H(n+1,m) \\ x^V(n,m+1) \end{pmatrix} = \begin{pmatrix} A_1 & A_2 \\ 0 & A_4 \end{pmatrix} \begin{pmatrix} x^H(n,m) \\ x^V(n,m) \end{pmatrix} + \begin{pmatrix} B_1 \\ B_2 \end{pmatrix} u(n,m) \quad (1.a)$$

$$y(n,m) = C_1 x^H(n,m) + C_2 x^V(n,m)$$

with boundary conditions :

$$\chi^H(0) = \{x^H(0,j),\ j \in \mathbb{N}\}$$
$$\chi^V(0) = \{x^V(i,0),\ i \in \mathbb{N}\} \quad (1.b)$$

$x^H(n,m)$ and $x^V(n,m)$ are respectively the horizontal and vertical local state vectors [10], of order N and M, $u(n,m)$ is the input to the system, $y(n,m)$ the output. A_i, B_i and C_i are matrices of proper order [8].
CRSD models proved most useful for such applications as filter design, model reduction [9,11] or image restoration [9]. For image modeling, the identification/approximation procedure of [8] can be applied to images defined by a Moving-Average (MA) equation. A second situation is described in [9], where the CRSD model parameters are related to a special class of covariance functions, defined as a sum of separable exponential kernels.
We now extend these results to a more general context. In Section 2, we propose to identify the CRSD model from the coefficients of a Quarter-Plane causal Auto-Regressive, Moving Average (ARMA) representation of the image. The CRSD model is first obtained in "canonical" form, and then reduced to its minimal balanced form using an eigenvalue decomposition of the system grammian matrices. In Section 3, we study the covariance structure of a stochastic process defined by a CRSD model, and propose an identification procedure from covariance functions. This approach generalizes the work by Faurre et al.[12] on Markov realizations of 1-D processes. For reasons mentionned later, it also appears preferable to the identification scheme of Attasi [4].

2. CRSD REALIZATION OF ARMA EQUATIONS

A large class of 2-D space invariant systems, with input $u(n,m)$ and output $y(n,m)$, can be modeled by difference equations, known as ARMA representations, as defined below :

$$\sum_{i,j} a(n-i,m-j)y(i,j) = \sum_{k,l} b(n-k,m-l)u(k,l) \tag{2}$$

Because of limiting factors inherent to the Roesser-CRSD model we consider, we shall restrict this study to ARMA equations which represent *strictly QP-causal* systems. If such is the case, the equation 2 can be brought into the form of equations 3,4.

$$\sum_{i,j \in D_y} a(i,j)y(n-i,m-j) = \sum_{i,j \in D_u} b(i,j)u(n-i,m-j) \tag{3}$$

where the domains D_y and D_u are finite, specified as :

$$\begin{aligned} D_y &= \{(i,j)/-\infty < i \le N \text{ and } -\infty < j < M\} \cup \{(i,j)/-\infty < i < N \text{ and } j=M\} \\ D_u &= \{(i,j)/-\infty < i < N \text{ and } -\infty < j < M\} \end{aligned} \tag{4}$$

Canonical CRSD Realization

In the 1-D case, a simple algebraic manipulation of such difference equation leads to a so-called *canonical state space realization*, obtained in a controller or observer form [13]. We extend this property with the following theorem and corollary.

Theorem 1 : A 2-D strictly QP-causal system defined by an ARMA equation such as 3 admits a CRSD model if and only if one can bring the ARMA relation into the form :

$$\sum_{i=0}^{N} \sum_{j=0}^{M} a_i b_j y(n-i,m-j) = \sum_{i=1}^{N} \sum_{j=1}^{M} c_{ij} u(n-i,m-j) \tag{5}$$

where we can assume without loss of generality $a_0=b_0=1$. ■

Proof : Assume a 2-D system is defined by a minimal CRSD model of order N,M (see [9] for a definition). Its transfer function can be written as :

$$H(z_1,z_2) = \frac{Y(z_1,z_2)}{U(z_1,z_2)} = \frac{N(z_1,z_2)}{D_1(z_1)D_2(z_2)} \tag{6}$$

where $N(z_1,z_2)$, $D_1(z_1)$ and $D_2(z_2)$ are polynomials in (z_1,z_2), $\deg\{D_1(z_1)\}=N$, $\deg\{D_2(z_2)\}=M$ and the degrees in z_1 and z_2 of $N(z_1,z_2)$ are strictly less than N and M [9]. Multiplying both sides of 6 by $z_1^{-N}z_2^{-M}$ and taking the inverse Z-transform yields an ARMA input/output relation of the desired format 5. Conversely, assume we have an equation such as 5. Then, a *"2-D canonical CRSD representation"* of the system can be simply obtained with the following choice for A_1, A_2, A_4, B_2 and C_1 [14].

$$A_1 = \begin{pmatrix} -a_1 & 1 & 0 & & \emptyset \\ -a_2 & 0 & \ddots & \ddots & \\ \vdots & & \ddots & \ddots & 0 \\ \vdots & \emptyset & & \ddots & 1 \\ -a_N & \cdots & \cdots & \cdots & 0 \end{pmatrix} \qquad A_4 = \begin{pmatrix} -b_1 & -b_2 & \cdots & -b_M \\ 1 & 0 & \emptyset & \vdots \\ 0 & \ddots & \ddots & \vdots \\ & \ddots & \ddots & 0 \\ \emptyset & & 0 & 1 \end{pmatrix} \tag{7.a}$$

$$A_2 = \begin{pmatrix} c_{11} & c_{12} & \cdots & c_{1M} \\ c_{21} & c_{22} & & \vdots \\ \vdots & & \ddots & \vdots \\ c_{N1} & \cdots & \cdots & c_{NM} \end{pmatrix} \qquad B_1 = 0, \quad C_2 = 0 \tag{7.b}$$

$$B_2^T = [1 \quad 0 \quad \ldots \quad 0], \qquad C_1 = [1 \quad 0 \quad \ldots \quad 0] \tag{7.c}$$

Corollary 2 : If a 2-D system is described by an equation of type 5, an "obvious" CRSD realization is obtained by choosing the model matrices as in equations 7. ■

In [8,9], the CRSD model 1 is shown to be related to a 1-D infinite-input system, with state vector $x^H(n,m)$, coupled with a 1-D infinite-output system, with state vector $x^V(n,m)$. For the canonical realization of equation 7, those 1-D systems, denoted R^H and R^V respectively, appear to be in canonical observer and controller forms -see [14] for details-. Thus, they are respectively observable and controllable. However, minimality of this realization, in the sense of [9], will be guaranteed only if the polynomials $P(z_1,z_2)$ and $Q(z_1,z_2)$, defined as :

$$P(z_1,z_2) = \sum_{i=1}^{N} \sum_{j=1}^{M} c_{ij} z_1^{N-i} z_2^{M-j} \tag{8.a}$$

$$Q(z_1,z_2) = [\sum_{i=0}^{N} a_i z_1^{N-i}][\sum_{j=0}^{M} b_j z_2^{M-j}] \tag{8.b}$$

have no common zeroes in z_1 and z_2. Indeed, since $P(z_1,z_2)/Q(z_1,z_2)$ represents the transfer function of the realization 7, irreducibility implies minimality of the realization [15,9]. However, this condition is impractical to check when only the ARMA coefficients are known. Although a minimal or reduced model could be obtained from [8], we propose a different method which will produce a minimal balanced realization of equation 5, based upon canonical factorization of a CRSD model [7].

Balancing of the canonical CRSD model

We assume that the original ARMA system is asymptotically stable, or equivalently, that the spectral norm of A_1 and A_4, defined in 7, is bounded :

$$||A_1||_S < 1, \qquad ||A_4|| < 1 \tag{9}$$

The controllability and observability of a CRSD model can be specified in term of the controllability and observability grammian of its two related 1-D systems R^H and R^V [8]. For an asymptotically stable CRSD system, the grammian matrices are obtained as the unique nonnegative solution of the Lyapunov system 10 :

$$W_c^H - A_1 W_c^H A_1^T = A_2 W_c^V A_2^T \tag{10.a}$$

$$W_c^V - A_4 W_c^V A_4^T = B_2 B_2^T \tag{10.b}$$

$$W_o^H - A_1^T W_o^H A_1 = C_1^T C_1 \tag{10.c}$$

$$W_o^V - A_4^T W_o^V A_4 = A_2^T W_o^H A_2 \tag{10.d}$$

and can be computed through various methods [16,18]. For the canonical realization 7, we can assert that W_o^H and W_c^V are positive definite matrices. However, W_c^H and W_o^V might not be full rank matrices. Assume that $\mathrm{Rank}\{W_c^H\}=K$ and $\mathrm{Rank}\{W_o^V\}=L$. We can obtain, through eigenvalue decomposition :

$$W_c^H = [U_1 | U_2] \left(\begin{array}{c|c} \Sigma_c & \\ \hline & \emptyset \end{array}\right) [U_1 | U_2]^T \tag{11.a}$$

$$W_o^V = [U_3 | U_4] \left(\begin{array}{c|c} \Sigma_o & \\ \hline & \emptyset \end{array}\right) [U_3 | U_4]^T \tag{11.b}$$

Following [7] the following properties are satisfied :

1. The unobservable subspace of R^H is the empty set.
2. The controllable subspace of R^V is the set R^M.
3. The column vectors of U_1 are in the controllable subspace of R^H.
4. The column vectors of U_4 are in the unobservable subspace of R^V.

The equivalence transformation [7,8]

$$T = \left(\begin{array}{c|c} T_1 & \emptyset \\ \hline \emptyset & T_4 \end{array}\right) \tag{12.a}$$

$$\text{where } T_1 = \left(\begin{array}{c|c} \Sigma_c^{-1/2} & \emptyset \\ \hline \emptyset & Id \end{array}\right) [U_1 | U_2]^{-1} \qquad (12.b)$$

$$\text{and } \quad T_4 = \left(\begin{array}{c|c} \Sigma_o^{1/2} & \emptyset \\ \hline \emptyset & Id \end{array}\right) [U_3 | U_4]^{-1} \qquad (12.c)$$

produces an "equivalent" CRSD model in "canonical" structure [7].

$$\hat{A}_1 = \begin{pmatrix} A_{1,11} & A_{1,12} \\ \emptyset & A_{1,22} \end{pmatrix} \qquad \hat{A}_4 = \begin{pmatrix} A_{4,11} & A_{4,12} \\ \emptyset & A_{4,22} \end{pmatrix} \qquad (13.a)$$

$$\hat{A}_2 = \begin{pmatrix} A_{2,11} & A_{2,12} \\ A_{2,21} & A_{2,22} \end{pmatrix} \qquad (13.b)$$

$$\hat{C}_1 = [C_{1,1} | 0] \qquad \hat{B}_2^T = [B_{2,1} | 0]^T \qquad (13.c)$$

where $A_{1,11}$ and $A_{4,11}$ are of order K and L respectively.
The CRSD subsystem $\{A_{1,11}, A_{2,11}, A_{4,11}, C_{1,1}, B_{2,1}\}$ is controllable, observable and realizes the same system impulse response, or ARMA equation, as the original system 7 [7].
The grammians of the subsystem are given by [14] :

$$\hat{W}_c^H = Id_{K\times K}, \quad \hat{W}_o^V = Id_{L\times L} \qquad (14.a)$$

$$\hat{W}_o^H - A_{1,11}^T \hat{W}_o^H A_{1,11} = C_{1,1}^T C_{1,1} \qquad (14.b)$$

$$\hat{W}_c^V - A_{4,11} \hat{W}_c^V A_{4,11}^T = B_{2,1} B_{2,1}^T \qquad (14.c)$$

If we perform the eigenvalue decompositions :

$$\hat{W}_o^H = \tilde{U} S_o \tilde{U}^T, \quad \text{and} \quad \hat{W}_c^V = \tilde{V} S_c \tilde{V}^T \qquad (15)$$

and if we introduce a second similarity transform :

$$\tilde{T}_i = S_o^{1/2} \tilde{U}^T, \qquad \tilde{T}_4 = S_c^{-1/2} \tilde{V}^T \qquad (16)$$

we obtain a CRSD model of order K,L, in balanced form, which is minimal and equivalent to the original system 7. Indeed, the grammians are now :

$$\tilde{W}_o^H = \tilde{T}_1^{-T}\hat{W}_o^H\tilde{T}_1^{-1} = S_o \tag{17.a}$$

$$\tilde{W}_c^V = \tilde{T}_4\hat{W}_c^V\tilde{T}_4^T = S_o \tag{17.b}$$

$$\tilde{W}_c^H = \tilde{T}_1 Id_{KxK}\tilde{T}_1^T = S_o \tag{17.c}$$

$$\tilde{W}_o^V = \tilde{T}_4^{-T} Id_{LxL}\tilde{T}_4^{-1} = S_c \tag{17.d}$$

In practical implementations, this approach amounts to the computation of four eigenvalue decompositions and solutions to Lyapunov equations. However, the computational cost and requirements of this approach will be much smaller than when using Lashgari's approach to model balancing and approximation. Typically, in image modeling, the order of the initial ARMA equation is a few rows and columns, as opposed to the computation on large Hankel matrices in [8].

3. CRSD MODELS FOR 2-D STOCHASTIC PROCESSES

Consider a 2-D stochastic process y(n,m), generated as the output of a strictly QP-causal CRSD system driven by white Gaussian noise :

$$\begin{pmatrix} x^H(n+1,m) \\ x^V(n,m+1) \end{pmatrix} = \begin{pmatrix} A_1 & A_2 \\ \emptyset & A_4 \end{pmatrix} \begin{pmatrix} x^H(n,m) \\ x^V(n,m) \end{pmatrix} + \begin{pmatrix} 0 \\ B_2 \end{pmatrix} u(n,m) \tag{18}$$

$$y(n,m) = C_1 x^H(n,m)$$

where u(n,m) is a zero-mean, white Gaussian noise process, with covariance matrix :

$$E\{u(n,m)u^T(n',m')\} = Id \times \delta_{n'-n,m'-m} \tag{19}$$

We assume that the matrices A_1 and A_4 are stable matrices, with eigenvalues within the unit circle.
The system output is given as [7] :

$$y(n,m) = \sum_{i>0,j>0} C_1 A_1^{i-1} A_2 A_4^{j-1} B_2 u(n-i,m-j) \tag{20}$$

Under the assumptions that the boundary conditions are mutually uncorrelated, zero-mean Gaussian process, uncorrelated with u(n,m), the stochastic processes y(n,m), $x^H(n,m)$ and $x^V(n,m)$ are Gaussian, zero-mean, stationary processes. We introduce the following notations :

$$\begin{cases} R_y(k,l) = E\{y(n,m)y^T(n+k,m+l)\} \\ \Lambda_H(k,l) = E\{x^H(n,m)x^{HT}(n+k,m+l) \\ \Lambda_V(k,l) = E\{x^V(n,m)x^{VT}(n+k,m+l) \end{cases} \quad (21)$$

We also define $P(m)=\Lambda_H(0,m)$ and $\Lambda=\Lambda_V(0,0)$. The second-order statistics of the processes $y(n,m)$, $x^H(n,m)$ and $x^V(n,m)$ are characterized in the next theorem.

Theorem 3 : (Proof in [14]) The covariance matrices defined in 21 are given by the set of equations :

$$R_y(n,m) = C_1\Lambda_H(n,m)C_1^T,\ R_y(-n,-m) = R_y^T(n,m),\ \forall\ n,m \quad (22.a)$$

For all $n>0$,

$$\Lambda_H(n,m) = A_1{}^nP(m),\ \forall\ m \text{ and } \Lambda_H(-n,m) = P(m)A_1{}^{nT}\ \forall\ m \quad (22.b)$$

and for all $m>0$

$$\begin{cases} \Lambda_V(0,m) = A_4{}^m\Lambda(0) \text{ and } \Lambda_V(0,-m) = \Lambda(0)A_4{}^{mT} \\ \Lambda_V(n,m) = 0\ \forall\ n \neq 0,\ \forall\ m. \end{cases} \quad (22.c)$$

$P(m)$ and $\Lambda(0)$ are the unique solutions of

$$\begin{cases} P(m) - A_1P(m)A_1{}^T = A_2\Lambda_V(0,m)A_2{}^T \text{ with } P(-m) = P^T(m) \\ \Lambda(0) - A_4\Lambda(0)A_4{}^T = B_2B_2{}^T \end{cases} \quad (22.d)$$

∎

Remark : If the CRSD model is balanced, the vectors $x^H(n,m)$ and $x^V(n,m)$ have diagonal variance, given by the grammians of R^H and R^V, respectively.

Remark : For stochastic processes defined from (18) or, equivalently, by the input/output relation :

$$y(n,m) = \sum_{i,j>0} K(i,j)u(n-i,m-j) \quad (23)$$

we have the following theorem [4].

Theorem 4 : The class of CRSD stochastic processes is dense in the class of Σ-sequences corresponding to $K(i,j)$ being square summable. ∎

Definition 5 : The spectrum of a 2-D stochastic process is defined as the 2-D Fourier transform of its covariance matrix, namely

$$S_y(z_1,z_2) = \sum_{-\infty < i,j < \infty} R_y(i,j)z_1^{-i}z_2^{-j} \quad ∎ \quad (24)$$

One major problem with 2-D stochastic processes lies in the non-uniqueness of the spectral factorization process. Typically, it has been shown that one may possibly factor the spectrum in two non-symmetric half-plane factors, or in four quarter plane factors [17]. This restriction justifies the various models considered for 2-D processes ; it is, in some way, analogous to the causality problem for deterministic processes. However, if a process is generated by a CRSD system, we can factor its spectrum in two quarter-plane factors, one being strictly causal, the other strictly anti-causal. In fact, we can give the following theorem [4,9].

Theorem 6 : A 2-D process $y(n,m)$ has a CRSD representation such as 18 if and only if its spectrum admits the following factorization :

$$\begin{cases} S(z_1,z_2) = H(z_1,z_2)\, H\,(z_1,z_2) \\ H(z_1,z_2) = C_1(z_1I-A_1)^{-1}A_2(z_2I-A_4)^{-1}B_2 \end{cases} \tag{25}$$

Remark : Since the impulse response of the system 18 is real, we also have :

$$H^*(z_1,z_2) = H^T(1/z_1,1/z_2)$$

Various techniques are available to obtain factorization of a complex function [1,17]. In his study, Attasi [4] showed that a realization can be directly obtained from the covariance matrix, without performing the spectral factorization. We now extend this method to the context of strictly QP-causal CRSD models.

Definition 7 : We shall say a 2-D stationary stochastic process $y(n,m)$ with covariance $R_y(n,m)$ is *CRSD-realizable* if we can find matrices A_1, A_2,A_4,B_2,C_1, $\{P(m), m \geq 0\}$ and $\Lambda(0)$ which satisfy the system 22. The statistics of the output of 18 are then equal to $R_y(n,m)$. ∎

The analog of Attasi's results for CRSD models is given in the following theorem. We assume the covariance $R_y(n,m)$ satisfy :

$$\sum_{n,k}\ \sum_{m,l}\ u^T(n,m)R_y(k-n,l-m)u(k,l) > 0\ \forall\ u(n,m). \tag{26}$$

Non-negativity of the above is obvious, since the left hand side represents the variance of the process $\sum_{n,m} y(n,m)u(n,m)$.

Theorem 8 : A stochastic process $y(n,m)$, with covariance $R_y(n,m)$ is CRSD-realizable if and only if

1. $R_y(n,m) = R_y^T(-n,-m)$ for all n,m; (27.a)

2. There exists a set of matrices H, F and G(m) such that

$$R_y(n,m) = HF^nG(m), \text{ for all } n \geq 0,\ \forall\ m \tag{27.b}$$

with $\{H,F\}$ observable;

3. There exist a sequence of matrices P(m) such that

$$\begin{cases} P(-m) = P^T(m),\ P(0) > 0 \\ G(m) = P(m)H^T \ \forall\ m \geq 0 \\ R_y(-n,m) = HP(m)F^{nT}H^T \text{ for all } n>0,\ m\varepsilon \mathbf{Z} \end{cases} \tag{27.c}$$

such that the operator defined as

$$Q(m) = P(m) - FP(m)F^T \tag{27.d}$$

is a strictly positive operator in the sense of [12], i.e.

$$\sum_{m,k} u^T(m)Q(k-m)u(k) \geq 0 \text{ for all } u(.)$$ ∎

The proof of this theorem is given in [14]. It closely follows a similar argument in [5]. It is also constructive in a sense that a practical method is given to compute the matrix sequence P(m) from the matrices H,F and G(m). However, as was the case in [5], the positivity of the associated Q(m) cannot be guaranteed when determining P(m), and the obtained solution might not be acceptable. However, for CRSD models, it is possible to show that an acceptable solution always exists, and can practically be computed. We first assume that a set of matrices {H,F,G(m) for mε**Z**} has been obtained, with the pair {H,F} observable, such that :

$$R_y(n,m) = HF^nG(m) \ \forall\ n \geq 0,\ m\varepsilon \mathbf{Z} \tag{28}$$

Equations 27.a and 28 imply :

$$R_y(-n,m) = G^T(-m)F^{nT}H^T \ \forall\ n > 0,\ m\varepsilon \mathbf{Z} \tag{29}$$

We then introduce the following notations :

$$<u,R_y u> = \sum_{n=-\infty}^{-1} \sum_{k=-\infty}^{-1} \sum_{m=-\infty}^{-1} \sum_{l=-\infty}^{-1} u^T(n,m)R_y(k-n,l-m)u(k,l) \tag{30.a}$$

$$u_n(z) = \sum_{m=-\infty}^{+\infty} u(n,m)z^{-m},\ |z| = 1 \tag{30.b}$$

$$\rho_n(z) = \sum_{m=-\infty}^{+\infty} R_y(n,m)z^{-m},\ |z| = 1 \tag{30.c}$$

$$<u,\rho u>(z) = \sum_{n=-\infty}^{-1} \sum_{k=-\infty}^{-1} u_n^T(z)\rho_{k-n}(z)u_k(1/z) \tag{30.d}$$

Observe that we can write :

$$\begin{cases} \rho_n(z) = HF^n G(z) \\ \rho_{-n}(z) = G^T(1/z) F^{nT} H^T \\ \rho_{-n}(z) = \rho_n(1/z)^T \end{cases} \tag{31}$$

where $G(z)$ is the Z-transform of $G(m)$. $<u,\rho u>(z)$ is an Hermitian form. Our goal is to find a sequence of matrices $P(m)$ and $Q(m)$ such that 27.c and 27.d holds. This is equivalent to finding functions $P(z)$ and $Q(z)$ for $|z|=1$ such that :

$$P(z)H^T = G(z),\ P(z) = P^T(1/z)$$
$$\text{which implies } HP(z) = G^T(1/z) \tag{32}$$
$$Q(z) = P(z) - FP(z)F^T \geq 0.$$

Lemma 9 : $<u,\rho u>(z)$ is non-negative for all z and $u_n(z)$.

Proof : From Parseval's theorem, we can write :

$$<u,R_y u> = \sum_{n,k} \frac{1}{2\pi} \int_{-\pi}^{+\pi} u_n{}^T(e^{j\omega}) \rho_{k-n}(e^{j\omega}) u_k(e^{-j\omega})\, d\omega$$

Assume there exists a sequence $v_n(z)$ defined on the unit circle and ω_0 such that $<v,\rho v>(e^{j\omega_0})$ is negative. By continuity, there is a neighborhood Ω_0 of ω_0 where this property holds. For the sequence $u(n,m)$ such that :

$$u_n(e^{j\omega}) = v_n(e^{j\omega}) \text{ if } \omega \in \Omega_0, \quad = 0 \text{ of } \omega \notin \Omega_0$$

the expression $<u,R_y u>$ is negative, which contradicts 30.

Lemma 10 : Assume $\rho_n(z)$ defined as in 30.c . If there exists $P(z)$ and $Q(z)$ such that 32 holds, then :

$$<u,\rho u> = x_0{}^T(z)P(z)x_0(1/z) + \sum_{n=-\infty}^{-1} x_n{}^T(z)Q(z)x_n(1/z) \tag{33.a}$$

$$\text{where } x_n(z) = F^T x_{n-1}(z) + H^T u_{n-1}(z),\ n \leq 0,\ |z|=1 \quad \blacksquare \tag{33.b}$$

Proof : The solution of 33.b is given as :

$$x_n(z) = \sum_{k=-\infty}^{n-1} F^{(n-k-1)\,T} H^T u_k(z)$$

We can write $<u,\rho u>(z)$ in the form :

$$<u,\rho u>(z) = \sum_{n=-\infty}^{-1} u_n{}^T(z) G^T(1/z) F^T x_n(1/z) + \sum_{n=-\infty}^{-1} x_n{}^T(z) F G(z) u_n(1/z) + \sum_{n=-\infty}^{-1} u_n{}^T(z) H G(z) u_n(1/z) \tag{34}$$

Using 32 , one can show that 34 can be rewritten as [14]:

$$<u,\rho u>(z) = \sum_{n=-\infty}^{-1} x_{n+1}^T(z)P(z)x^{n+1}(1/z) - x_n^T(z)P(z)x_n(1/z)$$

$$+x_n^T(z)Q(z)x_n(1/z) \qquad (35)$$

$$= x_0^T(z)P(z)x_0(1/z) + \sum_{n=-\infty}^{-1} x_n^T(z)Q(z)x_n(1/z)$$

which completes the proof.

Since the pair {H,F} is observable, the system defined in 33.b is controllable. For any function $\xi(z)$ defined on $|z|=1$, we can find a control law $u_n(z)$ which drives the system from $x_{-\infty}(.)=0$ to $x_0(.)=\xi(.)$. Define the set of such control laws as $\Psi(\xi)$.

Theorem 11 : The function $P^*(z)$ defined on the unit circle as :

$$x_0^T(z)P^*(z)x_0(1/z) = \inf <u,\rho u>(z), \text{ for } u_n(z)\varepsilon\Psi\{x_0(z)\} \qquad (36)$$

satisfy the conditions 32 , and yields a CRSD representation of the covariance $R_y(n,m)$. ■

Proof : The above definition is consistent. For each z, the expression $<u,\rho u>(z)$ is a Hermitian form in $u_n(z)$. Its minimum under the *linear* constraint $u_n(z)\varepsilon\Psi\{x_0(z)\}$ is a hermitian form in $x_0(z)$. $Q^*(z)$, defined from 32 is positive. Following [12], we notice that :

$$x_0^T(z)FP^*(z)F^Tx_0(1/z) = \inf <u,\rho u>(z) \text{ for } u_n(z)\varepsilon\Psi\{F^Tx_0(z)\}$$

If $u_n(z)\varepsilon\Psi\{x_0(z)\}$, then the control law $v_n(z)$:

$$v_n(z) = u_{n+1}(z) \text{ if } n<-1, =0 \text{ if } n=-1$$

is in the set $\Psi\{F^Tx_0(z)\}$. Also, from its definition 30.d , $<v,\rho v>(z)=<u,\rho u>(z)$, and therefore :

$$\inf <v,\rho v> \leq \inf <u,\rho u>$$

which implies that $Q^*(z)$ is non-negative. We also have $P^{*T}(z)=P(1/z)$ from its definition 29 . That $G(z)=P^*(z)H^T$ can be proven in a way similar to [12]. Once $P^*(z)$ has been found, we obtain $Q^*(z)$ as a nonnegative function, which satisfies :

$$Q^*(z) = Q^{*T}(1/z)$$

The operator $Q^*(m)$, for $m\varepsilon\mathbf{Z}$, obtained by inverse Z-transform is then positive, and standard 1-D identification techniques can then be applied. The CRSD model obtained from $\{H,F,Q^*(m)\}$ in the previous section will exactly realize the original covariance $R_y(n,m)$.

The last step towards a complete solution of this identification is a practical method to obtain $P^*(z)$ previously defined. This is done in the following theorem.

Theorem 12 : The function $P^*(z)$, for $|z|=1$, is given as solution of the following spectral Riccati equation :

$$\begin{cases} P^*(z) = [\lim_{N\to\infty} \Pi_N(z)]^{-1} \\ \Pi_{N+1}(z) = F^T\Pi_N(z)F + [H^T - F^T\Pi_N FG(z)]E(z)[H - G^T(1/z)F^T\Pi_N(z)F] \\ \text{with } E(z) = [\rho_0(z) - G^T(1/z)F^T\Pi_N FG(z)]^{-1} \\ \text{and } \Pi_0(z)=0 \end{cases} \quad (37)$$

■

Proof : If we introduce $<u,\rho u>_N(z)$ as :

$$<u,\rho u>_N(z) = \mathbf{U}_N^T(z)\mathbf{R}_N\mathbf{U}_N(1/z)$$

where $\mathbf{U}_N^T(z) = [u_{-1}^T(z)|u_{-2}^T(z)|......u_N^T(z)]$

$$\text{and } \mathbf{R}_N = \begin{bmatrix} \rho_0(z) & \rho_{-1}(z) & \cdots & \rho_{-N+1}(z) \\ \rho_1(z) & \rho_0(z) & \rho_{-1}(z) & \\ & \ddots & \ddots & \ddots \\ \rho_{N-1}(z) & & & \end{bmatrix} \quad (38)$$

then $P^*(z) = \lim_{N\to\infty} P_N(z)$ where :

$$x_0(z)P_N(z)x_0(1/z) = \inf <u,\rho u>_N(z) \text{ for } u_n(z)\varepsilon\Psi_N\{x_0(z)\}$$

$$\text{and } \underline{\mathbf{C}}_N\{x_0(z)\} = [H^T|F^TH^T|.......F^{(N-1)T}H^T]$$

Following an argument from [12], it can be shown that the solution of the minimization problem 36 , is given as :

$$P_N(z) = \underline{\mathbf{C}}_N\mathbf{R}_N^{-1}\underline{\mathbf{C}}_N^{T\ -1} \quad (39)$$

From the Toeplitz structure of R_N, we have :

$$\mathbf{R}_{N+1} = \left[\begin{array}{c|c} \rho_0(z) & G^T(1/z)F^T\mathbf{R}_N \\ \hline \mathbf{R}_N^T FG(z) & \mathbf{R}_N \end{array}\right] \quad (40)$$

and therefore

$$\mathbf{R}_{N+1}^{-1} = \left[\begin{array}{c|c} E(z) & A(z) \\ \hline B(z) & C(z) \end{array}\right] \quad (41)$$

$$\begin{cases} E(z) = \rho_0(z) - G^T(1/z)F^T\underline{\mathbf{C}}_N\mathbf{R}_N^{-1}\underline{\mathbf{C}}_N^{\,T}FG(z) \\ A(z) = -\,E(z)G^T(1/z)F^T\underline{\mathbf{C}}_N\mathbf{R}_N^{-1} \\ B(z) = -\,\mathbf{R}_N^{-1}\underline{\mathbf{C}}_N^{\,T}FG(z)E(z) \\ C(z) = \mathbf{R}_N^{-1} + \mathbf{R}_N^{-1}\underline{\mathbf{C}}_N^{\,T}FG(z)E(z)G^T(1/z)F^T\underline{\mathbf{C}}_N\mathbf{R}_N^{-1} \end{cases} \quad (42)$$

using a "well-known" matrix inversion rule. We finally have :

$$P_{N+1}(z) = \underline{\mathbf{C}}_{N+1}\mathbf{R}_{N+1}^{-1}\underline{\mathbf{C}}_{N+1}^{\,T\;-1} \quad (43)$$

$$\text{with } \underline{\mathbf{C}}_{N+1} = [\, H^T \mid F^T\underline{\mathbf{C}}_N \,]$$

An "adequate" algebraic manipulation of 43 using 42 yields the Riccati equation 37 .

Practically, the computation of the spectral Riccati equation can be made on equally spaced points of the unit circle. The coefficients of $P^*(m)$ and $Q^*(m)$ are then obtained using FFT algorithms.

The merit of this approach stems from the constructive manner in which a solution is always obtained. The spectral Riccati equation can also be compared to a similar equation introduced in [4] for estimation purpose. As in [12], one can then attempt to bridge the gap between the estimation filter and the stochastic identification problem, by defining Minimum-Variance representations of a 2-D process. One difficulty lies in the concept of duality, not introduced so far for CRSD models. The promising eventual extensions motivate further research in that domain.

4. CONCLUSION

We presented two different techniques to identify the parameters of a CRSD model for images from either an ARMA representation or a covariance function. Since most 2-D processes are usually described in terms of such statistical information, or possibly of Moving-Average equations, the combination of these two methods with the algorithm in [8] provides an efficient identification package for image modeling. Practical examples which illustrate these two procedures will be given at the conference.

REFERENCES

[1] J.W. Woods and C.H. Radewan, "Kalman Filtering in Two Dimensions", IEEE Transactions on Information Theory, Vol.IT-23, pp.473-482, July 1977.

[2] M.S. Murphy and L.M. Silverman, "Scene Model Representation and Restoration", Proc. Symposium on Image Science Mathematics, November 1976.

[3] L.M. Silverman and F.J. Clara, "Recent Results in Recursive and Nonlinear Image Restoration", in Analysis and Optimization of Systems, Springer-Verlag, pp.721-743, 1980.

[4] S. Attasi, "Modeling and Recursive Estimation for Double Indexed Sequences", in System Identification : Advances and Case Studies, Academic Press, 1976.

[5] A.O. Aboutalib and L.M. Silverman, "Restoration of Motion Degraded Images", IEEE Transactions on Circuits and Systems, Vol.CAS-22, March 1975.

[6] R.P. Roesser, "A Discrete State-Space Model for Linear Image Processing", IEEE Transactions on Automatic Control, Vol.AC-20, N° 1, pp.1-10, February 1975.

[7] T. Hinamoto, "Realizations of a State-Space Model from Two-Dimensional Input-Output Map", IEEE Transactions on Circuits and Systems, Vol.CAS-27, N° 1, pp.36-44, January 1980.

[8] B. Lashgari, L.M. Silverman, J-F. Abramatic, "Approximation of 2-D Separable in Denominator Filters", IEEE Transactions on Circuits and Systems, Vol.CAS-30, N° 2, pp.107-121, February 1983.

[9] B. Lashgari, Two-Dimensional Approximation, Model Reduction and Minimum Variance Estimation, PhD dissertation, University of Southern California, December 1981.

[10] S.Y. Kung, B.C. Levy, M. Morf, T. Kailath, "New Results in 2-D Systems Theory, Part II : 2-D State-Space Models, Realization and the Notions of Controllability, Observability and Minimality", Proceedings of the IEEE, Vol.65, N° 6, pp.945-961, June 1977.

[11] F.J. Clara and L.M. Silverman, "2-D Discrete Space Varying Systems : Identification, Balancing and Model Reduction", Sixth IFAC Symposium, Vol.2, pp.1304-1309, June 1982.

[12] P. Faurre, M. Clerget, F. Germain, Opérateurs Rationels Positifs, Dunod, Méthodes Mathématiques de l'Informatique, Vol.8, 1979.

[13] T. Kailath, Linear Systems, Prentice-Hall, Inc., Prentice-Hall Information and System Sciences Series, 1980.

[14] F.J. Clara, A state space approach to image modeling, restoration and identification, PhD Dissertation, University of Southern California, October 1983.

[15] R. Eising, 2-D Systems, an Algebraic Approach, PhD dissertation, Mathematisch Centrum, Amsterdam, March 1979.

[16] B.D.O. Anderson, J.B. Moore, Optimal Filtering, Prentice Hall, Inc. Information and System Sciences Series, 1979.

[17] M.P. Ekstrom and J.W. Woods, "Two-Dimensional Spectral Factorization with Applications in Recursive Digital Filtering", IEEE Transactions on Acoustics, Speech and Signal Processing, Vol. ASSP-24, N° 2, pp.115-128, April 1976.

[18] A.J. Laub, "A Schur Method for Solving Algebraic Riccati Equarions", IEEE Transactions on Automatic Control, Vol.AC-24, N° 6, pp.913-921, December 1979.

A SOLUTION OF AN INVERSE PROBLEM IN THE 1 D WAVE EQUATION APPLICATION TO THE INVERSION OF VERTICAL SEISMIC PROFILES

D. Macé and P. Lailly
Institut Français du Pétrole
Rueil-Malmaison - France

SUMMARY

We deal with the inversion of a vertical seismic profile in 1D. A seismic source being located at the vicinity of the earth surface we measure the vibratory state at different depths in a well. We have to find the distribution of acoustic impedance versus depth from these measurements. The excitation resulting from the seismic source is unknown. So we have to identify both the distributed parameter (acoustic impedance) in the 1D wave equation and the Neumann boundary condition at one edge of the domain from an observation of the vibratory state in a part of the domain.

The inverse problem is very close to the inversion of seismic surface data which was studied previously [2]. We shortly recall some mathematical results (uniqueness and stability of the solution) and the solution of the optimization problem which is here of large size (~ 1500 unknowns).

The numerical examples show the efficiency of the proposed solution and the interest of such an approach for the geophysicist: the redundancy available in the data allows a reliable inversion of strongly noise corrupted data provided that the proper mathematical constraints on the solution have been implemented to ensure stability.

INTRODUCTION

Reflection seismic surveys are one of the most important tools for oil exploration. The principle of a reflection seismic experiment is the following. A seismic source (explosive charge for instance) is fired at a point S near the surface of the earth (fig. 1). The

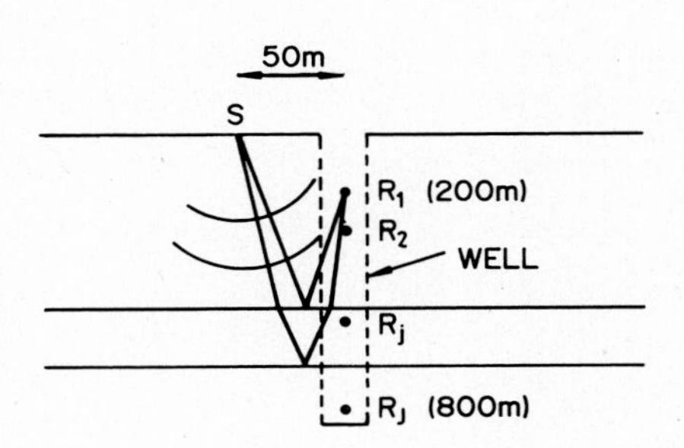

Figure 1

Principle of a VSP experiment

shot generates elastic waves which propagate into the earth and generate reflections, transmissions, and diffractions along their path of propagation. We record the velocity of the displacement at different points R_i as a function of time (seismogram). For surface land seismics, the measurement points R_i are located on the surface at different distances from the shot. An alternative is to perform a vertical seismic profile (VSP): the measurement points R_i are located at different depths in a well which has been drilled because oil is expected in this area. The geophysicist attempts to find out some information on the subsurface from these measurements. As will be seen, this problem can be formulated as an inverse problem. Different previous papers [1] [2] for instance) discuss the inversion of surface seismic data. In this paper, we deal with the inversion of VSP.

1 - MATHEMATICAL MODELING OF THE PROBLEM

1.1 - Physical assumptions

We deal with the simplest problem, which relies on a 1D model, and assume :

- the earth has no horizontal variations (the substratum depends only on the depth)

- the excitation (seismic source) does not vary horizontally (the seismic source generates a plane wave which propagates vertically). (*)

- each layer of the substratum is a linearly elastic isotropic solid.

1.2 - The forward problem

We introduce the following notations:

(1)
- t : time
- x : depth measured by the travel time from the surface
- $y(x,t)$: vertical displacement at depth x and time t
- $\sigma(x)$: acoustic impedance of the substratum
- $g(t)$: seismic pulse (vertical component of the traction on the surface resulting from the seismic source)
- T : duration of the observation

As a consequence of the previous assumptions, the 1D wave equation will model the wavefield, characterized by the displacement $y(x,t)$:

$$\sigma(x)\frac{\partial^2 y}{\partial t^2} - \frac{\partial}{\partial x}\left(\sigma(x)\frac{\partial y}{\partial x}\right) = 0 \text{ in } R^+ \times [0,T] \quad (2)$$

$$\sigma(0)\frac{\partial y}{\partial x}(0,t) = g(t) \qquad \text{(B.C.)} \quad (3)$$

$$y(x,t=0) = \frac{\partial y}{\partial t}(x,t=0) = 0 \quad (4)$$

(*) This assumption may seem to be unrealistic but geophysicists have some special techniques to transform the original problem (spherical wave) into a plane wave problem.

A vertical seismic profile consists of a set of J + 1 seismograms, each seismogram being the observation of $\frac{\partial y}{\partial t}$ at depth x_j ($j = 0, \ldots J$) as a function of time. Define the forward problem (computation of a surface synthetic seismogram or of a synthetic VSP) as :

$$\begin{cases} \text{given } \sigma(x) \text{ and } g(t), \text{ solve the 1D wave equation (2) (3) (4)} \\ \text{and obtain } \frac{\partial y}{\partial t}(x_j,t) \qquad j = 0, \ldots J \end{cases}$$

1.3 - Inversion of surface data : recall of some results

The problem is stated as follows : (see [1] [2] [3] [4])

$$\begin{cases} \text{find } \sigma(x), \text{ given } g(t), \sigma(0), \text{ and an observation of } \frac{\partial y}{\partial t}(0,t) \\ \text{on the time interval } [0,T]. \end{cases}$$

Before giving some mathematical results, we first present the physical quantities of interest. Observe a typical seismic pulse g(t) in figure 2. The mean frequency (25 Hz) characterizes the spectrum well but the pulse appears short in time. An actual impedance distribution (measured in a well) varies rapidly with the depth (figure 3) : the scale of variation of σ is smaller than the wavelength of the

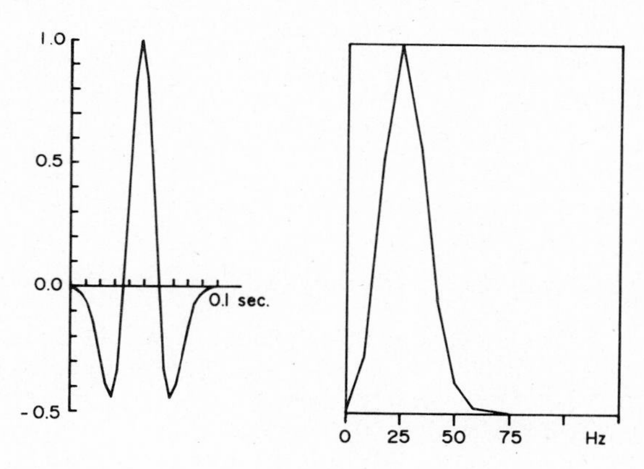

Figure 2

A classical seismic pulse (left) and its spectrum (right)

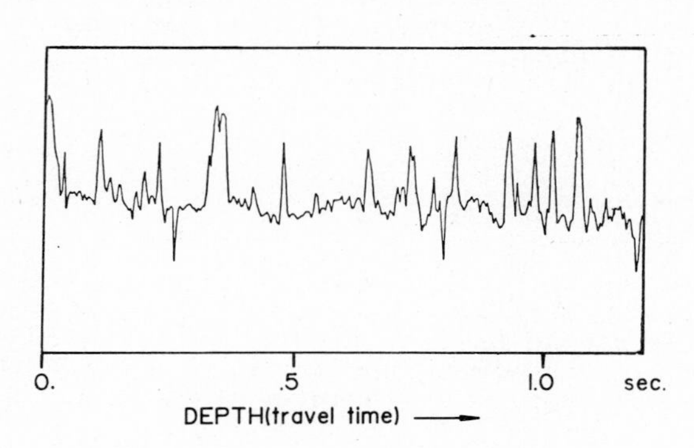

Figure 3

An actual impedance distribution

seismic pulse. The synthetic seismogram associated to the pulse $g(t)$ is displayed in figure 4 : even with the knowledge of $g(t)$, the complexity of the function $\sigma(x)$ makes the interpretation of this seismogram difficult by hand (*). This difficulty motivates a solution of the seismic inverse problem.

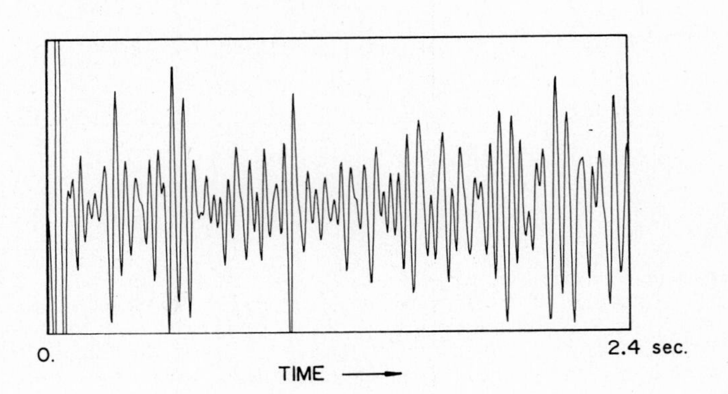

Figure 4

The surface seismogram obtained when the impedance distribution of fig. 3 is excited by the seismic pulse of fig. 2

Above-mentioned references discuss the injectivity of the mapping $\sigma \to \frac{\partial y}{\partial t}(0,t)$ when $g(t) = \delta(t)$. Of course the impedance can be determined only on $[0, \frac{T}{2}]$ because the waves propagate with a velocity equal to 1.

As shown in [1] a simple example illustrates the unstability of the inverse problem. Consider a regular seismic pulse (wavelength larger than the scale of variation of $\sigma(x)$) :

i) if we want to find $\sigma(x)$ in the set

$$(5) \quad \Sigma_b = \left\{\sigma(x) \text{ s.t. } 0 < \sigma^- \leq \sigma(x) \leq \sigma^+ < \infty \text{ for } x \in [0,\tfrac{T}{2}]\right\},$$

then the inverse problem is unstable : two very close seismograms can be generated by two very different impedance distributions, as can be seen on figure 5. One can find (see [1], theorem 5) a distance d on Σ_b which makes continuous the mapping :

$$\sigma \to \frac{\partial y}{\partial t}(0,t) \text{ (with } y \text{ solution of (2) (3) (4))}$$

but which is very weak with respect to classical distances.

(*) The principle of such an interpretation is the following : if we assume that the seismic pulse is localized near $t = 0$, a reflection visible on the seismogram at time τ means that there is a discontinuity in the impedance at depth $\tau/2$ and the amplitude of the discontinuity can be calculated from the amplitude of the reflection.

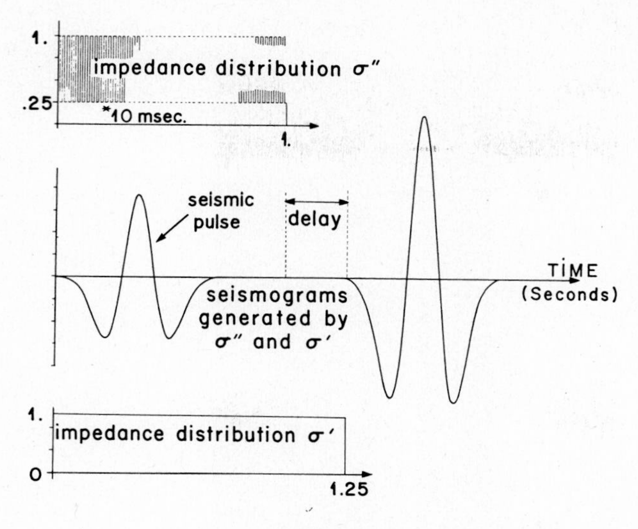

Figure 5

The problem of oscillating impedances
(unstability of the seismic inverse problem)

ii) If we want to find σ in the set

(6) $\Sigma_M = \left\{ \sigma \in \Sigma_b \text{ s.t. } \operatorname{var} \sigma \leq M \right\}$, where var σ is the total variation of σ(x) on the depth interval $[0,\frac{T}{2}]$ and M is some positive constant then the previous difficulty disappears : we have regularized the inverse problem by eliminating the unstable oscillating solutions, which do not, in general, interest the geophysicist. Mathematically because we have restricted the problem to a set that is compact in $L^2(]0,\frac{T}{2}[)$, two impedance distributions that are clos for the previous distance d will be close for the L^2 norm.

1.4 - Mathematical formulation of the inversion of VSP data

First let us describe a typical VSP record more precisely, in order to understand better the mathematical formulation of the inverse problem.

The distance between two observation points x_j and x_{j+1} is much shorter than the spatial wavelength, and the number of observation points of the order of one hundred. Our data (figure 6) have been recorded for 1 second at 66 observation points. The figure plots positive amplitudes in black and negative in white.

We can interpret some events that can be seen on the display:

- the direct downgoing wave
- some reflected upgoing waves
- some multiply reflected downgoing waves
- some guided waves (guided by the well), which differ from the previous waves by their greater amplitudes and by a lower velocity of propagation
- noise from outside sources.

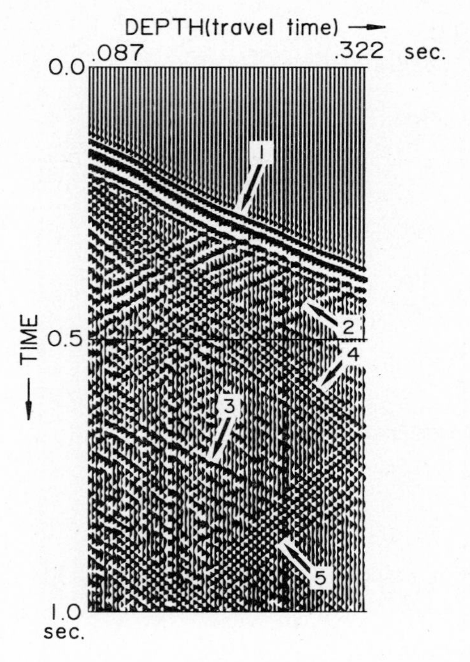

Figure 6

An actual VSP record (the distance (travel time) between observation points is not constant)

Geophysical interpretation : 1 - direct (downgoing) wave
2 - reflected (upgoing) waves
3 - multiply reflected (downgoing) wave
4 - well-guided wave
5 - reflection of the guided wave on the bottom of the well

Our 1D model cannot explain those guided waves ; so our observation is strongly contaminated by both correlated and uncorrelated noises. Fortunately the redundancy available in the leads us to expect a meaningful result.

To invert the data the geophysicist must find the function $\sigma(x)$ (*). The function $g(t)$ is unknown and must be determined in the inversion algorithm, even if this function is of low interest for the geophysicist. One can easily check that attempting to solve for $\sigma(x)$ beginning at zero depth will not yield a unique solution. However, as the observation points are close with respect to the wavelength, the VSP data contain the information about $\frac{\partial y}{\partial x}(x_o,t)$ and then, as it has been shown for the inversion of surface data, it seems possible to determine $\sigma(x)$ for $x \geq x_o$ and simultaneously the Neumann boundary condition.

(*) of course the most interesting region is below the bottom of the well that is $x \geq x_J$

So we change the origin of depth and set the new origin at $x = x_o$. We still denote by x_j the observation depths. The previous considerations lead us to the following mathematical formulation of the inversion of VSP data :

$$(7)\quad \begin{cases} \text{given } \sigma(0) = \sigma_o \text{ and observation } Y_d(x_j,t) \text{ (VSP data for } j = 0, \\ \ldots J \text{ and } t \in [0,T]), \text{ find } (\tilde{\sigma}(x), \tilde{g}(t)) \in P_{ad} \text{ that minimizes} \\ \text{over } P_{ad} \text{ the cost function :} \\ C(\sigma,g) = \sum_{j=0}^{J} \int_0^T (\frac{\partial y}{\partial t}(x_j,t) - Y_d(x_j,t))^2 \, dt = || \frac{\partial y}{\partial t} - Y_d ||^2 \end{cases}$$

where y is a function of σ and g through (2) (3) (4). We must choose the set of admissible parameters P_{ad} sufficiently wide to contain the sought actual parameters and sufficiently narrow to let us expect one (existence) stable solution of the inverse problem. As we have seen previously, the admissible impedances must be in the set :

$$(8)\qquad \Sigma_M = \{\sigma(x) \text{ defined on } [0,X] \text{ (*) s.t.}$$

i) $\sigma(0) = \sigma_o$

ii) $0 < \sigma^- \leq \sigma(x) \leq \sigma^+ < \infty$ for all x in $[0,X]$

iii) $\text{var}\ \sigma \leq M$

where var σ is the total variation of σ on $[0,X]\}$.

<u>Remark 1</u> : the Neumann boundary condition $\tilde{g}(t)$ we are trying to identify is no more the seismic pulse but the vertical component of the traction on the horizontal surface at the depth of the shallowest observation point. Then $\tilde{g}(t)$ is no more localized at the vicinity of $t = 0$ (difference with the original seismic pulse) and we must identify it on the whole interval $[0,T]$.

In other words, we want to find the function $\tilde{\sigma}(x)$ and $\tilde{g}(t)$ such that the associated synthetic VSP best fits the observed VSP. Of course, because of the importance of the noise, we can only expect a very rough fit.

It should be noticed that, mainly for the sake of simplicity, we have used a simple least-squares criterion. If we assume some statistical properties for the noise (which is mainly correlated), we may improve the previous formulation by using a generalized least-squares criterion with the (assumed) covariance matrix of the noise. It is also possible (as suggested by Tarantola and Valette [8]), to modify the cost function in order to introduce some information we can have on the solution.

As can be seen from (7), the inverse problem is formulated as an optimization problem with constraints.

(*) X which is the maximum depth that is investigable using our VSP data, can be easily calculated from the duration of observation T, the minimum and the maximum depths of observation. Note that X is much greater than x_j (fortunately for the geophysicist !) : our measurements are distributed in the very upper part of the domain.

2 - THE NUMERICAL METHOD

As the numerical method does not differ substantially from the algorithm which has been developed for the inversion of surface data (see [1] [2]), we specify here only the main features of the numerical solution and explain the choices that have been made, with emphasis on our specific problem.

2.1 - The discrete problem

In order to solve the problem on a computer, we need to discretize the inverse problem. We approximate the set Σ_M defined in (8) by the set $\Sigma_{h,M}$.

$$\Sigma_{h,M} = \left\{ \sigma \in \Sigma_M \text{ that are piece-wise constant on the intervals }]ih, (i+1)h[\; i \in \mathbb{N} \right\}. \tag{9}$$

Then the elements of $\Sigma_{h,M}$ will be characterized by a sequence of parameters σ_i which are the values of σ in the interval $]ih,(i+1)h[$.

Remark 2 (justification of the approximation) : for a given seismic pulse and a given accuracy, it is possible to find h (sufficiently small) such that, for any given impedance distribution σ^* in Σ_M there exists an impedance distribution in $\Sigma_{h,M}$ whose synthetic VSP fits the synthetic VSP associated to σ^* with the previous accuracy. h depends on the spectral content of the seismograms (the higher the frequencies, the smallest h). We shall see that h must be chosen small with respect to the wavelength so that the number of unknowns σ_i is large ($\sim$ 500).

Remark 3 : the set $\Sigma_{h,M}$ is not the only one to have the property mentioned in remark 2. The reasons of the choice $\Sigma_{h,M}$ (instead of the set of continuous piecewise linear functions for instance) are:

- we want to approximate actual impedance distributions which are irregular functions
- $\Sigma_{h,M}$ will allow very efficient computations (see § 2.3 hereafter).

Remark 4 : for the elements of $\Sigma_{h,M}$, the total variation var σ of the function σ on the interval $[0,X]$ has a very simple form :

$$\text{var } \sigma = \sum_{i=0}^{I} |\sigma_{i+1} - \sigma_i| \tag{10}$$

with I = X/h.

Then, the discrete inverse problem is formulated in the following way:

$$\left\{ \begin{array}{l} \text{given } \sigma_o \text{ and the observation } Y_d(x_j,t) \text{ for } j = 0, \dots J \text{ and} \\ t \in [0,T], \text{ find } \tilde{\sigma}_h (\text{in } \Sigma_{h,M}) \text{ and } \tilde{g} \text{ that minimize the cost} \\ \text{function (7).} \end{array} \right.$$

2.2 - The optimization method

The optimization problem appears to be very difficult:

i) large size (about 500 unknowns for the impedance σ, 1000 unknows for the seismic pulse $g(t)$ and the cost function as a function of σ appears to be flat

ii) the cost function $C(\sigma,g)$ is complicated (we must solve the 1D wave equation to evaluate $C(\sigma,g)$).

iii) the constraint var $\sigma \leq M$ is difficult to handle.

The points i) and ii) lead us to eliminate the explorative methods (Monte-Carlo for instance) as well as Newton like methods. Assuming a convex cost function, we have tried different gradient type methods : we have used successively and with increasing efficiency the simplest steepest-descent method, then a conjugate gradient method and finally an ε-subgradient method [6]. For the reasons explained in i), we absolutely need a rapidly converging method and nevertheless the number of iterations is large ($\sim$ 300).

The constraint on the variation was classically handled by penalization. Then the difficulty is that the penalized cost function is no more differentiable (cf. (10)). This was an other reason to use the ε-subgradient method. As other gradient types method, the ε-subgradient method requires at each iteration, the computation of $C(\sigma,g)$ and the gradient of C with respect to σ and g.

2.3 - The computation technique

Classically (cf. [7] for instance), we evaluate the gradients by introducing the adjoint problem ; then we have to compute other solutions of the wave equation.

As the optimization method requires a large number of computations of synthetic VSP, we need an efficient algorithm to solve the wave equation (2) (3) (4). Because we are dealing with impedance distributions in $\Sigma_{h,M}$, the method of characteristics provides us with an exact (*) solution of the wave equation.

The total computing time required to process the field record presented in section 3.2 hereafter ($\sim$ 1600 parameters to identify) is 3 min on a Cray computer. It has required 300 iterations of the gradient algorithm (**) that is 1800 solutions of the 1D wave equation for a medium constituted of 600 layers. Hence we can easily understand the crucial importance to dispose of an efficient optimization method and of a fast solver of the wave equation.

3 - NUMERICAL RESULTS

Before inverting field recorded VSP, we first applied the method on simulated VSP in order to test the algorithm and to study the stability of the result when noise corrupts the data.

(*) and consequently very efficient : we do not need, as in finite difference solutions, to discretize the problem with a thin sampling rate in order to make the numerical error small.

(**) starting from constant $\sigma(x)$ and $g(t)$. Of course, better initial guesses, especially for $g(t)$, could be used !

3.1 - Inversion of simulated VSP

a) Computation of the simulated VSP : given the impedance distribution $\sigma(x)$ (fig. 3) and the seismic pulse $g(t)$ (fig. 2), we solved the 1D wave equation (2) (3) (4) and computed the seismograms that would have been recorded at 51 depths in the well, equally distributed between the depths 0.2 sec and 0.6 sec.

The impedance sampling rate was chosen equal to 4 ms which appeared to give for the seismic pulse $g(t)$ of the figure 2, a good approximation (cf. remark 2). So, the number of layers used to compute the simulated VSP is 300. The duration of observation was 1.8 sec, the time sampling rate being 4 ms.

The synthetic VSP obtained is displayed on figure 7 a.

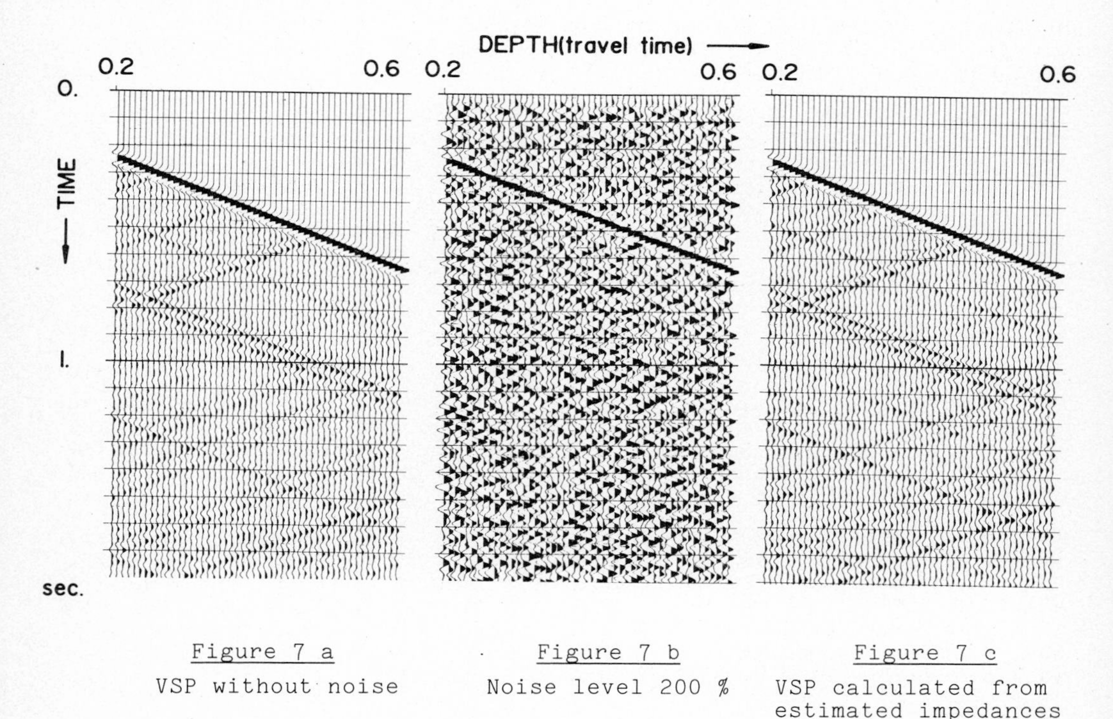

Figure 7 a
VSP without noise

Figure 7 b
Noise level 200 %

Figure 7 c
VSP calculated from estimated impedances

b) Inversion of noiseless VSP

As mentioned above, we are looking for :

$\sigma(x)$ from 0.2 sec to 1.2 sec (250 discrete unknowns)

and $g(t)$ with $0 \leq t \leq 1.8$ sec (450 discrete unknowns : Neumann condition at depth 0.2 sec)

So that the total number of unknowns is 700.

The dotted line on figure 8 (top) is the computed impedance obtained at iteration 500 (*) : by comparison with the exact impedance (full line) we can see that the high frequency information (shape of the impedance distribution) is very well recovered but not the low frequency. This is a consequence of the use of a zero mean seismic pulse as is well known from geophysicists. As can be seen on figure 9 the Neumann condition computed at iteration 500 fits very well with the exact Neumann condition.

c) Inversion of a noise corrupted VSP

Then we corrupted the first VSP with a noise which was computed by convolution of a white noise with the seismic pulse (so we are in the worst case where the spectra of the noise and that of the wavelet are the same).

The noise level is 200 %.

$$\text{with noise level} = \frac{\| \text{PSV with noise} - \text{PSV without noise} \|}{\| \text{PSV without noise} \|}$$

where $\| \ \|$ is defined in (7).

We thus obtained the noise corrupted VSP (fig. 7 b) that we now invert in two ways :

- first (as suggested previously) with a bound on the variation of the impedance (the bound was 29 000, the variation of the exact impedance being 25 000). The computed impedance distribution (fig. 8) is satisfactory for the geophysicist : the relative amplitude of the major peaks are well recovered. The Neumann boundary condition is also well recovered (fig. 9). If we generate the synthetic VSP associated to this computed impedance and to the computed Neumann condition at x_o (the other result of the inversion) we obtain a VSP (fig. 7 c) which is practically identical to the noiseless VSP. As expected previously and due to the redundancy of information in the VSP data and to the global inversion of all the seismograms, we thus have eliminated a very large part of the noise.

- second, without bound on variation. We can see in fig. 8, that parasitic oscillations of great amplitude spoil the result, the geological interpretation becoming very difficult.

The synthetic VSP computed from this bad result is infinitely close to the one computed from the result obtained with a bound on the variation.

This illustrates the unstability of the inverse problem mentioned in the first part of the paper, and the importance of the constraint on the admissible impedance distributions which has been used.

3.2 - Inversion of a field VSP record

We processed the inversion of the VSP record presented on figure 6. The depth interval of observation is 87 ms - 322 ms, the number of unknowns is :

(*) in all the numerical runs we have used for the iteration 0 of the gradient algorithm $\sigma(x) = \text{cte} = \sigma_o$ and $g(t) = 0$.

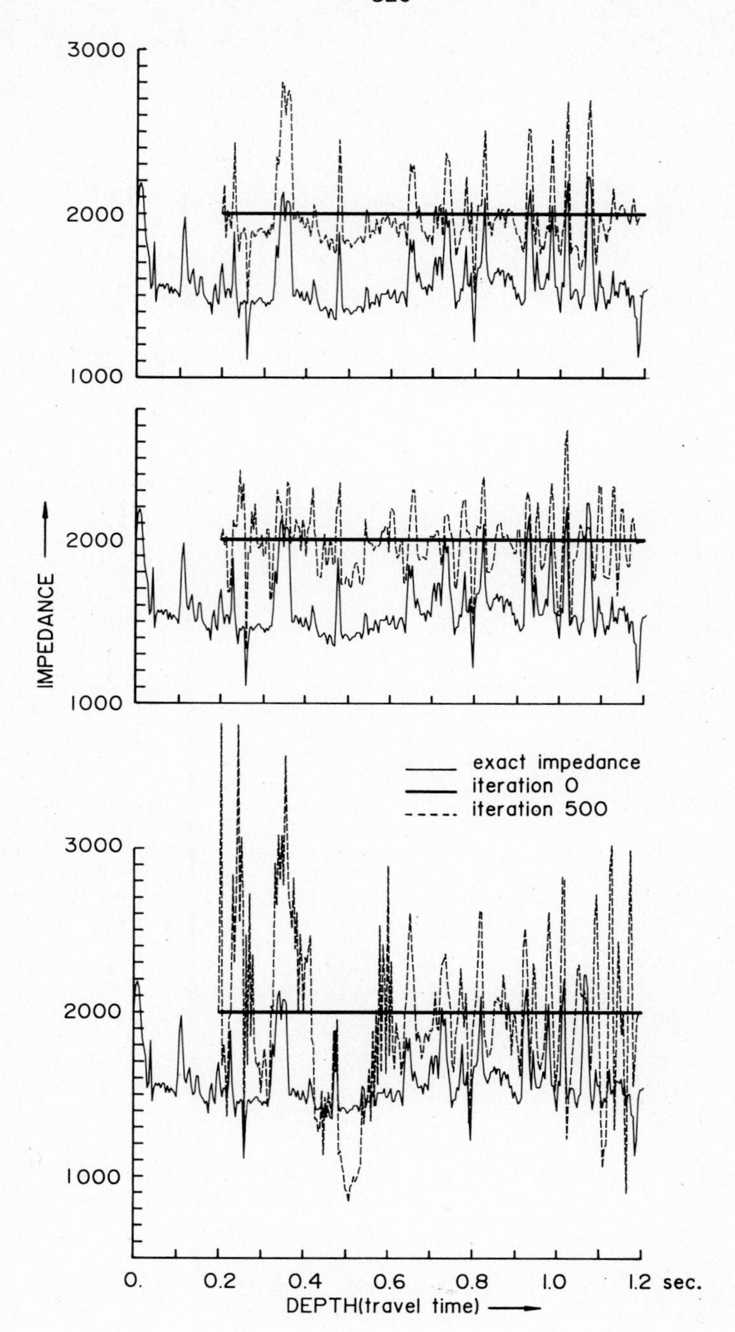

Figure 8

Impedance computed by inversion of the noiseless VSP (top), noise corrupted VSP with bound on the variation (middle), noise corrupted VSP without bound on the variation (bottom)

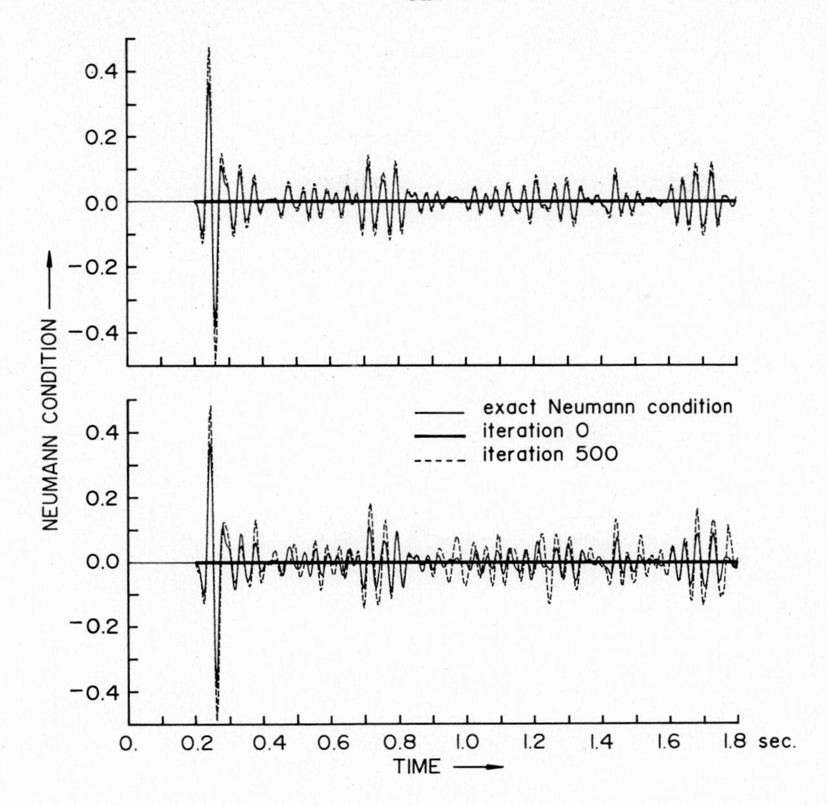

Figure 9

Neumann condition computed by inversion of the noiseless VSP (top), of the noise corrupted VSP with bound on the variation (middle)

- for $\sigma(x)$: 606 (sampling rate of 1 ms (*))
- for $g(t)$: 973 (time sampling rate of 1 ms)

The result obtained after 300 iterations and with a bounded variation constraint is on fig. 10.

We have compared the result with a measurement of the impedance in the well (impedance log) which is available from x = .150 sec to x = .315 sec (fig. 11). Again the shape of the impedance has been recovered which is satisfactory for the geophysicist.

Generating the synthetic VSP from the computed results at iteration 300, we can compare it (fig. 12) to the field VSP : on the computed VSP, we find all the major events that can be seen on the recorded VSP, and a large part of the noise which made it difficult to interpret, is eliminated.

(*) in this example the seismic pulse is centered on 60 Hertz ; we thus have been led to use a sampling rate for the impedance lower than the previous one.

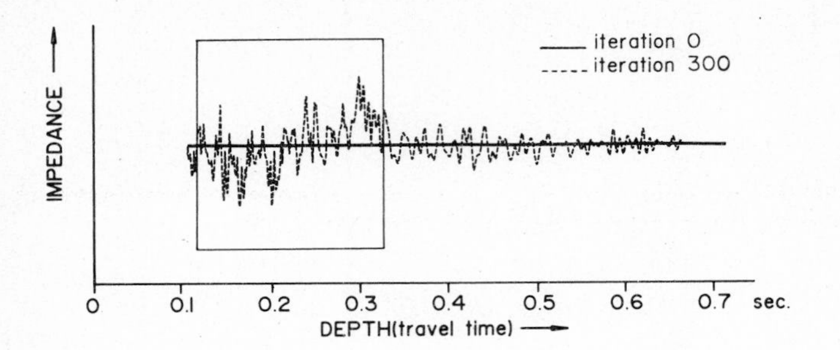

Figure 10

Inversion of the recorded VSP

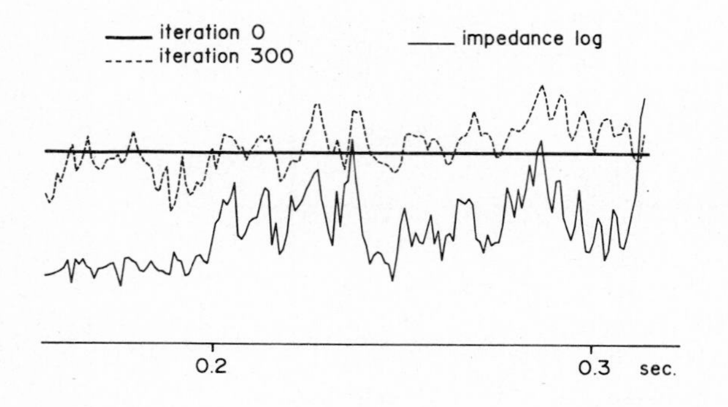

Figure 11

Comparison with the well log data (zoom of the square in fig. 10)

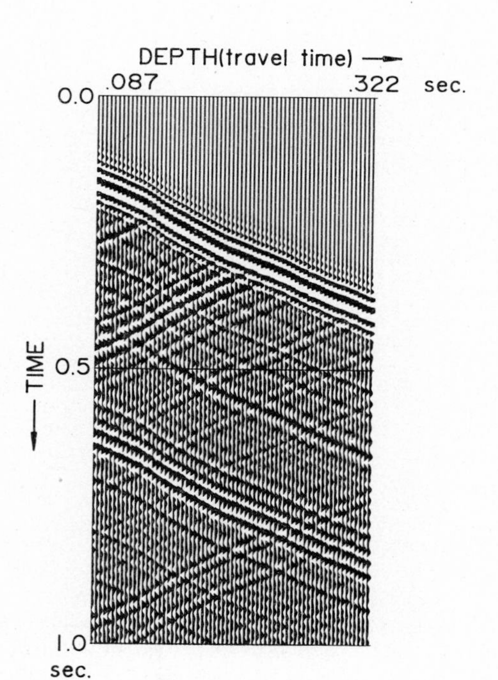

Figure 12

VSP calculated from estimated impedances

4 - CONCLUSIONS

We have seen an application of modern mathematical and numerical techniques to an industrial problem. The main features of the studied problem were :

- redundant but strongly noise corrupted data
- large number of unknowns
- complicated state equation (wave equation).

It has been illustrated that the redundancy available in the data allows a separation between the noise and the signal. But this does not ensure stability : despite of the redundancy, we needed to introduce some adequate constraint on the unknown parameter to regularize the inverse problem which was originally ill-posed. A previous mathematical study [1] was the key to define a proper regularization procedure.

The large size of the problem leads us to take great care to set the numerical algorithm (optimization scheme, solver of the wave equation). A tremendous effort will be necessary to apply these techniques to more complicated but more realistic problems (2 D - 3 D) which are submitted to the geophysicists for oil exploration.

ACKNOWLEDGEMENTS

This work has been done in the framework of a research contract between INRIA (Institut National de Recherche en Informatique et Automatique), S.N.E.A.(P) (Société Nationale Elf-Aquitaine) and I.F.P. (Institut Français du Pétrole).

We would like to thank especially Messrs Bamberger from Ecole Polytechnique, Chavent, Lemaréchal from INRIA, Komatitsch and Legrand from S.N.E.A.(P), and our colleagues from I.F.P. for the very fruitful advices we had from them.

REFERENCES

[1] BAMBERGER A., CHAVENT G., LAILLY P., About the stability of the inverse problem in the 1D wave equation. Application to the interpretation of seismic profiles, J. Appl. Math. Optim., Vol. 5, 1977, pp. 1-47.

[2] BAMBERGER A., CHAVENT G., HEMON C., LAILLY P., Inversion of normal incidence seismograms, Geophysics, Vol. 47, 1982, n° 5, pp. 757-770.

[3] GERVER M.L., The inverse problem for the vibrating string equation, Izv. Acad. Sci. URSS, Physic Solid Earth, 1970-1971.

[4] GOPINATH B., SONDHI M.M., Inversion of the telegraph equation and the synthesis of non uniform lines, Proc. I.E.E.E., Vol. 29, 1971, n° 3, pp. 383-392.

[5] KUNETZ G., Quelques exemples d'analyse d'enregistrements sismiques, Geophys. Prospect., Vol. 11, 1963, pp. 409-422.

[6] LEMARECHAL C., Nondifferentiable optimization subgradient and ε-subgradient methods, Lect. Notes in Econ. and Math. Systems,V.117, Springer-Verlag, Berlin, 1976.

[7] LIONS J.L., Contrôle optimal de systèmes gouvernés par des équations aux dérivées partielles, Dunod-Gauthier Villars, Paris, 1968.

[8] TARANTOLA A., VALETTE B., Generalized non linear inverse problems solved using the least-squares criterion, Reviews of Geophys. and Space Physics, Vol. 20, n° 2, 1982, pp. 219-232.

Session 16

NONLINEAR SYSTEMS I

SYSTÈMES NON LINÉAIRES I

APPLICATION D'UNE NOUVELLE METHODE DE COMMANDE DES SYSTEMES NON LINEAIRES - LA PSEUDO-LINEARISATION - A UN EXEMPLE INDUSTRIEL.

MOUYON P. - CHAMPETIER C. - REBOULET C.
Département d'Etudes et de Recherches
en Automatique du C.E.R.T. -
Boite Postale 4025
31055 TOULOUSE Cedex (France)

RESUME

Pour des systèmes non linéaires du type $\dot{x} = f(x,u)$ $x \in \mathbb{R}^n$, $u \in \mathbb{R}^m$, nous présentons une nouvelle méthode de commande : la pseudo-linéarisation. Cette méthode consiste à chercher des transformations $(z,v) = T(x,u)$ telles que dans l'espace des z, le modèle linéaire tangent soit indépendant du point de fonctionnement. La procédure permet de traiter un exemple industriel : la machine asynchrone.

I - INTRODUCTION

L'élaboration d'une commande boucle fermée d'un système non linéaire repose en général sur une étape préliminaire : la linéarisation. Plusieurs méthodes conceptuellement très différentes ont été proposées. Dans la pratique, la plus couramment utilisée est encore la linéarisation locale autour d'un ou de plusieurs points de fonctionnement. Toutefois, sa mise en oeuvre peut s'avérer délicate sur une grande plage de fonctionnement, en présence de transitoire rapide. Une autre méthode consiste en la linéarisation du comportement entrée-sortie du système avec, au surplus, la possibilité de découpler ce comportement (/2/, /3/, /5/, /8/, /9/). Simple et efficace, cette méthode présente cependant l'inconvénient de faire apparaître dans certains cas des inobservabilités. La linéarisation globale /6/, /7/, /11/ quant à elle a pour objet de linéariser la dynamique du système dans tout l'espace d'état. Les conditions d'existence d'une telle linéarisation sont toutefois extrêmement restrictives dès que la dimension du système est supérieure à deux.

Récemment, nous avons développé une alternative à ces trois approches /1/, /10/, à partir des considérations suivantes : il n'est pas besoin dans la pratique d'avoir une très bonne commande dans tout l'es-

* Cette recherche a été financée par la Direction des Recherches, Etudes et Techniques (D.R.E.T.)

pace d'état. De toute façon, les saturations sur les commandes empêchent généralement d'atteindre cet objectif. Par contre, au voisinage de tout point d'équilibre, on est en droit d'exiger de bonnes performances. C'est pourquoi, nous proposons de chercher des transformations dans l'espace état-commande (changement de variables, feedback) telles que le linéarisé du système après transformations soit indépendant du point de fonctionnement. Autrement dit, le système est équivalent à un système linéaire unique, auquel s'ajoute des termes non linéaires du second ordre au voisinage de l'ensemble des points d'équilibre. Après cette pseudo-linéarisation, la commande est synthétisée en ne considérant que la partie linéaire, la partie non linéaire n'intervenant que lors de brefs transitoires et n'affectant que peu les performances et la stabilité.

Dans ce papier, nous exposons le principe de la méthode et sa mise en oeuvre dans le cas multivariable. Nous traitons ensuite en détail l'application de la méthode à un exemple industriel : la machine asynchrone.

II - PRESENTATION DE LA METHODE

Considérons un système non linéaire de la forme :

$$\dot{x} = f(x,u) \tag{1}$$

avec $x \in \mathbb{R}^n$, $u \in \mathbb{R}^m$, f de classe C^2.

L'ensemble $\mathcal{V}_{x,u}$ des points de fonctionnement du système est donné par n équations non linéaires :

$$\mathcal{V}_{x,u} = \{(x_o,u_o) \; ; \; f(x_o,u_o) = 0\} \tag{2}$$

Nous noterons $\mathcal{V}_x$ la projection de $\mathcal{V}_{x,u}$ dans l'espace d'état :

$$\mathcal{V}_x = \{x_o, \text{t.q. } \exists \, u_o \text{ t.q. } f(x_o,u_o) = 0\} \tag{3}$$

Au voisinage d'un point $(x_o,u_o) \in \mathcal{V}_{x,u}$, le comportement du système peut être considéré comme linéaire, le modèle tangent s'écrivant :

$$\dot{\delta x} = F(x_o,u_o)\delta x + G(x_o,u_o)\delta u \tag{4}$$

avec

$$\delta x \in \mathbb{R}^n, \; \delta u \in \mathbb{R}^m \; ; \; F = \frac{\partial f}{\partial x} \; ; \; G = \frac{\partial f}{\partial u}$$

En général , ce modèle dépend du point de fonctionnement. Le but de la pseudo-linéarisation est de trouver des transformations non linéaires de la forme (changement d'état, feedback implicite) :

$$(z,v) = T(x,u) = (T_1(x),\ldots T_n(x), \; T_{n+1}(x,u),\ldots,T_{n+m}(x,u)) \tag{5}$$

avec

$$\det \frac{\partial z}{\partial x} \neq 0 \ ; \ \det \frac{\partial v}{\partial u} \neq 0$$

telles que dans l'espace des z, le modèle linéaire tangent soit indépendant du point de fonctionnement et s'écrive sous la forme canonique :

$$\begin{aligned} \delta\dot{z}_{\mu_i+1} &= \delta z_{\mu_i+2} \qquad i = 0,\ldots, m-1 \\ &\vdots \\ \delta\dot{z}_{\mu_i+\nu_{i+1}} &= \delta v_{i+1} \end{aligned} \tag{6}$$

où les ν_i sont les indices de commandabilité, les μ_i étant définis par :

$$\mu_o = 0 \ , \quad \mu_i = \sum_{k=1}^{i} \nu_k \qquad i = 1,\ldots, m$$

Deux remarques s'imposent : tout d'abord, la commandabilité du modèle linéaire tangent est invariante par les transformations (5). Donc pour pouvoir obtenir un linéarisé commandable (6), il faut imposer qu'en tout point de fonctionnement, le modèle tangent soit commandable, i.e. :

pour tout $(x_o,u_o) \in \mathcal{V}_{x,u}$, la paire $(F(x_o,u_o),G(x_o,u_o))$ est commandable (7)

D'autre part, la structure du linéarisé (c'est-à-dire les indices de commandabilité), est elle aussi invariante par les transformations considérées. On supposera que les ν_i restent constants sur l'ensemble des points de fonctionnement, et qu'en particulier rg G = cste = m.

Le linéarisé dans l'espace des z s'obtient aisément. Nous avons:

$$\delta\dot{z}_i = \frac{\partial T_i}{\partial x} f(x,u) \qquad i = 1,\ldots, n$$

D'où au voisinage d'un point (x_o,u_o) de $\mathcal{V}_{x,u}$:

$$\delta\dot{z}_i = \alpha_i \ F(x_o,u_o)\delta x + \alpha_i \ G(x_o,u_o)\delta u \quad i = 1,\ldots,n$$

où α_i est la 1-forme dT_i restreinte à $\mathcal{V}_x$: $\alpha_i = (dT_i)|\mathcal{V}_x$. (Un champ de forme α sur $\mathbb{R}^n$ s'identifie à une application de $\mathbb{R}^n$ dans $\mathbb{R}^n$. $\alpha|_{\mathcal{V}}$ est la restriction de cette application à un sous-ensemble $\mathcal{V}$).

On pose d'autre part :

$$\alpha_{n+i} = (d\ T_{n+i})\big|\mathcal{V}_{x,u} \qquad i = 1,\ldots, m$$

Pour obtenir (6), on montre facilement que les conditions suivantes doivent être satisfaites par les α_i (par commodité, on abandonne la référence au point de fonctionnement) :

pour i = 0,..., m-1

$$\alpha_{\mu_i+1} F^k G = 0 \qquad k = 0,\ldots, \nu_{i+1} - 2 \tag{8a}$$

$$\alpha_{\mu_i+k+1} = \alpha_{\mu_i+1} F^k \qquad k = 1,\dots, \nu_{i+1} - 1 \tag{8b}$$

$$\alpha_{n+k} = \alpha_{\mu_k} (F\ G) \qquad k = 1,\dots, m \tag{8c}$$

Le problème revient donc à trouver des 1-formes $\alpha_1,\dots,\alpha_n$ (resp $\alpha_{n+1},\dots,\alpha_{n+m}$) vérifiant les équations (8) ci-dessus en tout point de $\mathcal{V}_x$ (resp $\mathcal{V}_{x,u}$) et telles qu'il existe des transformations $T_1(x),\dots,$ $T_n(x)$ (resp $T_{n+1}(x,u),\dots,T_{n+m}(x,u)$) telles que :

$$\begin{aligned} \alpha_i &= (d\ T_i)|_{\mathcal{V}_x} \qquad && i = 1,\dots,n \\ \alpha_{n+i} &= (d\ T_{n+i})|_{\mathcal{V}_{x,u}} \qquad && i = 1,\dots,m \end{aligned} \tag{9}$$

La première étape de nature algébrique s'effectue en mettant le système linéaire (F,G) sous forme de Brunovski. C'est en fait l'intégration des 1-formes obtenues qui n'est pas immédiate. Il faut toutefois noter qu'en comparaison avec la linéarisation globale /6/, cette intégration ne doit s'effectuer que sur une sous-variété de dimension m et non pas dans tout l'espace ce qui est nettement moins restrictif. Les résultats théoriques concernant l'existence de transformations pseudo-linéarisantes dans le cas multivariable feront l'objet d'une publication ultérieure.

III - UN EXEMPLE INDUSTRIEL - LA MACHINE ASYNCHRONE

III.1 - Les équations

Nous présentons ici une application de la pseudo-linéarisation à la commande d'une machine asynchrone d'induction. Nous rappelons tout d'abord les équations décrivant le fonctionnement de la machine, ainsi que la signification physique de quelques grandeurs. Pour plus de précision, on se reportera à /4/.

Le système est de dimension trois, et possède deux entrées :

$$\dot{x}_1 = -\frac{x_1}{\tau} + (\omega_s - x_3)\ x_2$$

$$\dot{x}_2 = -\frac{x_2}{\tau} - (\omega_s - x_3)\ x_1 + \frac{k}{\tau} u$$

$$\dot{x}_3 = -\frac{x_3}{T} + K\ (g\ u\ x_1 - C_r)$$

x_1 et x_2 sont à un coefficient près les composantes du flux rotorique dans un système d'axes liés au courant statorique. x_3 est la vitesse de rotation. Les deux commandes sont ω_s, la pulsation statorique, et u l'intensité du courant statorique. La quantité gux_1 représente le couple électromagnétique. Le couple résistant C_r que nous envisageons ici est du type frottement visqueux (cas d'une hélice de bateau, d'un ventilateur, etc..) ; il est proportionnel au carré de la vitesse :

$$C_r = \alpha x_3 |x_3| = \alpha x_3^2 \quad \text{sur la plage de fonctionnement } x_3 \geq 0, \text{ seule considérée ici ;}$$

$x_1^2 + x_2^2$ représente à un coefficient près le carré du flux rotorique.

Nous supposons ici l'état entièrement mesurable. En pratique, on construit facilement un observateur de x_1 et x_2. L'observation de x_3 plus délicate peut être faite en suivant la méthode indiquée dans /4/ par exemple. Ceci nous permet de poser :

$$\omega_s = x_3 + \omega_r$$

Si, enfin, on néglige le frottement interne devant le couple résistant (ce qui paraît légitime) le système se met alors sous la forme :

$$\begin{aligned} \dot{x}_1 &= -\frac{x_1}{\tau} + \omega_r x_2 \\ \dot{x}_2 &= -\frac{x_2}{\tau} - \omega_r x_1 + \frac{k}{\tau} u \\ \dot{x}_3 &= K(g\,u\,x_1 - \alpha\,x_3^2) \end{aligned} \tag{10}$$

Les objectifs de la commande sont de deux types : on veut que la machine travaille à flux constant. Il faut donc amener rapidement le "flux" $\phi = x_1^2 + x_2^2$ à une valeur de consigne et l'y maintenir. Le deuxième objectif est la régulation de la vitesse de rotation ; x_3 doit atteindre sa valeur de consigne en 7 ou 8 secondes au maximum et ceci sur une grande plage de fonctionnement (de 0 à 157 rd/s).

III.2 - Recherche de la première transformation : $Z_1 = T_1(x)$

Le linéarisé du système s'écrit :

$$\begin{bmatrix} \delta\dot{x}_1 \\ \delta\dot{x}_2 \\ \delta\dot{x}_3 \end{bmatrix} = \begin{bmatrix} -\frac{1}{\tau} & \omega_r & 0 \\ -\omega_r & -\frac{1}{\tau} & 0 \\ Kgu & 0 & -2kx_3 \end{bmatrix} \begin{bmatrix} \delta x_1 \\ \delta x_2 \\ \delta x_3 \end{bmatrix} + \begin{bmatrix} x_2 & 0 \\ -x_1 & \frac{k}{\tau} \\ 0 & kgx_1 \end{bmatrix} \begin{bmatrix} \delta\omega_r \\ \delta u \end{bmatrix} \tag{11}$$

En posant :

$$\psi = \frac{x_1}{x_2} \qquad \phi = x_1^2 + x_2^2 ,$$

on a l'équilibre :

$$\omega_r = \frac{\psi}{\tau}$$

$$u = \frac{\phi}{kx_2} \tag{12}$$

$$\alpha x_3^2 = \frac{Kg\psi}{k}\ \phi$$

On peut vérifier facilement que le linéarisé est gouvernable, avec pour indices de gouvernabilité $\nu_1 = 2$, $\nu_2 = 1$. On cherchera donc à le mettre sous la forme :

$$\dot{\delta z} = \begin{bmatrix} 0 & 1 & 0 \\ 0 & 0 & 0 \\ 0 & 0 & 0 \end{bmatrix} \delta z + \begin{bmatrix} 0 & 0 \\ 1 & 0 \\ 0 & 1 \end{bmatrix} \delta w \tag{13}$$

La première transformation $\delta z_1 = \alpha_1 \delta x_1 + \alpha_2 \delta x_2 + \alpha_3 \delta x_3$ doit vérifier $\dot{\delta z}_1 = \delta z_2$. Il faut donc éliminer les termes de commande qui apparaissent dans $\dot{\delta z}_1$. Les équations à satisfaire sont donc :

$$\alpha_1 x_2 - \alpha_2 x_1 = 0 \tag{14}$$

$$\alpha_2 \frac{k}{\tau} + Kgx_1\alpha_3 = 0 \qquad \text{à l'équilibre}$$

La difficulté réside en l'intégration de (14) ; existe-t-il une transformation non linéaire $z_1 = T_1(x)$ telle que :

$$x_2 \frac{\partial T_1}{\partial x_1} - x_1 \frac{\partial T_1}{\partial x_2} = 0$$

$$\frac{k}{\tau} \frac{\partial T_1}{\partial x_2} + Kgx_1 \frac{\partial T_1}{\partial x_3} = 0 \qquad \text{à l'équilibre ?} \tag{15}$$

On peut encore écrire (15) sous la forme suivante :

$$dT_1 = \gamma \left(-\frac{K\tau g}{k} \frac{x_1^2}{x_2} , - \frac{K\tau g}{k} x_1, 1\right)$$

$$\gamma = \frac{\partial T_1}{\partial x_3} \qquad \text{à l'équilibre} \tag{16}$$

La recherche de T_1 se ramène à la recherche du facteur intégrant γ. On trouve facilement qu'il faut et suffit de poser :

$$\gamma = - \frac{2}{K\tau\alpha x_3^2}\ \bar{\gamma}$$

En effet, à l'équilibre, on a aussi

$$\gamma = - \frac{2k}{K\tau g\phi\psi}\ \bar{\gamma}$$

et le gradient de T_1 s'écrit :

$$dT_1 = \bar{\gamma}\left(\frac{2x_1}{\phi}, \frac{2x_2}{\phi}, \frac{-2}{K\tau\alpha x_3^2}\right)$$

où $\left(\frac{2x_1}{\phi}, \frac{2x_2}{\phi}, \frac{-2}{K\tau\alpha x_3^2}\right)$ est le gradient de Log $\phi + \frac{2}{K\tau\alpha x_3}$

La présence du facteur $\bar{\gamma}$ permet de choisir alors pour T_1, n'importe quelle fonction de

$$\text{Log } \phi + \frac{2}{K\tau\alpha x_3} \quad :$$

$$T_1(x) = F\left(\text{Log } \phi + \frac{2}{K\tau\alpha x_3}\right)$$

Remarque :

Pour trouver T_1, nous avons utilisé les équations d'équilibre (12). T_1 ne satisfait aux équations (15) qu'à l'équilibre. On montre facilement qu'il n'existe pas de solution à (15) dans tout l'espace d'état. En d'autres termes, la linéarisation globale est impossible.

III.3 - Les autres transformations : $T_2(x)$ et $T_3(x)$

Pour ne pas introduire de singularité, nous avons choisi :

$$z_1 = \frac{x_3}{x_3 \text{ Log } \phi + \frac{2}{K\tau\alpha}} \tag{17}$$

On pose pour simplifier les notations :

$$A(x) = x_3 \text{ Log } \phi + \frac{2}{K\tau\alpha}) \tag{18}$$

Les valeurs numériques des divers coefficients ainsi que les valeurs nominales de x_3 et ϕ permettent d'assurer la non nullité de $A(x)$ au voisinage de l'équilibre.

Pour trouver la transformation $z_2 = T_2(x)$, on peut calculer $\delta z_2 = \delta \dot{z}_1$ et intégrer ; on préfère ici calculer $\dot{z}_1$ et éliminer les termes de commande qui nécessairement pourront être mis sous forme de termes du second ordre au voisinage de l'équilibre. Les deux démarches sont équivalentes. La dérivée de $\dot{z}_1$ s'écrit :

$$\dot{z}_1 = \frac{-x_3^2}{A^2(x)}\left(\frac{\dot{\phi}}{\phi} - \frac{2}{K\tau\alpha}\frac{\dot{x}_3}{x_3^2}\right)$$

soit :

$$\dot{z}_1 = \frac{-2x_3^2}{\tau A^2(x)}\frac{k\ ux_2}{\phi}\left(1 - \frac{g\phi\psi}{k\alpha x_3^2}\right)$$

On fait alors apparaître le produit $(\frac{kux_2}{\phi} - 1)(1 - \frac{g\phi\psi}{k\alpha x_3^2})$ qui est du second ordre au voisinage de l'équilibre.

Pour avoir $\delta \dot{z}_1 = \delta z_2$, on peut prendre :

$$z_2 = -\frac{2}{\tau} \frac{(x_3^2 - \frac{g}{k\alpha}\phi\psi)}{A^2(x)} \tag{19}$$

La troisième transformation doit être indépendante des précédentes. On choisit :

$$z_3 = \phi = x_1^2 + x_2^2 \tag{20}$$

III.4 - Les commandes : w_1 et w_2

On pourrait calculer w_1 et w_2 tels que le linéarisé soit sous forme de Brunowsky au voisinage de l'équilibre, puis faire un retour d'état dans l'espace des z pour fixer les dynamiques désirées pour les z_i. Il est plus simple de tenir compte de la dynamique désirée dès le début.

En particulier, z_3 vérifie :

$$\dot{z}_3 = -\frac{2}{\tau} z_3 + \frac{2}{\tau} k\, u\, x_2$$

Dans la pratique, la dynamique naturelle du flux est suffisamment rapide. On ne recherchera pas à la modifier et on pose

$$u = \frac{w_2}{kx_2} \quad . \tag{21}$$

La dérivée de z_2 vaut au second ordre près :

$$\dot{z}_2 \cong -\frac{2g}{\tau k\alpha} \frac{1}{A^2(x)} (B(x) + C(x)\, w_2 + D(x)\, \omega_r)$$

avec

$$\begin{aligned} B(x) &= 2(\frac{\phi\psi}{\tau} - \frac{Kk\alpha^2}{g} x_3^2) \\ C(x) &= \psi(2K\alpha x_3 + \frac{1}{\tau}(\psi^2-1)) \\ D(x) &= -\phi(1+\psi^2) \end{aligned} \tag{22}$$

La dynamique désirée pour z_1 est :

$$\ddot{z}_1 + \alpha_1\dot{z}_1 + \beta_1 z_1 = \beta_1 w_1$$

soit encore :

$$\dot{z}_2 = -\alpha_1 z_2 - \beta_1 z_1 + \beta_1 w_1$$

Il suffit pour l'obtenir, de poser :

$$\omega_r = \frac{1}{D(x)} (-B(x) - C(x)w_2 + E(x) + G(x)w_1) \tag{23}$$

avec

$$E(x) = \frac{k\alpha}{g} x_3(-\alpha_1 x_3 + \frac{\tau}{2}\beta_1 A(x)) + \alpha_1 \phi\psi$$

$$G(x) = -\frac{k\alpha\tau}{2g} \beta_1 A^2(x) \qquad (24)$$

III.5 - Structure du système bouclé

Les transformations proposées en (17), (19, (20), (21), (23) conduisent à un système toujours non linéaire mais dont le linéarisé est indépendant du point de fonctionnement et s'écrit :

$$\delta\dot{z} = \begin{bmatrix} 0 & 1 & 0 \\ -\alpha_1 & -\beta_1 & 0 \\ 0 & 0 & -\frac{2}{\tau} \end{bmatrix} \delta z + \begin{bmatrix} 0 & 0 \\ \beta_1 & 0 \\ 0 & \frac{2}{\tau} \end{bmatrix} \delta w$$

Les commandes w_1 et w_2 agissent sur z_1 et z_3. A l'équilibre, on a $z_1 = w_1$ et $z_3 = w_2$. Pour les simulations, nous avons choisi d'appliquer des échelons $w_1 = z_{1_c}$ et $w_2 = z_{3_c}$. Les valeurs de consigne z_{1_c} et z_{3_c} sont calculées à partir des valeurs de consigne sur le flux et la vitesse de rotation ϕ_c et x_{3_c}. La figure suivante schématise la structure de commande complète :

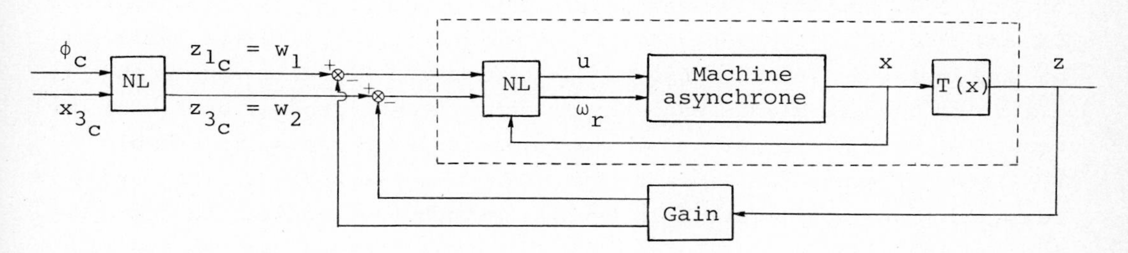

III.6 - Simulations

La machine que nous avons simulée a les caractéristiques suivantes :

$\tau = 0{,}568$ s

$T = 30{,}5$ s

$k = 0{,}972$

$K = 0{,}1036\ kg^{-1}\ m^{-2}$

$g = 0{,}042$ H

$\alpha = 0{,}0142\ Nm/(rd/s^2)$

Les valeurs nominales sont :

$u_N = 175$ A

$\omega_{r_N} = 5$ rd/s

$\omega_N = 157$ rd/s

$\phi_N = 3265$ A^2

Nous avons à tenir compte des contraintes suivantes :

$|\omega_r| \leq 10{,}5$ rd/s

$|u| \leq 350$ A

$|gux_1| \leq 600$ N.m

La dynamique du linéarisé en boucle fermée a été fixée en prenant $\alpha_1 = 2 = \beta_1$. La première simulation représente l'accélération de la machine de O à 157 rd/s, la seconde sa décélération de 157 rd/s à O rd/s.

VI - CONCLUSION

La plupart des systèmes non linéaires ont un modèle linéaire tangent dépendant du point de fonctionnement. Dans ce cas, une loi de commande linéaire ne permet en général pas d'obtenir de bonnes performances. L'idée forte de la pseudo-linéarisation est de chercher des transformations dans l'espace des états et des commandes, telle que le linéarisé du système devienne indépendant du point de fonctionnement. La commande est alors réalisée par un unique retour linéaire, efficace en tout point de fonctionnement. Les termes non linéaires du second ordre n'ont que peu d'influence sur le comportement général de la loi de commande. Pour illustrer la procédure, nous avons choisi un exemple réaliste fortement non linéaire sur lequel les méthodes classiques de commande sont inefficaces : la machine asynchrone. Une loi de commande pseudo-linéarisante suivie d'un retour linéaire classique ont été synthétisés puis validés dans le cadre d'une simulation sur la plage entière de fonctionnement de la machine.

VII - BIBLIOGRAPHIE

/1/ CHAMPETIER C., REBOULET C., MOUYON P. : "A new approach to linearize non linear systems : comparison with classical methods" - Soumis au congrès IFAC, Budapest, 1984

/2/ CLAUDE D. : "Découplage des systèmes : du linéaire ou non linéaire" Colloque CNRS, Septembre 1982, Belle-Ile

/3/ CLAUDE D., FLIESS M., ISIDORI A. : "Immersion directe et par bouclage d'un système non linéaire dans un linéaire" CR Acad. Sc., Paris, 1983, t 296, série I, pp.237-240

/4/ DE FORNEL B., REBOULET C., BOIDIN M. : "Speed control by microprocessor for an induction machine fed by a static convecter" 3rd IFAC Symposium Lausanne, Sept. 1983

/5/ FREUND E. : "The structure of decoupled nonlinear systems" Int. J. Control, 1975, vol.21, n°3, pp 443-450

/6/ HUNT L., SU R. : "Local transformations for multi-input nonlinear systems" Joint Automatic Control, June 1981, Charlottesville, VA

/7/ HUNT L., SU R., MEYER G. : "Global transformations of nonlinear systems" IEEE Trans. Automatic Control, vol.28, N°1, Jan. 1983, pp. 24-31

/8/ MERCIER O. : "Calcul de commandes non interactives pour les systèmes linéaires et non linéaires" Rapport ONERA n° 1981-6

/9/ PORTER W.A. : "Diagonalization and inverses for nonlinear systems" Int. J. Control, 1970, vol.11, n°1, pp. 67-76

/10/ REBOULET C., CHAMPETIER C. : "A new method for linearizing nonlinear systems : the operation-point-independent linearization" A paraître dans Int. J. Control

/11/ SU R. : "On the linear equivalents of non linear systems" Systems and Control Letters, vol. 2, n°1, July 1982, pp 48-52.

ACCELERATION

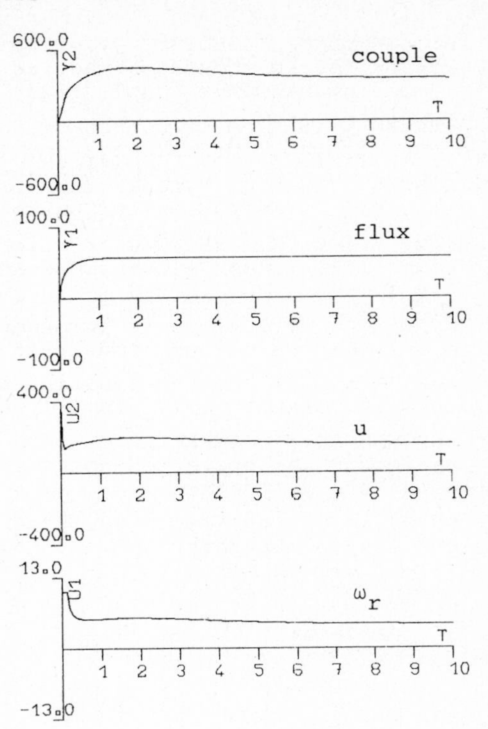

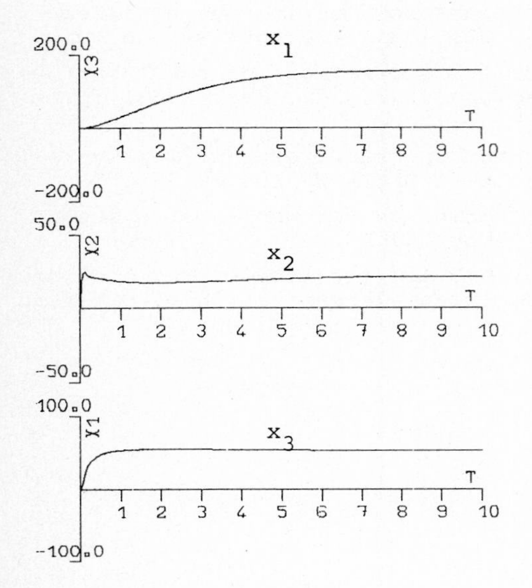

DECELERATION

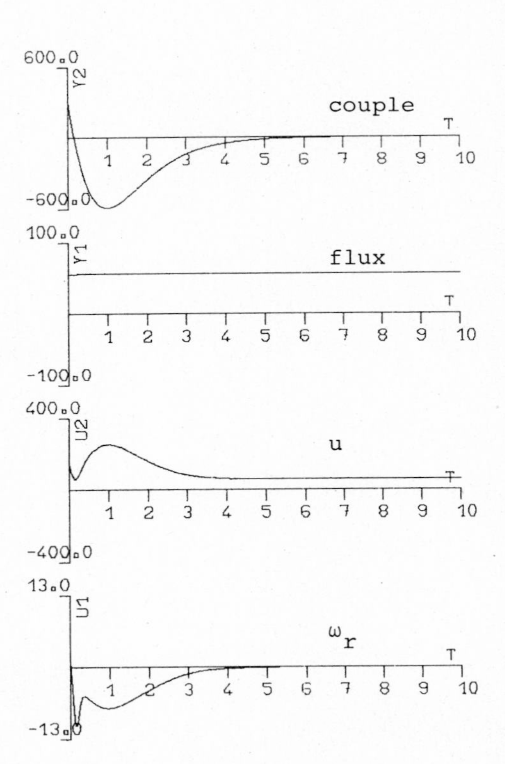

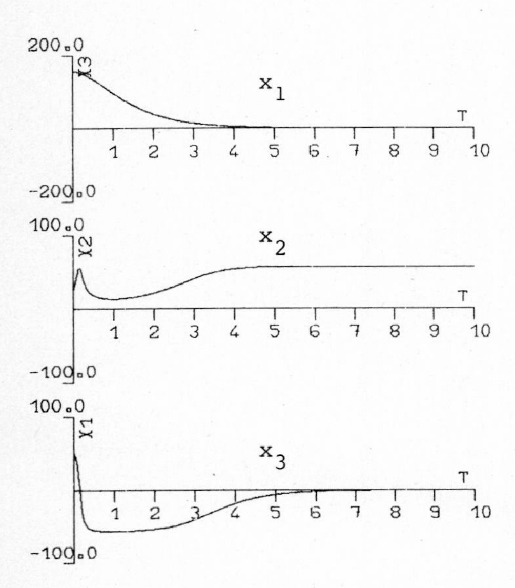

LINEARISATION PAR DIFFEOMORPHISME ET IMMERSION DES SYSTEMES

Daniel Claude
Laboratoire des Signaux et Systèmes, C.N.R.S. - E.S.E.
Plateau du Moulon
91190 Gif-sur-Yvette, France

ABSTRACT

In this paper we show that the linearization of a nonlinear dynamics via coordinate transformations is in fact a particular case of a linearization by "immersion" which was introduced recently (D. Claude, M. Fliess, A. Isidori, C.R. Acad. Sc. Paris, 296, Série I, 1983). We establish that the linearization by diffeomorphism is related directly to the "characteristic numbers" also known as "differential indices" that are used in nonlinear decoupling too.

I. INTRODUCTION

Par ses applications pratiques, la linéarisation d'un système non linéaire est d'une importance considérable.

Les méthodes par approximation sont bien connues (cf. [8]) et ont montré depuis longtemps leur efficacité. Cependant, leur domaine de validité est par nature très restreint. On est alors conduit à chercher comment un système non linéaire peut se ramener à un système linéaire.

La première approche a consisté à ne considérer que la dynamique du système, afin de mieux suivre ses trajectoires dans son espace d'état. Ainsi, Krener [16], donna en 1971 une condition pour que la dynamique soit difféomorphe, au moins localement, à celle d'un système linéaire, à coefficients dépendants du temps. A la suite de Brockett [1], Jakubczyk et Respondek [15] d'une part, Su [19] d'autre part ont donné des conditions d'existence de bouclage assurant un difféomorphisme entre la dynamique du système non linéaire et la dynamique d'un système linéaire à coefficients constants. Cette technique a été appliquée en aéronautique par Meyer, Su et Hunt [17].

La seconde approche à l'équivalence avec un système linéaire, prend les systèmes dans leur totalité, c'est-à-dire avec leurs sorties. Mais, là aussi, la recherche d'un changement de coordonnées sur l'espace d'état (Cf. Isidori et Krener [14]), conduit à des conditions difficilement compatibles avec les singularités les plus communes. Néanmoins, comme dans l'étude du découplage des systèmes non linéaires (Cf. [4], [5] et

[6]), la considération du seul comportement entrée-sortie permet de résoudre d'une nouvelle manière le problème par l'utilisation de la notion d'immersion (Cf. Fliess [11] et Fliess et Kupka [12]). On peut alors présenter des conditions acceptables amenant un système non linéaire à avoir par bouclage le même comportement entrée-sortie qu'un système linéaire aux agréables propriétés algébriques (systèmes c.i.f. ; Cf. Claude, Fliess et Isidori [7]). Cette méthode vient d'être employée par l'auteur pour la détermination de lois de commande pour le pilotage d'un hélicoptère, à l'occasion d'un contrat D.R.E.T.

Nous montrons dans cet article que la linéarisation par changement de coordonnées d'une dynamique analytique à n entrées, définie sur une variété de dimension N, est en fait un cas particulier de la linéarisation c.i.f. du système obtenu en adjoignant à cette dynamique les sorties adéquates. Plus précisément, nous mettons en évidence que la linéarisation par difféomorphisme d'une dynamique est liée au fait que la valeur de la "somme des nombres caractéristiques" (Cf. [4]) de n fonctions de sorties du système caractérisant le difféomorphisme est égale à N-n.

II. RAPPELS SUR L'IMMERSION DANS UN SYSTEME LINEAIRE

Nous prenons un système en temps continu, à entrées apparaissant linéairement. (Pour l'immersion des systèmes en temps discret voir Monaco et Normand-Cyrot [18]) :

$$\Sigma \left\{ \begin{array}{l} \dot{q}(t) = A^o(q) + \sum_{i=1}^{n} u_i(t) A^i(q) \\ y(t) = h(q) \end{array} \right.$$

L'état q appartient à une variété $\mathbb{R}$-analytique Q, de dimension finie N. Les champs de vecteurs $A^o, A^1, \ldots, A^n : Q \to TQ$ et la fonction de sortie $h = (h_1, \ldots, h_r) : Q \to \mathbb{R}^r$ $(r = 1,2,\ldots)$ sont analytiques.
Les entrées $u_1, \ldots, u_n$ à valeurs réelles, sont continues par morceaux.

On considère pour Σ les lois de bouclage par retour d'état, de la forme :

$$u_\ell = \alpha_\ell(q) + \sum_{i=1}^{n} \beta_\ell^i(q) v_i(t) \qquad (\ell = 1,\ldots,n) \tag{1}$$

Les fonctions α_ℓ, $\beta_\ell^i : Q \to \mathbb{R}$ sont analytiques et les v_i désignent de nouvelles entrées. On suppose la matrice $\Phi = (\beta_\ell^i)$ inversible. Σ devient:

$$\hat{\Sigma}\left\{\begin{array}{l}\dot{q}(t) = \hat{A}^o(q) + \sum_{i=1}^{n} v_i(t)\hat{A}^i(q) \\ y(t) = h(q)\end{array}\right.$$

où

$$\hat{A}^o = A^o + \sum_{\ell=1}^{n} \alpha_\ell A^\ell, \quad \hat{A}^i = \sum_{\ell=1}^{n} \beta^i_\ell A^\ell \ (i = 1,\dots,n)$$

Le problème est de trouver les lois de bouclage de type (1) de sorte que $\hat{\Sigma}$ ait même comportement entrée-sortie que le système linéaire Λ suivant :

$$\Lambda\left\{\begin{array}{l}\dot{\eta}(t) = F\eta(t) + \sum_{i=1}^{n} u_i(t)G^i \\ y(t) = H\eta(t)\end{array}\right.$$

L'état η appartient à un $\mathbb{R}$ espace vectoriel $\mathcal{Q}$ de dimension finie, dont $G^1,\dots,G^n$ sont des éléments. Les applications $F : \mathcal{Q} \to \mathcal{Q}$ et $H : \mathcal{Q} \to \mathbb{R}^r$ sont $\mathbb{R}$-linéaires.

Le concept de même comportement entrée-sortie, non évident en non linéaire, est formalisé par la notion d'immersion (Cf. Fliess [11] et Fliess et Kupka [12]) qui peut se définir pour les systèmes analytiques à l'aide de la notion de série génératrice d'un système (Cf. Fliess [10]).

Le système $\hat{\Sigma}$, initialisé en $q(0)$, a pour série génératrice $h|q(0) = (h_1|q(0),\dots,h_s|q(0)$ avec

$$h_s|q(0) = h_s|q(0).1 + \sum_{\nu\geq 0} \sum_{j_o,\dots,j_\nu=0}^{n} C^{j_o}\dots C^{j_\nu}.h_s|q(0).c_{j_\nu}\dots c_{j_o} \quad (s=1,\dots,r)$$

Chaque C^j appartient à l'ensemble des champs de vecteurs du système

$$\mathcal{C} = \{\hat{A}^o,\dots,\hat{A}^n\} = \{C^o, C^1,\dots,C^n\}$$

et chaque c_j appartient à l'alphabet $c = \{a_o, a_1,\dots,a_n\}$
la barre $|q(0)$ désigne l'évaluation en $q(0)$.
De même, le système Λ, initialisé en $\eta(0)$, a pour série génératrice :

$$f|\eta(0) = \Big(f_1|\eta(0),\dots,f_s|\eta(0)\Big) \text{ avec}$$

$$f_s|\eta(0) = H_s(1-Fa_o)^{-1}.\Big(\eta(0) + \sum_{i=1}^{n} G^i a_i\Big) \quad (s = 1,\dots,r).$$

où H_s désigne la forme $\mathbb{R}$-linéaire définie par la s-ième ligne de H.

Une immersion de $\hat{\Sigma}$ dans Λ est une application analytique $\tau : Q \to \mathcal{Q}$, telle que $\hat{\Sigma}$ et Λ, respectivement initialisés en $q(0)$ et $\eta(0) = \tau(q(0))$, aient même série génératrice.

En introduisant le $\mathbb{R}$-espace vectoriel engendré par les dérivés de Lie successives :

$$\hat{\mathcal{E}}_o = \{L_{\hat{A}o}^{\nu} h_s \mid s=1,\ldots,r;\nu\geq 0\} ,$$

on peut énoncer (Cf. [7]).

<u>Proposition 1</u> :

Pour que $\hat{\Sigma}$ puisse être immergé dans un système linéaire de type Λ, il faut et il suffit que les deux conditions suivantes soient satisfaites.

(i) $\hat{\mathcal{E}}_o$ est de dimension finie

(ii) pour tout $\psi \in \hat{\mathcal{E}}_o$, la dérivée de Lie $L_{\hat{A}i}\psi$ $(i = 1,\ldots,n)$ est une constante.

Ce résultat d'existence ne permet pas de donner immédiatement les lois de bouclages. Cependant, l'utilisation des méthodes apparues dans l'étude du découplage ([4], [5] et [6]), offre là aussi un calcul explicite de lois de bouclage intéressantes.

Rappelons que les nombres caractéristiques φ_s $(s = 1,\ldots,r)$ de Σ sont les nombres entiers tels que :

$$\forall \ell = 1,\ldots,n,\ \forall \nu,\ 0 \leq \nu < \varphi_s$$

$$A^{\ell}.L_{Ao}^{\nu} h_s \equiv 0 \quad \text{et} \quad \exists \ell A^{\ell}.L_{Ao}^{\varphi_s} h_s \not\equiv 0 .$$

On suppose qu'ils existent pour $s = 1,\ldots,r$. Ils sont alors strictement inférieurs à la dimension de Q (Cf. [5]).
D'autre part, on sait (Cf. [20]) que Λ vérifie $J \subset V^* \subset V_o^*$, où :

- J est l'espace vectoriel des états inobservables de Λ ;
- V^* est le plus grand (F,G) invariant de Λ, contenu dans KerH, avec $G = (G_1,\ldots,G_n)$;
- $V_o^* = \bigcap_{s=1}^{n} V_s^*$, où V_s^* est le plus grand (F,G) invariant contenu dans

$\mathrm{Ker}H_s$, H_s désignant la forme $\mathbb{R}$-linéaire définie par la s-ième ligne de H.

Un système linéaire est dit c.i.f. (comme il faut !) si tous ses nombres caractéristiques sont définis et si $J = V^* = V_o^*$. Notons γ_s $(s = 1,\dots,r)$ les nombres caractéristiques de Λ, supposés définis. Il vient :

Proposition 2 :

Une condition nécessaire et suffisante pour que Λ soit un système linéaire c.i.f. est que, pour tout $s = 1,\dots,r$, $H_s F^{\gamma_s+1}$ soit combinaison $\mathbb{R}$-linéaire de $H_1,\dots,H_1F^{\gamma_1},\dots,H_r,\dots,H_rF^{\gamma_r}$.

Comme dans l'étude du découplage [5], définissons :

- La matrice $r \times n$ $\Omega = (A^{\ell}.L_{A_o}^{\varphi_s} h_s)$;

- La matrice $r \times 1$ $\Delta = (\alpha_{\ell})$;

- La matrice $n \times n$ $\Phi = (\beta_{\ell}^{i})$;

- La matrice $r \times 1$ $\Gamma^o = (-L_{A_o}^{\varphi_s} h_s + g_s^o)$, où g_s^o est combinaison $\mathbb{R}$-linéaire des fonctions $h_1,\dots,L_{A_o}^{\varphi_1} h_1,\dots,h_r,\dots,L_{A_o}^{\varphi_r} h_r$;

- La matrice $r \times n$ $\Gamma^1 = (g_s^i)$, où les g_s^i sont des constantes réelles ;

Il vient alors :

Proposition 3 :

Si tous les nombres caractéristiques φ_s de Σ sont définis, une condition nécessaire et suffisante pour que Σ soit immergé par un bouclage de type (1) dans le système linéaire c.i.f. (3) ci-dessous est que les équations (2) suivantes admettent une solution avec Φ inversible :

$$\left.\begin{aligned} \Omega.\Delta &= \Gamma^o \\ \Omega.\Phi &= \Gamma^1 \end{aligned}\right\} \qquad (2)$$

Le système $\hat{\Sigma}$ s'immerge alors, par l'application $\tau : Q \to \mathcal{Q}$, définie par

$$\tau(q) = \left(h_1(q),\dots,L_{A_o}^{\varphi_1} h_1(q),\dots,h_r(q),\dots,L_{A_o}^{\varphi_r} h_r(q)\right),$$

dans le système linéaire c.i.f. et observable, d'espace d'état $\mathcal{Q} = \mathbb{R}^m$ avec $m = \sum_{s=1}^{r} \varphi_s + r$:

$$\left.\begin{array}{ll} \dot{\eta}_s^m = \eta_s^{m+1} & (s = 1,\dots,r \; ; \; 0 \leq m < \varphi_s) \; ; \\ \dot{\eta}_s^{\varphi_s} = F_s^o + \sum_{i=1}^{n} v_i(t) f_s^i & (s = 1,\dots,r) \; ; \\ y_s = \eta_s^o & \end{array}\right\} \quad (3)$$

où

$$f_s^o = \sum_{k=1}^{r} \sum_{j=0}^{\varphi_k} c_{s,j}^k \eta_k^j \quad \text{si} \quad g_s^o = \sum_{k=1}^{r} \sum_{j=0}^{\varphi_k} c_{s,j}^k L_{A^o}^j h_k$$

et

$$f_s^i = g_s^i \qquad (s = 1,\dots,r \; ; \; i = 1,\dots,n).$$

III. IMMERSION ET DIFFEOMORPHISME

Jakubczyk et Respondek [15] ainsi que Meyer, Su et Hunt [17] cherchent un difféomorphisme local $\tau = \mathbb{R}^N \to \mathbb{R}^N$ qui transforme, à l'aide d'un bouclage de type (1), la dynamique D d'un système de type Σ en la dynamique, mise sous forme canonique de Brunovsky (Cf. [2]), d'un système linéaire Λ contrôlable. (Si Σ est analytique, on peut, sans restriction, choisir τ analytique).

Cela signifie que si, $\kappa_1,\dots,\kappa_n$ désignent les invariants de seconde espèce de la paire (F,G) (invariants de Popov-Kronecker), on a pour le système Λ, écrit sous la forme :

$$\dot{\eta} = F\eta + Gu \quad \text{avec} \quad u = (u_1,\dots,u_n)^t \; ,$$

la décomposition par blocs suivante (Cf. [2] et [8]) :

$$F = (F_{ij}) \; ; \quad G = (G_i)$$

$$F_{ii} = \begin{bmatrix} 0 1 0 \dots 0 \\ \dots\dots\dots \\ 0 \dots\dots 1 \\ x \dots\dots x \end{bmatrix}_{\kappa_i \times \kappa_i} \qquad F_{ij} = \begin{bmatrix} \\ \\ x \dots\dots , x \end{bmatrix}_{\kappa_i \times \kappa_j} \qquad (4)$$

$$G_i = \begin{bmatrix} \\ \\ \underbrace{0 \dots}_{i} \; 1 x \dots x \end{bmatrix}_{\kappa_i \times n} \qquad \text{avec } i,j = 1,\dots,n$$

les x désignent des termes non déterminés a priori.

On décompose de même l'état η sous la forme :

$$\eta = (\eta_1,\dots,\eta_n)$$

avec $$\eta_s = (\eta_s^o,\dots,\eta_s^{\kappa_s - 1}), \quad s = 1,\dots,n. \tag{5}$$

En prenant alors les données de (4) et (5) et $y_s = \eta_s^o$ avec $s = 1,\dots,n$ on construit un système linéaire Π de type (3).

Le lien avec la linéarisation c.i.f. est maintenant clairement établi et la proposition 3 permet d'énoncer :

Proposition 4 :

Une condition nécessaire et suffisante pour que la dynamique du système Σ soit localement difféomorphe à une dynamique linéaire contrôlable mise sous la forme canonique de Brunovsky est que les conditions suivantes soient satisfaites :

(i) Il existe des fonctions analytiques $h_1,\dots,h_n$, définissant les sorties du système Σ pour lesquelles on a :

$$N = n + \sum_{s=1}^{n} \varphi_s \quad \text{avec} \quad \dim Q = N$$

(ii) Il existe une carte locale telle que l'application

$$\tau = \mathbb{R}^N \to \mathbb{R}^N, \quad \text{définie par :}$$

$$\tau(q) = \left(h_1(q),\dots,L_{A^o}^{\varphi_1} h_1,\dots,h_n(q),\dots,L_{A^o}^{\varphi_n} h_n(q)\right)$$

soit un difféomorphisme.

(iii) Le système (2) admet une solution avec

$$g_s^i = 0 \quad \text{pour} \quad i < s \; ; \quad g_s^s = 1 \quad (s = 1,\dots,n)$$

▲ Supposons qu'un difféomorphisme local τ et un bouclage de type (1) transforment la dynamique D de Σ en celle du système Π. On peut écrire :

$$\tau(q) = \left(\tau_1^o(q),\dots,\tau_1^{\kappa_1 - 1}(q),\dots,\tau_n^o(q),\dots,\tau_n^{\kappa_n - 1}(q)\right)$$

avec $q \in \mathbb{R}^N$ et supposer τ analytique.
Le système Π peut s'écrire :

$$\begin{cases} \dot{\eta} = B^o + \sum_{i=1}^{n} u_i B^i \\ y = f(\eta) \end{cases}$$

avec $y = (y_1,\dots,y_n)$; $f = (f_1,\dots,f_n)$ et $y_s = f_s(\eta) = \eta_s^o$ $(s = 1,\dots,n)$;
où

$$B^o = \sum_{s=1}^{n} \left(\sum_{m=0}^{\kappa_s - 2} \eta_s^{m+1} \frac{\partial}{\partial \eta_s^m} + f_s^o \frac{\partial}{\partial \eta_s^{\kappa_s - 1}} \right).$$

et

$$B^i = \sum_{s=1}^{n} f_s^i \frac{\partial}{\partial \eta_s^{\kappa_s - 1}}$$

avec f_s^o la forme linéaire sur $\mathbb{R}^N$ donnée par la ligne de la matrice F formée par la dernière ligne des matrices $F_{s,s}$ et $F_{s,j}$ et f_s^i le i-ième élément de la dernière ligne de G_s $(s = 1,\dots,n$; $j = 1,\dots,n)$ (Etant entendu que le terme $\sum_{m=0}^{\kappa_s - 2}$ disparaît quand $\kappa_s = 1$).

Un calcul élémentaire montre que $L_{B^o}^m f_s = \eta_s^m$ pour $0 \leq m \leq \kappa_s - 1$ et que $B^i . L_{B^o}^{\kappa_s - 1} f_s = f_s^i$. Ainsi, $\kappa_1 - 1,\dots,\kappa_n - 1$ sont les nombres caractéristiques du système Π.

D'autre part, comme τ est un difféomorphisme local, on peut trouver une carte sur laquelle (Cf. [3], p. 27) :

$$B^j . g = (\hat{A}^j . g \circ \tau) \circ \tau^{-1} \quad \text{avec} \quad g \in C^\omega(\mathbb{R}^N) \text{ et } j = 0,1,\dots,n$$

On a ainsi $L_{\hat{A}^o}^m \tau_s^o = \tau_s^m$ pour $0 \leq m \leq \kappa_s - 1$ et $\hat{A}^i . L_{\hat{A}^o}^{\kappa_s - 1} \tau_s^o = f_s^i$.

Si on prend maintenant le système Σ de dynamique D et de sorties $y_s = \tau_s^o$ $(s = 1,\dots,n)$, on trouve alors un système bouclé $\hat{\Sigma}$ dont les nombres caractéristiques sont $\kappa_1 - 1,\dots,\kappa_n - 1$.

Comme Φ est un inversible, il est facile de montrer (Cf. [6]) que Σ et $\hat{\Sigma}$ ont mêmes nombres caractéristiques et la condition (i) est bien vérifiée.

De même, le système Π, vérifiant :

$$L_{B^o}^{\kappa_s} f_s = f_s^o = \sum_{k=1}^{n} \sum_{j=0}^{\kappa_k - 1} c_{s,j}^k \eta_k^j$$

et, $$B^i.L_{B^o}^{\kappa_s - 1} f_s = f_s^i \qquad i = 1,\ldots,n \; ; \; s = 1,\ldots,n$$

on a, $$\hat{A}^o.L_{A^o}^{\kappa_s - 1} \tau_s^o = L_{\hat{A}^o}^{\kappa_s} \tau_s^o = g_s^o = \sum_{k=1}^{n} \sum_{j=0}^{\kappa_k - 1} c_{s,j}^k \tau_k^j$$

et, $$\hat{A}^i.L_{A^o}^{\kappa_s - 1} \tau_s^o = g_s^i = f_s^i \qquad i = 1,\ldots,n \; ; \; s = 1,\ldots,n.$$

Cela signifie, comme indiqué en [5] p. 546 que les données du système Σ, de dynamique D et de sorties $y_s = \tau_s^o$ $(s = 1,\ldots,n)$, vérifient le système (2) avec $g_s^i = 0$ si $i < s$ et $g_s^s = 1$ $(i = 1,\ldots,n)$.
Réciproquement, si les conditions (i), (ii) et (iii), sont satisfaites, on peut montrer comme dans [6] et [7] que les séries génératrices $h|q(0)$ et $f|\tau(q(0))$ sont égales et que l'application τ définit une immersion de $\hat{\Sigma}$ dans Π. La dynamique du système $\hat{\Sigma}$ est bien alors localement difféomorphe à une dynamique linéaire contrôlable mise sous la forme canonique de Brunovski ▲

Remarque : En fait comme nous le verrons sur les exemples ci-dessous, la seule condition forte est la condition (i).

IV. EXEMPLES

On va maintenant expliciter sur quelques exemples les résultats présentés précédemment.

a) Exemple 1

On considère la dynamique suivante étudiée par Su [19] :

$$\begin{pmatrix} \dot{x}_1 \\ \dot{x}_2 \end{pmatrix} = \begin{pmatrix} x_2 \\ 0 \end{pmatrix} + \begin{pmatrix} \cos x_1 \\ 1 \end{pmatrix} u \qquad (6)$$

et on applique la proposition 4.
On cherche une sortie $y(t) = h(x_1,x_2)$ telle que le nombre caractéristique du système ainsi formé soit égal à 1 :
On a :

$$A^o = x_2 \frac{\partial}{\partial x_1} \; ; \quad A^1 = \cos x_1 \frac{\partial}{\partial x_1} + \frac{\partial}{\partial x_2}$$

On doit donc avoir :

$$A^1.h \equiv 0 \quad \text{et} \quad A^1.L_{A^o} h \not\equiv 0$$

C'est bien le cas avec la fonction donnée par Su :

$$h = -x_2 + \mathrm{Log}\left|\mathrm{tg}\left(\frac{1}{2}x_1 + \frac{\pi}{4}\right)\right|$$

pour laquelle

$$L_{A^o}h = \frac{x_2}{\cos x_1} \quad ; \quad A^1.L_{A^o}h = x_2 \,\mathrm{tg}\, x_1 + \frac{1}{\cos x_1}$$

et

$$L^2_{A^o}h = x_2^2 \frac{\mathrm{tg}\, x_1}{\cos x_1} .$$

On prend alors des bouclages de la forme :

$$u = \frac{1}{A^1.L_{A^o}h}\left(- L^2_{A^o}h + \lambda_1 h + \lambda_2 L_{A^o}h + v\right)$$

L'application $\tau(x_1,x_2) = \left(h(x_1,x_2), L_{A^o}h(x_1,x_2)\right)$ est bien un difféomorphisme local qui permet d'immerger le système de dynamique (6) et de sortie

$$y(t) = h(x_1,x_2) = -x_2 + \mathrm{Log}\left|\mathrm{tg}\left(\frac{1}{2}x_1 + \frac{\pi}{4}\right)\right| ,$$

dans le système suivant :

$$\left.\begin{aligned} &\dot{\eta}_o = \eta^1 \\ &\dot{\eta}_1 = \lambda_1\eta^o + \lambda_2\eta^1 + v \quad \text{avec} \quad \lambda_1, \lambda_2 \in \mathbb{R} \\ &y(t) = \eta^o \end{aligned}\right\} \qquad (7)$$

En prenant $\lambda_1 = \lambda_2 = 0$ on retrouve bien les solutions de Su.

b) <u>Exemple 2</u> (Cf. Krener [16])

On a une dynamique à coefficients dépendants du temps présentée sous la forme usuelle :

$$\left.\begin{aligned} &\dot{x}_o = 1 \\ &\dot{x}_1 = u \\ &\dot{x}_2 = u.x_o \end{aligned}\right\} \qquad (8)$$

On a $A^o = \frac{\partial}{\partial x_o}$; $A^1 = \frac{\partial}{\partial x_1} + x_o \frac{\partial}{\partial x_2}$.

On cherche une sortie $y(t) = h(x_o,x_1,x_2)$ telle que le nombre caractéristique du système ainsi formé soit égal à 1. On doit donc avoir :

$$A^1.h \equiv 0 \quad \text{et} \quad A^1.L_{A^o}h \not\equiv 0$$

On trouve facilement que la fonction h suivante convient :

$$h(x_o,x_1,x_2) = x_o x_1 - x_2$$

Ainsi, on retrouve que l'application $d : \mathbb{R}^3 \to \mathbb{R}^3$, définie par :

$$d(x_o,x_1,x_2) = (x_o, x_o x_1 - x_2, x_1)$$

est un difféomorphisme qui permet de transformer la dynamique (8) en la dynamique :

$$\left.\begin{aligned} \dot\eta^o &= 1 \\ \dot\eta^1 &= \eta^2 \\ \dot\eta^2 &= u \end{aligned}\right\} \quad (9)$$

Elle a de plus même comportement que la dynamique linéaire

$$\left.\begin{aligned} \dot\eta^1 &= \eta^2 \\ \dot\eta^2 &= u \end{aligned}\right\} \quad (10)$$

et l'application $\tau : \mathbb{R}^2 \to \mathbb{R}^2$, définie par :

$$\tau(x_o,x_1,x_2) = (x_o x_1 - x_2, x_1)$$

est une immersion du système composé de la dynamique (8) et de la sortie $y = x_o x_1 - x_2$, dans le système de dynamique (10) et de sortie $y = \eta^1$.

c) <u>Exemple 3</u> (Cf. Hunt et Su [13])

On envisage le système suivant :

$$\left.\begin{aligned} \dot x_1 &= \tfrac{1}{2}x_1^2 + e^{x_2} + x_2 \\ \dot x_2 &= x_1^2 + u \\ y &= h(x_1,x_2) \end{aligned}\right\} \quad (11)$$

On a :

$$A^o = \left(\frac{1}{2}x_1^2 + e^{x_2} + x_2\right)\frac{\partial}{\partial x_1} + x_1^2 \frac{\partial}{\partial x_2} \quad ; \quad A^1 = \frac{\partial}{\partial x_2}$$

On cherche une fonction h telle que :

$A^1 h \equiv 0$ soit, h fonction de x_1 seul.

On trouve :

$$A^o h = \left(\frac{1}{2}x_1^2 + e^{x_2} + x_2\right)h'(x_1).$$

On veut ensuite

$$A^1 L_{A^o} h = \left(e^{x_2} + 1\right)h'(x_1) \not\equiv 0$$

afin que le nombre caractéristique du système (11) soit égal à 1.
On voit que des solutions simples sont données par les fonctions h telles que

$$h = Cx_1 \quad \text{avec} \quad C \neq 0 \ (C \in \mathbb{R})$$

Les bouclages de type (1), solutions du système (2) avec les conditions de la proposition 4 sont alors donnés par l'égalité :

$$C(e^{x_2}+1)u = -C\left[\left(\frac{1}{2}x_1^2 + e^{x_2} + x_2\right)x_1 + x_1^2\left(e^{x_2} + 1\right)\right]$$
$$+ \lambda C\left(\frac{1}{2}x_1^2 + e^{x_2} + x_2\right) + v \quad \text{avec } \lambda \in \mathbb{R}.$$

Ainsi, en prenant $C = -\frac{1}{2}$ et $\lambda = 0$, on retrouve les résultats de Hunt et Su. L'immersion $\tau : \mathbb{R}^2 \to \mathbb{R}^2$, définie par :

$$\tau(x_1,x_2) = \left(\frac{1}{2}x_1, \frac{1}{2}(\frac{1}{2}x_1^2 + e^{x_2} + x_2)\right)$$

permet d'immerger le système (11) dans le système linéaire :

$$\left.\begin{aligned} \dot{\eta}^o &= \eta^1 \\ \dot{\eta}^1 &= v \\ y &= \eta^o \end{aligned}\right\} \qquad (12)$$

L'application τ est de plus un difféomorphisme.

BIBLIOGRAPHIE

[1] R.W. BROCKETT, Feedback invariants for nonlinear systems, Proc. VIIth I.F.A.C. World Congr., Helsinski, 1115-1120 (1978)

[2] P. BRUNOVSKY, A classification of linear controllable systems, Kybernetica cislo 3, 173-188 (1970)

[3] Y. CHOQUET-BRUHAT, Géométrie différentielle et systèmes extérieurs, Dunod, Paris (1968)

[4] D. CLAUDE, Decoupling of nonlinear systems, Systems Control Lett., 1, 242-248 (1982)

[5] D. CLAUDE, Découplage des systèmes : du linéaire au non linéaire, dans "Développements et utilisations d'outils et modèles mathématiques en automatique, analyse de systèmes et traitement du signal", vol. 3, I.D. Landau ed., C.N.R.S., Paris, 533-555 (1983)

[6] D. CLAUDE, Découplage des systèmes non linéaires, séries génératrices non commutatives et algèbres de Lie, à paraître

[7] D. CLAUDE, M. FLIESS et A. ISIDORI, Immersion, directe et par bouclage, d'un système non linéaire dans un linéaire, C.R. Acad. Sc. Paris, 296, série I, 237-240 (1983)

[8] F. CSÁKI, Modern Control Theories, Akadémiai Kiadò, Budapest (1972)

[9] J. DESCUSSE, Les invariants fondamentaux d'une paire (A,B), leurs liens avec certaines formes canoniques, dans "Outils et modèles mathématiques pour l'automatique, l'analyse de systèmes et le traitement du signal, vol. 1, I.D. Landau ed., C.N.R.S., Paris, 65-86 (1981)

[10] M. FLIESS, Fonctionnelles causales non linéaires et indéterminées non commutatives, Bull. Soc. Math. France, 109, 3-40 (1981)

[11] M. FLIESS, Finite-dimensional observations-spaces for nonlinear systems, in Feedback Control of Linear and Nonlinear Systems, D. Hinrichsen et A. Isidori, ed., Lect. Notes Control Informat. Sc., 39, Springer-Verlag, 73-77 (1982)

[12] M. FLIESS et I. KUPKA, A finiteness criterion for nonlinear input-output differential systems, S.I.A.M. J. Control Optimiz., 21, 721-728 (1983)

[13] L.R. HUNT et R. SU, Global mapping of nonlinear Systems, Proc. Joint Automatic Control Conference, Charlottesville, VA, USA, série FA. 3C, 2, 1-5 (1981)

[14] A. ISIDORI et A.J. KRENER, On Feedback equivalence of nonlinear Systems, Systems Control Lett., 2, 118-121 (1982)

[15] B. JAKUBCZYK et W. RESPONDEK, On linearization of control systems, Bull. Acad. Polon. Sc. (Math.), 28, 517-522 (1980)

[16] A.J. KRENER, On the equivalence of control systems and the linearization of nonlinear systems, S.I.A.M. J. Control, 11, 670-676 (1973)

[17] G. MEYER, R. SU et L.R. HUNT, Applications to aeronautics of the theory of transformations of nonlinear systems, dans "Développements et utilisations d'outils et modèles mathématiques en automatique, analyse de systèmes et traitement du signal, vol. 3, I.D. Landau ed., C.N.R.S., Paris, 675-688 (1983)

[18] S. MONACO et D. NORMAND-CYROT, The immersion under feedback of a multidimensional discrete-time nonlinear system into a linear system, Int. J. Control, 38, 245-261 (1983)

[19] R. SU, On the linear equivalents of nonlinear systems, Systems Control Lett., 2, 48-52 (1982)

[20] W.M. WONHAM, Linear Multivariable Control : A Geometric Approach, Springer-Verlag (1977).

ON THE CONTROLLABILITY PROPERTIES OF ELASTIC ROBOTS

G. Cesareo and R. Marino
Seconda Università di Roma, Tor Vergata
Dipartimento di Ingegneria Elettronica
Via O. Raimondo, 00173 Roma, Italy

ABSTRACT: The note presents an integrated algorithm which, given the configuration (spatial geometry, mass distribution, structural parameters) of an industrial robot constituted by N+1 rigid links interconnected by N joints (elastic or rigid, controlled or uncontrolled) allows the symbolic generation of the equations of motion and the analysis of their controllability properties. The equations of motion are deterministic and highly nonlinear; their structural control properties are investigated using recently developed results from geometric nonlinear control theory. The algorithm is implemented via symbolic and algebraic manipulation (SAM) systems and is intended to be a guide in the choice of more suitable robot configurations and in the design of state feedback control laws. Two examples are reported. In the first one a controlled elastic shoulder is considered: the algorithm automatically establishes that it is controllable and that its nonlinearities can be compensated by state feedback. In the second one a planar robot with elastic controlled joints is considered: the algorithm automatically establishes the strong accessibility property.

1. INTRODUCTION

An industrial robot can be thought as N+1 rigid bodies (links) interconnected by N prismatic and/or rotary joints. Each joint may be controlled by a motor; it can be either rigid or characterized by a certain amount of elasticity, which may include the distributed elasticity of subsequent links. The aim of the dynamic control of an industrial robot is to force its end body to follow a given trajectory in terms of position and orientation, i.e. a trajectory $(y_1(t), y_2(t), y_3(t), \alpha(t), \beta(t), \gamma(t)$ in the output space.

This work was partly supported by MPI (fondi 40%).

The main contribution of this note is the presentation of an algorithm which, given the robot configuration, automatically investigates the controllability property and the linearizability properties.

Recall that if each joint is rigid and controlled by one motor the robot can be easily shown to be feedback linearizable and completely controllable. Therefore the algorithm presents some interest in the cases when not every joint is controlled and/or some joints are elastic.

For instance if any motor fails to work, one is interested in determining the portion of the output space which can be controlled by the remaining motors. In presence of elasticity at certain joints, simulation runs prove that control schemes developed for the rigid case may produce elastic oscillations or, if discrete controls are employed, resonant frequencies may be excited. Besides one could be interested not only in controlling positions and orientations of the end body but also in controlling the elastic forces themselves. In those cases the state space controllability needs to be studied.

The equations of motion for robot manipulators are cumbersome and highly nonlinear but linear in the inputs.

The starting point of this note is the possibility of the symbolic generation of such equations of motion for robots with elastic and/or rigid joints, given the robot configuration, i.e. types of joints, spatial geometry, mass distribution, control locations and so on [1]. The symbolic expression of the equations of motion enable us to apply the algorithm presented in [2] which, given a nonlinear control system linear in the inputs, automatically computes the accessible and strong accessible part of the system and establishes the possibility of compensating the nonlinearities by state feedback controls. The last property, which will be called feedback linearizability property, is the most interesting from a "practical" point of view, since allows us to control the system as easily as a linear system ([5], [6]). Besides the nonaccessible (strongly nonaccessible) directions represent the directions along which the system cannot be driven in a positive (arbitrarily finite) time.

The overall algorithm can be implemented by systems which allow symbolic and algebraic manipulations (SAM systems). The point of view presented in this note would not make sense if SAM systems were not available. The algorithm is tested in two examples: a controlled elastic shoulder and a planar robot with elastic actuated joints.

2. ROBOT MODELS AND NONLINEAR GEOMETRIC TECHNIQUES

The equations of motion of N+1 links interconnected by N rigid joints are in matrix form:

$$B(q)\,\ddot{q} + f(q,\dot{q}) + e(q) = m(t) \qquad (1)$$

where

q N-vector of joint variables, giving the relative displacements ot two adjacent links ;

$\dot{q}$ time derivative of q;

$B(q)$ the N x N matrix which produces the inertial forces $B(q)\ddot{q}$;

$f(q,\dot{q})$ N-vector of terms due to centrifugal and Coriolis forces ;

$e(q)$ N-vector of terms due to conservative forces ;

$m(t)$ N-vector of generalized forces delivered by the N motors, which represent the controls;

the components of $f(q,\dot{q})$ are given by:

$$f_i(q,\dot{q}) = \dot{q}^T F_i(q)\,\dot{q} \qquad (2)$$

where

$$F_i = \frac{1}{2}\left(\frac{\partial b_i}{\partial q} - \frac{\partial B}{\partial q_i} + \left(\frac{\partial b_i}{\partial q}\right)^T\right)$$

and b_i is the i-th column of B.

In the elastic case the state variables are 2N, two times as many as in the rigid case (N). Note that elasticity at joints usually occurs when rotary joints are used. In this case q_i is replaced by q^e_{2i-1} and q^e_{2i}, where q^e_{2i-1} denotes the displacement between the rotor and the stator and q^e_{2i} denotes the relative displacement of the two adjacent links. The dynamics of an elastic robot can be modeled by equations of the type (1)

$$B_e(q_e)\,\ddot{q}_e + f_e(q_e,\dot{q}_e) + e_e(q_e) = m_e(t) \qquad (3)$$

with the following peculiarities:

- q is replaced by q_e;
- $e_e(q_e)$ contains the elastic forces;
- the number of motors is still N and therefore the N even components of the 2N-vector $m_e(t)$ are zero.

Recalling that the inertia matrix $B(q)$ is nonsingular, (1) can be rewritten as

$$\ddot{q} = -B^{-1}(q)\ f(q,\dot{q}) - B^{-1}(q)\ e(q) + B^{-1}(q)\ m(t) \qquad (4)$$

One can define a state feedback and a state dependent linear change of coordinates in the control space R^N so that the transformed system becomes a linear controllable system with controllability indices (2,...,2). It is well-known that such a transformation does not affect the controllability property of the system; moreover in this case it holds globally, since the inertia matrix is nonsingular everywhere.

More precisely let us define the new controls $v(t)$

$$m(t) = f(q,\dot{q}) + e(q) + B(q)\ v(t)$$

system (4) becomes

$$\ddot{q}(t) = v(t)$$

which is completely controllable, being N chains of double integrators. $v(t)$ can be designed via linear control techniques. The preceding argument is a very simple application of a general theory on the linearization of nonlinear systems via feedback transformations ([4], [5]). Thus the complete controllability of the rigid robot can be established because of its *global* feedback equivalence to a linear controllable system; this fact is however a sufficient but not a necessary condition for controllability. Unfortunately such a simple argument does not apply if not every joint is controlled or if some joint is elastic.

Let us rewrite the elastic system (3)

$$\ddot{q}_e = -B_e^{-1}(q_e)\ f_e(q_e,\dot{q}_e) - B_e^{-1}(q_e)\ e_e(q_e) + B_e^{-1}(q_e)\ m_e(t)$$

in state space form as

$$\dot{x} = \overline{f}(x) + \overline{G}(x)\ u(t) \qquad (5)$$

where

$$x = \begin{pmatrix} q_e \\ \dot{q}_e \end{pmatrix} \qquad x \in \mathbb{R}^n\ , \quad n = 4N$$

$$\overline{f}(x) = \begin{pmatrix} \dot{q}_e \\ -B_e^{-1}(q_e)\ f_e(q_e,\dot{q}_e) - B_e^{-1}(q_e)\ e_e(q_e) \end{pmatrix}$$

$$\overline{G}(x) = \begin{pmatrix} 0 \\ H(q_e) \end{pmatrix}$$

$H(q_e)$ is the 2N x N matrix containing the odd columns of $B_e^{-1}(q_e)$. Let $\overline{g}_i(x)$ denote the i-th column of the matrix $\overline{G}(x)$, being $\overline{g}_i(x)$ a vector field. $u(t)$ is an N-vector which collects the nonzero components of $m_e(t)$.

We will now recall some notation, definitions and results from the differential geometric theory of nonlinear control systems.

Sussmann and Jurdjevic [7] introduced two basic distributions for nonlinear control theory:

- the strongly accessible distribution L_o, which is the smallest involutive distribution which contains

$$\text{span } \{ad_{\overline{f}}^{\ell}\overline{g}_i : \ell \geq 0,\ 1 \leq i \leq N\}$$

  dim $L_o(x_o)$ gives the dimension of the strongly accessible set from x_o, i.e. the set of points x for which, given any positive time t, there exists an admissible control which takes x_o into x in a time t ;
- the accessible distribution L, which is the smallest involutive distribution which contains

$$\text{span } \{\overline{f}, \overline{g}_1, \ldots, \overline{g}_N\}$$

  dim $L(x_o)$ gives the dimension of the accessible set from x_o, i.e. the set of points x for which there exists an admissible control which takes x_o into x in a positive time. Note that $L(x) = \text{span}\{L_o(x), \overline{f}(x)\}$.

Define

$$G^o = \text{span } \{\overline{g}_1, \ldots, \overline{g}_N\}$$

and inductively

$$G^j = \text{span } \{G^{j-1}, [G_{\overline{f}}, G^{j-1}]\} \qquad (6)$$

where

$$G_{\overline{f}} = \overline{f} + G^o$$

and

$$[G_{\overline{f}}, G^{j-1}] = \text{span}\,\{[X,Y] \; : \; X \in G_{\overline{f}}, \; Y \in {}^{j-1}\}$$

$[X, Y]$ denotes the Jacobi bracket, also denoted by $ad_X Y$. Define inductively

$$M^o(x) = G^o(x)$$

$$M^j(x) = \text{span}\,\{M^{j-1}(x), \; [\overline{f}, M^{j-1}](x)\} \tag{7}$$

After setting the preliminary notation let us recall some results of interest to us from the geometric nonlinear control theory. It is shown in [2] that if for some integer k, $G^k(x) = G^{k+1}(x)$ for every x in V, open subset in $\mathbb{R}^n$, then $G^k(x) = G^{k+\ell}(x)$ for any integer $\ell \geq 0$ and any $x \in V$. This guarantees the existence of an integer p, $p \leq n-1$, defined as the smallest integer such that $G^{p-1}(x) \subset G^p(x)$ and $G^p(x) = G^{p+\ell}(x)$ for every $x \in N_p$, open subset in $\mathbb{R}^n$ [12], and every integer $\ell \geq 0$. Moreover it is proved that $L_o(x) = G^p(x)$ for every $x \in N_p$. This allows the construction of an algorithm, presented in [2], which computers L_o in a finite number of steps.

It was recently proved (see [3], [4], [5], [6]) that a strongly accessible system can be locally transformed into a linear controllable system by a feedback-transformation ($\tilde{x} = T(x)$, T local diffeomorphism, $v = a(x) + S(x)\,u$, $S(x)$ locally nonsingular) if and only if the distributions M^j are locally involutive and locally of constant rank m_j for each j, $j = 1,\dots, n-1$. Recall that (see [12]) if M^j is involutive for each j, $0 \leq j \leq \ell$, then $M^j = G^j$ for each j, $0 \leq j \leq \ell$. If there exists a state x_o such that $\overline{f}(x_o) = 0$, the case in which only a state transformation $\tilde{x} = T(x)$ is required to transform (locally in U_{x_o}, a neighborhood of x_o) the strongly accessible system into a linear controllable one ([8], [9]) is characterized by a commutative structure of the nested involutive distributions M^j:

$$[ad^t_{\overline{f}}\overline{g}_i, \; ad^s_{\overline{f}}\overline{g}_j] = 0 \;, \quad 0 \leq t, s \leq n-1 \;, \quad 1 \leq i, j \leq N$$

or equivalently

$$[\overline{g}_i, \; ad^t_{\overline{f}}\overline{g}_j] = 0 \;, \quad 0 \leq t \leq 2(n-1) \;, \quad 1 \leq i, j \leq N$$

For feedback-linearizable (or in particular state linearizable) systems (5) the controllability indices $(k_1,\dots,k_N)$ are defined as follows: k_i

is equal to the number of $s_j \geq i, j \geq 0$, where $s_o = m_o$, $s_1 = m_1 - m_o, \ldots, s_i = m_i - m_{i-1}, \ldots$.

3. AN INTEGRATED ALGORITHM FOR THE CONTROLLABILITY OF ELASTIC ROBOTS

In this section we present an algorithm which, given the robot configuration, generates the equation of motion and analyzes their controllability and linearizability properties. The algorithm is divided into two parts: the first produces the equations of motion, the second computes $L_o(x)$, $L(x)$ and establishes the linearizability properties.

The whole algorithm integrates two more general programs. The former, called DYMIR [1] (Dynamical Models of Industrial Robot), allows the symbolic computation of dynamic models of robots with rigid and/or elastic joints. The latter was recently developed [2] as a tool for computer-aided analysis of nonlinear systems.

DYMIR computes respectively the matrices $B(q)$ and $B_e(q)$ and the vectors $e(q)$, $e_e(q_e)$, $f(q,\dot{q})$, $f_e(q_e,\dot{q}_e)$ which constitute equations (1) and (3). In both cases the input is the kinematic description of the robot (see the definition given in [11]).

In the elastic case DYMIR iterates the following five steps N times for the computation of $B_e(q_e)$ and $e_e(q_e)$ (the subscript r denotes quantities relative to the rotors).

Step 1. Computation of potential energy U

$$U_i = P_i + P_{r,i} + \frac{1}{2} k_i \, (q_{e,2i} - q_{e,2i-1}/n_i)^2$$

where P_i and $P_{r,i}$ are gravity potentials of link i and rotor i respectively.

Step 2. Computation of angular velocities ω^i, ω^i_r

$$\omega^i_r = C^i_r \, (\omega^{i-1} + (1-\sigma_i) \, g^i \, \dot{q}_{e,2i-1})$$

$$\omega^i = C^i \, (\omega^{i-1} + (1-\sigma_i) \, g^i \, \dot{q}_{e,2i})$$

where C^i, C^i_r are the rotational transformation matrices between frames R^{i-1} and R^i or R^i_r (Cartesian frames attached to the end of the arm or of the rotor)

g^i is the i-th joint versor (z-axis of R^i)

$$\sigma_i = \begin{cases} 1 \text{ if joint } i \text{ is a prismatic one} \\ 0 \text{ if joint } i \text{ is a rotary one} \end{cases}$$

Step.3 Computation of the velocities v_g^i, $v_{g,r}^i$ of the mass centres of link i and rotor i respectively

$$v_{g,r}^i = v_r^i + \omega_r^i \times d_r^{i,i}$$

$$v_g^i = v^i + \omega^i \times d^{i,i}$$

where $v_r^i = C_r^i(v^{i-1} + (\omega^{i-1} + (1-\sigma_i) g^i \dot{q}_{e,2i-1}) \times d_r^{i-1,i} + \sigma_i g^i \dot{q}_{e,2i-1}$

$$v^i = C^i(v^{i-1} + (\omega^{i-1} + (1-\sigma_i) g^i \dot{q}_{e,2i}) \times d^{i-1} + \sigma_i g^i \dot{q}_{e,2i}$$

$d^{i,i}$ vector of center of mass of link i relative to R^i;

$d_r^{i,i}$ vector of center of mass of rotor i relative to R_r^i;

$d_r^{i-1,i}$ vector of origin of R_r^i in R^{i-1} coordinates;

$d^{i-1,i}$ vector of origin of R^i in R^{i-1} coordinates.

Step 4. Computation of kinetic energy T_i

$$T_i = \tfrac{1}{2}((\omega^i)^T J^i \omega^i + m_i (v^i)^T) + ((\omega_r^i)^T J_r^i \omega_r^i + m_{r,i} (v_{g,r}^i)^T v_{g,r}^i))$$

where J^i is the inertia tensor of link i relative to the frame R_g^i ($R_g^i = R^i$ translated into the center of mass);

J_r^i is the inertia tensor of the rotor;

m_i is the mass of link i .

Step 5. Computation of the contributions B_{jk}^i, e_j^i of T_i and U_i to $B_e(q_e)$ and $e_e(q_e)$

$$B_{jk}^i = \text{coeff}(T_i; \dot{q}_{e,j} \dot{q}_{e,k}) \quad j,k=1,\ldots,i; \quad j \leq k$$

$$e_j^i = \frac{\partial U_i}{\partial q_j}$$

where $\text{coeff}(T_i; \dot{q}_{e,j} \dot{q}_{e,k})$ gives the coefficients of $\dot{q}_{e,j} \dot{q}_{e,k}$ In T_i and is a build-in function of REDUCE.

Step 6. Once the five steps are iterated varying i from 1 to N the matrices F_i are computed according to (2).

Remark. DYMIR was run for some different kinds of robots. For a three degrees of freedom robot the CPU time (UNIVAC 1100/80A) is between 20 and 35 seconds depending on the kinematic structure of the robot. For the elastic case CPU time is on average twice as much as in the rigid case.

Step 7. Computation of the vector fields $\overline{f}, \overline{g}_1, \ldots, \overline{g}_N$, given the matrices $B_e(q_e)$ and $F_{e,i}(q_e)$.

Remark. At this point the vector fields $\bar{f}(x),\bar{g}_1(x),\ldots,\bar{g}_N(x)$ are explicitely obtained and constitute the input required by the algorithm proposed in [2] which analyzes the controllability properties. Computing $L_o(x),L(x)$ and establishes the linearitability properties.

The overall agorithm procedes iterating the following step.

Step 8. Compute M^j (7)

until M^j is not involutive

OR dim M^j=dim M^{j-1}

OR dim M^j=n

If M^j is involutive and dim M^j=n then the robot (5) is feedback linearizable. In particular if (8) holds then the system is state linearizable.

If M^{j-1} is involutive, but $M^j=M^{j-1}$ then $L_o=G^j=M^j$.

If M^{j-1} is involutive, but M^j is not involutive set $M^j=G^j$ and iterate step 9.

Step 9. Which (dim $G^j<n$) AND ($G^j\neq G^{j-1}$) compute G^{j+1} (6).

Remark. Step 9 stops when j=p and $G^p=L_o$ as discussed before. The whole algorithm ends with the computation of L.

Step 10. $L=$ span$\{\bar{f},L_o\}$

4. IMPLEMENTATION AND EXAMPLES

The algorithm presented in section 3 can be implemented via SAM systems. The most suitable SAM systems for our purpose are REDUCE [14] and MACSYMA. Their appeal relis upon the following features:

a) integer and rational arithmetics with infinite precision;
b) manipulation of polynomials, rational and elementary functions;
c) algebraic manipulation of matrices whose entries are allowed to be polynomial, rational and elementary functions or combination of those: for instance computation of inverses and determinants, products and additions;
d) differential calculus: derivatives, partial derivatives;
e) as far as other types of manipulations not included in the previous four points are concerned, new or special rules for specific needs can be added via "pattern matching" techniques. The interactive use is also of interest.

SAM systems are available: REDUCE can be mounted on any machine of the size of an IBM/148: MACSYMA is available on DEC VAX-11.

We implemented the algorithm in REDUCE. REDUCE presents some problems in terms of feasibility and performance in handling long expressions. This imposes limitations on the number of joints and more generally on the kinematic configurations. If long expressions are involved, garbage collection and differentiation are highly CPU time consuming. In those cases relabeling parts of expressions as new variables is of some help.

In steps 8, 9 and 10 linear independence of vector fields is checked through a gaussian procedure developed in [2]. Involutivity can be checked on the basis of such a procedure and the possibility of performing Jacobi brackets by REDUCE [2].

We now report on the application of the algorithm to two simple robot structures: a shoulder and a planar robot.

Example 1. A shoulder is modeled by two rigid links interconnected by two elastic rotary joints whose axes are incident and orthogonal. Given the kinematic description, according to the definition given in [11], the output of the first seven steps is

$$\bar{G}_{51} = G_1$$
$$\bar{G}_{72} = G_2$$
$$\bar{f}_1 = x_5$$
$$\bar{f}_2 = x_6$$
$$\bar{f}_3 = x_7$$
$$\bar{f}_4 = x_8$$
$$\bar{f}_5 = G_1\ K_1\ (n_1x_2-x_1)/n_1^2$$
$$\bar{f}_6 = (A_1n_1x_6(x_8-x_6)\sin(2x_4)+K_1(n_1x_2-x_1))/(n_1A_1\cos^2x_4+A_2)$$
$$\bar{f}_7 = G_2\ K_2\ (n_2x_4-x_3)/n_2^2$$
$$\bar{f}_8 = -G_3\ K_2[(n_2x_4-x_3)\ +\ G_4n_2\cos x_4]/n_2$$

where only nonzero components are listed; n_i is the gear ratio, K_i the elastic constant of joint i. Step 8 is iterated four times and establishes that $\dim M^3(x)=8$ for every $x\in\mathbb{R}^8$. Then the elastic shoulder is feedback linearizable but not state linearizable since $[ad_f^2g_1,ad_f^3g_1](x)\neq 0$.

Example 2. A planar robot is constituted by two rigid links connected by two elastic joints with parallel axes; the link axis belong to the same vertical plane. The output of step 7 is

$$\bar{G}_{51} = G_1$$
$$\bar{G}_{61} = A_2/DT_1$$
$$\bar{G}_{72} = (A_3\cos^2x_4-A_1A_2)/DT_1$$

$$\overline{G}_{82} = -DT_2/DT_1$$

$$\overline{f}_1 = x_5$$

$$\overline{f}_2 = x_6$$

$$\overline{f}_3 = x_7$$

$$\overline{f}_4 = x_8$$

$$\overline{f}_5 = (n_1x_2-x_1)G_1K_1/n_1^2$$

$$\overline{f}_6 = -\Big[\sin x_4\big[x_6DT_2+A_2x_8(x_6+x_8)\big]A_3n_1n_2^2-\big[K_1(n_1x_2-x_1) + n_1A_5\cos x_2+n_1DT_3\big]n_2^2A_2-\big[K_2A_2(n_2x_4-x_3) + (DT_3n_2+K_2(n_2x_4-x_3))DT_2n_2\big]n_1\Big]/n_1n_2^2DT_1 +$$

$$\overline{f}_7 = \{\sin x_4\big[DT_2x_6+x_8(x_6+x_8)A_2\big]A_3JRZ_2n_1n_2^2 - \big[K_1(n_1x_2-x_1)+n_1A_5\cos x_2+DT_3n_1\big]JRZ_2n_2^2A_2 - \big[(A_3^2\cos^2x_4-A_1A_2)(x_3-n_2x_4)K_2+((x_3-x_4n_2)K_2 - DT_3n_2)DT_2JRZ_2n_2\big]n_1\}/JRZ_2n_1n_2^2DT_1$$

$$\overline{f}_8 = \{\sin x_4\big[(2DT_2+A_1-JRZ_2-A_2)x_6+x_8(x_6+x_8)DT_2\big]\cdot A_3n_1n_2^2-\big[(n_1x_2-x_1)K_1+n_1A_5\cos x_2+DT_3n_1\big]\cdot DT_2n_2^2-\big[(2DT_2+A_1-JRZ_2-A_2)((x_3-x_4n_2)K_2\cdot DT_3n_2)n_2-(x_3-n_2x_4)DT_2K_2\big]n_1\}/n_1n_2^2DT_1$$

where only nonzero components are listed and

$$DT_1 = A_3^2\cos^2 x_4 + A_2\,(JRZ_2-A_1)$$

$$DT_2 = A_3\cos x_4 + A_2$$

$$DT_3 = A_4\cos(x_2+x_4).$$

These substitutions are needed for better run time performance.

Step 8 is iterated two times and establishes that $M^1(x)$ is not involutive in any open subset of $\mathbb{R}^8$. Step 9 is initiated by $G^1=M^1$ and stops for p=3, that is

$$L_o(x)=G^3(x)=\text{span}\{\overline{g}_1,\overline{g}_2,\ ad_{\overline{f}}\overline{g}_1,\ ad_{\overline{f}}\overline{g}_2,\ ad_{\overline{f}}^2\overline{g}_1,\ ad_{\overline{f}}^2\overline{g}_2,\ [\overline{g}_2,\ ad_{\overline{f}}\overline{g}_2],ad_{\overline{f}}^3\overline{g}_1\}$$

in an open and dense subset of $\mathbb{R}^8$. Thus the system is strongly accessible. This example costs 300$ on UNIVAC 1100/80A.

5. AKNOWLEDGMENTS

Profitable discussions with Professors F. Nicolò and S. Nicosia are gratefully aknowledged.

6. REFERENCES

[1] G. Cesareo, F. Nicolò, S. Nicosia, *DYMIR: a code for generating dynamic models of robots*, Dipartimento di Informatica e Sistemistica, Università di Roma, Technical Report n.02.83, 1983.

[2] R. Marino, G. Cesareo, *Nonlinear control theory and symbolic algebraic manipulation*, presented at the Int. Symp. MTNS in Beer Sheva, Israel, June 1983.

[3] R.W. Brockett, *Feedback invariants for nonlinear systems*, Proc.IFAC Congress, Helsinki, 1978.

[4] B. Jakubczyk, W. Respondek, *On linearization of control systems*, Bull. Acad. Pol. Sci. Vol.XXVIII, n.9-10, 517-522.

[5] L.R. Hunt, R. Su, G. Meyer, *Design of multiinput nonlinear systems*, in Differential Geometric Control Theory, 268-298, R. Brockett... ed., Birkhäuser, 1983.

[6] R. Marino, *Feedback equivalence of nonlinear systems with applications to power systems*, Doctoral Dissertation, Washington University, St. Louis, 1982.

[7] H.J. Sussmann; V.J. Jurdjevic, *Controllability of nonlinear systems*, J. Diff. Eq., 12, 95-116, 1972.

[8] A.J. Krener, *On the equivalence of control systems and linearization of nonlinear systems*, SIAM J. Contr., Vol.11, 1973.

[9] W. Respondek, *Geometric methods in linearization of control systems*, in Banach Center Publications, Semester on Control Theory, Dec.1980.

[10] A.C. Hearn, *REDUCE-2 User's Manual*, Univ. of Utah, Rep. UCP-19, 1973.

[11] R.P. Paul, B. Shimano, *Kinematic control equations for simple manipulators*, IEEE, CDC Conference, 1398-1406, San Diego, 1979.

[12] R. Marino, W.M. Boothby, D.L. Elliott, *Geometric properties of linearizable control systems*, submitted to Int. J. of Math. System Theory.

[13] A. Ficola, R. Marino, S. Nicosia, *A singular perturbation approach to the dynamic control of elastic robots*, Proc. Allerton Conference, Monticello Illinois, 1983.

SUR LA COMMANDE NON INTERACTIVE DES SYSTEMES NON LINEAIRES EN TEMPS DISCRET

S. Monaco[*], D. Normand-Cyrot[**]

[*] Dipartimento di Informatica e Sistemistica
Università di Roma "La Sapienza",
Via Eudossiana, 18, 00184 Roma, ITALIE

[**] Laboratoire des Signaux et Systèmes, CNRS-ESE,
Plateau du Moulon, 91190 GIF-SUR-YVETTE, FRANCE

ABSTRACT

The noninteracting control problem of nonlinear discrete-time systems is studied. This approach, mostly algebraic, uses tools and techniques of proof recently introduced in solving related nonlinear control problems in the discrete-time case. Necessary and sufficient conditions for the construction of the appropriate feedback law are given. A parallel with the results actually known for nonlinear differential systems is also noted.

RESUME

On étudie le problème de la commande non interactive des systèmes non linéaires en temps discret. Cette approche, essentiellement algébrique, utilise des outils et des techniques de preuve récemment introduits pour la résolution de divers problèmes d'automatique non linéaire en temps discret. Des conditions, nécessaires et suffisantes, pour la construction de la loi de bouclage appropriée sont données. On note également un parallèle entre les résultats actuellement connus pour les systèmes différentiels en temps continu.

Ce travail a été réalisé pendant le séjour de S. Monaco au Laboratoire des Signaux et Systèmes avec le support financier du C.N.R. Italien.

1. INTRODUCTION

Etant donné un système linéaire analytique en temps discret de la forme:

$$\Sigma \begin{cases} x(t+1)=x(t)+f(x(t))+\sum_{i=1}^{q} u_i(t)g_i(x(t))=x(t)+f(x(t))+g(x(t))\underline{u}(t) \\ y(t) = h_i(x(t)) \quad , \; i = 1,\dots,q \end{cases}$$

où l'état $x(t) \in R^N$, les fonctions $f, g_1,\dots,g_q : R^N \to R^N$ sont analytiques sur tout R^N, les fonctions de sorties $h_1,\dots,h_q: R^N \to R$ sont analytiques, les entrées $\underline{u} = (u_1,\dots,u_q)$ sont à valeurs réelles; sous quelles conditions existe t'il une loi de bouclage statique telle que chaque sortie du système bouclé ne dépende que d'une seule entrée. Il s'agit du problème bien connu de la commande non interactive étudié dans le cas des systèmes linéaires [1] et récemment généralisé aux systèmes non linéaires différentiels [2], [3].

Dans cette communication le problème est posé dans le cas des systèmes non linéaires en temps discret. On obtient des critères analogues à ceux existant en temps continu et ce grâce à des outils introduits récemment, qui ont déjà permis aux auteurs eux-mêmes [4-5] d'aborder pour la première fois et de résoudre quelques problèmes classiques d'automatique non linéaire.

De façon précise, considérons une *loi de bouclage* linéaire analytique, β, de *rang plein*, de la forme:

$$u_i(x) = \alpha_i(x) + \sum_{j=1}^{q} \beta_{ij}(x)v_j \quad (i = 1,\dots,q) \tag{1}$$

défini pour $x \in V$, ouvert dense de $\mathbb{R}^N$, $v = (v_1,\dots,v_q) \in \Omega(x)$ l'ensemble des entrées admissibles, les fonctions α_i et $\beta_{ij}: \mathbb{R}^N \to \mathbb{R}^N$, $(i,j)\in\{1,\dots,q\}$ sont analytiques, si la matrice $\beta(x) = (\beta_{ij}(x))_{ij}$ est inversible, le bouclage est dit de rang plein.

Sous quelles conditions, le système bouclé

$$\Sigma_\beta \begin{cases} x(t+1) = x(t)+\tilde{f}(x(t)) + \sum_{i=1}^{q} v_i(t)\tilde{g}_i(x(t)) \\ y_i(t) = h_i(x(t)) \quad , \; (i = 1,\dots,q) \end{cases}$$

avec $\tilde{f} = f + \sum_{i=1}^{q} \alpha_i g_i ; \; \tilde{g}_i = \sum_{j=1}^{q} \beta_{ij} g_i \quad , \quad (i = 1,\dots,q)$

est t'il, à un changement de coordonnées près: $z(t) = T(x(t))$, un système de la forme:

$$
\begin{cases}
\underline{z}^1(t+1) = \begin{bmatrix} 010-0 \\ | \quad\; 1 \\ 0 \text{---} 0 \end{bmatrix} \underline{z}^1(t) + \begin{bmatrix} 0 \\ 0 \\ 1 \end{bmatrix} v_1(t) \\
\vdots \\
\underline{z}^q(t+1) = \begin{bmatrix} 010-0 \\ | \quad\; 1 \\ 0 \text{---} 0 \end{bmatrix} \underline{z}^q(t) + \begin{bmatrix} 0 \\ 0 \\ 1 \end{bmatrix} v_q(t) \\
\underline{z}^{q+1}(t+1) = \tilde{f}_{q+1}(z(t);\ \underline{v}(t)) \\
y_i(t) = [\,10\text{---}0\,]\,\underline{z}^i(t)\ , \quad i = 1,\dots q
\end{cases}
\tag{2}
$$

où les $\underline{z}^i \in R^{n_i}$, $(i = 1,\dots,q+1)$, $\sum_{i=1}^{q+1} n_i = N$ correspondent à une partition adéquate de l'état $z(t) = T(x(t))$?

REMARQUES. (i) Les sorties $y_1,\dots,y_q$ du système (2) représentent une cascade de retardateurs par rapport aux commandes $v_1,\dots,v_q$.

(ii) La notion d'entrée admissible permet de définir le bouclage pour tout x de V. Etant donné $x \in V$ on définit l'ensemble

$$\Omega(x) = \{\underline{v} = (v_1,\dots,v_q)^T \in \mathbb{R}^q \,|\, x+\tilde{f}(x) + \sum_{i=1}^{q} v_i \tilde{g}_i(x) \in V\}$$

Rappelons brièvement la formulation quelque peu différente posée en temps continu [2]. Etant donné un système linéaire analytique

$$
\begin{cases}
\dot{x}(t) = (\frac{dx(t)}{dt}) = f(x(t)) + \sum_{i=1}^{q} u_i(t) g_i(x(t)) \\
y(t) = h_i(x(t))\ , \quad (i = 1,\dots,q).
\end{cases}
\tag{3}
$$

où $x(t) \in M$, variété analytique sur R^N, $u = \{u_1,\dots,u_q\} \in R^q$, $y_i \in R$, où $f, g_1,\dots,g_q$ sont des champs de vecteurs analytiques définis sur M et $h_1,\dots,h_q$ sont des fonctions analytiques; sons quelles conditions existe-t'il un bouclage linéaire analytique de la forme (1), avec $\beta(x)$ inversible, tel que le système (3), bouclé, admette à un changement de coordonnées près la représentation suivante:

$$\begin{cases} \dot{\underline{z}}^1(t) = \tilde{f}_1(\underline{z}^1) + \tilde{g}_1(\underline{z}^1)v_1 \\ \vdots \qquad \vdots \qquad \vdots \\ \dot{\underline{z}}^q(t) = \tilde{f}_q(\underline{z}^q) + \tilde{g}_q(\underline{z}^q)v_q \\ \dot{\underline{z}}^{q+1}(t) = \tilde{f}_{q+1}(\underline{z}^1,\ldots,\underline{z}^{q+1}) + \sum_{j=1}^{q} v_j \tilde{g}_{q+1}(\underline{z}^1,\ldots,\underline{z}^{q+1}) \\ y_i = h_i(\underline{z}^i) \ , \ (i = 1,\ldots,q). \end{cases} \tag{4}$$

où les $\underline{z}^i(t)$, $(i = 1,\ldots,q+1)$ sont une partition adéquate de l'état.

On vérifie que chaque entrée v_i du système (4) n'affecte que la sortie y_i correspondence mais sans précision sur la dynamique du système (4) qui peut cette fois être non linéaire.

On distingue ainsi deux formulations du problème, la première aboutissant à un système de la forme (2), la seconde à un système de la forme (4).

En temps continu, ces deux formulations coïncident. En effet, on peut construire, de manière immédiate, une loi de bouclage linéaire analytique transformant le système (3) en une cascade d'intégrateurs et une partie non linéaire, analogue continu de la représentation (2).

En temps discret il en est de même à condition de considérer une extension polynomiale de la loi de bouclage linéaire analytique qui permet de transformer le système Σ en une cascade de retardateurs de la forme (2).

2. PRELIMINAIRES

Toute sortie $y_i(t)$, $t \geq 0$ du système non linéaire en temps discret Σ est une fonctionnelle dépendant des commandes $\underline{u}(o),\ldots,\underline{u}(t-1)$ dont on peut calculer les coefficients grâce à des opérateurs différentiels introduits récemment par l'étude du comportement entrée-sortie des systèmes en temps discret et rappelés brièvement ici (cf. [4] et [6] pour une présentation plus complète).

Si $f : R^N \to R^N$ est une fonction analytique sur tout R^N, définissons les opérateurs différentiels suivants

$$L_f^{\otimes 1} = \sum_{i=1}^{N} f_i \frac{\partial}{\partial x_i} \quad (f_i \ i^{\text{ème}} \text{ composante de } f)$$

$$L_f^{\otimes 2} = L_f \otimes L_f = \sum_{i,j=1}^{N} f_i f_j \frac{\partial^2}{\partial x_i \partial x_j}$$

$$\vdots \qquad\qquad \vdots$$

$$L_f^{\otimes n} = \sum_{i_1,\ldots,i_n=1}^{N} f_{i_1} \ldots f_{i_n} \frac{\partial^n}{\partial x_{i_1} \ldots \partial x_{i_n}}$$

ainsi de suite *notons* $exp_{\otimes} L_f$ *ou* Δ_f, la série différentielle

$$\Delta_f = I + L_f + \frac{1}{2!} L_f^{\otimes 2} + \ldots + \frac{1}{n!} L_f^{\otimes n} + \ldots$$

où $L_f^{\otimes o} = I$ est l'opérateur identité.

Si $h : R^N \rightarrow R$ est une fonction analytique, un développement Taylorien conduit à l'égalité suivante:

$$\forall x \in \mathbb{R}^N, \; ho(I+f)(x) = \Delta_f(h)|_x \qquad (|_x \text{ évaluation en } x)$$

Revenons au système en temps discret Σ, supposé initialisé en x_o, chaque sotie $y_i(t)$ vérifie l'expression:

$$y_i(t) = \Delta_{f+g\underline{u}(o)} \otimes \ldots \otimes \Delta_{f+g\underline{u}(t-1)}(h_i)|_{x_o}, \; (\forall i=1,\ldots,q) \tag{5}$$

où chaque opérateur $\Delta_{f+g\underline{u}}$ admet le développement en puissances de u suivant:

$$\Delta_{f+g\underline{u}} = \sum_{n \geq 0} \frac{u^n}{n!} \Delta_f \otimes L_g^{\otimes n} \tag{6}$$

En étendant à la série différentielle Δ_f (cf. [6]) la notion de dérivée de Lie L_f d'une forme différentielle ω [2], on peut montrer:

LEMME 1. $\Delta_f(dh) = d(\Delta_f h)$ où dh et $d\Delta_f h$ sont les différentielles des fonctions h et ho(I+f).

Rappelons les notations et résultats suivants (cf. [4],[5]).

A chaque sortie y_i, $i = 1,\ldots,q$, du système Σ est associé un *indice relatif* d_i qui est le plus petit entier tel que:

(i) $\Delta_f \otimes Lg_{i_\nu} \otimes \ldots \otimes Lg_{i_1} \Delta_f^r h_i|_x = 0, \; \forall x, \; \forall r < d_i, \; \forall \nu \geq 1, \; \forall i_\nu \ldots i_1 \in \{1,\ldots,q\}$

(ii) $\exists \bar{\nu} \in N^+$ et une suite d'indices $\bar{i}_{\bar{\nu}},\ldots,\bar{i}_1$ tels que
$\Delta_f \otimes Lg_{\bar{i}_{\bar{\nu}}} \otimes \ldots \otimes Lg_{\bar{i}_1} \Delta_f^{d_i} h_i|_x \neq 0 \quad \forall x \in V$

REMARQUE. On déduit aisément du développement (5) que les entrées à l'instant t = 0 agissent sur la sortie y_i pour la première fois à l'instant $t = d_i + 1$. Les entrées ultérieures n'agissant éventuellement sur y_i qu'à l'instant $t > d_i+1$. Supposons les *indices* d_i, $(i=1,\ldots,q)$ *du système* Σ *définis*, alors:

LEMME 2. $\forall k < d_i$, $\Delta_f^k h_i = \Delta_{\tilde{f}}^k h_i$, $(\forall i \in \{1,\ldots,q\})$ où $\tilde{f}$ est la dérive du système bouclé Σ_β.

Associons au système Σ *la matrice* A(x) définie par

$$A(x) = \{a_{ij}(x)\}_{\substack{i=1,\ldots,q \\ j=1,\ldots,q}}$$

où

$$a_{ij}(x) = \Delta_f \otimes Lg_j \Delta_f^{d_i} h_i |_x$$

On notera, cette fois, pour la commande non interactive, le rôle fondamental joué par la matrice A(x) pour la résolution de problèmes liés au comportement entrée-sortie d'un système non linéaire, en temps continu [2],[3], comme en temps discret [4],[5].

Si le problème de commande non interactive, admet une solution,on peut montrer que le système étudié reproduit après bouclage le comportement entrée-sortie d'un système linéaire convenablement initialisé. Cette propriété s'exprime en termes d'*immersion*, (ou de simulation) notion introduite en temps continu [7], puis étendue au temps discret [4],[5].

Soit Σ' une copie, définie sur $R^{N'}$, du système linéaire analytique Σ:

$$\Sigma' \begin{cases} x'(t+1) = x'(t)+f'(x'(t)) + \sum_{i=1}^{q} u_i(t) g_i'(x'(t)) \\ y'(t) = h_i'(x'(t)), \quad (i = 1,\ldots,q). \end{cases}$$

DEFINITION. Le système Σ (resp. Σ_β) simule (resp.simule par bouclage) le système Σ' s'il existe une application analytique $\tau : R^N \to R^{N'}$ (resp. une application et une loi de bouclage) telle que les systèmes Σ (resp. Σ_β) et Σ' respectivement initialisés en x et $\tau(x) = x'$ aient le même comportement entrée-sortie.

Etant donné un système linéaire:

$$\begin{cases} x(t+1) = A_o x(t) + \sum_{i=1}^{q} u_i(t) A_i \\ y_i(t) = C_i x(t), \quad (i = 1,\ldots,q) \end{cases}$$

où l'état $x(t) \in R^N$, $A_o, A_1, \ldots, A_q$, $C_1, \ldots, C_q$ sont des matrices de dimension appropriée, si les indices sont définis, on note

$$V_i = \bigcap_{j=0}^{d_i} C_i A_o^j \quad , \quad (i = 1,\ldots,q)$$

$$J = \bigcap_{i=1}^{q} \bigcap_{j=0}^{N-1} C_i A_o^j$$

l'ensemble des états inobservables.

LEMME 3. Si le système Σ simule pas bouclage linéaire analytique de la forme (1) un système linéaire vérifiant la condition,

$$J = \bigcap_{i=1}^{q} V_i \tag{7}$$

alors:

$$\Delta_f \otimes Lg_{i_\nu} \otimes \ldots \otimes Lg_{i_1} \Delta_f^{d_i} h_i = 0, \quad (\forall i=1,\ldots,q), \ \forall \nu \geq 2 \quad \forall i_\nu \ldots i_1 \in \{1,\ldots,q\} \tag{8}$$

REMARQUES. (i) Il a été montré [5] que la condition (8) n'est plus nécessaire pour la simulation par bouclage avec un système linéaire si l'on considère une extension polynomiale de la loi de bouclage. On constate ainsi que ce type de bouclage permet un affaiblissement non négligeable des conditions de simulation avec un linéaire. Il en est de même pour le problème de la commande non interactive.

(ii) Les systèmes bilinéaires vérifient toujours, de par leur structure linéaire en l'état, la condition (8).

3. RESULTAT PRINCIPAL

Les indices du système Σ sont supposés définis pour $i \in \{1,\ldots,q\}$, la dérive $(I+f) : R^N \to R^N$ est supposée inversible (i.e. de matrice jacobienne non nulle).

THEOREME. Les conditions:

(i) $A(x)$ inversible, $\forall x \in V$ ouvert dense de R^N

(ii) $\Delta_f \otimes Lg_{i_\nu} \otimes \ldots \otimes Lg_{i_1} \Delta_f^{d_i} h_i |_x = 0, \quad \forall x \in \mathbb{R}^N, \ \forall i \in \{1,\ldots,q\}, \ \forall \nu \geq 2, \ i_\nu \ldots i_1 \in \{1,\ldots,q\}$

sont nécessaires et suffisantes pour la résolution, avec bouclage linéaire analytique et $\beta(x)$ inversible, du problème posé de commande non interactive du système Σ.

Démontrons tout d'abord un lemme préliminaire, pendant discret d'un résultat connu en temps continu.

LEMME 4. Si $A(x)$ est non singulière, alors l'application $\tau : R^N \to R^{\sum_{i=1}^{q} d_i+q}$ définie par:

$$\tau(x) = \mathrm{col}(h_i|_x,\dots,\Delta_f^{d_1}h_1|_x,\dots,h_q|_x,\dots,\Delta_f^{d_q}h_q|_x)$$

est de rang plein égal à $\sum_{i=1}^{q} d_i + q \leq N$.

DEMONSTRATION DU LEMME 4. Supposons que la matrice jacobienne de $\tau(x)$ soit singulière. Alors il existe $\sum_{i=1}^{q} d_i+q$ fonctions analytiques $\gamma_o^1,\dots,\gamma_{d_1}^1,\dots,\gamma_o^q,\dots,\gamma_{d_q}^q$ telles que:

$$\sum_{i=1}^{q}\ \sum_{j=0}^{q_i} \gamma_j^i d\Delta_f^j h_i = 0 \tag{9}$$

Composons à gauche par l'opérateur Δ_f, il vient:

$$\sum_{i,j} \Delta_f(\gamma_j^i)\Delta_f(d\Delta_f^j h_i) = 0$$

et d'après le lemme 1:

$$\sum_{i,j} \Delta_f(\gamma_j^i)d(\Delta_f^{j+1}h_i) = 0 \tag{10}$$

L'opérateur différentiel:

$$P_k^1 = \Delta_f \otimes L g_k \Delta_f^{-1} \quad , \qquad (k = 1,\dots,q)$$

est un opérateur différentiel du premier ordre (cf. les auteurs [8]), il résulte si $h : R^N \to R$ est analitique:

$$P_k^1(h)|_x = L_{P_k^1(x)} h = dh|_x\ P_k^1(x)$$

Multiplions l'expression (10) par la fonction $P_k^1(x)$, il vient

$$\sum_{i,j} \Delta_f(\gamma_j^i)d(\Delta_f^{j+1}h_i)P_k^1(x) = 0$$

et ainsi

$$\sum_{i,j} \Delta_f(\gamma_j^i)\Delta_f \otimes Lg_k \Delta_f^{-1}\Delta_f(\Delta_f^j h_i) = 0$$

$$\sum_{i,j} \Delta_f(\gamma_j^i)\Delta_f \otimes Lg_k \Delta_f^j h_i = 0$$

D'après la définition de d_i, on obtient pour $j = d_i$ et $\forall k \in \{1,...,q\}$

$$\sum_{i=1}^{q} \Delta_f(\gamma_{d_i}^i)\Delta_f \otimes Lg_k \Delta_f^{d_i - 1} h_i = 0$$

qui s'écrit sour la forme matricielle:

$$[\Delta_f(\gamma_{d_1}^1),...,\Delta_f(\gamma_{d_q}^q)]A(x) = 0$$

Par hypothèse $A(x)$ est non singulière, $(I+f)$ est inversible, il en résulte:

$$\gamma_{d_i}^i = 0 \ , \quad \forall i \in \{1,...,q\}$$

Reprenons l'égalité (9) avec $\gamma_{d_i}^i = 0$, $\forall i \in \{1,...,q\}$, il vient,

$$\sum_{i=1}^{q} \sum_{j=0}^{d_i - 1} \gamma_j^i d(\Delta_f^j h_i) = 0$$

Composons à gauche par l'opérateur Δ_f^2, il vient,

$$\sum_{i,j} \Delta_f^2(\gamma_j^i) d(\Delta_f^{j+2} h_i) = 0$$

Multiplions à gauche par $P_k^1(x)$, $(\forall k = 1,...,q)$, il vient,

$$\sum_{i=1}^{q} \sum_{j=0}^{d_i - 1} \Delta_f^2(\gamma_j^i)\Delta_f \otimes Lg_k \Delta_f^{j+1} h_i = 0$$

qui s'écrit sous forme matricielle et pour $j = d_i - 1$

$$[\Delta_f^2(\gamma_{d_1 - 1}^1),...,\Delta_f^2(\gamma_{d_{q-1}}^q)]A(x) = 0$$

On en déduit $\gamma_{d_i - 1}^i = 0$, $\forall i \in \{1,...,q\}$ et ainsi de suite $\gamma_j^i = 0$, $\forall i \in \{1,...,q\}$, $\forall j \in \{0,...,d_i\}$.

PREUVE DU THEOREME. Notons $V = \{x \in \mathbb{R}^N$ t.q. $A(x)$ non singulière$\}$ V ouvert dense de R^N, et notons $\Omega(x)$, l'ensemble des commandes admissibles: $\Omega(x) = \{\underline{v} \in \mathbb{R}^q$ t.q. $x+\tilde{f}(x)+\tilde{g}(x)\underline{v} \in V\}$ où $\tilde{f}$ et $\tilde{g}$ définissent le système bouclé Σ_β.

CONDITION SUFFISANTE. Posons $\forall x \in V$

$$\alpha(x) = -[A(x)]^{-1} \begin{bmatrix} \Delta_f^{d_1+1} h_1|_x \\ \vdots \\ \Delta_f^{d_q+1} h_q|_x \end{bmatrix} \quad \beta(x) = [A(x)]^{-1}$$

En reprenant des techniques utilisées en [4], on montre aisément que pour tout $x \in V$, pour tout $\underline{v} \in \Omega(x)$ la loi de bouclage,

$$u(x) = \alpha(x) + \beta(x)\underline{v}$$

et la transformation de coordonnées,

$$T : V \longrightarrow W \text{ ouvert dense de } R^N$$

définie par:

$$T(x) = \mathrm{col}(h_1|_x, \ldots, \Delta_f^{d_1} h_1|_x, \ldots, h_q|_x, \ldots, \Delta_f^{d_q} h_q|_x, \phi_1(x), \ldots, \phi_\ell(x))$$

$$z = \mathrm{col}(z_o^1, \ldots, z_{d_1}^1, \ldots, z_o^q, \ldots, z_{d_q}^q, z_1^{q+1}, \ldots, z_\ell^{q+1})$$

où les fonctions ϕ_i sont choisies linéairement indépendantes des précédentes et

$$\sum_{i=1}^{q+1} n_i = N;$$

permettent de représenter le système Σ sous la forme voulue (2).

En effet, $\forall i \in \{1, \ldots, q\}$,

$$z_j^i(t+1) = \Delta_f^j h_i|_{x(t+1)} = \Delta_{\tilde{f}} \Delta_f^j h_i|_{x(t)}, \quad \forall j = 0, \ldots, d_{i-1}$$

$$z_j^i(t+1) = \Delta_f^{j+1} h_i|_{x(t)} \qquad \text{(définition de } d_i\text{)}$$

$$= z_{j+1}^i(t)$$

$$\vdots$$

$$z_{d_i}^i(t+1) = \Delta_{\tilde{f}} \Delta_f^{d_i} h_i|_{x(t)} + \sum_{j=1}^{q} \Delta_{\tilde{f}} \otimes L\tilde{g}_j \Delta_f^{d_i} h_i|_{x(t)}$$

$$\stackrel{\text{(ii)}}{=} \Delta_f^{d_i+1} h_i|_{x(t)} + \sum_{\ell=1}^{q} (\alpha_\ell + \sum_{j=1}^{q} \beta_{\ell j} v_j) \Delta_f \otimes L g_\ell \Delta_f^{d_i} h_i|_{x(t)}$$

$$= v_i(t) \qquad \text{(définition du bouclage).}$$

On obtient la représentation linéaire:

$$\left\{\begin{array}{l}
\underline{z}^1(t+1) = \begin{bmatrix} 010 & & 0 \\ & \ddots & 1 \\ 0 & \cdots & 0 \end{bmatrix} \underline{z}^1(t) + \begin{bmatrix} 0 \\ \vdots \\ 0 \\ 1 \end{bmatrix} v_1(t) \\
\vdots \\
\underline{z}^q(t+1) = \begin{bmatrix} 010 & & 0 \\ & \ddots & 1 \\ 0 & \cdots & 0 \end{bmatrix} \underline{z}^q(t) + \begin{bmatrix} 0 \\ \vdots \\ 0 \\ 1 \end{bmatrix} v_q(t) \\
\\
y_i(t) = [1, 0 \cdots 0]\,\underline{z}^i(t) \\
\\
\underline{z}^i(t) \in \mathbb{R}^{n_i}, \quad (\forall i \in \{1,\ldots,q\})
\end{array}\right. \tag{11}$$

complétée d'une partie non linéaire inobservable de dimension n_{q+1}

$$\underline{z}^{q+1}(t+1) = \tilde{f}_{q+1}(z(t);\underline{v}(t))$$

où

$$\forall i \in \{1,\ldots,P\},\quad z_i^{q+1}(t+1) = \Delta_{\tilde{f}+\tilde{g}\underline{v}}(\phi_i)\Big|_{T^{-1}(z(t))}$$

$$= \Delta_{T^{-1}-I}\,\Delta_{\tilde{f}+\tilde{g}\underline{v}}(\phi_i)\Big|_{z(t)}$$

CONDITION NECESSAIRE. Par hypothèse le système Σ admet après bouclage linéaire analytique avec $\beta(x)$ inversible, une représentation de la forme (2). On en déduit immédiatement qu'il est simulé par bouclage et avec l'application

$$\tau(x) = \mathrm{col}(h_1|_x,\ldots,\Delta_f^{n_1-1} h_1|_x,\ldots,h_q|_x,\ldots,\Delta_f^{n_q-1} h_q|_x)$$

par le système linéaire (11) de dimension $\sum_{i=1}^{q} n_i$.

Notons de façon préliminaire l'identité:

$$d_i = n_i - 1$$

Pour celà, supposons $d_i > n_i - 1$, alors d'après la définition de d_i, $\forall k < n_i - 1$, $\forall j \in \{1,\ldots,q\}$:

$$\Delta_f \otimes Lg_j \Delta_f^k h_i = 0$$

et d'après le lemme 2, il suit:

$$\Delta_f \otimes Lg_j \Delta^k_{\tilde f} h_i = 0$$

On en déduit:

$$\Delta_{\tilde f} \otimes L\tilde g_i \Delta^k_{\tilde f} h_i = \sum_{j=1}^{q} \beta_{ji} \Delta_{\tilde f} \otimes Lg_j \Delta^k_{\tilde f} h_i = 0 \ , \quad \forall k < n_i$$

Pour $k = n_i-1$ la structure du système (2) conduit à la contradiction:

$$\Delta_{\tilde f} \otimes L\tilde g_i \Delta_{\tilde f}^{n_i-1} h_i = 1$$

qui entraîne $d_i \leq n_i-1$.

Supposons $d_i < n_i-1$, alors d'après la structure du système (2)

$$\Delta_{\tilde f} \otimes L\tilde g_j \Delta_{\tilde f}^{d_i} h_i = 0, \quad \forall i \in \{1,\dots,q\}, \quad \forall j \in \{1,\dots,q\}$$

ce qui entraîne avec $\beta(x)$ inversible

$$\Delta_f \otimes Lg_j \Delta_f^{d_i} h_i = 0, \quad \forall i \in \{1,\dots,q\}, \forall j \in \{1,\dots,q\}$$

ceci est impossible d'après la définition de d_i, il en résulte $d_i = n_i-1$.

Le système linéaire (11) ainsi construit avec $d_i = n_i-1$ satisfait de manière évidente la condition (7) du lemme 3 et la condition (ii) du théorème est ainsi vérifiée.

Il reste à montrer la non singularité de la matrice $A(x)$. Par hypothèse chaque sortie y_j $(j = 1,\dots,q)$, du système bouclé Σ_β ne dépend pas des commandes v_i, $(i = 1,\dots,q)$ pour $j \neq i$, on a:

$$\Delta_{\tilde f} \otimes L\tilde g_i \Delta^k_{\tilde f} h_j = 0 \ , \quad \forall k \geq 0, \quad i \neq j$$

Pour $k = d_j$ et d'après (ii) du théorème, ceci implique:

$$\Delta_f \otimes L\tilde g_i \Delta_f^{d_i} h_i = \Delta_{\tilde f} \otimes L\tilde g_i \Delta_f^{d_i} h_i = 1$$

et l'on obtient l'égalité matricielle:

$$A(x)\beta(x) = \begin{bmatrix} \Delta_f \otimes L\tilde g_1 \Delta_f^{d_1} h_1 |_x & & 0 \\ & \ddots & \\ 0 & & \Delta_f \otimes L\tilde g_q \Delta_f^{d_q} h_q |_x \end{bmatrix} = I, \ \forall x \in V$$

avec $\beta(x)$ non singulière; il en résulte que $A(x)$ est non songulière.

REMARQUE. Si l'on considère un bouclage polynomial analytique de la forme:

$$u(x) = \alpha(x)+\beta_1(x)\underline{v}+\beta_2(x)\underline{v}\cdot\underline{v}+ \ldots$$

où

$$\underline{v}\cdot\underline{v} \in \mathbb{R}^{q\times q} = (v_1v_1,\ldots,v_1v_q,\ldots,v_qv_1,\ldots,v_qv_q)$$

et $\beta_1(x)$ inversible, la condition algébrique:

$$A(x) \text{ non singulière}$$

est nécessaire et suffisante pour la résolution du problème de commande non interactive avec bouclage polynomial analytique. De plus, les deux formulations du problème posées dans l'introduction coïncident. En effet, il est possible, avec ce type de bouclage, de passer d'un système non linéaire Σ à un système linéaire de la forme (2). On calcule alors la loi de bouclage en généralisant à partir des formules d'inversion de Lagrange de séries formelles multivariables [9] des techniques proposées par les auteurs [5] dans le cas d'une entrée scalaire.

Soit, à titre d'exemple, le système bilinéaire:

$$\begin{cases} x(t+1) = A_o x(t) + \sum_{i=1}^{q} u_i(t)A_i x(t) \\ y(t) = C_i x(t) \;, \quad (i = 1,\ldots,q) \end{cases}$$

Supposons les indices relatifs définis pour tout $i \in \{1,\ldots,q\}$ d_i est le plus petit entier tel que,

$$C_i A_o^k A_j = 0, \quad \forall k < d_i \;, \quad \forall i \in \{1,\ldots,q\}$$

et

$$\exists j \in \{1,\ldots,q\} \text{ t.q. } C_i A_o^{d_i} A_j \neq 0.$$

Ces équations d'état étant linéaires en x, la condition (ii) du théorème est toujours vérifiée. La condition

$$A(x) \text{ non singulière}$$

$$A(x) = \begin{bmatrix} C_1 A_o^{d_1} A_1 x, \ldots, C_1 A_o^{d_1} A_q x \\ \vdots \qquad\qquad \vdots \\ C_q A_o^{d_q} A_1 x, \ldots, C_q A_o^{d_q} A_q x \end{bmatrix}$$

est une condition nécessaire et suffisante pour la résolution du problème avec la loi de bouclage:

$$\alpha(x) = -[A(x)]^{-1} \begin{bmatrix} C_1 A_o^{d_1} x \\ \vdots \\ C_q A_o^{d_q} x \end{bmatrix}$$

$$\beta(x) = [A(x)]^{-1}.$$

REFERENCES

[1] F.L. FALB et W.A. WOLOVICH. Decoupling in the design and synthesis of multivariable control systems. IEEE Trans. Aut. Cont., 12, 1967, 651-659.

[2] A. ISIDORI, A.J. KRENER, C. GORI-GIORGI et S. MONACO. Nonlinear decoupling via feedback: a differential geometric approach. IEEE Trans. Aut. Cont., 26, 1981, 331-345.

[3] D. CLAUDE. Découplage des systèmes du linéaire au non linéaire dans "Outils et Modèles Mathématiques pour l'Automatique, l'Analyse des Systèmes et le Traitement du Signal", (I.D. Landau ed.), ed. CNRS Paris, 3, 1983.

[4] S. MONACO et D. NORMAND-CYROT. The immersion under feedback of a multidimensional discrete-time nonlinear system into a linear system. Int. J. Cont., 28, 1983, 245-261.

[5] S. MONACO et D. NORMAND-CYROT. Formal power series and input-output linearization of nonlinear discrete time systems. CDC 1983, San Antonio.

[6] D. NORMAND-CYROT. Théorie et pratique des systèmes non linéaires en temps discret, Thèse d'Etat, Université Paris Sud, 1983.

[7] M. FLIESS et I. KUPKA. A finiteness criterion for nonlinear input output differential systems. SIAM J. on Cont. and Optimi., 21,1983, 721-728.

[8] S. MONACO et D. NORMAND-CYROT. On the realization of nonlinear discrete time systems. Rapport Université de Rome "La Sapienza" soumis pour publication.

[9] S.A. JONI. Lagrange inversion in higher dimensions and umbral operators. Linear and Multilinear Algebra. 6, 111-121, 1978.

A FAST ALGORITHM FOR SYSTEMS DECOUPLING USING FORMAL CALCULUS

F. GEROMEL*, J. LEVINE**, P. WILLIS*

ABSTRACT : The feedback decoupling problem of nonlinear systems is actually well understood in a theoretic point of view. However, to compute the decoupling feedbacks, apart of [9] the only method known by the authors, consists in using a formal derivation program to check if differential expressions are null [3]. We firstly recall the generic interpretation of these expressions in terms of the graph of the system and recall the algorithm of [9] using the minimal length of the paths joining one of the inputs to the ith output. Secondly, we describe the program, and give an application to the control of robot arms.

(*) Ecole Polytechnique
91128 PALAISEAU

(**) Centre d'Automatique et d'Informatique
Ecole Nationale Supérieure des Mines de Paris
35, Rue Saint-Honoré
77305 FONTAINEBLEAU - FRANCE

A - THEORY

I - The feedback decoupling problem :

We consider a linear-analytic system, given in local coordinates, by :

$$(\Sigma) \quad \begin{cases} \dot{x} = f_o(x) + \sum_{i=1}^{N} u^i f_i(x) + \sum_{j=1}^{N} w^j g_j(x) \\ y_k = h_k(x), \quad k = 1,\dots,p \end{cases}$$

where x belongs to a connected n-dimensional analytic manifold X, $u = (u^1,\dots,u^N)^T$ are the input functions, $h_1,\dots,h_p$ are the output functions, analytic on X and where :

$$\begin{cases} F_i(x) = \sum_{j=1}^{n} f_i^j(x) \frac{\partial}{\partial x_j}, \quad i = o,\dots,N \\ G_j(x) = \sum_{k=1}^{n} g_j^k(x) \frac{\partial}{\partial x_k}, \quad j = 1,\dots,M \end{cases} \qquad (1)$$

are analytic vector fields on X.

The <u>feedback decoupling problem</u> consists in finding analytic functions (α^i, β_i^j), $i=1,\dots,N$, $j=1,\dots,N$, eventually defined on an open subset $\mathcal{O}$ of X such that the feedback control :

$$u^i(x) = \alpha^i(x) + \sum_{j=1}^{N} \beta_i^j(x) v_j, \quad i = 1,\dots,N \tag{2}$$

makes the p outputs $y_1,\dots,y_p$ locally independent of w^i, $i=1,\dots,M$. We shall denote $\hat{F}_i, i=0,\dots,N$, the vector fields obtained by the feedback (2) :

$$\begin{aligned} \hat{f}_o(x) &= f_o(x) + \sum_{i=1}^{N} \alpha^i(x) f_i(x), \quad \hat{F}_o(x) = \sum_{j=1}^{N} \hat{f}_o^j(x) \frac{\partial}{\partial x_j} \\ \hat{f}_i(x) &= \sum_{j=1}^{N} \beta_i^j(x) f_j(x), \quad \hat{F}_i(x) = \sum_{j=i}^{n} \hat{f}_i^j(x) \frac{\partial}{\partial x_j}, \quad i=1,\dots,N. \end{aligned} \tag{3}$$

The problem is actually well understood and the differential geometric methods [4] together with the algebraic ones [1] draw an almost complete picture of the theoretic solution. In the geometric approach of the "structural" decoupling, we introduce the maximal involutive distribution $\mathcal{D}$ of constant rank, which is $(F_o, F_1,\dots,F_N)$-invariant. Isidori, Krener, Gori-Giorgi and Monaco [4] have proved the following :

<u>Theorem 1</u> : The structural decoupling problem has a local solution if and only if :

$$\text{span}\,\{G_1,\dots,G_M\} \subset \mathcal{D} \subset \bigcap_{i=1}^{P} \ker dh_i. \tag{4}$$

Furthermore, $\mathcal{D}$ can be obtained by the following induction (see [5]) :

$$\mathcal{D}_o = \text{span}\,\{dh_1,\dots,dh_p\} \tag{5}$$

$$\mathcal{D}_k = \sum_{i=0}^{N} L_{F_i} \mathcal{D}_{k-1} + \mathcal{D}_{k-1} \tag{6}$$

where L_{F_i} is the Lie derivative with respect to the vector field F_i, and

$$\mathcal{D} = (\bigcup_{k \geq 0} \mathcal{D}_k)^{\perp}. \quad \blacksquare \tag{7}$$

The algebraic methods, using Fliess' input-output map representation, give a "functional" point of view : in place of a distribution, one looks for a module $\mathcal{M}$ of vector fields, playing the same role as the distribution $\mathcal{D}$ but eventually with a non constant rank (see [1]). Claude [1] has proved the following :

<u>Theorem 2</u> : The outputs $y_1,\dots,y_p$ are decoupled with respect to $w^1,\dots,w^M$, if and only if there exists an analytic module $\mathcal{M}$ which is also a Lie subalgebra of vector fields on X such that :

$$\forall i : [\hat{F}_i, \mathcal{M}] \subset \mathcal{M}, \text{ and } \{G_1,\dots,G_M\} \subset \mathcal{M} \subset \mathcal{H} \tag{8}$$

with $\hat{F}_i$ defined by (3), and $\mathcal{H} = \{\varphi$: vector field on $X \mid \varphi(h_i) = 0 \quad \forall i = 1,\dots,p\}$. ■

Furthermore, α and β can be computed in a purely algebraic way (that is to say without solving differential or partial differential equations) by the procedure described hereafter.

For this purpose, we need the :

Definition 1 : The characteristic number ρ_i of order i is the unique integer satisfying :

$$\exists j \in \{1,\dots,N\} : F_j F_o^{\rho_i} h_i \not\equiv 0, \text{ and :} \tag{9}$$

$$\forall j \in \{1,\dots,N\},\ \forall m \in \{0,\dots,\rho_i - 1\},\quad F_j F_o^m h_i \equiv 0. \tag{10}$$

If $F_j F_o^m h_i \equiv 0 \quad \forall j, \forall m$, we set $\rho_i = +\infty$, and if $\exists j : F_j h_i \not\equiv 0$, $\rho_i = 0$. ■

Remark that $F_o^m h_i = F_o(F_o^{m-1} h_i)$ is a polynomial of differentials of h_i up to the order m, and that $F_o^0 h_i \equiv h_i$. ρ_i can be interpreted as the minimal number of integrations such that y_i is affected by one of the u_j.

To compute α and β, we introduce the following quantities :

$$\Delta_i^j(x) = F_j(x) F_o^{\rho_i}(x) h_i(x), \quad i = 1,\dots,p, \quad j = 1,\dots,N \tag{11}$$

$$\varphi_i(x) = \tilde{\varphi}_i(h_i(x), F_o(x)h_i(x), \dots, F_o^{\rho_i}(x)h_i(x)) - F_o^{\rho_i+1}(x)h_i(x), i=1,\dots,p \tag{12}$$

$$\psi_i^j(x) = \tilde{\psi}_i^j(h_i(x), F_o(x)h_i(x), \dots, F_o^{\rho_i}(x)h_i(x)), \quad i=1,\dots,p, \quad j=1,\dots,N \tag{13}$$

with $\tilde{\varphi}_i$ and $\tilde{\psi}_i^j$ arbitrary analytic functions.

Let us call : Δ the $p \times N$ matrix-valued analytic function whose $(i,j)^{th}$ element is Δ_i^j, $\varphi = (\varphi_1,\dots,\varphi_p)^T$, and ψ the $p \times N$ matrix-valued analytic function whose $(i,j)^{th}$ element is ψ_i^j.

Theorem 3 : If $G_k F_o^m h_i \equiv 0 \quad \forall k \in \{1,\dots,M\}$, $\forall m \leq \rho_i$, a necessary and sufficient condition for (α,β) to realize the local functional decoupling of (Σ), is that (α,β) locally solve the system :

$$\begin{aligned} \Delta\alpha &= \varphi \\ \Delta\beta &= \psi \end{aligned} \tag{14}$$

In this case, the change of variables :

$$X_o^i = h_i, \dots, X_{\rho_i}^i = F_o^{\rho_i} h_i, \quad i = 1,\dots,p, \tag{15}$$

Puts the system (Σ) locally into the form :

$$\left\{\begin{array}{l} \dot{X}^i_o = X^i_1 \\ \vdots \\ \dot{X}^i_{\rho_i-1} = X^i_{\rho_i} \\ \dot{X}^i_{\rho_i} = \tilde{\varphi}_i(X^i_o,\dots,X^i_{\rho_i}) + \sum_{j=1}^{N} \tilde{\varphi}^j_i(X^i_o,\dots,X^i_{\rho_i})v_j \\ y_i = X^i_o \end{array}\right. \tag{16}$$

$i = 1,\dots,p.$ ∎

Clearly, this procedure involves a huge amount of formal calculus, especially to determine the characteristic numbers ρ_i , $i = 1,\dots,p$: one must differentiate ρ_i times the expressions h_i, $F_o h_i$, etc..., whose complexity is growing very fast, and then check if $F_j F_o^m h_i$ is null or not. A program has been developed by Claude and Dufresne [3], using the language MACSYMA, to compute these formal expressions.

The aim of this paper is to introduce a faster method to compute ρ_i with the minimal number of formal differentiations : for this purpose, we shall prove that the numbers ρ_i can generically be very easily obtained on the system's graph. We shall also give a lower bound ν_i for ρ_i in the non-generic case, still obtained from the graph, and prove that either $\nu_i \leq n-1$ or $\nu_i = \rho_i = +\infty$. These results are finally synthesized in an algorithm to compute (α,β).

II - The system's graph :

As in [6], we introduce the following system's graph :

Definition 2 : We call Γ the system's graph of Σ in a given open subset $\mathcal{O}$ of X with given local coordinates, the oriented graph whose input-nodes are $(u^1,\dots,u^N,w^1,\dots,w^M)$, whose intermediate-nodes are the state variables $(x_1,\dots,x_n)$, and whose output-nodes are $(y_1,\dots,y_p)$. The oriented arcs of Γ are obtained as follows :

- There exists an oriented arc joigning u^i to x_k iff $f^k_i(x) \not\equiv 0$ in $\mathcal{O}$, $i = 1,\dots,N$, $k = 1,\dots,n$, and joinging w^i to x_k iff $g^k_i(x) \not\equiv 0$ in $\mathcal{O}$, $i = 1,\dots,M$, $k = 1,\dots,n$.
- There exists an oriented arc joining x_k to x_j iff $\frac{\partial f^j_o}{\partial x_k}(x) \not\equiv 0$ in $\mathcal{O}$, $j, k = 1,\dots,n$.
- There exists an oriented arc joining x_k to y_i iff $\frac{\partial h_i}{\partial x_k}(x) \not\equiv 0$ in $\mathcal{O}$, $i = 1,\dots,p$, $k = 1,\dots,n$. ∎

Definition 3 : We call $d(u^j,y_i)$ the minimal number of oriented arcs of Γ forming an oriented path joining u^j to y_i, and $d_i = \min_{1\leq j\leq N} d(u^j,y_i)$. ∎

An introductory example : $n = 3$, $N = 1$, $M = 1$, $p = 1$, $f_1(x) \not\equiv 0$, $\frac{\partial f_o^2}{\partial x_1}(x_1,x_2) \not\equiv 0$.

$$(17)\quad \begin{cases} \dot{x}_1 = f_o^1(x_1,x_2,x_3) + u\, f_1(x_1,x_2,x_3) \\ \dot{x}_2 = f_o^2(x_1,x_2) \\ \dot{x}_3 = f_o^3(x_1,x_2,x_3) + w g_1(x_1,x_2,x_3) \\ y_1 = h(x_2) \end{cases}$$

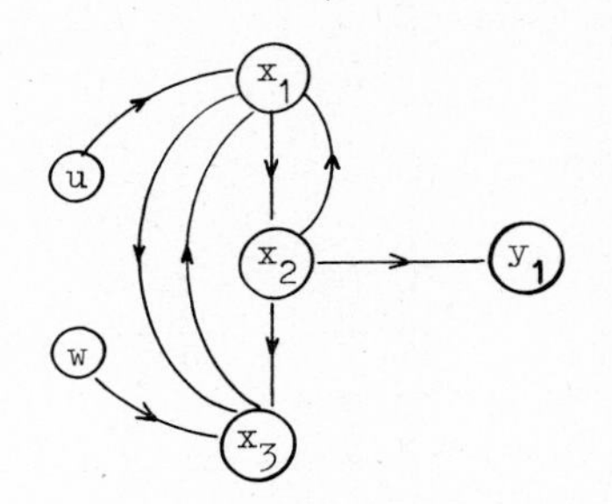

The system's graph Γ

It can be easily seen that $d_1 = d(u,y_1) = 3$, $d(w,y_1) = 4$.
We shall prove that we can predict that, generically, $\rho_1 = d_1 - 2 = 1$, and that $d(w,y_1) > d_1$ implies $G_1 h \equiv 0$ and $G_1 F_o h \equiv 0$. To check our assertion, let us go back to (9) and (10), and compute ρ_1. We first check that $(F_1 h)(x) = f_1(x)\frac{\partial h}{\partial x_1}(x_2) \equiv 0$, $(G_1 h)(x) = g_1(x)\frac{\partial h}{\partial x_3}(x_2) \equiv 0$. Then : $(F_o h)(x) = f_o^1(x)\frac{\partial h}{\partial x_1}(x_2) + f_o^2(x_1,x_2)\frac{\partial h}{\partial x_2}(x_2) + f_o^3(x)\frac{\partial h}{\partial x_3}(x_2) = f_o^2(x_1,x_2)\frac{\partial h}{\partial x_2}(x_2)$, $(F_1 F_o h)(x) = f_1(x)\frac{\partial f_o}{\partial x_1}(x_1,x_2)\frac{\partial h}{\partial x_2}(x_2) \not\equiv 0$, and thus $\rho = 1$;
Finally, we also have $G_1 F_o h = g_1\left(\frac{\partial f_o^2}{\partial x_3}\frac{\partial h}{\partial x_2} + f_o^2 \frac{\partial^2 h}{\partial x_3 \partial x_2}\right) \equiv 0$, as claimed above.

Thus, almost without computations, ρ_1 and the relations $G_1 h \equiv 0$ and $G_1 F_o h \equiv 0$, can be deduced from the system's graph (we only need to compute $F_1 F_o h$!). Clearly, the system's graph synthesizes the structure of the interactions of the input and output variables versus integration of the state variables. Thus, it is not surprising that, in general, but generically only, the minimal length d_i represents the minimum number of integrations for the inputs to affect y_i, namely ρ_i up to a constant equal to 2 since the first and last arcs do not represent integrations.

Remark : in Γ, we do not take into account the fact that $f_1,\dots,f_N,g_1,\dots,g_M$ depend on $x_1,\dots,x_n$ or not. For our purpose these interactions do not play any role in generic situations and, if they play a role in non-generic cases, the profit of the graph's method vanishes, as will be seen after.

III - The characteristic numbers ρ_i, their lower bounds ν_i, and the system's graph

Besides the characteristic numbers ρ_i, we shall introduce the numbers ν_i defined as follows :

Definition 4 : The number ν_i, $i = 1,\dots,p$, is the unique integer satisfying :

$$\exists j \in \{1,\dots,N\},\quad \exists k_{\nu_i},\dots,k_o \in \{1,\dots,n\} \quad \text{such that :}$$

$$f_j^{k_{\nu_i}} \frac{\partial f_o^{k_{\nu_{i-1}}}}{\partial x_{k_{\nu_i}}} \cdots \frac{\partial f_o^{k_o}}{\partial x_{k_1}} \frac{\partial h_i}{\partial x_{k_o}} \not\equiv 0 \text{ in } \mathcal{O}, \tag{18}$$

and : $\forall j \in \{1,\dots,N\}$, $\forall r < \nu_i$, $\forall k_r,\dots,k_o \in \{1,\dots,n\}$, we have :

$$f_j^{k_r} \frac{\partial f_o^{k_{r-1}}}{\partial x_{k_r}} \cdots \frac{\partial f_o^{k_o}}{\partial x_{k_1}} \frac{\partial h_i}{\partial x_{k_o}} \equiv 0 \text{ in } \mathcal{O} . \quad \blacksquare \tag{19}$$

Now we can state the main result :

Theorem 4 : • $\nu_i = d_i - 2$, $i = 1,\dots,p$. (20)

• $\nu_i \leqslant \rho_i$ and $\nu_i = \rho_i$ generically, $i = 1,\dots,p$. (21)

By generically, we mean : for every system Σ whose coefficients $f_o, f_1,\dots,f_N, g_1,\dots,g_M, h_1,\dots,h_p$, lie outside a closed subset, with empty interior, of the space of analytic vector-valued functions on $\mathcal{O} \subset X$, the functions f_o, f_i, g_i, h_i, depending locally on the same variables as those of the original system .

Corollary : If $\nu_i > n-1$, then $\rho_i = \nu_i = +\infty$

Remark 1 : From the corollary, we conclude that ν_i is computed in at most n-1 steps, and, generically, the same holds for ρ_i . The result for ρ_i was proved in [2]. However, it is remarkable that one can have $\nu_i < n-1$ whereas $\rho_i = +\infty$ as the following example proves :

$$\begin{cases} \dot{x}_1 = u x_1 \\ \dot{x}_2 = - u x_2 \\ y = x_1 x_2 \end{cases}$$

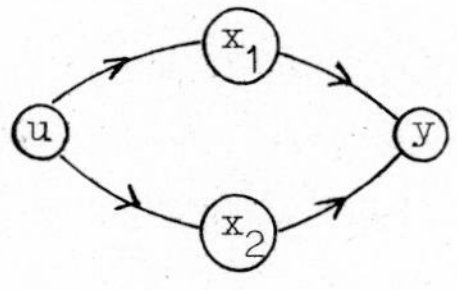

- Γ -

It is very easy to see that $\nu = 0$, but $\rho = +\infty$ since

$$F_1 h = x_1 \frac{\partial(x_1 x_2)}{\partial x_1} - x_2 \frac{\partial(x_1 x_2)}{\partial x_2} \equiv 0 ,$$

and $F_1 F_o h \equiv 0$ since $F_o \equiv 0$. Finally, this suffices to prove that $\rho = +\infty$ since $n = 2$. ■

Remark 2 : It would be a nice result, if $\rho_i > \nu_i$, that there exists a (non minimal) oriented path from one of the u^j to y_i of length $\rho_i + 2$.
Unfortunately, this is only true for linear systems. A counterexample in the non linear case :

$$\begin{cases} \dot{x}_1 = x_4 + ux_1 \\ \dot{x}_2 = x_2 w - x_2 u \\ \dot{x}_3 = -x_3 w \\ \dot{x}_4 = x_5 \\ \dot{x}_5 = u \\ y = x_1 x_2 x_3 \end{cases}$$

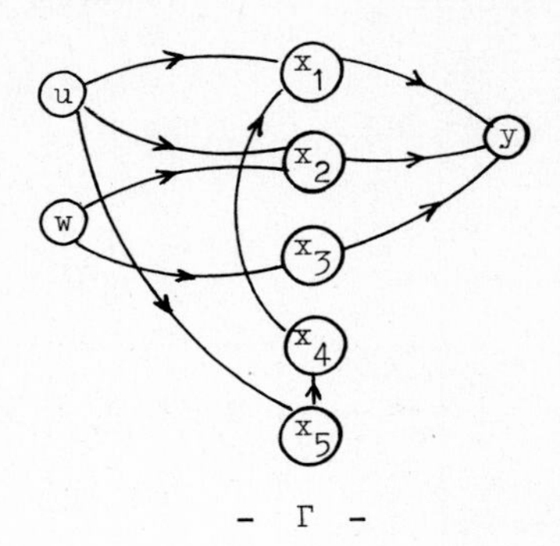

- Γ -

We have $\nu = 0$, whereas $\rho = 1$:

$$F_1 h = x_1 \frac{\partial}{\partial x_1}(x_1x_2x_3) - x_2 \frac{\partial}{\partial x_2}(x_1x_2x_3) \equiv 0, \quad G_1 h = x_2 \frac{\partial}{\partial x_2}(x_1x_2x_3) - x_3 \frac{\partial}{\partial x_3}(x_1x_2x_3) \equiv 0,$$

$$F_o h = x_2x_3x_4 \ , \ F_1F_o h = - x_2x_3x_4 \not\equiv 0.$$

But it can be seen that, in Γ, there is no oriented path joining u to y with length equal to $\rho + 2 = 3$. The only path of length larger than 2 is (u, x_5, x_4, x_1, y) of length 4. Thus, if $\rho_i > \nu_i$, we see that the graph does not give anymore information on ρ_i. However, to compute $F_1 F_o^r h_i$ with $r > \nu_i$, and if there is no path of length $r + 2$ in Γ, it is no need to compute the terms of the form (18) (with r in place of ν_i) since if there were a non zero expression in these terms, there should exist a path of length $r + 2$, which contradicts our assumption. ∎

Remark 3 : the two preceding examples give a good illustration of non-generic systems : in both there were orthogonality relations between F_1 and h, so that the expressions (18) are $\not\equiv 0$, but their sum is 0. Of course, this is non generic, for if we change, for example in Remark 1, ux_1 in $(1+\varepsilon)ux_1$, we obtain : $F_1 h = (1+\varepsilon)x_1x_2 - x_1x_2 = \varepsilon x_1x_2 \not\equiv 0$. ∎

Remark 4 : It is worth noting that if $r < \nu_i$, we necessarily have $F_j F_o^r h_i \equiv 0$ $\forall j = 1, \ldots, N$. In the same way, going back to the system (17) of the introductory example, we have $d(w, y_1) = 4$, and thus $G_1 F_o^r h \equiv 0 \quad \forall r < 4-2 = 2$. Also, this remark is useful to avoid computing a number of formal expressions : if ν_i or, more precisely d_i, is obtained only for paths joining $u^{j_1}, \ldots, u^{j_r}$ to y_i, one can be sure that $F_k F_o^{\nu_i} h_i \equiv 0 \quad \forall k \neq j_1, \ldots, j_r$, and one needs to check only those expressions $F_{j_1} F_o^{\nu_i} h_i, \ldots, F_{j_r} F_o^{\nu_i} h_i$ for minimal paths. ∎

IV - Description of the algorithm :

All the following computations must be done formally, for example with the languages MACSYMA or REDUCE.

1. The graph Γ

To avoid a complete construction of Γ with a number of useless nodes and arcs, one can determine $d_i = \nu_i + 2$, and $\mathcal{U}_i$ the subset of the $(u^1,\dots,u^N)$ corresponding to the minimal paths, directly from the data of Σ, and by a dynamic programming method :

- Starting from $y_i(i = 1,\dots,p)$, we build every incident arc with : $0 \not\equiv \frac{\partial h_i}{\partial x_{k_o}}$
  Then, for every x_{k_o} such that $(x_{k_o},y_i) \in \Gamma$, we test if there is an arc (u^j,x_{k_o}) in Γ by $f_j^{k_o} \not\equiv 0$. If $(u^j,x_{k_o}) \in \Gamma$, then $d_i = 2$, $\nu_i = 0$ and $u^j \in \mathcal{U}_i$.
- If $(u^j,x_{k_o}) \notin \Gamma$ $\forall j$, we change y_i into x_{k_o}, and build every incident arc to x_{k_o} by $\frac{\partial f_o^{k_o}}{\partial x_{k_1}} \not\equiv 0$; then again, for every x_{k_1} such that $(x_{k_1},x_{k_o}) \in \Gamma$,

we test if there is an arc (u^j,x_{k_1}) in Γ by $f_j^{k_1} \not\equiv 0$, and so on. If there is no arc from u^j, $\forall j$, to every path of length $\leqslant n-1$, then $\nu_i = +\infty$.

The same procedure can be done in parallel to determine $\min_{1\leqslant j\leqslant M} d(w^j,y_i) = \mu_i$, and $\mathcal{W}_i$ the subset of the $(w^1,\dots,w^M)$ corresponding to the $(\mu_i + 2)$ length in Γ.

2. Computation of ρ_i and the matrix Δ.

- We first compute $F_j F_o^{\nu_i} h_i$ $\forall j$ such that $u^j \in \mathcal{U}_i$.
  Two cases can happen :
  - either $F_j F_o^{\nu_i} h_i \not\equiv 0$ for at least one j with $u^j \in \mathcal{U}_i$.
    Then $\rho_i = \nu_i$, $\Delta_i^j = F_j F_o^{\nu_i} h_i$ $\forall j$ such that $u^j \in \mathcal{U}_i$
    $= 0$ $\forall j$ such that $u^j \notin \mathcal{U}_i$.
    If $\nu_i = +\infty$, then $\rho_i = +\infty$ and the ith line of Δ can be deleted.
  - or $F_j F_o^{\nu_i} h_i \equiv 0$, $\forall j$ such that $u^j \in \mathcal{U}_i$
    Then $\rho_i > \nu_i$ and one must compute $F_j F_o^r h_i$ $\forall r > \nu_i$, $\forall j = 1,\dots,N$, until the moment when one of these expressions becomes non 0 (ρ_i is then equal to the corresponding r) or until $r = n-1$ if every expression is null (then $\rho_i = +\infty$).
    If ρ_i is finite, the ith line of the matrix Δ is obtained by computing every expression (11) for $j = 1,\dots,N$.
    If $\rho_i = +\infty$, one can delete the ith line of Δ.

3. The comparison between ρ_i and μ_i.

- If $\rho_i < \mu_i$, we have $G_j F_o^m h_i \equiv 0$ $\forall m \leqslant \rho_i$, $\forall j$.
- If $\rho_i \geqslant \mu_i$, we have to look further if

$$G_j F_o^{\mu_i} h_i \equiv 0 \quad \forall j \text{ such that } w^j \in \mathcal{W}_i ,$$

and after if $G_j F_o^m h_i \equiv 0 \quad \forall m = \mu_i + 1, \ldots, \rho_i , \quad \forall j$.

Two cases can happen :

- $G_j F_o^m h_i \equiv 0 \quad \forall m \leq \rho_i, \forall j,$ then the decoupling problem has a local solution iff the system(14) has a local solution (α,β).
- $\exists j_o \in \{1,\ldots,M\}, \exists m_o < \rho_i$ such that $G_{j_o} F_o^{m_o} h_i \not\equiv 0$, then the decoupling problem has no solution, and the system is finitely decoupled up to the order m_o , $\forall(\alpha,\beta)$. (see [2]).

4. Inversion of the system (14). Same as in [3]. ■

Remark 5 : If $\nu_i < \rho_i$, and if Σ has a large dimension, it can be useful, in the evaluation of $F_j F_o^m h_i$ with $m > \nu_i$, to remark that if there is no path of length $m + 2$ joining u^j to y_i in Γ, every expression (18) with m in place of ν_i is necessarily null. Thus, we eliminate this way n formal differentiations in $F_j F_o^m h_i$. ■

Remark 6 : It is clear that this method is more efficient for larger ν_i's and larger n, N, M, p. If $\nu_i = \rho_i$ and if $\mathcal{U}_i$ does not contain too many elements, we need a very low number of formal derivations and the efficiency of this method is the highest. On the other hand, if $\nu_i < \rho_i$, since a minimal length in Γ is computed much faster than a formal derivation, the economy of time grows with ν_i. ■

Remark 7 : For linear systems, the graph's method can be significantly improved since ρ_i can be completely obtained from the graph : in place of step 2 of the algorithm, we have :

If $\nu_i < \rho_i$: delete every path of length $\nu_i + 2$, in the graph, and find the new minimal length $\nu_i' > \nu_i$. Check if $\exists j$ such that $F_j F_o^{\nu_i'} h_o \not\equiv 0$.

If yes, $\nu_i' = \rho_i$. If not, delete again every path of length $\nu_i' + 2$ and so on, until every path of length $\leq n + 2$ is deleted, then $\rho_i = + \infty$.

B - THE PROGRAM

I - Organization of the program

The programming language is MACSYMA.

The programming is made of :

1) The main program : prob ( )

It asks questions to the user and decides which subroutines to run.

2) The subroutines :

a) expli( ) : gives informations, if needed, on the program's use.

b) mit( ) : memorizes the formal equations of the system .

c) calnu( ) : computation of ν by the graph's method

d) calro( ) : computation of ρ

e) pmutcal( ) - trig( ) - resoud( ) : solves the Δ - system after reorganization of lines and columns of Δ

f) feedback( ) : gives the final result on the feedbacks.

For further informations and examples of sessions, see [8]

II - Session's display

At each step of the session, the user may choose between different tasks :

1) at the beginning : the inputs can be checked and corrected, and the user can ask for further informations (expli( )).

2) during the session : the user must answer the questions of recognizing null expressions. For example the program cannot check the nullity of an expression such as $x_1 \frac{\partial f}{\partial x_1} - x_2 \frac{\partial f}{\partial x_2}$ when f is not specified.

3) at the end : the user may help the program to simplify the results, for example, by giving rules of trigonometric simplifications. The user must be avare of the fact that some simplificiations automatically done by MACSYMA may be sometimes worse than no simplification at all.

Remark : In order to protect the intermediate results from manipulation's errors, the main program saves them step by step in auxiliary files.

C - EXAMPLE : THE ROBOT ARM

We study the decoupling problem for a 3 degrees of freedom robot arm. It is composed of three segments of length l_1, l_2, l_3. The links have relative angles noted x_1, x_2, x_3 and x_4, x_5 and x_6 are the respective angular velocities. The cartesian coordinates of the extremity are y_1, y_2, y_3 and we wish to control its motion along the y_1 - axes - such a problem arises in automatic sizing.
The motion's equations ([7]) are :

$$\dot{x}_1 = x_4$$
$$\dot{x}_2 = x_5$$
$$\dot{x}_3 = x_6$$
$$\dot{x}_4 = (u_1 - f_1(x_2,x_3,x_4,x_5,x_6)) \cdot \frac{1}{b_{11}(x_2,x_3,x_4)}$$
$$\dot{x}_5 = (b_{22}(x_3,x_5,x_6)u_2 + b_{32}(x_3,x_5,x_6\ u_3 + + b_{22}(x_3,x_5,x_6)f_2(x_2,x_3,x_4,x_5)) \cdot \frac{1}{\det(x_3,x_5,x_6)}$$

$$\dot{x}_6 = (b_{23}(x_3,x_5,x_6)u_2 + b_{33}(x_3,x_5,x_6)u_3 + \\ + b_{32}(x_3,x_5,x_6)f_3(x_3,x_4,x_5,x_6))\frac{1}{\det(x_3,x_5,x_6)}$$

with $\det(x_3,x_5,x_6) = (b_{22}b_{33} - b_{23}b_{32})(x_3,x_5,x_6)$.

The outputs are :

$$y_1 = \cos x_1\,(l_3\sin(x_3+x_2)+l_2\sin x_2)$$

$$y_2 = \sin x_1\,(l_3\sin(x_3+x_2)+l_2\sin x_2)$$

$$y_3 = l_3\cos(x_3+x_2)+l_2\cos x_2+l_1$$

The program finds that : $\nu_1 = \nu_2 = \nu_3 = 1$ and $\rho_1 = \rho_2 = \rho_3 = 1$ and that :

$$\det\Delta = \frac{l_2 l_3 \sin x_3}{b_{11}(x_2,x_3,x_4)\det(x_3,x_5,x_6)}\,(l_3\sin(x_3+x_2) + l_2\sin x_2)$$

The expressions of α_i and β_{ij}, $1\leq i,\ j\leq 3$, expressed as functions of $\varphi_1,\varphi_2,\varphi_3$ and $\psi_{11},\psi_{12},\psi_{33}$ as in (12),(13) are :

$$\alpha_1 = -\,[2l_3b_{11}\cos(x_2+x_3)x_4(x_6+x_5)+2l_2\cos(x_2)b_{11}\;x_4x_5 - l_3f_1\sin(x_3+x_2) - l_2f_1\sin x_2 + \\ +(\varphi_1\sin(x_1) - \varphi_2\cos(x_1))b_{11}] \times \frac{1}{l_2\sin(x_3+x_2)+l_2\sin x_2}$$

$$\alpha_2 = [(l_3^2b_{33}+(l_2l_3\cos x_3+l_3^2)b_{32})x_6^2 + (2l_3^2b_{33}+(2l_2l_3\cos x_3+2l_3^2)b_{32})x_5x_6 + \\ + ((l_2l_3\cos(x_3)+l_3^2)b_{33}+(2l_2l_3\cos(x_3)+l_3^2+l_2^2)b_{32})x_5^2 \\ + ((l_3^2b_{33}+l_3^2b_{32})\sin^2(x_3+x_2)+ \\ (l_2l_3\sin(x_2)b_{33}+2l_2l_3\sin(x_2)b_{32})\sin(x_2+x_3)+l_2^2\sin^2(x_2)b_{32})x_4^2 \\ + ((\varphi_2l_3\sin(x_1)+\varphi_1l_3\cos(x_1))b_{33} + (\varphi_2l_3\sin(x_1)+\varphi_1l_3\cos(x_1))b_{32})\sin(x_3+x_2) \\ + \varphi_3l_3(b_{33}+b_{32})\cos(x_3+x_2) \\ + ((\varphi_2l_2\sin(x_1)+\varphi_1l_2\cos x_1)\sin x_2+\varphi_3l_2\cos x_2)b_{32} - l_2l_3f_2\sin x_3] \times \frac{1}{l_2l_3\sin x_3} \\ + \frac{f_3b_{32}^2 - f_2b_{23}b_{32}}{\det(x_3,x_5,x_6)}$$

$$\alpha_3 = [f_2b_{22}b_{23}-f_3b_{22}b_{32}]/\det(x_3,x_5,x_6)$$

$$- [(l_3^2b_{23}+(l_2l_3\cos x_3+l_3^2)b_{22})x_6^2 + (2l_3^2b_{23}+(2l_2l_3\cos x_3+2l_3^2)b_{22})x_5x_6$$

$$+ ((l_2l_3\cos x_3+l_3^2)b_{23}+(2l_2l_3\cos(x_3)+l_3^2+l_2^2)b_{22})x_5^2$$

$$+ ((l_3^2b_{23}+l_3^2b_{22})\sin^2(x_3+x_2)+(l_2l_3\sin(x_2)b_{23}$$

$$+2l_2l_3\sin(x_2)b_{22})\sin(x_3+x_2)+l_2^2\sin^2x_2b_{22})x_4^2$$

$$+ ((\varphi_2l_3\sin(x_1)+\varphi_1l_3\cos(x_1))b_{23}+(\varphi_2l_3\sin(x_1)+\varphi_2l_3\cos(x_1))b_{22})\sin(x_3+x_2)$$

$$+ \varphi_3l_3(b_{23}+b_{22})\cos(x_3+x_2)$$

$$+ ((\varphi_2l_2\sin(x_1)+\varphi_1l_2\cos(x_1))\sin x_2+(\varphi_3l_2\cos(x_2))b_{22}] \times \frac{1}{l_2l_3\sin x_3}$$

$$\beta_{11} = -\frac{\psi_{11}\sin(x_1)b_{11}}{l_3\sin(x_3+x_2)+l_2\sin(x_2)}$$

$$\beta_{12} = \frac{\psi_{22}\cos(x_1)b_{11}}{l_3\sin(x_3+x_2)+l_2\sin(x_2)}$$

$$\beta_{13} = 0$$

$$\beta_{21} = \psi_{11}\cos x_1(l_3(b_{33}+b_{32})\sin(x_3+x_2)+l_2\sin(x_2)b_{32}) \times \frac{1}{l_2l_3\sin x_3}$$

$$\beta_{22} = \psi_{22}[l_3(b_{33}+b_{32})\sin(x_3+x_2)+l_2\sin x_2b_{32}] \times \frac{\sin x_1}{l_2l_3\sin x_3}$$

$$\beta_{23} = \psi_{33}(l_3(b_{33}+b_{32})\cos(x_3+x_2)+l_2\cos(x_2)b_{32}) \times \frac{1}{l_2l_3\sin x_3}$$

$$\beta_{31} = -\psi_{11}\cos x_1(l_3(b_{23}+b_{22})\sin(x_3+x_2)+l_2\sin x_2b_{22}) \times \frac{1}{l_2l_3\sin x_3}$$

$$\beta_{32} = -\psi_{22}\sin x_1(l_3(b_{23}+b_{22})\sin(x_3+x_2)+l_2\sin x_2b_{22}) \times \frac{1}{l_2l_3\sin x_3}$$

$$\beta_{33} = -\psi_{33}(l_3(b_{23}+b_{22})\cos(x_3+x_2)+l_2\cos x_2b_{22}) \times \frac{1}{l_2l_3\sin x_3}$$

Remark : In this example a direct computation would need 198 partial derivations and 198 multiplications, whereas with the graph's method, only 85 partial derivations and 56 multiplications have been executed. For other examples see [8].

CONCLUSION

We have proved that the feedback decoupling method of Claude and Dufresne [3] can be significantly simplified by the introduction of the system's graph. This graph has the property that the minimal length d_i between the ith output and the inputs $(u^1,\dots,u^N)$, is generically equal to the ith characteristic number ρ_i plus 2, and in general smaller or equal to $\rho_i + 2$. This property can be used to avoid a number of formal computations and is all the more efficient as d_i is large. This method requires the use of MACSYMA because of the formal manipulations, and gives an efficient CAO tool for non linear systems decoupling.

REFERENCES

[1] D. CLAUDE. Decoupling of nonlinear systems. Syst. and Contr. Letters. Vol.1, n°4 (1982), 242-248.

[2] D. CLAUDE. Decouplage des systèmes : du linéaire au nonlinéaire, in : Developpement et utilisation d'outils et modèles mathématiques en automatique, analyse des systèmes et traitement du signal. Vol.3, I.D. Landau ed., CNRS, Paris, 1983, 533-555.

[3] D. CLAUDE, P. DUFRESNE. An application of Macsyma to nonlinear systems decoupling Lecture Notes in Computer Sciences, Vol.144, Springer, 1982, 294-301.

[4] A. ISIDORI, A. KRENER, C. GORI-GIORGI, S. MONACO. Nonlinear decoupling via feedback. IEEE Trans. AC. Vol. AC26, n°2 (1981), 331-345.

[5] A. ISIDORI. The geometric approach to nonlinear feedback control : a survey. Analysis and Optimization of Systems. Lecture Notes in Control and information sciences n°44, Springer, 1982, 517-531.

[6] D. SILJAK. On reachability of dynamic systems. Int. J. Syst. Sc. Vol.8, n°3, (1977), 321-338.

[7] S. NICOSIA, F. NICOLO, D. LENTINI. Dynamical control of industrial robots with elastic and dissipative joints. 8th IFAC World Congress - Kyoto - 1981.

[8] F. GEROMEL, P. WILLIS. Algorithme de graphe pour le découplage de systèmes non-linéaires. Option Automatique. Ecole Polytechnique. Promotion 80. Juin 83.

[9] A. KASINSKY, J. LEVINE. A fast graph theoretic algorithm for the feedback decoupling problem of nonlinear systems. 8th MTNS conference. Beersheva. June 1983.

INPUT-OUTPUT DECOUPLING OF NONLINEAR SYSTEMS WITH AN APPLICATION TO ROBOTICS

H. Nijmeijer
Twente University of Technology
P.O. Box 217
7500 AE Enschede,
The Netherlands

J.M. Schumacher
Econometric Inst., Erasmus Univ.
P.O. Box 1738
3000 DR Rotterdam
The Netherlands

1. Control of robot motion

A major challenge that now faces robot technology is the creation of a generation of robots that are able to perform a variety of tasks in an environment that is subject to changes and uncertainty. Standard techniques are already available for programming robots that are made to repeat the same motion over and over again, under carefully controlled external conditions. For instance, one may use "teaching by showing": the robot arm is guided through its movements, and the corresponding joint motions are recorded so that they can be played back in the actual operation. Such methods, however, soon become unwieldy even if only slight variations occur in the task that has to be performed. Moreover, it may be very costly or outright impossible to control the working conditions of the robot in such a way that simple repetition of movements leads to satisfactory results. In these cases, some form of sensory feedback has to be used.
To get ahead in this direction, one needs to develop more sophisticated techniques of motion specification. Paths for most existing robots are specified by defining a sequence of intermediate points ("via points") which lead from the initial point to the end point of some desired trajectory. More advanced methods [21] call for "functionally defined motion", in which the trajectory is specified as a function from time to space coordinates. This specification can be done by the programmer, but possibly also (via a control algorithm) by the signals that are received from sensors.
The approach using functionally defined motion appears to be the natural way to deal with varying robot tasks and to incorporate sensory feedback. However, the difficulty of this approach is in the amount of computing time that is needed to translate the desired trajectory

into servo signals. At the moment, this computing time is too long to allow real-time applications of the method in any generality [10]. The problem is due to the complexity of the translation process, which we shall now describe in some detail.

The following figure views the robot as an input-output system with the signals to the servos as inputs, and the Cartesian coordinators of the arm tip (or whatever object one is interested in) as outputs.

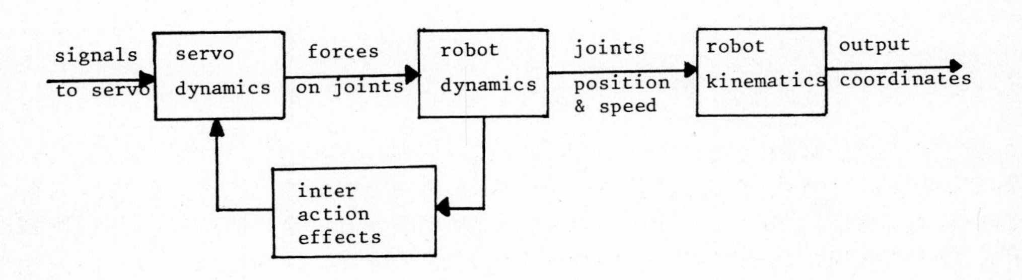

- diagram 1 -

If one wants to translate a desired trajectory into corresponding input signals, one may attempt to read this diagram from right to left. First, the trajectory (which is specified as a function of time into coordinates, so that the derivatives of the motion are also determined) has to be transformed into corresponding time functions that indicate position and speed of the joints. Although this is purely a kinematic problem, the amount of computation involved may already be very considerable; cf. [21]. Recently, Brockett [1] has proposed an interesting method for doing the kinematic transform via a differential equation; this may be helpful, in particular to deal with situations where desired trajectories are fed in in real time.

The second step consists of translating the motion of the joints into forces (either rotational or translational, depending on the nature of the joints) that have to be exerted in order to effect the desired motion. To do this, one needs a dynamic model of the robot. There are several ways to obtain such models, and we won't review these here. Quite generally, the equations to which these methods lead can be manipulated into the form

$$(1.1) \qquad \ddot{\Theta} = F(\Theta,\dot{\Theta}) + G(\Theta,\dot{\Theta})u$$

where Θ is a vector of joint positions, u is a vector of input forces, and G is a square invertible matrix depending on Θ and Θ. However, one also has to take into account the fact that the forces acting on the joints are produced by servo mechanisms that have dynamics of their own, which are moreover influenced by inherent feedback mechanisms (e.g., due to friction) from the robot dynamics. A specific model that is often used for servos is the first order model of the form

$$(1.2) \qquad \dot{u} = A_o(u) + B_o(u)w + E_o(\Theta,\dot{\Theta})$$

where w is a vector of signals to the servos, $B_o(u)$ is an invertible matrix, and $E_o(\Theta,\Theta)$ represents the interaction effects back from the robot dynamics into the servo dynamics.

The problem of finding the input signals that will produce a desired trajectory is, of course, of a feedforward nature. We won't discuss this problem directly in the present paper; rather, our attention is directed to ways in which feedback can be used to alleviate the complexity of the feedforward problem. In particular, we are interested in methods to eliminate the coupling that exists between the inputs and the outputs of the dynamical system that is represented by the robot. If we would be able to achieve a situation of noninteraction, so that every input affects just one output and none of the others, than the problem of trajectory following, in particular for trajectories that can be easily expressed in terms of the preferred coordinates corresponding to the selected outputs, would be greatly simplified.

A general theory for noninteracting control of nonlinear systems, based on state feedback, has been developed by the authors in [17,18]; cf. also [6,24]. A major problem in the application of this theory is the amount of calculation that is involved in the computation of the decoupling feedback. The development of specialized software seems to be necessary in order to make this approach practically feasible. Of course, the computational difficulty is a fast increasing function of the order of the system one is trying to decouple. Therefore, the fact that one should work with the combined robot and servo dynamics (1.1-2) rather than just the robot dynamics (1.1) is a cause for concern. In the present paper, we aim to show that - under certain conditions which are often fulfilled in practice - it is possible to neglect the servo dynamics first and then to derive a decoupling feedback law for the complete system in a relatively simple way.

Our approach in this is the following. First, assume that the servo dynamics can be written as a cascade of systems of the form (1.2), so that it is in fact sufficient to consider a single stage of this form. Next, note that feedback may be used to transform the dynamics (1.2) into a simple integration:

(1.3) $$w = (B_o(u))^{-1}(\hat{w}-A_o(u)-E_o(\Theta,\dot{\Theta}))$$

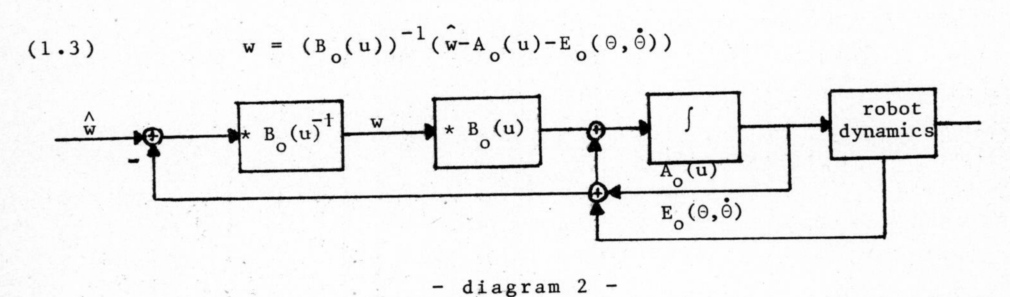

- diagram 2 -

This means that we can replace, without loss of generality, the equation (1.2) by

(1.4) $$\dot{u} = w$$

We consider the effect of adding this equation to the dynamics (1.1) with respect to the output equations

(1.5) $$z = H(\Theta,\dot{\Theta})$$

that represent the kinematics of the robot. In particular, we compare the zeros at infinity as defined in [18] for the two systems (1.1-5) and (1.1-4-5), and we derive conclusions for the decoupling problem. After presenting the necessary definitions and results in section 2, we do this in section 3. The theory is illustrated with an example in section 4.

2. Input-output decoupling.

We consider the nonlinear control system

(2.1) $$\dot{x} = A(x) + \sum_{i=1}^{m} B_i(x)u_i$$

(2.2) $$z_i = C_i(x), \quad i \in \underline{k}$$

where x is a vector of local coordinates for a smooth n-dimensional manifold M; $A, B_1, \ldots, B_m$ are smooth vector fields on M; $u_1, \ldots, u_m$ are piecewise smooth input functions; and for each $i \in \underline{k}$, $C_i : M \to N_i$ is a smooth surjective submersion into a p_i-dimensional manifold N_i ($p_i > 0$).

In this section, we explain the noninteracting control problem and mention some of the results that have been obtained for this problem. We shall work under the following standard assumptions.

Assumption 1. Writing $\Delta_o = \text{span}\{B_1, \ldots, B_m\}$, we have

(2.3) $$\dim \Delta_o = m.$$

Assumption 2. The rank of the map $C = (C_1, \ldots, C_k): M \to N_1 \times \ldots \times N_k$ equals $p_1 + + \ldots + p_k$.

Assumption 3. The system (2.1) satisfies the strong accessibility rank condition [27].

The control laws that we consider are of the form (often called "static state feedback")

(2.4) $$u = \alpha(x) + \beta(x)v$$

where α and β are smooth functions from M to $\mathbb{R}^n$ and $\mathbb{R}^{m \times m}$, respectively, $\beta(x)$ is nonsingular for all $x \in M$, and v is an $\mathbb{R}^m$-valued - function of time that acts as the new input. If we insert (2.4) into (2.1) we obtain the new dynamics

(2.5) $$\dot{x} = \tilde{A}(x) + \sum_{i=1}^{m} \tilde{B}_i(x)v_i$$

with

(2.6) $$\tilde{A}(x) = A(x) + \sum_{i=1}^{m} B_i(x)\alpha_i(x)$$

$$(2.7)\qquad \tilde{B}_i(x) = \sum_{j=1}^{m} B_j(x)\beta_{ji}(x)$$

We say that noninteraction has been achieved in the system (2.5)-(2.2) if the following holds. The inputs are divided into k groups,

$$(2.8)\qquad \underline{m} = I_1 \cup \ldots \cup I_k$$

(overlappings are not excluded). We write $\underline{v}_j$ for the vector with components v_i, $i \in I_j$. Corresponding to the division (2.8), we form the distributions R_i $(i \in \underline{k})$ defined by

$$(2.9)\qquad R_i = \text{span}[\text{inv.clos}\{ad^k_{\tilde{A}}\tilde{B}_j, ad^k_{\tilde{B}_\ell}\tilde{B}_j \;\; k\varepsilon Z^+, j\varepsilon I_i, \ell\varepsilon\underline{m}\}]$$

These are <u>regular local controllability distributions</u> ([13]).
Now, the input $\underline{v}_i$ does not affect the outputs z_j, for $j \neq i$, if and only if

$$(2.10)\qquad R_i \subset \bigcap_{j\neq 1} \ker C_j^*.$$

(see [6]).
Moreover, the reachable set of output values in N_i by applying arbitrary inputs $\underline{v}_i$ (and for fixed arbitrary other inputs $\underline{v}_j, j \neq i$) has nonempty interior in N_i ("$\underline{v}_i$ controls z_i") if and only if (cf. [14,15])

$$(2.11)\qquad R_i + \ker C_{i*} = TM$$

We say that noninteraction has been achieved, if (2.10) and (2.11) hold for all $i \in \underline{k}$. So the <u>static state feedback noninteracting control problem</u> can be stated as follows:

Given the system (2.1-2), find, if possible, a feedback law (2.4) and a partitioning (2.8) such that the corresponding regular local controllability distributions as defined in (2.9) satisfy (2.10) and (2.11) for all $i \in \underline{k}$.

A regular local controllability distribution can always be written in the form appearing in (2.9) for a suitable choice of feedback (2.4) and a suitable selection of input vector fields $\tilde{B}_j$. So instead of looking for a feedback directly, we can also try to look first for a set of regular local controllability distributions. A particular set of interest is the one defined as follows.

Definition 2.1 [13]. For each $i \in \underline{k}$, we let R_i^* denote the maximal regular local controllability distribution contained in $\bigcap_{j \neq i} \ker C_{j*}$.

Although each of these distributions separately can be written in the form (2.9) for some feedback (2.4), the question is whether there exists a single feedback which will do for all of the R_i^*.
This is called the compatibility problem. The following theorem, proved in [18], tells us exactly when this problem can be solved (cf. also the original linear version in [11]).

Theorem 2.2. Consider the system (2.1-2) and assume that the distributions $\sum_{i \in I} R_i^*$ have constant dimension for all subsets I in $\underline{k}$.

Then the static state feedback noninteracting control problem can be solved locally if and only if

$$(2.12) \qquad \Delta_o = \sum_{i \in \underline{k}} \Delta_o \cap R_i^*$$

So the condition (2.12) guarantees that the distributions R_i^* are compatible. It is immediate (cf. [13]) that (2.12) is equivalent to

$$(2.13) \qquad \Delta_o = \sum_{i \in \underline{k}} \Delta_o \cap D_i^*$$

where D_i^* denotes the largest local controlled invariant distribution in $\bigcap_{j \neq i} \ker C_{j*}$ ([6,12]). Because the D_i^* are easier to compute than the R_i^*, the form (2.13) is more convenient in applications. Another equivalent formulation can be given that is based on the theory of "zeros at infinity" for nonlinear systems, as developed in [18]. To formulate the definition, consider a general affine system given, in local coordinates, by

$$(2.14) \qquad \dot{x} = A(x) + \sum_{i=1}^{m} B_i(x) u_i$$

$$(2.15) \qquad y = C(x)$$

Recall the "V^*-algorithm" [12]:

(2.16) $$V^o = TM$$

(2.17) $$V^{\mu+1} = \ker C_* \cap \{X \varepsilon V(M) \mid [A,X] \;\varepsilon\; V^{\mu} + \Delta_o \text{ and } [B_i,X] \;\varepsilon\; V^{\mu} + \Delta_o, i \varepsilon \underline{m}\}.$$

Let a sequence of integers $\{p^{\mu}\}(\mu \;\varepsilon\; \mathbb{N})$ be defined by

(2.18) $$p^{\mu} = \dim(\Delta_o \cap V^{\mu-1}) - \dim(\Delta_o \cap V^*)$$

where V^* is the limit value of the non-increasing sequence of distributions defined by (2.16-17). We can now define, as in [18], assuming that the dimensions appearing in (2.18) are constant:

Definition 2.3. The orders of the zeros at infinity of the system (2.14-15) are given by the nonzero members of the sequence $\{n^{\mu}\}$ $(\mu \;\varepsilon\; \mathbb{N})$ defined by

(2.19) $$n^{\mu} = \{j \mid p^j \geq \mu\}.$$

The sequence $\{p^{\mu}\}$ can be recovered from the sequence $\{n^{\mu}\}$ by the rule

(2.20) $$p^{\mu} = \{j \mid n^j \geq \mu\}.$$

It turns out that, in the noninteraction problem, the integers p^{μ} are somewhat easier to work with than the integers n^{μ}.
Let, for $i \;\varepsilon\; \underline{k}$, $\{p_i^{\mu}\}$ be the sequence of integers defined in (2.18), taken with respect to the system (2.1) with output $z_i(x) = C_i(x)$. Also, let $\{p^{\mu}\}$ be the corresponding sequence for the system (2.1) with output $z(x)$ consisting of all components
$z_i(x) = C_i(x)$, $i \;\varepsilon\; \underline{k}$.
Denote by V_I^{μ} $(I \subset \underline{k})$ the result of performing the algorithm (2.16-17) on the system (2.1) with output $z_I(x)$ defined as the vector with components $z_i(x)$, $i \;\varepsilon\; I$.
The following result, which will be crucial below, was proved in [18] (Thm 5.1).

Theorem 2.4. Consider the system (2.1-2), and assume that the distributions V_I^{μ} and $V_I^* \cap \Delta_o$ have fixed dimension for all $\mu > 0$ and $I \subset \underline{k}$.

Then the condition (2.13) is equivalent to

(2.22) $$p^{\mu} = \sum_{i \varepsilon \underline{k}} p_i^{\mu} \qquad \text{for all } \mu > 0$$

Remarks.

(i) A specific open problem is to investigate what happens at singularity points of $\sum_{i \varepsilon I} R_i^*$, $I \subset \underline{k}$. The purpose of the present paper is not, however, to pursue the consequences of "singularities" or even to make a classification os singularities. The present authors believe that essential singularities will appear in any general constructive method to solve the decoupling problem and these need to be investigated separately.

(ii) For 1-dimensional outputs $y_i = C_i(x)$, $i \varepsilon \underline{k}$, i.e. $\dim N_i = 1$, $i \varepsilon \underline{k}$, the above result has also been obtained by an appropriate generalization of the Falb-Wolovich criterion (cf. [3]), see e.g. [28,29]. For a modern algebraic treatment using noncummutative generating power series, the reader may consult e.g. [30]. If rank $C_i = 1$, $i \varepsilon \underline{k}$, one can introduce numbers ρ_i, $i \varepsilon k$, the so-called characteristic numbers (sometimes called relative orders) defined by
$\rho_i = \min_{\rho \varepsilon \mathbb{N}} \{ad_A^{\rho} B_j C_i \neq 0 \text{ for an } j \varepsilon \underline{m}\}$.

For each $i \varepsilon \underline{k}$, the number ρ_i is the same as the order of the zero at infinity of the system $\dot{x} = A(x) + \sum_{i=1}^{m} B_i(x)u_i$, $y_i = C_i(x)$. But for multivariable output maps $C_i : M \rightarrow N_i$, i.e. $\dim N_i > 1$, the characteristic numbers are not well defined. A treatment of the general block decoupling problem as studied here from an algebraic point of view as in [30] does not seem to be available at present.

3. Actuator dynamics - the nonlinear case

Suppose that we add "first-order" actuator dynamics of the form

(3.1) $$\dot{u} = \gamma(x,u) + \delta(x,u)w$$

to the affine nonlinear system described by (2.1-2). In (3.1), we suppose that $\gamma(x,u)$ and $\delta(x,u)$ are smooth and that $\delta(x,u)$ is a square matrix which is invertible for all x,u. The new dynamics are given by

(2.1) and (3.1), and from the feedback transformation

(3.2) $$w = (\delta(x,u))^{-1}(\tilde{w}-\gamma(x,u))$$

it is clear that these dynamics are feedback equivalent to those given by (2.1) and

(3.3) $$\dot{u} = w$$

So we shall from now on work with the simpler form (3.3), remembering that the feedback laws that we obtain for the system (2.1-3.3) can be translated into corresponding laws for (2.1-3.1) via the transformation (3.2).

Let us complement the system (2.1) with a general output equation

(3.4) $$z = C(x)$$

We shall later apply the theory developed below to the various output mappings that play a role in the decoupling problem. After the actuator dynamics (3.3) have been added, we obtain a new system with output equation

(3.5) $$z = \tilde{C}(x,u) = C(x)$$

As a shorthand notation, let us write Σ for the system described by (2.1) and (3.4), and denote the system (2.1-3.3-3.5) by Σ^e.

Our first goal is to establish a direct correspondence between the structures at infinity, as defined in section 2, of Σ and of Σ^e. So we have to consider the relation between the $V^i(\Sigma)$'s and the $V^i(\Sigma^e)$'s (see the algorithm (2.16-17)).

The state space B of the system Σ^e can be considered as a fiber bundle π: B $\rightarrow$ M with the original input space $\mathbb{R}^m$ serving as the standard fiber, so that locally B can be identified with M $\times$ $\mathbb{R}^m$.

Using the local coordinates x for M that appear in (2.1), we can choose local coordinates for B as (x,u). Thus, in the fiber bundle π : B $\rightarrow$ M, the zero section is prescribed by gluing all these local charts together. A more sophisticated exposition of these arguments is given in [16]. We may conclude that any distribution D on M may also be considered as a distribution on B, given locally (on U M) as $D_U \times 0 \quad TU \times T\mathbb{R}^m \cong T(\pi^{-1}(U))$.

It is not our aim in this paper to consider singularities, so we shall take it as a <u>standing assumption</u> that the distributions $V^i(\Sigma)$ and $V^i(\Sigma^e)$ as well as their intersections with the input distributions of Σ and Σ^e (to be denoted by Δ_o and Δ_o^e, cf. Assumption 1 in Section 2) are of constant dimension.

Lemma 3.1. With the above notations and assumptions, we have, for $X \in V(M)$ and for all $k = 0,1,2, \ldots$:

$$(3.6) \qquad X \in V^k(\Sigma) \iff \exists\, U(x,u) \text{ s.t. } \begin{pmatrix} X(x) \\ U(x,u) \end{pmatrix} \in V^k(\Sigma^e).$$

Proof. The proof is by induction on k. For k = 0, the statement is trivially true. Now suppose it holds at k. To show that the implication holds from left to right also at k + 1, take $X \in V^{k+1}(\Sigma)$. By definition, we have

$$(3.7) \qquad C_* X = 0$$

$$(3.8) \qquad [A,X] \in V^k(\Sigma) + \Delta_o$$

$$(3.9) \qquad [B_i,X] \in V^k(\Sigma) + \Delta_o, \; i \in \underline{m}.$$

From (3.8), it follows that there exist functions $\alpha_j(x)$ such that

$$(3.10) \qquad [A,X] + \sum_{j=1}^{m} B_j \alpha_j \in V^k(\Sigma).$$

Also, it follows from (3.9) that for all $i \in \underline{m}$ there exists functions $\beta_{ij}(x)$ such that

$$(3.11) \qquad [B_i,X] + \sum_{j=1}^{m} B_j \beta_{ij} \in V^k(\Sigma).$$

We now define an $\mathbb{R}^m$-valued function U(x,u) with components

$$(3.12) \qquad U_j(x,u) = \alpha_j(x) + \sum_{i=1}^{m} \beta_{ij}(x) u_i$$

Computation shows that

$$(3.13) \qquad \left[\begin{pmatrix} A(x)+B(x)u \\ 0 \end{pmatrix}, \begin{pmatrix} X(x) \\ U(x,u) \end{pmatrix}\right] = \begin{pmatrix} [A,X] + \sum_{i=1}^{m} [B_i,X]u_i - BU \\ * \end{pmatrix}$$

where we left the second component of the vector at the right hand side unspecified, because it is not important. The first component is in $V^k(\Sigma)$ by construction (for each value of u) so that it follows from the induction assumption that

$$(3.14)\qquad [\begin{pmatrix} A(x)+B(x)u \\ 0 \end{pmatrix}, \begin{pmatrix} X(x) \\ U(x,u) \end{pmatrix}] \in V^k(\Sigma^e) + \Delta_o^e.$$

Noting also that

$$(3.15)\qquad \tilde{C}_*\begin{pmatrix} X \\ U \end{pmatrix} = (C_* \quad 0)\begin{pmatrix} X \\ U \end{pmatrix} = C_* X = 0,$$

we have proved

$$(3.16)\qquad \begin{pmatrix} X \\ U \end{pmatrix} \in V^{k+1}(\Sigma^e)$$

For the converse, suppose that (3.16) holds for some U(x,u). Then it follows that $C_*X = 0$ as in (3.15). Computing as in (3.13), we find that

$$(3.17)\qquad \begin{pmatrix} [A,X](x)+\sum_{i=1}^{m}[B_i,X](x)u_i-\sum_{i=1}^{m}B_i(x)U_i(x,u) \\ * \end{pmatrix} \in V^k(\Sigma^e)+\Delta_o^e.$$

Since this should hold for all u, we find in particular that (setting u = 0)

$$(3.18)\qquad \begin{pmatrix} [A,X](x)-\sum_{i=1}^{m}B_i(x)U_i(x,0) \\ ** \end{pmatrix} \in V^k(\Sigma^e).$$

By the induction assumption, this implies

$$(3.19)\qquad [A,X] \in V^k(\Sigma) + \Delta_o.$$

From (3.17) and (3.18), it follows that

$$(3.20)\qquad \left(\sum_{i=1}^{m}[B_i,X](x)u_i-\sum_{i=1}^{m}B_i(x)\big(U_i(x,u)-U_i(x,0)\big)\right) \in V^k(\Sigma^e)$$

Setting u equal to each of the standard unit vectors in $\mathbb{R}^m$ and using the induction assumption again, we get

$$(3.21)\qquad [B_i,X] \in V^k(\Sigma) + \Delta_o,\ i \in \underline{m}.$$

It follows that $X \in V^{k+1}(\Sigma)$.

We note from the proof that it is always possible to let U be an affine function of u. The lemma is now applied to obtain the following result.

Theorem 3.2. With the above notations and assumptions, we have

$$\dim\left(V^{k+1}(\Sigma^e) \cap \Delta_o^e\right) = \dim(V^k(\Sigma) \cap \Delta_o) \tag{3.22}$$

for all $k = 0,1,2,\ldots$

Proof. We start by taking a basis of vector fields $\{X_1,\ldots,X_s\}$ for $V^k(\Sigma) \cap \Delta_o$:

$$X_j(x) = \sum_{i=1}^{m} B_i(x)\beta_{ij}(x). \tag{3.23}$$

Let $U_1, \ldots, U_s$ be $\mathbb{R}^m$-valued functions with components $(U_j)_i(x) = \beta_{ij}(x)$ ($j \in \underline{s}$). We claim that

$$\begin{pmatrix} 0 \\ U_j(x) \end{pmatrix} \in V^{k+1}(\Sigma^e) \cap \Delta_o^e \qquad (j \in \underline{s}). \tag{3.24}$$

This is immediately verified. First of all, we certainly have

$$\tilde{C}_*\begin{pmatrix} 0 \\ U_j \end{pmatrix} = 0 \tag{3.25}$$

Moreover,

$$\left[\begin{pmatrix} A(x)+B(x)u \\ 0 \end{pmatrix}, \begin{pmatrix} 0 \\ U_j(x) \end{pmatrix}\right] = \begin{pmatrix} X_j(x) \\ * \end{pmatrix} \in V^j(\Sigma^e) + \Delta_o^e \tag{3.26}$$

according to lemma 2.1. Finally, one has

$$\left[\begin{pmatrix} 0 \\ Z(x,u) \end{pmatrix}, \begin{pmatrix} 0 \\ U_j(x) \end{pmatrix}\right] = \begin{pmatrix} 0 \\ \frac{\partial Z}{\partial u}(x,u)U_j(x) \end{pmatrix} \in \Delta_o^e \tag{3.27}$$

for all smooth $\mathbb{R}^m$-valued functions Z. By the definition of $V^{k+1}(\Sigma^e)$, our claim is proved. Since the vector fields appearing in (3.24) are independent, it follows that

$$\dim(V^{k+1}(\Sigma^e) \cap \Delta_o^e) \geq \dim(V^k(\Sigma) \cap \Delta_o). \tag{3.28}$$

To prove the reverse inequality, we first use an extension of the theorem of Frobenius ([26]; cf. [7]) to show that there exists a basis of vector fields depending only on x for $V^{k+1}(\Sigma^e) \cap \Delta_o^e$. Both $V^{k+1}(\Sigma^e)$ (see [18]) and Δ_o^e are involutive, and so their intersection is involutive as well. Moreover, we observe that

$$(3.29) \qquad [\begin{pmatrix} 0 \\ U_1(x,u) \end{pmatrix}, \begin{pmatrix} 0 \\ U_2(x,u) \end{pmatrix}] = \begin{pmatrix} 0 \\ \frac{\partial U_2}{\partial u}(x,u)U_1(x,u) - \frac{\partial U_1}{\partial u}(x,u)U_2(x,u) \end{pmatrix}$$

so that all distributions on $\mathbb{R}^m$ obtained from $V^{k+1}(\Sigma^e) \cap \Delta_o^e$ by fixing x are involutive. Choosing straight coordinates by the ordinary Frobenius' theorem for every x, one is able to obtain a basis for $V^{k+1}(\Sigma^e) \cap \Delta_o^e$ of the form

$$(3.30) \qquad \{\begin{pmatrix} 0 \\ U_1(x) \end{pmatrix}, \ldots, \begin{pmatrix} 0 \\ U_\ell(x) \end{pmatrix}\}$$

Now, one has

$$(3.31) \qquad \begin{pmatrix} B(x)U_j(x) \\ 0 \end{pmatrix} = -[\begin{pmatrix} A(x)+B(x)u \\ 0 \end{pmatrix}, \begin{pmatrix} 0 \\ U_j(x) \end{pmatrix}] \in V^k(\Sigma^e) + \Delta_o^e.$$

By lemma 2.1, it follows that

$$(3.32) \qquad B(x)U_j(x) \in V^k(\Sigma)$$

for all j. Since the vector fields in (3.32) are obviously also in Δ_o and are independent, we see that the reverse in equality in (3.28) also holds, so that the proof is complete.

The theorem allows us to establish the connection between the orders of the zeros at infinity of Σ and those of Σ^e.

<u>Corollary 3.8</u>. With the above notations and assumptions, and recalling the definition of the zeros at infinity in Section 2, we have

$$(3.33) \qquad n^\mu(\Sigma^e) = n^\mu(\Sigma) + 1 \qquad (\mu \in \underline{m}).$$

<u>Proof</u>. It is immediate from the definitions and from the preceding theorem that (see (2.18))

$$(3.34) \qquad p^1(\Sigma^e) = p^1(\Sigma) = m$$

(3.35) $\quad p^{\mu+1}(\Sigma^e) = p^{\mu}(\Sigma) \quad (\mu=1,2,\ldots).$

The result (3.33) then follows from the definition (2.19).

Remark. In [18], it has been suggested to define zeros at infinity for a general nonlinear system, given locally by $\dot{x} = f(x,u)$, $y = h(x,u)$, using the following procedure. First, introduce input dynamics $\dot{u} = v$ so that an affine system with input v is obtained. Compute the orders of the zeros at infinity for this affine system, and finally subtract 1 in order to compensate for the integration that had been added. The above result shows that this proposed definition is consistent in the sense that, when applied to an affine system, it leads to the same orders of zeros at infinity as would have been obtained from the original (direct) definition.
We now have an immediate corollary for the decoupling problem.

Corollary 3.4. Consider the system (2.1-2) and also the extended system obtained by adding the input dynamics

(3.36) $\quad \dot{u} = v$

(or a feedback equivalent version of this). Assuming that the conditions of Thm. 2.4 hold for the original system as well as for the extended system, the static state feedback noninteracting control problem can be solved locally for the extended system if and only if this can be done for the original system (2.1-2).
The next question we want to address is, given to a decoupling feedback law for the system (2.1-2), how to obtain from this a decoupling law for the system with input dynamics added. Indeed, it turns out that such a law can be obtained by differentiation, be it that an inversion is also necessary to get the law in a feedback form.

Proposition 3.5. Let the system (2.1-2) be given, and suppose that

(3.37) $\quad u = \alpha(x) + \beta(x)w$

is a decoupling feedback law for it. Then a decoupling feedback law for (2.1-2) with input dynamics (3.36) added is given by

$$v = (L_A\alpha)(x) + \sum_{i=1}^{m} (L_{B_i}\alpha)(x)u_i +$$

(3.38)

$$+ ((L_A\beta)(x) + \sum_{i=1}^{m} (L_{B_i}\beta)(x)u_i)\beta^{-1}(x)(u-\alpha(x)) +$$

$$+ \beta(x)\tilde{w},$$

where $\tilde{w}$ is the new input.

Proof. The extended system is given by

(3.39) $$\binom{\dot{x}}{\dot{u}} = \binom{A(x)+B(x)u}{0} + \binom{0}{I}v; \quad \binom{x}{u}(0) = \binom{x_o}{u_o}$$

It can be verified by direct computation that the solution of (3.38-39)(which is unique, at least for small t) can be found by solving

(3.40) $$\dot{x} = A(x) + B(x)\alpha(x) + B(x)\beta(x)w; \; x(0) = x_o$$

(3.41) $$\dot{w} = \tilde{w}; \; w(0) = \beta^{-1}(x_o)(u_o-\alpha(x_o))$$

and setting

(3.42) $$u = \alpha(x) + \beta(x)w.$$

Now, we know that (3.40) is decoupled with respect to the outputs (2.2) and the input w. It follows that (3.40-41) is also decoupled with respect to the outputs (2.2) and the input $\tilde{w}$.
Since the outputs depend only on x, the equations (3.40-41;2.2) describe the same input-output behaviour as the equations (3.38-39;2.2). We conclude that the system (3.38-39;2.2) is decoupled.

An example of the differentiation procedure will be shown in the next section.

4. An example.

We present here an example of a two by two decoupling problem, intended to illustrate the kind of computations that one has to do according to the theory above, rather than to show the power of the theory in handling block decoupling problems (which would take more space to treat). One of the simplest nontrivial examples of a robotic mechanism is the double pendulum sketched below.

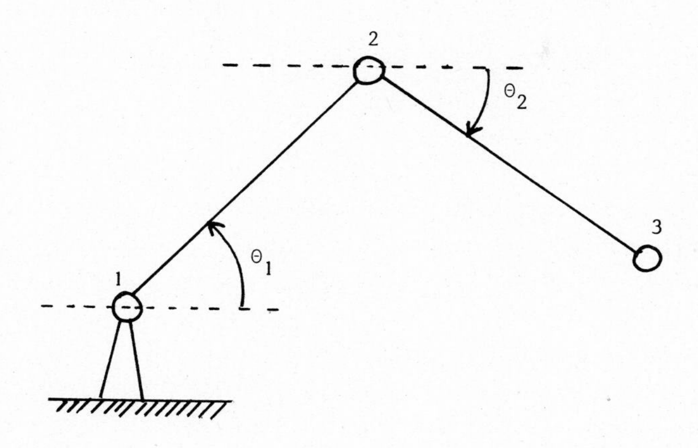

- diagram 3 -

Several models for this two-degrees-of-freedom structure have been derived in the literature, based on various assumptions about flexibility and various modeling methodologies. With the use of a finite element technique and under the assumption of rigid links, a model of the following form was obtained in [25].

$$(4.1)\qquad \begin{bmatrix} \alpha & \beta\cos\Delta\Theta \\ \beta\cos\Delta\Theta & \gamma \end{bmatrix}\begin{bmatrix} \ddot{\Theta}_1 \\ \ddot{\Theta}_2 \end{bmatrix} = -\begin{bmatrix} \beta\dot{\Theta}_2^2 \sin\Delta\Theta \\ \beta\dot{\Theta}_1^2 \sin\Delta\Theta \end{bmatrix} + \begin{bmatrix} u_1 \\ u_2 \end{bmatrix}$$

Here, Θ_1 and Θ_2 denote the angles that are indicated in the figure above; α, β and γ are constants depending on the weights and dimensions of various parts of the double pendulum, such that the matrix appearing on the left hand side of (4.1) is invertible for all Θ_1, Θ_2; $\Delta\Theta$ denotes $\Theta_2 - \Theta_1$; and u_1 and u_2 are the torques acting on the joints of the double pendulum. To keep the discussion as simple as possible, assume that the outputs are just the angles Θ_1 and Θ_2:

$$(4.2)\qquad z_1 = \Theta_1;\; z_2 = \Theta_2$$

It is immediately seen that the following is a decoupling control law for the system (4.1-2):

$$(4.3)\qquad \begin{bmatrix} u_1 \\ u_2 \end{bmatrix} = \begin{bmatrix} \beta\dot{\Theta}_2^2 & \sin\Delta\Theta \\ \beta\dot{\Theta}_1^2 & \sin\Delta\Theta \end{bmatrix} + \begin{bmatrix} \alpha & \beta\cos\Delta\Theta \\ \beta\cos\Delta\Theta & \gamma \end{bmatrix}\begin{bmatrix} v_1 \\ v_2 \end{bmatrix}.$$

This leads to the very simply input-output relation

$$(4.4)\qquad \ddot{z}_1 = v_1;\ \ddot{z}_2 = v_2$$

Let us now consider first-order actuator dynamics as in (3.1). As explained in the previous section, it is not a loss of generality to assume that we simply have a pure integration in both input channels:

$$(4.5)\qquad \dot{u}_1 = w_2;\ \dot{u}_2 = w_2.$$

To find a decoupling control law for the system (4.1-2-5) with w_1 and w_2 as inputs, we differentiate (4.3). This leads to

$$(4.6)\qquad \begin{bmatrix} \dot{u}_1 \\ \dot{u}_2 \end{bmatrix} = \begin{bmatrix} 2\beta\dot{\Theta}_2\cdot\ddot{\Theta}_2\sin\Delta\Theta + \beta\dot{\Theta}_2^2\cdot\Delta\dot{\Theta}\cdot\cos\Delta\Theta \\ 2\beta\dot{\Theta}_1\cdot\ddot{\Theta}_1\sin\Delta\Theta + \beta\dot{\Theta}_1^2\cdot\Delta\dot{\Theta}\cdot\cos\Delta\Theta \end{bmatrix} +$$

$$+ \begin{bmatrix} 0 & \beta\cdot\Delta\dot{\Theta}\cdot\sin\Delta\Theta \\ -\beta\cdot\Delta\dot{\Theta}\cdot\sin\Delta\Theta & 0 \end{bmatrix}\begin{bmatrix} v_1 \\ v_2 \end{bmatrix} +$$

$$+ \begin{bmatrix} \alpha & \beta\cos\Delta\Theta \\ \beta\cos\Delta\Theta & \gamma \end{bmatrix}\begin{bmatrix} \dot{v}_1 \\ \dot{v}_2 \end{bmatrix}.$$

Since $v_1 = \dot{\Theta}_1$ and $v_2 = \dot{\Theta}_2$, this simplifies to

$$(4.7)\qquad \begin{aligned} \begin{bmatrix} \dot{u}_1 \\ \dot{u}_2 \end{bmatrix} &= \begin{bmatrix} \beta\dot{\Theta}_2^2\cdot\Delta\dot{\Theta}\cdot\cos\Delta\Theta \\ \beta\dot{\Theta}_1^2\cdot\Delta\dot{\Theta}\cdot\cos\Delta\Theta \end{bmatrix} + \begin{bmatrix} \alpha & \beta\cos\Delta\Theta \\ \beta\cos\Delta\Theta & \gamma \end{bmatrix}\begin{bmatrix} \dot{v}_1 \\ \dot{v}_2 \end{bmatrix} + \\ &+ \begin{bmatrix} 0 & \beta(\dot{\Theta}_1+\dot{\Theta}_2)\sin\Delta\Theta \\ \beta(\dot{\Theta}_1+\dot{\Theta}_2)\sin\Delta\Theta & 0 \end{bmatrix}\begin{bmatrix} v_1 \\ v_2 \end{bmatrix}. \end{aligned}$$

In order to turn this into a decoupling control law for (4.1-2-5), we write w for $\dot{u}$ and $\tilde{w}$ for $\dot{v}$, and we solve for v from (4.3). We get

$$(4.8)\qquad \begin{bmatrix} w_1 \\ w_2 \end{bmatrix} = \begin{bmatrix} \beta\dot{\Theta}_2^2\cdot\Delta\dot{\Theta}\cdot\cos\Delta\Theta \\ \beta\dot{\Theta}_1^2\cdot\Delta\dot{\Theta}\cdot\cos\Delta\Theta \end{bmatrix} + \frac{\beta(\dot{\Theta}_1+\dot{\Theta}_2)\sin\Delta\Theta}{\alpha\gamma-\beta^2\cos^2\Delta\Theta}\cdot$$

$$\begin{bmatrix} -\beta\cos\Delta\Theta(u_1-\beta\dot{\Theta}_2^2\sin\Delta\Theta)+\alpha(u_2-\beta\dot{\Theta}_1^2\sin\Delta\Theta) \\ \gamma(u_1-\beta\dot{\Theta}_2^2\sin\Delta\Theta)-\beta\cos\Delta\Theta(u_2-\beta\dot{\Theta}_1^2\sin\Delta\Theta) \end{bmatrix} +$$

$$+ \begin{bmatrix} \alpha & \beta\cos\Delta\Theta \\ \beta\cos\Delta\Theta & \gamma \end{bmatrix}\begin{bmatrix} \tilde{w}_1 \\ \tilde{w}_2 \end{bmatrix}.$$

Recall that now u acts as the state of the actuator dynamics, which we assume also to be available for feedback. According to the theory of section 3, the input-output relation of the compensated system (4.1-2-5-8) is given by

$$(4.9)\qquad z_1^{(3)} = \tilde{w}_1;\; z_2^{(3)} = \tilde{w}_2$$

This may be verified by direct computation.
Another way to arrive at the same result would consist in differentiating (4.1) using (4.5), and then writing down the decoupling law 'by inspection' as in (4.3). In this simple example where the output equations are given by (4.2), the amount of calculation required by this method is exactly the same as in the method given above. If one has more complicated output equations, however, it is advantageous to compute the decoupling feedbacklaw in the fourth-order model (4.1) and then to differentiate the feedback law as above, rather than first differentiating the equations of motion and being faced with the task of computing a decoupling law in a sixth-order model. Still another method to derive the same results for this particular example would consist in using the agebraic approach of [30], or even to apply feedback immersion into a linear system in [31].

Acknowledgements.
We would like to thank Arjan van der Schaft and Frank Sperling for useful discussions.

References.

1. R.W. Brockett, paper presented at the 1983 MTNS Conference, Beer Sheva, Israel.
2. J.M. Dion, "Feedback block decoupling and infinite structure of linear systems", Int. J. Control, Vol. 37, pp. 521-533, 1983.
3. P.L. Falb and W.A. Wolovich, "Decoupling in the design and synthesis of multivariable control systems", IEEE Trans. Automat. Contr., Vol. AC-12, pp. 651-659, 1967.
4. M.L.J. Hautus, "The formal Laplace transform for smooth linear systems", Lecture Notes in Economics and Mathematical systems, Vol. 131, pp. 29-47, Springer, New York, 1976.
5. R.M. Hirschorn, "(A,B)-invariant distributions and disturbance decoupling of nonlinear systems", SIAM J. Contr. Optimiz., Vol. 19, pp. 1-19, 1981.
6. A. Isidori, A.J. Krener, C. Gori-Giorgi and S. Monaco, "Nonlinear decoupling via feedback: A differential geometric approach", IEEE Trans. Autom. Contr., Vol. AC-26, pp. 331-345, 1981.
7. B. Jakubczyk and W. Respondek, "On linearization of control systems", Bull. Acad. Polon. Sci., Ser. Sci. Math., Vol. XXVIII, pp. 517-522, 1980.
8. T. Kailath, <u>Linear systems</u>, Prentice-Hall, Englewood Cliffs, New Jersey, 1980.
10. T. Lozano-Pérez, "Robot programming", Proc. IEEE, Vol. 71, pp. 821-841, 1983.
11. A.S. Morse and W.M. Wonham, "Status of noninteracting control", IEEE Trans. Automat. Contr., Vol. AC-16, pp. 568-581, 1971.
12. H. Nijmeijer, "Controlled invariance for affine control systems", Int. J. Contr., Vol. 34, pp. 824-833, 1981.
13. H. Nijmeijer, "Controllability distributions for nonlinear control systems", Systems Control Lett., Vol. 2, pp. 122-129, 1982.
14. H. Nijmeijer, "Feedback decomposition of nonlinear control systems", IEEE Trans. Automat. Contr., Vol. AC-28, pp. 861-862, 1983.
15. H. Nijmeijer, "The triangular decoupling problem for nonlinear control systems", Nonlinear Anal. Appl., to appear.
16. H. Nijmeijer and A.J. van der Schaft, "Controlled invariance for nonlinear systems", IEEE Trans. Automat. Contr., Vol. AC-27, pp. 904-914, 1982.
17. H. Nijmeijer and J.M. Schumacher, "The noninteracting control problem for nonlinear control systems", Memo 427, Dept. Appl. Math., Twente Univ. of Technology, May 1983.
18. H. Nijmeijer and J.M. Schumacher, "Zeros at infinity for affine nonlinear control systems", Memo 441, Dept. Appl. Math., Twente Univ. of Technology, September 1983.
19. R.E. O'Malley, Jr., and A. Jameson, "Singular perturbations and singular arcs", Part I, IEEE Trans. Autom. Contr., Vol. AC-20, pp. 218-226, 1975; Part II, ibid., Vol. AC-22, pp. 328-337, 1977.
20. D.H. Owens, <u>Feedback and Multivariable Systems</u>, Peter Peregriners London, 1978
21. R.P. Paul, <u>Robot Manipulations: Mathematics, Programming, and Control</u>, MIT Press, Cambridge, Mass., 1981.
24. P.K. Sinha, "State feedback decoupling of nonlinear systems", IEEE Trans. Automat. Contr., Vol. AC-22, pp. 487-489, 1977.
25. F.B. Sperling, "Nonlinear modeling of robotic mechanisms", Ir. thesis, Lab. for Measurement and Control, Delft Univ. of Technology, 1983.
26. M. Spivak, <u>A Comprehensive Introduction to Differential Geometry</u>, Publish or Perish, Boston, Mass., 1970.

27. H.J. Sussman and V. Jurdjevic, "Controllability of nonlinear systems", J. Diff. Eq., Vol. 12, pp. 95-116, 1972.
28. W.A. Porter, "Diagonalisation and inverses for nonlinear systems", Int. J. Control, 11, pp. 67-76, 1970.
29. E. Freund, "The structure of decoupled nonlinear systems", Int. J. Control, 21, pp. 443-450, 1975.
30. D. Claude, "Découplage des Systèmes: du linéaire au nom linéaire", in "Developpements et Utilisations d'Outils et Modèles Mathématiques en Automatique, Analyse de Systèmes et Traitement du Signal", vol. 3 I.D. Landau (ed.) CNRS, Paris, pp. 533-555, 1983.
31. D. Claude, M. Fliess & A. Isidori, "Immersion, directe et par bouclage, d'un système non lineaire dans un linéaire", C+R+ Acad. Sc. Paris, 296, série I, pp. 237-240, 1983.

Session 17

BIOTECHNOLOGICAL SYSTEMS AND BIOENGINEERING

GÉNIE BIOMÉDICAL ET SYSTÈMES BIOTECHNOLOGIQUES

OPTIMAL SENSOR ALLOCATION FOR IDENTIFICATION OF UNKNOWN PARAMETERS IN A BUBBLE-COLUMN LOOP BIOREACTOR

Axel Munack

Institut f. Regelungstechnik
Universität Hannover
Appelstr. 11, D-3000 Hannover 1

ABSTRACT

Applications of some recently implemented control concepts to batch SCP-cultivation processes in tower loop reactors are considered. These complex distributed parameter processes show some peculiarities which suggest to use adaptive control algorithms. Due to the complex models, decomposition/coordination methods are proposed for parameter identification. After a brief description of the principles of the used control and identification algorithms the results of computations of optimal sensor allocations in the various cases are demonstrated and discussed.

INTRODUCTION

A fundamental step in control of complex biotechnological systems is system modelling. This is by far not a trivial task, since the processes are not understood in detail, and, in cases where a detailed study has been performed, the models are too complicated to be used for on-line control. So usually a model reduction is performed which results in system equations with time-varying parameters, since the complex biological models in general exhibit various nonlinearities. In that sense, reduction of complex models leads to the same situation of time-varying system parameters as in the case, where simple formal models are used. Particular problems arise in batch operation of processes, where simple models may not be valid in the whole region of operation. So, on-line identification techniques have to be used in order to identify time-variant system parameters. Although biotechnological processes are relatively slow, the complexity of the models necessitates the use of fast algorithms for parameter estimation, particularly in case of systems which exhibit spatial concentration profiles and therefore have to be modelled by partial differential equations (PDEs). - The bubble column loop fermenter used as application example in the following has to be modelled as such a distributed parameter system (DPS). - Following a short description of this system, the used methods for adaptive control and identification of unknown system parameters will be explained in brief. Then the used approach for optimal sensor allocation will be presented and the results will be demonstrated and discussed.

DESCRIPTION OF THE PLANT

A schematic diagram of the used double-reactor system (column and loop) is shown in fig.1. Since only the column is gassed from the bottom, gas phase (G) balances have only to be formulated in this part, while liquid phase balances appear in the column (F) and the loop (B). In modelling the fermentation process, only mass balances are taken into account since isothermal conditions are assured by control, and momentum is negligible. The main task of the optimization is to guarantee maximal SCP production with minimal cost, which means low substrate and oxygen consumption.

A general model, only based on the nutrients O_2 and S, the products X and CO_2, and the spatially varying gas velocity u_G leads to a nonlinear system with six parabolic PDEs and five plug flow equations, cf. Luttmann(1980); Luttmann/Munack/Thoma(1984). For control purposes in a limited area of operating conditions, several simplifications can be introduced in order to reduce this large number of system equations; these are:

1) biomass and substrate are well mixed in the liquid phase,
2) the substrate concentration is nowhere growth-limiting (extended culture),
3) the oxygen mole fraction drops linearly in the gas phase of the reactor,
4) the respiration quotient is rq=0, (RQ=1) ; therefore the velocity

SYSTEM VARIABLES

longitudinal coordinate	(column)	x
longitudinal coordinate	(loop)	x^*
time		t
dissolved oxygen concentration	(column)	$O_F(x,t)$
	(loop)	$O_B(x^*,t)$
substrate concentration	(column)	$S_F(x,t)$
	(loop)	$S_B(x^*,t)$
biomass concentration	(column)	$X_F(x,t)$
	(loop)	$X_B(x^*,t)$
oxygen gas phase mole fraction		$x_{OG}(x,t)$
linear gas velocity		$u_G(x,t)$

SYSTEM PARAMETERS

longitudinal pressure profil		$P(x,t)$
linear liquid velocity	(column)	$u_F(t)$
	(loop)	$u_B(t)$
relative liquid hold up		$\varepsilon_F(t)$
relative gas hold up		$\varepsilon_G(t)$
longitudinal dispersion coefficient	(liquid)	$D_F(t)$
	(gas)	$D_G(t)$
volumetric O_2-mass transfer coefficient		$k_L \cdot a(x,t)$
respiratory quotient		$RQ(t)$
dilution (feed) rate		$D(t)$
maximum specific growth rate		$\mu_{max}(t)$
specific death rate (including maintenance)		μ_T
oxygen yield coefficient		$Y_{X/O}(t)$
substrate yield coefficient		$Y_{X/S}(t)$
oxygen saturation constant		K_O
substrate saturation constant		K_S

OPERATING VARIATIONS

aeration rate	recirculation ratio
$0.083\ VVM_N < \beta^N < 1.06\ VVM_N$	$0 \leq \gamma \leq \gamma_{max}(u_{Fomax}, D)$
dilution rate	substrate reservoir concentration
$0\ h^{-1} < D < D_{crit}\ \ (\approx 0.5\ h^{-1})$	$0 < S_R < S_{Rcrit}\ \ (= 45.0\ g/l)$

Fig.1: Bubble column loop fermenter; system variables, system parameters, and control parameters

of the gas phase is reciprocal to the pressure profile in the reactor,

5) residence time in the loop is small in comparison with that one in the column. Conditions in the loop are assumed to be quasistationary.

So the model is drastically reduced to a quasilinear PDE of parabolic type, describing the dissolved oxygen concentration in liquid phase of the column, an implicit algebraic equation for the corresponding concentration in the loop, and an ordinary differential equation for the biomass. The simplified model equations are given below. They are the basis for the computations carried out in the following.

Normalized dissolved oxygen concentration in the liquid phase of the reactor

$$\text{PDE:}\ \frac{\partial c_{OF}(z,\tau)}{\partial \tau} = \frac{1}{Bo_F(\tau)} \cdot \frac{\partial^2 c_{OF}(z,\tau)}{\partial z^2} - v_F(\tau) \cdot \frac{\partial c_{OF}(z,\tau)}{\partial z} + \tag{1a}$$

$$+ St_F^E(\tau) \cdot \Psi(K_{St}(\tau),z) \cdot [p(z,\tau) \cdot c_{OG}(z,\tau) - c_{OF}(z,\tau)]$$

$$- q_{O/Xm}(\tau) \cdot \frac{c_{OF}(z,\tau)}{k_O + c_{OF}(z,\tau)} \cdot c_X(\tau) \quad , \quad \tau \in]0,\tau_e[, \quad z \in]0,1[\; ;$$

$$\text{IC:}\ c_{OF}(z,0) = p(z,0) \; , \quad z \in]0,1[\; ; \tag{1b}$$

$$\text{BC:}\ \left. \frac{\partial c_{OF}(z,\tau)}{\partial z} \right|_{z=0} = Bo_F(\tau) \cdot v_F(\tau) \cdot [c_{OF}(0,\tau) - c_{OB}(1,\tau)] \; , \tag{1c}$$

$$\left. \frac{\partial c_{OF}(z,\tau)}{\partial z} \right|_{z=1} = 0 \quad , \quad \tau \in]0,\tau_e[\; .$$

Normalized dissolved oxygen concentration in the loop

$$c_{OB}(z^*,\tau) - c_{OB}(0,\tau) + k_O \cdot \ln \frac{c_{OB}(z^*,\tau)}{c_{OB}(0,\tau)} = -q_{O/Xm}(\tau) \cdot \frac{z^*}{v_B(\tau)} \cdot c_X(\tau), \tag{2}$$

$$\text{BC:}\ c_{OB}(0,\tau) = c_{OF}(1,\tau) \quad , \quad \tau \in]0,\tau_e[\quad , \quad z^* \in [0,1].$$

Normalized concentration of cells

$$\frac{dc_X(\tau)}{d\tau} = y_{X/O}(\tau)\cdot\left[\frac{v_B(\tau)}{v_B(\tau)+v_F(\tau)}\cdot q_{O/Xm}(\tau)\cdot c_X(\tau)\cdot\int_0^1 \frac{c_{OF}(z,\tau)}{k_O+c_{OF}(z,\tau)}\,dz + \right. \tag{3}$$

$$\left. + \frac{v_B(\tau)\cdot v_F(\tau)}{v_B(\tau)+v_F(\tau)}\cdot\left(c_{OB}(0,\tau)-c_{OB}(1,\tau)\right) - \frac{\mu_t}{y_{X/O}(\tau)}\cdot c_X(\tau)\right]$$

$$\tau \in \,]0,\tau_e[\quad ;$$

$$\text{IC:} \quad c_X(0) = 1 \quad .$$

In (1)-(3), the time variable is denoted as τ in order to emphasize that usually a normalization of the physical time is introduced. - The system equations are coupled via the states; so C_X enters into the dissolved oxygen equations, and on the other hand, C_{OF} enters into the cells equation. The physical coupling of column and loop is modelled via coupling terms in the boundary conditions. Four parameters have turned out to be unknown; these are two fluiddynamical parameters, k_La (included in St_F^E) as well as K_{St}, both describing oxygen transfer from gas phase into liquid phase, and two biological parameters, the metabolic quotient $q_{O/Xm}$ and the yield coefficient $y_{X/O}$. Other parameters, like Bo_F, k_O and μ_t can be identified off-line.

ADAPTIVE CONTROL

In order to perform an optimal operation of this plant, adaptive control has been considered, using the above formulated system equations. The algorithms used are based on the heuristical OLFO-concept, which means Open Loop Feedback Optimal control. The structure of this algorithm is demonstrated in fig. 2.

The plant is driven by the control input, and furthermore influenced by external disturbances. It should be mentioned again, that the model of the plant is not entirely known; its structure, however, is fixed by physical, chemical, and biological considerations.

The OLFO-controller can be divided into the two parts identification and optimization. At first, an identification algorithm with a model of the system is used to identify the unknown system parameters by means of an output error minimization. These parameters are then used

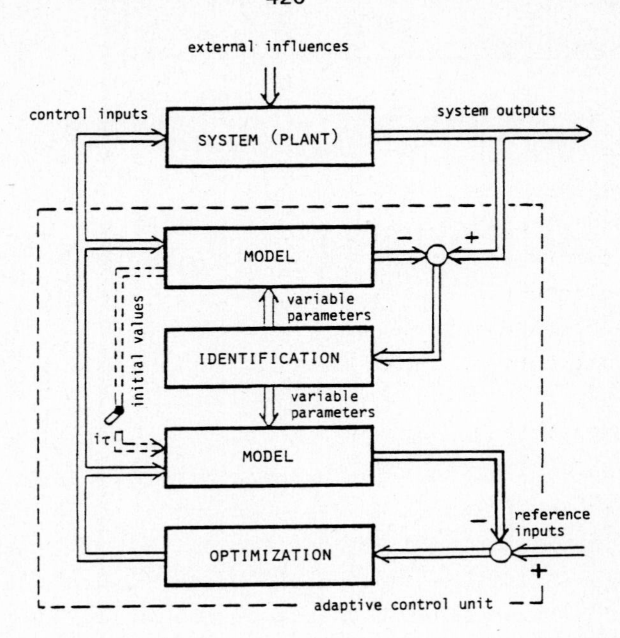

Fig.2: Structure of Open Loop Feedback Optimal control (OLFO)

to compute optimal control functions by means of a further model (computationally : the same) and an optimization algorithm. The model used for optimization has to be set to proper initial values for each optimization cycle. These values can be taken from the identification part of the procedure, since in parallel to the parameter identification also a state estimation is performed. By application of the computed control functions to the plant, the loop is closed. However, the OLFO-controller cannot be looked at as a continuous process, since a great deal of computation is involved; therefore a cyclic operation is performed, and between the adaption time-instants the loop is open, as shown in the following.

The finite control interval $]0,T[$, which includes the whole batch operation time of the plant, is divided into N subintervals, the so-called adaption intervals, each of length τ (not to be confused with the normalized time in (1)-(3)). During each interval, say $]k\tau,(k+1)\tau[$, estimates of the system's parameters and system's states are computed by the identification procedure using measurements of the preceding interval $](k-1)\tau,k\tau[$. These parameters and the prediction of their courses in the future, as well as the state estimation at time $k\tau$, form the basis for calculations of the optimal control function in the remaining interval $](k+1)\tau,T[$. It is imposed on the system from time $t=(k+1)\tau$, but is valid only until $t=(k+2)\tau$, since then new measurement data will have led to updated parameter estimates and, consequently, to new optimal control functions. Looking once again on the structure of the OLFO-controller, two feedback paths are to be seen: one path, cyclically closed, is a signal path containing the initial values for optimization, and the other path, also cyclically closed, but at different time-instants, is a higher-level path containing the estimated parameters of the system. So, in spite of the pure feedforward structure of the optimal control algorithm, there are two feedback paths to be found in this adaptively controlled system.

As already demonstrated, e.g. in Luttmann/Munack(1983), the results of Lions(1971) and Chavent(1974) can be used to formulate algorithms for the optimization, respectively the identification part of the OLFO-controller. These will not be discussed here since both algorithms have been proven to work reliably and the simulation results have been published already - Munack(1980).

With respect to identification, particularly sensor allocation, the use of OLFO-controllers brings up a new problem. In these algorithms, parameter identification is not carried out for the complete operation time, but the data used in each identification run are measurements only of the preceding adaption interval. So the sensors must be allocated in order to guarantee a good identifiability in each of the intervals.

DECOMPOSITION/COORDINATION METHODS

Methods of decomposition/coordination have already been applied to control of distributed-parameter systems, cf. Bensoussan/Glowinski/Lions(1973), Cambon/LeLetty(1973), Pradin/Titli(1975), or Pradin (1979). However, simulations have shown that very often the gain in computational speed is not very high, if no parallel processor system is used. But application of these ideas to the parameter identification problem seems to yield very promising results, cf. Munack(1983). Since the formulae used can be found in the paper cited, here only a short outline of the principle will be given.

The treated systems are described by a set of N subsystems, which means N coupled parabolic partial differential equations of the form ($i = 1,2,...,N$)

$$\text{PDE:}\quad \frac{\partial y_i}{\partial t} - \frac{\partial}{\partial x}\left[a_{2i}(x,\underline{P})\,\frac{\partial y_i}{\partial x}\right] + a_{1i}(x,\underline{P})\,\frac{\partial y_i}{\partial x} + a_{oi}(x,\underline{P})\,y_i =$$

$$= \frac{\partial y_i}{\partial t} + A_{ii}(x,\underline{P})\,y_i = -\sum_{k\neq i}^{N} a_{oik}(x,\underline{P})\,y_k + b_i(x,\underline{P})\,u_i + f_i(x,\underline{P})$$

$$\text{in }]0,1[\times]0,T[\;;$$

$$a_{2i}(x),\ a_{1i}(x),\ a_{oi}(x),\ a_{oik}(x) \in L^\infty(0,1) \;;\ a_{2i}(x) > 0 \ \forall x \in]0,1[\,. \tag{4}$$

$$\text{IC:}\quad y_i(0) = y_{io} \qquad \text{in }]0,1[\;;$$

$$\text{BC:}\quad \left.\begin{aligned} -a_{2i}(0,\underline{P})\left.\frac{\partial y_i}{\partial x}\right|_{x=0} + c_{oi}\left.y_i\right|_{x=0} &= c_{oi}\,y_{eoi} \\ a_{2i}(1,\underline{P})\left.\frac{\partial y_i}{\partial x}\right|_{x=1} + c_{1i}\left.y_i\right|_{x=1} &= c_{1i}\,y_{eli} \end{aligned}\right\} \text{ in }]0,T[\,.$$

Arguments of the dependent variables are only written, if this clarifies the notation. $\underline{P}$ is a parameter vector which is partly to be identified during the course of the batch fermentation process; another part of $\underline{P}$ is considered to be identified once for a complete cultivation and then fixed for all others in order to keep the number of parameters low which must be identified on-line; and a third part of $\underline{P}$ may be known from the beginning. Each of the N equations as well as the whole system of equations is assumed to admit a unique solution in an appropriate solution space over $]\emptyset,1[\times]0,T[$, cf. Lions(1971).

M sensors may be located to measure the state of each subsystem, which means that we have installed in total MxN sensors, yielding measurements

$$s_i^j = \int_0^1 \chi_i^j(x)\,y_i\,dx \;, \qquad i = 1,2,...,N \;;\ j = 1,2,...,M \;, \tag{5}$$

with χ_i^j denoting the spatial characteristic of the j-th sensor, located at the i-th subsystem. Assuming the χ_i^j are known, errors

$$e_i^j(\hat{\underline{P}}) = \int_0^1 \chi_i^j(x)\,\hat{y}_i(\hat{\underline{P}})\,dx - s_i^j \tag{6}$$

are defined, where $y_i(\hat{\underline{P}})$ means the state of subsystem i of the identification model with parameter vector $\hat{\underline{P}}$.

In order to minimize the error, a functional

$$J^I(\underline{\hat{P}}) = \sum_{i=1}^{N} J_i^I(\underline{\hat{P}}) = \sum_{i=1}^{N} \int_0^T \sum_{j=1}^{M} e_i^j W_i^j e_i^j \, dt \quad , \quad W_i^j \geq 0 \tag{7}$$

is introduced, which is used to formulate the parameter identification problem as optimization problem. The objective is to find $\underline{\hat{P}}_{opt}$ with

$$J^I(\underline{\hat{P}}_{opt}) \leq J^I(\underline{\hat{P}}) \quad , \quad \underline{\hat{P}} \in P_{ad} \quad , \quad \underline{\hat{P}}_{opt} \in P_{ad} \, .$$

P_{ad} is the set of admissible (physically, biologically meaningful) parameters. The couplings via the states of the subsystems prevent the direct decomposition of the problem (if not $a_{0ik}=\emptyset$); in other words, a parameter identification, separately performed in each subsystem, does not yield correct results, because of the fact that the states of the other subsystems enter into the state equations and are subject to change during the identification of the other subsystems.

However, several methods are known to decompose coupled optimization problems into decoupled subproblems, where the couplings are taken into account by means of a so-called coordinator. Among these are the introduction of Lagrange multipliers, the penalty function method, and the re-injection method developed by Takahara. In some practical tests, the latter has proven to be a very simple and reliable tool. It roughly consists of a parametric decomposition and separation of the optimality system for the coupled problem. This optimality system is formed by the model equation and the corresponding adjoint equation which is given by

$$- \frac{\partial p_i}{\partial t} - \frac{\partial}{\partial x}\left[a_{2i}(x,\underline{\hat{P}}) \frac{\partial p_i}{\partial x} \right] - \frac{\partial}{\partial x}\left[a_{1i}(x,\underline{\hat{P}}) \, p_i \right] + a_{oi}(x,\underline{\hat{P}}) \, p_i =$$

$$- \frac{\partial p_i}{\partial t} + A'_{ii}(x,\underline{\hat{P}}) \, p_i = - \sum_{k \neq i}^{N} a_{oki}(x,\underline{\hat{P}}) \, p_k + \sum_{j=1}^{M} x_i^j W_i^j e_i^j \quad , \text{ in }]0,1[\times]0,T[\quad ,$$

$$\text{FC:} \quad p_i(T) = 0 \ \text{ in }]0,1[\ ; \qquad \text{BC: homogeneous.} \tag{8}$$

In case of $\underline{\hat{P}}_{opt}=\underline{P}$ and no measurement noise, the solution of the adjoint equations is identically zero (in contrast to the adjoint equations in optimal control problems). If $\underline{\hat{P}} \neq \underline{P}_{opt}$, the solution of (4) with $\underline{P}=\underline{\hat{P}}$ and (8) gives the opportunity to calculate the gradient of the identification functional with respect to the unknown parameters, such that iteratively the optimum can be found. For decomposition of the problem, so-called coordination variables $\underline{v}$ and $\underline{q}$ are introduced into the model equations and the adjoint equations to yield

$$\text{PDE:} \quad \frac{\partial \hat{y}_i}{\partial t} + A_{ii}(x,\underline{\hat{P}})\, \hat{y}_i = - \sum_{k \neq i}^{N} a_{oik}(x,\underline{\hat{P}})\, v_k + b_i(x,\underline{\hat{P}})\, u_i + f_i(x,\underline{\hat{P}}) \quad ,$$
$$y_i(0) = y_{io} \quad ; \tag{9}$$

$$\text{ADJ:} - \frac{\partial p_i}{\partial t} + A'_{ii}(x,\underline{\hat{P}})\, p_i = - \sum_{k \neq i}^{N} a_{oki}(x,\underline{\hat{P}})\, q_k + \sum_{j=1}^{M} X_i^j W_i^j e_i^j(\underline{\hat{P}}) \quad ,$$
$$p_i(T) = 0 \quad . \tag{10}$$

In this way, the problem is decoupled. However, it can be shown, that on each subsystem a modified sub-functional has to be optimized.

Coordination is performed by simply setting

$$v_i^{l+1} = \hat{y}_i^l(\underline{\hat{P}}^{*l}) \; ; \; q_i^{l+1} = p_i^l(\underline{\hat{P}}^{*l}) \tag{11}$$

- where $\underline{\hat{P}}^{*l}$ is the optimal parameter estimate of the preceding step no. l - and then iterating until the differences between consecutive iterations are small enough.

If an unknown parameter enters into several subsystems, it has to be guaranteed, that the identified values coincide. This can be taken into account by means of an augmented Lagrange modification of the subfunctionals, cf. Bertsekas(1976) and Munack/Thoma(1983).

Although the decomposition/coordination method has been presented here for a system of partial differential equations, of course an application to problems with mixed partial and ordinary differential equations is straightforward. Also the case of quasilinear equations can be treated by the methods described above.

OPTIMAL SENSOR ALLOCATION

From the literature, several methods are known for optimal allocation of sensors in DPS. Instead of reviewing the different approaches here, the reader is referred to the survey by Kubrusly/Malebranche(1983). A relatively simple approach is used here, which was developed by Qureshi/Ng/Goodwin(1980). It is based on the Fisher information matrix - cf. e.g. Goodwin/Payne(1977) - and will be outlined in the following.

The measurements are corrupted by noise,

$$s_i^j = \int_0^1 \chi_i^j(x)\cdot y_i(\underline{P})dx + w_i^j(t) \quad , \tag{12}$$

where $w_i^j(t)$ denotes the measurement noise of the j-th sensor, located at the i-th subsystem. Spatially uncorrelated white noise is considered with covariance

$$E\left\{\underline{w}^{jT}(t)\ \underline{w}^{l}(s)\right\} = \mathrm{diag}_i\ [(\sigma_i^j)^2\delta_{jl}\ \delta(t-s)] = \underline{C} \quad . \tag{13}$$

Then the information matrix is given by

$$F = \sum_{j=1}^{M} \int_0^T \left(\frac{\partial \underline{s}^j}{\partial \underline{P}}\right) \quad \underline{C}^{-1} \left(\frac{\partial \underline{s}^j}{\partial \underline{P}}\right) dt \tag{14a}$$

where

$$\frac{\partial \underline{s}^j}{\partial \underline{P}} = \begin{bmatrix} \frac{\partial s_1^j}{\partial P_1} & \frac{\partial s_1^j}{\partial P_2} & \cdots & \\ \frac{\partial s_2^j}{\partial P_1} & \ddots & & \\ \vdots & & \ddots & \\ \vdots & & & \frac{\partial s_N^j}{\partial P_L} \end{bmatrix} \tag{14b}$$

and $\underline{P}$ has L unknown components $P_1 \ldots P_L$.

Since

$$\frac{\partial s_i^j}{\partial P_\ell} = \chi_i^j(x)\ \frac{\partial y_i(\underline{P})}{\partial P_\ell} \quad , \tag{14c}$$

$\underline{F}$ can be computed by the state sensitivity functions, which satisfy the following coupled set of PDEs:

$$\frac{\partial}{\partial t}\left[\frac{\partial y_i(\underline{P})}{\partial P_\ell}\right] - \frac{\partial}{\partial x}\left[a_{2i}(x,\underline{P})\cdot\frac{\partial}{\partial x}\left[\frac{\partial y_i(\underline{P})}{\partial P_\ell}\right]\right] + a_{1i}(x,\underline{P})\cdot\frac{\partial}{\partial x}\left[\frac{\partial y_i(\underline{P})}{\partial P_\ell}\right] + a_{0i}(x,\underline{P})\cdot\left[\frac{\partial y_i(\underline{P})}{\partial P_\ell}\right] =$$

$$= -\sum_{k=i}^{N} a_{0ik}(x,\underline{P})\cdot\left[\frac{\partial y_k(\underline{P})}{\partial P_\ell}\right] + \frac{\partial}{\partial x}\left[\left[\frac{\partial a_{2i}(x,\underline{P})}{\partial P_\ell}\right]\frac{\partial y_i}{\partial x}\right] - \frac{\partial a_1(x,\underline{P})}{\partial P_\ell}\cdot\frac{\partial y_i}{\partial x} - \frac{\partial a_{0i}(x,\underline{P})}{\partial P_\ell} +$$

$$- \sum_{k=i}^{N} \frac{\partial a_{0ik}(x,\underline{P})}{\partial P_\ell}\cdot y_k + \frac{\partial b_i(x,\underline{P})}{\partial P_\ell}\cdot u_i + \frac{\partial f_i(x,\underline{P})}{\partial P_\ell} \tag{15a}$$

$i = 1,\ldots,N \quad , \quad \ell = 1,\ldots,L$

with initial conditions

$$\left.\frac{\partial y_i(x,\underline{P})}{\partial P_\ell}\right|_{t=0} = 0 \quad , \quad x \in]0,1[\quad , \tag{15b}$$

and boundary conditions

$$- a_{2i}(0,\underline{P})\cdot\frac{\partial}{\partial x}\left[\frac{\partial y_i(x,\underline{P})}{\partial \underline{P}_\ell}\right]\Bigg|_{x=0} + c_{0i}\cdot\frac{\partial y_i(0,\underline{P})}{\partial \underline{P}_\ell} = c_{0i}\cdot y_{e0i} + \frac{\partial a_{2i}(0,\underline{P})}{\partial \underline{P}_\ell}\cdot\frac{\partial y_i(x,\underline{P})}{\partial x}\Bigg|_{x=0}$$

(15c)

$$a_{2i}(1,\underline{P})\cdot\frac{\partial}{\partial x}\left[\frac{\partial y_i(x,\underline{P})}{\partial \underline{P}_\ell}\right]\Bigg|_{x=1} + c_{1i}\cdot\frac{\partial y_i(1,\underline{P})}{\partial \underline{P}_\ell} = c_{1i}\cdot y_{e1i} + \frac{\partial a_{2i}(1,\underline{P})}{\partial \underline{P}_\ell}\cdot\frac{\partial y_i(x,\underline{P})}{\partial x}\Bigg|_{x=1}$$

For fixed inputs, $\underline{F}$ is only a function of the sensor characteristics, which consist of the local characteristics and the allocation of the sensors. In practical cases, the local characteristics are usually fixed for a certain type of sensors. - The estimation accuracy may be evaluated by computing the determinant of $\underline{F}$, which means that we have to find

$$\max_{X_i^j(x)} \det F \tag{16}$$

in order to determine the sensor positions which give best parameter estimation accuracy.

Using the adaptive control algorithm, however, results in a different measure, since the cyclical structure of the algorithm has to be taken into account. Best accuracy is given here, if

$$\max_{X_i^j(x)} \min_{i=2,\frac{T}{\tau}} \det F\,[(i-1)\tau\,,\,i\tau] \tag{17}$$

where the times in brackets denote the integration bounds in (14a).

APPLICATION TO THE BUBBLE COLUMN BIOREACTOR

As indicated above, mainly four parameters have turned out to be time-varying, such that they have to be determined on-line during the cultivation. These are the fluiddynamical parameters k_La and K_{St} as well as the biological parameters $q_{O/Xm}$ and $y_{X/O}$. Measurements can be taken of the dissolved oxygen concentration in the liquid phase of the column, C_{OF}, the overall oxygen transfer rate OTR, calculated by exhaust gas analysis, and the cells concentration C_X.

Since the control input (the air flow rate) is restricted by biological considerations, no research has been carried out in order to optimize this input function w.r.t. parameter identification. So it must be pointed out that the numerical results presented in the following only apply to the treated reactor, the considered process, and the used operating conditions of the plant. However, corresponding results may be computed for other processes and other conditions as well using the same ideas and methods.

According to the various demands on the identification procedure, the following sensor allocation problems are solved:

Determination of the entire set of four parameters, by using measurements of a) C_{OF}, b) C_{OF},OTR, c) C_{OF},C_X, d) C_{OF},OTR,C_X during the whole cultivation time (off-line identification), where 1...4 sensors for C_{OF} are taken into account each.

Determination as above, but using criterion (17) in order to compute corresponding results in case of OLFO-control.

A possible (and reasonable) decomposition of the system equations can be made by treating (1) and (2), i.e. the dissolved oxygen equations as one subsystem and by taking the biological growth equation (3) as the second subsystem. Then the measurements of C_{OF} and OTR are affected by the first subsystem, while C_X clearly belongs to the second. Three parameters may be grouped undoubtedly, these are k_La and K_{St} (to the first subsystem) and $y_{X/O}$ (to the second subsystem). The parameter $q_{O/Xm}$, however, enters into both subproblems. - Here only the simple case is considered, where this parameter is identified either in the first or in the second system. A complete treatment by identification in both subsystems and then ensuring convergence of the estimates by augmented Lagrange modification of the identification functionals is under progress.

So, in the decomposed case, the corresponding allocation problems are:

Determination of k_La, K_{St}, and $q_{O/Xm}$ by using measurements of a) C_{OF}, b) C_{OF},OTR , while determining $y_{X/O}$ by measurements of C_X, where 1...4 sensors for C_{OF} are taken into account each, in case of off-line identification (data of the entire cultivation time interval are used).

Determination of k_La and K_{St} by using measurements of a) C_{OF}, b) C_{OF},OTR while determining $q_{O/Xm}$ and $y_{X/O}$ by measurements of C_X, where 1...4 sensors for C_{OF} are taken into account each, in case of off-line identification.

As above, but using criterion (17) in order to compute corresponding

results in case of OLFO-control.

In order to give some ideas about the system treated, in the following the states $C_{OF}(x,t)$ and $C_X(t)$ and the measurement OTR(t) are shown as well as the corresponding sensitivities for coupled identification. The plots are valid for a treatment of the whole fermentation interval; sensitivities in case of OLFO-control look different, since here a state-adaption is performed at the beginning of each adaption interval. Sensitivities for decomposed identification look (slightly) different, too, since the coordination variables have to be fixed during identification in the subsystems.

The plots show a high dominance of the biological parameters, particularly $q_{O/Xm}$, compared with the fluiddynamical parameters. A quite unexpected decrease of the sensitivities at time t=11..12h at the bottom of the column (x=0) -cf. fig. 4c,d- is caused by the nonlinear behaviour of the plant; under the operating conditions actually

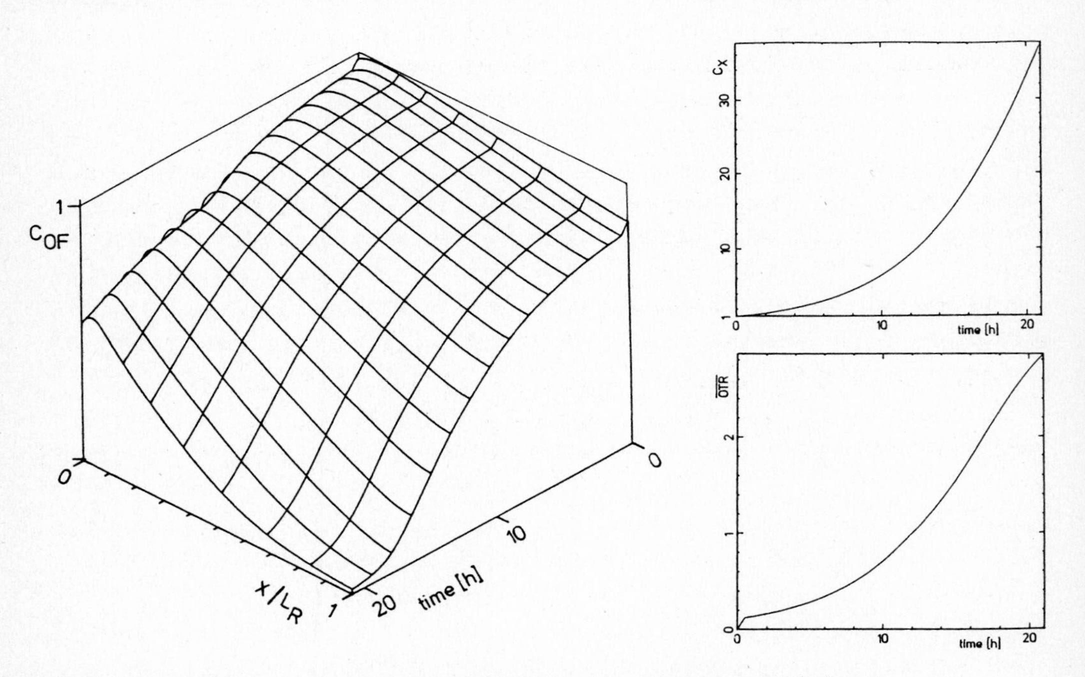

Fig.3a-c: States $C_{OF}(x,t)$ and $C_X(t)$; measurements OTR(t)

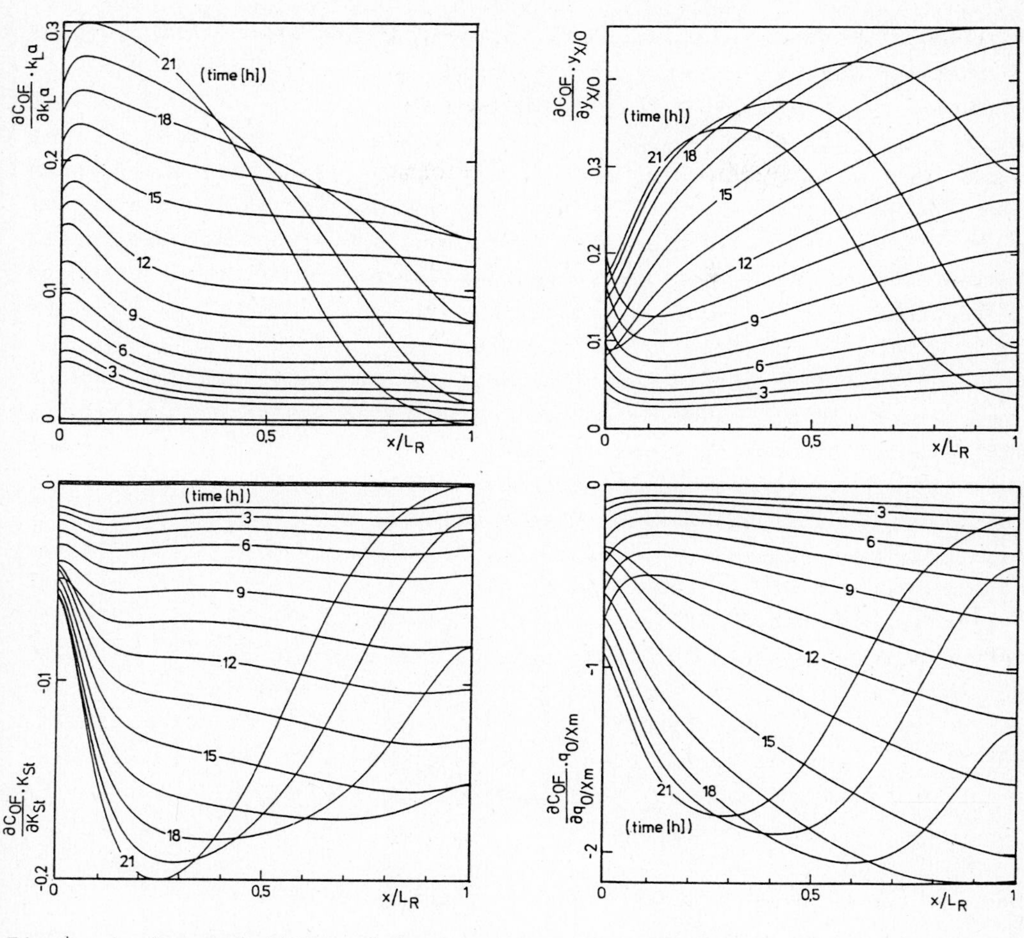

Fig.4a-d: Sensitivities of $C_{OF}(x,t)$ w.r.t. k_La, K_{St}, $q_{O/Xm}$, and $y_{X/O}$

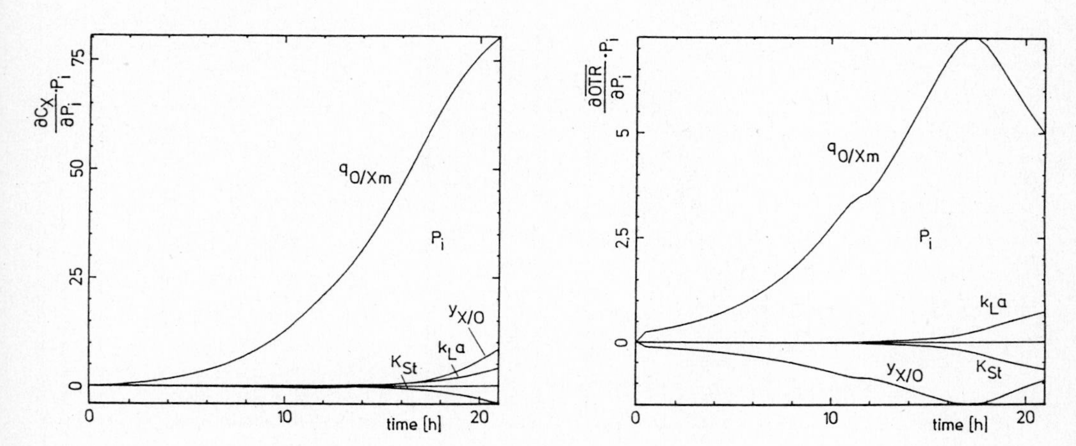

Fig.5a,b: Sensitivities of $C_X(t)$ and OTR(t) w.r.t. the unknown parameters

applied, at this time the dissolved oxygen concentration at the bottom of the loop practically goes down to zero.

For the calculations of optimal sensor allocations, variances of the noise components for each sensor have to be specified. It is known that OTR is a comparatively precise measurement, compared with relatively high errors in dissolved oxygen and cells concentration measurements. This was taken into account by a variance of 2% for OTR and 5% for the other measurement values.

A very valuable result is the following: The OTR-measurement is most significant in all cases, where $q_{O/Xm}$ has to be estimated; in detail: adding this measurement gives an increase of det $\underline{F}$ in order of magnitude of 2...7 decades! In fact, measuring C_{OF} with only one sensor and, additionally, OTR gives a higher value of det $\underline{F}$ than by taking eight C_{OF} measurements. This holds for off-line identification. In the case of cyclic identification it can be stated that taking two optimally allocated C_{OF}-sensors plus OTR-measurements is as good as taking eleven C_{OF}-sensors. These results are not only due to the lower variance in the OTR measurements, since tests with taking the same variances for all measurements also yield a considerable increase in det $\underline{F}$. So one can draw the conclusion, that, even for the treated distributed parameter system, the 'lumped' OTR-measurement should be taken into account in any case.

Some of the results are summarized in fig.6.

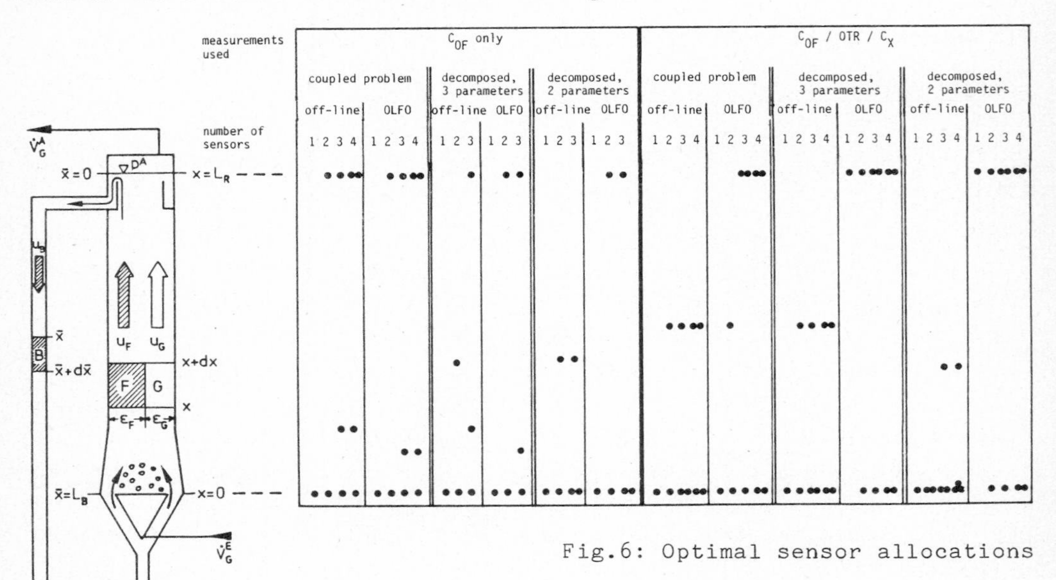

Fig.6: Optimal sensor allocations

Here it is quite unexpected, that taking more and more sensors, in most cases the new locations are the same as some already existing. This means, that installing a further sensor at the same height of the column adds more information to parameter estimation purposes than installing it at another place elsewhere in the column. However, the sensors then have to be separated over the cross-section of the reactor in order to fulfil the assumed uncorrelated noise condition.

Only in case of pure C_{OF}-measurements the positions are somewhat distributed over the column, while in case of coupled identification and of decomposed identification with three parameters to be identified the optimal positions are at the bottom and in the middle of the column, if the whole time-interval is used. Identification with the OLFO-algorithm would need sensors at the bottom and at the top of the column for best performance as well in the coupled as in the decoupled case.

The close relationship between optimal allocations in the decomposed and the coupled case indicates that couplings are not too strong in the system under consideration.

It can be deduced from these results, that for a practical installation it would be advisable to pose three sensors at the bottom, the top, and at 45% of height of the column. This would give best identification results using the different approaches.

CONCLUSIONS

For a relevant biotechnological system - the bubble column loop fermenter - the optimal sensor allocation problem for parameter identification has been considered. The problem has been solved by determining the Fisher information matrix, which is based on sensitivity calculations. By this, the advantages of sensitivity analyses are underlined again, which have been emphasized earlier by other authors, e.g. Goodson/Polis(1974).

Besides parameter identifications using measurement data of the whole batch fermentation interval, also the case of adaptive (OLFO-) control has been considered. Furthermore, decomposition/coordination techniques have been applied to solve the identification problem at a high speed.

The results emphasize the great importance of OTR measurements for this plant and show, that different sensor locations should be used in the adaptive case compared with off-line identifications using the complete available data set. Since the interconnections between oxygen and cellmass balances are not very strong, the sensor positions are essentially the same in the coupled as in the decomposed case.

Further research should be done in the area of defining measures of the strength of interconnection in these problems, such that best separation into subproblems for more complex plants can be performed computationally.

REFERENCES

Bensoussan,A.; Glowinski,R.; Lions,J.L.: Methode de décomposition appliquée au contrôle optimal de systèmes distribués. 5th Conf. on Optimization Techniques, Rom. LNCS 3, Springer, Berlin, Heidelberg,New York, 1973.

Bertsekas,D.P.: Multiplier methods: a survey. Automatica, pp. 133-145, 1976.

Cambon,P.; LeLetty,L.: Applications of decomposition and multi-level techniques to the optimization of distributed parameter systems. 5th IFIP Conf. on Optimization Techniques. LNCS 3, Springer, Berlin,Heidelberg,New York, 1973.

Chavent.G.: Identification of functional parameters in partial differential equations. In: Goodson,R.E.; Polis,M.P.(eds.): 'Identification of Parameters in Distributed Systems'. ASME, New York, 1974.

Goodson,R.E.; Polis,M.P.: Parameter identification in distributed systems: a synthesizing overview. In: Goodson,R.E.;Polis,M.P. (eds.): 'Identification of Parameters in Distributed Systems'. ASME, New York, 1974.

Kubrusly,C.S.; Malebranche.H.: A survey on optimal sensors and controllers location in DPS. In: Babary,J.P.; LeLetty,L.:'3rd IFAC Symposium on Control of Distributed Parameter Systems'. Pergamon, Oxford, 1983.

Lions,J.L.: Optimal Control of Systems Governed by Partial Differential Equations. Springer, Berlin,Heidelberg,New York, 1971.

Luttmann,R.: Modellbildung und Simulation von SCP-Prozessen in Blasensaulenschlaufenfermentern. Dr.-Ing. Dissertation, Universität Hannover. 1980.

Luttmann,R.; Munack,A.: Economic optimization of distributed parameter fermentation processes (a case study). In: Halme,A.(ed.): 'Modelling and Control of Biotechnical Processes'. Pergamon, Oxford, 1983.

Luttmann,R.; Munack,A.; Thoma,M.: Mathematical modelling, parameter identification, and adaptive control of single cell protein processes in tower loop bioreactors. In: Advances in Biochemical Engineering, Springer, Berlin,Heidelberg,New York, to appear 1984.

Munack,A.: Application of adaptive control to a bubble-column fermenter. In: Bensoussan,A.;Lions,J.L.(eds.): 'Analysis and Optimization of Systems'. LNCIS 28 ,Springer, Berlin,Heidelberg,New York, 1980.

Munack,A.; Thoma,M.: Application of decomposition/coordination methods to control and identification of interconnected distributed parameter processes. In: Babary,J.P.;LeLetty,L.(eds.): '3 rd IFAC Symposium on Control of Distributed Parameter Systems'. Pergamon. Oxford, 1983.

Pradin,B.: Calcul hiérarchisé pour la commande en boucle ouverte de systèmes a paramètres répartis. Thèse Docteur d'Etat, Université Paul Sabatier, Toulouse, 1979.

Pradin,B.; Titli,A.: Methods of decomposition-coordination for the optimization of interconnected, distributed parameter systems. IFAC Boston, part IA, paper 15.3, 1975.

Qureshi.Z.H.; Ng,T.S.; Goodwin,G.C.: Optimum experimental design for identification of distributed parameter systems. Int. J. Control, pp. 21-29, 1980.

EXTRACTION OF WEAK BIOELECTRICAL SIGNALS BY MEANS OF SINGULAR VALUE DECOMPOSITION

J.Vanderschoot*, J.Vandewalle**, J.Janssens*, W.Sansen**, G.Vantrappen*
*Laboratory of G.I.Motility, Department of Medical Research,
**ESAT Laboratory, Department of Electrical Engineering,
University of Leuven, Belgium

Abstract - Most measurements of human electrical activity contain large amounts of electrical heart activity (electrocardiogram, ECG). Whenever other sources of lower energy are of interest, a need arises to eliminate this ECG. Some applications allow a frequency domain operation (e.g. gastro-intestinal slow wave detection), or even a time domain operation (e.g. blanking of QRS complex of an ECG). Usually none of these methods are adequate. The proposed method uses a totally different approach, in this sense that it manipulates geometrically all measurements at the same time. It decomposes the measurements on the basis of an oriented energy, by means of the singular value decomposition. After an introduction to the problem and a definition of some useful concepts, the basic idea of the method is presented in a low dimensional geometrical example, and generalized to higher dimensions. It is shown that the method can easily be applied under realistic clinical conditions. A discussion is given about : the influence of noise; considerations on correlations between source signals; number and location of electrodes; certain dynamical problems like interference. Results on real data are given, in order to illustrate and verify the main features of the method : the multidimensional approach to the estimation of equivalent dipole vector sources; the insensitivity of ECG elimination quality to actual electrode positions; the minimal rank representation of the source signals and the resulting signal to noise ratio improvement; 50 Hz or 60 Hz interference elimination.

I Introduction

The problem of reduction of ECG and noise in multiple bioelectrical measurements has received a lot of attention in the literature. The proposed methods usually split up the problem in (1) ECG elimination, and (2) noise elimination. Most of these publications deal with the detection of the fetal electrocardiogram (FECG) in abdominal recordings. This is no coincidence, since (1) the FECG signal has a frequency spectrum intertwined with the maternal ECG spectrum, making a simple filtering inadequate, and (2) the measurement noise is relatively large. Furthermore, all techniques try to obtain only one signal with good ECG and noise reduction. However, many applications require estimates not of one measured signal, but of more than one source signal. Indeed, the information on most sources can only be described and analyzed adequately by vector signals, e.g. the electrical heart activity.

The method that will be presented takes into account that all electrical sources in the body have an equivalent dipole vector model, and that potential measurements are linear static combinations of these dipole signals and noise |5|. It tries to reconstruct these source signals with an inverse linear static combination. This reconstruction is optimal in the sense of discrimination between different sources and at the same time optimal for the S/N ratios.

The method can be depicted as in fig.1. **T** represents the transfer from sources to electrodes, and **P** represents the estimate of the inverse transfer. The identification block, which estimates the **T** transfer on the basis of the electrode signals, uses the singular value decomposition (SVD) |1,2,3|, a numerically very reliable technique. If the transfer **T** is time invariant, only one SVD has to be executed. Then the computationally simple **P** projection can be applied continuously, and even abrupt changes in the source signals are tracked immediately. If the transfer is time variant, the SVD and adjustment of **P** should be executed in an on line, adaptive fashion. For that, an iterative SVD algorithm could be used, see e.g. |4|.

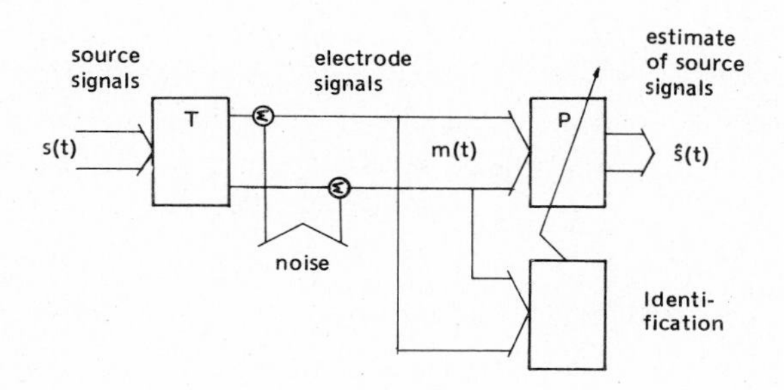

Fig.1. General overview of the method.

DEFINITIONS For further discussions, some concepts should be defined. "Real space" is the term used for the 3 dimensional physical space. In this space, electrical sources in the human body can be represented by equivalent current source dipole vectors with fixed positions |5|. The magnitude and direction of these vectors can be defined by their x,y,z components. These components are generally functions of time, they will be called "source signals". If there are n such vectors at different fixed positions in real space, a 3n dimensional "source space" can be defined. In this space, a point is completely defined by each of the real space components of each source. A curve in this space describes completely the evolution in time of all source signals. The "source signal vector" is defined by :

$$\mathbf{s}^T(t) = [\mathbf{s}_1{}^T(t) \;\ldots\ldots\; \mathbf{s}_n{}^T(t)] = [s_1(t) \;\ldots\ldots\; s_r(t)] \tag{1}$$

where each $\mathbf{s}_i{}^T(t)$ describes the dipole vector at the i-th position , and r=3n.The aim of the presented method will be to extract this source signal vector, entirely or partially. Now, if p potential differences $m_i(t)$ are recorded, these measurement signals can also be arranged in a "measurement vector" :

$$\mathbf{m}^T(t) = [m_1(t) \;\ldots\ldots\; m_p(t)] \tag{2}$$

This is a vector in a p dimensional "measurement space". As has been proven elsewhere |5|, the transfer from sources to electrodes can be well represented by a linear static combination :

$$\mathbf{m}(t) = \mathbf{T}\,\mathbf{s}(t) \tag{3}$$

where **T** is a pxr matrix, called the "transfer matrix". This matrix contains the classically known lead vectors. A lead vector is a vector in real space, associated with one measurement signal, and one dipole position in real space. It is defined such that the contribution to that measurement of the dipole source at that position is given by the orthogonal projection of the dipole vector onto the lead vector. Therefore :

$$m_i(t) = \sum_{j=1}^{n} \mathbf{c}_{ij}^T \mathbf{s}_j(t) \tag{4}$$

where $\mathbf{c}_{ij}$ is the lead vector of measurement signal i for the dipole at position j. On the other hand, $\mathbf{T}$ can also be written as :

$$\mathbf{T} = [\mathbf{t}_1 \; \ldots\ldots \; \mathbf{t}_r] \tag{5}$$

where each $\mathbf{t}_i$ is a column of $\mathbf{T}$, representing a vector in the p dimensional measurement space. Each of these r vectors is associated with one source signal. Physically, we can say that, if only one source signal were non zero, all measurements would fall on the direction of its associated vector. Therefore $\mathbf{t}_i$ will be called the "source vector" of the i-th source signal. If q measurement vector samples are taken they can be put into a "measurement matrix" :

$$\mathbf{M}_{pxq} = [\mathbf{m}_1 \; \ldots\ldots \; \mathbf{m}_q] \tag{6}$$

where $\mathbf{m}_i$ is the measurement vector at sampling time t_i. The source signal vector samples at these sampling times can be arranged into a "source matrix" :

$$\mathbf{S}_{rxq} = [\mathbf{s}_1 \; \ldots\ldots \; \mathbf{s}_q] \tag{7}$$

Measurement and source matrices are then related by :

$$\mathbf{M}_{pxq} = \mathbf{T}_{pxr} \, \mathbf{S}_{rxq} \tag{8}$$

In section II, the principal idea of the method and its formalization will be given. Section III will deal with the problem of how to satisfy the requirements of the method for practical bioelectrical applications. Experimental results on real data will be discussed in section IV.

II Oriented energy and source selection

Suppose that $\mathbf{A}_{pxq}$ is a matrix representing a set of q vector samples $\mathbf{a}$ from a p dimensional signal space, with $q \gg p$:

$$\mathbf{A} = [\mathbf{a}_1 \; \ldots\ldots \; \mathbf{a}_q] \tag{9}$$

Then the energy of this set in the direction of a unit vector $\mathbf{e}$ can be defined as |4| :

$$E_{\mathbf{e}}[\mathbf{a}] = \sum_{i=1}^{q} (\mathbf{e}^T \mathbf{a}_i)^2 \tag{10}$$

It is the sum of the squared orthogonal projections of all $\mathbf{a}_i$ onto $\mathbf{e}$. For example, fig.2. and 3. give polar plots of the oriented energy of sets of vectors in a 2, resp. 3 dimensional signal space. In general, for higher dimensions, polar plots of the oriented energy all have the same properties : orthogonal directions of extremal energy, and a maximum, a minimum, saddlepoints. A numerically very reliable technique to find these directions and values of extremal energy |4|, is the singular value decomposition of the matrix **A** |1,2,3| :

$$\mathbf{A}_{pxq} = \mathbf{U}_{pxp}\, \mathbf{\Sigma}_{pxq}\, \mathbf{V}_{qxq} = [\mathbf{u}_1 \dots \mathbf{u}_p]\, [\mathrm{diag}(\sigma_1 \dots \sigma_p) \mid \mathbf{O}\,]\, \mathbf{V}_{qxq} \qquad (11)$$

in which **U** and **V** are orthonormal matrices, and $\mathbf{\Sigma}$ is a real, pseudodiagonal, nonnegative definite matrix. The singular spectrum is the set of diagonal elements (singular values) of $\mathbf{\Sigma}$. The number of non zero singular values is equal to the rank of matrix **A** . It can easily be proven |4|, that each column $\mathbf{u}_i$ of **U** represents a unit vector in the p dimensional signal space in a direction of extremal energy of **A**, and

$$E_{\mathbf{u}_i}[\mathbf{a}] = \sigma_i^2 \qquad (12)$$

where σ_i is the diagonal element of $\mathbf{\Sigma}$, corresponding to $\mathbf{u}_i$ (also called the i-th left singular basis vector). The eq.(12) means that the extremal energy in the direction of $\mathbf{u}_i$ equals σ_i^2. In further discussions, it will be assumed that columns of **U**, diagonal elements of $\mathbf{\Sigma}$,and rows of **V** are permuted so that the singular values are ordered along the diagonal of $\mathbf{\Sigma}$ in decreasing order of magnitude. In the example of fig.3., this implies that $\mathbf{u}_1=\mathbf{x}$, $\mathbf{u}_2=\mathbf{y}$, $\mathbf{u}_3=\mathbf{z}$.

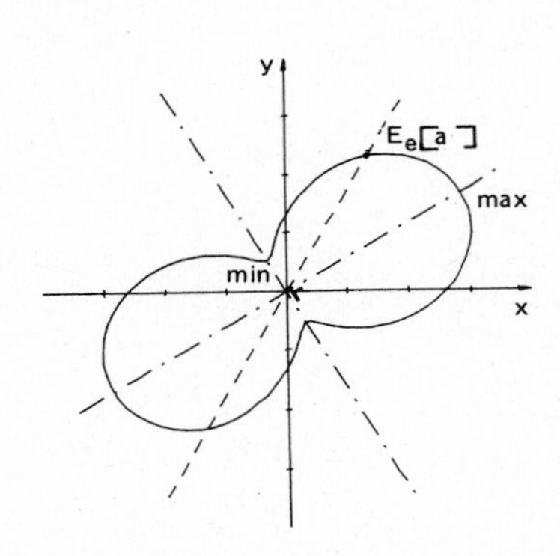

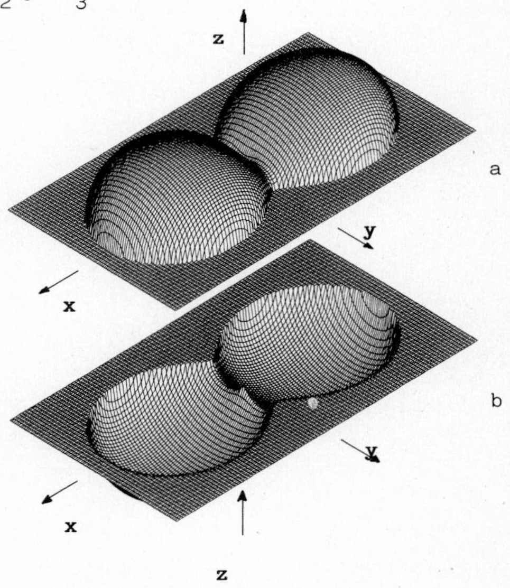

Fig.2. (above) Oriented energy in a 2 dimensional signal space.

Fig.3. (right) Oriented energy in a 3 dimensional signal space.

The basic idea of the method will now be demonstrated on the basis of fig.4. and formalized in theorem 1 afterwards. Fig.4.a. represents a source matrix **S**, see eq.(7), for a 2 dimensional source space. The two source signals are zero mean Gaussian, not correlated, and they have different energies. Fig.4.b. represents the measurement matrix **M**, the **T** mapping of **S** into a 3 dimensional measurement space. There will be no samples outside the plane $(\mathbf{t}_1,\mathbf{t}_2)$, so that $\mathbf{u}_3$ will be orthogonal to this plane. Therefore, the plane$(\mathbf{u}_1,\mathbf{u}_2)$ will coincide with the plane $(\mathbf{t}_1,\mathbf{t}_2)$. Due to the non orthogonality of $\mathbf{t}_1$ and $\mathbf{t}_2$ in fig.4.b., directions of extremal energy in the plane $(\mathbf{t}_1,\mathbf{t}_2)$ will not coincide with $\mathbf{t}_1$ and $\mathbf{t}_2$ respectively. However, if $\mathbf{t}_1$ can be made orthogonal to $\mathbf{t}_2$, by proper location of the electrodes, $\mathbf{u}_1$ resp. $\mathbf{u}_2$ will coincide with $\mathbf{t}_1$ resp. $\mathbf{t}_2$. Projections of the measurement vector signal onto the $\mathbf{u}_1$ and $\mathbf{u}_2$ vectors will then result in the respective source signals. This idea can now be formalized, and generalized to higher dimensional problems.

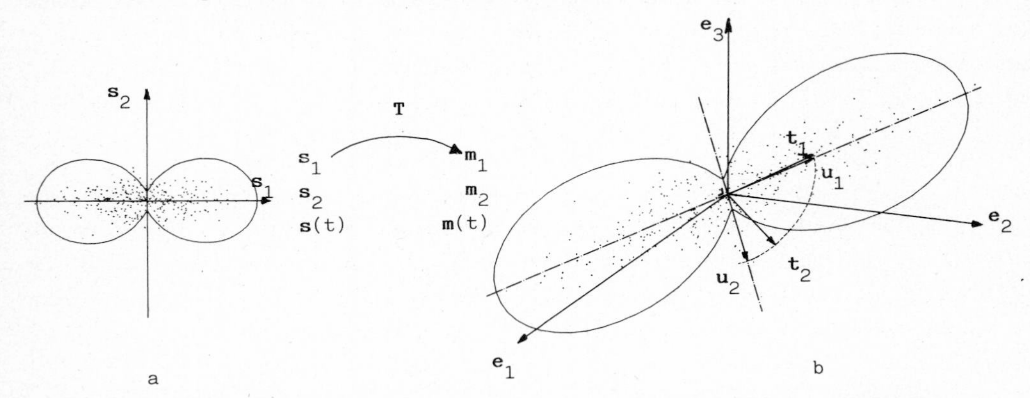

Fig.4. 300 samples and their oriented energy in (a) source,(b) measurement space.

Theorem 1

For a matrix $\mathbf{M}$, given by :

$$\mathbf{M}_{pxq} = \mathbf{T}_{pxr}\,\mathbf{S}_{rxq} \qquad \text{with } r \leqslant p \leqslant q \tag{13}$$

with a singular value decomposition given by :

$$\mathbf{M}_{pxq} = \mathbf{U}_{pxp}\,\mathbf{\Sigma}_{pxq}\,\mathbf{V}_{qxq} \tag{14}$$

if (1) $\mathbf{SS}^T = \text{diag}(\sigma_{s1}^2 \;\ldots.\; \sigma_{sr}^2)$ i.e. the source signals are not correlated for zero shift in time,

(2) $\sigma_{s1} > \sigma_{s2} > \ldots. > \sigma_{sr} > 0$, i.e. the source signals have different non zero energies,

(3) $\mathbf{T}^T\mathbf{T} = \mathbf{I}_{rxr}$, i.e. columns of $\mathbf{T}$ are orthonormal, and rank($\mathbf{T}$) = r,

then for $1 \leqslant i \leqslant r$:

(1) $\mathbf{u}_i = \pm\, \mathbf{t}_i$, with $\mathbf{u}_i$ resp. $\mathbf{t}_i$ the i-th column of $\mathbf{U}$ resp. $\mathbf{T}$,

(2) $\sigma_i = \sigma_{si}$

(3) $\mathbf{u}_i^T\,\mathbf{M} = \pm\, \mathbf{s}^i$, with $\mathbf{s}^i$ the i-th row of $\mathbf{S}$. □

The proof of this theorem is based on the uniqueness of the SVD. The first statement of the theorem means that the first r left singular basis vectors $\mathbf{u}_i$ are, apart from their sign, equal to the respective source vector. The second statement says that the extremal oriented energy values are equal to the respective source energies. Remark that the last p-r singular values are zero since rank($\mathbf{S}$) = r. The third statement says that a source signal can be selected by projecting the measurements onto the corresponding left singular basis vector. The fact that the sign of the source vectors is not found, is not very important in a practical situation. E.g. in ECG, inspection of the projections is enough to know the sign if needed.

III Application under practical biomedical conditions

MEASUREMENT NOISE Theorem 1 does not take into account the influence of additive measurement noise, merely for the sake of its notational clarity. However, as will be shown, commonly occuring noise conditions will only have a small influence on the estimation of the source vectors. This will be demonstrated first by a low dimensional example.

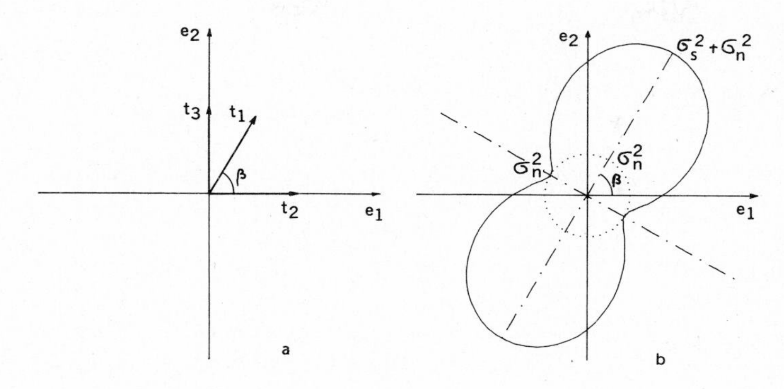

Fig.5. (a) source vectors for example 1, (b) oriented energy if $\sigma_n = \sigma_{n1} = \sigma_{n2}$.

Example 1

See fig.5.a.. Suppose that there is one signal source in the body, and measurements are done with 2 noise generating electrode-amplifier combinations. It is as if there are 3 sources. The first source has a source vector $\mathbf{t}_1$ making an angle ß with $\mathbf{e}_1$, a basis vector of the measurement space. The noise sources have a source vector ($\mathbf{t}_1$ and $\mathbf{t}_2$) coinciding with one basis vector each. Suppose now that these 3 source signals are not correlated (a quite realistic assumption) and that their resp. energies are given by σ_s^2, σ_{n1}^2, σ_{n2}^2. It can then be verified that extremal energy directions will be given by :

$$\theta_i = \tfrac{1}{2} \arctan(\sin 2\beta /(\cos 2\beta + \varepsilon)) \quad \text{with } \varepsilon = (\sigma_{n1}^2 - \sigma_{n2}^2) / \sigma_s^2 \tag{15}$$

Where θ_i is the angle between $\mathbf{u}_i$ and $\mathbf{e}_1$. Now, if the difference in energy between the two noise sources is much smaller than the energy of the source signal in the body, ε is a very small quantity. It is clear then that $\theta_1 \simeq \beta$. This means that $\mathbf{u}_1$ of the SVD of **M** will be almost the same as the source vector of the body source. And the accuracy does not depend on the S/N ratio in the measurements, but on the ratio of the difference in noise energy between the electrodes to the body source energy. Under the above assumption that ε is a very small quantity, the polar plot of the oriented energy of the measurements will be given by fig.5.b. In all directions the energy is enlarged by the same amount. Remark that the measurement matrix **M** will now be of full rank. The same conclusions can be made for more dimensional problems.

Theorem 2

Let $\mathbf{M}$, $\mathbf{N}$, $\mathbf{T}$, and $\mathbf{S}$ be resp. the measurement, the noise, the transfer, and the source matrix, satisfying :

$$\mathbf{M}_{pxq} = \mathbf{T}_{pxr}\,\mathbf{S}_{rxq} + \mathbf{N}_{pxq} \tag{16}$$

If (1) $\mathbf{N}\mathbf{N}^T = \sigma_N^2\,\mathbf{I}_{pxp}$, meaning that noise in the measurement signals is not correlated, and all the noise generators have the same energy,

(2) $\mathbf{N}\mathbf{S}^T = \mathbf{O}_{pxr}$, meaning that the additive noise is not correlated with the sources in the body,

then the SVD of $\mathbf{M}=\mathbf{U}_M\mathbf{\Sigma}_M\mathbf{V}_M$ and $\mathbf{TS}=\mathbf{U\Sigma V}$ are related by :

for $1 \leqslant i \leqslant r$:

(1) $\mathbf{u}_{Mi} = \pm\,\mathbf{u}_i$

(2) $\sigma_{Mi}^2 = \sigma_i^2 + \sigma_N^2$

and for $r+1 \leqslant i \leqslant p$:

(3) $\sigma_{Mi}^2 = \sigma_N^2$ □

The proof of this theorem is based on the fact that any orthonormal matrix $\mathbf{U}_N$ satisfies the definition of the SVD of $\mathbf{N}$, since all its singular values are equal. The most important result of theorem 2 is that under the given conditions on $\mathbf{N}$, combined with those of theorem 1, directions of extremal energy will still coincide with the source vectors. Then for $1 \leqslant i \leqslant r$:

$$\mathbf{u}_{Mi}^T\mathbf{M} = \pm\mathbf{s}^i + \mathbf{u}_{Mi}^T\mathbf{N} \tag{17}$$

Another result is that $\mathbf{M}$ in eq.(16) is now of full rank, since it has only non zero singular values. However, the numerical rank can be estimated by using results (2) and (3) of theorem 2. Indeed, there will be r singular values which are larger than the other, equal p-r singular values.

SOURCE SIGNALS Condition (1) of theorem 1 requires that $\mathbf{S}\mathbf{S}^T$ is a diagonal matrix. Suppose that the crosscorrelation functions of the source signals :

$$R_{ij}(\tau) = \lim_{T\to\infty} \frac{1}{T}\int_0^T s_i(t)\; s_j(t+\tau)\; dt \tag{18}$$

satisfy :

$$R_{ij}(0) = \delta_{ij}\;, \quad 1 \leqslant i,j \leqslant r \tag{19}$$

with δ_{ij} the Kronecker delta. Then $\mathbf{S}\mathbf{S}^T$ will be an unbiased estimator of :

$$q\begin{bmatrix} R_{11}(0) & \cdots\cdots & R_{1r}(0) \\ \vdots & & \vdots \\ R_{r1}(0) & \cdots\cdots & R_{rr}(0) \end{bmatrix} \tag{20}$$

The estimation error will depend on the number of samples, the sampling frequency, and the actual source waveforms. So, even if all source signals satisfy eq.(19), attention has to be paid to the mentioned parameters. The same arguments can be applied to conditions (1) and (2) of theorem 2.

Now, usually even eq.(19) is not satisfied. E.g. the x,y,z source signals of the heart are correlated in general. Then $\mathbf{S}\mathbf{S}^T$ is not a 3x3 diagonal matrix. However, this is no serious problem. Indeed, suppose that the SVD of **S** is given by :

$$\mathbf{S} = \mathbf{U}_S \boldsymbol{\Sigma}_S \mathbf{V}_S \tag{21}$$

We then define :

$$\mathbf{T}_2 = \mathbf{T}\,\mathbf{U}_S \quad \text{and} \quad \mathbf{S}_2 = \boldsymbol{\Sigma}_S \mathbf{V}_S \tag{22}$$

so that :

$$\mathbf{M} = \mathbf{T}\,\mathbf{S} = \mathbf{T}_2\,\mathbf{S}_2 \tag{23}$$

Then it can be easily verified that :

$$\mathbf{T}_2^T \mathbf{T}_2 = \mathbf{I} \quad \text{if} \quad \mathbf{T}^T\mathbf{T} = \mathbf{I} \tag{24}$$

$$\mathbf{S}_2 \mathbf{S}_2^T = \boldsymbol{\Sigma}_S^2 \tag{25}$$

The physical meaning of this is that :

(1) the original source signal vector samples are first projected onto an orthonormal set of basis vectors $\mathbf{U}_S$ in real space, where $\mathbf{U}_S$ is chosen in such a way that these projections are not correlated,

(2) the lead vectors (the rows of **T**) are rotated in real space in such a way that exactly the same measurement signals result.

The result of applying the method will then be that not the x,y,z components are estimated, but rather the source signal vector represented in this $\mathbf{U}_S$ basis. This can be a considerable advantage in clinical practice. E.g. in vector cardiography, the resulting signals will not depend on the physical rotation of the heart, which should facilitate interpretation of the recordings. One could call this a normalized ECG. For the description and the clinical value of another, manual normalization technique see |6|. Also in fetal ECG's, the resulting signals will not depend on a rotation of the fetus.

If the source matrix can be partitioned in uncorrelated blocks, then the same reasoning as in eq.(21)-(25) can be applied for each block, after partitioning **T** in a corresponding way. Another conclusion from eq.(21)-(25) is the following. Suppose **S** is not of full rank. E.g. if there is a dipole vector which stays in a plane. Then the singular values in $\boldsymbol{\Sigma}_S$ of eq.(21) will not all be non zero. The result will be that the method only "sees" the minimal representation of the sources. This is a clear advantage, since the signal to noise ratios will then be optimal.

<u>TRANSFER MATRIX</u> Condition (3) of theorem 1 says that columns of **T** should be orthonormal. It is certainly not evident that this can always be achieved. First it is shown that for many applications this condition can approximately be satisfied in an easy way. Secondly that small deviations of orthogonality have a limited influence on the resulting signals.

First of all, columns of **T** do not have to be of unit length. Indeed, if they are only orthogonal, they can be supposed to be of unit length by scaling the corresponding source signal. From this it can be seen that, under the conditions of theorem 2, an extra electrode which picks up one or more sources, will improve the signal to noise ratios of the projections for these sources. Indeed, in this case, in eq.(17), the term $\mathbf{s}^i$ will be scaled up, while the term $\mathbf{u}_{Mi}^T\mathbf{N}$ will still have the same energy.

However the best S/N ratio improvement that can be obtained by using N_e electrodes instead of 1, will be by a factor N_e .
Further on, in some situations, the condition of orthogonality of the transfer matrix **T** is quite easy to satisfy. Suppose there are 2 sources and 2 electrode signals. One of the electrodes ($\mathbf{e}_1$) picks up contributions from both sources, and the other electrode ($\mathbf{e}_2$) only from one source. See fig.6.. It is clear that the larger the relative contribution of source 2 to measurement signal 2, the more the orthogonality will be improved. This conclusion can readily be generalized to more dimensional sources. An interesting situation is the following. Suppose that a weak source, not too close to the heart, has to be measured. Electrodes in the neighbourhood of the weak source (like $\mathbf{e}_1$ above) will also pick up serious amounts of the ECG (like $\mathbf{t}_1$ and $\mathbf{t}_2$ above). However, electrodes near the heart (like $\mathbf{e}_2$), pick up almost only ECG (like $\mathbf{t}_2$). If enough electrodes are used in the neighbourhood of the heart, to span all dimensions of it (normally 3), then the 3 dimensional ECG subspace in the measurement space will be almost orthogonal to the subspace of the weak source. And projections onto the left singular vectors $\mathbf{u}_4$, $\mathbf{u}_5$, ... will be free of ECG. Remark that the positions of the three heart leads is not critical at all. This procedure turned out to be very effective to remove maternal ECG from abdominal fetal ECG recordings, see section IV and |7|.

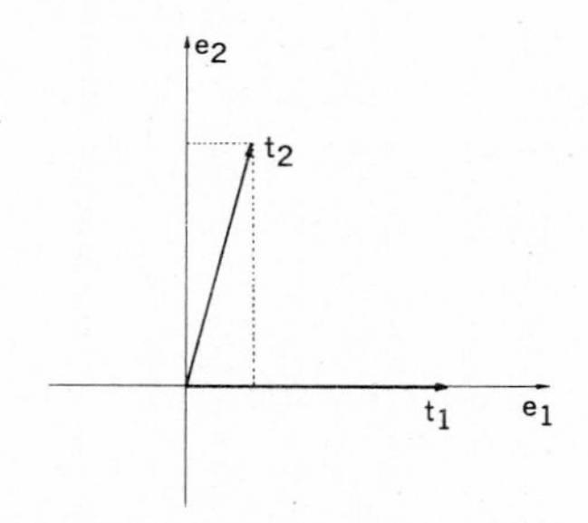

Fig.6. Orthogonalization by picking up source 2 seperately.

For an accurate determination of good electrode position, certainly for more than one dipole, numerical simulation of the transfer may be needed. Generally, the orthogonality of the source vectors can be achieved by a multitude of electrode arrangements, since there are many degrees of freedom.
The consequence of non orthogonality of the source vectors will be illustrated by the following low dimensional example.

Example 2

Suppose that there are 2 uncorrelated source signals, with energies of resp. σ_{s1}^2, σ_{s2}^2. Let the angle between $\mathbf{t}_1$ and $\mathbf{t}_2$ be α. Then it can be found that the angle between $\mathbf{t}_1$ and the direction of maximal energy will be given by :

$$\theta_1 = \tfrac{1}{2} \arctan(\sin 2\alpha / (\cos 2\alpha + s^2)) \quad \text{with } s^2 = \sigma_{s1}^2 / \sigma_{s2}^2 \qquad (26)$$

Fig.7.a. shows θ_1 vs. α for some values of s^2. An important conclusion from these curves is that even for moderate ratios of source singular values, the first singular vector of the measurements is not very far from the strongest source vector. E.g. for s=2, the worst case is about 7°, at an angle of 51° between the source vectors. So a projection of measurement vectors onto a direction orthogonal to $\mathbf{u}_1$ will contain only a small portion of the first source signal. Now a signal_2 to signal_1 ratio can be defined for this projection, in which "signal_2" means weak source contribution, and "signal_1" means strong source contribution. This ratio becomes :

$$S_2/S_1 = \sin^2(\theta_1 - \alpha) \; / \; (s^2 \sin^2\theta_1) \tag{27}$$

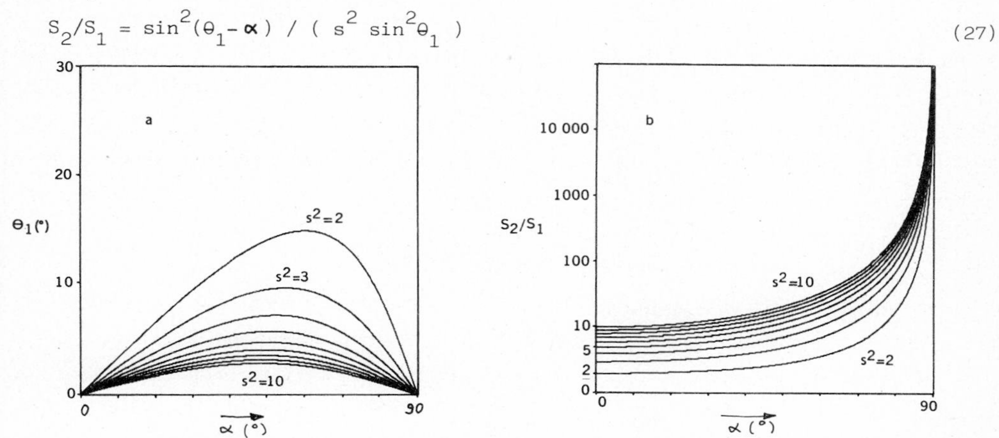

Fig.7. Influence of angle between source vectors ( ) and energy ratio of sources on (a) angle between $\mathbf{u}_1$ and $\mathbf{t}_1$,(b) S_2/S_1 ratio.

Fig.7.b. gives this ratio vs. α, for some values of s^2. First of all, in the neighbourhood of α=90°, signal_2 contributes more to the projected signal than signal_1, hence the ratio is high. Decreasing the angle between the source vectors will make the S_2/S_1 ratio less good. However, this ratio will never decrease below a limiting value, for every $|\alpha|>0$. This limiting value can be found to be :

$$\lim_{\alpha \to 0} S_2/S_1 = s^2 \tag{28}$$

This is in fact a very remarkable result : the stronger the disturbing source, the better it will be rejected. For the generalization to more than two sources, only a qualitative assertion will be given. The analysis in a formal, general way is quite involved. Indeed, the number of angles between n source vectors equals n(n-1)/2. In general, for any number of sources, a projection of the measurement vectors onto a certain left singular basis vector, may contain a large portion of lower energy sources, but it contains only a small portion of the higher energy sources.

<u>INTERFERENCE</u> A frequently occuring problem when measuring weak signals is interference from the power distribution network (50 or 60 Hz). This interference can be seen as resulting from an extra source, and a dynamic transfer. Although the presentation has been exclusively on static transfers, it will be shown now that at

least this dynamical problem can also be covered by the method. Indeed, suppose that the interferences in the measurement signals contain only the first harmonic of the interfering source, and that the transfer from this source is linear. Then it is possible to reconstruct any interference signal by a linear combination of 2 signals, i.e. 2 sinusoidal signals with the interfering frequency, but delayed relative to each other. These two sinusoidal signals can earily be introduced as 2 extra measurement signals, if the power network signal is available. We can look at it as if there are 2 extra source signals in the body (2 sinusoids in quadrature), for which the source vectors can be made almost orthogonal to all other source vectors. Another way of formulation is that we have constructed a set of p 2-tap adaptive FIR filters, having as input a signal, highly correlated to the interfering source. The advantage of this method of reducing interference, is that there is no need for additional signal processing (like e.g. a notch filter), and that no distortion is introduced.

IV Experimental results

The practical value of the approach has been verified in a number of experiments to record the fetal electrocardiogram (FECG) from electrodes on the skin. All these electrodes pick up maternal ECG (MECG). The results that will be presented focus the following :

- how to orthogonalize the MECG subspace to the FECG subspace for MECG elimination,
- the insensitivity of the MECG elimination quality on actual positions of the MECG electrodes,
- the numerical dimensionality of the source space,
- improvement of the S/N ratio by adding an electrode,
- elimination of power network interference (50 Hz) from estimated source signals, in the multidimensional all-in approach.

For more information on experimental conditions, and other results see |7|. Some simulation results are given in |8|. For all presented recordings and results, the sampling frequency was 250 Hz, and the number of vector samples on which SVD was applied was 1000. Notations are :

$m_i(t)$ is the i-th measurement signal,

$\hat{s}_i(t) = \mathbf{u}_i^T\, \mathbf{m}(t)$ is the projection of the measurement vector signal onto the i-th direction of extremal energy (where i gives the order of the corresponding singular value).

A result of one of the first experiments is shown in fig.8. The sophism that each measurement signal should contain a contribution from the weak source (the fetal heart), led to the application of abdominal electrodes only. Five measurement signals were recorded (fig.8.a.). Projections 1 to 3 are given in fig.8.b., projections 4 to 5 contain only noise. From the figure it is clear that the numerical rank of $\mathbf{M}$ should be estimated to be 2 or 3. The fact that $\hat{s}_3(t)$ contains only a very small FECG contribution, relative to the FECG contribution in $\hat{s}_2(t)$, shows that $\mathbf{t}_3$ should be very close to the plane $(\mathbf{t}_1,\mathbf{t}_2)$. This means that the required orthogonality is far from achieved. For better orthogonality, a numerical simulation of the transfer could have been performed, as mentioned in section III. However, further experiments were carried out by applying the simple orthogonalization technique of separate electrodes

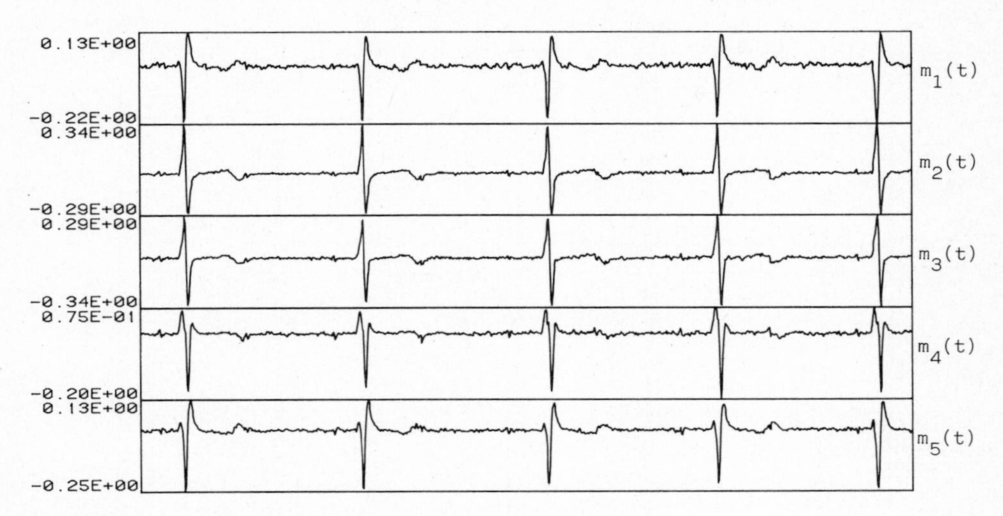

Fig.8.a. 5 abdominal recordings for FECG detection.

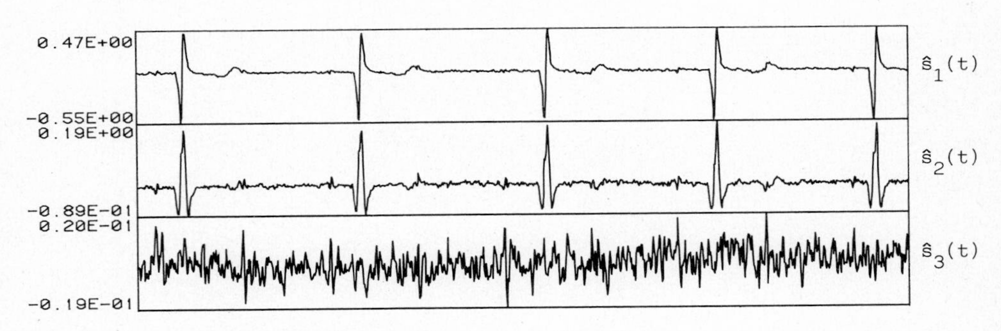

Fig.8.b. Projections onto $\mathbf{u}_1$ to $\mathbf{u}_3$.

for the MECG. This proved to be very efficient. E.g. fig.9.a. shows the 5 measurement signals of one of the subsequent experiments. Signals 1 to 3 are from thoracic electrodes, 4 and 5 from abdominal electrodes. The projection signal $\hat{s}_4(t)$ is shown in fig.9.b. It is very clear that this is a signal containing almost no contribution from the MECG. Even FECG peaks coinciding with MECG peaks are clearly discernable. Due to the fact that the 3 thoracic electrodes are so close to the maternal heart, the numerical dimension of the MECG subspace in the measurement space became 3 instead of 2 in the former experiment. Therefore the first non MECG extremal energy direction is $\mathbf{u}_4$. For all the experiments that have been done, the dimension of the

FECG subspace had to be estimated to be 1. This suggests that the equivalent fetal heart dipole does not change direction in real space.

To verify the insensitivity of the MECG elimination on positions of the thoracic electrodes, the next experiment has been carried out. On the same pregnant subject, the same electrode positions are applied as in the former experiment, except for $m_1(t)$ and $m_3(t)$. See fig.9.c. The signal $\hat{s}_4(t)$, shown in fig.9.d., resulting in that case is indeed very comparable to the former result (fig.9.b.).

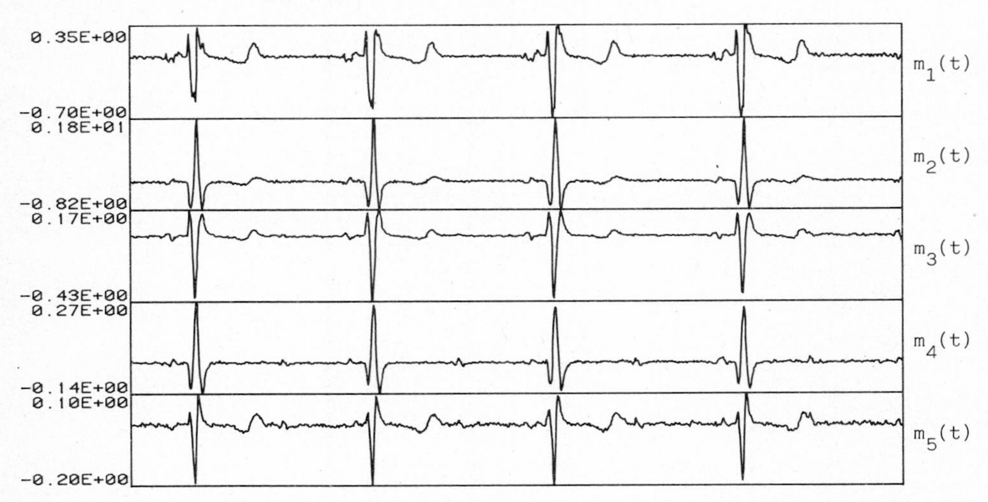

Fig.9.a. 2 abdominal and 3 thoracic electrodes to orthogonalize the MECG subspace to the FECG subspace.

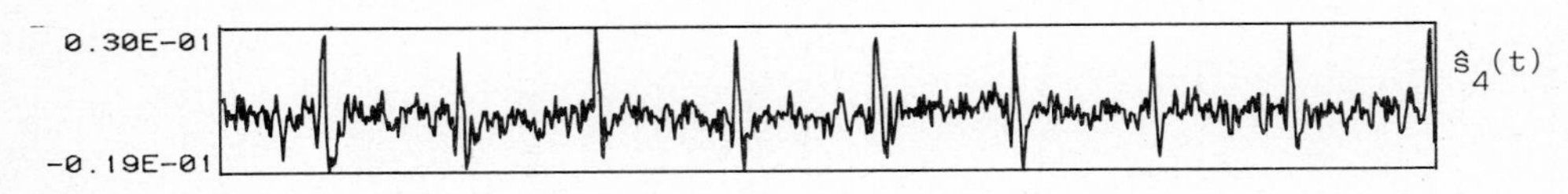

Fig.9.b. Projection onto $\mathbf{u}_4$.

Fig.10. shows $\hat{s}_4(t)$ if the same recordings as in fig.9.c. are used, but with $m_5(t)=0$. It can be seen that the S/N ratio is slightly lower now. It can also be seen that the absolute MECG residue is about the same. This confirms the statement that an extra electrode picking up the weak source, but not decreasing the orthogonality between source vectors (the same MECG residue) improves the S/N ratio.

The next experiment shows how 50 Hz interference can be eliminated in the same operation. Two measurement signals came from abdominal electrodes (one of them is shown in fig.11.a.), two others came from thoracic electrodes. Fig 11.b. gives $\hat{s}_3(t)$ if only these 4 measurement signals are applied. Fig 11.c. gives $\hat{s}_5(t)$ if also 2 other signals (the power network signal undelayed, as well as delayed over 1 sampling interval) are applied. For fig.11.b. there were 2 directions of higher extremal energy (MECG), and for fig.11.c. there were 4 directions of higher extremal energy (MECG and 50 Hz). So resp. $\mathbf{u}_3$ and $\mathbf{u}_5$ correspond to the fetal source vector. The improvement is obvious from the figures. Remark also that in this case it happened that a left singular basis vector changed sign, as discussed in connection with theorem 1.

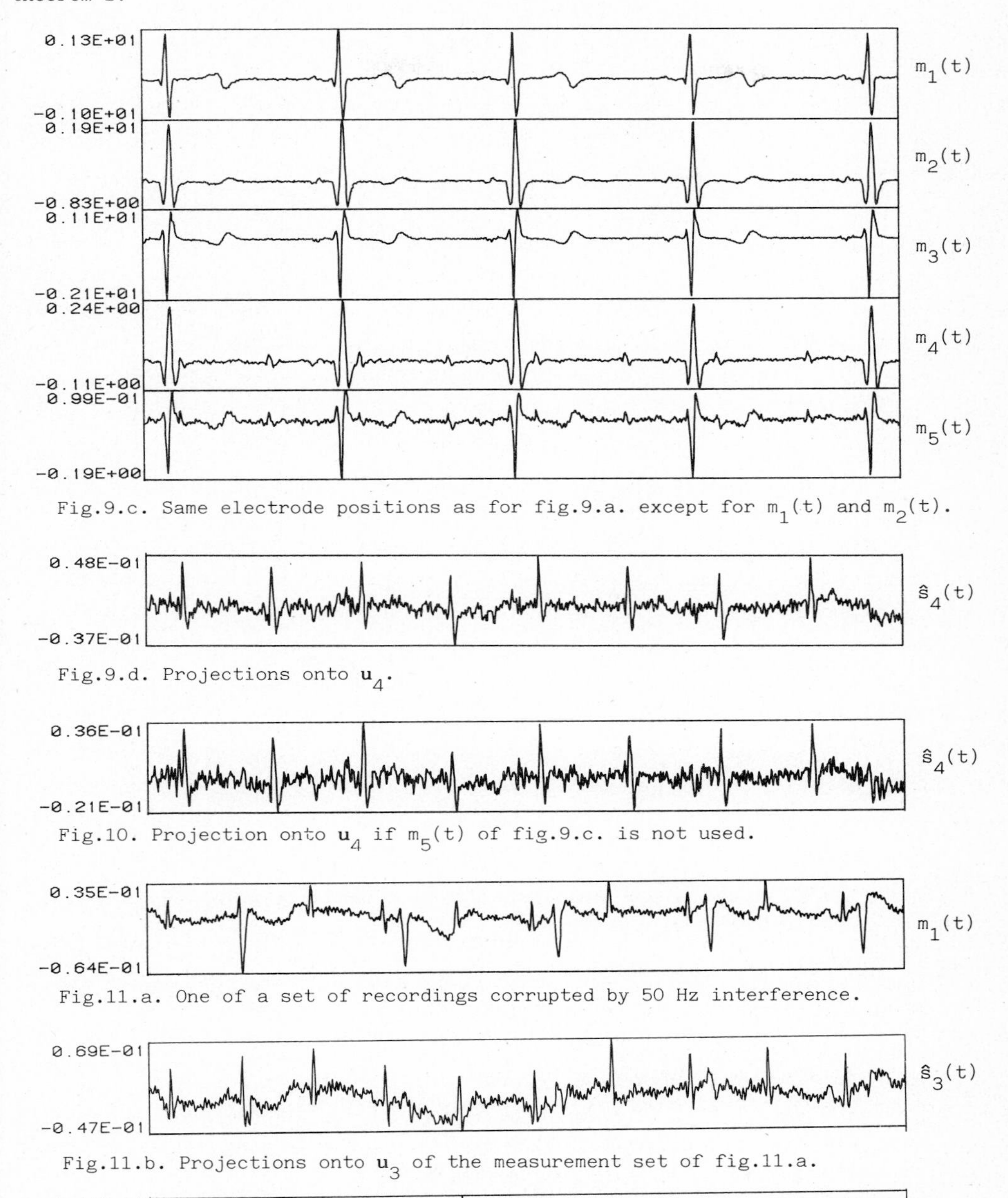

Fig.9.c. Same electrode positions as for fig.9.a. except for $m_1(t)$ and $m_2(t)$.

Fig.9.d. Projections onto $\mathbf{u}_4$.

Fig.10. Projection onto $\mathbf{u}_4$ if $m_5(t)$ of fig.9.c. is not used.

Fig.11.a. One of a set of recordings corrupted by 50 Hz interference.

Fig.11.b. Projections onto $\mathbf{u}_3$ of the measurement set of fig.11.a.

Fig.11.c. Projections onto $\mathbf{u}_5$ if also the power network signal is applied.

V Conclusions

A new technique has been presented for the solution of the so called inverse problem for current dipole sources at fixed positions in the human body, based on a simple geometrical, vector space interpretation of this problem. If electrode positions are chosen so that source vectors in the measurement space are nearly orthogonal, then the different source signals can be extracted from the measurements. Indeed, it has been shown that in this case directions of extremal energy correspond to directions spanned by individual sources. The numerical technique to find these directions accurately is the singular value decomposition (SVD). This technique also gives an energy spectrum, on the basis of which the number of sources can be estimated. For a time invariant transfer from sources to electrodes, this SVD has to be carried out only once, for a limited number of vector samples. Further measurements can be decomposed by a simple projection onto the orthonormal basis vectors correponding to the sources of interest. If a certain more dimensional stronger source (e.g. the heart) is of no interest, then only the subspace spanned by this source (not each individual source vector of it) has to be orthogonal to the other source vectors. A condition which can easily be satisfied for a practical problem, most of the time. It is shown that application of the method allows to achieve in one step : optimal discrimination between multidimensional sources in the body, optimal S/N ratios in the estimates of the source signals, and elimination of 50 or 60 Hz interference. Other, general advantages are : the method does not depend on the actual waveform of the signals; uniform sampling is not required (allowing optimization of calculation effort); conceptually fitting into the notions of bioelectricity; extendable to certain slowly varying or dynamical transfers; efficient : on line detection of the FECG with 8 electrodes and at a sampling rate of 250 Hz would require a multiply-add time of about 20 μs.

References

|1| C.Eckhart and G.Young, "A principal axis transformation for non Hermitean matrices," Bull.Amer.Math.Soc., vol.45, pp.118-121, 1939.

|2| G.H.Golub and C.Reinsch, "Singular value decomposition and least squares solutions," Numer.Math., vol.14, pp.403-420, 1970.

|3| C.L.Lawson and R.J.Hanson, "Solving least squares problems," Prentice Hall Series in Automatic Computation, Englewood Cliffs, 1974.

|4| J.Staar, "Concepts for reliable modelling of linear systems with application to on line identification of multivariable state space descriptions," Doct.Th., Kat.Univ.Leuven, Belgium, Jun.1982.

|5| R.Plonsey, "Bioelectric Phenomena," McGraw-Hill, New-York, 1969.

|6| R.McFee, R.S.Wilkinson and J.A.Abildskov, "On the normalization of the electrical orientation of the heart and the representation of electrical axis by means of an axis map," Am.Heart.J., vol.62, no.3, pp.391-397, 1961.

|7| J.Vanderschoot, G.Vantrappen, J.Janssens, J.Vandewalle and W.Sansen, "A reliable method for fetal ECG extraction from abdominal recordings," Submitted to 5-th Congr. Eur. Fed. Medical Informatics, Brussels, 1984.

|8| J.Vanderschoot, J.Vandewalle, G.Vantrappen, W.Sansen and J.Janssens, "Application of singular value decomposition to vector signal processing," Submitted to Electronic Letters.

A SYSTEM-ANALYTICAL APPROACH TO THE PROCESS OF FEVER

J. Werner and R. Graener
Institut für Physiologie, AG Elektrophysiologie,
Ruhr-Universität, D 4630 Bochum, FRG

SUMMARY

Pyrogen-induced fever is a strictly dynamic process without steady-states before reaching again the normal temperature level. Therefore former models of central control of metabolic heat production (Stitt, 1980; Simon, 1981; Werner, 1981) do not enable an adequate computation of the dynamics of fever, as the equations are formulated on the basis of steady-state relations. The purpose of this study is to derive from experimental results a simple model describing the febrile process.

Female rabbits were exposed to air-temperatures $< 20°C$ in a climatic chamber. After the animals had reached their thermal equilibrium, bacterial pyrogen was injected intravenously.

The dependence of metabolic heat production on the rate of change of core temperature dT_C/dt in the course of the experiment is a quasi-linear one, independently of the type of experiment. This is a property of the passive system due to instationary heat flow. The properties of the closed control-loop imply a strictly proportional controller. A pure change of controller-gain in fever is excluded, as this would not evoke changes of core temperature observed in the experiment. Fever is primarily due to a parallel shift of the controller characteristic via the pyrogen. The set of equations enables the determination of the central controller gains and of the fever-signal controlling directly metabolic heat production.

OUTLINE OF THE MODEL

Structure and dynamics of temperature regulation have been analyzed by system-analytical methods since many years (for review see Werner, 1975, 1981, 1983), whereas the process of pyrogen-induced fever up to now has been looked at only in steady-state terms. As a first approach we propose a lumped two-compartment-model (fig. 1), i.e. a mathematical description of the heat-transfer-processes in the body core (c) and in the body shell (s). This leads to the following set of equations:

$$(1) \quad m_c \cdot c_c \frac{dT_c}{dt} = M - k\,(T_c - T_s) - R$$

$$(2) \quad m_s \cdot c_s \frac{dT_s}{dt} = k\,(T_c - T_s) - hA\,(T_s - T_A) - E$$

with

- m = mass of the compartment c or s
- c = specific heat of the tissue
- T = temperature
- t = time
- M = metabolic heat production
- h = transfer coefficient skin/environment
- k = heat conductivity index
- A = skin area
- R = respiratory evaporative heat loss
- E = evaporative heat loss via skin
- T_A = air temperature

Because there is experimental evidence, that in the febrile process the rate of change of mean skin temperature dT_s/dt is minimal at a constant air-temperature, we assume $dT_s/dt = 0$. In the cold we neglect evaporative heat loss R and E. This delivers an approximative equation for the passive system:

$$(3) \quad \frac{dT_c}{dt} = c_1 M - c_2\,(T_s - T_A)$$

with

$$c_1 = \frac{1}{m_c\, c_c}$$

$$c_2 = \frac{hA}{m_c\, c_c}$$

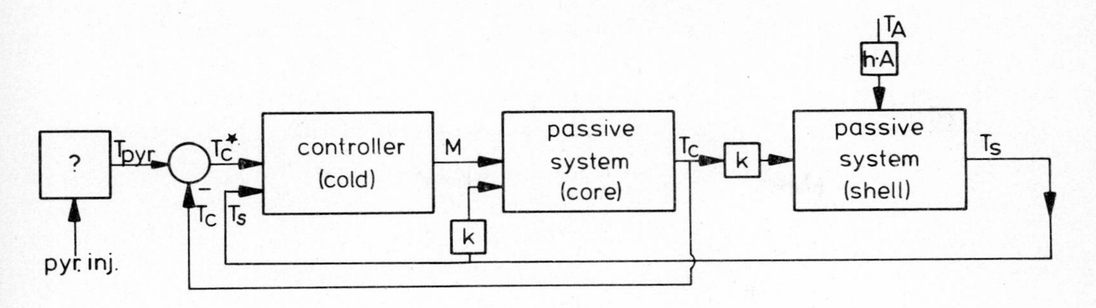

Fig. 1: Pyrogen-input in the thermoregulatory system in the cold (evaporative heat neglected)

For the controller no essential dynamic properties are assumed:

$$(4) \quad M = K_1 T_c^* - K_2 T_s$$

The controller input T_c^* may be imagined as follows (cf. fig. 1):

$$(5) \quad T_c^* = T_{pyr} - T_c,$$

where T_{pyr} is the pyrogen induced parallel shift of the central controller equation, although this does not mean that there exists an explicit computation of this difference in the central nervous system (Werner, 1981).

EXPERIMENTAL EVALUATION

This model first was evaluated by carrying out experiments on rabbits. Female rabbits of about 5 kg weight were exposed to air temperatures < 20°C in a climatic chamber. After the animals had reached their thermal equilibrium, bacterial pyrogen was injected intravenously (0,1 µg/kg). Hypothalamic or oesophageal temperature were considered to represent core temperature T_c. They were measured, as well as 9 skin temperatures delivering a mean skin temperature T_s, by use of thermocouples. Hypothalamic temperature was recorded by a chronically implanted thermocouple (diameter 300 µm). A recently developed amplifier circuit enabled an accuracy of 3/100°C. Furthermore metabolic heat production and respiratory evaporative heat loss were determinded by an open mask system.

Fig. 2 shows a typical experimental time course. It is obvious that oesophageal temperature is controlled by the increase of metabolic heat production. On the other hand, we do not know the time-course of the fever-input T_{pyr} controlling directly metabolic heat production.

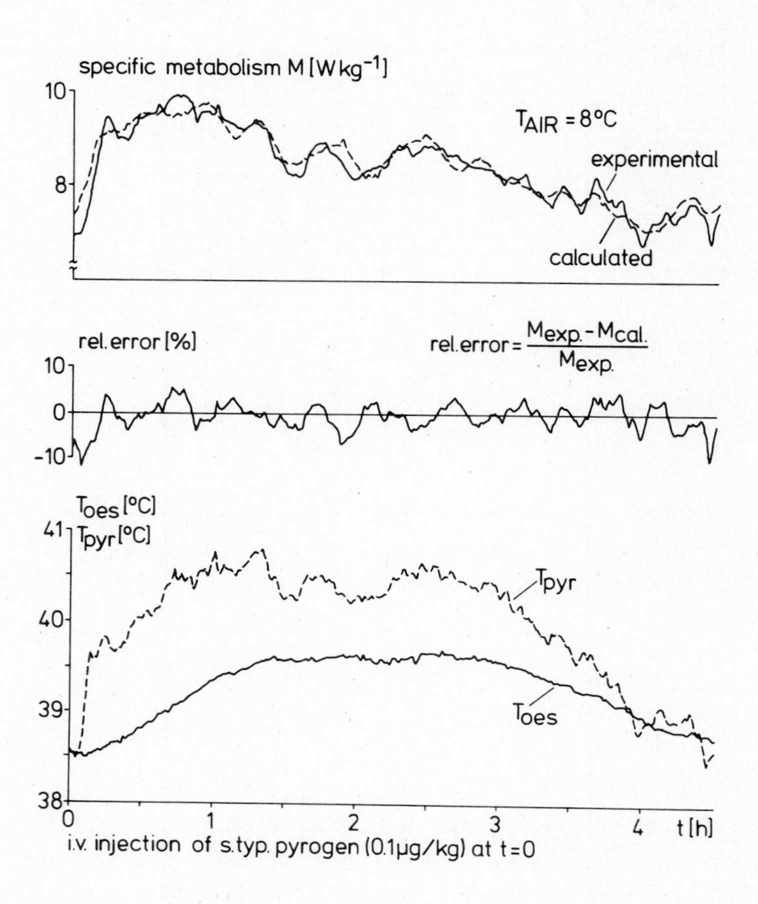

Fig. 2: The febrile process in experiment and computation. Upper part: comparison of experimental and calculated metabolism. Lower part: experimentally recorded oesophageal temperature T_{oes} and calculated fever signal T_{pyr}

To evaluate this, we eliminate T_s and T_c^* from equations (3) - (5) and get the following linear relationship

$$(6) \quad M = f^{\cdot}(T_c, dT_c/dt, T_A, T_{pyr}).$$

The parameters c_1 and c_2 are determined by using the physical properties of the body. The controller gains K_1 and K_2 are estimated by carrying out a multiple linear regression neglecting the input T_{pyr} in a first step. Then T_{pyr} is calculated by using the experimental time courses of T_c, T_s and M according to relation (4) and (5). Then the multiple linear regression is carried out again on equation (6). Now K_1, K_2 are varied, the corresponding T_{pyr} signal is calculated until the correlation coefficient of equation (6) becomes maximal.

The K_1, K_2 - values of this last iteration step are regarded as the true controller gains of equation (4), by which the final T_{pyr}-signal is calculated.

The fever signal T_{pyr} shown in fig. 2 was computed by choosing the values $K_1 = 5\ W\,kg^{-1}\,°C^{-1}$ and $K_2 = 0.26\ W\,kg^{-1}\,°C^{-1}$. If now metabolic heat production is calculated, we obtain a relatively small error between experiment and calculation (fig. 2). However, we should be capable of predicting metabolic heat production in quite different types of experiments. This is demonstrated in fig. 3 where two succeeding pyrogen injections were applied. The overall tendency is determined quite well, whereas there are some differences in the detailed time courses.

The experimental results demonstrate, corresponding with equation (6), a strong linear relationship between metabolic heat production and the first derivative of core-temperature dT_c/dt (cf. fig. 4), which is valid for all types of fever experiments carried out in our laboratory. This brings about two consequences:

1. Former models (Stitt, 1980; Werner, 1981; Simon, 1981) which do not take into account this relationship , but only infer a relation $M = f(T_c)$ are inadequate in predicting any time course.
2. The relation $M = f(dT_c/dt)$ may be used to get a first estimation of the time course of metabolic heat production.

An additional question is answered by the good compatibility between the experimental results and the model: changes of the central controller gains are not able to rise body temperature, as observed in fever, although this is suggested by some neurophysiological experiments. There must be a controller reference, which we expressed in our model as T_{pyr}.

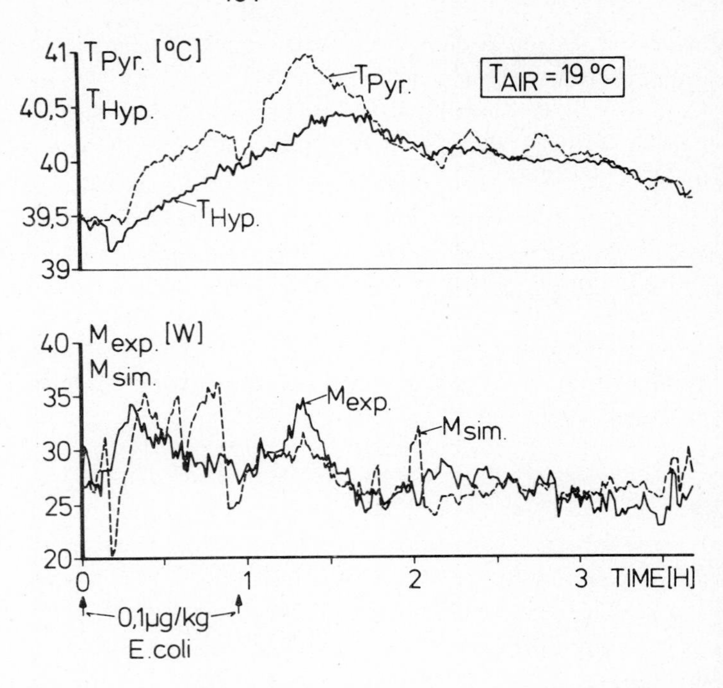

Fig. 3: Febrile process after two succeeding injections
T_{hyp} = hypothalamic temperature (exp.),
T_{pyr} = fever signal (calc.), M = metabolism

The next step not yet done is to evaluate a pharmacokinetic process delivering the transfer-function between fever signal T_{pyr} and the pyrogen-injections (cf. fig. 1).

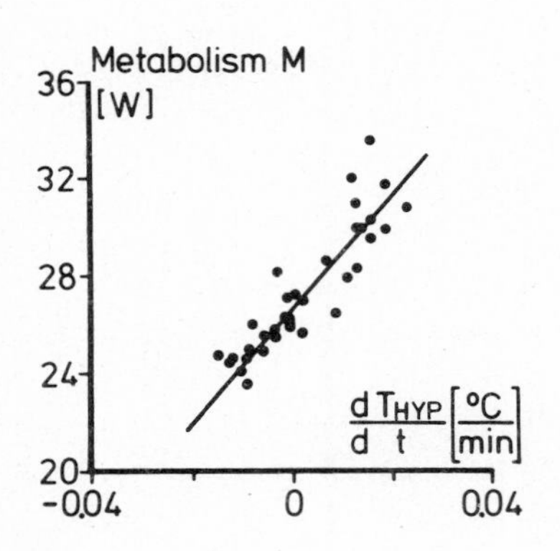

Fig. 4: Experimentally determined relation between metabolic heat production M and rate of change of core temperature dT_c/dt

CONCLUSIONS

It could be demonstrated that a mathematical analysis of the febrile process requires to take into account the dependence of metabolic heat production on the rate of change of core temperature. This is simply due to the physical properties of the passive system. A simple two compartment - model may be derived which enables the determination of the fever-signal controlling directly metabolic heat production and the computation of the central controller gains. However, pure changes of these gains cannot be responsible for temperature increases as observed in the experiments during the febrile process. Therefore this study also confirms the hypothesis that fever must be primarily due to a parallel shift of the controller characteristic.

REFERENCES

Simon, E. (1981): Pflügers Arch. 392, 79-88
Stitt, J.T. (1980): J. Appl. Physiol. 48, 494-499
Werner, J. (1975): Biol. Cybernetics 17, 53-63
Werner, J. (1981): Automatica 17, 351-362
Werner, J., Graener, R. (1981): Proc. of IFAC, XXI, Kyoto, 107-112
Werner, J. (1983): Proc. of IASTED: ACI 83, Copenhagen 14: 10-15

(Supported by the Deutsche Forschungsgemeinschaft, SFB 114)

ESTIMATION OF BIOMASS AND SECONDARY PRODUCT IN BATCH FERMENTATION

PROFESSOR JAMES RONALD LEIGH
MR MAN HEE NG

POLYTECHNIC OF CENTRAL LONDON
115 NEW CAVENDISH STREET
LONDON W1M 8JS TEL.: 01-486-5811

ABSTRACT

The work described in the paper represents a stage in the development of an industrially realistic, adaptive control strategy for those many batch fermentation processes where biomass and secondary product concentrations are difficult to measure accurately, reliably and rapidly enough to allow good on-line control to be achieved.

The paper describes results achieved in the estimation of biomass and secondary product formation in batch fermentation processes. The first results show that the most serious obstacle to producing a reliable and practicable on-line estimation algorithm is the presence and effect of batch-to-batch variations in the fermentation process. The problem is investigated on data from Sorbose fermentation and it is shown that estimation accuracy is most affected by two particular process parameters, one of which is found to be time-varying. The incorporation of this varying parameter into the on-line estimation scheme compensates well for batch-to-batch variations and good estimation accuracy is obtained even with an incorrect process model. The paper includes a description of the prototype hardware for on-line estimation.

Keywords Biocontrol, fermentation process, modelling, nonlinear estimation, Kalman filter application, computer application.

INTRODUCTION

The aims of the project are to produce reliable on-line estimates of biomass and secondary product formation by using the Extended Kalman filter operating on available measurements from fermentation processes.

Compared with other papers in the same field, the emphasis here is on the development of methods that are suited to an industrial, rather than a laboratory, environment. In particular, it has been assumed that significant batch to batch variation will occur between nominally identically fermentations and that only a minimal set of on-line measurements will be available to drive the estimation algorithms.

A previous paper (1) has explained how alternative variants of the Extended Kalman filter have been applied to the estimation of biomass in cellulose fermentation.

This paper, after briefly summarising the earlier work, describes the progress achieved in understanding and compensating for the effects of (unmodelled) batch-to-batch process variations on the estimation accuracy.

Early work was on cellulose fermentation but most of the results quoted here are based on data from sorbose fermentation, and it is for the latter process that a reliable and accurate estimation procedure is proposed and tested.

The work described forms a part of the research program of the Biotechnology Institute of University College London (UCL), Polytechnic of Central London (PCL) and University of Kent with advice and support from the Glaxo Research Laboratories, Greenford, U.K.

The estimates, when displayed graphically continuously in real-time throughout the progress of a fermentation batch, will form a new and valuable guide to assist process operators to obtain efficient and repeatable operation.

A later stage in the project will be to integrate the developed methods into industrially realistic adaptive control strategies for batch fermentation processes.

EQUIPMENT

The fermentation data on which the estimations are based have been obtained from a 10L fermenter instrumented as shown in Figure 1. and from 14L and 100L vessels with comparable instrumentation.

The prototype on-line estimation system uses the Motorola 68000 hardware shown in Figure 2.

MODELLING AND ESTIMATION OF THE MICROBIAL

GROWTH MECHANISM IN A CELLULOSE FEEDSTOCK

The batch is inoculated with cells that first produce an enzyme that hydrolises cellulose into glucose and other products. A proportion of the glucose is consumed by the bacteria. This growth is accompanied by the evolution of carbon dioxide, and the production of acids that will be neutralised by the pH control system.

The process has been modelled by the following stochastic non-linear differential equations.

$$\left.\begin{aligned} \frac{dx_1(t)}{dt} &= \frac{\mu_m x_1(t)\, x_2(t)}{K_s + x_2(t)} - K_D x_1 + w_1(t) \\ \frac{dx_2(t)}{dt} &= \frac{\mu_m x_1(t)\, x_2(t)}{Y_x (K_s + x_2(t))} + w_2(t) \\ y(t) &= \frac{1}{Y_c} \frac{\mu_m x_1(t)\, x_2(t)}{(K_s + x_2(t))} + v(t) \end{aligned}\right\} \quad 1.$$

where

$x_1(t)$	is the biomass concentration (state variable)
$x_2(t)$	is the substrate concentration (state variable)
$y(t)$	is the output (carbon dioxide evolution rate)
Y_c	is the observation yield coefficient
μ_m	is the maximum growth rate
K_s	is the substrate saturation coefficient
Y_x	is the substrate yield coefficient
K_D	is the death rate coefficient
$w(t)$ =	$(w_1(t), w_2(t))^T$ is the state noise vector
$v(t)$	is the output noise

Further equations were obtained from the literature for heat, power, carbon-dioxide and oxygen balance. These equations were used to augment the basic dynamic model.

Problems encountered

The fitting of this model to cellulose fermentation and the subsequent use of the model within an Extended Kalman filter algorithm have been fully described in (1).

We concentrate here on the most fundamental and practically important problem that was encountered in those stages of the work - the problem of batch-to-batch variations. Briefly, although the model (1) was able, after parameter fitting, to represent well any particular batch, no universal set of parameters could be found to give good agreement over a set of (nominally identical) fermentations. This meant, in effect, that in a typical industrial environment, unmodelled batch-to-batch deviations would prevent the realisation of the aim to produce an accurate on-line Kalman estimate of biomass.

The problem described was seen as the main obstacle to achieving success in the project. Possible actions to overcome the problem were considered to be:

(i) To use a combined state and parameter estimating Kalman filter.
(ii) To introduce some degree of adaptivity in the model without moving so far as in (i).
(iii) To continue with model refinement in the hope that the extent of the problem could be diminished.
(iv) To lay down a rigorous but industrially feasible regime for the operation of the fermenter with due attention to the procedures for culture preparation to inocculation of the batch.

We have investigated all four options in parallel and have achieved considerable success as will be seen in the results section of the paper.

MODELLING AND STATE ESTIMATION OF SORBOSE FERMENTATION

Another type of process, on which available data were made from the UCL fermenter, is the bacterial oxidation of sorbitol. Sorbose was produced from sorbitol conversion by the organism grown on sorbitol. The combination of Leudeking & Piret's(references 2, 3) and Koga's (4, 5) models has been used to describe such a process. The substrate consumption consists of two parts. The first part of the substrate consumption is used in bacterial growth and the second one is consumed to the production of sorbose. Besides equations for biomass and substrate concentration, the product formation is also modelled within the dynamic equations. The model is then modified to become:

$$\left.\begin{aligned}
\frac{d\,x_1(t)}{dt} &= \frac{\mu_m\, x_1(t)\, x_2(t)}{K_s + x_2(t)} - K_d\, x_1(t) + W_1(t) \\
\frac{d\,x_2(t)}{dt} &= -\frac{\mu_m\, x_1(t)\, x_2(t)}{Y_x\,(K_s + x_2(t))} - \frac{\mu'_m\, x_1(t)\, x_2(t)}{Y_x'\,(K'_s + x_2(t))} + W_2(t) \\
\frac{d\,x_3(t)}{dt} &= \frac{\mu'_m\, x_1(t)\, x_2(t)}{K'_s + x_2(t)} + M_p\, x_1(t) + W_3(t)
\end{aligned}\right\}$$

Measurement equation

$$Y(t) = \frac{1}{Yc}\,\frac{\mu_m\, x_1(t)\, x_2(t)}{(K_s + x_2(t))} + M_c x_1(t) + V(t)$$

where $x_3(t)$ is the sorbose concentration (product)

μ'_m is the maximum specific growth rate for the product formation

K'_s is the substrate saturation coefficient for the product formation

Y_x' is the substrate yield coefficient for the product formation

M_p, M_c are maintenance coefficients for the product formation and the remaining symbols are defined below equation 1.

BATCH-TO-BATCH VARIATIONS IN PROCESS PARAMETERS AND THEIR EFFECT ON ESTIMATION ACCURACY

Figure 3. shows the estimation performance on sorbose fermentation when correct model parameters are used in the Kalman filter. Figure 4 shows the estimation performance when the Kalman filter contains the model parameters applicable to a different (but nominally identical) batch. The problem of unmodelled batch-to-batch variation in the process is seen to have serious consequences for the estimation accuracy. The effect is similar to that noted prevoously for cellulose fermentation.

The next stage was to investigate the variability of particular parameters from batch-to-batch and the relative sensitivity of the model to parameter changes . Given this information it was hoped to choose a subset of parameters that might be identified and updated within each batch - thus effectively compensating the model for batch-to-batch variations.

An analytical sensitivity analysis was excluded because of the non-linearities in the equations and therefore the sensitivity analysis was performed numerically as follows:

The Kalman filter containing a correct model was run on typical sorbose data to provide a datum against which to measure deviations. Then each of the model parameters was varied by -50%, the Kalman filter was run again and the accuracy of estimation was recorded. The results are shown below:

	Mean square errors in the estimation of:		
Parameter subject to -50% change	Biomass concentration (x_1)	Secondary product concentration (x_2)	Carbon dioxide evolution rate (Y)
none	0.13139E-02	0.35121E-04	0.46934E-01
μ_m	0.11443E+00	0.11971E-01	0.60751E-01
K_s	0.14943E-02	0.36945E-04	0.46421E-01
μ_m'	0.13157E-02	0.56367E-02	0.46928E-01
K_s'	0.13136E-02	0.33464E-03	0.46934E-01
K_d	0.13141E-02	0.35061E-04	0.46931E-01
Y_x	0.13033E-02	0.34659E-04	0.46982E-01
m_p	0.13139E-02	0.35167E-03	0.46934E-01
Y_c	0.72951E-01	0.60472E-02	0.31013E-01
m_c	0.11709E-01	0.10743E-02	0.50737E-01

The table shows that the model fit is most affected by the two parameters μ_m and Y_c.

The estimations with these two parameters 50% in error are shown in Figure 5, 6 respectively - note that the datum against which they are to be compared is that of Figure 3.

INTRODUCTION OF A TIME VARYING PARAMETER - ESTIMATED ON-LINE TO OVERCOME THE EFFECT OF BATCH-TO-BATCH VARIATIONS IN PROCESS PARAMETERS

It has been demonstrated in the above section that two model parameters (μ_m and Y_c) have the most significant effect on estimation accuracy. Further, a more detailed analysis of specific results showed that the parameter Y_c varied significantly not only from batch-to-batch but also during the progress of a sorbose fermentation. (This type of information was found from a detailed examination of previous parameter estimation records).

Accordingly Y_c was re-designated as a time varying process parameter $Y_c(t)$ that would be estimated on-line, along with the process states.

Figure 7 shows the results obtained when the estimation, otherwise identical to that of Figure 4 (i.e. with an incorrectly specified set of parameters) was re-run with the new facility of the one time varying parameter, $Y_c(t)$, estimated on line.

The introduction of the additional facility of one time varying parameter has clearly allowed the effect of the batch-to-batch variation to be largely compensated for in a practical way.

Further tests on fermentations involving E.coli have confirmed the value of the approach in compensating for batch-to-batch deviations.

CONCLUSIONS

The practicability of using the Extended Kalman Filter for estimation of biomass and secondary product fermentation has been demonstrated. The major problem of inaccurate estimation due to batch-to-batch deviations has been overcome by the inclusion of one time varying parameter into the estimation algorithm. Further work will concentrate on extensive testing of the approach on the prototype laboratory system.

REFERENCES AND BIBLIOGRAPHY

1. Swiniarski, R. et al (1982)

   Progress towards estimation of biomass in a batch fermentation process. Proc. IFAC Workshop, Modelling and Control of Biotechnical Processes, Helsinki, Finland.

2. Leudeking, R. and Piret, E.L. (1959)

   A Kinetic Study of the Lactic Acid Fermentation, Batch Process at Controlled pH.
   J.Biochem. Microbiol. Technol. Eng., 1, pp.393-412.

3. Roels, J.A. and Kossen, N.W.F. (1978)

   On the Modelling of Microbiol Metabolism, Progress in Industrial Microbiology, 14, pp.95 - 203.

4. Koga, S. et al (1967)

   Computer Simulation of Fermentation Systems.
   App. Microbiol. 15, pp.683-689.

5. Fredrickson, A.G. et al (1971)

   Mathematical Models for Fermentation Processes.
   Advances in Applied Microbiology, 13, pp.419-465.

6. Nelligan, I. and Calam, C.T. (1980)

   Experience with the Computer Control of Yeast Growth.
   Biotechnology Letters, 2, No.12. pp.531-536.

7. Svrek, W.Y. et al (1974)

   The Extended Kalman Filter Applied to a Continuous Culture Model. Biotech. and Bioeng. 16, pp.827-846.

8. Yousefpour, P. and Williams, D. (1981)

   Real-time Optimisation of a Fermentation Process.
   Biotechnology Letters, 3, No.9. pp. 519-524.

9. Katoh, H. et al (1981)

   Analysis and State Estimation of Enzyme Kinetic Model for Microbial Growth Processes.
   Proc. of IFAC 8th Triennial World Congress, pp.2803-2808.

10. Waterworth, G. and Swanick, B.H. (1981)

    Investigation into the Adaptive Control of a Fermentation Process.
    Trans. Inst. M C, 3, No.1. p.56.

11. Nelligan, I. and Calam, C.T. (1983)

Optimal Control of Penicillin Production, Using a Mini-Computer. Biotechnology Letters, Vol.5. No.8. pp.561-566.

ACKNOWLEDGMENTS

The advice of Dr Fish, UCL, and the provision of data from the Department of Chemical and Biochemical Engineering, UCL is gratefully acknowledged.

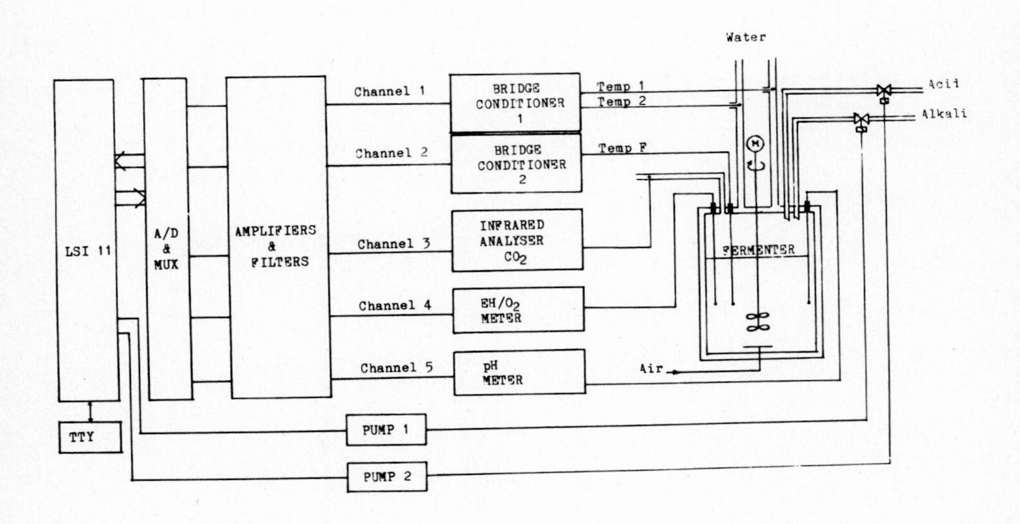

Fig.1. Configuration of Data Logging System hardware

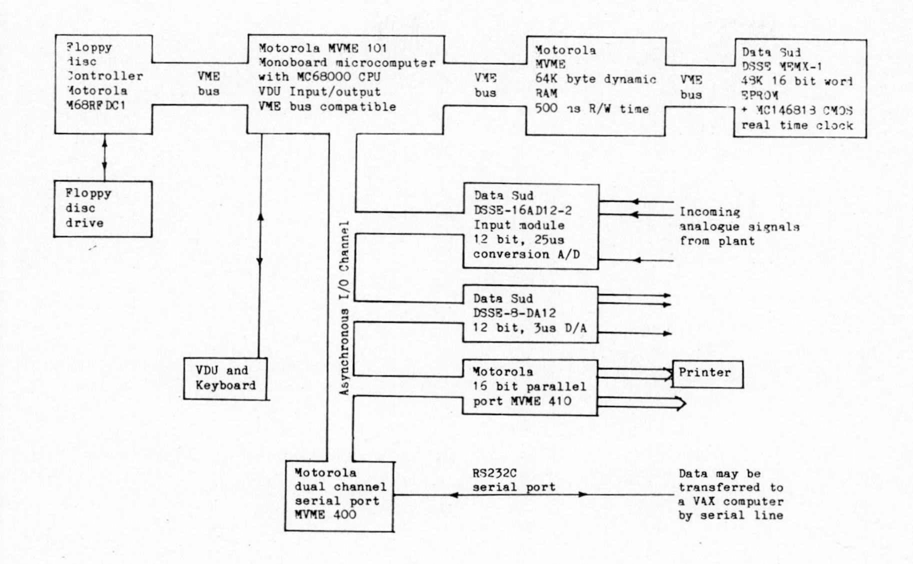

Fig. 2. Configuration of M68000 System

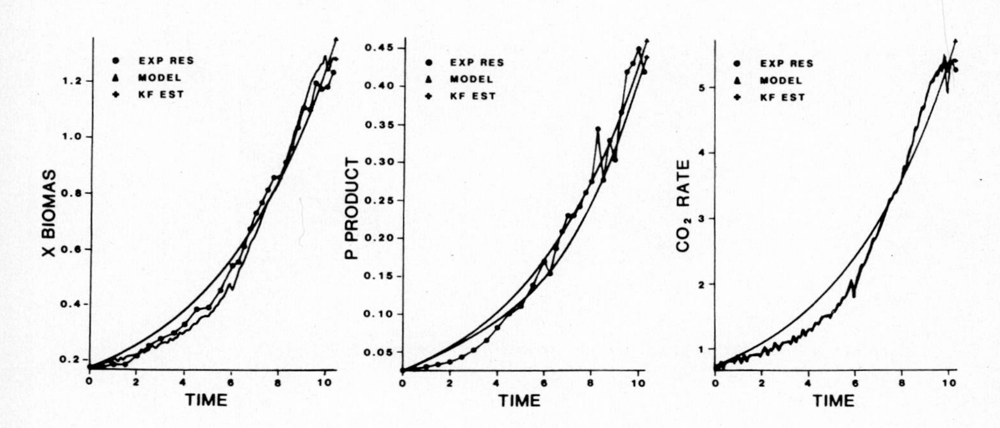

State Estimation on Sorbose Fermentation - Correct Model

Fig.3.

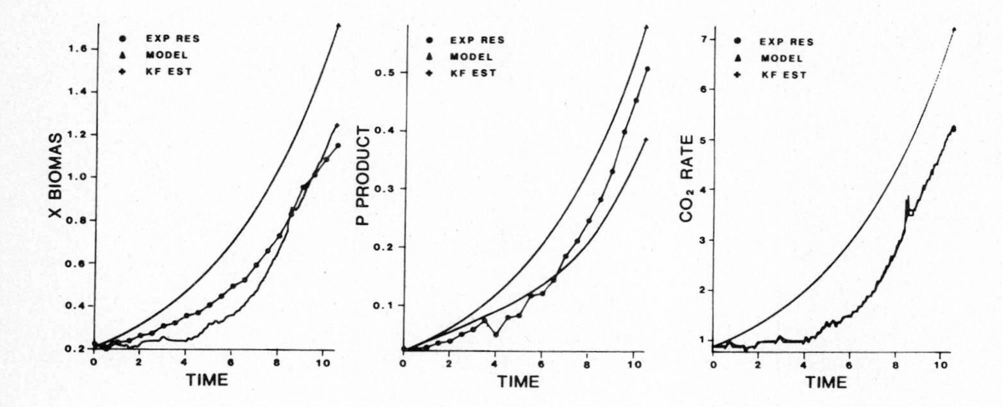

State Estimation on Sorbose Fermentation - Using the Model of an Earlier Batch

Fig.4.

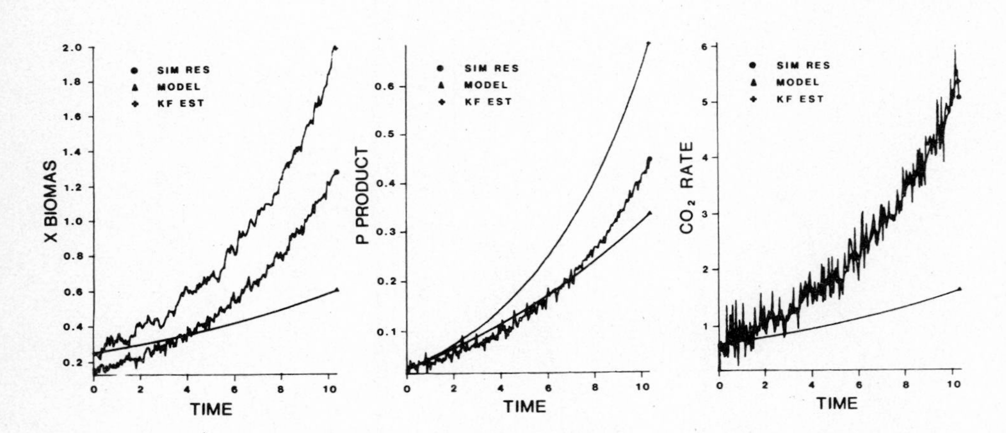

Effect of a Change in Parameter μm on Simulated Sorbose Fermentation

Fig.5.

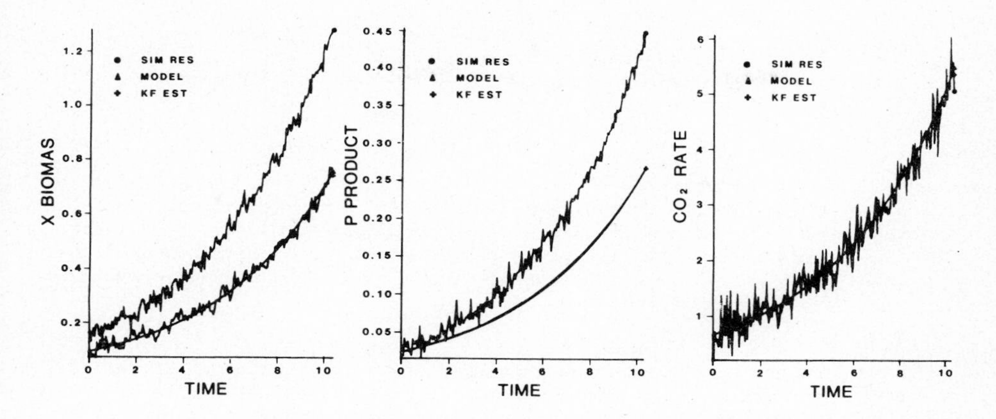

Effect of a Change in Parameter Yc on Simulated Sorbose Fermentation

Fig.6.

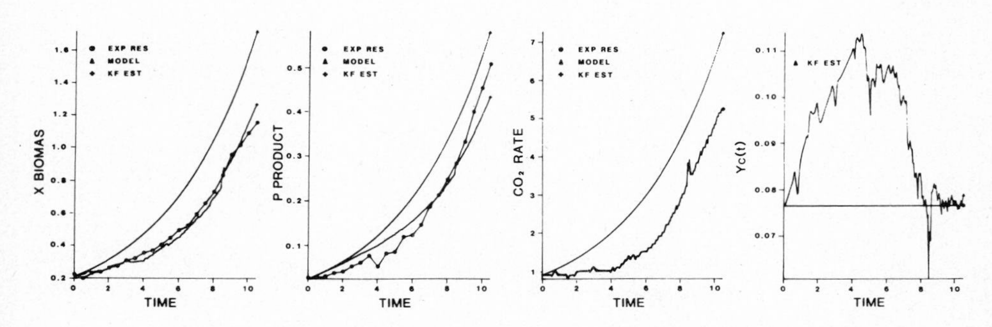

Estimation Performance Under the Same Conditions as for figure 4 Except that now the Adaptive Parameter Yc(t) is Identified During the Batch and Used to Improve the Model

Fig.7.

STABILITY AND CONVERGENCE ANALYSIS OF A CONTINUOUS-TIME ADAPTIVE CONTROL ALGORITHM FOR FERMENTATION PROCESSES.

D. Dochain and G. Bastin

Laboratoire d'Automatique, de Dynamique et d'Analyse des Systèmes
Université de Louvain, Bâtiment Maxwell, B-1348 LOUVAIN-LA-NEUVE.

ABSTRACT.

In this paper, a continuous-time non linear adaptive controller is designed for the regulation of the substrate concentration of non linear fermentation processes. The design is based on a time varying non linear model obtained from material balance equations. The convergence and the stability of the algorithm are demonstrated, using Liapunov's techniques, both in the deterministic case and in the bounded disturbance case.

1. INTRODUCTION.

In recent years, control of fermentation processes has become a subject of increasing importance (see e.g. Halme, 1983). Since fermentation processes are highly non linear and some of their parameters are time varying and/or badly defined, adaptive control should be of interest.

A commonly used approach to the control of non linear systems is to consider them as time varying linear systems and to use black-box linear approximate models to implement the control law.

In this paper, an alternative approach is considered, in which the non linear structure of the system is explicitely used for the design of the adaptive control scheme. This approach has already been used by the authors in the implementation of discrete-time adaptive control of both substrate concentration control and production rate control of fermentation processes (Dochain and Bastin, 1984).

Here a continuous time non linear adaptive controller is proposed (section 3). Particular convergence and stability properties are emphasized in both the deterministic (section 3) and the bounded disturbance (section 4) cases by using Liapunov's techniques.

2. DESCRIPTION OF THE SYSTEM.

We consider the usual state-space representation of fermentation systems deduced from material balance equations :

$$\dot{X} = (\mu - k_d - U)X \qquad (1a)$$

$$\dot{S} = -k_1 \mu X + U(V-S) \qquad (1b)$$

$$Y = k_2 \mu X \qquad (1c)$$

with state variables : X : bacterial concentration
S : substrate concentration

inputs : U : dilution rate (i.e. influent flow rate)
V : influent substrate concentration

outputs : S : substrate concentration
Y : production rate of the reaction product

parameters : $\mu(X,S)$: bacterial growth rate
k_1, k_2 : yield coefficients
k_d : bacterial decay rate

Throughout this paper, we shall assume that :

a) the dilution rate U is the control input
b) the influent substrate concentration V is an external measurable disturbance input.
c) the substrate concentration S and the production rate Y are measurable outputs.

A typical example : the anaerobic digestion process.

Model (1) is well-suited to describe the last stage (methanization stage) in an anaerobic digestion process. Anaerobic digestion can be used, for instance, for the treatment of wastes in sugar industries : V is the influent acetic acid concentration (i.e. the input pollution level), S is the output pollution level and Y is the methane gas flow rate. One of the main advantages of such a waste water treatment system is methane gas production, which can be used as an auxiliary energy supply. Further details on the anaerobic digestion process can be found in Bastin et al. (1983), Antunes and Installé (1981), Van den Heuvel and Zoetmeyer (1982).

3. SUBSTRATE CONCENTRATION CONTROL : THE DETERMINISTIC CASE.

3.1. The adaptive algorithm.

We consider now the problem of regulating the substrate concentration S at a prescribed level S^* despite the disturbance input V, by acting on the dilution rate U. In the biomethanization example mentioned above, this is a depollution control problem, with V and S as the input and the output pollution levels respectively. The model (1) can be rewritten as follows :

$$\dot{S} = -KY + U(V-S) \qquad (2)$$

$$\text{with } K = \frac{k_1}{k_2}$$

Let us define

$$\overline{U}(t) = \frac{F(S^*-S)+\hat{K}Y}{V-S} \quad , F > 0 \qquad (3)$$

where $\hat{K}$ is an estimate of the parameter K.

Obviously, realistic operating conditions impose bounds on the command input values. An evident lower bound is zero, since the input flow rate cannot be negative. Therefore, we calculate the control input as follows :

$$U(t) \begin{cases} = \overline{U}(t) & \text{if } 0 \leqslant \overline{U}(t) \leqslant U_{max} \\ = 0 & \text{if } \overline{U}(t) < 0 \\ = U_{max} & \text{if } \overline{U}(t) > U_{max} \end{cases} \qquad (4)$$

Note the parallel between control law (4) and a minimum variance control strategy. The parameter $\hat{K}$ is estimated by using the following adaptive law :

$$\dot{\hat{K}} \begin{cases} = 0 & \text{if } S > S^* \text{ and } \hat{K} = 0 \\ & \text{if } S < S^* \text{ and } \overline{U} > U_{max} \qquad (5.a) \\ = CY(S^*-S) & \text{otherwise} \qquad (5.b) \end{cases}$$

with $\hat{K}(0) \geqslant 0$

We can notice the similarity between the algorithm (5) and the so-called "MIT-rule" estimation algorithm (see Aström, 1983).

3.2. Stability and convergence properties.

In this section, we first demonstrate the BIBO stability of the system. Then, asymptotic stability and convergence properties of the closed-loop system (1)(3)(4)(5) are analyzed by Liapunov's techniques.

3.2.1. BIBO stability.

BIBO stability of system (1) can be proven under the following assumptions :

Assumptions

The growth-rate $\mu(S,X)$ is a continuous differentiable function of X and S and fulfills the following assumptions for $X \geq 0$ and $S \geq 0$:

A.1. $0 \leq \mu(S,X) \leq \mu^*$

A.2. $\mu(0,X) = 0$

A.3. $\mu(S,0) > k_d$

A.4. $\frac{\delta\mu}{\delta X} < 0$

A.5. $\lim_{X\to\infty} \mu(S,X) < k_d$

Theorem 3.1.

If (i) $U \geq 0$
(ii) $0 \leq V \leq V_{max}$
(iii) $0 \leq S(0) \leq V_{max}$ and $X(0) \geq 0$

Then, there exist maximum values X_{max} and Y_{max} such that, if $X(0) \leq X_{max}$,

a) $0 \leq S \leq V_{max}$

b) $0 \leq X \leq X_{max}$

c) $0 \leq Y \leq Y_{max}$, $\forall t > 0$

Proof.

1) $X \geq 0$ and $Y \geq 0$; straightforward by using (1a), (1c) and (iii)
2) For S = 0, we have, using (i), (ii), (1b) and A.2 :

$\dot{S} \geq 0$

The conclusion $S \geq 0$ for all S follows.

3) For $S = V_{max}$, we have, using (1b), (1c) and (ii)

$$\dot{S} = -\frac{k_1}{k_2} Y + U(V-V_{max}) \leq 0 \quad \text{since } Y \geq 0$$

The conclusion $S \leq V_{max}$ for all S follows.

4) Write $\mu_{max}(X) = \max_{0\leq S\leq V_{max}} \mu(S,X)$

Consider X_{max}, the value of X which fulfills the following relationship $\mu_{max}(X_{max}) = k_d$. X_{max} exists since assumptions A.3, A.4, A.5, hold.

For $X = X_{max}$, we have, using (i) :

$\dot{X} \leq 0$

The conclusion $X \leq X_{max}$ for all X follows.

5) It is obvious that, by using (1.c) and A.1 :

$$Y_{max} = k_2\mu^* X_{max}$$

and $Y \leqslant Y_{max}$

Q.E.D.

3.2.2. Asymptotic Stability.

Asymptotic stability of the closed-loop system (2)(3)(4)(5) is demonstrated under the following assumptions :

Assumptions :

B.1 V is constant
B.2 S^* is constant, with $S^* < V$
B.3 $0 \leqslant X(0) \leqslant X_{max}$
B.4 $0 \leqslant S(0) \leqslant V$
B.5 U is calculated by (4) with $U_{max} \geqslant \frac{KY_{max}}{V-S^*}$

Theorem 3.2.

Let us consider the following positive definite function :

$$W(t,S,\hat{K}) = \frac{1}{2}(S^*-S)^2 + \frac{1}{2C}(K-\hat{K})^2 \qquad (6)$$

Then the time derivative of $W(t,S,\hat{K})$ computed along the solution of (2)(3)(4)(5) is semi-negative definite.

Proof.

A. if $U = \overline{U}$

$$\dot{W}(t,S,\hat{K}) = -(S^*-S)\dot{S} - \frac{1}{C}(K-\hat{K})\dot{\hat{K}}$$

$$= -F(S^*-S)^2 \leqslant 0$$

B. if $U = 0$

Since $\hat{K} \geqslant 0$, $U = 0$ only when $S > S^*$

if $\dot{\hat{K}} \neq 0$ and $\hat{K} \neq 0$

$$\dot{W}(t,S,\hat{K}) = -(S^*-S)(-KY) - \frac{1}{C}(K-\hat{K})CY(S^*-S)$$

$$= (S^*-S)\hat{K}Y \leqslant 0$$

if $\dot{\hat{K}} = \hat{K} = 0$

$$\dot{W}(t,S,\hat{K}) = (S^*-S)KY \leqslant 0$$

C. if $U = U_{max}$

if $S > S^*$

$$\dot{W}(t,S,\hat{K}) = -(S^*-S)(-KY+U_{max}[U-S]) - \frac{1}{C}(K-\hat{K})CY(S^*-S)$$

$$= -(S^*-S)(-\hat{K}Y+U_{max}[V-S])$$

Since $U_{max} < \frac{F(S^{*}-S)+K\hat{Y}}{V-S} < \frac{K\hat{Y}}{V-S}$,

$\dot{W}(t,S,\hat{K}) \leqslant 0$

if $S < S^{*}$

Since $\dot{\hat{K}} = 0$ when $S < S^{*}$ and $U(t) = U_{max}$,

$\dot{W}(t,S,\hat{K}) = -(S^{*}-S)(-KY + U_{max}[V-S])$

Since $U_{max} \geqslant \frac{KY_{max}}{V-S^{*}} > \frac{KY}{V-S}$

$\dot{W}(t,S,\hat{K}) \leqslant 0.$

So, we can conclude that W is a Liapunov function (W is a positive definite function and its time derivative along the solution of (2)(3)(4)(5) is semi-negative definite) Therefore, the closed-loop system(2),(3),(4),(5) is stable, and theorem 3.1 is demonstrated.

Q.E.D.

3.2.3. Convergence.

It is straightforward from the above paragraph that $\dot{W}$ is negative definite in $(S-S^{*})$ ($\dot{W} = 0$ if and only if $S = S^{*}$)

Lemma 3.1.

The time derivative of S is bounded :

$|\dot{S}| < L$

Proof.

By using lemma 3.1 and assumption B.5, we know that :

$0 \leqslant U \leqslant U_{max}$
$0 \leqslant S \leqslant V$
$0 \leqslant Y \leqslant Y_{max}$

Since $\dot{S} = -KY + U(V-S)$, we deduce :

$-KY_{max} - U_{max}V \leqslant \dot{S} \leqslant U_{max}V$

By choosing $L > KY_{max} + U_{max}V$, it is straightforward that :

$|\dot{S}| < L$

and lemma 3.1. is demonstrated.

Therefore, convergence of the substrate concentration control algorithm can be deduced from the following theorem.

Theorem 3.3.

If $W(t,S,\hat{K})$ is a Liapunov's function such that $\dot{W}$ is negative definite in $(S-S^{*})$ and if there exists a constant $L > 0$ such that $|\dot{S}| < L$,
then

$$\lim_{t\to\infty} S = S^{*}$$

Proof : the proof of this theorem can be found in Peiffer and Rouche (1969).

4. SUBSTRATE CONCENTRATION CONTROL : THE BOUNDED DISTURBANCE CASE.

We consider now that a bounded non measured disturbance is added to the state-space equation (2) :

$$\dot{S} = -KY + U(V-S) + d \qquad (7)$$

with $\sup_t |d(t)| = \delta$

Then, with the adaptive law (5), the boundedness of the system is not guaranteed anymore. Therefore, in order to achieve boundedness in the presence of disturbances, we use an idea proposed by Ioannou and Kokotovic (1983), and we modify the adaptive law (5) by adding a decay term : $-\sigma C\hat{K}$:

$$\dot{\hat{K}} \begin{cases} = 0 & \text{if } S > S^* \text{ and } \hat{K} = 0 \\ & \text{if } S < S^* \text{ and } \overline{U}(t) > U_{max} \\ = -\sigma C\hat{K} + CY(S^*-S) & \text{otherwise} \end{cases} \qquad (8)$$

with $\sigma > 0$ and $\hat{K}(0) \geqslant 0$

In order to prove boundedness properties, we first consider the following lemma :

Lemma 4.1.

The time derivative of the function $W(t,S,\hat{K})$ (6) computed along the solutions of (7)(8)(3)(4) is bounded as follows :

$$\dot{W}(t,S,\hat{K}) \leqslant -\frac{F}{2}(S^*-S)^2 - \frac{\sigma}{2}(K-\hat{K})^2 + \gamma$$

with $\gamma = \sup\{\frac{\delta^2}{2F^2} + \frac{\sigma}{2}K^2,\ \gamma_1\delta^2 + \frac{\sigma}{2}K^2,\ \gamma_2\delta^2 + \frac{\sigma}{2}K^2,\ \gamma_2\delta^2 + \frac{1}{2\sigma}Y^2_{max}S^{*2}\}$

and $\gamma_1 = \{\frac{1}{2} + \frac{2+F}{4\delta}(V-S^*)\}^2$

$\gamma_2 = \{\frac{1}{2} + \frac{2+F}{4\delta}S^*\}^2$

The technical details of the proof are not given in the present paper and can be found in Dochain (1984).

Now, in order to establish the stability properties of (7)(8)(3)(4), we can apply the following theorem.

Theorem 4.1.

Given any $\sigma > 0$, the solution $S,\hat{K}$ of (7)(8)(3)(4) is bounded for every bounded initial condition S_o, K_o.
Furthermore, $(S-S^*)$, $(K-\hat{K})$ converge to the residual set :

$$D = \{S,\hat{K} : \frac{F}{2}(S^*-S)^2 + \frac{\sigma}{2}(K-\hat{K})^2 \leqslant \gamma\}$$

Proof.

Since σ is positive, $\dot{W}$ is strictly negative outside D.
Therefore, W is strictly decreasing outside D and any solution $S,\hat{K}$ starting outside D enters it after a finite time T. Boundedness of the solution $(S-S^*)$, $(K-\hat{K})$ follows from boundedness of the disturbance term d ($\sup |d(t)| = \delta$) and of the residual set D.

Q.E.D.

It is worth noting that the adaptive law (8) does not guarantee convergence to the substrate concentration S to its prescribed level S^*. It only achieves boundedness in the presence of disturbances. But, an appropriate choice of σ can improve the convergence of the closed loop system.

5. CONCLUSIONS.

This paper has dealt with the substrate concentration control of fermentation processes. A continuous-time adaptive scheme has been proposed. The convergence and the stability properties of the algorithm have been analyzed in the deterministic case. In the bounded disturbance case, the adaptive scheme has been modified so as to ensure boundedness and convergence to a residual set.

6. REFERENCES.

ANTUNES S., INSTALLE M. (1981), The Use of Phase-Plane Analysis in the Modelling and the Control of a Biomethanisation Process, Proc. VIIIth IFAC World Congress, Kyoto, Japan, Vol. XXII, pp.165-170.

ÅSTROM K.J. (1983), Theory and Applications of Adaptive Control - A Survey, Automatica, September, pp. 471-468.

BASTIN G., DOCHAIN D., HAEST M., INSTALLE M., OPDENACKER Ph. (1983a), Modelling and Adaptive Control of a Continuous Anaerobic Fermentation Process, Modelling and Control of Biotechnical Processes, A. Halme Ed., Pergamon Press.

BASTIN G., DOCHAIN D., HAEST M., INSTALLE M., OPDENACKER Ph., (1983b), Identification and Adaptive Control of a Biomethanization Process, Modelling and Data Analysis in Biotechnology and Medical Engineering, Vansteenkiste G.C. & Young P.C., Ed., North-Holland Publ. Cy.

DOCHAIN D. (1984) , Stability Analysis of a Continuous Time Adaptive Controller for Fermentation Processes, Internal Report, Laboratoire d'Automatique et d'Analyse des Systèmes, Université de Louvain, Louvain-la-Neuve.

DOCHAIN D., BASTIN G. (1984), Adaptive Identification and Control Algorithms for non linear Bacterial Growth Systems, to be published (Automatica).

HALME A. Editor (1983), Modelling and Control of Biotechnical Processes. Proceedings of the first IFAC Workshop, Helsinki, Finland, August 17-19, 1982. Published by Pergamon Press.

IOANNOU P.A., KOKOTOVIC P.V. (1983), Adaptive Systems with Reduced Models, Springer Verlag.

PEIFFER K., ROUCHE N. (1969), Liapunov's Second Method Applied to Partial Stability, Journal de Mécanique, Juin, vol. 8, n°2.

Session 18

NONLINEAR SYSTEMS II

SYSTÈMES NON LINÉAIRES II

SUPERVISORY CONTROL OF A CLASS OF DISCRETE EVENT PROCESSES

P.J. Ramadge and W.M. Wonham[(0)]

Systems Control Group
Dept. of Electrical Engineering
University of Toronto
Toronto, Ont. M5S 1A4
CANADA

ABSTRACT

This paper studies the control of a class of discrete event processes, i.e. processes that are discrete, asynchronous and possibly nondeterministic. The controlled process is described as the generator of a formal language, while the controller, or supervisor, is constructed from the grammar of a specified target language that incorporates the desired closed-loop system behavior. The existence problem for a supervisor is reduced to finding the largest controllable language contained in a given legal language. Two examples are provided.

1. INTRODUCTION

In this paper we study the control of a class of systems broadly known as discrete event processes. The principal features of such processes are that they are discrete, asynchronous and (possibly) nondeterministic. Typical examples include computer networks, flexible manufacturing systems, and the start-up and shut-down procedures of industrial plants.

At the present time there is little unifying theory for the control of discrete event processes. Nor is it entirely clear what such a theory ought to encompass. Numerous approaches have appeared in the literature. A representative sampling of these could include boolean models [Aveyard, 1974]; Petri nets [Peterson, 1981]; formal languages [Beauquier and Nivat, 1980], [Park, 1981]; temporal logic [Pnueli, 1979], [Hailpern and Owicki, 1983]; and port automata and flow networks [Milne and Milner, 1979], [Steenstrup, Arbib and Manes, 1981]. All of this work is concerned, in one way or another, with the problem of how to ensure, by control, the orderly flow of events; and to this end how to bring together ideas from logic, language and control theory itself. The variety of approaches reflects the diversity of areas in which discrete event processes play an important role. It also indicates that to date no dominant paradigm has emerged upon which a broad and detailed theory of control might be based.

In this article we investigate a simple abstract model of a controlled discrete event process, our main objective being to determine qualitative structural features of the relevant basic control problems. Specifically we take the controlled process to be the generator of a formal language, and study how the grammar of a specified (target) language may be employed as a controller. Our approach is similar in spirit to some qualitative theories of multivariable control synthesis that have emerged over the last decade in the context of standard dynamic systems (e.g. [Wonham, 1979], [Nijmeijer, 1983]). The present article is based

(0) This research was partially supported by NSERC (Canada), Grant No. A-7399.

mainly on [Ramadge, 1983], while earlier versions were summarized in [Ramadge and Wonham, 1982a,b].

The paper is organized as follows. In Sect. 2 we define the class of controlled processes and controllers (supervisors), of interest; and in Sect. 3 we discuss various associated formal languages. Sects. 4 and 5 develop criteria for the existence of a supervisor for which the corresponding closed-loop controlled system satisfies given linguistic requirements; the main idea here is that of a controllable language. Sect. 6 introduces the notion of a supervisor that is proper, namely nonblocking and nonrejecting. In Sect. 7 we pose two problems of supervisor synthesis: the Supervisory Marking Problem (SMP) and the Supervisory Control Problem (SCP). Each of these is then shown to be solvable in a minimally restrictive, or 'optimal', fashion in the class of proper supervisors, the 'optimality' depending on a semilattice property of the relevant classes of languages. Sect. 8 defines a congruence (or simplification) of supervisors. The latter, combined with some notions of reduction of languages and grammars in Sect. 9, leads in Sect. 10 to our main result, the Quotient Structure Theorem. According to this, every efficiently constructed supervisor is structurally equivalent to a quotient (i.e. high-level, or lumped, model) of a grammar of the desired closed-loop generated language. We conclude in Sects. 11 and 12 with two simple but practical illustrations.

2. CONTROLLED DISCRETE-EVENT PROCESSES

2.1 Generators

To establish notation we first recall various standard ideas from automaton and language theory (cf. [Hopcroft and Ullman, 1979]). We define a generator to be a deterministic automaton

$$G = (Q, \Sigma, \delta, q_0, Q_m)$$

where Q is the set of states q, Σ is the alphabet or set of output symbols σ, $\delta:\Sigma\times Q \to Q$ is the transition function, $q_0 \in Q$ is the initial state and $Q_m \subset Q$ is a subset of states to be called marker states[1]. We always assume that Σ, but not necessarily Q or Q_m, is finite. In general, δ is only a partial function (pfn), meaning that, for each fixed $q \in Q$, $\delta(\sigma,q)$ is defined only for some subset $\Sigma(q) \subset \Sigma$ that may depend on q. Formally G is equivalent to a directed graph with node set Q and an edge $q \to q'$ labeled σ for each triple (σ,q,q') such that $q' = \delta(\sigma,q)$. Such an edge, or state transition, will be called an event.

We interpret G as a device that starts in q_0 and executes state transitions, i.e. generates a sequence of events, by following its graph. Events are considered to occur spontaneously (no auxiliary forcing mechanism is postulated), asynchronously (i.e. without reference to a clock) and instantaneously. An event is thought of as signaled (to an outside observer, say) by its label σ.

Let Σ^* denote the set of all finite strings s of elements of Σ, including the empty string 1[2]. In standard fashion we construct the extended transition function

---

[1] The terms generator and marker are nonstandard, but better suited to our interpretation than e.g. 'machine' and 'final'.

[2] 1 plays the role of identity of string concatenation, i.e. $1s = s1 = s$.

$$\delta : \Sigma^* \times Q \to Q \quad \text{(pfn)}$$

according to

$$\delta(1,q) = q \quad q \in Q,$$

and

$$\delta(s\sigma,q) = \delta(\sigma,\delta(s,q))$$

whenever $q' = \delta(s,q)$ and $\delta(\sigma,q')$ are both defined. Any subset of Σ^* is a language over Σ. The strings of a language are often called words. The language generated by G is

$$L(G) = \{w : w \in \Sigma^* \ \& \ \delta(w,q_0) \text{ is defined}\}$$

The language marked by G is

$$L_m(G) = \{w : w \in L(G) \ \& \ \delta(w,q_0) \in Q_m\}$$

We interpret $L(G)$ as the set of all possible finite sequences of events that can occur; while $L_m(G) \subset L(G)$ is a distinguished subset of these sequences that may be 'marked', or recorded, perhaps representing completed 'tasks' (or sequences of tasks) carried out by the physical process that G is intended to model.

To conclude this subsection we remark that it is usually convenient to eliminate states of G that can never by reached (or 'accessed') from q_0. Namely let

$$Q_{ac} = \{q : \exists\, w \in \Sigma^*,\ \delta(w,q_0) = q\}$$

$$Q_{ac,m} = Q_{ac} \cap Q_m$$

$$\delta_{ac} = \delta \,|\, (\Sigma \times Q_{ac})$$

The accessible component of G, denoted by $Ac(G)$, is then defined to be

$$Ac(G) = (Q_{ac},\Sigma,\delta_{ac},q_0,Q_{ac,m}).$$

A generator G is accessible if $G = Ac(G)$.

We say that G is co-accessible if every string in $L(G)$ can be completed to one in $L_m(G)$, i.e.

$$(\forall w)\, w \in L(G) \implies (\exists s)\, s \in \Sigma^* \quad \& \quad ws \in L_m(G)$$

If G is both accessible and co-accessible it is said to be trim [Eilenberg, 1974]. It is well known (cf. [Eilenberg, 1974] Sect. III.5) that to every language (i.e. subset of Σ^*) there corresponds a trim generator; if required to be minimal, it is essentially unique.

2.2 Controlled discrete-event processes

To a generator $G = (Q,\Sigma,\delta,q_0,Q_m)$ we now adjoin a means of control. For this let $\Sigma_c \subset \Sigma$ be a distinguished subset of the alphabet; we say that an event (σ,q,q') is a controlled event if $\sigma \in \Sigma_c$. Let

$$\Gamma = \{0,1\}^{\Sigma_c}$$

be the set of all binary assignments to the elements of Σ_c. Each assignment $\gamma \in \Gamma$, i.e. each function

$$\gamma : \Sigma_c \to \{0,1\} ,$$

is a control pattern. An event (with label) σ is said to be enabled by γ if $\gamma(\sigma) = 1$, or disabled by γ if $\gamma(\sigma) = 0$. It is convenient to extend each $\gamma \in \Gamma$ to a map $\gamma : \Sigma \to \{0,1\}$ by defining $\gamma(\sigma) = 1$ for each $\sigma \in \Sigma - \Sigma_c$. If $\delta : \Sigma \times Q \to Q$ is the transition function of G, we define an augmented transition function

$$\delta_c : \Gamma \times \Sigma \times Q \to Q \quad \text{(pfn)}$$

according to

$$\delta_c(\gamma,\sigma,q) = \begin{cases} \delta(\sigma,q), & \text{if } \delta(\sigma,q) \text{ is defined and } \gamma(\sigma) = 1 \\ \text{undefined}, & \text{otherwise .} \end{cases}$$

Formally, the object

$$G_c = (Q, \Gamma \times \Sigma, \delta_c, q_0, Q_m)$$

is just another generator, constructed from G by a specification of Σ_c. However, we interpret G_c as a version of G that admits external control, as follows. For brevity call 'an event labeled σ' simply 'an event σ'. For each fixed $\gamma \in \Gamma$ there is a generator $G(\gamma)$ formed by deleting from the graph of G those events σ with $\gamma(\sigma) = 0$, i.e. those events that the control pattern γ disables. Then external control action would consist simply in switching the control pattern through a sequence of elements $\gamma, \gamma', \gamma''$... in Γ, like switching the pattern of red and green lights in a traffic network. Observe that such control is 'permissive' (cf. [Peterson, 1981]): while disabled events are certainly prevented from occurring, enabled events are not necessarily forced to occur.

A structure G_c as described above will be called a controlled discrete-event process (CDEP).

2.3 Example - a primitive CDEP

A user of a resource may be modeled as a CDEP with three states I (idle), R (request) and U (use), and with transitions as shown. Here we take (with some change of notation)

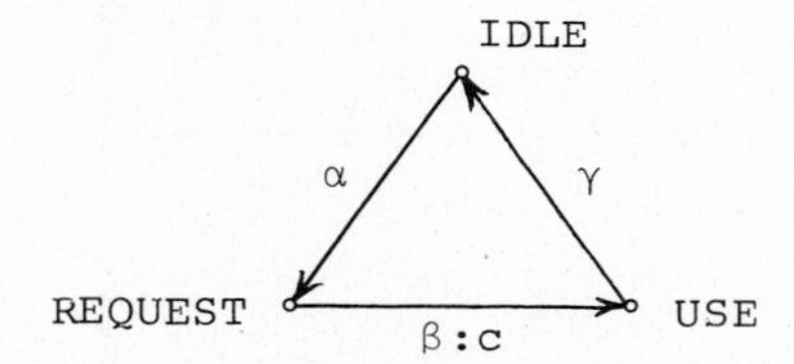

$\Sigma = \{\alpha, \beta, \gamma\}$ and $\Sigma_c = \{\beta\}$. The (two) control patterns correspond to evaluations $c = 0$ or $c = 1$ of the control variable c. A transition $R \to U$ may occur only when $c = 1$.

2.4 Supervisors

Our objective will be to design a controller that switches control patterns in such a way that a given CDEP G_c, as described in Sect. 2.2, behaves in obedience to various constraints. Such a controller will be called a supervisor. Formally a supervisor S is a pair

$$S = (S,\phi) .$$

Here

$$S = (X,\Sigma,\xi,x_0,X_m)$$

is a deterministic automaton with state set X, input alphabet Σ, transition (partial) function $\xi:\Sigma\times X \to X$, initial state x_0 and marker subset $X_m \subset X$; while

$$\phi:X \to \Gamma$$

is a (total) function that maps supervisor states x into control patterns γ. Thus for each $x \in X$,

$$\gamma := \phi(x) \in \{0,1\}^{\Sigma_c}.$$

(As before we extend $\phi(x)$ to a map $\phi(x):\Sigma \to \{0,1\}$ with $\phi(x)(\sigma) = 1$ for each $\sigma \in \Sigma - \Sigma_c$.) S will always be assumed to be accessible. We call ϕ the <u>state feedback map</u>.

We interpret S conventionally, as a device that executes a sequence of state transitions (according to ξ) in response to an appropriate input string $w \in \Sigma^*$. Thus we may couple S to G_c in a feedback loop by allowing the state transitions of S to be forced by G_c, and requiring G_c to be constrained by the successive control patterns determined by the states of S. Formally, define the partial function

$$\xi\times\delta_c:\Sigma\times X\times Q \to X\times Q \quad \text{(pfn)}$$

according to

$$(\sigma,x,q) \longmapsto (\xi(\sigma,x),\delta_c(\phi(x),\sigma,q)) .$$

Thus $(\xi\times\delta_c)(\sigma,x,q)$ is defined iff $\delta(\sigma,q)$ is defined, $\phi(x)(\sigma) = 1$, and $\xi(\sigma,x)$ is defined. This yields the generator

$$(X\times Q,\Sigma,\xi\times\delta_c,(x_0,q_0),X_m\times Q_m) .$$

We define the <u>supervised discrete event process (SDEP)</u>, denoted by S/G_c, to be the accessible generator

$$S/G_c = Ac(X\times Q,\Sigma,\xi\times\delta_c,(x_0,q_0),X_m\times Q_m) \qquad (2.1)$$

From now on we shall assume that $\xi\times\delta_c$ has been extended to strings of Σ^* in the way described in Sect. 2.1 for δ. Of course, so far there is nothing to guarantee that $(X\times Q)_{ac}$ is anything more than the singleton $\{(x_0,q_0)\}$, or that $L(S/G_c)$ is any larger than the singleton $\{1\}$ consisting of the empty string alone.

In analogy to the case of G itself, we wish to interpret the language $L(S/G_c)$ generated by S/G_c as the set of all possible finite sequences of events that can occur when S is coupled to G_c as just described. For this it is necessary to ensure that transitions of S are actually defined whenever they can occur in G and are enabled by ϕ. To formalize this relationship we shall say that S is <u>complete with respect to</u> G_c provided the following is true: for all $s \in \Sigma^*$, $\sigma \in \Sigma$ the three conditions

(i) $s \in L(S/G_c)$

(ii) $s\sigma \in L(G)$ {i.e. $\delta(s\sigma,q_0)$ is defined}

(iii) $[\phi \circ \xi(s,x_0)](\sigma) = 1$ {i.e. σ is enabled at $\xi(s,x_0)$}

together imply that

(iv) $s\sigma \in L(S/G_c)$ {i.e. $\xi(s\sigma,x_0)$ is defined}

While the definition (2.1) is admissible as it stands, it will be of real interest only when it is physically meaningful, namely when S is complete with respect to G_c.

3. LANGUAGES OF S/G_c

3.1 Definitions

Let $L \subset \Sigma^*$. The closure of L, denoted by $\bar{L}$, is the set of all strings that are prefixes of words of L, i.e.

$$\bar{L} = \{s : s \in \Sigma^* \ \& \ (\exists t)t \in \Sigma^*,\ st \in L\}$$

For instance, if $L = \emptyset$ then $\bar{L} = \emptyset$ and if $L \neq \emptyset$ then $1 \in \bar{L}$. A language L is closed if $L = \bar{L}$. If G is any generator then $L(G)$ is closed; if in addition G is trim then

$$L(G) = \bar{L}_m(G) .$$

Let G_c be a CDEP constructed from a generator G. For simplicity we shall denote G_c simply by its underlying generator G. We refer to $L_m(G)$ as the uncontrolled (discrete-event) process language. Let S be a supervisor for G, $L(S/G)$ the language generated by S/G and $L_m(S/G)$ the language marked by S/G. Define the language controlled by S in G to be

$$L_c(S/G) := L(S/G) \cap L_m(G) \tag{3.1}$$

In other words, $L_c(S/G)$ consists of those (marked) strings of the uncontrolled process language that 'survive' in the presence of supervision.

It is clear from the definitions that

$$L_m(S/G) \subset L_c(S/G) \subset L_m(G) \tag{3.2}$$

and, if G is trim,

$$\bar{L}_m(S/G) \subset \bar{L}_c(S/G) \subset \bar{L}(S/G)\,[= L(S/G)] \subset \bar{L}_m(G) \tag{3.3}$$

3.2 Examples

Let G be the generator over $\Sigma = \{\alpha,\beta,\gamma\}$ displayed below.

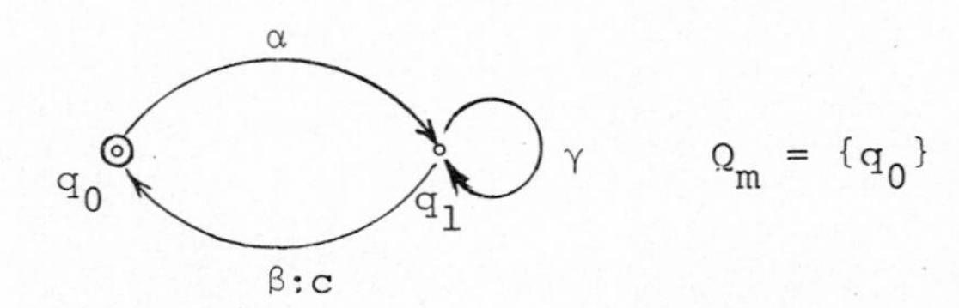

Then [3]

$$L_m(G) = (\alpha\gamma^*\beta)^*$$

We shall consider two different supervisors, each specified by its transition graph, as follows.

(i)

S $\quad x_0 \xrightarrow{\alpha} x_1$, $x_1 \xrightarrow{\beta} x_0$, $x_1 \xrightarrow{\gamma} x_2$, $x_2 \xrightarrow{\gamma} x_2$ $\qquad X_m = \{x_0\}$

$c=0 \qquad c=1 \qquad c=0$

This gives for S/G the transition graph

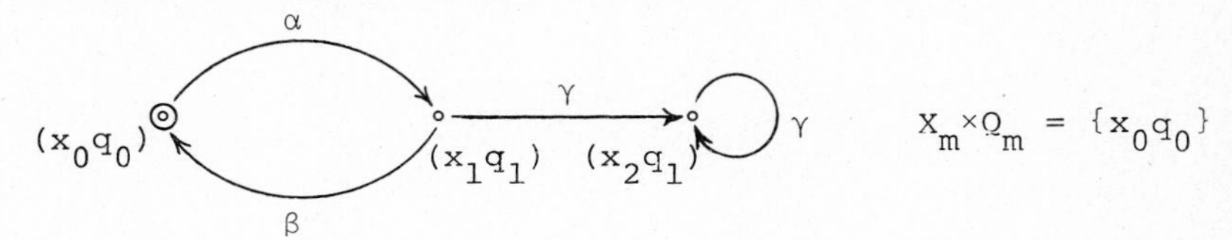

It is seen that

$$L(S/G) = (\alpha\beta)^*(1+\alpha\gamma^*)$$

$$L_c(S/G) = L(S/G) \cap L_m(G) = (\alpha\beta)^* = L_m(S/G)$$

For future reference we note, however, that

$$\bar{L}_c(S/G) = (\alpha\beta)^*(1+\alpha) \subsetneq L(S/G)$$

(ii)

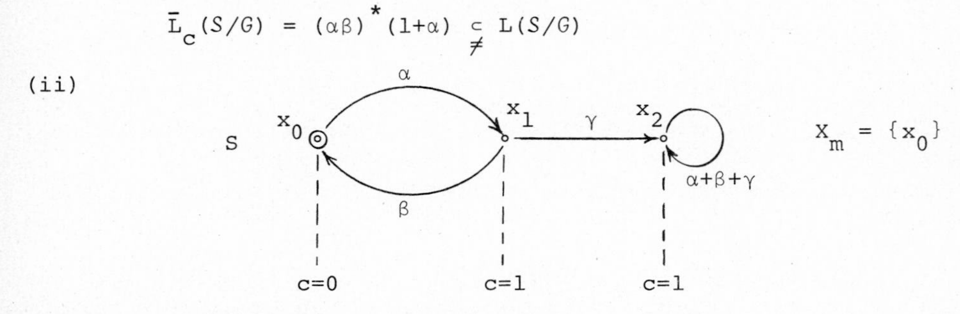

This gives for S/G the transition graph

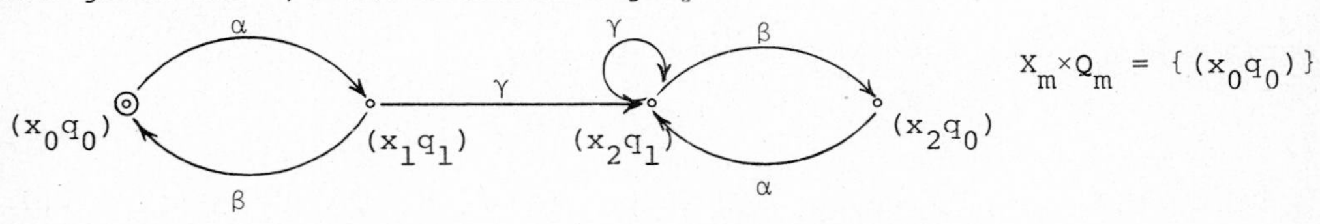

It is seen that

---

[3] For the notation of regular expressions used here and below see e.g. [Hopcroft and Ullman, 1979].

$$L_c(S/G) = L_m(G) = (\alpha\gamma^*\beta)^*$$

Again for future reference, we note that

$$L_m(S/G) = (\alpha\beta)^* \subsetneq L_c(S/G)$$

and

$$\bar{L}_m(S/G) \subsetneq \bar{L}_c(S/G)$$

4. MARKING AND CONTROL

A supervisor S performs two, essentially independent tasks: marking (as described by $L_m(S/G)$) and control (as described by $L_c(S/G)$, $L(S/G)$). If a given controlled behavior is achievable, then any marking task is simultaneously achievable that is consistent with the controlled behavior.

Without essential loss of generality we assume that the generator G is trim, namely

$$L(G) = \bar{L}_m(G) .$$

Proposition 4.1

(i) For each sublanguage $K \subset L_m(G)$ there exists a complete supervisor S such that, for S/G, we have

$$L(S/G) = L(G), \qquad L_m(S/G) = K .$$

(ii) Let $L \subset L(G)$. If there exists a complete supervisor S for which $L(S/G) = L$ then for every sublanguage K of $L \cap L_m(G)$ there exists a complete supervisor S_K such that, correspondingly,

$$L(S_K/G) = L, \qquad L_m(S_K/G) = K . \qquad \square$$

In later work we shall make use of the following definition: if $L \subset \Sigma^*$, a grammar M for L is an accessible generator G such that $L_m(G) = L$. While a grammar and an accessible generator are formally no different, we interpret a grammar $M = (Q,\Sigma,\delta,q_0,Q_m)$ as a device which, like a supervisor, is forced externally by strings in Σ^*; its action is thus to 'mark' or 'recognize' precisely the words of L, regarded as input strings to M.

In the proof of Prop. 4.1 our construction merely installs a grammar that acts as a marking device 'in parallel' with the original supervisor (vacuous in part (i), S in part (ii)). This does nothing to change the control action, but might be thought of as a means of recording when words in K have been completed.

5. CONTROLLABILITY

Let $G = (Q,\Sigma,\delta,q_0,Q_m)$ be a fixed CDEP. We assume that G is trim, i.e. $L(G) = \bar{L}_m(G)$. Write $\Sigma_u = \Sigma - \Sigma_c$, i.e. Σ_u is the set of (labels of) events that cannot be disabled. Let $K \subset \Sigma^*$, $L \subset \Sigma^*$ be arbitrary languages. We say that K is

(i) L-closed if $K = \bar{K} \cap L$

(ii) (Σ_u, L)-invariant if

$$\bar{K}\Sigma_u \cap L \subset \bar{K}$$

(iii) controllable if $K \subset L(G)$ and K is $(\Sigma_u, L(G))$-invariant i.e.

$$\bar{K}\Sigma_u \cap L(G) \subset \bar{K} .$$

Recall that $\bar{K}$ is the language consisting of K together with all the prefixes (including the empty word) of words in K. Thus K is L-closed iff any prefix of K that is a word of L is also a word of K.

The language $\bar{K}\Sigma_u \cap L$ consists of all strings $s' = s\sigma$ where $s' \in L$, $s \in \bar{K}$ and $\sigma \in \Sigma_u$. If we think of L as representing 'physically possible behavior', and $\bar{K}$ as 'legally admissible behavior', then the string $s\sigma$ is a legally admissible string s followed by an uncontrolled symbol σ such that $s\sigma$ is physically possible. K is (Σ_u, L)-invariant precisely when all such strings are legally admissible, i.e. certain instances of uncontrolled behavior are nonetheless legal.

Finally, thinking of $L(G)$ as the uncontrolled process language, i.e. the physically possible uncontrolled behavior of our CDEP, we have that K is controllable if every prefix $s \in \bar{K}$ is physically possible, and every physically possible string $s\sigma$, with $s \in \bar{K}$ and σ uncontrolled, is nevertheless in $\bar{K}$.

Proposition 5.1

Let $K_1 \subset L_m(G)$, $K_2 \subset L_m(G)$ and $K_3 \subset L(G)$ with $K_3 \neq \emptyset$. There exists a complete supervisor S such that, for S/G,

$$L_m(S/G) = K_1$$
$$L_c(S/G) = K_2 \qquad (5.1)$$
$$L(S/G) = K_3$$

iff

(i) $K_1 \subset K_2$

and

(ii) $K_2 = K_3 \cap L_m(G)$

and

(iii) K_3 is closed and controllable. □

6. PROPER SUPERVISORS

To specify controlled behavior in a way that is intuitively satisfying, more stringent conditions must be placed on the three languages

$$L_m(S/G), \quad L_c(S/G), \quad L(S/G)$$

that describe the closed-loop system S/G. We shall say that S is nonblocking if

$$\bar{L}_c(S/G) \ [= \overline{L(S/G) \cap L_m(G)}] = L(S/G)$$

and that S is <u>nonrejecting</u> if

$$\bar{L}_c(S/G) = \bar{L}_m(S/G).$$

By definition we always have $\bar{L}_c(S/G) \subset L(S/G)$. If S blocks, i.e. fails to be nonblocking, then there exists a string s generated by S/G [i.e. $s \in L(S/G)$] that can never be completed to a word $st \in L_c(S/G)$, i.e. $s \notin \bar{L}_c(S/G)$. In this sense the CDEP may be blocked from ever completing a 'task'. This undesirable situation is illustrated by supervisor (i) of Sect. 3.2. Here, for instance, the string $\alpha\gamma \in L(S/G) - \bar{L}_c(S/G)$.

If S rejects, i.e. fails to be nonrejecting, then there exists a string $s \in \bar{L}_c(S/G)$ that can be completed to a 'task' in $L_c(S/G)$ but never to a task that is marked, i.e. (say) recorded. By contrast, if $\bar{L}_c(S/G) = \bar{L}_m(S/G)$, so that

$$L_m(S/G) \subset L_c(S/G) \subset \bar{L}_m(S/G)$$

then for every $s \in L_c(S/G)$ there is some t such that $st \in L_m(S/G)$, and then $st \in L_c(S/G)$ as well. In Sect. 3.2 the supervisor (ii) rejects: only strings of the form $(\alpha\beta)^*$ are marked, while $L_c(S/G) = (\alpha\gamma^*\beta)^*$ represents the complete set of tasks that may be performed.

A supervisor S is <u>proper</u> if it is nonblocking and nonrejecting, i.e.

$$\bar{L}_m(S/G) = \bar{L}_c(S/G) = L(S/G).$$

<u>Theorem 6.1</u>

Let $K \subset L_m(G)$.

(i) There exists a proper supervisor S such that $L_m(S/G) = K$ iff K is controllable. In that case,

$$L_c(S/G) = L_m(G) \cap \bar{K}.$$

(ii) There exists a proper supervisor S such that $L_c(S/G) = K$ iff K is controllable and $L_m(G)$- closed.

□

7. SUPERVISOR SYNTHESIS PROBLEMS

Let languages L_a, $L_g \subset \Sigma^*$ be given, with

$$L_a \subset L_g \subset L_m(G)$$

We interpret L_g as 'legal behavior', i.e. each word of L_g is a 'legal task'; and L_a as 'minimal acceptable behavior', i.e. control of the CDEP G in such a way that a language smaller than L_a is generated is considered inadequate. We now introduce the

<u>Supervisory Marking Problem (SMP)</u>:

Construct a proper supervisor S for G such that

$$L_a \subset L_m(S/G) \subset L_g$$

Similarly we define the

Supervisory Control Problem (SCP):

Construct a proper supervisor S for G such that

$$L_a \subset L_c(S/G) \subset L_g$$

If SCP is solvable then (as shown in the proof of Theorem 6.1(ii)) we can always arrange that $L_m(S/G) = L_c(S/G)$, so that automatically SMP is solvable as well. For a converse to this statement, consider the special but interesting case where L_g is $L_m(G)$-closed, i.e.

$$L_g = \bar{L}_g \cap L_m(G)$$

Then L_g is a sublanguage of $L_m(G)$ with the property that if a string $st \in L_g$ and $s \in L_m(G)$ then also $s \in L_g$. Now if SMP is solvable, the language $L_m(S/G)$ satisfies

$$L_a \subset L_m(S/G) \subset L_g,$$

so that

$$L_a \subset L_m(S/G) \subset L_c(S/G) .$$

Also, since S is proper,

$$\bar{L}_m(S/G) = \bar{L}_c(S/G)$$

so that

$$L_c(S/G) \subset \bar{L}_c(S/G) = \bar{L}_m(S/G) \subset \bar{L}_g$$

But $L_c(S/G) \subset L_m(G)$ by definition, i.e.

$$L_c(S/G) \subset \bar{L}_g \cap L_m(G) = L_g$$

Hence

$$L_a \subset L_c(S/G) \subset L_g$$

and so SCP is solvable as well.

When SMP or SCP is solvable, it may be considered desirable that the solution be <u>minimally restrictive</u> in the sense that $L_m(S/G)$ or $L_c(S/G)$, considered as sublanguages of $L_m(G)$, be as large as possible, subject to the constraint that it is a sublanguage of L_g. The fact that minimally restrictive solutions are possible in principle is due to a certain semilattice property that we now describe. For this, let $L \subset L(G)$ be an arbitrary sublanguage of $L(G)$. Let

$$\underline{C}_G(L) := \{K: K \subset L \ \ \& \ \ K \text{ is controllable}\}$$

$$\underline{F}_G(L) := \{K: K \subset L \ \ \& \ \ K = \bar{K} \cap L_m(G)\}$$

Thus $\underline{C}_G(L)$ (resp. $\underline{F}_G(L)$) are the controllable (resp. $L_m(G)$-closed) sublanguages of L.

<u>Proposition 7.1</u>

$\underline{C}_G(L)$ and $\underline{F}_G(L)$ are nonempty classes of languages that are closed under arbitrary unions.

□

By Prop. 7.1 each of $\underline{C}_G(L)$, $\underline{F}_G(L)$ contains a unique supremal element with respect to inclusion, which we denote by

$$\sup \underline{C}_G(L), \qquad \sup \underline{F}_G(L)$$

respectively. In fact, $\underline{C}_G(L)$ and $\underline{F}_G(L)$ are complete subsemilattices of the semilattice of all sublanguages of L, partially ordered by inclusion, and with join operation the union of languages.

On the basis of Theorem 6.1 and Prop. 7.1 we immediately conclude the following.

Theorem 7.1

(i) SMP is solvable iff

$$\sup \underline{C}_G(L_g) \supset L_a$$

(ii) SCP is solvable iff

$$\sup\{\underline{C}_G(L_g) \cap \underline{F}_G(L_g)\} \supset L_a$$

In each case the corresponding supervisor is minimally restrictive. □

8. CONGRUENCES OF SUPERVISORS

Let $\mathcal{S} = (S,\phi)$ and $\hat{\mathcal{S}} = (\hat{S},\hat{\phi})$ each be supervisors for G, where as usual

$$S = (X,\Sigma,\xi,x_0,X_m), \qquad \phi : X \to \{0,1\}^\Sigma$$

$$S = (\hat{X},\Sigma,\hat{\xi},\hat{x}_0,\hat{X}_m), \qquad \hat{\phi} : \hat{X} \to \{0,1\}^\Sigma$$

We shall say that a (total) function $\pi : X \to \hat{X}$ is a congruence from $\mathcal{S}$ to $\hat{\mathcal{S}}$, and write $\pi : \mathcal{S} \to \hat{\mathcal{S}}$, provided

(i) $\pi : X \to \hat{X}$ is surjective

(ii) $\pi(x_0) = \hat{x}_0$ and $X_m = \pi^{-1}(\hat{X}_m)$

(iii)[4] $\hat{\xi} \circ (id_\Sigma \times \pi)(\sigma,x) = \pi \circ \xi(\sigma,x)$ for all (σ,x) where $\xi(\sigma,x)$ is defined.

(iv) $\hat{\phi} \circ \pi = \phi$

The situation is displayed in the diagrams[5] below.

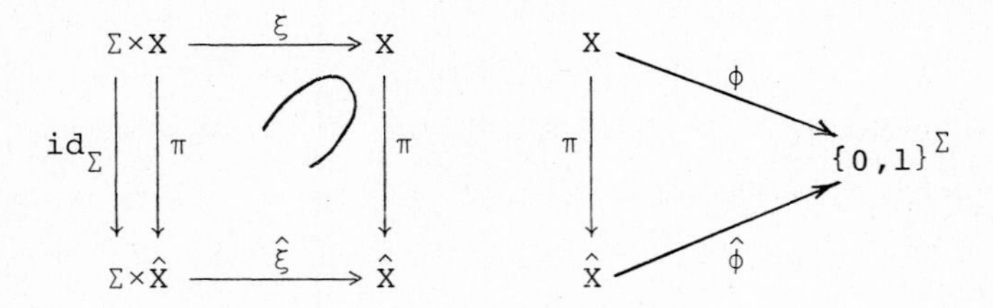

[4] By definition $id_\Sigma \times \pi : \Sigma \times X \to \Sigma \times \hat{X} : (\sigma,x) \longmapsto (\sigma,\pi(x))$.

[5] The symbol ⊃ means that the left hand diagram is only 'partially commutative', in the sense of (iii).

Congruences represent very close relationships between supervisors, as expressed in the following.

Proposition 8.1

Let S be complete with respect to G, and let $\pi : S \to \hat{S}$ be a congruence. Then

(i) π is unique

(ii) $(L_m, L_c, L)(S/G) = (L_m, L_c, L)(\hat{S}/G)$

(iii) S is nonblocking (resp. nonrejecting, proper) iff $\hat{S}$ is nonblocking (resp. nonrejecting, proper).

(iv) $\hat{S}$ is complete with respect to G.

□

9. REDUCED LANGUAGES AND GRAMMARS

In this section we introduce some definitions needed for the main result in the section to follow.

Let $L \subset \Sigma^*$. For $s, s' \in \Sigma^*$ write

$$s \equiv_L s'$$

if

$$\{t : t \in \Sigma^* \;\; \& \;\; st \in L\} = \{t' : t' \in \Sigma^* \;\; \& \;\; s't' \in L\}$$

Thus s and s' are 'equivalent mod L' if the corresponding sets of postfixes, leading to completed words of L, are identical[6].

Let $K \subset L \subset \Sigma^*$. We say that L is K-reduced if, whenever, $s, s' \in \bar{K}$ and $s \equiv_K s'$, then $s \equiv_L s'$. Similarly, if

$$M = (X, \Sigma, \xi, x_0, X_m)$$

is a grammar for L, then M is K-reduced if a counterpart condition holds, namely whenever $s, s' \in \bar{K}$ and $s \equiv_K s'$, then

$$\xi(s, x_0) = \xi(s', x_0) .$$

For example, let $\Sigma = \{\alpha, \beta, \gamma, \mu, \nu\}$. Suppose

$$L = \alpha\mu + \beta\nu + (\alpha+\beta)\gamma$$

$$K = (\alpha+\beta)\gamma$$

Here L is not K-reduced, since $\alpha \equiv_K \beta$ but it is not true that $\alpha \equiv_L \beta$. On the other hand, suppose

$$L = (\alpha+\beta)(\gamma+\mu+\nu)$$

$$K = (\alpha+\beta)\gamma$$

---

(6) This is just standard Nerode equivalence: see e.g. [Hopcroft and Ullman, 1979, p. 65].

Clearly L is K-reduced.

As above, let M be a grammar for L and let $K \subset L$. M is K-trim if every state of M is visited by some word of K, i.e. if $x \in X$ then there are strings s and $t \in \Sigma^*$ such that $\delta(s,x_0) = x$ and $st \in K$. If M is trim, i.e. every state is both accessible and co-accessible under the action of strings of L, then clearly M is L-trim; conversely if M is L-trim then it is trim. Clearly M is L-trim if it is K-trim. In linguistic terms, if M is any grammar for L and M is K-trim, then every equivalence class of $\equiv_L$ in $\bar{L} \subset \Sigma^*$ has a nonempty intersection with $\bar{K}$.

10. QUOTIENT STRUCTURE THEOREM

The following is the main result of this paper. It states, roughly, that 'every efficiently constructed supervisor is a quotient (high-level, or lumped, model) of the desired closed-loop behavior'.

Theorem 10.1

Let $S = (S,\phi)$ be a complete supervisor for G. Let $K_1 := L_m(S/G)$, $K_3 := L(S/G)$ and assume that S is K_3-reduced and K_3-trim. Let

$$\hat{S}^0 = (X^0,\Sigma,\xi^0,x_0^0,X^0)$$

be a trim grammar for K_3. Then there exist a subset $X_m^0 \subset X^0$ and a state feedback map $\phi^0: X^0 \to \{0,1\}^\Sigma$ with the following properties.

(i) Define

$$S^0 := (S^0,\phi^0), \qquad S^0 := (X^0,\Sigma,\xi^0,x_0^0,X_m^0)$$

Then S^0 is a complete supervisor for G with

$$L_m(S^0/G) = K_1, \qquad L(S^0/G) = K_3$$

(ii) There is a congruence $\pi: S^0 \to S$

(iii) If S is proper then so is S^0. □

11. EXAMPLE 1

We consider two users of a single resource, each modeled as in Subsect. 2.3, giving the state transition graphs G_1, G_2 of Fig. 11.1. For G we take the 'shuffle' of G_1, G_2, namely the process determined by the concurrent actions of G_1 and G_2 under the assumption that these actions are asynchronous and independent. This assumption rules out the simultaneous occurrence of an event in G_1 with an event in G_2, but otherwise places no constraint on their joint behavior. The graph of G is thus as shown in Fig. 11.2. Here the state ◎ is both q_0 and (as a singleton) Q_m, while $L_m(G)$ consists of all words over the alphabet

$$\Sigma = \{\alpha_i \; \beta_i \; \gamma_i \; ; \; i = 1,2\}$$

corresponding to paths in the graph that begin and end at ◎ .

The objective of supervisory control is to manipulate the binary controls c_1, c_2 in order to satisfy the following synchronization requirements.

(i) Mutual exclusion: G_1, G_2 never simultaneously occupy their respective USE states.

(ii) Fair usage: The USE states of G_1, G_2 are occupied according to first-come-first-served discipline, namely the index sequence of events β_i must coincide with the index sequence of events α_j.

In practical terms this standard problem would, of course, be solved by a queue; but instead we shall approach it via the ideas of previous sections. However, we defer the question of how conditions like (i) and (ii) may be formalized, simply taking it for granted that from them the 'legal' behavior $L_g \subset L_m(G)$ can be explicitly determined. In fact the reader may convince himself that L_g is described by the grammer displayed in Fig. 11.3[7].

By inspection of Fig. 11.3 it is easy to see that L_g is both controllable and $L_m(G)$-closed. That is,

$$L_g = \sup\{\underline{C}_G(L_g) \cap \underline{F}_G(L_g)\} .$$

By Theorem 6.1(ii) there exists a proper supervisor $S = (S, \phi)$ such that $L_c(S/G) = L_g$. As demonstrated in the proof of Prop. 5.1, the state transition diagram for L_g (Fig. 11.3) can serve to define S; it just remains to identify the state feedback map ϕ. For each state x of S, $\phi(x)$ is a map

$$\phi(x):\{c_1, c_2\} \to \{0,1\} \qquad ,$$

i.e. a binary evaluation of each of the controls c_1, c_2. So, with reference to Fig. 11.3, it is enough to define

$$\phi(x)(c_1) = \begin{cases} 1 & \text{if an edge labeled } \beta_1 \text{ issues from x} \\ 0 & \text{otherwise} \end{cases}$$

and similarly for $\phi(x)(c_2)$. The resulting control patterns are tabulated in Fig. 11.4. The supervisor $S = (S, \phi)$ then certainly determines

$$L(S/G) = \bar{L}_g$$

$$L_m(S/G) = L_g$$

We remark that in this example there exists an alternative supervisor

$$S^\circ = (S^\circ, \phi^\circ)$$

defined by setting $S^\circ = S$, and with ϕ° as tabulated in Fig. 11.4. This determines exactly the same language controlled in G as S does, namely

$$L(S^\circ/G) = L(S/G) = \bar{L}_g \qquad .$$

---

[7] Alternatively the grammer of Fig. 11.3 could be taken as the definition of L_g.

To verify this fact note that, for instance, states x_1, x_3 of S are entered only on the occurrence of the event β_1; but since β_1 can be immediately followed in G_1 only by γ_1, the enablement of β_1 by $\phi°$ in x_1 and x_3 can have no effect on the language controlled in G.

It may be left to the reader to verify that $S°$ is complete with respect to G. From $S°$ we construct a new supervisor $S' = (S',\phi')$ and a congruence $\pi: S° \rightarrow S'$ as tabulated in Fig. 11.4; the result is displayed in Fig. 11.5. By Prop. 8.1

$$(L_m,L_c,L)(S'/G) = (L_m,L_c,L)(S°/G) \quad ,$$

namely control and marking action are preserved. The simplified supervisor S' has just 5 states and is equivalent, in fact, to a queue (of maximum length 2) that stores events α in order of occurrence and is popped by the corresponding events γ. Intuitively it is fairly obvious that no simpler supervisor could yield the control action required.

12. EXAMPLE 2

In a manufacturing system we consider two machines M_1, M_2 connected in tandem and separated by a buffer B (Fig. 12.1). Each machine M_i is modeled as a CDEP over the alphabet $\{\alpha_i\ \beta_i\ \lambda_i\ \mu_i\}$ and having binary-valued controls $\{u_i,v_i\}$ (Fig. 12.2). The machine states are IDLE (I), WORKING (W) and DOWN (D). The control u enables/disables the transition for I to W (u = 1 allows M to 'acept a workpiece'); while v enables/disables the transition from D to I (v = 1 means, when M is in state D, that M is 'under repair'). The buffer B has one slot, i.e. is EMPTY (E) or FULL (F); it is not a CDEP but simply an automaton driven by M_1 and M_2 (Fig. 12.3). The system operates as follows. Machine M_1 takes a workpiece (event α_1), and either successfully completes processing and passes the workpiece to the buffer (event β_1); or breaks down and discards the workpiece (event λ_1), but in that case may later be repaired (event μ_1). Machine M_2 operates in the same way, but takes its workpiece from the buffer B, provided one is there.

The problem is to manipulate the controls in order to satisfy the four requirements stated informally below.

(i) M_1 executes α_1 only if B is in E.

(ii) M_2 executes α_2 only if B is in F (thereby driving B to E).

(iii) M_1 cannot execute α_1 while M_2 is in D_2.

(iv) If M_1 is in D_1 and M_2 is in D_2 then $v_1 = 0$.

Condition (iv) means that if both machines are down then M_2 must be repaired before M_1.

We shall not formalize these requirements or present the details of how the legal language L_g is derived from them: a systematic procedure for doing so will be reported elsewhere. Here we shall merely display the result. The language L_g that incorporates requirements (i) - (iv) with the system constraints is generated by the grammar shown in Fig. 12.4. This grammar defines a supervisor $S°$ such that $L(S°/G) = \bar{L}_g$; the control patterns are tabulated in Fig. 12.6. It can be verified that $S°$ admits the quotient S displayed in Fig. 12.5; the congruence is also tabulated in Fig. 12.6. The quotient represents a reduction from 12 states to 6. As will be shown elsewhere, it can actually be obtained directly from two modular 'subsupervisors', of which one is modeled on the buffer, and the other incorporates the logic of breakdown and repair.

13. CONCLUSION

In this paper we have introduced a broad class of controlled discrete-event processes together with some general concepts and results relating to their control or 'supervision'. Our main conclusion, the Quotient Structure Theorem, is similar in spirit to the Internal Model Principle of regulator theory; it may be roughly paraphrased by saying that 'supervisors must be modeled on the task to be accomplished'.

In future articles we shall discuss constructive methods for computing the supremal controllable (or closed controllable) sublanguage of a given language, as well as concrete methods for system specification and supervisor synthesis.

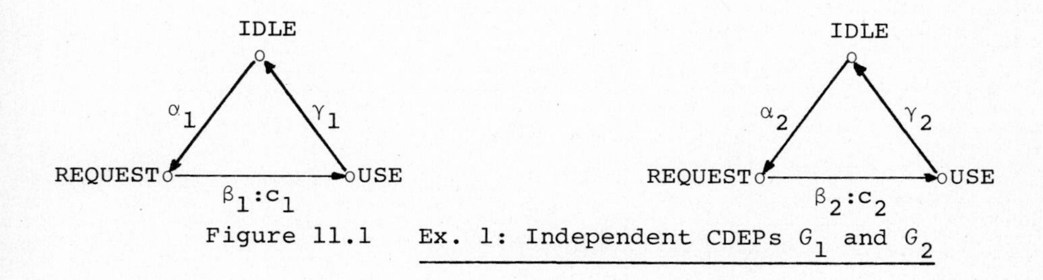

Figure 11.1 Ex. 1: Independent CDEPs G_1 and G_2

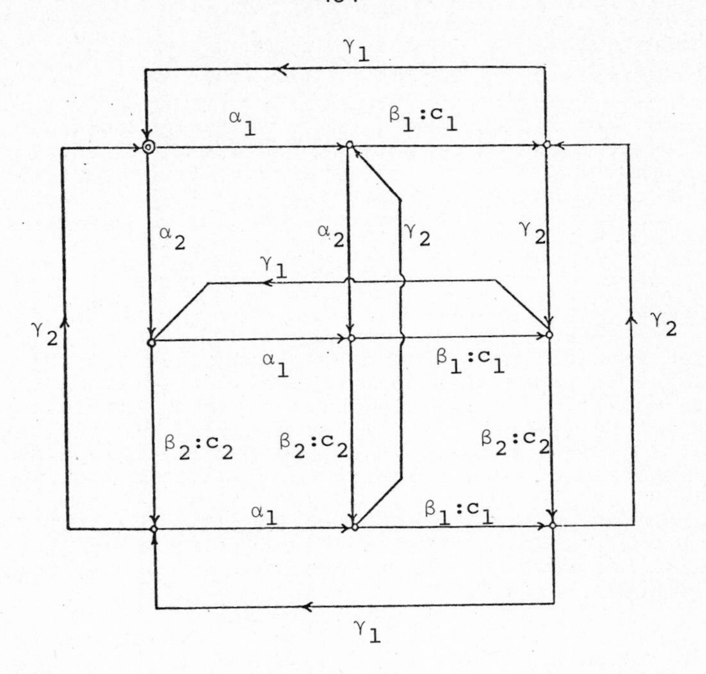

Figure 11.2 Ex.1: Shuffle Grammar G of G_1 and G_2

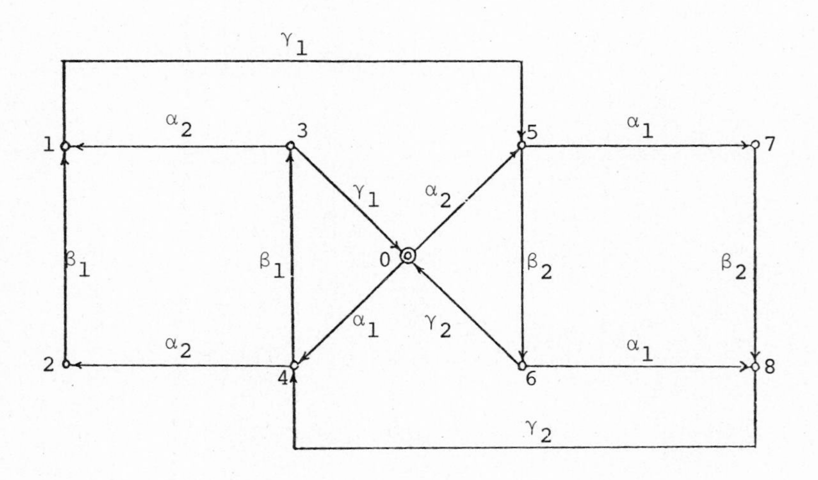

Figure 11.3 Ex.1: Grammar for L_g

State	x_0	x_1	x_2	x_3	x_4	x_5	x_6	x_7	x_8
ϕ	00	00	10	00	10	01	00	01	00
$\phi°$	00	10	10	10	10	01	01	01	01
π	x'_0	x'_1	x'_1	x'_2	x'_2	x'_3	x'_3	x'_4	x'_4

Figure 11.4 Ex.1: Control Data for S, $S°$ and S'

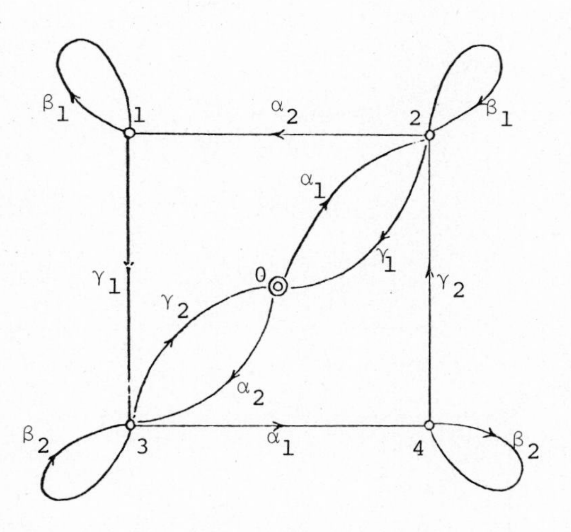

Figure 11.5 Ex.1: Quotient Supervisor S'

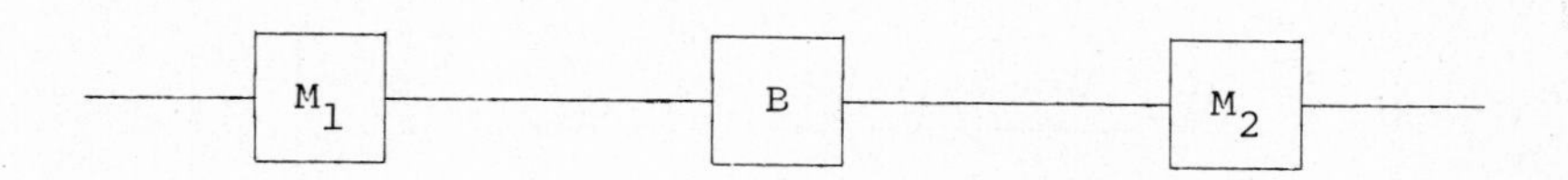

Figure 12.1 Ex.2: Machines Coupled by a Buffer

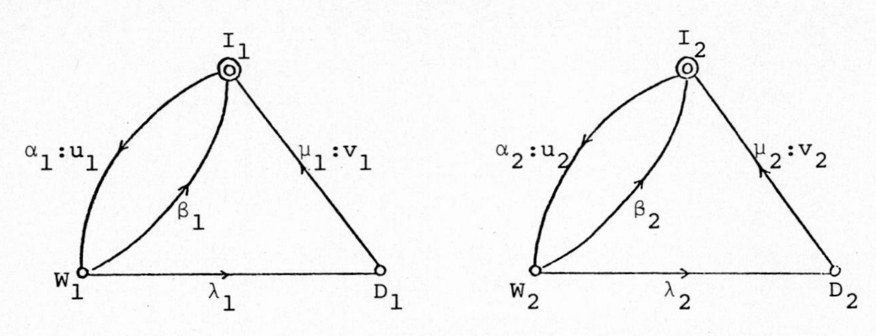

Figure 12.2 Ex.2: State Diagrams of Machines

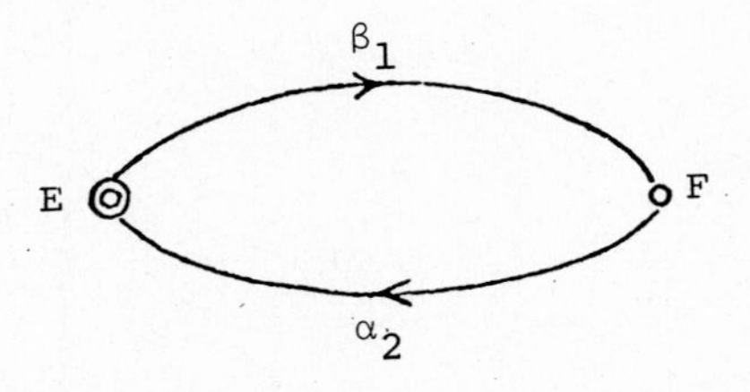

Figure 12.3 Ex.2: State Diagram of Buffer

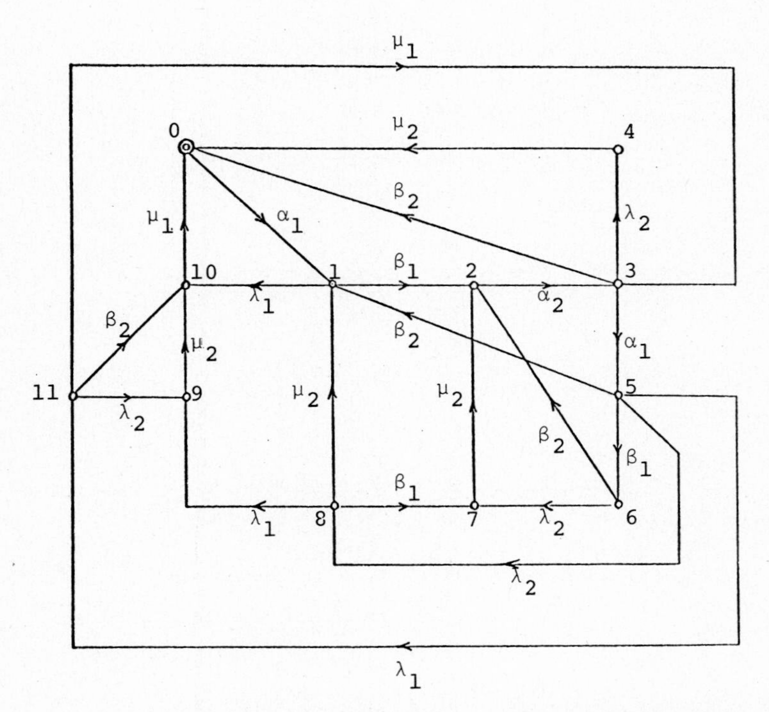

Figure 12.4 Ex.2: Grammar for L_g

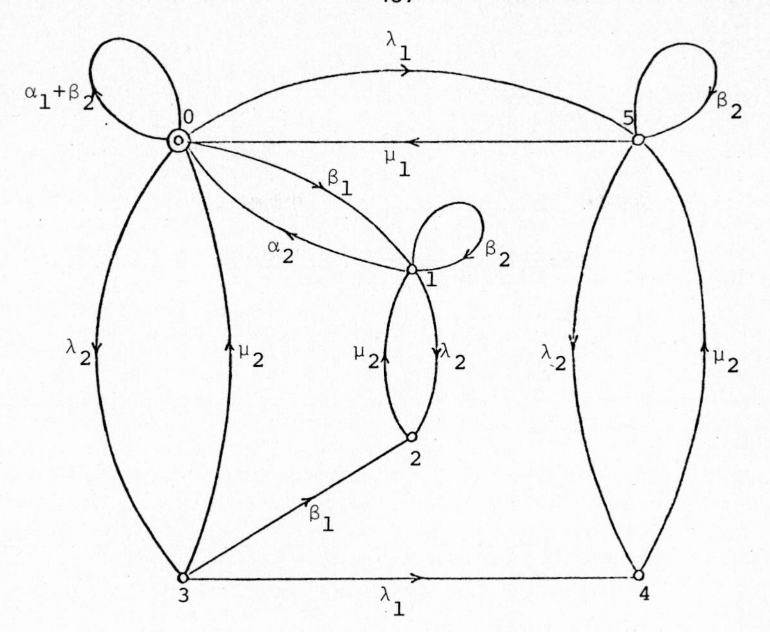

Figure 12.5 Ex.2: Quotient Supervisor S

x^0	x	u_1	v_1	u_2	v_2
0	0	1	-	0	-
1	0	1*	-	0	-
2	1	0	-	1	-
3	0	1	-	0	-
4	3	0	-	0	1
5	0	1*	-	0	-
6	1	0	-	1*	-
7	2	0	-	-	1
8	3	0	-	0	1
9	4	0	0	0	1
10	5	-	1	0	-
11	5	-	1	0	-

Figure 12.6 Ex.2: Control Data for S and S^0

Assignment (*) are determined by consistency for the quotient; entries (-) may be assigned arbitrarily, consistent with the quotient.

13. REFERENCES

Aveyard, R., 1974. A boolean model for a class of discrete event systems. IEEE Trans. Sys. Man and Cyb. SMC-4, pp. 249-258.

Beauquier, J., and Nivat, M., 1980. Application of formal language theory to problems of security and synchronization. In R.V. Book (Ed.), Formal Language Theory - Perspective and Open Problems, Academic Press, New York; pp. 407-454.

Eilenberg, S., 1974. Automata, Languages, and Machines, Vol. A. Academic Press, New York.

Hailpern, B.T., and Owicki, S.S., 1983. Modular verification of computer communication protocols. IEEE Trans. Commun. COM-31, pp. 56-68.

Hopcroft, J.E., and Ullman, J.D., 1979. Introduction to Automata Theory, Languages, and Computation. Addison-Wesley Pub. Co., Reading.

Milne, G., and Milner, R., 1979. Concurrent processes and their syntax. J. Assoc. Comp. Mach. 26, pp. 302-321.

Nijmeijer, H., 1983. Nonlinear Multivariable Control: A Differential Geometric Approach. Thesis, Rijksuniv. te Groningen.

Park, D., 1981. Concurrency and automata on infinite sequences. In Theoretical Computer Science, Lecture Notes in Computer Science 104, Springer-Verlag, New York; pp. 167-183.

Peterson, J.L., 1981. Petri Net Theory and the Modeling of Systems. Prentice-Hall, Inc., Englewood Cliffs.

Pnueli, A., 1979. The temporal semantics of concurrent programs. In Semantics of Concurrent Computation, Lecture Notes in Computer Science 70, Springer-Verlag, New York; pp. 1-20.

Ramadge, P.J., 1983. Control and Supervision of Discrete Event Processes. Ph.D. Thesis, Dept. of Electl. Engrg., University of Toronto.

Ramadge, P.J., and Wonham, W.M., 1982. Supervisory control of discrete event processes. In: Feedback Control of Linear and Nonlinear Systems, Lecture Notes in Control and Information Sciences No. 39. Springer-Verlag, Berlin; pp. 202-214.

Ramadge, P.J., and Wonham, W.M., 1982. Supervision of discrete event processes. Proc. 21st IEEE Conf. on Decision & Control, December; pp. 1228-1229.

Steenstrup, M., Arbib, M.A., and Manes, E.G., 1981. Port automata and the algebra of concurrent processes. Computer and Information Science Tech. Rpt. 81-25, University of Massachusetts (Amherst).

Wonham, W.M., 1979. Linear Multivariable Control: A Geometric Approach, Sec. ed., Springer-Verlag, New York.

QUELQUES REMARQUES ELEMENTAIRES SUR LE CALCUL DES LOIS DE BOUCLAGE EN COMMANDE OPTIMALE NON LINEAIRE(*)

Michel FLIESS
et
Houria BOURDACHE-SIGUERDIDJANE
Laboratoire des Signaux et Systèmes
C.N.R.S. - E.S.E.
Plateau du Moulon
91190 Gif-sur-Yvette, France.

RESUME

On considère des problèmes de commande optimale où les dimensions de l'état et du vecteur de commande sont égales. Des calculs simples à partir des crochets de Lie de champs de vecteurs montrent que la loi de bouclage optimal vérifie un système d'équations aux dérivées partielles quasi-linéaires du premier ordre qui, lorsque la commande n'apparaît pas dans le critère, dégénère en équations algébriques. On illustre les avantages numériques de l'approche par des exemples empruntés à la littérature et dont certains ont une origine concrète.

ABSTRACT

Problems of optimal control are considered where the dimensions of the state-space and of the control-vector are the same. Simple computations using Lie brackets of vector fields show that the optimal feedback law satisfies a system of quasi-linear first order partial differential equations. When the control does not appear in the criterion, this system degenerates into algebraic equations. We give numerical illustrations of our method using several examples from the literature, some from realistic models.

INTRODUCTION

Notre but est une première évaluation pratique de travaux actuels [3,4] montrant que, pour des problèmes de commande optimale sans contraintes, à temps fixe et extrémité libre, les lois de bouclage vérifient des équations aux dérivées partielles. Avec la forme de Bolza et la dimension du vecteur de commande égale à celle de l'état, on aboutit à un système quasi-linéaire du premier ordre. Nous en rappelons la démarche à partir du formalisme hamiltonien plus usuel que les séries de Volterra également présentées en [3,4]. Des dérivations successives du pseudo-hamiltonien font apparaître les crochets de Lie des champs de vecteurs associés. L'élimination du vecteur adjoint conduit aux équations. La contribution théorique originale de cette communication apporte la loi de bouclage lorsque la commande ne figure pas dans le critère sous forme de Lagrange : elle est solution d'équations algébriques, c'est-à-dire où n'apparaissent pas les dérivées des fonctions inconnues.

Notre point de vue, en fait très élémentaire, participe du développement durant ces quinze dernières années des méthodes géométriques puis algébriques en non-linéaire. Nos résultats, aussi simples qu'ils soient, ouvrent bien des perspectives quant au calcul numérique de la régulation optimale, certainement l'un des sujets les plus

(*)Travail effectué dans le cadre de l'A.T.P. "Thermique du Bâtiment" du C.N.R.S.

importants de l'automatique, qui a connu une multitude de publications dont il est impossible de rendre compte ici. Nous en donnons un avant-goût en empruntant des exemples monodimensionnels à Ozgoren, Longman et Cooper [12], Willemstein [14], Bell et Ye [1], Jacobson et Mayne [6], Intriligator [5]. Le cas algébrique est illustré par la commande d'un système endocrinien bidimensionnel (Kaulgud, Sentis et Bernard-Weil [8]). A plusieurs reprises, nous obtenons l'expression analytique de la boucle optimale, ce qui, à notre connaissance, est une première !

Nous remercions F. Lamnabhi-Lagarrigue pour d'utiles conversations.

A. LOIS DE BOUCLAGE OPTIMAL

1. Problème posé et formalisme hamiltonien

Considérons le système

$$\begin{cases} \dot{q}(t) = F(t, q(t), u(t)) \\ y(t) = h(q(t)). \end{cases} \tag{1}$$

L'état $q = (q^1,\ldots,q^N)$ appartient à R^N (1). La fonction $F = (F^1,\ldots,F^N) : R^{1+N+m} \to R^N$ est indéfiniment dérivable, c'est-à-dire C^∞, en ses arguments, à savoir t, $q^1,\ldots q^N$, et le vecteur de commande $u = (u_1,\ldots,u_m)$. La fonction de sortie $h : R^N \to R$ est aussi C^∞. On s'attaque au problème suivant de commande optimale sans contraintes, sous forme de Mayer, à temps et état initial fixes et à extrémité libre : optimiser, c'est-à-dire minimiser ou maximiser, $y(T) = h(q(T))$. Nous recherchons la commande optimale sous forme d'une loi de bouclage (*feedback law*) $u : R^{1+N} \to R^m$, fonction C^∞ de t et q.

Le pseudo-hamiltonien correspondant est :

$$H = \langle p,F(t,q,u)\rangle = \sum_{k=1}^{N} p_k F^k(t,q,u) \ , \tag{2}$$

où $p = (p_1,\ldots,p_N)$ est le vecteur adjoint. Les équations de Hamilton sont :

$$\begin{cases} \dot{q} = \dfrac{\partial H}{\partial p} \\ \\ \dot{p} = -\dfrac{\partial H}{\partial q} \ , \end{cases} \tag{3}$$

avec les conditions aux bouts $q(o)=q_o$ (état inital), $p(T)= - h_q(q(T))$. Le Principe du Maximum, équivalent, ici, au calcul des variations ordinaire, donne :

$$\frac{\partial H}{\partial u} = 0 \qquad (0 \leqq t \leqq T) \ ,$$

soit

$$H_{u_i} = \langle p, F_{u_i}(t,q,u)\rangle = 0 \qquad (i=1,\ldots,m). \tag{4}$$

---

(1) En automatique non linéaire, il est souvent avantageux de prendre un état évoluant sur une *variété différentiable* [9]. Une telle présentation, faite en [4], change peu les calculs.

Remarque : Comme on désire la commande optimale sous forme d'une loi de bouclage $u(t,q)$, il faut, dans l'écriture $\frac{\partial H}{\partial q}$, en tenir compte :

$$\frac{\partial H}{\partial q^k} = \sum_{i=1}^{N} p_i [F^i_{q^k}(t,q,u(t,q)) + \sum_{j=1}^{m} F^i_{u_j}(t,q,u(t,q)) \frac{\partial u_j}{\partial q^k}] . \qquad (5)$$

Pour se convaincre de ce fait parfois oublié dans la littérature, il suffit de remplacer en (1) u par $u_{op}(t,q) + v(t)$, où :

- $u_{op}(t,q)$ est la loi de bouclage optimal , supposée exister;
- $v(t)$ est une nouvelle commande en boucle ouverte.

Comme $v(t) = 0$, $0 \leq t \leq T$, est optimale, l'application des méthodes habituelles redonne (2), (3), (4) avec l'interprétation (5)

2. Crochets de Lie et conditions d'optimalité

Les crochets de Lie de champs de vecteurs apparaissent dans les dérivées totales par rapport au temps $\frac{d^\nu}{dt^\nu} \frac{\partial H}{\partial u}$. Peut-être convient-il d'en redonner brièvement la définition[(2)]. Toute fonction $R^N \to R^N$ sera appelée champ de vecteurs. Le crochet de Lie de deux champs de vecteurs dérivables $X = (X^1,\ldots,X^N)$, $Y = (Y^1,\ldots,Y^N)$: $R^N \to R^N$ est le champ de vecteurs $Z = (Z^1,\ldots,Z^N) : R^N \to R^N$ tel que

$$Z^k = \sum_{i=1}^{N} X^i \frac{\partial Y^k}{\partial q^i} - Y^i \frac{\partial X^k}{\partial q^i} .$$

On écrit $Z = [X,Y]$. Il est parfois nécessaire d'itérer ce crochet : on introduit la notation $ad^\nu_X Y$, définie par récurrence sur ν, $ad^o_X Y = Y$, $ad_X Y = [X,Y]$, $ad^{\nu+1}_X Y = [X, ad^\nu_X Y] = [X,\ldots[X,Y]\ldots]$ ($\nu+1$ fois). Il est classique d'envisager un champ de vecteurs $X : R^N \to R^N$ comme opérateur différentiel linéaire du premier ordre

$$X = \sum_{k=1}^{N} X^k \frac{\partial}{\partial q^k} .$$

En (1), F étant considéré comme un champ de vecteurs de R^N pouvant dépendre du temps, associons-lui le champ de vecteurs $A : R^{1+N} \to R^{1+N}$:

$$A = \frac{\partial}{\partial t} + \sum_{k=1}^{N} F^k(t,q,u) \frac{\partial}{\partial q^k} = \frac{\partial}{\partial t} + F.$$

Soit $G = (G^1,\ldots,G^N) : R^{1+N} \to R^N$ une fonction C^∞ de t et q, à laquelle on associe le champ de vecteurs $G = \sum G^k \frac{\partial}{\partial q^k}$. Avec les notations du paragraphe A.1, il vient :

Lemme 1 : $\frac{d^\nu}{dt^\nu} \langle p,G(t,q)\rangle = \langle p,ad^\nu_A G\rangle$.

---

(2) Pour plus de détails, renvoyons aux nombreux cours de géométrie différentielle, par exemple [9].

Démonstration. On écrit

$$\frac{d}{dt} \langle p,G\rangle = \langle \dot{p},G\rangle + \langle p,\dot{G}\rangle$$

et on applique (2).

Avec $H_{u_i} = \langle p,F_{u_i}\rangle$, on a :

Corollaire 2 : $\dfrac{d^\nu}{dt^\nu} H_{u_i} = \langle p,\mathrm{ad}_A^\nu F_{u_i}\rangle$.

Le Principe du Maximum, exprimé ici par (4), implique, pour $\nu = 0,1,2,\ldots$, $\dfrac{d^\nu}{dt^\nu} H_{u_i} = 0$, soit une hiérarchie infinie de conditions nécessaires.

Proposition 3. Une loi de bouclage optimal $u(t,q)$ satisfait la hiérarchie infinie de conditions nécessaires

$$\langle p,\mathrm{ad}_A^\nu F_{u_i}\rangle = 0 \qquad (i=1,\ldots, m;\ \nu= 0,1,2,\ldots). \tag{6}$$

Remarques : (i) L'apparition de crochets de Lie est bien connue en mécanique analytique, comme dans les problèmes *singuliers* de commande optimale[(3)].

(ii) $\mathrm{ad}_A^\nu F_{u_i}$ fait intervenir des dérivées partielles $\dfrac{\partial}{\partial q^k}$ pour lesquelles la remarque finale du paragraphe A.1 s'applique. La loi $u(t,q)$ pouvant aussi dépendre de t, une constatation analogue est vraie pour $\dfrac{\partial}{\partial t}$. Si l'on cherchait la commande optimale en boucle ouverte, c'est-à-dire fonction uniquement de t, la hiérarchie (6) resterait valide avec une modification évidente du rôle de $\dfrac{\partial}{\partial q^k}$.

(iii) C'est pour écrire la hiérarchie infinie (6) que nous travaillons avec des fonctions C^∞.

3. Equations aux dérivées partielles

L'élimination du vecteur adjoint p des conditions (6) conduit à des équations aux dérivées partielles quasi-linéaires [3,4]. Avec un problème sous forme de Bolza, tel que les dimensions de la commande et de l'état soient égales, on aboutit à un système d'équations du premier ordre.

On cherche à optimiser J en boucle fermée :

$$\begin{cases} \dot{q}(t) = F(t,q,u) \\ J = \phi(q(T)) + \displaystyle\int_0^T F^0(t,q,u)\, dt. \end{cases} \tag{7}$$

La première ligne de (7) est identique à celle de (1) avec, ici, $m = N$. Les fonctions $\phi : R^N \to R$, $F^0 : R^{1+2N} \to R$ sont C^∞. Comme à l'accoutumée, on se ramène à un

(3) Voir [4] pour des compléments et des références.

problème de Mayer en ajoutant une coordonnée q^o :

$$\begin{cases} \dot{q}^o(t) = F^o(t,q,u) & (q^o(0) = 0) \\ \dot{q}(t) = F(t,q,u) & \\ y(t) = \phi(q) + q^o & \end{cases}$$

Notons cependant que la loi de bouclage $u(t,q)$ dépend de q et non de q^o. Le rôle du champ de vecteurs A du paragraphe A.2 est tenu par

$$\tilde{A} = \frac{\partial}{\partial t} + F^o(t,q,u)\,\frac{\partial}{\partial q^o} + \sum_{k=1}^{N} F^k(t,q,u)\,\frac{\partial}{\partial q^k} = \frac{\partial}{\partial t} + \tilde{F}\ ,$$

où $$\tilde{F} = \sum_{\ell=o}^{N} F^{\ell}\,\frac{\partial}{\partial q^{\ell}}\ .$$

Le nouveau pseudo-hamiltonien est

$$\tilde{H} = <\tilde{p},\tilde{F}> = \sum_{\ell=o}^{N} p_{\ell}\, F^{\ell}(t,q,u),$$

où $\tilde{p} = (p_o,p_1,\ldots,p_N) = (p_o,p)$.

Pour $\nu = 0,1$, les conditions (6) deviennent

$$\sum_{\ell=o}^{N} p_{\ell}\, F^{\ell}_{u_i}(t,q,u) = 0 \quad \text{(Principe du Maximum)},$$

$$\sum_{\ell=o}^{N} p_{\ell}\, [\tilde{A},\tilde{F}_{u_i}]^{\ell} = 0\ ,$$

où $[\ \ ]^{\ell}$ dénote la $\ell^{\text{ème}}$ composante du crochet de Lie. Remarquons que dans $[\tilde{A},\tilde{F}_{u_i}]^{\ell}$ les dérivées $\frac{\partial u_i}{\partial t}$, $\frac{\partial u_i}{\partial q^k}$ figurent aux degrés un ou zéro. On est en présence d'un système de 2N équations linéaires homogènes en N+1 inconnues $p_o, p_1,\ldots,P_N$. La solution nulle étant exclue puisque l'on aurait pu poser $p_o \equiv 1$, les conditions déterminantales de compatibilité conduisent à l'énoncé suivant :

Théorème 4 : La loi de bouclage optimal du problème (7) satisfait nécessairement un système de N équations aux dérivées partielles quasi-linéaires du premier ordre exprimé par les déterminants :

$$\begin{vmatrix} F^o_{u_1} & F^1_{u_1} \cdots\cdots & F^N_{u_1} \\ \vdots & \vdots & \vdots \\ F^o_{u_N} & F^1_{u_N} \cdots\cdots & F^N_{u_N} \\ [\tilde{A},\tilde{F}_{u_i}]^o & [\tilde{A},\tilde{F}_{u_i}]^1 \cdots\cdots & [\tilde{A},\tilde{F}_{u_i}]^N \end{vmatrix} = 0 \quad (i=1,\ldots,N)\ . \qquad (8)$$

Si N = 1, on obtient l'unique équation

$$(F_u\, F^o_{(u)2} - F^o_u\, F_{(u)2})\left(\frac{\partial u}{\partial t} + F\,\frac{\partial u}{\partial q}\right) = F^o_u\, F_{ut} - F_u\, F^o_{ut} + F(F^o_u\, F_{uq} - F_u\, F^o_{uq}) - F_u(F^o_u\, F_q - F_u\, F^o_q)\ . \qquad (9)$$

Remarques : (i) La condition de Cauchy découle de la connaissance de $u(T,q)$ qui se calcule grâce à $p(T) = -\frac{\partial\phi}{\partial q}(q(T))$.

(ii) Il n'était point besoin ici de travailler dans la catégorie C^∞. Les fonctions F, F^o doivent être C^2 et la loi de bouclage C^1.

4. Horizon infini

On rencontre souvent des problèmes à horizon infini, c'est-à-dire où le T de (7) est infini. Il convient de se restreindre à la forme de Lagrange, c'est-à-dire de supposer $\phi \equiv 0$. Les équations (8) et (9) demeurent évidemment valables. Une simplification importante se produit si (7) est autonome, c'est-à-dire si F et F^o sont indépendantes de t :

$$\begin{cases} \dot{q}(t) = F(q,u) \\ J = \int_o^\infty F^o(q,u)\, dt. \end{cases}$$

Comme d'habitude et pour des raisons évidentes, nous cherchons la loi de bouclage comme fonction $u(q)$ du seul état et non du temps. Les calculs précédents restent vrais à condition d'ôter $\frac{\partial}{\partial t}$ dans $\tilde{A}$, c'est-à-dire de remplacer $\tilde{A}$ par $\tilde{F} = \sum_{\ell=o}^{N} F^\ell(q,u)\frac{\partial}{\partial q^\ell}$. En particulier, (9) dégénère en équation différentielle ordinaire :

$$F(F_u F^o_{(u)2} - F^o_u F_{(u)2})\frac{du}{dq} = F(F^o_u F_{uq} - F_u F^o_{uq}) - F_u(F^o_u F_q - F_u F^o_q). \quad (10)$$

5. Quatre exemples

La situation du théorème 4, où la dimension de la commande égale celle de l'état, n'a rien d'exceptionnel. Dans cette première prise de contact avec la pratique, nous nous contenterons de traiter des problèmes monodimensionnels restés sans solution acceptable d'un point de vue numérique. Rappelons que l'intégration d'une équation quasi-linéaire du premier ordre se ramène par la théorie des caractéristiques à un système d'équations différentielles ordinaires (cf. Courant et Hilbert [2]).

a) Une approche tentée par divers auteurs est le calcul du développement taylorien de la boucle optimale. Dans cet esprit, Willemstein [14] à étudié l'exemple suivant (voir aussi Ozgoren, Longman, Cooper [12], Nihtilä [11])

$$\begin{cases} \dot{q}(t) = (q(t))^3 + u(t) \\ J = \int_o^T [(q(t))^2 + (u(t))^2]\, dt. \end{cases}$$

L'équation (9) est ici

$$\frac{\partial u}{\partial t} + [(q)^3 + u]\frac{\partial u}{\partial q} = q - 3u(q)^2.$$

Puisque $p(T) = 0$, la condition de Cauchy est $u(T,q) = 0$. La boucle optimale, qui est analytique, admet, au voisinage de $q = 0$, le développement de Taylor en q, $u(t,q) = \sum_{\nu \geq 0} c_\nu(t)\,(q)^\nu$. En $q = 0$, la commande optimale est évidemment $u = 0$, d'où $c_0(t) \equiv 0$. Pour $\nu \geqq 1$, les c_ν sont déterminés par la hiérarchie infinie d'équations différentielles ordinaires :

$$\dot{c}_1(t) + (c_1(t))^2 - 1 = 0$$

$$\dot{c}_2(t) + 3\,c_1(t)\,c_2(t) = 0$$

$$\dot{c}_\nu(t) + (\nu+1)\,c_{\nu-2}(t) + \sum_{\alpha=1}^{\nu} \alpha c_{\nu-\alpha+1}(t)\,c_\alpha(t) = 0 \qquad (\nu \geqq 3)\,.$$

Comme $c_\nu(T) = 0$, on en déduit $c_1(t) = -\,\mathrm{th}(T-t)$, $c_2(t) = 0$, $c_3(t) = -1 + \dfrac{1}{[\mathrm{ch}(T-t)]^4}$, etc ... On retrouve très simplement les résultats de [14].

En dépit de quelques essais, il semble douteux que de tels développements soient pratiquement exploitables. De toute façon, il n'y a guère de mal à intégrer numériquement l'équation aux dérivées partielles.
La surface intégrale $u(t,q)$, pour $T=1$, est représentée par la figure suivante :

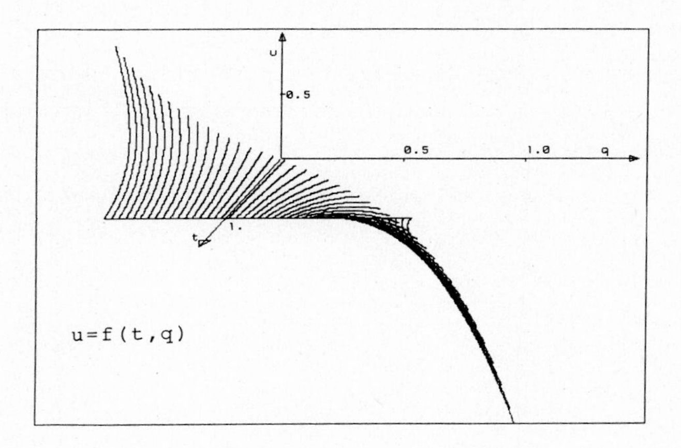

L'intégration analytique donne

$$p^{-1}\left[\frac{-2}{3((q)^2-(u)^2-2u(q)^3)}\,;\ g_2,\ g_3\right] -$$

$$p^{-1}\left[-\frac{1}{(q)^2} + \frac{1}{3((q)^2-(u)^2-2u(q)^3)}\,;\ g_2,\ g_3\right] = \pm\,(T-t)\,\sqrt{(q)^2-(u)^2-2u(q)^3}\ ,$$

où
$$g_2 = \frac{4}{3((q)^2-(u)^2-2u(q)^3)^2}$$

$$g_3 = -4\left[\frac{2}{27((q)^2-(u)^2-2u(q)^3)^3} + \frac{1}{((q)^2-(u)^2-2u(q)^3)}\right].$$

L'inverse $\wp^{-1}$ de la fonction elliptique de Weierstrass est fourni par l'intégrale (cf. Jordan [7])

$$\wp^{-1}(z;g_2,g_3) = \int_z^\infty \frac{dx}{\sqrt{4(x)^3-g_2x-g_3}} .$$

Avec T infini, l'équation (10) donne

$$[(q)^3 + u] \frac{du}{dq} = q - 3u(q)^2.$$

Comme $u(o) = 0$, il vient $(u)^2 + 2u(q)^3 - (q)^2 = 0$, soit

$$u(q) = -q\ [(q)^2 \pm \sqrt{1+(q)^4}\ ] ,$$

résultat que l'on aurait pu obtenir directement à partir du formalisme hamiltonien(4).
La commande

$$u(q) = -q\ [(q)^2 + \sqrt{1+(q)^4}\ ],$$

qui, pour q voisin de zéro, vaut approximativement $-q-(q)^3$, apparaît comme la limite pour T grand de la boucle calculée plus haut. Avec Nihtilä [11], remarquons qu'elle est stabilisante à l'origine.

b) Avec des buts analogues mais selon des méthodes différentes, Ozgoren, Longman et Cooper [12] d'une part, Bell et Ye [1] de l'autre, ont proposé l'exemple suivant

$$\begin{cases} \dot{q}(t) = \varepsilon\ u(t)\ q(t) \\ J = \frac{1}{2} \int_o^T [\alpha(q(t))^2 + (u(t))^2\ ]dt \qquad (\alpha > 0). \end{cases}$$

(9) devient

$$\frac{\partial u}{\partial t} + \varepsilon\ qu\ \frac{\partial u}{\partial q} = \alpha\ \varepsilon(q)^2.$$

La condition de Cauchy est aussi $u(T,q) = 0$. Les procédés classiques d'intégration donnent l'expression analytique de la solution; dans l'espace tridimensionnel (t,q,u) la surface intégrale a pour équation

$$u + q\sqrt{\alpha}\ \sin\ [\varepsilon(T-t)\ \sqrt{\alpha(q)^2-(u)^2}\ ] = 0.$$

(4) Pour une approche par développement de Taylor, voir Lukes [10].

On détermine le développement de Taylor comme précédemment :

$$u(t,q) = -\alpha\varepsilon(q)^2 \left[(T-t) - \frac{2}{3}\alpha\varepsilon^2 (T-t)^3 (q)^2 + \dots\right] .$$

La représentation graphique suivante résume l'intégration numérique :

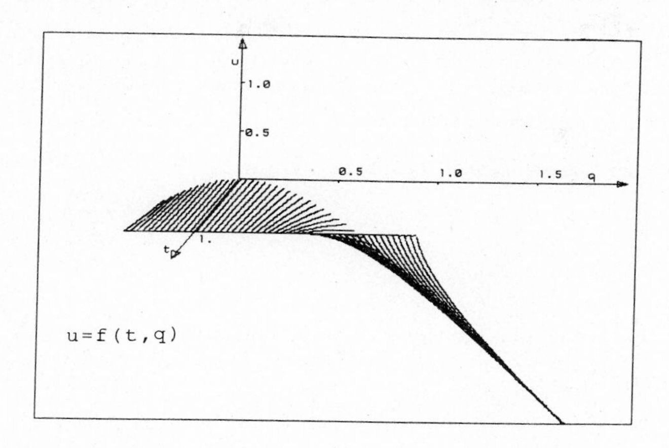

c) Quelques auteurs ont étudié l'exemple suivant, analysé par Jacobson et Mayne [6],

$$\begin{cases} \dot{q}(t) = -0{,}2\ q(t) + 10\ \text{th}\ u(t) \\ J = 10\ (q(T))^2 + \int_0^T [10(q(t))^2 + (u(t))^2]\ dt, \end{cases}$$

où $T = 0{,}5$. Ici , (9) devient

$$\frac{\partial u}{\partial t} + (10\ \text{th}\ u - 0{,}2\ q)\ \frac{\partial u}{\partial q} = \frac{100\ q + 0{,}2\ u\ \text{ch}^2\ u}{\text{ch}^2\ u + u\ \text{sh}2\ u} .$$

La condition de Cauchy est $u(T,q)\ [\text{ch}u(T,q)]^2 = 100\ q(T)$.

L'intégration numérique donne :

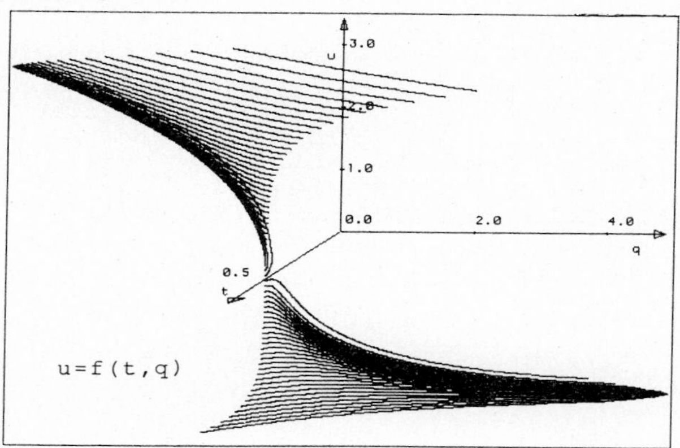

d) L'exemple suivant, à horizon infini et non autonome, a une origine économétrique (Intriligator [5]) :

$$\begin{cases} \dot{q}(t) = f(q(t)) - \lambda q(t) - u(t) \\ J = \int_0^{\infty} e^{-\delta t} \; I(u(t)) \, dt. \end{cases}$$

f et I sont des fonctions strictement concaves, monotones croissantes, les paramètres δ et λ sont des constantes positives.

Résoudre le problème en boucle fermée exige la résolution de l'équation (9) qui est ici :

$$\frac{\partial u}{\partial t} + (f(q) - \lambda q - u) \frac{\partial u}{\partial q} = (\lambda + \delta - f_q(q)) \frac{I_u(u)}{I_{(u)^2}(u)} \quad .$$

L'intégration numérique donne la représentation suivante,

avec $I(u) = (u - 0{,}05)^{0{,}4}$, $f(q) = (q)^{0{,}3}$:

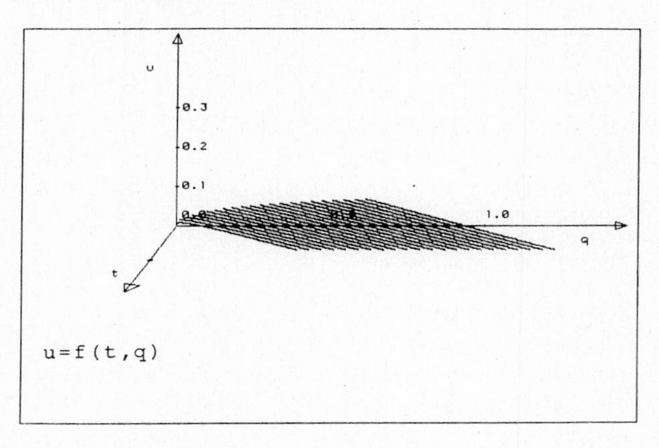

B. DEGENERESCENCE EN EQUATIONS ALGEBRIQUES

1. Théorie

Supposons que, dans le problème (7), F^o soit indépendante de la commande u, c'est-à-dire $F^o_{u_i} \equiv 0$ $(i = 1,\dots,N)$. En (8) et (9), les coefficients des dérivées $\frac{\partial u_i}{\partial t}$, $\frac{\partial u_i}{\partial q^k}$ sont alors nuls. Le théorème fournit des équations sans dérivées de la fonction inconnue que, par abus de langage, nous dirons "algébriques".

Nous considérons le problème de Lagrange :

$$\begin{cases} \dot{q}(t) = F(t,q,u) \\ J = \int_0^{T} F^o(t,q) \, dt, \end{cases} \tag{11}$$

différant de (7) par les conditions $\phi \equiv F^o_{u_i} \equiv 0$.

Dans la dérivée $\frac{d}{dt} H_{u_i}$, calculée à partir du corollaire 2, les coefficients de $p_1,\ldots,p_N$ sont compliqués; celui de p_o, qui est $\sum_{k=1}^{N} F^o_{q^k} F^k_{u_i}$, a une interprétation variationnelle remarquable. Comme au paragraphe A.1, remplaçons u par $u_{op}(t,q)+v(t)$, où $v(t) = (v_1(t),\ldots,v_N(t))$ est nulle hors de l'intervalle $[t_o,t_o+\Delta t]$, constante, égale à $(\varepsilon_1,\ldots,\varepsilon_N)$ à l'intérieur, les ε_i étant de "petits" paramètres. Un développement limité usuel montre que

$$\lim_{\Delta t\to o} \frac{\Delta J}{(\Delta t)^2} = \sum_{i=1}^{N} \frac{1}{2} \left[\sum_{k=1}^{N} F^o_{q^k} F^k_{u_i}\right] \varepsilon_i .$$

Il vient :

Proposition 5 : La loi de bouclage optimal du problème (11) satisfait nécessairement le système d'équations algébriques :

$$\sum_{k=1}^{N} F^o_{q^k} F^k_{u_i} = 0 \qquad (i=1,\ldots,N) . \tag{12}$$

Remarques : (i) De $F^o_{q^k} = 0$ $(k = 1,\ldots,N)$, il découle en général des trajectoires incompatibles avec la condition initiale. L'exclusion de ce cas entraîne la nullité du déterminant $\det(F^k_{u_i})_{i,k=1,\ldots,N}$, que l'on retrouve aussi à travers

$$H_{u_i} = \sum_{k=1}^{N} p_k F^k_{u_i} = 0.$$

(ii) La proposition 5 et sa démonstration restent valables en horizon infini.

2. Exemple

Considérons, avec Kaulgud, Sentis et Bernard-Weil [8], le problème bidimensionnel

$$\begin{cases} \dot q^1(t) = k_1 (q^1 - q^2 + u_1 - u_2) + c_1 (q^1 + q^2 - m + u_1 + u_2) \\ \qquad + k_2 (q^1 - q^2 + u_1 - u_2)^2 + c_2 (q^1 + q^2 - m + u_1 + u_2)^2 \\ \dot q^2(t) = k_3 (q^1 - q^2 + u_1 - u_2) + c_3 (q^1 + q^2 - m + u_1 + u_2) \\ \qquad + k_4 (q^1 - q^2 + u_1 - u_2)^2 + c_4 (q^1 + q^2 - m + u_1 + u_2)^2 \\ J = \int_o^\infty e^{-\alpha t} [(q^1(t))^2 + (q^2(t))^2] \, dt \qquad (\alpha > 0), \end{cases}$$

où k_1, k_2, k_3, k_4, c_1, c_2, c_3, c_4, m et α sont des constantes. Il est supposé fournir le choix d'un traitement permettant de rétablir certains équilibres hormonaux dans un système endocrinien[5].

[5] En [8], on impose des bornes à u_1 et u_2 que nos méthodes ne peuvent prendre en compte.

Des équations (12) on tire la loi de bouclage optimal

$$u_1 = -\frac{1}{4}\left[\frac{k_1q^1 + k_3q^2}{k_2q^1 + k_4q^2} + \frac{c_1q^1 + c_3q^2}{c_2q^1 + c_4q^2}\right] - q^1 + \frac{m}{2}$$

$$u_2 = \frac{1}{4}\left[\frac{k_1q^1 + k_3q^2}{k_2q^1 + k_4q^2} - \frac{c_1q^1 + c_3q^2}{c_2q^1 + c_4q^2}\right] - q^2 + \frac{m}{2} .$$

Les calculs numériques sont effectués avec des valeurs des constantes empruntées à la thèse de Sellam [13]. Voici d'abord u_1, u_2 en fonction de q^1 et q^2 :

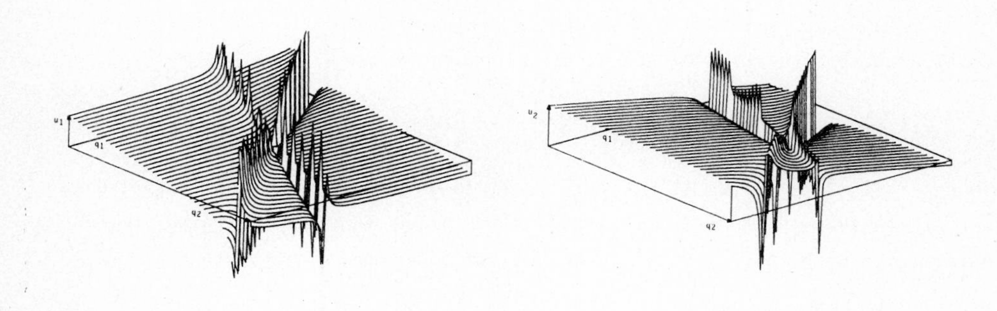

puis q^1, q^2, u_1, u_2 en fonction du temps

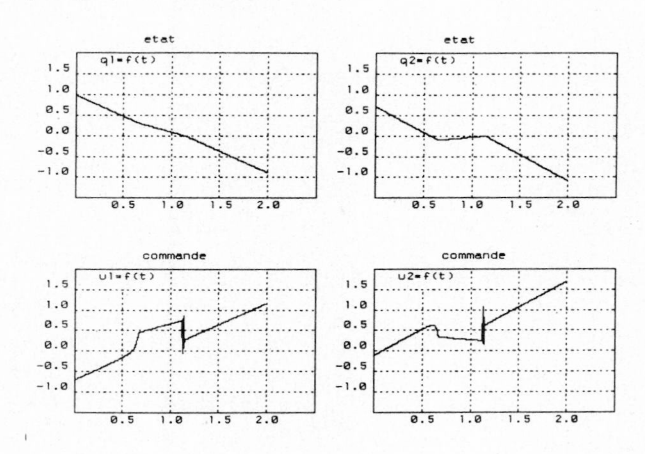

En modifiant les constantes k_1, k_2, k_3, k_4 on obtient les courbes suivantes :

u_1, u_2 en fonction de q^1, q^2,

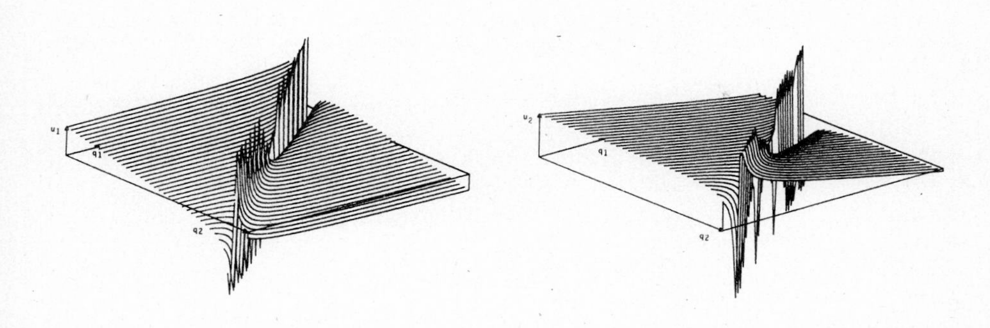

et enfin,

q^1, q^2, u_1, u_2 en fonction du temps.

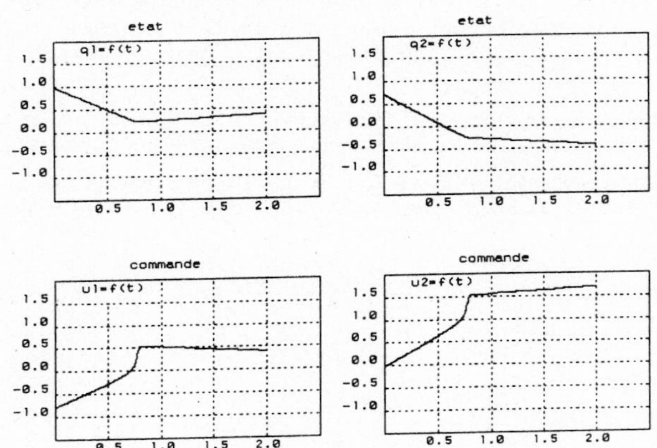

CONCLUSION

L'approche présentée dans cette communication est en plein développement théorique et pratique. Divers exemples issus de la régulation thermique ou de la technologie aérospatiale sont en cours d'étude et seront bientôt publiés.

BIBLIOGRAPHIE

[1] BELL (D.J.) et YE (Q.). A perturbation method for sub-optimal feedback control of bilinear systems, Internat. J. Systems Sci., 12, 1981, p. 1157-1168.

[2] COURANT (R.) et HILBERT (D.). Methoden der mathematischen Physik II, Springer, Berlin, 1937 (Traduction anglaise : Methods of Mathematical Physics, Vol. II, Interscience, New York, 1962).

[3] FLIESS (M.). On a possible connection between Volterra series and nonlinear optimal control, Proc. 7th Conf. Informat. Sci. Systems, p. 402-407, Baltimore, 1983.

[4] FLIESS (M.). Lie brackets and optimal nonlinear feedback regulation, Proc. IXth IFAC World Congress, Budapest, July 1984.

[5] INTRILIGATOR (M.D.). Economic systems, in "Control and Dynamic Systems", C.T. Leondes ed., Vol. 13, p. 135-160, Academic Press, New York, 1977.

[6] JACOBSON (D.H.) et MAYNE (D.Q.). Differential Dynamic Programming, American Elsevier, New York, 1970.

[7] JORDAN (C.). Cours d'analyse, t.2, Gauthiers-Villars, Paris, 1959 (nouveau tirage de la 3ème édition).

[8] KAULGUD (N.), SENTIS (R.) et BERNARD-WEIL (E.). Regulation of an endrocrinal system, Rapp. Rech. n° 180, INRIA, Le Chesnay, 1982.

[9] LEBORGNE (D.). Calcul différentiel et géométrie, Presses Universitaires de France, Paris, 1982.

[10] LUKES (D.L.). Optimal regulation of nonlinear dynamical systems, SIAM J. Control, 7, 1969, p. 75-100.

[11] NIHTILÄ (M.T.). An approach to state-feedback control of non-linear differential systems, Systems Sci., 6, 1980, p. 211-223.

[12] OZGOREN (M.K.), LONGMAN (R.W.) et COOPER (C.A.). Application of Lie transform based canonical perturbation methods to the optimal control of bilinear systems, Proc. AAS-AIAA Astrodynamics Specialist Conf., Nassau, Bahamas, 1975, ASS Publicat. Office, Tarzana, CA, 1975.

[13] SELLAM (S.). Etude mathématique d'un système endocrinien : identification, stabilité, contrôle optimal, Thèse 3ème Cycle, Université Paris VI, Paris, 1978.

[14] WILLEMSTEIN (A.P.). Optimal regulation of nonlinear dynamical systems on a finite interval, SIAM J. Control Optimiz., 15, 1977, p. 1050-1069.

OPTIMIZATION OF SYSTEMS POSSESSING SYMMETRIES

J. W. Grizzle and S. I. Marcus
Department of Electrical Engineering
The University of Texas at Austin
Austin, Texas 78712
U.S.A.

ABSTRACT

It is shown that a symmetry in an optimization problem induces a decomposition of the optimal feedback control law into two factors. One factor can be calculated algebraically and depends only on the symmetry; the other factor corresponds to a lower dimensional optimization problem. This gives *a priori* information about the structure of the optimal feedback control law and indicates a possibly more efficient method for optimizing such systems.

I. INTRODUCTION

In a previous paper [1], the authors showed that a symmetry in a control system could be used to deduce structural information about the system. In particular, it was shown that the system could be decomposed into a cascade of lower dimensional subsystems and a feedback loop. In this paper, the problem of optimizing systems which possess symmetries will be addressed. It will be shown that a symmetry in an optimization problem gives a decomposition of the optimal feedback control law into two parts: one part can be calculated *algebraically* and depends only on the symmetry, and the other part corresponds to a lower dimensional optimization problem. This means that the partial differential equation which must be solved to determine the optimizing controller can be replaced with one of lower dimension. This provides obvious computational advantages. Another implication is that one has some *a priori* information about the optimal controller's structure and the resulting signal/flow patterns in the closed-loop system.

A problem very similar to this one was considered in [2]. In his thesis [2], van der Schaft shows that optimization problems give rise to Hamiltonian control systems [3]. He shows that if the resulting Hamiltonian control system possesses a generalized conserved quantity satisfying a certain condition, then one can use it to deduce information about the optimal trajectory. In previous work [3], he had established the relationship between symmetries and generalized conserved quantities in Hamiltonian control systems. His result is essentially an application of "Noether's Theorem" for Hamiltonian control systems.

The approach taken here is totally different; it capitalizes upon the structural results obtained in [1], which were alluded to at the beginning of the introduction.

In brief, the contrasts and parallels of the two results are analogous to those which exist between Noether's Theorem and the reduction procedure in classical mechanics [4]; in Section V this will be made more precise.

The remainder of this paper is structured as follows. Section II contains the definitions needed to define an optimization problem and a symmetry therein; also, it summarizes some results from [1] which are needed for this paper. Section III states the decomposition result and gives its proof. Section IV contains an illustrative example. Section V compares the results obtained here to those of [2] and Section VI contains the conclusions.

II. DEFINITIONS AND PRELIMINARIES

This section gives the definitions needed to describe a symmetry in an optimization problem. In addition, it summarizes some of the recent work of the authors which will be needed later.

Definition 2.1 (Nonlinear Control System) [3]: A *nonlinear control system* Σ is a 3-tuple $\Sigma(B,M,f)$ where $\pi:B\to M$ is a smooth fiber bundle and f is a smooth map such that the diagram

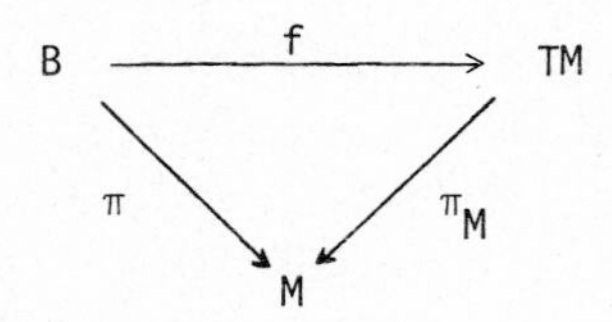

commutes, where π_M is the natural projection of TM on M.

In the above, M is to be interpreted as the state space and the fibers of B as the (possibly state-dependent) input spaces. If one chooses fiber respecting coordinates (x,u) for B, then locally this definition reduces to (with the abuse of notation $f:(x,u)\to(x,f(x,u))$

$$\dot{x} = f(x,u).$$

For this paper, B will always be assumed to be trivial; i.e., there exists a manifold U such that $B = M\times U$.

Definition 2.2 [4]: Let M be a smooth manifold. A left *action* (or G-action) of a Lie group G on M is a smooth mapping $\Phi:G\times M\to M$ such that (i) for all $x\in M$, $\Phi(e,x) = x$ and (ii) for every $g,h\in G$, $\Phi(g,\Phi(h,x)) = \Phi(gh,x)$ for all $x\in M$.

At various times, it will be useful to hold one variable fixed and consider an action Φ as a function of the remaining variable; $\Phi_g:M\to M$ will denote the function $x\mapsto\Phi(g,x)$ and $\Phi_x:G\to M$ will denote the function $g\mapsto\Phi(g,x)$. Note that because $(\Phi_g)^{-1} = \Phi_{g^{-1}}$, Φ_g is a diffeomorphism.

Example 2.1: (a) Let X be a complete vector field on M and let X_t denote its flow (i.e., $\dot{X}_t = X(X_t)$). Then $X_t: M \to M$ by $x \mapsto X_t(x)$ is an $\mathbb{R}$-action on M.

(b) Let $M = \mathbb{R}^2$ and $G = SO(2)$, the group of 2×2 orthogonal matrices with positive determinant. Then $\Phi: G\times M \to M$ by $(g,x) \mapsto gx$ is an SO(2) action on $\mathbb{R}^2$.

The following terminology regarding actions is useful.

Definition 2.3 [4]: Let Φ be an action of G on M. For $x\in M$, the *orbit* (or Φ-orbit) of x is given by

$$G\cdot x = \{\Phi_g(x) \mid g\in G\}.$$

An action is *free at* x if $g \mapsto \Phi_g(x)$ is one-to-one. It is *free* if it is free at x for all $x\in M$. Φ is said to be *proper* (or act properly) if $(g,m) \mapsto (m,\Phi(g,m))$ is a proper map (i.e., the inverse images of compact sets are compact).

Remark 2.1: If G is compact, then Φ is necessarily proper [4].

Example 2.2: Let M, Φ and G be as in Example 2.1(b). Then the orbits consist of the origin and circles about the origin. The action is not free since $g \mapsto \Phi_g(0) = 0$ for all g. However, if the origin is removed, the action is then free. Since SO(2) is compact, the action is proper.

It is now possible to define, in a global manner, a symmetry in a control system.

Definition 2.4 (based on [3]): Let $\Sigma(B,M,f)$ be a nonlinear control system and let θ and Φ be actions of G on B and M respectively. Then Σ has *symmetry* (G,θ,Φ) if the diagram

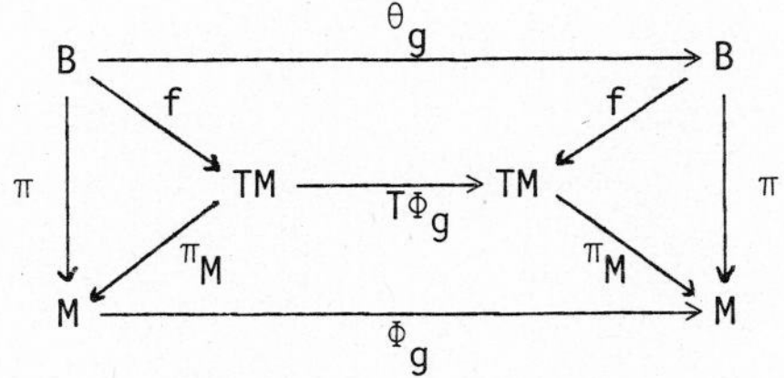

commutes for all $g\in G$, where $T\Phi_g$ is the tangent map of Φ [4].

An important special case of the above occurs when the symmetry lies "entirely in the state space".

Definition 2.5: Let $B = M\times U$ for some manifold U. (G,Φ) is a *state space symmetry* of $\Sigma(B,M,f)$ if (G,θ,Φ) is a symmetry of Σ for $\theta_g = (\Phi_g, Id_U)$: $(x,u) \mapsto (\Phi_g(x),u)$.

Note that state space symmetries can be defined globally only for systems in which B is a trivial bundle since, otherwise, the input spaces are state dependent.

Definition 2.6 [5]: A system $\Sigma(B,M,f)$ is *feedback equivalent* to a system $\tilde{\Sigma}(B,M,\tilde{f})$ if there exists a bundle isomorphism γ: $B \to B$ such that the diagram

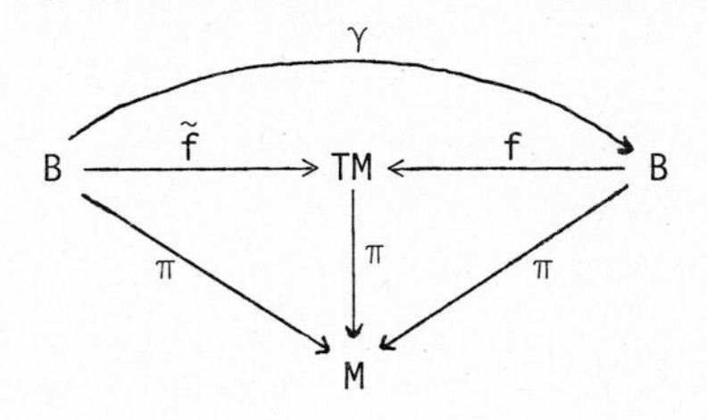

commutes. γ is called the *feedback function.*

In local coordinates, the abuse of notation $(x,u) \mapsto (x,\gamma(x,u))$ will be used. γ being an isomorphism implies that for each $x \in M$, viewing γ_x as a map from the fiber over x to the fiber over x, γ_x is a diffeomorphism. Hence this corresponds to "feedback with full control" as in [6].

Next, a general optimization problem will be defined, some preliminary results will be proved about optimizing systems which are related by feedback, and finally a symmetry in an optimization problem will be defined.

Definition 2.7: A *nonlinear optimization problem* $(\Sigma,K,L,\mathcal{U})$ consists of a nonlinear control system $\Sigma(M\times U,M,f)$, a pair of smooth functions $K: M \to \mathbb{R}$ and $L: M\times U \to \mathbb{R}$, a class of functions $\mathcal{U}$ (possibly depending on M) with domain $[0,T]$ and range in U, and an objective functional $J: M\times\mathcal{U} \to \overline{\mathbb{R}}$ (the real line plus the points at infinity) given by

$$J(x_0,u(\cdot)) = K(x(T)) + \int_0^T L(x(t),u(t))dt,$$

where $\dot{x}(t) = f(x(t),u(t))$, $x(0) = x_0$.

The following is assumed to hold throughout this paper.

Assumption 2.1: (a) $\mathcal{U}(x_0) = \{u: [0,T] \to U \mid u(\cdot)$ is Borel measurable and $\dot{x} = f(x,u)$ has a well-defined solution for all $t \in [0,T]$, $x(0) = x_0\}$.

(b) For each $x_0 \in M$, there exists a $u^*(\cdot) \in \mathcal{U}(x_0)$ such that $J(x_0,u^*(\cdot)) = \min_{u(\cdot)\in\mathcal{U}} J(x_0,u(\cdot))$.

The above simplifying assumptions are made because this paper is concerned with the structure of the optimal controller, not its existence. It does not seem desirable to obscure the essential geometric nature of the following constructions with tedious functional analytic details.

Definition 2.8: Let $(\Sigma,K,L,\mathcal{U})$ be a nonlinear optimization problem and let $\gamma: M\times U \to M\times U$ be a feedback function (recall that γ is an isomorphism and hence invertible).

(a) Define a function Γ on $M\times\mathcal{U}$ by $\Gamma(x_0,u(\cdot))(t) = (x_0,\gamma_{x(t)}(u(t)))$, where $\dot{x}(t) = f(x(t),u(t))$ and $x(0) = x_0$ (note that Γ is well-defined by Assumption 2.1).

(b) Let $(\Sigma',K',L',\mathcal{U}')$ be the nonlinear optimization problem given by $\Sigma' = \Sigma(B,M,f\circ\gamma^{-1})$, $K' = K$, $L' = L\circ\gamma^{-1}$, so that $J'(x_0,u(\cdot)) = K'(x'(T)) + \int_0^T L\circ\gamma^{-1}(x'(t),u(t))dt$, where $\dot{x}'(t) = f\circ\gamma^{-1}(x'(t),u(t))$, $x'(0) = x_0$, and $\mathcal{U}' = \Gamma(\mathcal{U})$.

Proposition 2.1:

$$J = J'\circ\Gamma.$$

Proof: Let $u(\cdot) \in \mathcal{U}(x_0)$ and define $x(t)$ by $\dot{x}(t) = f(x(t),u(t))$ for $t \in [0,T]$ and $x(0) = x_0$. Define $v(t) = \gamma_{x(t)}(u(t))$. Then corresponding to $J'\circ\Gamma$, one has $\dot{x}'(t) = f\circ\gamma^{-1}(x'(t),v(t)) = f(x'(t),\gamma^{-1}_{x'(t)}\circ\gamma_{x(t)}(u(t)))$, where $x'(0) = x_0$. Since clearly $\dot{x}(t) = f(x(t),\gamma^{-1}_{x(t)}\gamma_{x(t)}(u(t)))$, one must have $x'(t) = x(t)$ (solutions to differential equations are unique!). Hence, $J'\circ\Gamma(x_0,u(\cdot)) = K(x'(T)) + \int_0^T L\circ\gamma^{-1}(x'(t),\gamma_{x(t)}(u(t)))dt = K(x(T)) + \int_0^T L(x(t),u(t))dt = J(x_0,u(\cdot))$. □

Actually, the same type of argument can be used to establish the following.

Proposition 2.2: Γ is a bijection (onto its range), and Γ^{-1} is given by $\Gamma^{-1}(x_0,u(\cdot))(t) = (x_0,\gamma^{-1}_{x'(t)}(u(t)))$.

Next, a symmetry in an optimization problem is defined.

Definition 2.9: $(\Sigma,K,L,\mathcal{U})$ is said to have *symmetry* (G,θ,Φ) if (i) Σ has symmetry (G,θ,Φ), (ii) $K\circ\Phi_g(x) = K(x)$ for all $g\in G$, and (iii) $L\circ\theta_g(x,u) = L(x,u)$ for all $g\in G$. If $\theta_g(x,u) = (\Phi_g(x),u)$, then (G,θ,Φ) is said to be a *state space symmetry*; as before, the notation (G,Φ) will be used.

Since they will be needed later on, the following results from [1] are stated.

Theorem 2.1: Suppose $\Sigma(M\times U,M,f)$ is a control system with state space symmetry (G,Φ). Then if Φ is free and proper and $p\colon M\to M/G$ admits a cross-section σ, Σ is isomorphic to the system

$$\dot{y} = \hat{f}(y,u)$$

$$\dot{g} = (T_eL_g)(T_e\hat{\Phi}_{\sigma(y)})^{-1}\,[f(\sigma(y),u) - (T_y\sigma)\hat{f}(y,u)]$$

which evolves on $M/G\times G$, where L_g is the left translation operator on G (which should not be confused with the running cost $L(x,u)$).

Theorem 2.2: Suppose $\Sigma(M\times U,M,f)$ has symmetry (G,θ,Φ). Then if Φ is free and proper and if $p\colon M\to M/G$ admits a cross-section, there exists a system Σ' with state space symmetry (G,Φ) to which Σ is feedback equivalent.

III. MAIN RESULTS

Suppose $(\Sigma,K,L,\mathcal{U})$ has state space symmetry (G,Φ) and Φ acts freely and properly. Then, it is known that $p\colon M\to M/G$, by $m\to G\cdot m$, is a (principal) fiber bundle [4] and that Σ projects to a control system $\hat{\Sigma}(M/G\times U,M/G,\hat{f})$. It also follows that L projects to a smooth function $\hat{L}$ on $M/G\times U$ and that K projects to a smooth function $\hat{K}$ on M/G. Even more, the following holds.

Proposition 3.1: Assume $p\colon M\to M/G$ admits a cross-section and that $(\Sigma,K,L,\mathcal{U})$ is an optimization problem which satisfies Assumption 2.1 and which has state space symmetry (G,Φ). Then $\mathcal{U}(x) = \mathcal{U}(\Phi_g(x))$ for all $x\in M$ and $g\in G$.

Proof: Let $u(\cdot)\in\mathcal{U}(x_0)$ and let $x(t)$ be the solution of Σ corresponding to $u(\cdot)$ and $x(0) = x_0$. From Theorem 2.1, it follows that the solution of Σ corresponding to $u(\cdot)$ and $x(0) = \Phi_{g_0}(x_0)$ is $\Phi_{g_0}(x(t))$. Hence, $u(\cdot)$ is a member of $\mathcal{U}(\Phi_{g_0}(x_0))$. □

From this proposition, one has that $\mathcal{U}(p(x)) = \mathcal{U}(x)$ is well-defined, and hence, one can consider the nonlinear optimization problem $(\hat{\Sigma},\hat{K},\hat{L},\mathcal{U})$.

Proposition 3.2:

$$J(x_0,u(\cdot)) = \hat{J}(p(x_0),u(\cdot)) \text{ for all } (x_0,u(\cdot))\in M\times\mathcal{U}(x_0).$$

Proof: From the definition of $\hat{L}$ and $\hat{K}$, one has

$$J(x_0,u(\cdot)) = K(x(T)) + \int_0^T L(x(t),u(t))dt = \hat{K}(p(x(T)) + \int_0^T \hat{L}(p(x(t)),u(t))dt,$$

where $\dot{x}(t) = f(x(t),u(t))$ and $x(0) = x_0$. But Theorem 2.1 gives that $y(t) = p(x(t))$ satisfies the equation

$$\dot{y}(t) = \hat{f}(y(t),u(t)),\ y(0) = p(x_0).$$

Hence the result follows. □

This proposition implies that the optimal control function $u^*(\cdot)$ for $(\Sigma,K,L,\mathcal{U})$ can be calculated from $(\hat{\Sigma},\hat{K},\hat{L},\mathcal{U})$, which is a lower dimensional problem. Note that if the solution to $(\hat{\Sigma},\hat{K},\hat{L},\mathcal{U})$ is given in closed loop form, that is, $u^*(\cdot) = \hat{\gamma}^*(y(\cdot))$, then $\gamma^*(x(\cdot)) = \hat{\gamma}^*(p(x(\cdot)))$ is the optimal feedback solution for $(\Sigma,K,L,\mathcal{U})$.

Now suppose that $(\Sigma,K,L,\mathcal{U})$ has symmetry (G,θ,Φ), that Φ is free and proper, and that $p\colon M \to M/G$ admits a cross-section. Then Theorem 2.1 guarantees the existence of a feedback function $\gamma\colon M\times U \to M\times U$ such that $\Sigma' = \Sigma(M\times U,M,f\circ\gamma^{-1})$ has state space symmetry (G,Φ). It is straightforward to check that $(\Sigma',K',L',\mathcal{U}')$, as defined in Definition 2.9, also has state space symmetry (G,Φ). Thus the optimal control function $u'^*(\cdot)$ for $(\Sigma',K',L',\mathcal{U}')$ can be calculated from $(\hat{\Sigma}',\hat{K}',\hat{L}',\mathcal{U}')$, its lower dimensional projection. Then by Proposition 2.2, $u^*(\cdot) = \Gamma^{-1}(u'^*(\cdot))$ is the optimal solution to the original problem. The latter, of course, has the feedback implementation given in Proposition 2.2. In summary, the following has been established.

Theorem 3.1: Let $(\Sigma,K,L,\mathcal{U})$ be a nonlinear optimization problem which has symmetry (G,θ,Φ). Suppose that Φ acts freely and properly, and that $p\colon M \to M/G$ admits a cross-section. Then the optimal control function $u^*(\cdot)$ can be determined in the following manner:

(i) calculate the feedback function $\gamma\colon M\times U \to M\times U$ such that $(\Sigma',K',L',\mathcal{U}')$ has state space symmetry (G,Φ) (note that this is a purely algebraic operation);

(ii) calculate either $\hat{u}'^*$ or $\hat{\gamma}'^*$ for the quotient problem $(\hat{\Sigma}',\hat{K}',\hat{L}',\mathcal{U}')$;

(iii) set $u^*(\cdot) = \Gamma^{-1}(\hat{u}'^*(\cdot))$, or $\gamma^* = \gamma^{-1}\circ\hat{\gamma}'^*$.

Thus the optimal feedback controller has the structure depicted in Figure 3.1.

For step (ii) of the above, one would normally use the Maximum Principle [7] to calculate the open loop control $\hat{u}'^*$, or the Hamilton-Jacobi-Bellman [7] equation (i.e., dynamic programming) to calculate the closed loop control law $\hat{\gamma}'^*$. In the latter technique, one has eliminated $k = \dim G$ variables by solving $(\hat{\Sigma}',\hat{L}',\hat{K}',\mathcal{U}')$ instead of $(\Sigma,L,K,\mathcal{U})$, whereas in the former, one has eliminated $2k$ variables: k state variables and k co-state variables.

IV. AN EXAMPLE

To exemplify the use of Theorem 3.1, consider a particle of unit mass, in a planar inverse-square-law gravitational field, which has thrusters in the "x-y" directions. The equations of motion in rectangular coordinates are

$$\begin{bmatrix} \dot{q}_1 \\ \dot{q}_2 \\ \dot{p}_1 \\ \dot{p}_2 \end{bmatrix} = \begin{bmatrix} p_1 \\ p_2 \\ \dfrac{-q_1}{(q_1^2+q_2^2)^{3/2}} + u_1 \\ \dfrac{-q_2}{(q_1^2+q_2^2)^{3/2}} + u_2 \end{bmatrix} \triangleq f(x,u), \tag{4.1}$$

which are defined on $M = (\mathbb{R}^2 - \{0\}) \times \mathbb{R}^2$ and $U = \mathbb{R}^2$. It is straightforward to check (see [1]) that the system has symmetry (G,θ,Φ), for $G = SO(2)$,

$$\theta_g(q,p,u) = \begin{bmatrix} g & 0 & 0 \\ 0 & g & 0 \\ 0 & 0 & g \end{bmatrix} \begin{bmatrix} q \\ p \\ u \end{bmatrix}, \text{ and}$$

$$\Phi_g(q,p) = \begin{bmatrix} g & 0 \\ 0 & g \end{bmatrix} \begin{bmatrix} q \\ p \end{bmatrix}.$$

The objective will be to drive the particle to a given circular orbit in T units of time with minimum energy. Hence, reasonable choices of K and L are $K(q(T),p(T)) = (\sqrt{q_1^2(T)+q_2^2(T)} - a)^2$, for a equal to the radius of the desired orbit, and $L(q(t),p(t),u(t)) = \frac{1}{2}(u_1^2(t)+u_2^2(t))$. Clearly the nonlinear optimization problem $(\Sigma,K,L,\mathcal{U})$ has symmetry (G,θ,Φ). From [1] it is known that

$$\gamma(q,p,u) = \frac{1}{(q_1^2+q_2^2)^{1/2}} \begin{bmatrix} q_1 & -q_2 \\ q_2 & q_1 \end{bmatrix} \begin{bmatrix} u_1 \\ u_2 \end{bmatrix}$$

transforms Σ into a system with state space symmetry (G,Φ). Furthermore, the quotient system $\hat{\Sigma}'$ is given by

$$\begin{bmatrix} \dot{r} \\ \dot{p}_r \\ \dot{p}_\theta \end{bmatrix} = \begin{bmatrix} p_r \\ p_\theta^2/r^3 - 1/r^2 + u_1 \\ u_2 \end{bmatrix},$$

$$\hat{L}'(r,p_r,p_\theta,u_1,u_2) = \frac{1}{2}(u_1^2+u_2^2), \text{ and}$$

$$\hat{K}'(r(T),p_r(T),p_\theta(T)) = (r(T) - a)^2.$$

To calculate the optimal closed loop controller, let

$$\overline{V}(r,p_r,p_\theta,t) = (r(T) - a)^2 + \frac{1}{2}\int_0^T (u_1^2(\tau)+u_2^2(\tau))d\tau.$$

Then dynamic programming [7] gives

$$\hat{\gamma}'^*(r,p_r,p_\theta) = -\begin{bmatrix} \dfrac{\partial \overline{V}}{\partial p_r} \\ \dfrac{\partial \overline{V}}{\partial p_\theta} \end{bmatrix},$$

where $\overline{V}$ satisfies the equation

$$\frac{\partial \overline{V}}{\partial t} - \frac{1}{2}\left(\frac{\partial \overline{V}}{\partial p_r}\right)^2 - \frac{1}{2}\left(\frac{\partial \overline{V}}{\partial p_\theta}\right)^2 + \frac{\partial \overline{V}}{\partial p_r}\left[\frac{p_\theta^2}{r^3} - \frac{1}{r^2}\right] + \frac{\partial \overline{V}}{\partial r} p_r = 0,$$

$$\overline{V}(r(T),p_r(T),p_\theta(T),T) = (r(T) - a)^2;$$

the solution to the original problem is then

$$\gamma^*(q,p) = \frac{1}{(q_1^2+q_2^2)^{1/2}}\begin{bmatrix} q_1 & -q_2 \\ q_2 & q_1 \end{bmatrix}\begin{bmatrix} \dfrac{\partial \overline{V}}{p_r} \\ \dfrac{\partial \overline{V}}{\partial p_\theta} \end{bmatrix}(p(q_1,q_2,p_1,p_2)).$$

On the other hand, applying the Hamilton-Jacobi-Bellman equation directly entails the following.

Define

$$V(q,p,t) = \left(\sqrt{q_1^2(T)+q_2^2(T)} - a\right)^2 + \frac{1}{2}\int_0^T (u_1^2(\tau) + u_2^2(\tau))\,d\tau.$$

Then,

$$\gamma^*(q,p) = -\begin{bmatrix} \dfrac{\partial V}{\partial p_1} \\ \dfrac{\partial V}{\partial p_2} \end{bmatrix}$$

and V satisfies the equation

$$\frac{\partial V}{\partial t} - \frac{1}{2}\left(\frac{\partial V}{\partial p_1}\right)^2 - \frac{1}{2}\left(\frac{\partial V}{\partial p_2}\right)^2 - \frac{q_1}{(q_1^2+q_2^2)^{3/2}}\frac{\partial V}{\partial p_1} - \frac{q_2}{(q_1^2+q_2^2)^{3/2}}\frac{\partial V}{\partial p_2} + p_1\frac{\partial V}{\partial q_1} + p_2\frac{\partial V}{\partial q_2} = 0,$$

$$V(q(T),p(T),T) = \left(\sqrt{q_1^2(T)+q_2^2(T)} - a\right)^2.$$

V. RELATION TO PREVIOUS WORK

This section establishes some of the relationships between Theorem 3.1 and the result contained in [2]. The presentation will be intentionally brief.

Let $\Sigma(X\times U,X,g)$ be a control system and let $(\Sigma,L,\mathcal{U})$ (i.e., $K\equiv 0$) be an associated nonlinear optimization problem. To $(\Sigma,L,\mathcal{U})$, one can associate a Hamiltonian control system [2,3] as follows. Let $(M,\omega) = T^*X$ equipped with its canonical symplectic form

[4], and similarly, let $(W,\omega_e) = T^*U$. Define a Hamiltonian $H\colon M\times U \to \mathbb{R}$ by

$$H(x,\lambda,u) = L(x,u) + \lambda^T g(x,u)$$

where (x,λ) are canonical coordinates for M. Then

$$\begin{aligned} \dot{x}_i &= \frac{\partial H}{\partial \lambda_i} \\ \dot{\lambda}_i &= -\frac{\partial H}{\partial x_i} \\ y_j &= \frac{\partial H}{\partial u_j} \end{aligned} \tag{5.1}$$

defines a Hamiltonian control system which will be denoted Σ_H.

In [2], van der Schaft considers the symmetries admitted by (5.1), and not those admitted by the nonlinear optimization problem, to be of fundamental importance. In general, a symmetry in one does not imply the presence of a symmetry in the other. However, if $(\Sigma,L,\mathcal{U})$ has a symmetry (G,θ,Φ) and θ satisfies a certain condition, then one can show that Σ_H also possesses a symmetry. Towards this end, write $\theta_g(x,u) = (\Phi_g(x),\theta_g^2(x,u))$, and assume that $\theta_g^2(x,u) = \theta_g^2(u)$ (i.e., θ^2 is in fact itself an action on U). Define $\tilde{\Phi}\colon G\times M \to M$, $\tilde{\theta}\colon G\times M\times U \to M\times U$ and $\tilde{\Psi}\colon G\times T^*U \to T^*U$ by

$$\tilde{\Phi}_g(x,\lambda) = (\Phi_g(x),(T^*\Phi_{g^{-1}})\lambda) \quad \text{(see [4], page 283)}$$

$$\tilde{\Psi}_g(u,y) = (\theta_g^2(u),(T^*\theta^2_{g^{-1}})y)$$

$$\tilde{\theta}_g(x,\lambda,u) = (\tilde{\Phi}_g(x,\lambda),\theta_g^2(u)).$$

Then one has that

$$\begin{aligned} T\tilde{\Phi}_g f_H(x,\lambda,u) &= f_H(\tilde{\theta}_g(x,\lambda,u)) \\ h_H(\tilde{\theta}_g(x,\lambda,u)) &= \tilde{\Psi}_g h_H(x,\lambda,u), \end{aligned} \tag{5.2}$$

for $f_H = \begin{bmatrix} \frac{\partial H}{\partial \lambda} \\ -\frac{\partial H}{\partial x} \end{bmatrix}$ and $h_H = \begin{bmatrix} u \\ \frac{\partial H}{\partial u} \end{bmatrix}$, which gives that Σ_H has symmetry $(G,\tilde{\theta},\tilde{\Phi},\tilde{\Psi})$. Now, in preparation for applying the results of [2], assume that G is one dimensional, and that S and T are the infinitesimal generators [4] of the actions Φ and θ^2 respectively. Then from [4], there exist momentum maps for $\tilde{\Phi}$ and $\tilde{\Psi}$ given by

$$\begin{aligned} F(x,\lambda) &= \lambda^T S(x) \\ F_e(u,y) &= y^T T(u). \end{aligned} \tag{5.3}$$

From [2], it follows that (F,F_e) is a generalized conservation law for Σ_H:

$$\frac{d}{dt} F\Big|_{\Sigma_H} = F_e \circ h_H; \tag{5.4}$$

i.e., the derivative of F along the trajectories of Σ_H is a function of the external variables. Noting that $F_e(u,0) \equiv 0$ for all $u \in U$ and that $y = \frac{\partial H}{\partial u} \equiv 0$ is the first order necessary condition for optimality, one arrives at

$$\frac{d}{dt} F \Big|_{\Sigma_H} \equiv 0 \tag{5.5}$$

for trajectories of Σ_H corresponding to minimizing u^*, or

$$\frac{d}{dt} F \Big|_{\Sigma_H} (x^*,\lambda^*,u^*) \equiv 0. \tag{5.6}$$

The observation that this gives information about the optimal trajectory x^* is the result of [2, page 242] (for this special case). However, it should be noted that this is rather indirect information. To see this, note that the example considered in the previous section satisfies the assumptions which were made on θ. Let $f(x,u)$ denote the dynamics (4.1) expressed in polar coordinates, and let $\lambda = (\lambda_1,\lambda_2,\lambda_3,\lambda_4)$ be the corresponding canonical coordinates on the cotangent bundle. Then a straightforward calculation gives that

$$F(x,\lambda) = \lambda_2$$

and (5.6) then yields

$$\dot{\lambda}_2 = 0;$$

and finally one obtains $\lambda_2^* \equiv 0$ since $\lambda_2^*(0) = 0$. This in turn yields information about x^* since x^* depends on the co-adjoint variables λ^*, but this sort of information is significantly different than that given by Theorem 3.1. In particular, it does not give the structure of the optimal closed loop system.

VI. CONCLUSIONS

Based on the authors' previous work in [1], a notion of symmetry has been defined in nonlinear optimization problems. It was shown that a symmetry gives rise to a decomposition of the optimal feedback controller into two factors: one factor is calculated algebraically and depends only on the symmetry, and the other factor corresponds to a lower dimensional optimization problem. The existence of such a decomposition has implications for constructing more efficient numerical optimization algorithms, as well as providing *a priori* information about the structure of the optimal controller.

ACKNOWLEDGEMENTS

This research was supported in part by the Air Force Office of Scientific Research under Grant AFOSR-79-0025, in part by the National Science Foundation under Grant ECS-8022033, and in part by the Joint Services Electronics Program under Contract

F49620-82-C-0033.

REFERENCES

[1] J. W. Grizzle and S. I. Marcus, "The Structure of Nonlinear Control Systems Possessing Symmetries," to appear in IEEE Transactions on Automatic Control.

[2] A. J. van der Schaft, System Theoretic Descriptions of Physical Systems, Ph.D. Dissertation, Mathematics Centrum, Amsterdam, June 1983.

[3] A. J. van der Schaft, "Symmetries and Conservation Laws for Hamiltonian Systems with Inputs and Outputs: A Generalization of Noether's Theorem," Systems and Control Letters, Vol. 1, No. 2, August 1981, pp. 108-115.

[4] R. Abraham and J. Marsden, Foundations of Mechanics, 2nd Ed., Massachusetts, The Benjamin/Cummings Publishing Co., 1978.

[5] H. Nijmeijer and A. J. van der Schaft, "Controlled Invariance for Nonlinear Systems," IEEE Transactions on Automatic Control, Vol. AC-27, No. 4, August 1982, pp. 904-914.

[6] A. Isidori, A. J. Krener, C. Gori-Giorgi and S. Monaco, "Nonlinear Decoupling via Feedback: A Differential Geometric Approach," IEEE Transactions on Automatic Control, Vol. AC-26, No. 2, April 1981, pp. 331-345.

[7] E. Lee and L. Markus, Foundations of Optimal Control Theory, John Wiley & Sons, Inc., New York, 1967.

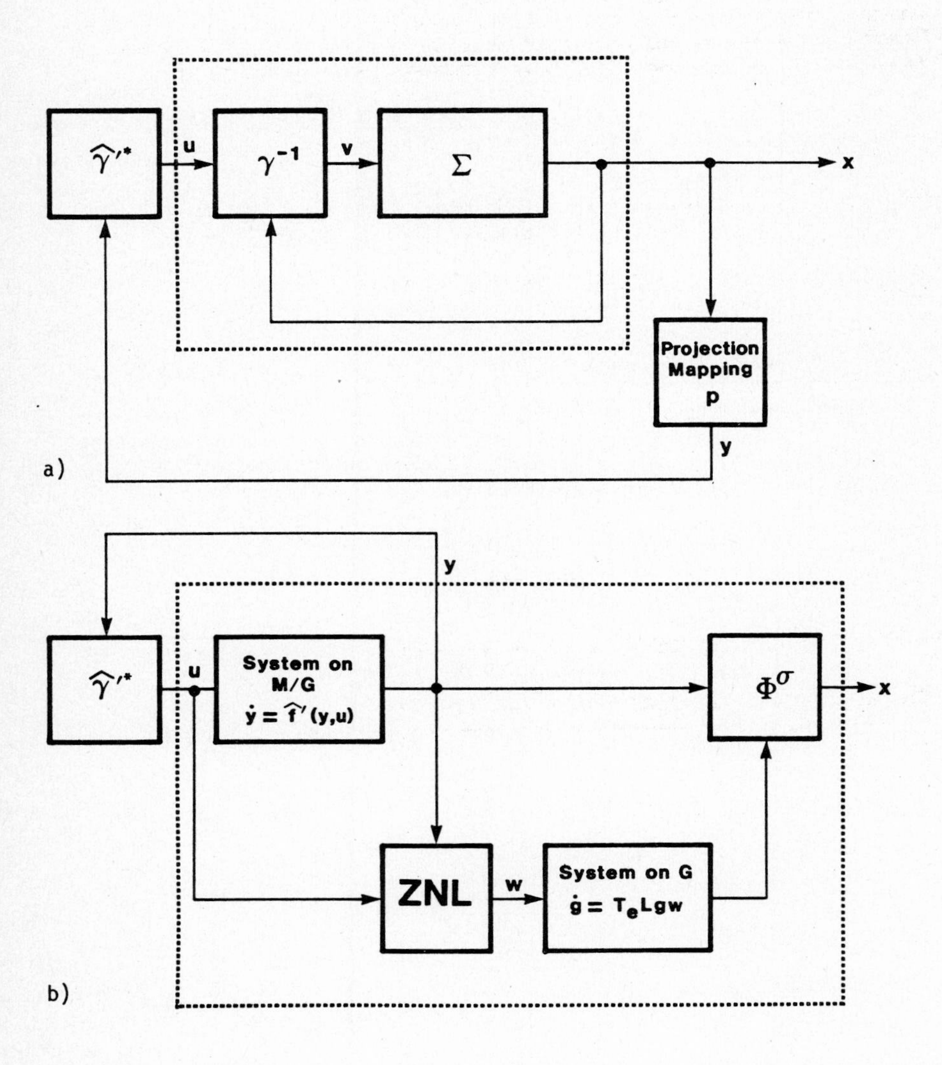

Figure 3.1. Structure of the optimal feedback controller. a) Representation in terms of Σ; b) representation in terms of Σ'.

SUR LES CONDITIONS NECESSAIRES D'OPTIMALITE DU DEUXIEME ET TROISIEME ORDRE DANS LES PROBLEMES DE COMMANDE OPTIMALE SINGULIERE(*)

F. Lamnabhi-Lagarrigue
E.S.E. - L.S.S.
Plateau du Moulon
91190 Gif-sur-Yvette, France

INTRODUCTION

D'une façon générale, un problème de commande optimale est dit singulier dès que le Principe du Maximum est trivialement satisfait sur une portion de trajectoire. Les cas les plus communs, comportant des arcs singuliers, apparaissent lorsque l'hamiltonien associé au problème de commande dépend linéairement d'une entrée. Considérons par exemple le problème en dimension N suivant :

$$\dot{x} = A_o(x) + u(t)A_1(x), \qquad t \in [0,T] \tag{1}$$

où les champs de vecteurs A_o et A_1 sont supposés analytiques ; $x(0) \in \mathbb{R}^N$ est donné, l'instant final T est fixé. En utilisant des fonctions analytiques par morceaux, prenant leurs valeurs dans un ensemble borné U, le problème est de minimiser la fonctionnelle

$$y(T) = h\Big(x(T)\Big) \tag{2}$$

h étant une fonction analytique. L'hamiltonien associé s'écrit :

$$H = \lambda\Big(A_o(x) + u(t)A_1(x)\Big)$$

avec la fonction de commutation

$$\Phi(t) = \lambda A_1(x)$$

λ étant le vecteur adjoint. Le problème est dit singulier si la fonction Φ s'annule sur des sous-intervalles $(t_a,t_b) \subset [0,T]$; les arcs extrémaux sont dits singuliers sur (t_a,t_b) et non-singuliers ailleurs. Il est clair que le Principe du Maximum est trivialement satisfait sur les arcs singuliers.

(*) Travail effectué dans le cadre de l'ATP "Thermique du Bâtiment" du CNRS.

Ceci a stimulé la recherche, à la fin des années soixante, de nouveaux tests d'optimalité ; citons la condition (du deuxième ordre) de Legendre-Clebsch généralisée, encore appelée, test de Kelley-Contensou (voir [5], [8], [11] et la bibliographie citée). Notons toutefois que l'on trouve très peu de formulations des conditions nécessaires d'optimalité du troisième ordre [13]. Pourtant, mis à part le problème de distinguer si une trajectoire est minimale ou non lorsque les critères du second ordre sont trivialement satisfaits, ces tests, qui sont des critères égalités, peuvent aussi donner des indications précises sur la commande.

Le but est ici de proposer une nouvelle interprétation des conditions nécessaires d'optimalité d'ordre supérieur (deuxième et troisième) pour les problèmes de commande optimale (1)-(2). On utilisera un nouveau développement fonctionnel ([2],[4]) sous la forme d'une série de Volterra et une variante de la formule de Baker-Campbell-Hausdorff. Ce nouveau formalisme algébrique semble, d'une part, plus rigoureux que le formalisme hamiltonien classique ; il évite, en effet, les notations

$$\frac{\partial^j}{\partial u^j}\,\frac{d^i}{dt^i}\,\frac{\partial H}{\partial u}\,, \qquad i,j > 1$$

qui peuvent devenir confuses, surtout lorsqu'il s'agit de tester une commande singulière en boucle fermée. D'autre part, la connaissance précise des noyaux de Volterra permet d'accéder aisément aux conditions nécessaires d'optimalité d'ordre supérieur à deux. Enfin, ces techniques se généralisent aux problèmes de commande optimale de la forme

$$\begin{cases} \dot{x} = f(t,x;u) \\ y(T) = h\big(x(T)\big) \qquad t \in [0,T] \end{cases} \tag{3}$$

où la fonction f est supposée analytique par rapport à t, x et u.

I. UN NOUVEAU DEVELOPPEMENT FONCTIONNEL

Considérons le système non linéaire

$$\begin{cases} \dot{x} = A_0(x) + u(t)A_1(x) \\ y(t) = h\big(x(t)\big), \qquad t \in [0,T] \end{cases} \tag{4}$$

La fonction de sortie $h : \mathbb{R}^N \longrightarrow \mathbb{R}$, et les champs de vecteurs A_j, $j = 0,1$, sont analytiques et définis dans un voisinage de $x(0) = a$

(si nous écrivons le champ de vecteurs A_j

$$A_j = \sum_{k=1}^{N} \theta_j^k(x_1,\dots,x_N) \frac{\partial}{\partial x_k}, \qquad j = 0,1,$$

rappelons que la première ligne de (4) est équivalente à

$$\dot{x}_k(t) = \theta_o^k(x_1,\dots,x_N) + u(t)\theta_1^k(x_1,\dots,x_N), \quad (k=1,\dots,N)\Big)$$

On peut montrer [2], [4] que la sortie y peut être développée en série de Volterra sous forme triangulaire :

$$\begin{aligned} y(t) = {} & w_o(t,a) + \int_o^t w_1(t,\sigma_1,a)u(\sigma_1)d\sigma_1 \\ & + \int_o^t \int_o^{\sigma_2} w_2(t,\sigma_2,\sigma_1,a)u(\sigma_2)u(\sigma_1)d\sigma_2 d\sigma_1 \\ & + \int_o^t \int_o^{\sigma_3} \int_o^{\sigma_2} w_3(t,\sigma_3,\sigma_2,\sigma_1,a)u(\sigma_3)u(\sigma_2)u(\sigma_1)d\sigma_3 d\sigma_2 d\sigma_i \\ & + \dots \end{aligned}$$

où $t \geq \sigma_3 \geq \sigma_2 \geq \sigma_1 \geq 0$, avec les noyaux suivants :

$$\begin{aligned} w_o(t,a) &= \sum_{\nu\geq 0} A_o^\nu . h\Big|_{x(0)} \frac{t^\nu}{\nu!} = e^{tA_o}.h\Big|_{x=a} \\ w_1(t,\sigma_1,a) &= e^{\sigma_1 A_o} A_1 e^{(t-\sigma_1)A_o}.h\Big|_{x=a} \qquad (5) \\ w_2(t,\sigma_1,\sigma_2,a) &= e^{\sigma_1 A_o} A_1 e^{(\sigma_2-\sigma_1)A_o} A_1 e^{(t-\sigma_2)A_o}.h\Big|_{x=a} \\ w_3(t,\sigma_3,\sigma_2,\sigma_1,a) &= e^{\sigma_1 A_o} A_1 e^{(\sigma_2-\sigma_1)A_o} A_1 e^{(\sigma_3-\sigma_2)A_o} A_1 e^{(t-\sigma_3)}.h\Big|_{x=a} \end{aligned}$$

Les membres de droite de (5) sont les développements de Taylor à l'origine des noyaux de Volterra correspondants. On peut encore écrire :

$$\begin{aligned} w_1(t,\sigma_1,a) &= e^{\sigma_1 A_o} A_1 e^{-\sigma_1 A_o} w_o(t,x)\Big|_{x=a} \\ w_2(t,\sigma_2,\sigma_1,a) &= e^{\sigma_1 A_o} e^{(\sigma_2-\sigma_1)A_o} A_1 e^{-\sigma_2 A_o} w_o(t,x)\Big|_{x=a} \\ w_3(t,\sigma_3,\sigma_2,\sigma_1,a) &= e^{\sigma_1 A_o} A_1 e^{(\sigma_2-\sigma_1)A_o} A_1 e^{(\sigma_3-\sigma_2)A_o} A_1 e^{-\sigma_3 A_o} w_o(t,x)\Big|_{x=a} \end{aligned}$$

Remarque : On ne s'intéresse pas ici au problème de la convergence de tels développements (voir Lésiak et Krener [10], Rugh [12]).

II. CONDITION NECESSAIRE D'OPTIMALITE

Soit le système non-linéaire (4).

Le problème peut être formulé ainsi : T étant fixé, trouver des conditions nécessaires satisfaites par une commande optimale u (c'est-à-dire la commande qui minimise $y(T) = h\big(x(T)\big)$, supposée analytique (*) sur [0,T]). Dans la suite, on notera y(T) par $W_u(T,\tau,b)$ où, la commande u étant donnée, le temps initial et le point initial sont respectivement $\tau \leq T$ et $x(\tau) = b$, $\tau \in [0,T]$. Décrivons maintenant les principales étapes de la méthode.

II.1. Variation de la commande

Faisons agir sur le système (4) la nouvelle entrée $\tilde{u}$:

$$\tilde{u} = u + \delta u(t)$$

où $\delta u(t)$ est une perturbation arbitraire. On obtient le nouveau système :

$$\begin{cases} \dot{x} = A_o(x) + uA_1(x) + \delta u(t)A_1(x) \\ \dot{x}_o = 1 \\ y(t) = h\big(x(t)\big), \qquad t \in [0,T] \end{cases}$$

Dans la suite, X dénotera l'état augmenté (x_o,x) et a, le point initial $(\tau,x(\tau)=b)$.

En utilisant la représentation entrée/sortie du paragraphe précédent, on peut établir que la série de Volterra de la variation de la fonctionnelle y(T), c'est-à-dire

$$\delta W = W_{u+\delta u}(T,\tau,b) - W_u(T,\tau,b)$$

s'écrit :

$$\delta W = \delta_1 W + \delta_2 W + \delta_3 W + o(|\delta u|^3) \tag{6}$$

avec :

$$\delta_1 W = \int_\tau^T e^{(\sigma_1-\tau)B_o} A_1 e^{-(\sigma_1-\tau)B_o} W_u(T,\tau,X)\Big|_{X=a} \delta u(\sigma_1)\, d\sigma_1$$

$$\delta_2 W = \int_\tau^T \int_\tau^{\sigma_2} e^{(\sigma_1-\tau)B_o} A_1 e^{(\sigma_2-\sigma_1)B_o} A_1 e^{-(\sigma_2-\tau)B_o} W_u(T,\tau,X)\Big|_{X=a} \times \delta u(\sigma_1)\delta u(\sigma_2)\, d\sigma_1 d\sigma_2$$

(*) On peut aussi supposer, avec quelques simples modifications [9], que la commande est analytique par morceaux avec un nombre fini de points de commutation.

et

$$\delta_3 W = \int_\tau^T \int_\tau^{\sigma_3} \int_\tau^{\sigma_2} e^{(\sigma_1-\tau)B_o} A_1 e^{(\sigma_2-\sigma_1)B_o} A_1 e^{(\sigma_3-\sigma_2)B_o} A_1 e^{-(\sigma_3-\tau)B_o}$$

$$W_u(T,\tau,X)\Big|_{X=a} \delta u(\sigma_1)\delta u(\sigma_2)\delta u(\sigma_3) d\sigma_1 d\sigma_2 d\sigma_3$$

où

$$A_o = \sum_{k=1}^{N} \theta_o^k(x) \frac{\partial}{\partial x_k} ,$$

$$A_1 = \sum_{k=1}^{N} \theta_1^k(x) \frac{\partial}{\partial x_k} \quad \text{et} \quad B_o = A_o + uA_1 + \frac{\partial}{\partial x_o}$$

$\delta_i W$ est le terme d'ordre i par rapport à u dans le développement de δW. Il est encore appelé *i-ème variation de la fonctionnelle y*.

II.2. Variation de la commande "concentrée" en un point

Plus précisément, choisissons la variation $\delta u(t)$ "concentrée" en un point $\theta \in [0,T]$ c'est-à-dire telle que :

$$\delta u(t) = 0 \quad \text{si} \quad t \notin [\theta, \omega(\varepsilon)]$$

où
$$\omega(\varepsilon) \geq \theta \quad \text{et} \quad \lim_{\varepsilon \to 0} \omega(\varepsilon) = \theta$$

Pour cette variation, (6) devient :

$$\delta W = e^{(\theta-\tau)B_o} V(\varepsilon) e^{-(\theta-\tau)B_o} W_U(T,\tau,X)\Big|_{X=a}$$

où
$$V(\varepsilon) = \delta_1 V + \delta_2 V + \delta_3 V + o(|\delta u|^3) \tag{7}$$

avec

$$\delta_1 V = \int_\theta^{\theta+\varepsilon} e^{(\sigma_1-\theta)B_o} A_1 e^{-(\sigma_1-\theta)B_o} \delta u(\sigma_1) d\sigma_1$$

$$\delta_2 V = \int_\theta^{\theta+\varepsilon} \int_\theta^{\sigma_2} \left(e^{(\sigma_1-\theta)B_o} A_1 e^{-(\sigma_1-\theta)B_o}\right)\left(e^{(\sigma_2-\theta)B_o} A_1 e^{-(\sigma_2-\theta)B_o}\right) \times \delta u(\sigma_1)\delta u(\sigma_2) d\sigma_1 d\sigma_2$$

et

$$\delta_3 V = \int_\theta^{\theta+\varepsilon} \int_\theta^{\sigma_3} \int_\theta^{\sigma_2} \left(e^{(\sigma_1-\theta)B_o} A_1 e^{-(\sigma_1-\theta)B_o}\right)\left(e^{(\sigma_2-\theta)B_o} A_1 e^{-(\sigma_2-\theta)B_o}\right) \times \left(e^{(\sigma_3-\theta)B_o} A_1 e^{-(\sigma_3-\theta)B_o}\right) \delta u(\sigma_1)\delta u(\sigma_2)\delta u(\sigma_3) d\sigma_1 d\sigma_2 d\sigma_3 .$$

II.3. Condition nécessaire d'optimalité

Soit $V(\varepsilon) = \sum_{k \geq 0} V_k \varepsilon^k$, le développement en puissance de ε de $V(\varepsilon)$;

on obtient alors le lemme suivant,

Lemme II.1 : Si u est minimizante sur [O,T] alors

$$e^{(\theta-\tau)B_o} V_h e^{-(\theta-\tau)B_o} W_u(T,\tau,X)\Big|_{X=a} \geq 0$$

$\forall\tau \in [0,T]$ et $\forall\theta \in [\tau,T]$,

où V_h est le terme dans le développement de $V(\varepsilon)$ tel que

$$V_k W_u(T,\tau,X)\Big|_{X=a} = 0 \qquad \forall\tau \in [0,T], \quad k \leq h-1$$

(l'inégalité est inversée pour un maximum).

Appliquons une variante de la formule Baker-Campbell-Hansdorff :

$$e^x y e^{-x} = \sum_{\ell\geq 0} \frac{1}{\ell!} ad_x^\ell . y \tag{8}$$

où $ad_x^\ell y$ dénote $[x, ad_x^{\ell-1} y]$ avec $ad_x^o y = y$;

Le lemme II.1 peut encore s'écrire :

Lemme II.2 : Si u est minimizante sur [O,T] alors :

$$\sum_{\ell\geq 0} \frac{(\theta-\tau)^\ell}{\ell!} ad_{B_o}^\ell V_h W_u(T,\tau,X)\Big|_{X=a} \geq 0$$

$\forall\tau \in [0,T]$ et $\forall\theta \in [\tau,T]$.

Dans la suite, pour simplifier les notations, nous nous intéressons aux conditions nécessaires d'optimalité pour que $u = 0$ soit minimisante sur [O,T]. A partir du lemme précédent on obtient :

Lemme II.3 : Si $u = 0$ est minimisante sur [O,T] alors :

$$\sum_{\ell\geq 0} \frac{(\theta-\tau)^\ell}{\ell!} ad_{A_o}^\ell V_h^o W_o(T,\tau,x)\Big|_{x=b} \geq 0$$

où $W_o(T,\tau,b)$ étant la réponse libre de (4), c'est-à-dire :

$$W_o(T,\tau,b) = e^{(T-\tau)A_o} . h\Big|_{x=b} ,$$

et où V_h^o est l'expression de V_h pour $u = 0$,

$$\forall\tau \in [0,T] \quad \text{et} \quad \forall\theta \in [\tau,T].$$

III. CHOIX DE LA PERTURBATION δu

Il est clair que la formulation des conditions nécessaires d'optimalité dépend du choix de la variation $\delta u(t)$. Dans la littérature, l'obtention des conditions du second ordre fait appel à de nombreux types de variations, citons "les variations spéciales" [7], les "paquets de variations" [5], "les paquets de perturbations" [1], etc... Nous utiliserons ici celles de [13] et [14] parce qu'elles nous semblent bien adaptées aux formulations des conditions nécessaires d'optimalité d'ordre supérieur à deux.

Soit donc la perturbation "concentrée" en θ telle que, k étant un entier *arbitraire*

$$\int_{\theta}^{\theta+\varepsilon} (\sigma-\theta)^{i}\delta u(\sigma)d\sigma = 0 \ , \quad i \leq k-1$$

Une solution [14] de ces équations intégrales est :

$$\delta u(t) = a \sum_{\ell=0}^{k} (-1)^{\ell} C_k^{\ell} C_{k+\ell}^{\ell} \frac{1}{\varepsilon^{\ell}}(t-\theta)^{\ell}, \quad a = \delta u(\theta), \qquad t \in [\theta,\theta+\varepsilon] \tag{9}$$

et de plus

$$\int_{\theta}^{\theta+\varepsilon} (\sigma-\theta)^{m}\delta u(\sigma)d\sigma = (-1)^{k+1} a \frac{(m!)^2}{(m-k)!(m+k+1)!} \varepsilon^{m+1}, \quad m \geq k \tag{10}$$

III.1. Deuxième variation

Considérons la deuxième variation $\delta_2 V$ dans (7), avec u = 0,

$$\delta_2 V = \int_{\theta}^{\theta+\varepsilon}\int_{\theta}^{\sigma_2}\left(e^{(\sigma_1-\theta)A_o}A_1 e^{-(\sigma_1-\theta)A_o}\right)\left(e^{(\sigma_2-\theta)A_o}A_1 e^{-(\sigma_2-\theta)A_o}\right) \times \delta u(\sigma_1)\delta u(\sigma_2)d\sigma_1 d\sigma_2$$

Après avoir intégré par parties on obtient :

$$\delta_2 V = \frac{1}{2}\left(\int_{\theta}^{\theta+\varepsilon} e^{(\sigma_1-\theta)A_o}A_1 e^{-(\sigma_1-\theta)A_o}\delta u(\sigma_1)d\sigma_1\right)^2 + \frac{1}{2}\int_{\theta}^{\theta+\varepsilon}\int_{\theta}^{\sigma_2}\left[e^{(\sigma_1-\theta)A_o}A_1 e^{-(\sigma_1-\theta)A_o}, e^{(\sigma_2-\theta)A_o}A_1 e^{-(\sigma_2-\theta)A_o}\right] \times \delta u(\sigma_1)\delta u(\sigma_2)d\sigma_1 d\sigma_2$$

En appliquant (8), $\delta_2 V$ peut encore s'écrire :

$$\delta_2 V = \frac{1}{2}\Big(\sum_{\alpha\geq 0} \frac{1}{\alpha!}\, ad^{\alpha}_{A_o} A_1 \int_{\theta}^{\theta+\varepsilon} (\sigma-\theta)^{\alpha}\, \delta u(\sigma)\, d\sigma\Big)^2$$

$$+ \frac{1}{2} \sum_{\alpha,\beta\geq 0} \frac{1}{\alpha!\beta!}\Big[ad^{\alpha}_{A_o} A_1, ad^{\beta}_{A_o} A_1\Big] \int_{\theta}^{\theta+\varepsilon}\int_{\theta}^{\sigma_2} (\sigma_1-\theta)^{\alpha}(\sigma_2-\theta)^{\beta}$$

$$\times\ \delta u(\sigma_1)\,\delta u(\sigma_2)\, d\sigma_1 d\sigma_2$$

Pour k *quelconque*, en utilisant (9) et (10), on trouve :

$$\delta_2 V = a^2(-1)^k \sum_{\substack{\alpha,\beta\geq 0\\ \alpha+\beta\leq k-1}} \frac{1}{\alpha!\beta!}\Big[ad^{\alpha}_{A_o}.A_1, ad^{\beta}_{A_o}.A_1\Big] \sum_{\ell=k-(\alpha+\beta)}^{k} (-1)^{\ell}$$

$$\times\ C^{\ell}_k C^{\ell}_{k+\ell}\ \frac{(\alpha+\beta+\ell+1)!^2}{(\alpha+\beta+\ell+1-k)!(\alpha+\beta+\ell+k+2!)}\ \varepsilon^{\alpha+\beta+2} + o(\varepsilon^{k+1})$$

Choisissons k = 2, alors :

$$\delta_2 V = a^2\mu\Big[A_1, [A_o, A_1]\Big]\varepsilon^3 + o(\varepsilon^3), \quad \text{avec } \mu < 0$$

D'après le lemme II.3, on obtient :

Proposition III.1 : Si u = 0 est minimisante sur [0,T] pour le problème de commande (1)-(2) alors

$$\sum_{\ell\geq 0} ad^{\ell}_{A_o}\Big[A_1, [A_o, A_1]\Big] W_o(T,\tau,x)\Big|_{x=b} \leq 0$$

$\forall\tau \in [\theta,T]$ et $\forall\theta \in [\tau,T]$.

Si le champ de vecteurs $\Big[A_1, [A_o, A_1]\Big] \in \mathcal{L}_1$, l'espace vectoriel engendré par $ad^{\ell}_{A_o}.A_1$, $\ell \geq 0$, cette condition est trivialement satisfaite ; en effet, si u = 0 est optimal sur [0,T] on sait que la première variation est nulle, ce qui s'exprime [3] ici par

$$ad^{\ell}_{A_o} A_1 W_o(T,\tau,x)\Big|_{x=b} = 0, \quad \forall\tau \in [0,T],\ \forall\ell \geq 0 \tag{11}$$

D'une façon générale, on obtient :

Proposition III.2 : Si u = 0 est minimisante sur [0,T] et si

$$[A_1, ad^{\ell}_{A_o} A_1] \in \mathcal{L}_1 \quad \text{pour } \ell \leq k-2$$

alors i) k est pair

ii) $(-1)^{k/2} \sum_{\ell \geq 0} \frac{(\theta-\tau)^\ell}{\ell!} \, ad^\ell_{A_o} [A_1, ad^{k-1}_{A_o} A_1] \, w_o(T,\tau,x) \Big|_{x=b} \leq 0$

$\forall \tau \in [0,T]$ et $\forall \theta \in [\tau,T]$.

Remarques : (1) Cette proposition se montre en partie grâce à la formule :

$$[ad^\alpha_{A_o}.A_1, ad^\beta_{A_o}.A_1] = \sum_{\nu=0}^{\alpha} (-1)^{\alpha+\nu} C^\nu_\alpha ad^\nu_{A_o}.[A_1, ad^{\alpha+\beta-\nu}_{A_o}.A_1]$$

Il est facile de voir, par exemple, que k ne peut pas être impair sinon

$$[ad^{k-1}_{A_o}.A_1, A_1] = (-1)^{k-1}[A_1, ad^{k-1}_{A_o}.A_1] + B \quad \text{avec } B \in \mathscr{L}_1$$

impliquerait $\quad 2[A_1, ad^{k-1}_{A_o}.A_1] \in \mathscr{L}_1$.

(2) Si on choisit $\theta = \tau$, il vient :

$$(-1)^{k/2}[A_1, ad^{k-1}_{A_o}.A_1] w_o(T,\tau,x)\Big|_{x=b} \leq 0, \ \forall \tau \in [0,T]$$

On peut montrer que cette condition n'est autre que la condition de Legendre-Clebsch généralisée [5], [11]

$$(-1)^{k/2} \frac{\partial}{\partial u} \frac{d^k}{dt^k} \frac{\partial H}{\partial u}\Big|_{\substack{u=0\\x=b}} \leq 0, \quad \forall \tau \in [0,T],$$

H étant l'hamiltonien associé au problème de commande (4)

$$H = \sum_{k=1}^{N} (\theta^k_o + u\theta^k_1)\lambda_k$$

avec $\lambda = d\omega_o = \left(\frac{\partial \omega_o}{\partial x_k}\right)_{k=1,N}$.

III.2. Troisième variation

Intéressons-nous aux nouvelles conditions basées sur la troisième variation

$$\delta_3 V = \int_\theta^{\theta+\varepsilon}\int_\theta^{\sigma_3}\int_\theta^{\sigma_2} \left(e^{(\sigma_1-\theta)A_o} A_1 e^{-(\sigma_1-\theta)A_o}\right)\left(e^{(\sigma_2-\theta)A_o} A_1 e^{-(\sigma_2-\theta)A_o}\right)$$

$$\times \left(e^{(\sigma_3-\theta)A_o} A_1 e^{-(\sigma_3-\theta)A_o}\right) \delta u(\sigma_1)\delta u(\sigma_2)\delta u(\sigma_3) d\sigma_1 d\sigma_2 d\sigma_3 \qquad (11)$$

k étant fixé, ces nouvelles formulations proviendront, comme pour la deuxième variation, du développement en puissance de ε jusqu'à l'ordre (k+1) de $\delta_3 V$

$$\delta_3 V = a^3 \sum_{\nu=1}^{k} G_\nu \varepsilon^{\nu+1} + o(\varepsilon^{k+1}) \tag{13}$$

En intégrant par parties (12), on montre que le seul terme intervenant dans la détermination des G_ν est :

$$\frac{1}{3}\int_\theta^{\theta+\varepsilon}\int_\theta^{\sigma_3}\int_\theta^{\sigma_2}\Big[\Big(e^{(\sigma_1-\theta)A_o}A_1 e^{-(\sigma_1-\theta)A_o},\ e^{(\sigma_2-\theta)A_o}A_1 e^{-(\sigma_2-\theta)A_o}\Big],\ e^{(\sigma_3-\theta)A_o}A_1 e^{-(\sigma_3-\theta)A_o}\Big]\delta u(\sigma_1)\delta u(\sigma_2)\delta u(\sigma_3)d\sigma_1 d\sigma_2 d\sigma_3$$

En utilisant (8) et les propriétés (9) et (10) de u(t), on trouve :

$$G = \sum_{\substack{\ell=0 \\ s=0 \\ \ell+s\geq k-\nu}}^{k} \Phi(\nu,k,\ell,s)V(\nu,\ell,s)$$

avec

$$V(\nu,\ell,s) = \frac{1}{6}\sum_{\substack{\alpha\ \beta\ \gamma \\ \alpha+\beta+\gamma=\nu-2}} \frac{1}{\alpha!\beta!\gamma!}\ \frac{1}{(\alpha+\ell+1)(\beta+\alpha+\ell+s+1)}\Big[\Big[ad^\alpha_{A_o}.A_1, ad^\beta_{A_o}.A_1\Big], ad^\gamma_{A_o}.A_1\Big]$$

et

$$\Phi(\nu,k,\ell,s) = (-1)^{k+1}\frac{C^\ell_k C^\ell_{k+\ell} C^s_k C^s_{k+s}(\nu+\ell+s)!^2}{(\nu-k+\ell+s)!(\ell+\nu+k+s+1)!}$$

Comme le paramètre a dans (13) est *arbitraire*, a^3 peut être choisi soit positif, soit négatif. Le lemme II.3 permet alors d'écrire :

<u>Proposition III.3</u> : Si <u>u = 0</u> est minimisante sur [0,T] et si

$$[A_1, ad^{2\ell-1}_{A_o}.A_1] \in \mathcal{L}_1 \quad \text{pour} \quad \ell \leq k-1$$

alors $\displaystyle\sum_{\ell\geq 0}\frac{(\theta-\tau)^\ell}{\ell!}\, ad_{A_o} G_\nu W_o(T,\tau,x)\Big|_{x=b} = 0, \quad \forall\nu\leq k,\ \forall\tau\in[0,T],\ \forall\theta\in[\tau,T]$

Prenons <u>k = 2</u>, on obtient :

<u>Proposition III.4</u> : Si <u>u = 0</u> est minimisante sur [0,T] et si

$$\Big[A_1,[A_o,A_1]\Big] \in \mathcal{L}_1$$

alors $\displaystyle\sum_{\ell\geq 0}\frac{(\theta-\tau)^\ell}{\ell!}\, ad^\ell_{A_o}.\Big[A_1,[A_1,[A_o,A_1]]\Big]W_o(T,\tau,x)\Big|_{x=b} = 0$

$\forall\tau\in[0,T]$ et $\forall\theta\in[\tau,T]$.

Cette condition peut aussi s'exprimer sous la forme :

$$\boxed{ad^{\ell}_{A_o} \cdot \left[A_1,[A_1,[A_o.A_1]]\right] W_o(T,\tau,x)\Big|_{x=b} = 0, \ \forall \ell \geq 0, \forall \tau \in [0,T]} \tag{14}$$

Remarques : (1) La condition (14) doit être comparée à celle obtenue par Skorodinskii [13]

$$\frac{\partial^2}{\partial u^2} \frac{d^3}{dt^3} \frac{\partial H}{\partial u}\Bigg|_{\substack{u=0 \\ x=b}} = 0, \qquad \forall \tau \in [0,T].$$

(2) Ces critères du troisième ordre, contenant des égalités, pourront augmenter considérablement l'ensemble des conditions nécessaires disponibles pour déterminer la commande.

IV. GENERALISATION

Les arcs singuliers peuvent aussi apparaître dans des problèmes de commande optimale plus généraux que (4), soit :

$$\begin{cases} \dot{x} = f(t,x;u) \\ y(t) = h\big(x(t)\big), \qquad t \in [0,T], \end{cases} \tag{15}$$

où f est une fonction analytique par rapport à toutes les variables t, x et u.

Dans la suite, nous proposons pour ce type de problème, une nouvelle formulation des conditions nécessaires d'optimalité, obtenue, ici encore, à partir du développement fonctionnel du § I. Il faut cependant noter qu'il ne s'agit, ici, que d'une approximation du problème ; le formalisme rigoureux, utilisant un développement fonctionnel convenable, sera donné ultérieurement.

Comme précédemment, effectuons une perturbation sur l'entrée u du système (15)

$$\tilde{u} = u + \delta u(t).$$

On obtient le nouveau système :

$$\begin{cases} \dot{x} = f(t,x;u) + \sum\limits_{i=1}^{3} \frac{1}{i!} f_{u^{(i)}}(t,x;u)\,\delta \overset{i}{u}(t) + o(|\delta u|^3) \\ y(t) = h\big(x(t)\big) \end{cases}$$

où les nouvelles entrées $u_1(t) = \delta u(t)$, $u_2(t) = \delta u^2(t)$ et $u_3(t) = \delta u^3(t)$ sont ici considérées "indépendantes" et où $f_{u(i)}$ dénote la i-ème dérivée partielle par rapport à u de f.

Si X dénote, ici aussi, l'état augmenté (x_o, x) et a le point initial $(\tau, x(\tau))$, les variations $\delta_1 W$, $\delta_2 W$ et $\delta_3 W$ deviennent, pour ce problème :

$$\delta_1 W = \int_\tau^T e^{(\sigma_1-\tau)A_o} A_1 e^{-(\sigma_1-\tau)A_o} W_u(T,\tau,X)\Big|_{X=a} \delta u(\sigma_1)\, d\sigma_1$$

$$\delta_2 W = \frac{1}{2}\int_\tau^T e^{(\sigma_1-\tau)A_o} A_2 e^{-(\sigma_1-\tau)A_o} W_u(T,\tau,X)\Big|_{X=a} \delta u^2(\sigma_1)\, d\sigma_1$$

$$+ \int^T\int^{\sigma_2} e^{(\sigma_1-\tau)A_o} A_1 e^{(\sigma_2-\sigma_1)A_o} A_1 e^{-(\sigma_2-\tau)A_o} W_u(T,\tau,X)\Big|_{X=a} \delta u(\sigma_1)\delta u(\sigma_2)\, d\sigma_1 d\sigma_2$$

et

$$\delta_3 W = \frac{1}{6}\int_\tau^T e^{(\sigma_1-\tau)A_o} A_3 e^{-(\sigma_1-\tau)A_o} W_u(T,\tau,X)\Big|_{X=a} \delta u^3(\sigma_1)\, d\sigma_1$$

$$+ \frac{1}{2}\int_\tau^T\int_\tau^{\sigma_2} e^{(\sigma_1-\tau)A_o} A_1 e^{(\sigma_2-\sigma_1)A_o} A_2 e^{-(\sigma_2-\tau)A_o} W_u(T,\tau,X)\Big|_{X=a} \delta u(\sigma_1)\delta u^2(\sigma_2)\, d\sigma_1 d\sigma_2$$

$$+ \frac{1}{2}\int_\tau^T\int_\tau^{\sigma_2} e^{(\sigma_1-\tau)A_o} A_2 e^{(\sigma_2-\sigma_1)A_o} A_1 e^{-(\sigma_2-\tau)A_o} W_u(T,\tau,X)\Big|_{X=a} \delta^2 u(\sigma_1)\delta u(\sigma_2)\, d\sigma_1 d\sigma_2$$

$$+ \int_\tau^T\int_\tau^{\sigma_3}\int_\tau^{\sigma_2} e^{(\sigma_1-\tau)A_o} A_1 e^{(\sigma_2-\sigma_1)A_o} A_1 e^{(\sigma_3-\sigma_2)A_o} A_1 e^{-(\sigma_3-\tau)A_o}$$

$$\times W_u(T,\tau,X)\Big|_{X=a} \delta u(\sigma_1)\delta u(\sigma_2)\delta u(\sigma_3)\, d\sigma_1 d\sigma_2 d\sigma_3$$

où
$$A_o = \sum_{k=1}^{N} f^k(x_o,x;u)\frac{\partial}{\partial x_k} + \frac{\partial}{\partial x_o}$$

et
$$A_i = \sum_{k=1}^{N} f^k_{u(i)}(x_o,x;u)\frac{\partial}{\partial x_k} \; ; \quad i = 1,2,3 \, .$$

Pour une variation $\delta u(t)$ "concentrée" en un point $\theta \in [0,T]$, δW peut encore s'écrire :

$$\delta W = \sum_{\ell \geq 0} \frac{(\theta-\tau)^\ell}{\ell!}\, ad^\ell_{A_o} V(\varepsilon)\, W_u(T,\tau,X)\Big|_{X=a}$$

avec

$$V(\varepsilon) = \delta_1 V + \delta_2 V + \delta_3 V + o(|\delta u|^3) \tag{16}$$

$$\delta_1 V = \int_\theta^{\theta+\varepsilon} e^{(\sigma_1-\theta)A_o} A_1 e^{-(\sigma_1-\theta)A_o} \delta u(\sigma_1)\, d\sigma_1$$

$$\delta_2 V = \frac{1}{2}\int_\theta^{\theta+\varepsilon} e^{(\sigma_1-\theta)A_o} A_2 e^{-(\sigma_1-\theta)A_o} \delta u^2(\sigma_1)\, d\sigma_1$$

$$+ \int_\theta^{\theta+\varepsilon}\int_\theta^{\sigma_2} \left(e^{(\sigma_1-\theta)A_o} A_1 e^{-(\sigma_1-\theta)A_o}\right)\left(e^{(\sigma_2-\theta)A_o} A_1 e^{-(\sigma_2-\theta)A_o}\right) \delta u(\sigma_1)\delta u(\sigma_2) d\sigma_1 d\sigma_2$$

et

$$\delta_3 V = \frac{1}{6}\int_\theta^{\theta+\varepsilon} e^{(\sigma_1-\theta)A_o} A_3 e^{-(\sigma_1-\theta)A_o} \delta u^3(\sigma_1)\, d\sigma_1$$

$$+ \frac{1}{2}\int_\theta^{\theta+\varepsilon}\int_\theta^{\sigma_2} \left(e^{(\sigma_1-\theta)A_o} A_1 e^{-(\sigma_1-\theta)A_o}\right)\left(e^{(\sigma_2-\theta)A_o} A_2 e^{-(\sigma_2-\theta)A_o}\right) \delta u(\sigma_1)\delta u^2(\sigma_2) d\sigma_1 d\sigma_2$$

$$+ \frac{1}{2}\int_\theta^{\theta+\varepsilon}\int_\theta^{\sigma_2} \left(e^{(\sigma_1-\theta)A_o} A_2 e^{-(\sigma_1-\theta)A_o}\right)\left(e^{(\sigma_2-\theta)A_o} A_1 e^{-(\sigma_2-\theta)A_o}\right) \delta u^2(\sigma_1)\delta u(\sigma_2) d\sigma_1 d\sigma_2$$

$$+ \int_\theta^{\theta+\varepsilon}\int_\theta^{\sigma_3}\int_\theta^{\sigma_2} \left(e^{(\sigma_1-\theta)A_o} A_1 e^{-(\sigma_1-\theta)A_o}\right)\left(e^{(\sigma_2-\theta)A_o} A_1 e^{-(\sigma_2-\theta)A_o}\right)$$

$$\times \left(e^{(\sigma_3-\theta)A_o} A_1 e^{(\sigma_3-\theta)A_o}\right) \delta u(\sigma_1)\delta u(\sigma_2)\delta u(\sigma_3) d\sigma_1 d\sigma_2 d\sigma_3$$

IV.1. Deuxième variation

En utilisant (8) $\delta_2 V$ devient :

$$\delta_2 V = \frac{1}{2}\sum_{\alpha\geq 0}\frac{1}{\alpha!}\, ad^\alpha_{A_o}.A_2 \int_\theta^{\theta+\varepsilon} (\sigma-\theta)^\alpha \delta^2 u(\sigma)\, d\sigma$$

$$+ \frac{1}{2}\left(\sum_{\alpha\geq 0}\frac{1}{\alpha!}\, ad^\alpha_{A_o}.A_1 \int_\theta^{\theta+\varepsilon} (\sigma-\theta)^\alpha \delta u(\sigma)\, d\sigma\right)^2$$

$$+ \frac{1}{2}\sum_{\alpha,\beta\geq 0}\frac{1}{\alpha!\beta!}\left[ad^\alpha_{A_o}.A_1, ad^\beta_{A_o}.A_1\right]\int_\theta^{\theta+\varepsilon}\int_\theta^{\sigma_2} (\sigma_1-\theta)^\alpha(\sigma_2-\theta)^\beta \delta u(\sigma_1)\delta u(\sigma_2) d\sigma_1 d\sigma_2$$

Choisissons la même variation qu'au § III.

Si $k = 1$,

$$\delta u(t) = \begin{cases} a\left(1 - \frac{2}{\varepsilon}(t-\theta)\right), & t \in [\theta,\theta+\varepsilon] \\ 0 & \text{ailleurs} \end{cases}$$

alors,

$$\delta_2 V = -\frac{a^2}{3} A_2 \varepsilon + o(\varepsilon)$$

D'après le lemme II.2, une condition nécessaire d'optimalité est donc :

$$\sum_{\ell \geq 0} \frac{(\theta-\tau)^{\ell}}{!} \mathrm{ad}_{A_o}^{\ell} A_2 w_o(T,\tau,X)\Big|_{X=a} \leq 0, \ \forall \tau \in [0,T], \ \forall \theta \in [\tau,T] \quad (17)$$

Le membre de gauche de cette dernière expression est le développement de Taylor au point τ de la fonction $\frac{\partial^2 H}{\partial u^2}$, H étant l'hamiltonien associé au problème (15)

$$H = \lambda^T f$$

avec $\lambda = dw_o = \left(\frac{\partial w_o}{\partial x_k}\right)_{1 \leq k \leq N}$. La condition (17) est une formulation équivalente de la condition de Legendre-Clebsch.

Supposons que, sur tout l'intervalle [0,T], cette condition soit trivialement remplie (le problème est alors singulier) :

$$\sum_{\ell \geq 0} \frac{(\theta-\tau)^{\ell}}{\ell !} \mathrm{ad}_{A_o}^{\ell} A_2 w_u(T,\tau,X)\Big|_{X=a} = 0 \quad \forall \tau \in [0,T], \forall \theta \in [\tau,T]$$

ou encore :

$$\mathrm{ad}_{A_o}^{\ell} \cdot A_2 w_u(T,\tau,X)\Big|_{X=a} = 0, \quad \ell \geq 0, \ \forall \tau \in [0,T] \ ; \quad (18)$$

Si <u>k = 3</u>

$$\delta_2 V = -a^2\Big(\mu_o A_2 \varepsilon + \mu_1 \mathrm{ad}_{A_o} \cdot A_2 \varepsilon^2 + \mu_2 \mathrm{ad}_{A_o}^2 \cdot A_2 \varepsilon^3$$
$$+ \mu_3\Big[A_1,[A_o,A_1]\Big]\varepsilon^3\Big) + o(\varepsilon^3) \ , \quad \mu_3 > 0$$

D'après (18) et le lemme II.2 on obtient :

<u>Proposition IV.1</u> : Si u est minimisante sur [0,T] et si le Principe du Maximum est trivialement satisfait (i.e. les relations (18) sont vérifiées), alors :

$$\sum_{\ell \leq 0} \frac{(\theta-\tau)^{\ell}}{\ell !} \mathrm{ad}_{A_o}^{\ell} \Big[A_1,[A_o,A_1]\Big] w_u(T,\tau,X)\Big|_{X=a} \leq 0,$$

$$\forall \tau \in [0,T] \quad \text{et} \quad \forall \theta \in [\tau,T].$$

Remarque : Si H désigne l'hamiltonien associé au problème (15), un calcul simple montre que :

$$\frac{\partial}{\partial u} \frac{d^2}{dt^2} \frac{\partial H}{\partial u} = \Big(\Big[A_1,[A_o,A_1]\Big] + \Big[A_o,[A_o,A_2]\Big]\Big) w_u \ ;$$

Comme le deuxième terme du membre de droite est nul pour un problème singulier, on voit clairement que le formalisme hamiltonien est dans ce cas plus lourd que celui basé sur les champs de vecteurs.

Désignons par $\mathcal{L}_2$ l'espace vectoriel engendré par $ad^{\ell}_{A_o}.A_2$, $\ell \geq 0$; la proposition III.2 devient, pour le problème de commande (15) :

Proposition IV.2 : Si u est minimisante sur [O,T], si le Principe du Maximum est trivialement satisfait et si

$$[A_1, ad^{\ell}_{A_o}.A_1] \in \mathcal{L}_1 \cup \mathcal{L}_2 \quad \text{pour } \ell \leq k-2$$

alors i) k est pair

$$\text{ii)} \quad (-1)^{k/2} \sum_{\ell \geq 0} \frac{(\theta-\tau)^{\ell}}{\ell!} \, ad^{\ell}_{A_o}\left[A_1, ad^{k-1}_{A_o}.A_1\right] W_u(T,\tau,X)\Big|_{X=a} \geq 0.$$

$\forall \tau \in [O,T]$ et $\forall \theta \in [\tau,T]$.

IV.2. Troisième variation

Considérons le développement en puissance de ε de $\delta_3 V$ jusqu'à l'ordre k+1,

$$\delta_3 V = a^3 \sum_{\nu=1}^{k} G_{\nu} \varepsilon^{\nu+1} + O(\varepsilon^{k+1})$$

En intégrant par parties certains termes de $\delta_3 V$ dans (16), on montre que les seuls termes intervenant dans la détermination des coefficients G_{ν} sont :

$$\frac{1}{6} \int_{\theta}^{\theta+\varepsilon} e^{(\sigma-\theta)A_o} A_3 e^{-(\sigma-\theta)A_o} \delta u^3(\sigma)\, d\sigma,$$

$$\frac{1}{4} \int_{\theta}^{\theta+\varepsilon} \int_{\theta}^{\sigma_2} \left[e^{(\sigma_1-\theta)A_o} A_1 e^{-(\sigma_1-\theta)A_o},\ e^{(\sigma_2-\theta)A_o} A_2 e^{-(\sigma_2-\theta)A_o} \right] \delta u(\sigma_1) \delta u^2(\sigma_2) d\sigma_1 d\sigma_2$$

$$\frac{1}{4} \int_{\theta}^{\theta+\varepsilon} \int_{\theta}^{\sigma_2} \left[e^{(\sigma_1-\theta)A_o} A_2 e^{-(\sigma_1-\theta)A_o},\ e^{(\sigma_2-\theta)A_o} A_1 e^{-(\sigma_2-\theta)A_o} \right] \delta u^2(\sigma_1) \delta u(\sigma_2) d\sigma_1 d\sigma_2$$

et

$$\frac{1}{3} \int_{\theta}^{\theta+\varepsilon} \int_{\theta}^{\sigma_3} \int_{\theta}^{\sigma_2} \left[\left(e^{(\sigma_1-\theta)A_o} A_1 e^{-(\sigma_1-\theta)A_o},\ e^{(\sigma_2-\theta)A_o} A_1 e^{-(\sigma_2-\theta)A_o} \right),\ e^{(\sigma_3-\theta)A_o} A_1 e^{-(\sigma_3-\theta)A_o} \right] \delta u(\sigma_1) \delta u(\sigma_2) \delta u(\sigma_3) d\sigma_1 d\sigma_2 d\sigma_3$$

Après quelques calculs simples utilisant les propriétés (9) et (10) de $\delta u(t)$ et la formule (8) on obtient :

$$G_\nu = \sum_{\substack{\ell=0 \\ s=0 \\ \ell+s\leq k-\nu}}^{k} \Phi(\nu,k,\ell,s)\, V(\nu,\ell,s)$$

où $V(v,\ell,s)$ est le *champ de vecteurs*

$$\frac{1}{6}\frac{1}{\nu!}\,\mathrm{ad}^{\nu}_{A_o}.A_3 + \frac{1}{4}\sum_{\substack{\alpha,\beta\geq 0 \\ \alpha+\beta=\nu-1}} \frac{1}{\alpha!\beta!}\,\frac{1}{\alpha+\ell+s+1}\left[\mathrm{ad}^{\alpha}_{A_o}.A_2,\mathrm{ad}^{\beta}_{A_o}.A_1\right]$$

$$+\frac{1}{4}\sum_{\substack{\alpha,\beta\geq 0 \\ \alpha+\beta=\nu-1}} \frac{1}{\alpha!\beta!}\,\frac{1}{\alpha+\ell+1}\left[\mathrm{ad}^{\alpha}_{A_o}.A_1,\mathrm{ad}^{\beta}_{A_o}.A_1\right]$$

$$+\frac{1}{6}\sum_{\substack{\alpha,\beta,\gamma\geq 0 \\ \alpha+\beta+\gamma=\nu-2}} \frac{1}{\alpha!\beta!\gamma!}\,\frac{1}{(\alpha+\ell+1)(\beta+\ell+\alpha+s+1)}\left[[\mathrm{ad}^{\alpha}_{A_o}.A_1,\mathrm{ad}^{\beta}_{A_o}.A_1],\mathrm{ad}^{\gamma}_{A_o}.A_1\right]$$

et où

$$\Phi(\nu,k,\ell,s) = (-1)^k\,\frac{C^{\ell}_{k}C^{\ell}_{k+\ell}\,C^{s}_{k}C^{s}_{k+s}\,(\nu+\ell+s)!^2}{(\nu-k+\ell+s)!\,(\ell+\nu+k+s+1)!}$$

D'où la proposition :

<u>Proposition IV.3</u> : Si u est minimisante sur $[0,T]$, si le Principe du Maximum est trivialement satisfait et si

$$\left[A_1,\mathrm{ad}^{2\ell-1}_{A_o}A_1\right] \in \mathcal{L}_1 \cup \mathcal{L}_2 \quad \text{pour } \ell \leq k-1$$

alors

$$\sum_{\ell\geq 0}\frac{(\theta-\tau)^\ell}{\ell!}\,\mathrm{ad}^{\ell}_{A_o}G_\nu W_u(T,\tau,X)\Big|_{X=a} = 0 \quad \forall\nu\geq k \quad \forall\tau\in[0,T],\ \forall\theta\in[\tau,T]$$

ou encore

$$\boxed{\mathrm{ad}^{\ell}_{A_o}.G_\nu W_u(T,\tau,X)\Big|_{X=a} = 0,\ \forall\nu\leq k,\ \forall\ell\geq 0,\ \forall\tau\in[0,T]}\ .$$

Prenons par exemple <u>$k = 2$</u>, on obtient la condition :

$$\mathrm{ad}^{\ell}_{A_o}[A_1,A_2]W_u(T,\tau,X)\Big|_{X=a} = 0,\ \forall\ell\geq 0,\ \forall\tau\in[0,T] \text{ et } \forall\theta\in[\tau,T]$$

qui doit être comparée à celle obtenue par Skorodinskii [13]

$$\frac{\partial^2}{\partial u^2}\frac{d}{dt}\frac{\partial H}{\partial u}\Big|_{X=a} = 0,\quad \forall\tau\in[0,T].$$

BIBLIOGRAPHIE

[1] A.A. AGRACEV and R. GAMKRELIDZE. A second order optimality principle for a time optimal problem. Math. USSR Sbornik, 29, 1976, pp. 547-576.

[2] M. FLIESS. Fonctionnelles causales non linéaires et indéterminées non commutatives. Bull. Soc. Math. France, 109, 1981, pp. 3-40.

[3] M. FLIESS. On a possible connection between Volterra series and nonlinear optimal control. Proc. 7th Conf. Informat. Sci. Systems, pp. 402-407, Baltimore, 1983.

[4] M. FLIESS, M. LAMNABHI and F. LAMNABHI-LAGARRIGUE. An algebraic approach to nonlinear functional expansions. IEEE Trans. Circuits Systems, 30, 1983, pp. 554-570.

[5] R. GABASOV and F.M. KIRILLOVA. High order necessary conditions for optimality. SIAM J. Contr., 10, 1972, pp. 127-168.

[6] W. GRÖBNER. Die Lie-Reihen un ihre Anwendungen, (2nd edition), Berlin : VEB Deutscher Verlag der Wissons-chafter, 1967.

[7] H.J. KELLEY, R.E. KOPP and H.G. MOYER. Singular extremals. In Topics and Optimization, (G. Leitman ed.), Academic Press, New York, 1967.

[8] H.W. KNOBLOCH. Higher order necessary conditions in optimal control theory. Lect. Notes Contr. and Inf. Sc. n° 34, Springer-Verlag, Berlin, 1981.

[9] F. LAMNABHI-LAGARRIGUE. A Volterra series interpretation of some higher order conditions in optimal control, Proc. 1983 MTNS Beer-Sheva, to appear in Lect. Notes Contr. and Inf. Sc., Springer-Verlag, Berlin.

[10] C. LESIAK and A.J. KRENER. The existence and uniqueness of Volterra series for nonlinear systems, IEEE Trans. Automat. Contr., 23, 1978, pp. 1090-1095.

[11] C. MARCHALL and P. CONTENSOU. Singularities in optimization of deterministic dynamic systems. J. Guidance and Control, 4, 1980, pp. 240-252.

[12] W.J. RUGH. Nonlinear System Theory, Baltimore : The Johns Hopkins University Press, 1981.

[13] I.T. SKORODINSKII. Third variation of a functional on singular controls. Diff. Equations, 16, 1980, pp. 923-928.

[14] V.A. SROCHKO. Investigation of the second variation on singular controls. Diff. Equations, 10, 1974, pp. 809-822.

CONTROLLABILITY FOR POLYNOMIAL SYSTEMS

Dirk AEYELS

Dept of Systems Dynamics
State University of Gent
Grote Steenweg Noord 2
9710 Gent(Zwijnaarde)
België

ABSTRACT

Sufficient conditions for controllability of a class of nonlinear systems defined on $\mathbb{R}^2$ are presented. The approach can be extended to a class of systems defined on $\mathbb{R}^n$.

1. Introduction

Let $\dot{x} = f(x) + u.g(x)$, $x \in \mathbb{R}^n$, $u \in \mathbb{R}$, f and g smooth and $\mathbb{R}^n$-valued, be a nonlinear control system. At first we are interested in deriving sufficient conditions for local controllability at a point $x_o \in \mathbb{R}^n$. *Local controllability at* x_o means that for any $t > 0$, there exists a neighborhood of x_o that can be reached along the trajectories of the system in no more than t units of time. It is remarked that no a priori bounds are imposed on the control values.

The vectorfields f and g defining the system have the following restrictions : the drift term f is assumed to be polynomial on $\mathbb{R}^n$ and the vectorfield g is a constant vector b. *The derived results generalize the classical Kalman-controllability condition for linear systems.* Rather than exhibiting the theory on $\mathbb{R}^n$, we will be satisfied with illustrating our approach by means of systems defined on $\mathbb{R}^2$. *It is stressed that the theory is extendable to* $\mathbb{R}^n$. For the general result and for proofs we refer to a forthcoming paper [1] .

2. Controllability of $\dot{x} = f(x) + ub$, $u \in \mathbb{R}$, $x \in \mathbb{R}^2$, $b \in \mathbb{R}^2$

Consider the system $\dot{x} = f(x) + ub$, with the specifications given above. For the moment we will allow Dirac-impulses as inputs. Later on we will indicate how the results obtained, can also be derived when Dirac -impulses are not allowed. Dirac -impulses are being used because they have the property that, when $u = \alpha\delta(t)$ at $t = 0$ is applied to the differential equation above with initial condition $x(0^-) = x_o$, then $x(0^+) = x_o + b\alpha$.

Starting from x_o, Dirac-impulses with different intensities are applied, alternated with the control value $u = 0$. In other words, the trajectories corresponding to the constant vectorfield b and the vectorfield f(x) are being followed alternately. Since the flow corresponding to b is a translation on $\mathbb{R}^n$, the expression of x(t) is manageable - at least transparent enough so as to give rise to some original results.

Consider the system above, and apply $\alpha\delta(.)$ at time zero, then set $u = 0$ on the interval $(0,t)$ and finally apply $\beta\delta(.)$ at time t. Then $x(t) = \Phi(t,x_o+b\alpha) + b\beta$ when starting in x_o. Here $\Phi(t,x)$ is the flow corresponding to the vectorfield f which associates to x the point $\Phi(t,x)$ at time t.

A Taylor expansion of x(t) at t = 0 gives

$$x(t) = (x_o+b\alpha+t.\dot{\Phi}_{t=0}(t,x_o+b\alpha)+\frac{t^2}{2!}\ddot{\Phi}_{t=0}(t,x_o+b\alpha)+\ldots)+b\beta,$$

$$= x_o+b(\alpha+\beta)+t.f(x_o+b\alpha)+\frac{t^2}{2!}(\frac{\partial f}{\partial x}.f)(x_o+b\alpha)+\ldots$$

$$= x_o+b(\alpha+\beta)+t.f(x_o+b\alpha)+O(t^2),$$

where $O(t^2)$ is a term of order t^2, containing α but not β. Since f is polynomial, $f(x_o+b\alpha) = f(x_o) + \frac{\partial f}{\partial x}\Big|_{x=x_o}.b\alpha+ \frac{1}{2!}\frac{\partial}{\partial x}(\frac{\partial f}{\partial x}.b\alpha)\Big|_{x_o}.b\alpha+\ldots,$

for *all* $\alpha \in \mathbb{R}$ (the series reduces to a *finite sum!*), and therefore

$$x(t)=x_o+(\alpha+\beta)b+t(f(x_o)+\frac{\partial f}{\partial x}\Big|_{x_o}.b\alpha+\frac{1}{2!}\frac{\partial}{\partial x}(\frac{\partial f}{\partial x}.b)\Big|_{x_o}.b\alpha^2+\ldots)+O(t^2).$$

Let γ be such that $\beta = -\alpha+\gamma.t$ and consider the expression

$$t(\gamma b+(f(x_o)+\frac{\partial f}{\partial x}\Big|_{x_o}.b\alpha+\frac{1}{2!}\frac{\partial}{\partial x}\frac{\partial f}{\partial x}.b)\Big|_{x_o}.b\alpha^2+\ldots))$$

The coefficiënt of t in this expression defines a function F of two variables α,γ into $\mathbb{R}^2$.

Sufficient conditions for local controllability at x_o can now be derived as follows. *Suppose there exist real numbers* α^*, γ^* *such that* $F(\alpha^*,\gamma^*) = 0$ *and such that* $DF(\alpha^*,\gamma^*)$ *is invertible.* Then F is one to one in a neighborhood of α^*,γ^*, by the inverse function theorem. Also the function (tF) (with the obvious definition $(tF)(\alpha,\gamma):=t.F(\alpha,\gamma)$) maps a neighborhood of α^*,γ^* onto a neighborhood of the origin, and this is true for any $t > 0$. Therefore there exists a set C of real numbers α,β such that for all $t > 0$

$\{x_o+(\alpha+\beta)b+t(f(x_o)+\ldots),\alpha,\beta \in C\}$ contains a neighborhood of x_o. This implies controllability at x_o if the additional term $O(t^2)$ does not destroy the properties attached to the F-function so far described. We will not be explicit about this matter but one can show, by using perturbation theory of mappings, that *if $t > 0$ is taken small enough* local controllability at x_o is maintained under the conditions stipulated above.

3. Main result

The following theorem is an immediate consequence of the considerations above. Let d denote the highest degree appearing in the polynomials defining f.

Theorem. If d is odd, and $\frac{\partial}{\partial x}\left(\ldots \frac{\partial f}{\partial x}.b\right)\Big|_{x_o}.b$ (taking derivatives d times) is not parallel to b, then the system $\dot{x} = f(x) + bu$, $x \in \mathbb{R}^2$, $u \in \mathbb{R}$, is locally controllable at x_o.

For an anlysis of the case "d is even", one is referred to [1] .

4. Remarks

1. When Dirac-impulses are not allowed, the results derived above remain valid. This is done by replacing for example $\alpha\delta(.)$ at $t = 0$ by an input function equal to α/ε on the interval $(0;\varepsilon)$ and equal to zero elsewhere, and then letting ε go to zero. Again perturbation theory of functions can be invoked to show that, if ε is taken small enough the results on controllability derived above, remain true.
2. Notice that local controllability of each point $x \in \mathbb{R}^n$ implies global controllability. This will be discussed to a greater extent in [1] .
3. As to the relation of this theory to the existing literature we mention the following :
   i) the local controllability criterion reduces to the Kalman-criterion(as in the Lee-Markus [2] result for nonlinear systems) when x_o is an equilibrium point for the field f.
   ii) the major new characteristic seems to be that we are able to provide conditions for local controllability at points $x_o \in \mathbb{R}^n$ which are not equilibrium points. This should be compared to the Hermes-Sussmann result [3] where the set of attainability from x_o at time t is investigated when x_o is an equilibrium point.
   iii) It should be mentioned that we are able to incorporate multiple inputs (if the system is defined on $\mathbb{R}^n$). For this we also refer to [1] .

iv) In the sense indicated in ii) and iii) our result is stronger than the Hermes-Sussmann result. Our theory however seems to be heavily dependent on the constancy of the control vector b - a feature not present in the Hermes-Sussmann theory.

5. Bibliography

[1] D. Aeyels, Global Controllability for Polynomial Systems.

[2] E.B. Lee and L. Markus, Foundations of Optimal Control Theory, John Wiley, New York, 1967.

[3] H.J. Sussmann, Lie Brackets and Local Controllability : a Sufficient Condition for Scalar-Input Systems, SIAM J. Control and Optimization, Vol.21, No 5, Sept 83.

Session 19

COMPUTER AIDED CONTROL SYSTEM DESIGN II

CAO EN AUTOMATIQUE II

COMPUTER AIDED DESIGN OF CONTROL SYSTEMS

K J Åström

Department of Automatic Control
Lund Institute of Technology
S-220 07 Lund 7, Sweden

1. INTRODUCTION

Computer aided engineering is now finding extensive use in a wide range of disciplines like architecture, mechanical design, drafting, VLSI design, circuit board lay-out, solution of PDE system dynamics, and control system design. The techniques used in the different disciplines vary considerably depending on the applications. In some cases CAE is mostly graphical drafting, in other cases it is mostly numerics. There are however also many common elements in the different applications. This paper discusses some uses of computer aided engineering for design of control systems.

Solution to a control problem involves modeling, analysis, control law design and implementation. The rapid development of control theory in the fifties and sixties has been followed by computer tools for aiding the design work. The development has typically followed the evolution from algorithms and software libraries to interactive program packages for modeling, analysis and design. Several packages are now available. The functions of the packages depend critically on the available hardware which is currently in a stage of rapid development. In spite of several shortcomings of current packages they are undoubtedly very useful in terms of productivity gains.

The paper is organized as follows. Some characteristics of current packages are discussed in Section 2. This discussion naturally leads to desirable new features. The viewpoint that a CAE package may be regarded as a high level problem solving language is elaborated in Section 3. This gives a unified approach to design and implementation of CAE packages. Section 4 deals with data structures. It is suggested that considerable attention should be devoted to find appropriate ways to characterize a dynamical system which is the key concept in control systems. A few aspects on implementation issues are briefly covered in Section 5.

2. SOME CHARACTERISTICS OF CURRENT AND FUTURE DESIGN PACKAGES

The field of computer aided engineering of control systems is still in its early stages of development. An overview of some packages are found in Atherton (1981), Edgar (1981), Edmunds (1979), Furuta and Kajiwara (1979), Hashimoto and Takamatsu (1981), Lemmens and van den Boom (1979), Munro (1979), Rosenbrock (1974), Tyssoe (1981) and Wieslander (1979a,1979b) and Åström (1983). More references are also found in these papers. Special workshops and symposia devoted to CAD for control systems have been organized by IFAC, see Mansour (1979) and Leininger (1982), by GE-RPI, see Spang and Gerhart (1981), and by IEEE, see Herget and Laub (1982) and Strunce (1983). Computer aided tools are also popular in many other fields e.g. mechanical design and VLSI design. The seminal work on computer graphics by Newman and Sproull (1979) and the text Foley and van Dam (1982) contain much material and many rererences. The field is in a state of rapid development due to an increased understanding of the technology and the drastic development of computer and graphics hardware.

Some of the packages like Matrix-X, CNTRL-C and INTOPS are based on the package MATLAB, which was developed by Moler (1980) for matrix computations. Other packages are based on existing languages with interactive implementations like APL. Other packages like those developed in Lund (Åström 1983) and their predecessor RATTLES-DELIGHT are new language designs. Typical tasks for packages for control design are computation of frequency and time responses for linear systems, control systems design based on polynomial and matrix operations, simulation of linear and nonlinear systems, data analysis, spectral factorization, parameter estimation and optimization. The packages normally includes graphics and standard numerical packages like Eispack and Linpack. Few packages include formula manipulation and other forms of symbolic computations.

Computer hardware

Most current design packages were designed to run on main frames or minicomputers. A few of them are also running on personal computers. The personal computers which are projected to appear within a few years have specifications like: a primary memory of 2 Mbytes, a secondary memory 100 Mbytes, a computing speed of one megaflop/s and a price less than 20k$. See Dertouzos and Moses (1980). These computers are also expected to have a high resolution bit mapped color graphics display. With computers like this it is possible to have single user work-stations with packages which are much more sophisticated than all our current packages. The existence of computers like Apollo, Lisa, PERQ and Sun make the predictions quite credible.

The renaissance of graphics

Graphics has played a major role in engineering. The first books used in engineering education were books of drawings of machines by Leonardo da Vinci. Graphical representations have been used extensively ever since. Graphics in the forms of Bode diagrams, Nichols charts, root loci, block diagrams and signal flow diagrams are important tools in classical control theory. Modern control theory has however not been much influenced by graphics. This can partly be explained by lack of proper tools for graphics.

There has been a drastic development of the computer output devices. A teletype is capable of writing at a speed of 10 ch/s (110 Baud). A regular terminal connected to a 19.2 kBaud channel can write a screen i.e. 80 x 24 ch in a second. A good vector graphics terminal can refresh up to 100 000 long vectors or a million short vectors per second. A high resolution bit mapped display may refresh 512 x 512 pixel frames at rates of 60 frames/s (15 Mbit/s).

The input devices have unfortunately not developed at the same rate. We still have ordinary keyboards. A very good typist may type at a rate of 8 ch/s. A normal engineer types considerably slower. Pointing devices like roll balls, mouses and touch panels have been invented. These devices may perhaps be used to increase the input rate indirectly by combining the rapid output rate with feedback via the picking device (dynamic menus). Speech input is another possibility. There are however no indications of a more drastic increase in the input rate.

The possibility of using graphics will be improved considerably because of the emerging graphics standards like the Graphical Kernel System (GKS) or raster graphics extensions of SIGGRAPH Core. See Foley and van Dam (1982) and Anon.(1982).

The man-machine interface

A high bandwidth information transmission is required for an efficient man-machine communication. This implies a high rate of transmission of symbols and a high information content in each symbol. The graphics hardware, which is now becoming available, is fast and flexible. Individual picture elements may be changed instantaneously. It is possible to zoom, scroll and pan a picture. Color and animation add extra dimensions. Imaginative use of color graphics is still in its infancy in CAD packages for control systems. Interesting ideas have been proposed by Polak (1982) in connection with applications of optimization techniques. Animation has not been used much. It is clear that a lot can be learned from designers of video games. Interesting ideas on the use of graphics in this direction are demonstrated in Elmqvist (1982).

3. HIGH LEVEL PROBLEM SOLVING LANGUAGES

In this section it is argued that a CAE package may conveniently be viewed as an implementation of a high level problem solving language. To arrive at this we will start to discuss the user interface. It is important to realize that there is a wide range of users, from novices to experts, with different abilities and demands. For a novice who needs a lot of guidance it is natural to have a system where the computer has the initiative and the user is gently led towards a solution of his problem. For an expert user it is preferable to have a system where the user keeps the initiative and where he gets advice and and help on request only. Attempts of guidance and control by the computer can lead to frustration and inefficiency. It is highly desirable to design a system so that it will accomodate a wide range of users. This makes it more universal. It also makes it possible to gradually shift the initiative from the computer to the user as he becomes more proficient.

A menu driven dialog is a good solution for the novice user, in particular if good pop-up menus are combined with a good pointing device. A command driven dialog is more flexible for the experienced user. A command may have the following form

```
NAME LARG1 LARG2... ← RARG1 RARG2...
```

It has a name and it may also have left arguments and right arguments. The arguments may be numbers or names of objects in a data base. In packages developed in Lund the objects are implemented as files because this is a simple way to deal with objects having different types. This is illustrated by a few examples.

The command

```
INSI U 100
   >PRBS 4 7
   >EXIT
```

generates an input signal of length 100 called U. The command has options to generate several input signals. The options are selected by additional subcommands. PRBS is a subcommand which selects a PRBS signal. The optional arguments 4 and 7 indicate that the PRBS signal should change at most every fourth sampling period and that its period should be 2^7-1. The subcommand EXIT denotes the end of the subcommands.

The command

```
DETER Y ← SYST U
```

generates the response of the linear system called SYST to the input signal U.

The command

```
ML PAR ← DAT N
```

fits an ARMAX model of order N to the data in the file called DAT and stores the parameters in a file called PAR.

The command

```
OPTFB L CLSYS ← LOSS SYS
```

computes the optimal feedback gain L and the corresponding closed loop system CLSYS for the system SYS and the loss function LOSS.

Short form commands and default values

It is highly desirable to have simple commands in a command dialog. This is in conflict with the requirement that commands should be explicit and that it may sometimes be desirable to have variants of the commands. These opposite requirements may be resolved by allowing short forms of the commands. The standard form for the simulation command is SIMU. If no other command starts with the letter S it is, however, sufficient to type S alone. It may also be useful to have a simple way of renaming the commands.

A similar mechanism may be used for commands which use arguments by introducing a default mechanism so that previous values of the arguments are used unless new values are specified explicitly.

Problem solving languages

With a menu driven dialog the choices available to the user at each situation are those which were anticipated by the system designer. A command driven dialog is more flexible because the user can apply the system in ways that were not anticipated by the designer of the system. The two approaches can be combined by having a command driven dialog as the base line design and to have facilities in the command dialog to generate dynamic menus. The inclusion of a rule based expert system offers interesting possibilities. See Barr and Feigenbaum (1982). Ideas in this direction have been persued by Gale and Pregibon (1983) who have tried to construct an expert interface called REX (Regression EXpert) to a program for regression analysis.

The design of a command driven package may be viewed as a construction of a high level language for solving control problems. From this viewpoint design of a CAD system involves the selection of vocabulary, grammar and semantics. The vocabulary defines the basic language elements i.e. the data structures and the operators. The grammar tells how the basic language elements may be combined into new language elements and the semantics tells how the language elements should be interpreted. The language should be rich enough to solve many problems. It should also be simple so that it is easily learned. A CAD program is simply an interpreter for the design language.

The commands are part of the grammar, they determine how useful a package is and how easy it is to learn. The commands should be complete in the sense to cover a wide range of techniques. Otherwise the designer will only try those approaches for which commands are available. Commands should also have a considerable expression power so that a control system designer can do what he wants with a few commands. The commands should also reflect the natural concepts from a theoretical point of view. This would make it easy for a user well versed in control theory to use a package. The commands should also be few and simple so that they are easy to learn and remember. This is of course in conflict with requirements on completeness and expression power. Selection of commands is thus a good exercise in engineering design.

Macros

The commands are normally read from a terminal in a command driven system. It is, however, useful to have the option of reading a sequence of commands from a file in storage instead. Since this is analogous to a macro facility in an ordinary programming language the same nomenclature is adopted. The construction

```
MACRO NAME
    Command 1
    Command 2
    Command 3
END
```

thus indicates that the Command 1, Command 2 and Command 3 are not executed but stored in memory. The command sequence is then activated simply by typing NAME.

Macros are convenient for simplification of a dialog. Command sequences that are commonly used may be defined as macros. A simple macro call will then activate a whole sequence of commands. The macro facility is also useful in order to generate new commands. Macros may also be used to rename commands. This is useful in order to tailor a system to the needs of a particular user. The usefulness of macros may be extended considerably by introducing commands to control the program flow in a macro, facilities for handling local and global variables and by allowing macros to have arguments. By having commands for reading the keyboard and for writing on the terminal it is also possible to implement menu driven dialogs using macros.

An interactive CAD program based on a command dialog with a macro-facility may be viewed as an extendable high level problem solving language. A set of basic commands which correspond to the elements of the theory and which allow coverage of a certain problem area are first determined. Simplifications and extensions are then generated using the macro facility. Error checking is an important aspect at interactive computing. Realising that a CAD system may be viewed as a high level program solving language the writing of a package simply becomes construction of an interpreter for the language.

4. DATA STRUCTURES

Accepting the viewpoint that a CAE program may be viewed as a high level problem solving language the design of a vocabulary is one of the key issues. A wide range of data structures are required to deal naturally with control problems. Apart from common mathematical objects like integers, real and complex numbers, it is desirable to have polynomials, rational functions, matrices and matrix fractions. Such objects are conveniently viewed as abstract data types. They can be conveniently handled as packages in Ada or similar constructs in other modern computer languages. The overloading facility in Ada offers a convenient way to use natural notations. There is also a need to have signals and systems. Signals are conveniently represented as arrays. When working with experimental data there is also a need to tag the data with verbal information about experimental conditions. This can be handled by making signal a record where the pure signal is an array which is part of a record.

Descriptions of control systems problems require flexible data structures. Many problems may be characterized in terms of arrays only. Arrays will go a long way to describe linear systems in state space form and to describe signals. Many problems can be solved using a matrix language like MATLAB, Moler (1980) and one of its extension Matrix_X, Walker et al. (1982). It is, however, clear, that it is not sufficient to only have matrices.

For simple systems with only one data type, like matrices, all data may be stored in a stack or in a simple array. A more sophisticated data structure was used in the Lund packages. Our experiences indicate that it would be very useful to have a more flexible system. It is probably a good idea to build a system around some general database system. The need for multiple descriptions of a system is one special problem which is conveniently solved using databases. A typical example is when a system is represented both as a transfer function and as a state equation. Small systems are not much of a problem because it is easy to transform from one form to another. Such computations may however be extensive for large systems. To obtain a reasonable efficiency it is then necessary to store the different descriptions. It may also be desirable to have models of different complexity for the same physical object as well as linearized models for different operating conditions. Since it is very difficult to visualize all possible combinations a priori it is a useful to have a database system which admits modifications of the structure of the data.

System descriptions

Since dynamical systems is a fundamental notion, its representation becomes a key issue. Many different representations of systems are used in control theory. The ordinary differential equation model

$$\begin{cases} \dfrac{dx}{dt} = f(x,u,t) \\ y = g(x,u,t) \end{cases} \tag{1}$$

where x is the state vector, u the input vector and y the output vector, is a common case. Often the fundamental form of the equations is not (1), where the derivative is solved explicitly but rather

$$\begin{cases} F\left[\dfrac{dx}{dt}, x, u, t\right] = 0 \\ G(x,y,u,t) = 0 \end{cases} \tag{2}$$

The following discussion is restricted to systems of type (1). Other issues arise when operating with models of type (2). This is treated in depth in Elmqvist (1978, 1979a, 1979b). Partial differential equations and differential equations with time delay are also common. In this paper the discussion is, however, restricted to differential equation models. Linear systems where the functions f and g take the form

$$\begin{cases} f = A(t)x + B(t)u \\ g = C(t)x + D(t)u \end{cases} \tag{3}$$

is an important special case. For linear systems it is also possible to use other representations like input-output models of the form

$$\frac{d^n y}{dt^n} + A_1 \frac{d^{n-1} y}{dt^{n-1}} + \ldots + A_n y = B_1 \frac{d^{n-1} y}{dt} + B_2 \frac{d^{n-2} y}{dt^{n-2}} + \ldots + B_n u \tag{4}$$

which also can be represented by the matrix fraction

$$G(s) = A^{-1}(s)\ B(s) \tag{5}$$

where A and B are the polynomials

$$\begin{cases} A(s) = s^n + A_1 s^{n-1} + \ldots + A_n \\ B(s) = B_1 s^{n-1} + B_2 s^{n-2} + \ldots + B_n \end{cases}$$

The discrete time versions of (1), (2), (3), (4) and (5) are also needed. It is a key issue to find suitable computer representations of systems. Some operations to be performed on systems will be discussed before treating this.

System interconnections

Interconnections of systems is a fundamental issue. The elementary connections of systems are series, parallel and feedback connections. They can be represented graphically as is shown in Fig. 1 or algebraically as

$$S_p = S_A + S_B$$ "Parallel connection

$$S_c = S_A \cdot S_B$$ "Series connection

$$S_c = [I+S_B \cdot S_A]^{-1} S_A = [I+S_B \cdot S_A] \setminus S_A$$ "Feedback connection

where "\" is the notation introduced in MATLAB to denote the solution X = A\B of the linear equation

AX = B.

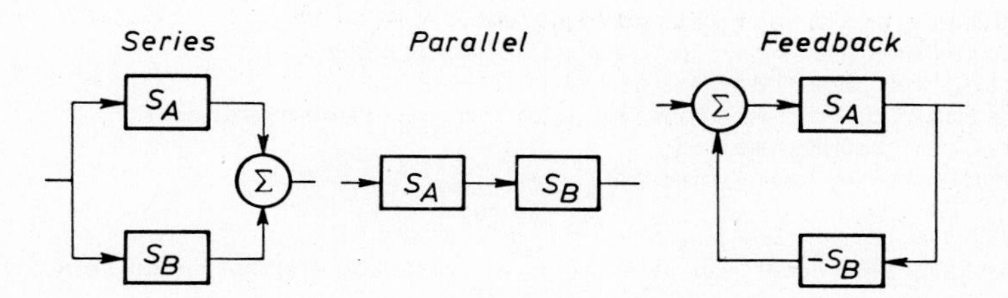

Fig. 1 - The basic system interconnection.

For more complex systems it is desirable to have appropriate notations for interconnected hierarchical systems. These notations should be such that details of the subsystem can be hidden and that signals and variables at the lower levels can be accessed in a well controlled fashion.

The system description introduced by Elmqvist (1977) in the simulation language Simnon has been very easy to operate with and very easy to teach. Elmqvist introduced the classes of continuous and discrete time systems defined as follows.

```
CONTINUOUS SYSTEM    <system identifier>
[INPUT    <simple variable>*]
[OUTPUT   <simple variable>*]
[STATE    <simple variable>*]
[DER      <simple variable>*]
[TIME     <simple variable>*]

[INITIAL                                           ]
[Computation of initial values for state variables]

[Computation of auxiliary variables]
[Computation of output variables]
[Computation of derivatives]
[Parameter assignment]
[Initial value assignment]
END
```

```
DISCRETE SYSTEM  <system identifier>
[INPUT   <simple variable>*]
[OUTPUT  <simple variable>*]
[STATE   <simple variable>*]
[NEW     <simple variable>*]
[TIME    <simple variable>*]
TSAMP    <simple variable>*]

[INITIAL                                                    ]
[[Computation of initial values for state variables]        ]
[[Computation of initial values for output variables]       ]
[[Computation of initial values for the TSAMP-variable]     ]
[SORT                                                       ]

[Computation of auxiliary variables]
[Computation of output variables]
[Computation of new values of the states]
Updating of the TSAMP-variable
[Modification of states in continuous subsystems]
[Parameter assignment]
[Initial value assignment]
END
```

where the standard BNF notations of <...> as a syntactic element, * as repetition, and [ ] an optimal element are used. Elmqvist allowed connections of systems at one level using connecting systems defined as

```
CONNECTING SYSTEM  <system identifier>
[TIME   <simple variable>]
[Computation of auxiliary variables]
[Computation of input variables]
[Parameter assignment]
END
```

The variables in each system description are local. The notation <variable name>[<system name>] is used in the interconnecting system and at the interaction level to separate variables with the same names.

Elmqvists notation is very natural. Long experience of using it has shown that it is very easy to teach and use. It therefore seems attractive to make marginal extensions of it to allow hierarchical interconnections with controlled access to parameters and variables.

Hierarchical system connections can be obtained by adding definitions of inputs and outputs in the connecting system and adding the new language element INCLUDE to give the names of the subsystems that are connected. The syntax of the connecting system thus becomes

```
CONNECTING SYSTEM  <system identifier>
[INPUT     <variable>*]
[OUTPUT    <variable>*]
[INCLUDE   <system identifier>*]
[TIME      <simple variable>]
[Computation of input variables]
[Parameter assignment]
END
```

The variables in Simnon have only one type reals. This should be extended to allow other datatypes like arrays, matrices and polynomials. When using hierarchically connected systems it is also useful to replace the notation for referencing variables to the common dot notation. A variable would thus be referenced as

```
<system name>.<system name>...<variable name>
```

Controlled access to variables and parameters can be obtained by introducing a heading

```
EXPORT  <variable>*
```

to list those variables which are exported up to the next level. In many teaching situations it is desirable to have access to all variables. This can be achieved by using EXPORT ALL or some similar construction.

Since system interconnections are often visualized graphically there should be facilities for representing and manipulating system interconnections graphically as well as textually. Interesting ideas in this direction have been proposed by Elmqvist (1982).

It would also be desirable to have the notion of system type and facilities for creating instances of the type. This would give a simple way of generating special classes of systems. Linear systems can then be defined as

```
type LINEAR_STATE_SPACE_SYSTEM
INPUT  u: vector
OUTPUT y: vector
STATE  x: vector
DERIVATIVE  dx: vector
A B C D : matrix
y  = C*x + D*y
dx = A*x + B*u
END
```

A similar construction can be used for linear polynomial systems. Instance of linear systems can then be created by

```
S: LINEAR SYSTEM
```

The parameters can be accessed as

```
S.A = matrix (1 2 ; 3 4)
```

It is a nontrivial design issue to decide when and how dimension compatibility should be checked. This has to do with how arrays are implemented. From the user point of view it would, however, be desirable to define a linear system as was done above without a need for specifying the dimensions.

In some cases it is also desirable to be able to hide a system description so that a user of the system can only make operations like simulation. An example from teaching is in courses on system identification, where it is desirable for students to find the properties of an unknown system, or in courses on adaptive control, when it is desirable to check that an algorithm works on an unknown system. The possibility to hide details of a system description would also be a possibility to get controlled access to industrial models. This can be achieved by using the mechanisms introduced in Ada, where the declarations and a body of a procedure are separated. See DOD (1983).

System operations

Apart from interconnections there are many other operations that are desirable to perform on systems, e.g. computation of equilibrium values, simulation, linearization, system inversion. For linear systems it is also natural to be able to transform coordinates, compute poles and zeros, determine observability and controllability, and perform Kalman decomposition. Some of these operations are conveniently done numerically. Others require formula manipulation.

5. IMPLEMENTATION ISSUES

A CAE package represents a substantial software development. Looking at the details there are needs for good subroutine packages like Eispack, Garbow et al. (1977) and Smith et al. (1976), and Linpack, Dongarra et al. (1979), which are now available in the public domain. Similar packages for the numerical calculations required for analysis and design of control systems are also needed. The numerical problems that arise in automatic control are however starting to receive attention from numerical analysts. See van Doren (1981), Hammarling (1982) and Laub (1980). Packages for graphics are also needed.

Most data processing in current packages is inspired from numerical analysis. The powers of non-numeric data processing have not been exploited. It would be highly desirable to have facilities for symbolic manipulation. This can e.g. be used for model simplification, generation of code for computing equilibrium points, generation of simulation code, linearization, etc. If symbolic manipulations are included it is also possible to generate code for realization of the control laws. Symbolic calculations are likely to be included in future packages.

Since the basic subroutine libraries are likely to be written in Fortran and programs for symbolic manipulation are written in Lisp, it is likely that future systems will include code written in several languages.

6. CONCLUSIONS

Interactive computing is a powerful tool for problem solving. An engineer can come to the work station with a problem and he can leave with a complete solution after a few hours. The results are well documented in terms of listings, text and graphs. The problem solver can obtain the solution by himself without relying on programmers as intermediaries. Our projects have shown that the productivity in analysing and designing control systems can be increased substantially by using these tools. We believe that interactive computer aided design tools is one possibility to make modern control theory cost effective.

Computer aided design of control systems is still in its infancy. A small number of systems have been implemented in a few places. There are many possible future developments which are mainly driven by the computer development. Packages of the type we have been experimenting with can easily be fitted into the personal computers or work stations that will be available in a few years time. The bit mapped high resolution color displays that will be available on these computers offer new possibilities for an efficient man-machine dialog. With the drastic increase in computer capacity, that is forth coming, it is also possible to make much more ambitious projects. Applications of computer aided design also appear in many other branches of engineering. Cross fertilization between the fields will most likely lead to a rapid development.

7. REFERENCES

Anonymous (1982): Graphical Kernel System (GKS) - Functional Description. Draft International Standard ISO/DIS 7942 Version 7.02, August 9, 1982. Available through American National Standards Institute Inc. New York, N.Y.

Åström, K.J. (1983): Computer aided modeling, analysis and design of control systems - A perspective. IEEE Control Systems Magazine 3:2, 4-16.

Atherton, D.P.(1981): The role of CAD in education and research. IFAC Congress VIII, Kyoto, Japan.

Barr, A. and E.A. Feigenbaum (1982): The Handbook of Artificial Intelligence. Vol II. W. Kaufmann Inc. Los Altos, Calif.

Dertouzos, M.L. and J. Moses (1980): The Computer Age: A twenty year view. MIT Press Cambridge, Mass.

DOD (1983): Reference Manual for the Ada Programming Language. ANSI/MIL-STD-1815A, United States Department of Defense, Washington, D.C.

Dongarra, J.J., C.B. Moler, J.R. Bunch, and G.W. Stewart (1979): LINPACK - Users' guide. SIAM, Philadelphia.

Edgar, T.F. (1981): New results and the status of computer-aided process control system design in North America. Engineering Foundation Conference on Chemical Process Control-II, Sea Island, Georgia.

Edmunds, J.M. (1979): Cambridge linear analysis and design programs. IFAC Symposium on Computer Aided Design of Control Systems, Zurich, 253-258.

Elmqvist, H. (1977): SIMNON - An Interactive Simulation Program for Nonlinear Systems. Simulation '77, Montreux, Switzerland, June 1977.

Elmqvist, H. (1978): A Structured Model Language for Large Continuous Systems. Ph.D. Thesis. Dept of Automatic Control, Lund Institute of Technology, Lund, Sweden, Report CODEN: LUTFD2/(TFRT-1015)/1-226/(1978).

Elmqvist, H. (1979a): Dymola - A Structured Model Language for Large Continuous Systems. Summer Computer Simulation Conference, Toronto, Canada, July 1979.

Elmqvist, H. (1979b): Manipulation of Continuous Models Based on Equations to Assignment Statements. Simulation of Systems '79. Sorrento, Italy, September 1979.

Elmqvist, H. (1982): A graphical approach to documentation and implementation of control systems. Proc. 3rd IFAC/IFIP Symposium on Software for Compurter Control, SOCOCO 82. Madrid, Spain.

Foley, J.D. and A. van Dam (1982): Fundamentals of interactive computer graphics. Addison Wesley, Reading, Mass.

Furuta, K. and H. Kajiwara (1979): CAD system for control system design. J of the Society of Instrument and Control Engineers, Japan, 18 (9). (In Japanese).

Gale, W.A. and D. Pregiborn (1983): Using expert systems for developing statistical strategy. Proc. Joint Statistical Meetings, Toronto, Canada.

Garbow, B.S., et al. (1977): Matrix eigensystem routines - Eispack Guide Extension. Lecture Notes in Computer Science, Vol. 51, Springer-Verlag, New York.

Hammarling, S. (1982): Some notes on the use of the orthogonal similarity transformations in control. NPL Report DITC.

Hashimoto, I. and Y. Takamatsu (1981): New results and the status of computer aided process control systems design in Japan. Engineering Foundation Conference on Chemical Process Control-II, Sea Island, Georgia.

Herget, C.J. and A.J. Laub (Eds.)(1982): Proc IEEE CSS Workshop on Computer Aided Control System Design. Berkeley, Calif. IEEE Control Systems Magazine 2:4. Special Issue on Computer-Aided Design of Control Systems.

Laub, A.J. (1980): Survey of computational methods in control theory. In A.M. Erisman et al. (Eds.), Electric Power Problems. The mathematical challenge, SIAM, Philadelphia, pp 231-260.

Leininger, G. (Ed.)(1982): Computer aided design of multivariable technological systems. Preprints second IFAC symposium on Computer Aided Design of Multivariable Technological systems. West Lafayette, Indiana, USA.

Lemmens, W.J.M. and A.J.W. Van den Boom (1979): Interactive computer programs for education and research: a survey. Automatica 15, 113-121.

Mansour, M. (Ed.)(1979): Preprints first IFAC Symposium on CAD of Control systems. Zurich. Pergamon.

Moler, C. (1980): Matlab users' guide. Report Department of Computer Science, University of New Mexico.

Munro, N. (1979): The UMIST control system design and synthesis suites. IFAC Symposium on Computer Aided Design of Control Systems, Zurich, 343-348.

Newman, W.M. and R.F. Sproull (1979): Principles of interactive computer graphics. McGraw-Hill, New York.

Polak, E. (1981): Optimization-based computer-aided-design of control systems. Proc JACC. University of Virginia.

Rosenbrock, H.H. (1974): Computer-aided control system design. Academic Press, New York.

Smith, B.T. et al. (1976): Matrix eigensystem routines - Eispack guide. 2nd ed., Lecture Notes in Computer Science, Vol. 6, Springer-Verlag, New York.

Spang, H.A., III, and L. Gerhart (Eds.) (1981): Preprints GE-RPI, Workshop on control design. Schenectady, N.Y.

Strunce, R. (Ed.)(1983): Preprints CACSD '83 IEEE Control Systems Society Symposium on Computer-Aided Control System Design, Cambridge, Mass., September 28-30.

Tyssoe, A. (1981): New results and the status of computer aided process control systems design in Europe. Engineering Foundation Conference on Chemical Process Control-II, Sea Island, Georgia.

Van Doren, P. (1981): A generalized eigenvalue approach for solving Riccati equations. SIAM J Sci. Stat. Comput. 2, 121-135.

Walker, R., C. Gregory, and S. Shah (1982): $Matrix_X$ - A data analysis, system identification, control design and simulation package. IEEE Control Systems Magazine 2:4, 30-37.

Wieslander, J. (1979a): Interaction in computer aided analysis and design of control systems. PhD thesis, Dept of Automatic Control, Lund Institute of Technology, Lund, Sweden, Report CODEN: LUTFD2/(TFRT-1019)/1-222/(1979).

Wieslander, J. (1979b): Design principles for computer aided design software. Preprints, IFAC Symposium on CAD of Control Systems, Zurich, 493.

TOWARDS AN EXPERT SYSTEM IN STOCHASTIC CONTROL :

THE HAMILTON-JACOBI EQUATION PART

C. GOMEZ - J.P. QUADRAT - A. SULEM

INRIA
Domaine de Voluceau
BP 105 - Rocquencourt
78153 Le Chesnay Cédex (FRANCE)

I - INTRODUCTION

Stochastic control problemscan be solved completely or approximatively by different kinds of approaches :

- dynamic programming,
- decoupling technique,
- stochastic gradient,
- perturbation method.

These methods are described in Goursat-Quadrat [G]. We are designing an expert system which will be able to manage all this approaches to solve a particular application.

In this context a set of tools to solve the dynamic programming equation at a numerical or theoretical level is discussed here.

In the first part a class of dynamic programming equations which can be solved automatically by a generator of fortran program is described.

In the second part a program which is able to make some reasoning on Partial-differential equation in general and on Hamilton-Jacobi equation in particular is described.

II - A DYNAMIC PROGRAMMING SOLVER

We would like the system to be able to solve a large class of stochastic control problems of diffusion type. The generality is associated to the structure of the state and control space. Here we discuss only a general admissible set of controls, which may be a union of sets which are cartesian product of intervals. This generality is necessary if we want the same system to be able to manage continuous control, finite set of controls,stopping-time problem and every mixing of them.

To define precisely the structure of this kind of set we introduce the :

a) Notations

Let $E = \{\ell_0, \ell_1, \ldots, \ell_m\}$ be a finite set where ℓ_i denotes a set of $\mathbb{R}$. In the following $m \leq 2$, ℓ_0 denotes $\mathbb{R}$ and ℓ_1 the interval $[0,1]$.

E^* is the commutative monoïd obtained by commutative concatenation elements of E^*. The words w of E are $\prod_{i=1}^{m} \ell_i^{n_i}$ where $\forall i = 1, \ldots, m$, $n_i \in \mathbb{N}$. The degree of $w \in E^*$ denoted $|w|$ is $\sum_{i=1}^{m} m_i$. The neutral element for the concatenation is denoted 1. The product ww' of $w = \prod_{i=1}^{m} \ell_i^{n_i}$ and $w' = \prod_{i=1}^{m'} \ell_i^{n'_i}$ is $\prod_{i=1}^{m} \ell_i^{n_i+n'_i}$.

$\mathbb{N}\langle E\rangle$ denotes the set of polynomials with integer constants and variables belonging to E. $p \in \mathbb{N}\langle E\rangle$ is written :

$$(1.1) \qquad p = \sum_{w\in E^*} (p,w)w$$

where $(p,w) \in \mathbb{N}$. p has a finite number of non zero coefficients. p is the notation for the name of the set :

$$(1.2) \qquad \sum_{w\in E^*} \sum_{i=1}^{(p,w)} \prod_{i=1}^{m} \ell_i^{m_i}$$

where Σ is the union of sets. Each connected subset of p is completely defined by the couple (w,j) where $w \in E^*$, $1 < j \leq (p,w)$. Thus (p,w,j) denotes the jth connected subset of structure w of the set p.

b) The stochastic control problem

Defined on some probability space (Ω, F_t, F, p) we consider the controlled diffusion processes :

$$(1.3) \qquad dX_t = b(X_t, S(X_t))dt + \sigma(X_t, S(X_t))dB_t + \eta(X_t)d\xi_t$$

where

t denotes the time and belongs to ℓ_1 ;

X_t denotes the state and belongs to $w = \ell_1^n$;

$p \in \mathbb{N}\langle E\rangle$ is the admissible set of controls ;

$S : w_x \to p_u$ denotes a feedback ;

$b : wp \to w$ is the drift term ;
$(x,u) \; b(x,u)$

$\sigma : wp \to \mathcal{L}(w,w)$ is the diffusion term where
$(x,u) \quad \sigma(x,u)$

$\mathcal{L}(w,w)$ denotes the set of matrices of dimension $|w|$;

$B_t \in \ell_1^n$ is a n dimensional independent brownian motion ;

Γ is the boundary of w ;

$\eta : \Gamma \to w$ the inside boundary normal of w ;

ξ_t is an increasing process strictly increasing only when X_t belongs to Γ.

Under some hypotheses, that we do not precise here, the processes X_t and ξ_t are well defined, see for example A.Bensoussan [B], P.L.Lions [L1] and all their references.

Given an open set $\mathcal{O} \supset w$ such that the boundary of $\mathcal{O} \cap w$ is included in Γ, and a finite family $\mathcal{Q}$ of open sets included in $\mathcal{O}$ we define the stopping time τ and the stochastic index i by $(\tau(\omega), i(\omega)) = \arg\min_{\substack{t \in \mathbb{R}^+ \\ q \in \mathcal{Q} \text{ or } \mathcal{O}}} (X_t(\omega) \notin q)$, then denoting :

$c : wp \to \mathbb{R}^+$ an instantaneous cost ;
$xu \quad c(x,u)$

$f : w \times \mathcal{Q} \to \mathbb{R}^+$ a final cost function ;
$w \; q \quad f(x,q)$

$\lambda : wp \to \mathbb{R}^+$ a discount factor ;
$xu \quad \lambda(x,u)$

we want to solve the stochastic control problem :

$$V(x) = \operatorname*{Min}_{\substack{S \\ \mathcal{Q}}} \mathbb{E} \int_0^\tau e^{-\int_0^t \lambda(X_s, u_s)ds} c(X_t, u_t)dt + e^{-\int_0^\tau \lambda(X_s, u_s)ds} f(X_\tau, i)$$

<u>Remark</u>

In this formulation of the problem the dependency of the defined functions with the time is not explicit, but can be introduced at least at a theoretic level by adding a new state y of dynamic $\dot{y} = 1$. Then, for example, defining $\mathcal{O} = \{y | y<T\}$, $\mathcal{Q} = \{\emptyset\}$ the classical finite horizon stochastic control problem :

$$\operatorname*{Min}_{S} \mathbb{E} \int_0^T c(s, X_s, S(s, X_s))ds$$

enters in this formalism.

c) <u>A class of Hamilton-Jacobi equation</u>

The dynamic programming methods give the optimality conditions that satisfy the function V, that is the Hamilton-Jacobi equation. The class of Hamilton-Jacobi equation associated to our stochastic control problem is defined by a grammar. To define precisely this grammar we use the Backus -Naur notation. The grammar is the following:

<stochastic-control-problem> ::= <domain>, <inside-condition>, <boundary-conditions>

<domain> ::= $[0,1]^n$ | $[0,1]^n \times [0,T]$

<boundary-conditions> ::= Σ <boundary-condition>, <boundary-element>
<boundary-element>

<boundary-element> ::= $\{\langle y\rangle : \langle y\rangle \in \langle domain\rangle, \langle x_i\rangle = 1\}$ | $\{\langle y\rangle, \langle y\rangle \in \langle domain\rangle, \langle x_i\rangle = 0\}$ | $\{(x,1),\ x \in [0,1]^n\}$

$\langle x_i\rangle$ ::= $x_1 | x_2 | \dots | x_n$

<y> ::= x | (x,t)

<boundary-condition> ::= V=f | $\frac{\partial V}{\partial \vec{n}} = f$

<operator> ::= <evolution-operator> + <space-operator>

<evolution-operator> ::= ␣ | $\frac{\partial V}{\partial t}$

<operators> ::= <operator> | <operator> + <operators> | Min(<operator>, <operators>)

<inside-condition> ::= <operators> = 0

<space-operator> ::= $-\lambda(\langle y\rangle)\ V + \sum_{i=1}^{n} b_i(\langle y\rangle) \frac{\partial}{\partial x_i} V + \frac{\partial^2}{\partial x_i^2} V + C(\langle y\rangle)$ |

$$\min_{u \in \mathbb{R}^m} \Big[-\lambda(\langle y\rangle,u)V + \sum_{i=1}^{n} \Big(b_i(\langle y\rangle,u) \frac{\partial}{\partial x_i} V + a_i(\langle y\rangle,u) \frac{\partial^2}{\partial x_i^2} V\Big) + C(\langle y\rangle,u)\Big]$$

Moreover we have semantic constraints which define a sublanguage of this grammar. These constraints impose an appropriate use of <y>. We do not discuss this point here.

Examples of Hamilton-Jacobi equations described by this grammar :

$$\begin{cases} -\lambda V + \mathrm{Min}\Big(b_1(x)\frac{\partial V}{\partial x} + C_1(x),\ b_2(x)\frac{\partial V}{\partial x} + C_2(x)\Big) + \frac{\partial^2 V}{\partial x^2} = 0 \ \text{for}\ x \in\ ]0,1[\\ V(0) = V(1) = f(x) \end{cases}$$

$$\begin{cases} \frac{\partial V}{\partial t} + \min_{u \in \mathbb{R}} \Big[b_1(t,x,u)\frac{\partial V}{\partial x_1} + b_2(t,x,u)\frac{\partial V}{\partial x_2} + C(t,x,u)\Big] + \Delta V = 0 \ \text{for}\ x \in\ ]0,1[\ \times\ ]0,1[\\ V = f_1 \ \text{for}\ x_1 = 0 \\ V = f_2 \ \text{for}\ x_1 = 1 \\ \frac{\partial V}{\partial x_2} = g_1 \ \text{for}\ x_2 = 0 \\ \frac{\partial V}{\partial x_2} = g_2 \ \text{for}\ x_2 = 1 \end{cases}$$

$$\begin{cases} \text{Min } \{-\lambda V + \underset{u \in \mathbb{R}}{\text{Min}} \ (b(x,u) \dfrac{\partial V}{\partial x} + C(x,u)), \ f-V\} = 0 \text{ for } x \in \ ]0,1[\\ V = f_1 \text{ for } x = 0 \qquad\qquad \dfrac{\partial V}{\partial x} = f_2 \text{ for } x = 1 \end{cases}$$

For each partial differential equation accepted by this grammar we generate a fortran program which solves it numerically after discretization. The program generator is written in MACSYMA. Indeed we need some algebraïc manipulations (derivation, simplification) to generate it.

Moreover a semi-natural language interface based on a program given in Queinnec [Q] facilitates the specification of the problem.

d) Example

An example of a session is given in the following. The prompt character is "==>", thus after this symbol are written the input given by the user.

The user defines the statement of his stochastic control problem in semi-natural language. When the problem is well posed, the robot can generate the associate Dynamic Programming equation and the list calling the fortran program generator for numerical solution.

```
(c1) robot();
hello
we shall try to solve your stochastic control problem
please give the statement of the problem

==>  let [s1,s2] be a state variable denoting a water-level
state : [s1, s2]
physical meaning of [s1, s2] : water-level

==>  s1 belongs to [0,1]
domain of s1 : [0, 1]

==>  s2 belongs to [0,1]
domain of s2 : [0, 1]

==>  [v1,v2] is a control variable denoting the turbined-water
control : [v1, v2]
physical meaning of [v1, v2] : turbined-water

==>  v1 belongs to [0,1]
domain of v1 : [0, 1]

==>  v2 belongs to [0,1]
domain of v2 : [0, 1]

==>  d is a drift term equal to v1
drift : d
value of d : v1
the dimensions of the state and the drift must be the same

==>  delete d
drift : d
value of d : v1
deleted

==>  d is a drift term equal to [v1,v2]
drift : d
value of d : [v1, v2]
```

```
==>  l is a diffusion term equal to [1,2]
diffusion : l
value of l : [1, 2]

==>  what do you know
diffusion : l
value of l : [1, 2]

drift : d
value of d : [v1, v2]

control : [v1, v2]

physical meaning of [v1, v2] : turbined-water

state : [s1, s2]
physical meaning of [s1, s2] : water-level

state : [s1, s2]
physical meaning of [s1, s2] : water-level

==>  please cancel l
diffusion : l
value of l : [1, 2]
deleted

==>  l is a diffusion diagonal matrix equal to [1,1]
diffusion : l
value of l : [1, 1]

==>  the problem consists in the minimization of a cost function
the problem leads to a hamilton-jacobi equation

==>  f is the instantaneous-cost equal to v1**2+v2**2+
     s1*(1-s1)*s2*(1-s2)+((0.5-s1)*s2*(1-s2))**2+
     ((0.5-s2)*s1*(1-s1))**2+2*s2 *(1-s2)+2*s1*(1-s1)
instantaneous-cost : f
                2        2                2         2    2
value of f : v2   + v1   + (0.5 - s1)  (1 - s2)  s2  +

                                                      2    2          2
(1 - s1) s1 (1 - s2) s2 + 2 (1 - s2) s2 + (1 - s1)  s1  (0.5 - s2)  +

2 (1 - s1) s1

==>  what is d ?
drift : d
value of d : [v1, v2]

==>  for s1 = 0 the limit condition is of the dirichlet type and equal
                                                              to s1*s2

limit condition for s1 = 0 : s1 s2
 dirichlet type
domain of s1 : [0, 1]

==>  s2 = 1 stopping (1-s1)*(1-s2)
limit condition for s2 = 1 : (1 - s1) (1 - s2)
 dirichlet type .
domain of s2 : [0, 1]

==>  s2 = 0 dirichlet s1*s2
limit condition for s2 = 1 : (1 - s1) (1 - s2)
 dirichlet type .
limit condition for s2 = 0 : s1 s2
 dirichlet type
domain of s2 : [0, 1]

==>  stop
you must precise if the problem is static or parabolic
a limit condition is missing for s1 = 1

do you still wish to stop ?
no

==>  static
the problem is static
```

```
==>  s1 = 1 dirichlet (1-s1)*(1-s2)
limit condition for s1 = 1 : (1 - s1) (1 - s2)
 dirichlet type .
limit condition for s1 = 0 : s1 s2
 dirichlet type
domain of s1 : [0, 1]

==>  stop
a discount factor is missing

do you still wish to stop ?
no

==>  let a be a discount factor equal to 1
discount : a
value of a : 1

==>  what is the dynamic programming equation ?
Hamilton-Jacobi equation :
                dv        dv        2      2
- v + min (v1 --- + v2 --- + v2  + v1  +
               ds1       ds2

           2          2   2
 (0.5 - s1)  (1 - s2)  s2  + (1 - s1) s1 (1 - s2) s2

                                                                        2
                               2    2          2                       d v
 + 2 (1 - s2) s2 + (1 - s1)  s1  (0.5 - s2)  + 2 (1 - s1) s1 + ---  +
                                                                          2
                                                         2            ds1
                                                        d v
                                                        ---)
                                                           2
                                                        ds2

==>  please generate the fortran program
done
you will find it in the segment belman.fortran

==>  stop

good bye
```

belman.fortran

```
      subroutine prodyn(n1,n2,epsimp,impmax,v,ro,u,eps,nmax)
      dimension v(n1,n2),u(2,n1,n2)
c        Resolution de l equation de Bellman dans le cas ou:
c               Les parametres sont
c               L etats-temps est:  x1 x2
c               La dynamique du systeme est decrite par l operateur

c                            2        2   2
c       Minu( (0.5 - x1)  (1 - x2)  x2  + (1 - x1) x1 (1 - x2) x2
c
c                              2    2           2                          2
c + 2 (1 - x2) x2 + (1 - x1)  x1   (0.5 - x2)   + 2 (1 - x1) x1 + u2
                                                         + p2 u2
c
c        2
c + u1   + p1 u1 + q2 + q1 )
c               ou v designe le cout optimal
c               ou pi designe sa derivee premiere par rapport a xi
c               ou qi designe sa derivee seconde par rapport a xi
c               Le probleme est statique
c               Les conditions aux limites sont:
c                    x2 = 0 v= x1 x2
c                    x2 = 1 v= (1 - x1) (1 - x2)
c                    x1 = 0 v= x1 x2
c                    x1 = 1 v= (1 - x1) (1 - x2)
c        Les nombres de points de discretisation sont: n1 n2
c                    x2 = 1 correspond a i2 = n2
c                    x2 = 0 correspond a i2 = 1
c                    x1 = 1 correspond a i1 = n1
c                    x1 = 0 correspond a i1 = 1
```

```
c       Le taux d actualisation vaut: 1
c       impmax designe le nbre maxi d iterations du systeme implicite
c       epsimp designe l erreur de convergence du systeme implicite
c       ro designe le pas de la resolution du systeme implicite
c                                       par une methode iterative
c       Minimisation par la methode de Newton de l'Hamiltonien
c       L inversion de la Hessienne est faite formellement
c       nmax designe le nombre maxi d iteration de la methode de Newton
c       eps designe l erreur de convergence de la methode de Newton

      h2 = float(1)/(n2-1)
      h1 = float(1)/(n1-1)
      u2 = u(2,1,1)
      u1 = u(1,1,1)
      hih2 = h2**2
      hih1 = h1**2
      h22 = 2*h2
      h21 = 2*h1
      nm2 = n2-1
      nm1 = n1-1
      do  119   i2 = 1 , n2 , 1
      do  119   i1 = 1 , n1 , 1
      v(i1,i2) = 0.0
  119 continue
      imiter = 1
  113 continue
      erimp = 0
      do  111   i1 = 1 , n1 , 1
      x1 = h1*(i1-1)
      v(i1,n2) = 0
      v(i1,1) = 0
  111 continue
      do  110   i2 = 2 , nm2 , 1
      x2 = h2*(i2-1)

      v(n1,i2) = 0
      v(1,i2) = 0
  110 continue
      do  109   i1 = 2 , nm1 , 1
      x1 = h1*(i1-1)
      q2 = (v(i1,i2+1)-2*v(i1,i2)+v(i1,i2-1))/hih2
      q1 = (v(i1+1,i2)-2*v(i1,i2)+v(i1-1,i2))/hih1
      p2 = (v(i1,i2+1)-v(i1,i2-1))/h22
      p1 = (v(i1+1,i2)-v(i1-1,i2))/h21
      niter = 0
      w0 = -1.0e+20
  101 continue
      niter = niter+1
      if ( niter - nmax )  102 , 102 , 103
  103 continue
      write(8,901)i1,i2
  901 format(' newton n a pas converge', 2 i3)
      goto  104
  102 continue
      u1 = -p1/2.0
      u2 = -p2/2.0
      ww = (0.5-x1)**2*(1-x2)**2*x2**2+(1-x1)*x1*(1-x2)*x2+2*(1-x2)*x2+(
     1   1-x1)**2*x1**2*(0.5-x2)**2+2*(1-x1)*x1+u2**2+p2*u2+u1**2+p1*u1+
     2   q2+q1
      er = abs(ww-w0)
      if ( er - eps )  104 , 104 , 105
  105 continue
      w0 = ww
      goto  101
  104 continue
      u(1,i1,i2) = u1
      u(2,i1,i2) = u2
      w0 = ww
      w0 = w0-v(i1,i2)
      vnew = ro*w0+v(i1,i2)
      v(i1,i2) = vnew
      erimp = abs(w0)+erimp
  109 continue
      imiter = imiter+1
      if ( imiter - impmax )  116 , 115 , 115
  116 continue
      if ( epsimp - erimp )  113 , 112 , 112
  115 continue
      write(8,907)
  907 format(' schema implicite n a pas converge')
  112 continue
      do  117   i1 = 1 , n1 , 1
```

```
      do  117   i2 = 1 , n2 , 1
      write(8,900)i1,i2,v(i1,i2)
  900 format(' v[', (i3,','), i3,']:', e14.7,'$')
      write(8,902)i1,i2,u(1,i1,i2),u(2,i1,i2)
  902 format(' u[', (i3,','), i3,']:[', (e14.7,','), e14.7,']$')
  117 continue
      return
      end
```

III - AUTOMATIC PROVING OF THEORETICAL RESULTS ON PARTIAL DIFFERENTIAL EQUATIONS

We can prove existence results on the Hamilton-Jacobi equation by the monotony method of J.L. Lions [L2]. In this part we try to represent the theoretical informations and its use in such a way that the system is able to prove the existence of a class of partial differential equations. More precisely the type of problems we deal with is in its abstract form :

$$(P)\quad \begin{cases} Au = f \text{ in } \Omega \text{ bounded open of } \mathbb{R}^n \\ B_j u = g_j \quad j = 0, \ldots, m-1 \text{ on the boundary } \Gamma \text{ of } \Omega, \end{cases}$$

A and B_j are differential operators.

We find the existence of a solution for (P) and the functional space V to which this solution belongs. The demonstrations are made by computations in Banach spaces and it seems possible to perform them by using MACSYMA [M] as a formal calculus system. Moreover, the functional analysis theorems used by the demonstration can be encoded in rules in a PROLOG [C] system written in LISP named LOGIS [G].

a) Presentation of the system

The system is completely embedded in the MACSYMA environment. The activation of the system is done by MACSYMA functions calling other MACSYMA functions, LISP functions or LOGIS.

The general structure of the systems is described by the following figure ;

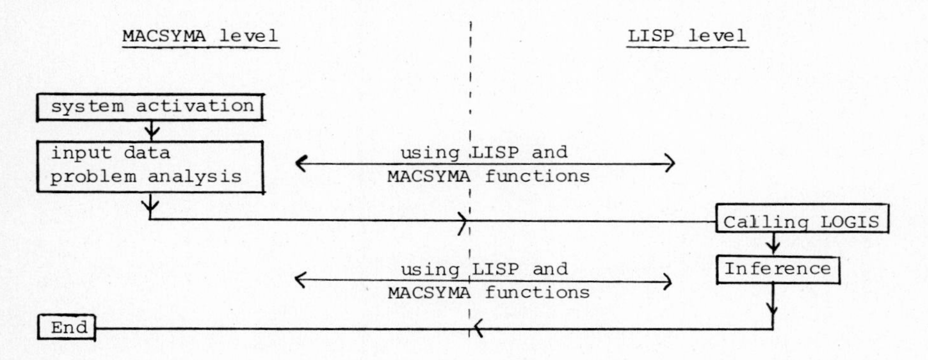

b) Data structure used for representing the mathematical objects

- 1) Functional spaces

The functional spaces used are Sobolev spaces [A] $W^{m,p}(\Omega)$ (functions belonging to $L^p(\Omega)$ with the m first derivatives in the weak sense belonging to $L^p(\Omega)$) and $W_0^{m,p}(\Omega)$ (the closure of $W^{m,p}(\Omega)$ in $\mathcal{D}(\Omega)$).

They are represented by MACSYMA lists.

space	internal representation
$W^{m,p}(\Omega)$	[p,m]
$H^m(\Omega)$ ($\equiv W^{m,2}(\Omega)$)	[2,m]
$L^p(\Omega)$ ($\equiv W^{0,p}(\Omega)$)	[p,0]
$\mathcal{D}(\Omega)$	[inf, inf]
$W_0^{m,p}(\Omega)$	[p, m, 0]

- 2 Functions

The mathematical functions appearing in the problem can possess various properties put in property lists.

As an example, for the equation

$$- \sum_{i=1}^{n} \frac{\partial}{\partial x_i} \left(a \frac{\partial u}{\partial x_i}\right) + |u|^{p-2} u = f$$

we have the following properties table :

function	nature	space	property
u	solution	unknown	nil
a	coefficient	[q, 0]	a > 0
q	constant	[r]	nil
p	constant	[r]	p ≥ 2
f	smember	unknown	nil

c) Inference

The part of the system using existence theorems and deciding what computations must be done to satisfy the hypotheses is encoded in Horn clauses with the PROLOG syntax of LOGIS. This is the core of the system determining what theorem and what method must be applied.

The production rules trigger the execution of MACSYMA functions or LISP functions. Moreover, all the results or informations obtained during the inference are kept in the fact data base to avoid useless computations and to increase the speed of the system.

d) Formal calculus

The main formal calculus parts (Green formula application, variational formulation obtention, ...) are written in LISP. Indeed, the major part of work consists in list handling (MACSYMA expressions) and it is much more efficient to do it in LISP. When necessary, MACSYMA functions (expand, diff, ...) are called.

The programming technique used has been data driven programming. This allows an easier knowledge updating.

e) Example

The main purpose of the system is to perform the following operations from the problem (P) :

(i) to obtain the variational formulation of (P) :

$$(vf) \quad \begin{cases} a(u,v) = (f,v) \qquad u,v \in V \\ V \text{ space to be found} \end{cases}$$

(ii) to verify hypothesis of various theorems (Lax-Milgram, monotony theorem,...) for the operator A of (P) or the variational formulation (vf).

Various variational formulations can be computed in order to apply a theorem.

In case of failure, the user will be able to do other computations by using, if he wants to, the functions defined by the system (Green formula application, functional space research, ...).

Only part (i) and (vf) has been completed. An example of execution is given below :

```
(c1) cogito();
dimension de l'espace r**n   :
3;
la dimension de l'espace est 3
est-ce-correct ?
oui;
```

```
equation a resoudre dans omega sous la forme "a(u)=f" :
sum(diff(diff(a*delta(u),x[i])+diff(u,x[i])*
          abs(diff(u,x[i]))^(p-2),x[i]),i,1,n)+u=f;
conditions sur la frontiere gamma sous la forme
          "[b[1](u)=0,...,b[m](u)=0]" :
[u=0,diff(u,nor)=0];
le probleme a resoudre est :
 n
====
\                         p - 2
 >     d    du   !du !              d
/     --- (--- !---!          +  --- (a delta(u))) + u = f dans omega
====  dx    dx   !dx !              dx
i = 1   i     i  !  i!                i

avec sur gamma :
u = 0
 du
---- = 0
dnor

est-ce-correct ?
oui;

nature de u :
solution;

fonction : u
**********

    nature : solution

    espace : inconnu

    proprietes : nil

est-ce-correct ?
oui;

nature de a :
coefficient;

espace de a :
[l,inf];

proprietes de a :
nil;

fonction : a
**********

    nature : coefficient

    espace : [l, inf]

    proprietes : nil

est-ce-correct ?
oui;

nature de p :
constante;

proprietes de p :
nil;

fonction : p
**********

    nature : constante

    espace : [r]

    proprietes : nil

est-ce-correct ?
oui;
```

```
proprietes de f :
nil;

fonction : f
**********

     nature : smembre

     espace : inconnu

     proprietes : nil

est-ce-correct ?
oui;

==>  ((solution *espace))
**** bien_pose ****

**** non_lineaire ****
op_principal
solution

**** formulation_variationnelle ****

op_principal
   3
  ====                      p - 2
  \          du   !du !           dv
-  >    %int(---  !---!           ---) + %int(a delta(u) delta(v)) + %int(u v)
  /          dx   !dx !           dx
  ====         i  !  i!             i
  i = 1
%int(f v)
[w 1 p] inter [h0 2]

==>  fin
Time= 13821 msec.
(d1)                                        true
```

CONCLUSION

Generalizations of the class of Hamilton-Jacobi equations are investigated : (extension of the state space and the optimization algorithms, control of jumps processes ...).

In this paper only the part of the system dealing with Dynamic Programming method has been described. Program generators are also available for stochastic gradient, decoupling and perturbation methods.

The natural language interface will be extended in particular for theorem prooving purposes (hypothesis specifications).

In the future, inference mechanism will take a more important part in the choice of resolution methods.

REFERENCES

[A] R.A. ADAMS : "Sobolev Spaces". Academic Press, 1975.

[B] A. BENSOUSSAN : "Stochastic control by functional analysis methods". North Holland, 1982.

[G1] M. GOURSAT - J.P. QUADRAT : "Numerical methods in optimal stochastic control" To appear in the Encyclopedia on the System Theory.

[G2] P. GLOESS : "Logis User's Manual". UTC/GI, BP 233, 60206 Compiègne Cédex, France, First Editions, January 1984.

[L1] P.L. LIONS : Thesis, Paris 9.

[L2] J.L. LIONS : "Quelques méthodes de résolution des problèmes aux limites non linéaires". Dunod, Gauthier-Villars, 1969.

[M] MATHLAB GROUP : "Macsyma Users' Manual". Laboratory for Computer Science, M.I.T., Version 10, January 1983.

[Q] C. QUEINNEC : "LISP : language d'un autre type". Eyrolles 1983.

IMPACT

Interactive Mathematical Program for Automatic Control Theory

Magnus Rimvall
François Cellier
Institute for Automatic Control
Swiss Federal Institute of Technology (ETH)
CH-8092 Zuerich, Switzerland
Tel. 01 / 256 28 42

Abstract.

IMPACT, a new CAD-program for Control Systems which is presently under development at our institute, is presented. The program will give access to algorithms useful in control systems theory in an interactive manner. It is aimed at inexperienced students as well as skilled control scientists for the analysis, synthesis and simulation of control systems. IMPACT is coded in ADA, portability is one of the main design goals.

A first section discusses the chosen mode of interaction, and compares it with other common methods. A second section presents the data structures available in IMPACT, and discusses the operations which can be performed on these structures. The IMPACT command language is thereafter presented, in particular are the very versatile macro-facilities explained. Finally, some implementational aspects are discussed.

1. INTRODUCTION

In the last decades, digital computers have thoroughly changed the computational tools used by control engineers. However, this revolution is not yet over, its thrust has just shifted from the point of raw computing power to the question of user friendly and adaptive systems.

Let us look at the (nowadays) simple problem of calculating the eigenvectors and inverse eigenmatrix of several 8*8 matrices. Forty years ago, you would need a lot of paper and almost unlimited patience to solve this problem.

Twenty years ago, you probably had a digital computer at your disposal. However, you would most likely have to write a program yourself, which calculated the eigenvectors and inverted the eigenvector-matrix. Only if you were extremely lucky, you might have had access to one of the first libraries containing general-purpose programs for mathematical operations (e.g. SSP /SSP68/).

Ten years ago, you most certainly had access to some library containing mathematical algorithms, e.g. IMSL /IMSL82/, EISPACK /GARB77/ /SMIT74/ or LINPACK /DUNG79/. Unfortunately, you still had to write a program which read the matrices, called the algorithm-routine(s) and printed the result. You were bound to loose a lot of time until the input format corresponded to the input data, all the parameters of the library calls were correct and in the right order, and so on.

Today, for most people, things are not that much different. New and better algorithms have emerged, but you still loose lots of time writing programs accessing these algorithms. Needed is a package adapted to control theory, which has not only a software interface, but also an interactive interface for easy access.

1.1 MATLAB

One of the first persons to realize the importance of an interactive interface to packages containing complex mathematical algorithms was C. Moler /MOLE80/. In his program MATLAB, a milestone in the history of interactive programs, an easy-to-use, interactive interface is provided to the LINPACK and EISPACK matrix manipulation libraries. Using a very natural input command language, it is possible to perform matrix operations in MATLAB with the same ease as one makes scalar computations on a pocket calculator. For example can the above mentioned problem be solved in a few lines of input:

```
A = <1, 4, 5, 0, 0, -1, 3, 1
      0, 1, 0, 1, 0,  0, 0, 0
        ...
     -3, 0, 0, 0, -1, 0, 0, 2>;
<P,DUM> = EIG(A);
INVP = INV(P)
```

Although MATLAB is, as we have seen, extremely easy to use, and moreover very versatile, many problems in control theory are not (and were never intended to be) solvable using MATLAB. One reason for this is of course the lack of suitable algorithms, another, more fundamental cause is the lack of data structures.

As MATLAB is written in a very well-structured manner, it is quite simple to add new algorithms. $MATRIX_X$ is one such extended MATLAB package for control scientists /WALK82/, another extended version of MATLAB is presently under development by the group of P. van Dooren (unpublished). Although these new products are definite upgradings of

MATLAB, they only partially provide the control engineer with an adequate tool. The reason for this is the mentioned lack of data structures adapted to control problems. MATLAB uses the complex matrix (with the scalar as a special case) as the only data structure. Control engineers often work with more complex structures, like polynomial matrices, transfer-function matrices and linear as well as non-linear system descriptions. Furthermore, although the input command language of MATLAB is well suited for smaller problems, a better structured command language is needed for more complex problems. In particular, some form of macro/procedure facility accepting parameters must be available. Finally, versatile graphical output and an interface to a data base should be present.

1.2 IMPACT

Seen through the eyes of the user, IMPACT appears to be just another extension to MATLAB, the IMPACT command language is similar to that of MATLAB. However, seen from an implementational view, IMPACT is only a conceptual superset of MATLAB. As IMPACT is implemented in ADA, not one single line of code has been taken from MATLAB. Furthermore, several new data structures are introduced.

The development of IMPACT is made with the objective of serving a very inhomogenous group of users. On one hand, IMPACT is aimed at being used by students with little experience in control theory and no experience at all in CAD. Using only the most basic structures of the input command language which are simple enough to be learnt in a few hours, these students will be able to access complex algorithms in order to solve control problems with a minimum of tutorial. An on-line HELP facility contains all needed information on the command language syntax as well as on the numerical algorithms, making self-tutorial possible. On the other hand, the experienced control scientist is provided with a full-fledged structured command language with all elements found in a higher computer language including WHILE and FOR loops, IF-THEN-ELSE statements, and so forth. Furthermore, a large selection of IMPACT functions and procedures gives the user access to a wide range of algorithms. To further enhance the structurability, four different macro facilities have been introduced. Access to a data-base will be provided to store away variables, plots and macros for later reuse.

2. MODE OF INTERACTION

When designing a new interactive system, one of the first decisions must be, in which form the man-machine interaction is to take place. This decision should not be taken lightheartedly, as this mode of interaction determines the user-friendlyness, and thereby also user acceptance, of the system; although this interface is not the brain of any CAD-system, it certainly serves as both eyes and mouth.

The mode of interaction also influences the structure of the kernel controlling the package. In particular, the data-structures of the kernel are very closely knitted to the user interface. As any late changes in the central data structures are the worst of all possible nightmares for any software developer, the design of the interactive interface should be done carefully, so that no later modifications need to be done /INFO79/.

Apart from some more exotic ways of communication, like speech input and natural language input, four basically different ways of interactive input exist:

- question-and-answer method
- menu-driven operation
- command-language communication
- graphical input.

Of these four, the first three work with alphanumeric information, which, at least in principle can work on any alphanumerical terminal. The graphical input requires special hardware in form of a graphical terminal for the graphic echo and some graphical input device (in form of e.g. a joystick or a lightpen). Although very interesting in the field of system documentation and modelling /ELMQ82/, where structures rather than numerical data are entered, the complexity of a program allowing such an input makes it unattainable for a CACSD system designed primarily to solve numerical problems. However, a later interface between IMPACT and a graphical input system would be an interesting and meaningful extension.

Of the three alphanumeric input modes, the menu and the question-and-answer methods let the computer be in charge of the conversation, whereas the command language method gives the user almost total control. In the last few years, several menu-driven interactive programs using a "mouse" or joystick as input device have emerged. These systems are extremely handy and speedy for dexterous users. However, until a hardware standard has been set, such systems are not very portable. On the other hand, the speed of portable menu-driven systems can be compared with that obtained by question-and-answer methods.

If the only design goals are minimal learning time and maximum accessibility by non-specialists, the question-and-answer method is most certainly the right answer. However, this method gets very tiresome after a while, as the user always can anticipate the next question, but cannot speed up the input. As an example, compare the conversation in INTOPS /AGAT79/ (a somewhat old-fashioned interactive program for control system operations) with the equivalent commands in IMPACT. Both examples produce a BODE diagram of the transfer function

$$G(s) = \frac{1}{s^3 + 9s^2 + 5s + 9}$$

INTOPS question-and-answer conversation:

```
P>      OP CODE = ENTER
P>      NAME = NUME
P>      COMMENT = NUMERATOR
P>      ORDER = 0
P>      P( 0) = 1
P>      NUME       NUMERATOR
P>      P( 0) =   0.10000E+01

P>      OP CODE = ENTER
P>      NAME = DENO
P>      COMMENT = DENOMINATOR
P>      ORDER = 3
P>      P( 0) = 9
P>      P( 1) = 5
P>      P( 2) = 9
P>      P( 3) = 1
P>      DENO       DENOMINATOR
P>      P( 0) =   0.90000E+01
P>      P( 1) =   0.50000E+01
P>      P( 2) =   0.90000E+01
P>      P( 3) =   0.10000E+01

P>      OP CODE = BODE
P>      GH NUMERATOR = NUME
P>      GH DENOMINATOR = DENO
P>      NUMBER OF FREQUENCY VALUE = 100
P>      OMEMAX = 1000.
P>      OMEMIN = .1
```

(user input is underlined).

Command language input of IMPACT:

```
BODE (1/<9^5^9^1> //DOMAIN=LOGDOM(.1,1000.,100) )
```

or a little more verbose but better readable:

```
S = <^1>;
G = 1 / (S**3 + 9*S*S + 5*S + 9);
FREQ = LOGDOM(.1,1000.,100);
BODE (G //DOMAIN=FREQ )
```

With right, the advocates of menu-driven and question-and-answer interaction claim that their methods are specially advantageous for users unfamiliar with the system. In our example, the user of INTOPS needed to know only the existence of the two commands ENTER and BODE, whereas the user of IMPACT must know how to form a transfer-function, and how to put a variable as parameter of a function. However, due to the very natural notation of the IMPACT command language (which will be described in more detail later), any inexperienced user will be able to use IMPACT after only a few hours.

On the other hand, anyone familiar with both systems will save a factor 10 in time as well as in number of input lines when he uses IMPACT to construct the BODE diagram. For an easy-to-learn system like IMPACT, this means that the command language pays off heavily already after a few hours of use.

In order not to discourage the beginner during these first few hours, an on-line help will provide answers to all questions in a structured manner, also making self-tutorial possible. The help-facility can of course also be accessed by the advanced user when he needs information, e.g. about a lesser used algorithm and the corresponding function call parameters.

3. DATA STRUCTURES IN IMPACT

Whereas MATLAB supports only the double-precision complex matrix, IMPACT provides the user with several other data structures. In this chapter, most of these will be presented, together with some of the operations which can be performed on these different structures.

3.1 Matrices

As in MATLAB, all matrices in IMPACT are stored away using complex elements of high precision. Any scalar can be stored away as a one by one matrix. Matrix input is done using a very natural notation:

```
A = <1,2,3
     4,5,6
     7,8,9>;
```

constructs a 3*3 matrix A. If the column vector B has been entered as

```
B = <1.5 ; 4.3 ; 1>;
```

the equation A*x = B can be solved e.g. through

```
X = INV(A)*B
```

Many matrix operations, as different eigenvector operations, inversions and transformations, have been included in IMPACT.

3.2 Polynomial matrices

A polynomial matrix is a matrix where each element is a polynomial with (complex) coefficients.

Polynomial matrices are entered into IMPACT using a notation similar to that used for normal matrices. For example, the input line

```
Q = < 2^3^1 ; 4^4^1 >
```

will result in the polynomial column vector

```
Q(p)        =
    2. + 3.*p + 1.*p**2
    4. + 4.*p + 1.*p**2
```

which is the IMPACT form to describe the polynomial matrix

$$\langle \begin{matrix} 2. + 3.* p + 1.* p^2 \\ 4. + 4.* p + 1.* p^2 \end{matrix} \rangle$$

An alternative way of entering the polynomial matrix Q might be to first define the variable P as

```
P = <^1>;
```

Thereafter the polynomial matrix Q can be entered as

```
Q  =  <2 + 3*P + P**2; 4 + 4*P + P**2>
```

The basic matrix operations addition, subtraction and multiplication may be used on polynomial matrices (using the symbols +, - and *) if the basic dimensional rules are fulfilled. For example, the input lines

```
P      =  <^1>;
Z      =  <1+1*P  , 2*P>;
WROW   =  < 1     , 2+2*P>;
WCOL   =  WROW';
XADD   =  Z + WROW  , XMULT = Z * WCOL
```

will result in the output

```
XADD(p)      =
    2. + 1.*p     2. + 4.*p
XMULT(p)     =
    1. + 5.*p + 4.*p**2
```

Until now, all polynomial matrices have been entered in a non-factorized manner, specified through all non-zero coefficients of the polynomial elements. To further enhance the flexibility of IMPACT, polynomials can also be given in factorized form. Example:

```
QF  = FACTOR (Q)
```

will transform the matrix Q to a factorized form, resulting in

```
QF(p)    =
   (p + 1.)*(p + 2.)
   (p + 2.)*(p + 2.)
```

It is of course possible to enter factorized polynomial matrices directly:

```
QF = <-1¦-2
      -2¦-2>
```

Due to an ill-conditioned re-factorization, operations on factored polynomial matrices, where at least part of the factors need to be de-factorized before the operation (like addition), can be extremely badly conditioned. In IMPACT, the user is responsible for the testing on the accuracy of the result. However, IMPACT provides the user with a few tools: you can simulate a smaller computer word-length in that you specify the accuracy to be used in each arithmetic operation. This enables you to test the error-propagation of the used algorithm. Furthermore, IMPACT will warn you each time an ill-conditioned operation is performed. Moreover, as neither of the two "classical" polynomial representations is optimal with respect to numerical behaviour, IMPACT offers yet another representation which shall be discussed in Section 3.5 of this paper.

Furthermore, IMPACT supports several complex polynomial operations. For instance have algorithms calculating the least common left/right denominators, the Smith-form and eigenvalues /KAIL80/ been included.

3.3 Transfer-function matrices

Only in special cases is the inverse of a polynomial matrix another polynomial matrix. However, the inverse of a polynomial matrix can (as long as the matrix is non-singular) always be defined as matrix with rational function elements, a so called transfer-function matrix.

Transfer-function matrices are entered in a manner similar to that used by polynomial matrix entry. The input sequences

```
S = <^1>;
G = < 1/S      , 1/(S+1)
      1/(S+1)  , 1/(S*(S+1)) >;
G = FACTOR(G)
```

and

```
G = ONES(2) ./ < ¦0, ¦-1; ¦-1, 0¦-1 >
```

(where ONES(2) returns a 2*2 matrix filled with ones and ./ denotes an element-by-element division) both result in the factored 2*2 transfer-function matrix

```
G(p)    =
            1.                 1.
         -------           ----------
            p               (p + 1.)
            1.                 1.
         --------          ----------
         (p + 1.)          p*(p + 1.)
```

In control theory, transfer-function matrices are used to describe systems in the frequency domain. Interesting enough, many mathematical operations on transfer-functions have a physical meaning. For example, the addition of two systems corresponds to the parallel connection:

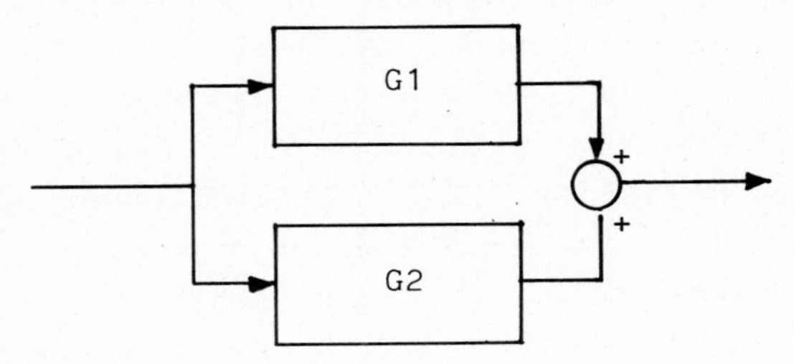

GTOT = G1 + G2

A cascading of two systems is mathematically described through the multiplication in reverse order of the two system components. A feedback can either be described directly:

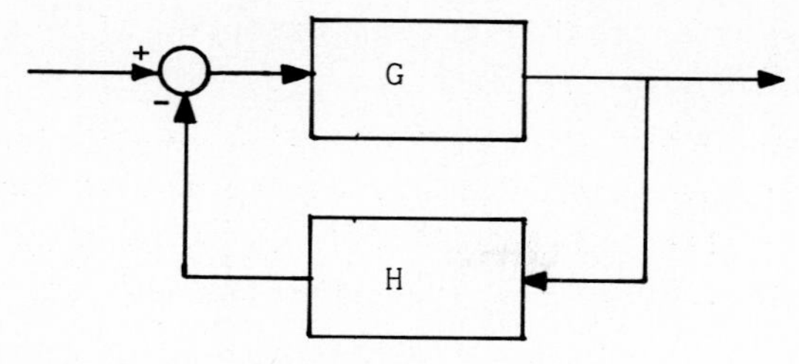

GTOT = G / (1 + G*H)

or through the use of the special feedback operator \\ (which does not correspond to any trivial mathematical operation):

GTOT = G \\ (-H)

3.4 System descriptions

In the time domain, a linear system is normally described by four different matrices:

$$\dot{x} = A*x + B*u$$
$$y = C*x + D*u$$

As this is a very common representation, IMPACT provides the user with a special data-structure, the linear system description. Given three matrices A, B, C of right dimensions, the function LCSYS will form a continuous linear system description out of these matrices,

CSYS1 = LCSYS(A,B,C)

whereas LDSYS will form a discrete linear system description with a sampling rate of DT:

DSYS1 = LDSYS(F,G,H,DT)

The D matrix was here assumed to be a null matrix of correct dimensions. However, if the user wants to define a D-matrix, this can be entered through the use of default redefinition:

CSYS2 = LCSYS(A,B,C //D=DD)

will include the matrix DD as the direct-path matrix.

Mathematical operations on system descriptions have been defined such that the physical meaning is the same as if the same operation were performed on transfer-function matrices. For example, if a system of 2nd order has been defined through the matrices

```
A = <1, 1
     0, 1>;
B = <0
     1>;
C = <1, 0>;
SIMPLE = LCSYS(A,B,C);
```

the operation

```
CASC = SIMPLE * SIMPLE
```

will result in a system of order 4 with the component matrices

```
CASC.A = <1, 1, 0, 0
          0, 1, 0, 0
          0, 0, 1, 1
          1, 0, 0, 1>
CASC.B = <0
          1
          0
          0>
CASC.C = <0, 0, 1, 0>
```

Note that the dimension of the system matrix is doubled, just as the order of the physical system.

A linear system given in the time domain by three (four) matrices can always be transformed into a frequency representation, and vice versa. The transformation from the time to the frequency domain is given through the formula

```
S = <^1>;
G = C * INV(S*EYE(A) - A) * B
```

This valid IMPACT statement is also available as a separate function

```
G = TRANS(LCSYS(A,B,C))
```

The such determined transfer function matrix is not unique, as each transfer-function component could have reducible factors. The function REDUCE will shorten any common factors of a transfer-function (using the machine tolerance, or any other given tolerance, to determine if two factors are equal or not).

As the transformation from the frequency to the time-domain is not unique, IMPACT provides the user with a range of transformations resulting in linear system descriptions in different canonical forms, Jordan form etc.

3.5 Domain and trajectory variables

A domain is a sequence of discrete, increasing values on the real axis which can be used to form the independent variable of a table.

```
TIME = LINDOM(0.,50.,0.1)
```

would thus define a sequence TIME with 501 elements, the first of which has the value 0 and the last the value 50, using an increment of 0.1. With the help of the '&'-operator, domains can be concatenated. For example would

```
PULSE_BASE = LINDOM(0.,1.,0.01) & LINDOM(1.1,10.,0.1) & 20.
```

be a non-equidistant domain with 202 points.

A trajectory is a table of function values which uses a domain as independent variable. Such a table results from a variety of operations performed on domains. E.g. would the operation

```
TRA = SIN(TIME)
```

result in a table where each entry contains an independent variable copied from the domain TIME and the sine-value thereof.

Mathematical operations are defined on trajectories using the same domain, e.g. would the operation

```
TRB = TRA + COS(TIME)
```

once again be a table with one row of values as function of the independent variable TIME, whereas

```
TRC = <TRA, COS(TIME), TRA + COS(TIME)>;
```

would be a table where each entry is a row-vector with three elements. Note that TRB = TRC(3).

All graphical functions return a trajectory as result, this trajectory can then be plotted with the command PLOT.

```
PL1 = BODE(G1)
```

will compute a bode-diagram and store this diagram as a trajectory with one complex element (containing amplitude as real and phase as imaginary part) per entry. If you want to compare the BODE-diagrams of two transfer-functions, combine the trajectories into one trajectory and plot this (the option //BODE will insure that the axes are logarithmically scaled and correctly labelled):

```
PL12 = <BODE(G1),BODE(G2)>;
PLOT(PL12//BODE)
```

On each plot, you will now find two different-colored/shaped curves from your two systems.

Furthermore, domains and trajectories can be used to <u>simulate</u> system behaviour /CELL83/. If SSYS is any system representation (e.g. a transfer-function matrix or a system description),

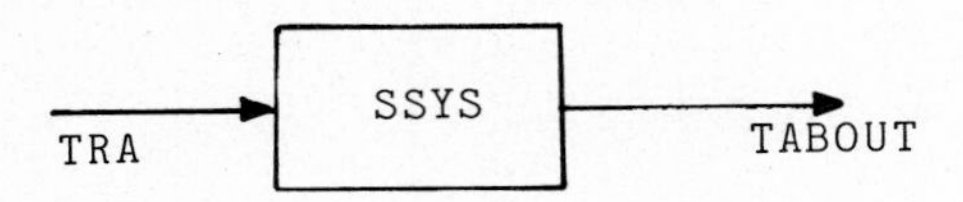


```
TABOUT = SSYS * TRA
```

will perform a simulation and store away the values of the output signal at the discrete times of the trajectory TRA, thus making TABOUT another trajectory variable specified over the same domain as TRA.

Finally, domains are useful for yet another purpose. As stated previously, polynomial operations may be numerically ill-conditioned. It has been found that, in many cases, a better behaviour results when polynomials (or rational functions) are represented by a set of supporting values rather than by coefficients or roots.

```
FREQ = LOGDOM(.1,1000.,100)
```

generates a domain consisting of 100 values distributed logarithmically over the interval from 0.1 to 1000.

```
PP1 = TRAJEC(P1,FREQ)
```

computes a trajectory (matrix) of the polynomial (matrix) P1 by evaluating each polynomial (or each transfer-function, resp.) at each of the supporting values of FREQ. Obviously, this gives rise to a third representation of polynomials which often exhibits better numerical properties. (All primitive polynomial operations such as addition, multiplication, and inversion become trivial.)

```
Q1 = ROOTS(PP1)
```

reestablishes the factored representation of PP1. This can be obtained by a numerically well-behaving Fast Fourier Transform (/GEIG73/).

4. COMMAND LANGUAGE

For IMPACT, a very versatile command language has been developed using the command language of MATLAB as a base. However, IMPACT allows for a more structured input, in form of more general statements and the availability of four kinds of macros.

The requirement that novices as well as very experienced users should be able to use IMPACT is reflected in the design of the IMPACT command language. On one hand, the basic commands are extremely easy to learn, but still powerful enough to make all kinds of operation possible, although not necessarily in the most optimal way. On the other hand, more complicated language elements can be used to perform complex operations. In particular, a hierarchical structuring of a problem is possible through the use of macros.

Due to their very natural notation, the basic commands can be mastered in a few hours time. The most essential statement is the assignment statement, which uses a notation similar to that of a normal mathematical formula. If we, for example, want to determine whether or not a certain linear system is controllable, we first enter the state and input matrices:

```
A = <1,0,1
     0,1,2
     0,0,1>
B = <0; 0; 1>
```

Thereafter we will get our answer by calling a procedure CONTR with the command

```
CONTR(A,B)
```

The procedure CONTR writes the wanted result on the terminal.

For more complex problems, a full-fledged, structured input language is available, including IF..THEN..ELSE, FOR/WHILE-loops, and so on. Example: The heat-diffusion in a long metal bar can be approximately modelled through a set of N differential equations (t(1). denotes the derivative of the variable t(1)):

```
t(1). = -2*t(1) + t(2)
t(N). = -2*t(N) + t(N-1)
t(i). = -3*t(i) + t(i-1) + t(i+1)    ,   1 < i < N
```

The state-matrix of this model can be obtained through the statements

```
FOR i = 1:n DO
  FOR j = 1:n DO
    IF (j = i) THEN IF (i = 1) OR (i = n) THEN a(i,j) = -2
                                          ELSE a(i,j) = -3
                    ENDIF
                ELSIF abs(j-i)=1 THEN a(i,j) = 1
                                 ELSE a(i,j) = 0
    ENDIF
  ENDFOR
ENDFOR
```

When a sequence of statements like these are to be performed several times, the user should use a macro to avoid typing errors and to save time.

4.1 Macros

IMPACT provides the user with four different types of macros.

4.1.1 Function macros

If the previously described model of a metal bar is to be used several times, each time with a different value for N, the user can save time by defining a function macro returning the wanted state matrix:

```
FUNCTION bar_matrix(n)
  FOR i = 1:n DO
    ::
  ENDFOR
  RETURN a
ENDFUNCTION
```

4.1.2 Procedural macros

Example: We want to write a procedure to add a new MACRO to our private MACRO library (PRILIB.INT) or replace an old one by a newer update. This can be performed by:

```
PROCEDURE ADDMAC (FILNAM)
  LOAD('PRILIB');
  READ(FILNAM);
  SAVE('PRILIB',MACRO);
ENDPROCEDURE
```

This procedure is executed by

```
ADDMAC('NEWMAC.IMP')
```

Upon call, only one variable is known within the procedure, namely the variable FILNAM which is of type text-string and contains the name of the file which the new MACRO is currently stored in. LOAD('PRILIB') loads all variables from file PRILIB.INT containing the MACRO library. READ(FILNAM) reads the new MACRO and converts it to its internal representation. If such a MACRO variable was already in the library, this variable is now overwritten by the new definition. SAVE('PRILIB',MACRO) saves all currently accessible MACRO's (but not the text-string variable FILNAM) in a new cycle of the file PRILIB.INT. Upon return from the procedure, the old context is reestablished, and all previously visible variables are accessible again.

4.1.3 String macros

Until now, we have considered macros which look and work as functions or procedures. In this chapter, we will extend our macro definitions to include a general macro concept.

When we allow ourself a slight simplification, a macro definition connects a string of characters (possibly divided onto several lines) with a macro name. Each time this macro is called, the corresponding string is inserted at the point of call with each formal parameter being replaced by its actual value. In the more general case, the string of a macro could be inserted not only as a factor within any type of expressions (function macros returning one variable) or as a statement (procedure macro changing the values of one or several variables), but anywhere in an IMPACT input.

As a trivial example, let us consider a user who, for estetical reasons, dislikes the element-by-element operations '.*' . Such a user could avoid this symbol through defining a new string macros :

```
MACRO ELMULT
  .*
ENDMACRO
```

which thereafter can be used in statements as

```
C =  A ELMULT B            -- Equivalent to C=A.*B
```

Generally, the string macro is a very versatile instrument. It can for example be used to dynamically define new functions, and will certainly be used to shorten other macros through the use of "tricky" string operations. However, inexperienced users are warned not to use the string macro.

4.1.4 System macros

As we live in an imperfect world, control scientists usually have to use non-linear models to describe a real system. IMPACT provides the so-called system macro for this modelling. Consider the following example describing a discrete PI-regulator :

```
SYSTEM discr_regulator(kp,ki,dt)
  DSTATE int
  INITIAL int0=0;
  INPUT err
  OUTPUT u
    NEXT.int = int + ki*err*dt;
    u = kp*err + int;
ENDSYSTEM
```

This definition of a discrete system with one difference equation can be used to create a variable of the same type discr_regulator:

```
REG1 = discr_regulator(1,1,0.1)
```

The thus created system variable can then be used in any mathematical operations to construct parallel and/or concatenated systems, perform simulations, and so on. Given the predefined systems SMP (sampler) and ZOH (Zero-order-hold), a sampled data system using REG1 as regulator can be created through the single statement:

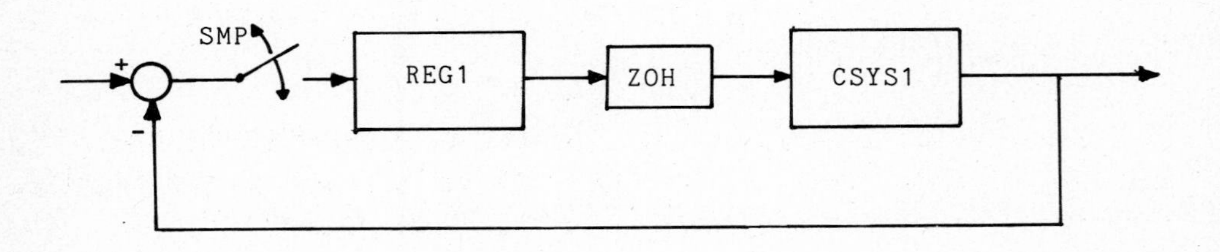


```
STOT = (CSYS1*ZOH*REG1*SMP(.1))\\(-1)
```

5. IMPLEMENTATION CONSIDERATIONS

Although MATLAB, as most other larger scientific software-projects of the sixties and seventies, is coded in FORTRAN, and although most available algorithm-libraries are FORTRAN-coded, it has been decided that IMPACT shall be coded in ADA. There are manifold reason for this /BIRD83/:

- ADA allows almost any types of data-structures to be directly defined, avoiding the hazzle of redefining all structures into arrays (the only structure available in FORTRAN). Furthermore, through the use of discriminants, ADA allows for the dynamic sizing of arrays, which means that no unnecessary space has to be reserved, as would be the case in a language like PASCAL.

- ADA, due to recursiveness, allows for a much more elegant coding of the IMPACT expression parser than FORTRAN would do. In this way, IMPACT shall be easier maintainable and updateable than MATLAB.

- ADA provides for a unique means of exception handling which shall prove very useful for our task.

- ADA is per definition portable, there may not exist any sub- and/or super-set of ADA with that name.

- ADA is highly structured, making a modular programming possible, resulting in reliable and easily maintainable code. Furthermore through the use of visibility rules, all system-dependencies can be hidden from the user as well as from most of the people involved in the development of IMPACT.

- ADA-libraries of algorithms are expected to emerge on the market in the near future. Therefore, IMPACT is to contain a well-defined interface for later incorporation of new algorithms.

At the present state of development, an IMPACT users' manual /RIMV83/ exists which describes the command language as well as each available function (algorithm).

To simplify the construction of the expression parser needed in IMPACT, the syntax of the IMPACT command language has been defined using an extended Backus-Naur form. This syntax has then been tested for consistency using a general purpose parser /BONG79/. LL(1) parsibility rules have been applied wherever applicable.

The actual coding of IMPACT has commenced late 1983 using one of the first, almost complete ADA compilers. It is expected that a first subset of IMPACT will be available during 1984.

6. REFERENCES

/AGAT79/ Agathoklis, P., _et alia_; "Educational Aspects of Using Computer-Aided Design in Automatic Control"; in Proc. of the IFAC Symposium on Computer-Aided Design of Control Systems, Zurich, Switzerland; Pergamon Press, London; 1979.

/ASTR83/ Åstrom, K.J.; Computer-Aided Modeling, Analysis and Design of Control Systems, A Perspective; Report CODEN: LUTFD2/(TFRT-7251), Department of Automatic Control, Lund Institute of Technology, Sweden; 1983.

/BIRD83/ Birdwell, J.D.; "Future Directions in Computer-Aided Control System Design, Software Development"; _IEEE Control Systems Magazine_, February 1983.

/BONG79/ Bongulielmi, A.P. and F.E. Cellier; "On the Usefulness of Using Deterministic Grammars for Simulation Languages"; Proc. of the SWISSL Workshop, St. Agata, Italy; to appear in Simuletter; 1979.

/CELL83/ Cellier, F.E. and M. Rimvall; "Computer Aided Control Systems Design"; _Proc. First European Simulation Conference ESC'83_, (W. Ameling, Ed.); Informatik Fachberichte, Springer Verlag; 1983.

/CUEN79/ Cuenod, M.A.; (Editor); _Proc. First IFAC Symposium on CAD of Control Systems_; Pergamon Press; 1979.

/DUNG79/ Dungorra, J.J, Bunch, J.R., Moler, C.B., Stewart, G.W.; _LINPACK Users' Guide_; Society for Industrial and Applied Mathematics; 1979.

/ELMQ82/ Elmqvist, H.; "A Graphical Approach to Documentation and Implementation of Control Systems"; Proc. 3rd IFAC/IFIP Symposium on Software for Computer Control, SOCOCO'82, Madrid, Spain; 1982.

/GEIG81/ Geiger, P.; Nullstellenbestimmung bei Polynomen und allgemeinen analythischen Funktionen als Anwendung der schnellen Fouriertransformation. Diss.Math.ETH 6759; 1981.

/GARB77/ Garbow, B.S., _et alia_; _Matrix Eigensystem Routines, EISPACK Guide Extensions_; Springer, Lecture Notes in Computer Science, _51_; 1977.

/HERG82/ Herget, C.J. and A.J. Laub; Special Issue on Computer-Aided Control System Design Programs; _IEEE Control Systems Magazine_, December 1982.

/IMSL82/ IMSL Library Reference Manual, Edition 9; IMSL, 1982.

/INFO79/ Infotech state of the art report : Man/Computer Communication, Vol 1-2, 1979.

/KAIL80/ Kailath, T.; Linear Systems; Prentice-Hall; 1980.

/MOLE80/ Moler, C.; MATLAB, Users' Guide; Department of Computer Science, University of New Mexico, Albuquerque, USA; 1980.

/RIMV83/ Rimvall, M.; IMPACT, Interactive Mathematical Program for Automatic Control Theory, A Preliminary User's Manual; Institute for Automatic Control, ETH Zurich, Switzerland; 1983.

/SMIT74/ Smith, B.T. et alia; Matrix Eigensystem Routines, EISPACK Guide; Springer, Lecture Notes in Computer Science, 6; 1974.

/SSP68/ System/360 Scientific Subroutine Package; Version III Programmers Manual, IBM, 1968.

/WALK82/ Walker, R., et alia; "MATRIX$_x$, A Data Analysis, System Identification, Control Design, and Simulation Package"; IEEE Control Systems Magazine, December 1982.

THE USE OF SYMBOLIC COMPUTATION FOR POWER SYSTEM STABILIZATION: AN EXAMPLE OF COMPUTER AIDED DESIGN

R. Marino and G. Cesareo
Seconda Università di Roma, Dipartimento di Ingegneria Elettronica
Via O. Raimondo I00173, Roma, Italy

Abstract. We consider the nonlinear model of a power system network in emergency for a stability crises. We present algorithms, to be implemented by symbolic and algebraic manipulation (SAM) systems, which, given the incidence matrix of a power system network, determine the controllability properties induced by the choice of the number and the location of power controllers, namely the strong accessibility and the feedback linearizability. If the number and the location of controls are to be chosen, one can determine the minimum number and the location of controls which induce the desired properties. The effect of structural or parameter perturbations can also be evaluated. These computer aided design techniques are applied for a network of five nodes and the state feedback stabilizing control laws are symbolically generated.

1. INTRODUCTION

We refer to the nonlinear control problem of power system networks in stability crises, following the geometric approach to nonlinear control theory. In particular we are interested into two properties, characterized by necessary and sufficient conditions: strong accessibility [11] and feedback linearizability ([7], [2]). The former guarantees the possibility of controlling in a finite time as many directions as the state space dimensions; the latter allows us to compensate nonlinearities by state feedback. While strong accessibility establishes the existence of admissible stabilizing controls, the stronger feedback linearizability property allows us to transform by state feedback the system into a linear controllable one and therefore to use linear control techniques.

In [4], [7] and [9] the geometric approach to power system stabilization is extensively discussed and it is shown that a power system network is always feedback linearizable with a number of controls

---

This work was partly supported by MPI (fondi 40%).

between 1 and N, if N is the number of nodes of the network reduced at its generating points. The problem of reducing the number of power controllers depends on the network structure: the stronger the sufficient conditions on the network structure are the smaller the number of controls is. In fact one control is enough if the network graph is a straight line. In this paper we adapt to power system equations the algorithms presented in [8] which allow us to check on computers the two aforementioned structural control properties. More precisely, given the incidence matrix of the power system network, at a preliminary stage one can decide the most convenient control locations for inducing strong accessibility or the stronger property of feedback linearizability. When the power controls are given or already established, the use of symbolic manipulation enable us to analyze the effect of parameter or structural perturbations on the aforementioned properties. This is a precious tool since stability problems for power systems arise because of the occurrence of perturbations. Subsequently, as far as feedback linearizable cases are concerned, the symbolic expressions of stabilizing state feedback control laws are automatically generated by computers: given the state, the controls and the frequency at which the power system is going to be stabilized are computed. The whole scheme can be considered as a sort of computer aided design as shown in the simple case of a five machine network.

The paper is organized as follows: in section 2 the stability problem is discussed and the nonlinear mathematical model is presented; in section 3 basic definitions, notation and results are recalled from nonlinear control theory; in section 4, after a brief discussion on the use of SAM systems, the algorithms are introduced and an example of computer aided design is reported for a five machine network.

2. POWER SYSTEM STABILIZATION

Consider a power system network reduced at its N generation nodes. Each node i is characterized by the voltage (E_i, δ_i); each a.c. line ij which connects the nodes i and j is characterized by the impedance (Z_{ij}, θ_{ij}). In nominal conditions each machine at the i-th node is supposed to rotate at synchronous speed $d\delta_i^o/dt=\omega_s$ and the mutual angle positions $\delta_i-\delta_j$ are supposed to be at their nominal values $\delta_i^o-\delta_j^o$ which, along with the nominal voltage magnitudes $(E_1^o,\ldots,E_N^o)$, constitute the stable load flow solution corresponding to a network structure and a set of power injections. If parameter perturbations (changes in the values of power injections or line impedances) or structural perturbations (changes in the network structure) occur, the stable load flow solution is in general no longer the same. The unperturbed load flow solution constitutes the initial point for a perturbed trajectory $(E_i(t),\delta_i(t))$. The dynamics of the mutual angle positions is much faster than the dynamics of the voltage magnitudes. There are physical bounds on the mutual angle positions $\delta_i-\delta_j$: when they are reached circuit breakers are supposed to disconnect the corresponding line ij . These bounds can be reached in less than 3 s . Thus the stabilization problem for power system network can be formulated as follows: find fast acting active power controls, their number, their location and the expression of the control law so that, whenever parameter or structural perturbations occur, the mutual angle positions are kept within the admissible bounded region, i.e. each synchronous machine is prevented from going out of step.

For stabilization purposes the following nonlinear model is considered:

$$\begin{bmatrix} \dot{\delta}_1 \\ \\ \dot{\delta}_N \\ \hline \dot{\omega}_i \\ \\ \\ \dot{\omega}_N \end{bmatrix} = \begin{bmatrix} \omega_1-\overline{\omega} \\ \cdots\cdots \\ \omega_N-\overline{\omega} \\ \hline P_1-\sum\limits_{\substack{j=1\\ j\neq 1}}^{N} k_{1j}\sin(\delta_1-\delta_j+\alpha_{1j}) \\ \cdots\cdots \\ P_N-\sum\limits_{\substack{j=1\\ j\neq N}}^{N} k_{Nj}\sin(\delta_N-\delta_j+\alpha_{Nj}) \end{bmatrix} + \begin{bmatrix} 0 \\ \cdots \\ 0 \\ \hline 1 \\ \cdots \\ 0 \end{bmatrix}\gamma_1 u_1(t) +$$

$$+ \begin{bmatrix} 0 \\ \cdots \\ 0 \\ \hline 0 \\ 1 \\ \cdots \\ 0 \end{bmatrix}\gamma_2 u_2(t) + \cdots\cdots + \begin{bmatrix} 0 \\ \cdots \\ 0 \\ \hline 0 \\ 0 \\ \cdots \\ 1 \end{bmatrix}\gamma_N u_N(t) \triangleq$$

$$\triangleq \quad f(x) + \sum_{i=1}^{N}\gamma_i u_i(t) g_i \tag{1}$$

Recall that the linearization around stable equilibrium points is a gross simplification since, due to perturbations, the state can be very far from any equilibrium point.

The notation is as follows:

γ_i is one only in the case when a power control is acting at the node i ; otherwise γ_i is zero;

$k_{ij} = C_{ij} \frac{1}{M_i} \frac{E_i E_j}{Z_{ij}}$;

$C_{ij} = \begin{cases} 1 & \text{if there is an a.c. line connecting nodes } i \text{ and } j \\ 0 & \text{otherwise;} \end{cases}$

$M_i = J_i \omega_s$, where ω_s is the synchronous speed and J_i the momentum of inertia;

$P_{mi} = P_i^M - P_i^L + \sum_{\substack{j=1 \\ j \neq i}}^{N} C_{ij} \frac{E_i^2}{Z_{ij}} \sin \alpha_{ij}$, where $\alpha_{ij} = \theta_{ij} - 90°$, P_i^M is the power delivered by the machine, P_i^L is the power absorbed by the loads; $P_i = P_{mi}/M_i$; $\bar{\omega}$ is the common angular speed at which all machines tend to rotate in presence of effective stabilizing control action.

As far as control locations are concerned, some control may already exist and restrictions or costs of various kind may be imposed on locating additional controls. Few results are available on the stabilization of nonlinear systems. Besides, since parameter and structural perturbations are likely to occur, one may think of using adaptive techniques or at least robust control schemes. In any case a preliminary property to be checked is the strong accessibility. Once the controls are enough and well located so that strong accessibility is guaranteed, one may use adaptive techniques or feedback linearizing techniques: both were successfully used in the dynamic control of robot arms ([12],[13]), whereas the last technique is currently used in designing autopilots for helicopters [2].

3. SOME FACTS FROM NONLINEAR SYSTEM THEORY

Consider the class of nonlinear systems which can be described by equations of type

$$\dot{x} = f(x) + \sum_{i=1}^{m} u_i(t) g_i(x) \triangleq f(x) + G(x)u(t) \qquad (2)$$

where: $x \in \mathbb{R}^n$; $f, g_1, \ldots, g_m$ are smooth vector fields defined on $\mathbb{R}^n$; $u_1(t), \ldots, u_m(t)$ are Lebesgue measurable functions and represent the inputs of the system. Many significant physical situations can be modeled by equations of type (2): for instance helicopters, robot arms and in general most of controlled mechanical systems, whenever the controls are forces or torques; in fact, according to Newton laws, those models are in general

$$\ddot{q} = a(q,\dot{q}) + \sum_{i=1}^{m} m_i(t)\, b_i(q,\dot{q})$$

and, in state space form, can be expressed as in (2), by defining

$$x = \begin{bmatrix} q \\ \dot{q} \end{bmatrix}, \quad f(x) = \begin{bmatrix} \dot{q}-\dot{q}_o \\ a(q,\dot{q}) \end{bmatrix}, \quad g_i(x) = \begin{bmatrix} 0 \\ b_i(q,\dot{q}) \end{bmatrix}$$

Let us introduce some notations and definitions (see also [3], [4]).

Notation 1. Let $f(x)$ be a smooth vector field on $\mathbb{R}^n$ and $T(x)$ a smooth real valued function on $\mathbb{R}^n$; one denotes the Lie derivative of T with respect to f as

$$L_f T(x) = \langle dT(x), f(x) \rangle = \sum_{i=1}^{n} \frac{\partial T(x)}{\partial x_i} f_i(x)$$

Notation 2. Let $f(x)$ and $g(x)$ be two vector fields defined on $\mathbb{R}^n$; $ad_f g$ or $L_f g$ or $[f,g]$ denotes the Lie bracket of two vector fields defined as

$$[f,g] = \left[\frac{dg}{dx}\right] f - \left[\frac{df}{dx}\right] g$$

where $\left[\frac{dg}{dx}\right]$ and $\left[\frac{df}{dx}\right]$ are jacobians. Also $ad_f^0 g = g$ and, inductively, $ad_f^{\ell+1} g = ad_f(ad_f^{\ell} g)$.

Definition 1. $F(x) = \text{span}\{f_1(x), \ldots, f_k(x)\}$ is an involutive distribution in U, open subset in $\mathbb{R}^n$, if $[f_i, f_j](x) \in F(x)$ for every i,j such that $1 \leq i,j \leq k$ and every $x \in U$.

Let us now define

$$G^0 = \text{span}\,\{g_1, \ldots, g_m\}$$

$$G_f = f + G^0$$

$$G^j = \text{span}\,\{G^{j-1}, [G_f, G^{j-1}]\}$$

In [11] the strongly accessible distribution L_o is introduced: it is defined as the smallest involutive distribution which contains

$$\text{span}\ \{ad_f^{\ell} g_i\ ;\quad \ell \geq 0\ ,\quad 1 \leq i \leq m\}$$

It is proved in [11] that the strongly accessible set from x_o, i.e. the set of points x for which, given any positive time t, there exists a control which takes x_o into x in a preassigned time t, is open and dense in the integral manifold of L_o through x_o.
It is proved in [5] that $L_o(x)=G^{n-1}(x)$ for every x where the distribution $G^{n-1}(x)$ is nonsingular. This allows the computation of L_o in a finite number of steps as shown in [8].

In favourable cases, characterized by necessary and sufficient conditions on the vector fields $f, g, \ldots, g_m$, strongly accessible systems of type (1) can be controlled through an equivalent, more precisely feedback equivalent, linear controllable system.

Definition 2. A system (2) is said to be feedback equivalent in U_{x_o}, open subset in $\mathbb{R}^n$, to the linear controllable system

$$\dot{y} = Ay + \sum_{i=1}^{m} v_i b_i = Ay + Bv \qquad y \in \mathbb{R}^n \tag{3}$$

if there exist a diffeomorphism $T: U \rightarrow T(U)$ so that $T(x_o)=0$ and an affine, state dependent, nonsingular transformation of the control space $\mathbb{R}^m$, which we call S-transformation

$$v = a(x) + S(x)\, u \tag{4}$$

$(a(x_o)=0)$ such that

$$f(x) = \left[\frac{dT}{dx}\right]^{-1} (AT(x) + a(x))$$

$$G(x) = \left[\frac{dT}{dx}\right]^{-1} S(x)$$

Theorem 1 ([1],[2],[5]): The system (2) is locally feedback equivalent in U_{x_o} a neighborhood of x_o, to the system (3) if and only if

(i) $\text{span}\ \{G(x), ad_f G(x), \ldots, ad_f^{n-1} G(x)\} = T_x U_{x_o} \quad x \in U_{x_o}$;

(ii) $\text{span}\ \{G(x), \ldots, ad_j^i G(x)\}$ is an involutive distribution of constant rank r_i in U_{x_o} ;

(iii) $f(x_o) \in \text{span}\ \{g_1(x_o), \ldots, g_m(x_o)\}$.

A set of integers $k_1 \geq \dots \geq k_m$ ($\sum_{i=1}^{m} k_i = n$), called controllability indices, can be uniquely associated to systems (2) which are feedback equivalent to linear controllable systems, i.e. <u>feedback linearizable</u>; k_i is equal to the number $s_j \geq i$, $j \geq 0$ where $s_o = r_o$, $s_i = r_i - r_{i-1}$; the controllability indices are invariant under feedback transformations.

Let $\sigma_o = 0$, $\sigma_1 = k_i, \dots,$ $\sigma_i = \sum_{j=1}^{i} k_i, \dots, \sigma_m = n$.

If the conditions of Theorem 1 are satisfied, there are procedures ([2],[6]) for the construction of the transformations T and S which take the system (2) into a linear system in Brunovsky canonical form with controllability indices $k_1 \geq \dots \geq k_m$:

$$
\begin{bmatrix} \dot y_1 \\ \\ \\ \\ \\ \\ \\ \\ \\ \\ \\ \\ \dot y_n \end{bmatrix}
=
\begin{array}{r}
k_1\left\{\begin{array}{c} \\ \\ \\ \\ \end{array}\right. \\
k_2\left\{\begin{array}{c} \\ \\ \\ \\ \end{array}\right. \\
\\
k_m\left\{\begin{array}{c} \\ \\ \\ \\ \end{array}\right.
\end{array}
\begin{bmatrix}
010\dots0 & & \\
001\dots0 & & \\
\dots\dots & & \\
000\dots0 & & \\
 & 010\dots0 & \\
 & 001\dots0 & \\
 & \dots\dots & \\
 & 000\dots0 & \\
\dots\dots & \dots\dots & \dots\dots \\
 & & 010\dots0 \\
 & & 001\dots0 \\
 & & \dots\dots \\
 & & 000\dots0
\end{bmatrix}
\begin{bmatrix} y_1 \\ \\ \\ \\ \\ \\ \\ \\ \\ \\ \\ \\ y_n \end{bmatrix}
+
\begin{bmatrix}
00\dots0 \\
00\dots0 \\
\dots\dots \\
10\dots0 \\
00\dots0 \\
00\dots0 \\
\dots\dots \\
01\dots0 \\
\dots\dots \\
00\dots0 \\
00\dots0 \\
\dots\dots \\
00\dots1
\end{bmatrix}
\begin{bmatrix} v_1 \\ \\ \\ \\ \\ \\ \\ \\ v_m \end{bmatrix}
\tag{5}
$$

The components of the diffeomorphism $y=T(x)$ which correspond to the first elements of each block in (4), i.e. $T_1, T_{\sigma_1+1}, \dots, T_{\sigma_i+1}, \dots, T_{\sigma_{m-1}+1}$ satisfy for each $i=1,\dots,m$

$$\langle dT_{\sigma_{i-1}+1}, X \rangle = 0 \tag{6}$$

for any $X \in \text{span}\,\{ad_f^{\ell}\, G(x):\ i=0,\dots,k_i-2\}$.

In order to construct the S part of the feedback transformation, one has to find, in general, a relabeling of g_i so that the set of vector fields

$$\{g_i, \dots, ad_f^{k_i-1}\, g_i \ :\ i=1,\dots,m\}$$

is a spanning set in U_{x_o}. This is certainly possible by the definition of controllability indices, if the condition (i) of Theorem 1 holds in U_{x_o}.

An effective construction iterates the following two steps for $i=1,\dots,m$:

1. a smooth function $T_{\sigma_{i-1}+1}$ whose differential satisfies (6) and is independent on $(dT_1, \dots, dT_{\sigma_{i-1}})$ is computed;

2. compute, by Lie differentiation, the additional components

$$T_{j+1}(x) = L_f T_j(x) \qquad j = \sigma_{i-1}+1, \ldots, \sigma_i - 1 \tag{7}$$

Once the iteration of the two steps is over, the entries of the matrix $S(x)$ and the components of the vector $a(x)$, which specify the S part of the feedback transformation, are given by

$$s_{ij}(x) = \langle dT_{\sigma_i}(x), g_j(x) \rangle \qquad 1 \le i, j \le m$$

$$a_i(x) = \langle dT_{\sigma_i}(x), f(x) \rangle \qquad 1 \le i \le m$$

Note that the conditions of Theorem 1 imply the existence of a relabeling of g_i so that the matrix $S(x)$ is nonsingular in U_{x_o} In particular the following corollary holds.

<u>Corollary 2</u>. The matrix $S(x)$ (8) is singular in x_o if and only if the vector fields $\{g_i, \ldots, ad_f^{k_i-1} g_i : i=1,\ldots,m\}$ are linearly dependent in x_o.

With reference to system (1) feedback linearizability allows the following control scheme

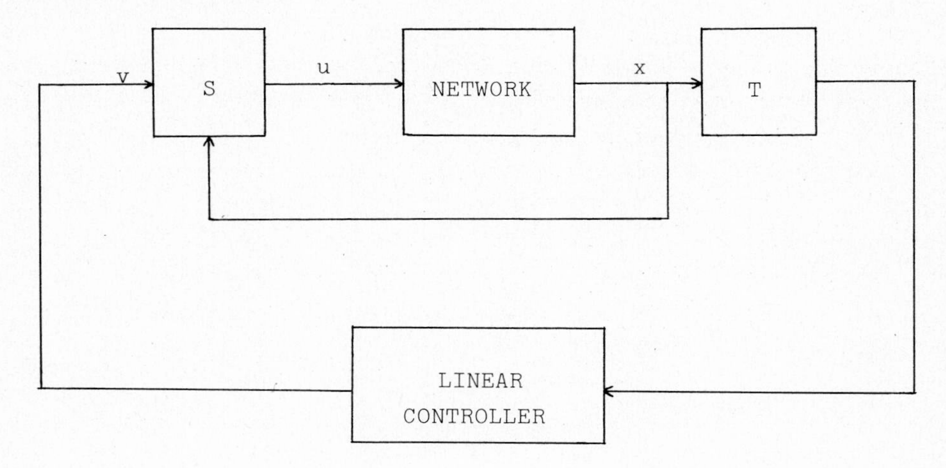

Figure 1

For instance a pole placement technique can be followed by the linear controller, i.e.

$$v_i = \sum_{j=1}^{k_i} c_j \, T_{\sigma_{i-1}+j}(x)$$

4. AN EXAMPLE OF COMPUTER AIDED DESIGN VIA COMPUTER ALGEBRA

In this section we adapt to the power system model (1) algorithms developed in [8] for nonlinear systems of type (2). Those algorithms, to be implemented by symbolic and algebraic manipulation systems, allow a computer to establish strong accessibility and feedback linearizability given the vector fields $f, g_1, \ldots, g_m$. They are based on the theory recalled in the previous paragraph and on two computer algebraic programs, reported in [8], which check the linear independence over the quotient field of analytic functions and the involutivity of a given set of vector fields. Due to the special structure of (1) the algorithms can be easily adapted so that the only inputs required are the incidence matrix of the power system network and the location of power controllers: the outputs are the distributions span$\{ad_f^{\ell} G(x): \ell=0,\ldots,i\}$, $i=0,\ldots,n-1$, whose involutivity in open subset of $\mathbb{R}^n$ is checked and the distribution $G^{n-1}(x)=L_o(x)$. The distributions are given via triangularized basis of vector fields, so that singularities are easily identified. In feedback linearizable cases the controllability indices are automatically computed; the construction of the transformations T and S reported in the previous paragraph can be implemented by symbolic systems (as shown in [15]) whenever the system of linear partial differential equations (6), which represents the crucial step of the procedure, can be easily solved. It is shown in [4] and [7] that this happens in correspondence to certain network structures. Thus, once feedback linearizability is established, in favourable cases we can ask the computer to generate the symbolic expressions of the state feedback control law $u(x)$ as shown in Figure 1.

In the process the computer also provides the symbolic expression of the angular speed $\bar{\omega}$ at which the power system is going to be stabilized; it is obtained by solving for $\bar{\omega}$ the equation $\Sigma\, u_i=0$; in [15] the solution is shown to be

$$\bar{\omega}(x) = \frac{\Sigma\,\Sigma\, s_{ij}^{inv}\,(v_j|_{\bar{\omega}=0} - a_j)}{\Sigma\,\Sigma\, s_{ij}^{inv}\, c_{\sigma_{j-1}+2}}$$

where s_{ij}^{inv} is the ij-th entry of S^{-1}. The parameter $\bar{\omega}(x)$ is needed to generate $u(x)$ and the input to the so called "viabilizing" control [14], whose task is to drive the difference $\bar{\omega}(x)-\omega_s$ to zero by injecting or withdrawing active power. According to the control strategy developed in [14] stabilizing and viabilizing controls lie on different hierarchical levels.

We now show the use of those design techniques in the case of the following network structure.

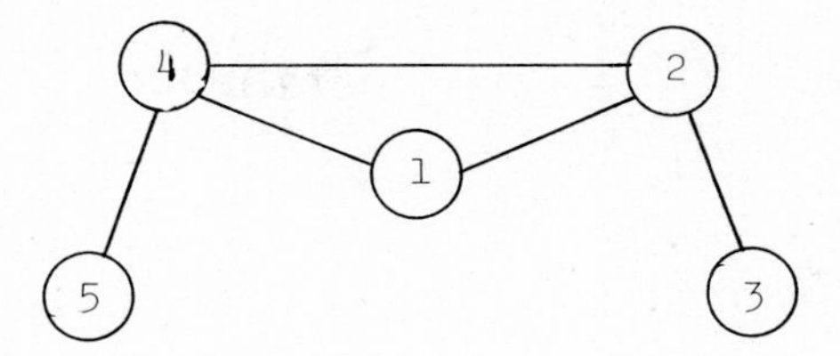

Figure 2

The problem is to design stabilizing power controllers. Suppose we choose to evaluate first the feasibility of feedback linearizing techniques. At a preliminary stage the minimum number of controls and their most convenient location (according to some cost) have to be established. In [1] it is shown that:

(i) one control induces the feedback linearizability property if and only if it is placed at the end of a straight chain of machines; (ii) under rather mild conditions (N+1)/2 controls induce the feedback linearizability if N is the (odd) number of machines. Thus, since one control is not enough whereas three controls are likely to be enough, we may analyze the case of two controls. To this purpose the "feedback linearizable" program needs to be applied 10 times, since there are $\binom{5}{2}$ possible ways of placing two controls. Due to symmetries in the network only six control configurations are to be tested. Asking the computer to do so, one obtains the results reported in the table where the results of the "strong accessibility" program are also reported.

CONTROL LOCATION	FEEDBACK LINEARIZABILITY AND CONTROLLABILITY INDEXES	STRONGLY ACCESSIBILITY
4&2	NO	YES
4&1 or 2&1	YES (6,4)	YES
4&5 or 2&3	NO	YES
3&5	YES (6,4)	YES
5&2 or 3&4	NO	YES
5&1 or 3&1	YES (6,4)	YES

As discussed in [4] at the actual state of the art fast acting controllers can be obtained connecting the nodes to be controlled via d.c. lines. In this case the distance between the controlled nodes can represent a cost to be minimized among the feedback linearizing control configurations. Let for instance the configuration 3&5 be the optimum solution. Then according to the algorithm and the control scheme reported in Section 3, the computer generates the stabilizing control law

$$\begin{bmatrix} u_3 \\ \\ u_5 \end{bmatrix} = S^{-1}(x) \begin{bmatrix} a_1(x) \\ \\ a_2(x) \end{bmatrix} + S^{-1}(x) \begin{bmatrix} \sum_{i=1}^{6} c_i T_i(x) \\ \\ \sum_{i=7}^{10} c_i T_i(x) \end{bmatrix}$$

where $(c_1,\ldots,c_6),(c_7,\ldots,c_{10})$ are the coefficients of the characteristic polynomials of the equivalent linear system to be designed according to linear control techniques and

$$\begin{aligned}
T_1(x) &= x_1 - x_1^o \\
T_2(x) &= f_1 \\
T_3(x) &= f_6 \\
T_4(x) &= K_{12}(f_1-f_2)+K_{14}(f_1-f_4) \\
T_5(x) &= H_{12}(f_1-f_2)^2+H_{14}(f_1-f_4)^2+K_{12}(f_6-f_7)+K_{14}(f_6-f_9) \\
T_6(x) &= 3H_{12}(f_1f_6-f_1f_7-f_2f_6+f_2f_7)+3H_{14}(f_1f_6-f_1f_9-f_4f_6+f_4f_9) \\
&\quad + K_{12}(f_2-f_1)^3+K_{14}(f_4-f_1)^3+f_1(K_{12}+K_{14})^2-2K_{14}^2f_4-2K_{12}^2f_2 \\
&\quad + f_1(K_{14}^2+K_{12}^2)+f_2(K_{14}K_{24}-K_{12}K_{24}-K_{12}K_{23}-K_{12}K_{14})+K_{12}K_{23}f_3 \\
&\quad + K_{14}K_{45}f_5+f_4(-K_{14}K_{45}-K_{14}K_{24}+K_{12}K_{24}-K_{12}K_{14}) \\
T_7(x) &= x_4 - x_4^o \\
T_8(x) &= f_4 \\
T_9(x) &= f_9 \\
T_{10}(x) &= -K_{14}f_1-K_{24}f_2+f_4(K_{45}+K_{24}+K_{14})-K_{45}f_5
\end{aligned}$$

$$\begin{aligned}
a_1(x) &= -6K_{14}f_4^2(f_6-f_9)+12f_1[K_{14}f_4(f_6-f_9) \\
&\quad + K_{12}f_2(f_6-f_7)]-6K_{12}f_2^2(f_6-f_7) \\
&\quad - K_{14}H_{45}(f_4-f_5)^2-H_{12}\{f_1^4-4f_1^3f_2+f_1^2(6f_2^2-9K_{12}-5K_{14})-2f_1[2f_2^3 \\
&\quad - f_2(2K_{24}+2K_{23}+9K_{12}+3K_{14})+2f_4(K_{24}-K_{14}) \\
&\quad + 2f_3K_{23}]+f_2^4-f_2^2(4K_{24}+4K_{23}+9K_{12}+K_{14}) \\
&\quad + 4f_2[f_4(K_{24}-K_{14})+f_3K_{23}]-3f_6^2+6f_6f_7-3f_7\} \\
&\quad - H_{14}\{f_1^4-4f_1^3f_4+f_1^2[6f_4^2-9K_{14}-5K_{12}]
\end{aligned}$$

$$+ 2f_1[f_4(2K_{45}+2K_{24}+4K_{14}+3K_{12}+5K_{14})-2f_2(K_{24}-K_{12})-2f_4^3-2K_{45}f_5\}$$
$$+ f_4^2(4K_{45}+4K_{24}+K_{14}+K_{12})+4f_2f_4(K_{24}-K_{12})+f_4^4+4f_4f_5K_{45}-3(f_6-f_9)^2\}$$
$$- K_{12}H_{12}(f_1-f_2)^2-K_{12}H_{23}(f_2-f_3)^2-K_{12}H_{24}(f_2-f_4)^2$$
$$- K_{14}H_{14}(f_1-f_4)^2-6f_1^2[f_6(K_{12}-K_{14})-K_{12}f_7-K_{14}f_9]$$
$$- K_{14}K_{45}(f_9-f_{10})+2f_6(K_{14}^2+K_{12}^2+ K_{12}K_{14})$$
$$+ f_7(K_{14}K_{24}-K_{12}K_{24}-K_{12}K_{23}-2K_{12}^2-K_{12}K_{14})$$
$$- K_{14}K_{24}f_9+K_{12}K_{23}f_8+K_{12}K_{24}f_9-K_{14}f_9(K_{12}+2K_{14})$$

$$a_2(x) = H_{45}(f_4-f_5)^2+H_{14}(f_1f_4)^2+H_{24}(f_2-f_4)^2$$
$$+ K_{45}(f_9-f_{10})+K_{14}(f_9-f_6)+K_{24}(f_7+f_9)$$

$$s_{11}^{inv} = 1/K_{23}K_{12}$$
$$s_{12}^{inv} = K_{14}/K_{23}K_{12}$$
$$s_{21}^{inv} = 0$$
$$s_{22}^{inv} = 1/K_{45}$$

s_{ij}^{inv} is the entry ij of the matrix $s^{-1}(x)$

$$\overline{\omega}(x) = \frac{\sum_{i=1}^{2}\sum_{j=1}^{2} s_{ij}^{inv}(v_j|_{\omega=0}-a_j)}{\sum_{i=1}^{2}\sum_{j=1}^{2} s_{ij}^{inv}c_{\sigma_{j-1}+2}}$$ is the angular speed at which the

system is going to be stabilized. $K_{ij}=k_{ij}\cos(\delta_i-\delta_j+\alpha_{ij})$ and $H_{ij}=$ $=k_{ij}\sin(\delta_i-\delta_j+\alpha_{ij})$; since it is assumed $M_1=\ldots=M_5$, $K_{ij}=K_{ji}$ and $H_{ij}=-H_{ij}$.

The power system network can be stabilized at any state belonging to the submanifold $\{x\in\mathbb{R}^{10}:f(x)\in \mathrm{span}\{g_1,g_2\}\}$, that is the components (x_1^o,x_4^o) of the induced equilibrium point can be arbitrarity chosen. Parameter perturbations on power injections and line impedances can be detected on line and the control law can be adjusted accordingly. Besides the linearizing technique is robust. Once the control locations are established, structural perturbations such as line removals could affect feedback linearizability or strong accessibility. In our case if we examine the determinant of $S(x)$ provided by the computer we realize that $S(x)$ becomes singular everywhere if either the line 1-2 or 2-3 or 4-5 is removed. According to Corollary 2 this implies that $\{g_1,\ldots,$ $ad_f^5g_1,g_2,\ldots,ad_f^3g_2\}$ is no longer a spanning set of vector fields. Actually if either the line 2-3 or 4-5 is removed the power system is no longer feedback linearizable with two controls whereas if the line 1-2 is removed the control law should be recomputed attributing the controllability indexes (6,4) to g_2 and g_1 respectively. If data on lines fault probability are available a reasonable cost in the choice of control locations could be the probability of preserving feedback linearitability (or strong accessibility). In our case for instance the removals of lines 1-4 and 2-4 preserve feedback linearizability and the control law can be adjusted by setting $k_{14}=0$ or $k_{24}=0$.

As far as strong accessibility is concerned we analyzed on computer the case when the line 2-3 is removed: it turns out that strong accessibility is preserved even if, in this case, the machines 1,2 and 4 are controlled only through machine 5.

One can conclude that strong accessibility is more affected by structural perturbations than feedback linearizability.

5. REFERENCES

[1] B. Jacubczyk, W. Respondek, *On linearization of control systems*, Bull. Acad. Polon. Sci. Ser. Sci. Math., Vol.28, 9-10, 517-522, 1980.

[2] L.R. Hunt, R. Su, G. Meyer, *Design for multiinput nonlinear systems*, in Differential Geometric Control Theory, R. Brockett... ed., 268-298, Birkhäuser, 1983.

[3] W.M. Boothby, *An Introduction to Differentiable Manifolds and Riemannian Geometry*, Academic Press, N.Y., 1979.

[4] R. Marino, *Feedback equivalence of nonlinear systems with applications to power system equations*, Dc. Sc. Dissertation, Washington University, St. Louis, Missouri, 1982.

[5] R. Marino, W.M. Boothby, D.L. Elliott, *Geometric properties of linearizable control systems*, submitted to Int. J. Math. System Theory.

[6] W.M. Boothby, R. Marino, *Some remarks on feedback linearizable systems*, PREPRINT.

[7] R. Marino, *On the stabilization of power systems with a reduced number of controls*, Sixth Int. Conf. on Analysis and Optimization of Systems, INRIA, Nice, 1984.

[8] R. Marino, G. Cesareo, *Nonlinear control theory and symbolic algebraic manipulation*, presented at 6-th Int. Symp. MTNS, Beer Sheva, Israel, 1983.

[9] R. Marino, *A geometric approach for state feedback stabilization of power system networks*, Int. Conf. on Modelling, Identification

and Control, IASTED, Innsbruck, 1984.

[10] A.C. Hearn, *REDUCE-2 User's Manual*, University of Utah, Symb. Comp. Group, Technical Report n. UCP-19, 1973.

[11] H.J. Sussmann, V. Jurdjevic, *Controllability of nonlinear systems*, J. Diff. Eqs., Vol.12, 95-116, 1972.

[12] E. Freund, *Fast nonlinear control with arbitrary pole placement for industrial robots and manipulators*, Int. J. Robotics Research, 1.1., 65-78, 1982.

[13] M. Brady, J.M. Hollerback, T.L. Johnson, T. Lozano-Perez, M.T. Mason eds., *Robot Motion: Planning and Control*, MIT Press, 1982.

[14] J. Zaborszky, K.W. Whang, K.V. Prasad, *Operation of the large interconnected power system by decision and control in emergencies*, Report SSM 7907, Dept. of System Science and Math., Washington University, St. Louis, Missouri, 1979.

[15] G. Cesareo, *Symbolic algebraic manipulation and nonlinear control theory*, Technical Report no.82-14, Istituto di Automatica, Università di Roma, 1982.

Session 20

PRODUCTION AUTOMATION

AUTOMATISATION DE LA PRODUCTION

SHORT TERM PRODUCTION SCHEDULING OF AN AUTOMATED MANUFACTURING FACILITY

Stanley B. Gershwin, Ramakrishna Akella, Yong Choong, and Sanjoy K. Mitter

Laboratory for Information and Decision Systems
Massachusetts Institute of Technology
77 Massachusetts Avenue
Cambridge, Massachusetts 02139

ABSTRACT

We describe extensions to the on-line hierarchical scheduling scheme for flexible manufacturing systems of Kimemia and Gershwin. Major improvements to all levels of the algorithm are reported, including algorithm simplification, substantial reductions of off-line and on-line computation time, and improvement of performance. Simulation results based on a detailed model of an IBM printed circuit card assembly facility are presented.

1. INTRODUCTION

This paper describes extensions to the work reported by Kimemia (1982) and Kimemia and Gershwin (1983) on the on-line scheduling of flexible manufacturing systems. Major improvements to all levels of the hierarchical algorithm are reported and simulation results are presented. The results indicate that the approach is practical, well-behaved, and robust. A full description of the results appears in Gershwin, Akella, and Choong (1984) and Akella, Choong, and Gershwin (1984).

A flexible manufacturing system (FMS) is one in which a family of related parts can be made simultaneously. It consists of a set of computer-controlled machines and transportation elements. The changeover time between different operations at a machine is small compared with operation times.

Processing a mix of parts makes it possible to utilize the machines more fully than otherwise. This is because different parts spend different amounts of time at the machines. Each part type may use some machines heavily and others very little or not at all. If complementary part types are selected for simultaneous production, the machines that are lightly used by some parts can be loaded with others that do require them.

In principle, therefore, line balancing can keep several machines busy at the same time. However, scheduling such a system is difficult because there are several machines, several part types, and many parts. In addition, like all manufacturing systems, a FMS is subject to random disturbances in the form of machine failures and repairs, material unavailability, "hot" items or batches, and other phenomena. These effects further complicate an already difficult optimization problem.

Gershwin, Akella, and Choong present a hierarchical description of the manufacturing scheduling problem. At the top of the hierarchy are the long term decisions, such as what capital equipment to acquire. At the bottom is the decision of which part to load into an existing FMS, and when to load it. This is an extension to Kimemia and Gershwin's short term hierarchy (Figure 1).

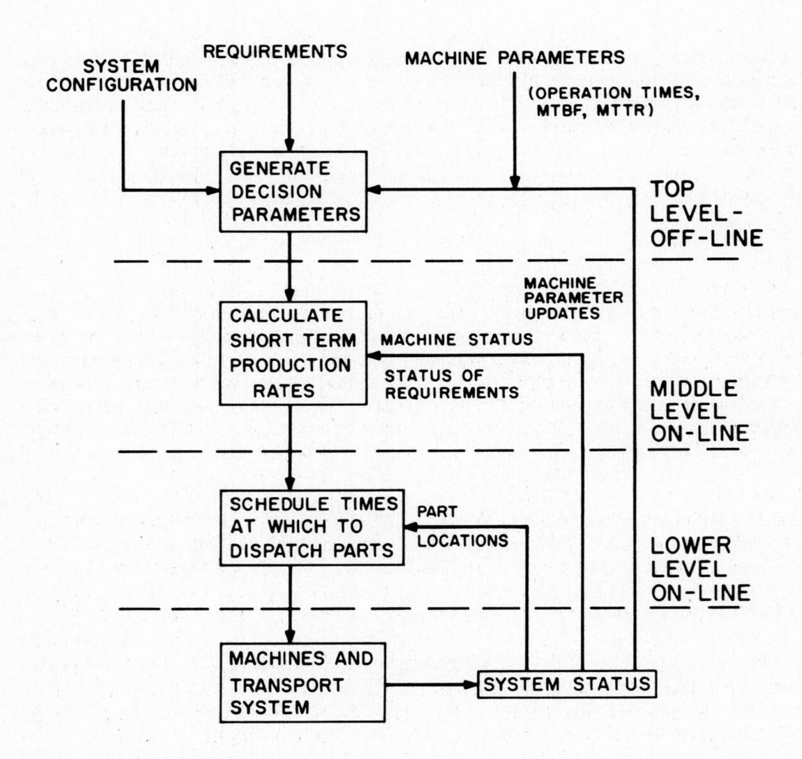

Figure 1. Hierarchical Production Algorithm.

2. CONTINUOUS FORMULATION

The purpose of the short-term FMS scheduling algorithm is to solve the following problem: when should parts (whose operation times at machines are on the order of seconds or minutes) be dispatched into an FMS whose machines are unreliable (with mean times between failures and mean times to repair on the order of hours) to satisfy production requirements that are specified for a week? Kimemia and Gershwin's approach decomposed the problem into two parts: a continuous dynamic programming problem to determine the instantaneous production rates and a combinatorial algorithm to determine the dispatch times.

The continuous part is further divided into the top and middle levels. The top level calculates a value or cost-to-go function and is executed off-line. The middle level uses the cost-to-go function to determine instantaneous flow rates and part mixes.

Assume that the production requirements are stated in the form of a demand rate vector d(t). Let the instanteous production rate vector be denoted u(t). Define x(t) to be production surplus. It is the cumulative difference between production and demand and satisfies

$$\frac{dx}{dt} = u(t) - d(t). \tag{1}$$

If x(t) is positive, more material has been produced than is currently required. This surplus or safety stock is helpful to insure that material is always available over the planning horizon. However, it has a cost. Expensive floor space and material handling systems must be devoted to storage. In addition, working capital has been expended in the acquisition and processing of stored materials. This capital is not recovered until the processing is complete and the inventory is sold.

If x(t) is negative, there is a backlog, which is also costly. Backlog represents either starved machines downstream or unsatisfied customers. In the former case, valuable capital is underutilized; in the latter, sales and good will may be lost.

The production rate vector u is limited by the capabilities of the machines. Let part type j require time τ_{ij} on machine i for all of its operations. (Note that the order in which parts go to machines is not relevant for this calculation. Nor is the number of times a part visits a machine. For simplicity, we assume here that there is only one path for each part.) Then

$$\sum_j \tau_{ij} \, u_j(t) \le \alpha_i(t) \tag{2a}$$

where $\alpha_i(t)$ is 1 if machine i is operational and 0 if it is down. More generally, if there is a set of identical type i machines, $\alpha_i(t)$ is the number of these that are operational at time t. Note also that

$$u_j \geq 0. \qquad (2b)$$

Inequalities (2a) and (2b) can also be written as

$$u(t) \in \Omega(\alpha(t)). \qquad (2)$$

3. TOP-LEVEL COMPUTATION (GENERATE DECISION TABLES)

Costs are incurred when x is far from zero. Kimemia and Gershwin describe the following dynamic optimization problem:

$$\text{minimize } E \int g(x(t))\, dt$$

$$\text{subject to (1), (2),} \qquad (3)$$
$$\text{and initial conditions } x(0) \text{ and } \alpha(0).$$

The optimal value of the cost of this problem is called $J(x(0), \alpha(0))$.

Kimemia and Gershwin suggest a decomposition by which the n'th order Bellman partial differential equation for $J(x,\alpha)$ is replaced by n first order Bellman ordinary differential equations (where n is the number of part types, ie, the dimensionality of x, u, and d).

Kimemia further suggests approximating the solution to each one-dimensional dynamic programming problem with a quadratic cost function. Not only does this reduce data requirements, but it also simplifies the middle-level computation. As a result, the cost function is then written

$$J(x, \alpha) = \tfrac{1}{2}\, x^T A(\alpha)\, x + b(\alpha)^T\, x + c(\alpha) \qquad (5)$$

where $A(\alpha)$ is a diagonal matrix, $b(\alpha)$ is a vector, and $c(\alpha)$ is a scalar whose value is not important. In this section, we propose an alternate technique for obtaining approximate values of the coefficients of (5).

The function $J(x(t),\alpha)$ is a decreasing function of t when α remains constant. The hedging point, given by

$$H_i(\alpha) = -b_i(\alpha)/A_{ii}(\alpha) \qquad (6)$$

is the minimum value of $J(x,\alpha)$ for α fixed. It is the value that x reaches if α stays constant for a long time and if d is feasible, ie if $d \in \Omega(\alpha)$.

In order to calculate the hedging point, consider Figure 2 which demonstrates a typical trajectory of $x_i(t)$. Assume x_i has reached $H_i(\alpha)$, the hedging point corresponding to the machine state before the failure. Then u_i is chosen to be d_i

and x_i remains constant.

A failure occurs at time t_0 that forces u_i to be 0. This causes x_i to decrease at rate $-d_i$. In fact, if the failure lasts for a length of time T_r, then the minimum value of x_i is

$$H_i - d_i T_r. \tag{7}$$

Just after the repair (at time $t_0 + T_r$), u_i is assigned the value U_i. Assuming that this value is greater than demand d_i, x_i increases at rate $U_i - d_i$ until it reaches the hedging point H_i (at time t_3). At that time, u_i resumes its old value of d_i and x_i stays constant until the next failure, at time $t_0 + T_r + T_f$.

To simplify the analysis, we make several assumptions:

1. u_i is constant between the repair ($t_0 + T_r$) and when x_i reaches H_i (t_3).

2. T_r and T_f can be replaced by their expected values, the MTTR and MTBF.

3. The cost function g( ) in (3) penalizes positive areas in Figure 2 with weight a and negative areas with weight b, where a and b are positive scalars.

The positive area between t_0 and $t_0 + T_r + T_f$ is the area between t_0 and t_1 plus the area between t_2 and $t_0 + T_r + T_f$, where

$$t_1 = t_0 + H_i / d_i,$$

$$t_2 = t_0 + T_r - (H_i - d_i T_r) / (U_i - d_i)$$

and

$$t_3 = t_0 + T_r + d_i T_r / (U_i - d_i).$$

The positive area is

$$PA = \frac{1}{2} \frac{H_i^2 U_i}{d_i (U_i - d_i)} + H_i T_f - \frac{d_i T_r}{U_i - d_i} .$$

The absolute value of the negative area is

$$NA = \frac{1}{2} \frac{(H_i - d_i T_r)^2 U_i}{d_i (U_i - d_i)} .$$

Both terms are always positive.

The cost function, according to assumption 3, is then

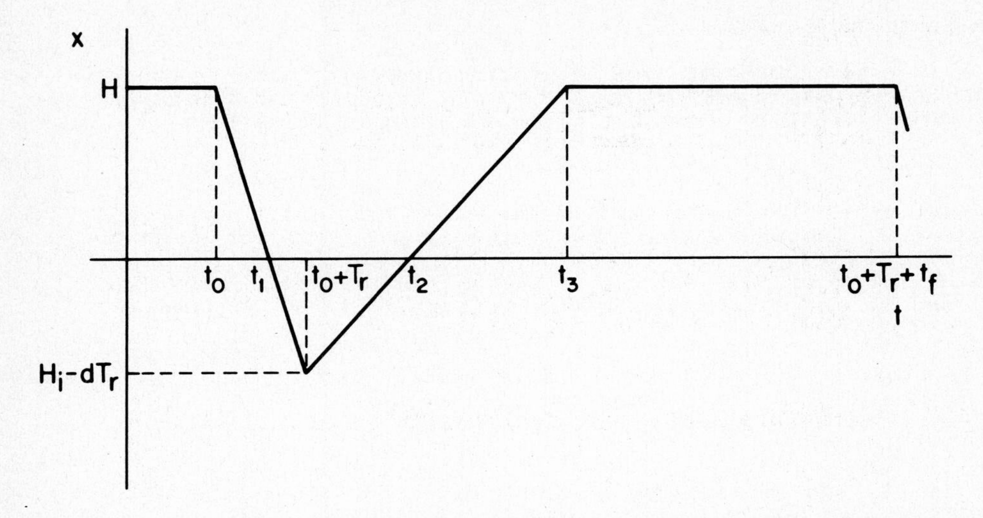

Figure 2. Typical x Trajectory During a Repair-Failure Cycle.

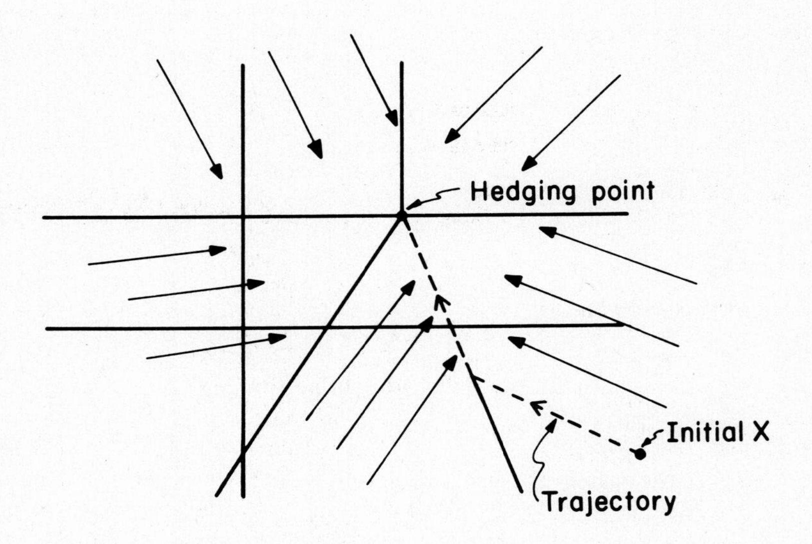

Figure 3. Regions of x-Space, and a Trajectory.

a PA + b NA

This quantity is the cost incurred per average repair and failure cycle of a machine. To find H_i, we must minimize it. This is not difficult because the cost is quadratic in H.

The minimizing H_i is

$$\frac{T_r\ d_i\ (b\ U_i\ -\ a\ d_i)\ -\ T_f\ a\ d_i\ (U_i\ -\ d_i)}{(a\ +\ b)\ U_i}\ . \qquad (8)$$

For machine states in which the demand is not feasible, this approach does not apply. The hedging point for such states must be larger than (8).

This calculation does not consider the fact that the failure may occur before the state reaches the hedging point or that the repair may occur before the state becomes negative. It assumes a specific form of g. These considerations and others should be the subject of future research, but it is important to observe that this method produced very satisfactory results.

$A_{ii}(\alpha)$ must be positive in order for J to be convex. Its value reflects the relative priority of part type i. Parts that have great value, or that would cause great difficulty if backlogged, or that pass through relatively unreliable machines should have large values of A_{ii}.

4. MIDDLE LEVEL (CALCULATE SHORT TERM PRODUCTION RATES)

Chattering

The optimal production rate vector u(t) satisfies the following linear programming problem.

$$\text{minimize} \quad \frac{\partial J(x,\alpha)}{\partial x}\, u \qquad (9)$$

subject to (2).

This is a feedback law since the problem is only specified when x and α are determined. The numerical solution of (9) is implemented on-line at the middle level of the hierarchical algorithm.

For every α, x-space is divided into a set of regions (open, connected sets) and the boundaries between them. Each region is associated with a corner of $\Omega(\alpha)$. When x is in the interior of region R_i, the value of u that satisfies (9) is the corresponding corner P_i.

Kimemia and Gershwin implemented (9) in a simulation by solving it every time step (one minute). This worked well while x was in the interior of a region R_i. However, when x crossed certain boundaries between regions, this approach worked poorly. After x(t) crossed such an attractive boundary, the value of u corresponding to the new region R_j was such that the derivative (1) pointed toward R_i. When x(t) crossed the boundary back into R_i, the derivative pointed again to R_j. Thus, u(t) jumped between adjacent corners P_i and P_j of $\Omega(\alpha)$.

This chattering behavior is undesirable. It allows the flow rate to change more frequently than parts are loaded into the system. The flow rates are such that, very often, at least one of the machines is fully utilized or totally unutilized (since u(t) is at an extreme point of $\Omega(\alpha(t))$. When u jumps frequently from one corner of $\Omega(\alpha)$ to another, the algorithm is trying to switch rapidly from keeping one set of machines fully loaded or unloaded to keeping another fully loaded or unloaded.

It cannot do this successfully: neither set of machines is fully loaded or unloaded. As a result, if the demands on the system are near its capacity, it will fail to meet the demands. This behavior was observed by Kimemia (1982).

Planar Boundaries

Kimemia shows that the optimal $J(x,\alpha)$ is convex in x for each α. He also shows that J decreases when u satisfies (9) and d is feasible. The minimal value of J is achieved when x is at the hedging point. When J is given by (5), its minimum is reached at (6). Gershwin, Akella, and Choong show that when J is quadratic, the regions of x-space (in which the solution u of (9) is constant at a corner of $\Omega(\alpha)$) are cones.

Linear program (9) can now be written

$$\begin{aligned} \text{minimize} \quad & c(x)^T u \\ \text{subject to} \quad & D u = e \\ & u \geq 0 \end{aligned} \tag{10}$$

where u has been expanded with slack variables so that inequality constraint (2a) can be written as an equality, and

$$c(x) = A x + b.$$

(Note that arguments α and t are suppressed.) The standard solution of (10) (Luenberger, 1977) breaks u into basic (u_B) and non-basic (u_N) parts, with c(x) and D broken up correspondingly. The basic part of D is a square, invertible matrix. By using the equality in (10), u_B can be eliminated, and the problem becomes

$$\text{minimize } c_R(x)^T u_N \qquad (11)$$
$$\text{subject to } \quad u_N \geq 0$$

where the constraint on u_B has been suppressed, and where

$$c_R(x)^T = c_N(x)^T - c_B(x)^T D_B^{-1} D_N$$

is the reduced cost. If all components of c_R are positive, then there is a solution to (11): $u_N = 0$. This and the corresponding u_B form an optimal solution to (10). Otherwise, (11) does not have a bounded optimal solution and (11) is not equivalent to (10).

It is important to note that since c is a function of x, the basic/non-basic breakup of this problem depends on x. That is, the set of components of u that are treated as basic varies as a function of x.

At every x in region R_i, corner P_i is the optimal value of u for (10). In each region, then, there must be a basic/non-basic break-up of (10) which is constant. Consequently, $c_R(x)$ must be positive everywhere in its own region and it must have some negative components elsewhere. The boundaries of the regions are determined by some components of $c_R(x)$ being equal to zero.

The boundaries of the regions are therefore portions of hyperplanes. This is because c(x) is linear in x. Consequently $c_N(x)^T$ and $c_B(x)^T$ and therefore $c_R(x)$ are also linear in x.

Qualitative Behavior of Trajectories

Since u is constant throughout a region, dx/dt is also constant. The buffer state x travels along a straight line in the interior of each region. As indicated in Figure 3, such lines may intersect with one or more boundaries of the region. When x reaches a boundary, u and therefore dx/dt changes.

Some boundaries are such that when they are reached, the trajectory continues, after changing direction, into the adjacent region. Others, that we call attractive boundaries, are different. The trajectories on both sides of such boundaries point toward them. Consequently, the trajectories tend to move along the boundaries.

We can now qualitatively describe the trajectory. After a machine state change, x(t) is almost always in the interior of a region. It moves in the characteristic direction of that region (which corresponds to a corner of the $\Omega(\alpha(t))$ polyhedron) until it reaches a boundary. If the boundary is not attractive, x(t) moves in the interior of the next region until it reaches the next boundary. The production rate vector u jumps to an adjacent corner. This behavior continues until x(t) encounters an attractive boundary. At this time, the trajectory begins to move along the boundary and u(t) jumps to a point on the edge of $\Omega(\alpha)$ between the corners corresponding to the regions on either side of the boundary.

The trajectory continues until it hits the next attractive boundary. After that, x(t) moves along the intersection of three regions. The production rate vector is on the surface determined by the three corners corresponding to these regions.

This behavior continues: x(t) moves to lower dimensional boundaries and u(t) jumps to higher dimensional faces. It stops when either the machine state changes (that is, a repair or failure takes place) or u(t) becomes constant. If the demand is feasible (that is, if $d \in \Omega(\alpha)$) then the constant value for u is d. When that happens, x also becomes constant and its value is the hedging point. If the demand is not feasible, x does not become constant. Instead, some or all of its components decrease without limit.

Consequently, for a constant machine state, the future behavior of x(t) would be determined from its current value. We call this the "conditional future trajectory" or the "projected trajectory".

4.2 Calculation of the Conditional Future Trajectory

Assume that the conditional future trajectory is to be calculated at time t_0. This may be due to a machine state change.

As soon as the machine state change occurs, linear program (10) is solved. Thus the basic/non-basic split is determined and the $c_R(x)$ function is known.

The production rate vector at $t=t_0$ is denoted u_0. The production rate remains constant at this value until $t=t_1$, which is to be determined. In $[t_0,t_1]$, x is given by

$$x(t) = x(t_0) + (u_0 - d)(t - t_0)$$

where $x(t_1)$ is on a boundary. Then t_1 is the smallest value of t for which some component of $c_R(x(t))$ is zero. It is easy to calculate this quantity since c_R is linear in x and x is linear in t. Once t_1 is found, $x(t_1)$ is known. Define h(x(t)) to be the component of $c_R(x(t))$ that reaches zero at $t=t_1$. Because h is a linear scalar function of x, we can write

$$h(x(t)) = f^T (x(t) - x(t_1)).$$

For $t > t_1$, there are two possibilities. The trajectory may enter the neighboring region and travel in the interior until it reaches the next boundary. Alternatively, it may move along the boundary it has just reached. To determine whether or not the boundary is attractive, we must consider the behavior of

$h(x(t))$ in its neighborhood.

We know that $h(x)$ is negative in the region across the boundary since this is how the regions are defined. We must determine whether h is increasing or decreasing on trajectories inside that region. If h is decreasing, x moves away from the boundary (where h is zero) into the interior. If h is increasing, trajectories move toward the boundary which must therefore be attractive.

One value of x which is just across the boundary is

$$x'' = x(t_0) + (u_0 - d)(t_1 + \epsilon - t_0)$$

$$= x(t_1) + (u_0 - d)\epsilon.$$

This is the value x would have if u were allowed to be u_0 until $t_1 + \epsilon$.

Let u" be the solution to (10) in the adjacent region. That is, (10) is solved with x given by x". (This can be performed efficiently.) Let x^* be the value of x at $t_1+\epsilon$ if u" were used after t_1. That is,

$$x^* = x(t_1) + (u'' - d)\epsilon.$$

Then

$$h(x'') = f^T (u'' - d)\epsilon.$$

Therefore h is increasing and the boundary is attractive if and only if

$$f^T (u'' - d) > 0.$$

If the boundary is not attractive, define $u_1 = u''$. Then the process is repeated to find t_2, $x(t_2)$, t_3, $x(t_3)$, and so forth until an attractive boundary is encountered. (It should be remembered that this is an on-line computation that is taking place at time t_0. The future trajectory is being planned.)

If the boundary is attractive, a value of u must be determined which will keep the trajectory on it. Otherwise chattering will occur. For the trajectory to stay on the boundary,

$$h(x(t)) = 0$$

or, since $h(x(t_1)) = 0$,

$$\frac{d}{dt} h(x(t)) = f^T(u - d) = 0. \qquad (12)$$

Although u is an optimal solution to (10), it is no longer

determined by this linear program. In fact u_0, u", and any convex combination of them are optimal. This is because one or more of the reduced costs is zero while x is on a boundary. Consequently, the new scalar condition (12) is required to determine the solution. The linear program is modified as follows:

$$\begin{aligned} &\text{minimize} \quad c(x)^T u \\ &\text{subject to } D u = e \\ &\qquad u \geq 0 \\ &\qquad f^T u = f^T d \end{aligned} \tag{13}$$

By adding equation (12) to (13), we are requiring that the solution keeps x(t) on the boundary. We are also replacing the reduced cost which has become zero with a new equation, so that the new problem has a unique solution.

The solution to (13) is the value of u that keeps the trajectory on the boundary. As before, this value is maintained until a new boundary is encountered.

New boundaries may still be attractive or unattractive. The same tests are performed: x is allowed to move slightly into the next region to determine the value of u. The time derivative of the component of the reduced cost that first reaches zero (h) is examined. If it is negative, the boundary is unattractive and the trajectory enters the new region. If it is positive, a new constraint is added to linear program (13).

Constraints, when added to (13), are not deleted. As the number of constraints increases, the surfaces that u is found on in $\Omega(\alpha)$) increase in dimension. That is, u is first on a corner. When the fist attractive boundary is encountered, u is on the edge formed by the convex combination of the corners corresponding to the regions adjacent to the boundary. When the next attractive boundary is reached, u is a convex combination of three corners, and so forth.

At the same time, x is found in regions of decreasing dimension. After a machine state change, $x(t_0)$ is in the interior of a region of full dimensionality. The first attractive boundary x(t) reaches is a hyperplane separating regions of full dimensionality, so its dimensionality is one less than full. The next boundary is the intersection of two such boundaries and thus has dimensionality one smaller.

Since this is a finite dimensional system, this process must terminate. There are two cases. If the demand is feasible, ie if d is a feasible solution of (10), then d is a feasible solution of (13). This is because d satisfies (12) for all f. As new constaints of the form (12) are added to (13), d remains feasible. Finally, if enough linearly independent constraints are added, there is only a single feasible solution to (13) and that is $u=d$.

Since the dynamics of x are given by (1), x remains constant when $u=d$. The value of this constant is the hedging point, discussed above, which is the minimum of $J(x, \alpha)$ for the current value of α.

If the demand is not feasible, u cannot be equal to d and thus x cannot become constant. Instead, the process described above terminates with u satisfying linear program (13) including one or more constraints of the form (12). The vector $x(t)$ is eventually of the form

$$x(t) = x(t_j) + (u - d)\ t.$$

Since d is not feasible, some or all of the components of $u - d$ are negative. The corresponding components of x decrease without limit.

4.3 COMPUTATIONAL CONSIDERATIONS

The conditional future trajectory is calculated whenever the machine state changes, either due to a failure or a repair. It may also be calculated under other conditions: periodically, to ensure that the actual trajectory is close to the projected trajectory; or after unanticipated events such as parts not being loaded into the system in the prescribed manner.

To begin the computation, a linear programming problem (10) must be solved. The number of variables (production rates and slack variables) is the number of part types plus the number of distinct machines.

As each boundary is reached, one (if unattractive) or two (if attractive) additional programs are solved. The numerical effort is very small, however, since each starting basic feasible solution is the solution of the previous problem. We expect that no more than a few pivots of the simplex method will be required to find each new solution.

5. LOWER LEVEL

Lower Level (SCHEDULE TIMES AT WHICH TO DISPATCH PARTS)

The new part loading scheme is based on the conditional future trajectory $(x(s),\ s>t)$. Define the actual surplus of part type i at time t to be

$$x^A_i(t) = [\text{number of parts of type } i \text{ loaded during } [0,t]] - d_i\ t.$$

Note that $x^A_i(t)$ is an irregular sawtooth function of time. It jumps by 1 each time a part is loaded. At other times, it decreases at rate d_i.

The loading strategy insures that $x^A(t)$ is near $x(t)$. The strategy is: at each time step t, load a part of type i if

$$x^A_i(t) < x_i(t). \tag{14}$$

Do not load a part of type i otherwise. A rule is required to resolve conflicts; it probably does not matter what that rule is since conflicts will not arise very often.

Behavior of the New Strategy

Figure 4 demonstrates the behaviors of projected and actual trajectories. It shows a portion of a projected trajectory and of an actual trajectory that was determined by this method.

The actual trajectory remains as close as possible to the projected trajectory. (There are five other such trajectories because six part types are being produced in the simulated system). No difficulty is experienced at the time (about 9740) when the loading rate changes.

6. SIMULATION RESULTS

A detailed simulation of an IBM flexible manufacturing system was written to test the hierarchical scheduling policy and to compare it with other reasonable ploicies. The simulation is described in Akella, Bevans, and Choong (1984). A full description of the results appears in Akella, Choong, and Gershwin (1984).

Six part types are being made simultaneously. Failures and repairs of machines take place at random times. Each MTBF is 10 hours and each MTTR is 1 hour, so that the efficiencies are all 91%. Demands are chosen so that the machines are utilized 98%, 91%, 96%, and 97% of the expected available time.

A variety of alternative policies was formulated to compare with the hierarchical policy. Policy X was: If more than N parts are in the system, do not load a part. If N or fewer parts are in the system, load the part that is furthest behind or least ahead of demand. Do not allow any part type to get more than K parts ahead of demand.

Parameter N must be chosen. Little's law (Little, 1961) gives some guidance, but simulation experience indicates that behavior can depend critically on its value.

Figure 5 displays the results of four runs of the hierarchical policy and four runs of Policy X. All runs were performed with the same seed for the random number generator. That is, each had the same sequence of repairs and failures. The horizontal axis displays the average number of parts in the system. The

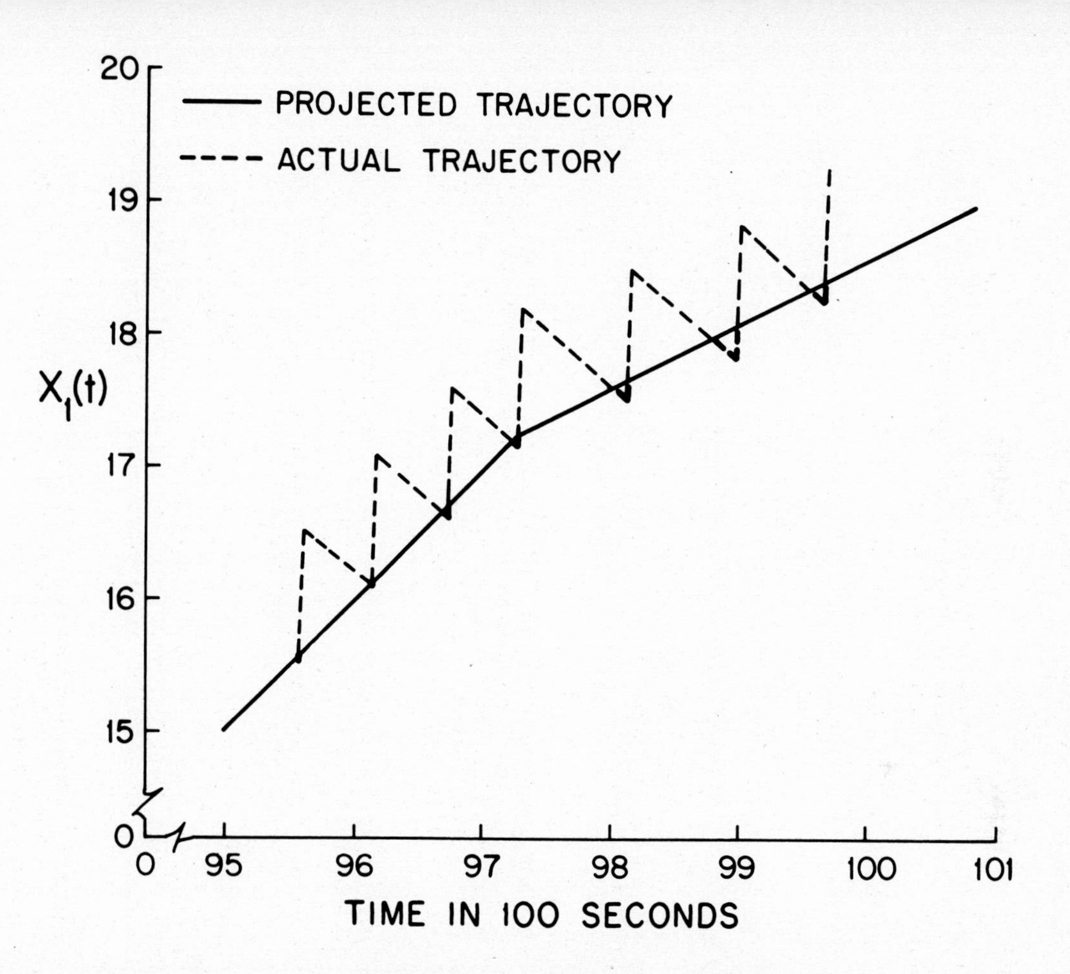

Figure 4. Lower Level Behavior.

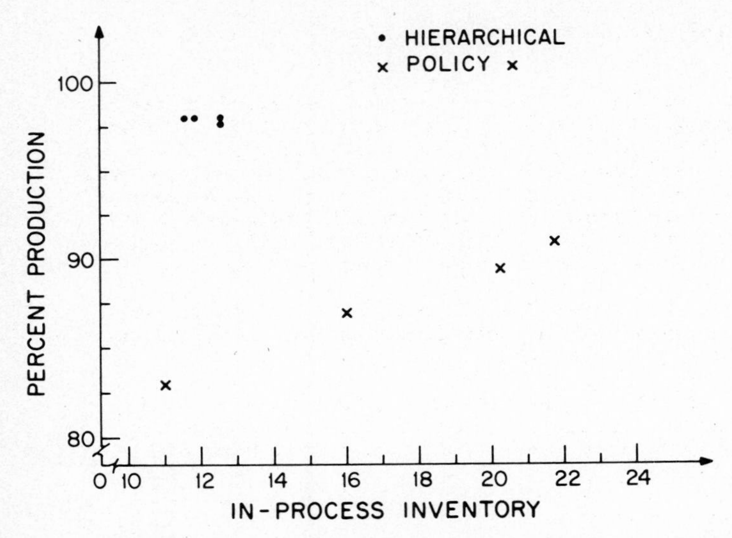

Figure 5. Production and Inventory Comparisons.

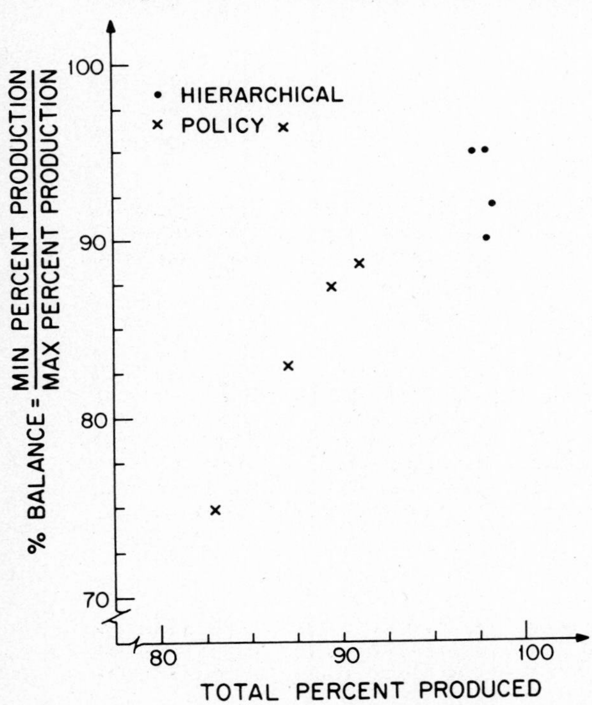

Figure 6. Balance Comparisons.

vertical axis shows the percentage of the total requirements that was actually produced.

The four hierarchical runs had different values of A and b. The Policy X runs had different values of N. Figure 5 indicates that the hierarchical strategy produced superior results. The in-process inventory was lower and the production percentage was greater: in fact, over 98%. In addition, although the values of the A and b parameters differed considerably, the four hierarchical points are clustered quite close together. This indicates that the policy is not sensitive to these parameters.

Figure 6 demonstrates how the policies satisfy balance requirements. The horizontal axis is the same as the vertical axis of Figure 5. To define balance, let

$$Z_i = \frac{\text{number of type i parts produced during the run}}{\text{number of type i parts required during the run}} .$$

Then balance is defined as

$$\frac{\min Z_i}{\max Z_i} .$$

It is important that balance be near 100% to ensure that only what is required is produced.

Akella, Choong, and Gershwin (1984) present a fuller comparison of these and other policies. The more sophisticated policies performed better than Policy X, but not as well as the hierarchical policy.

7. SUMMARY AND CONCLUSIONS

The hierarchical scheduling policy devised by Kimemia and Gershwin for flexible manufacturing systems has been further developed and tested. This policy is designed to respond to random disruptions of the production process. In its current formulation, it treats unpredictable changes in the operational states of the machines: repairs and failures. All levels of the policy have been improved, and the policy shows great promise for practical application.

REFERENCES

R. Akella, J. P. Bevans, and Y. Choong (1984), "Simulation of a Flexible Electronic Assembly System," Massachusetts Institute of Technology Laboratory for Information and Decision Systems Report to appear.

R. Akella, Y. Choong, and S. B. Gershwin (1984), "Performance of Hierarchical Production Scheduling Policy," Massachusetts Institute of Technology Laboratory for Information and Decision Systems Report LIDS-FR-1357.

S. B. Gershwin, R. Akella, and Y. C. Choong, "Short Term Production Scheduling of an Automated Manufacturing Facility," Massachusetts Institute of Technology Laboratory for Information and Decision Systems Report LIDS-FR-1356.

J. G. Kimemia (1982), "Hierarchical Control of Production in Flexible Manufacturing Systems, Massachusetts Institute of Technology Laboratory for Information and Decision Systems Report LIDS-TH-1215.

J. G. Kimemia and S. B. Gershwin (1983), "An Algorithm for the Computer Control of Production in Flexible Manufacturing Systems, IEE Transactions, Volume 15, No. 4, December, 1983, pp. 353-362.

D. Luenberger (1977) Introduction to Linear and Nonlinear Programming, Addison-Wesley.

J. D. C. Little (1961) "A Proof for the Queuing Formula: $L = \lambda W$," Operations Research, Volume 9, Number 3, pp. 383-387.

ACKNOWLEDGMENTS

Research support has been provided by the Manufacturing Research Center of the Thomas J. Watson Research Center of the International Business Machines Corporation; and by the U. S. Army Human Engineering Laboratory under Contract DAAK11-82-K-0018.

OPTIMISATION DE LA REPARTITION DES PALETTES DANS UN ATELIER FLEXIBLE

Yves Dallery
Laboratoire d'Automatique de Grenoble (L.A. 228,CNRS)
Institut National Polytechnique de Grenoble
B.P. 46, 38042 Saint-Martin d'Hères, France
et
Automatique Industrielle S.A.
9 Rue Benoît Malon, 92150 Suresnes, France

ABSTRACT

In a flexible manufacturing system, the pallet, which is the physical system on which the part is fixed, is a limited resource. Moreover, for each part type, a specific pallet type is needed. The number of pallets of each type is an important parameter because the performances of the flexible manufacturing system will be dependent on it.

In this paper, an algorithm for the optimization of the pallet numbers is proposed. It uses an analytical method which enables to evaluate the performances of a flexible manufacturing system ensuring a prescribed product mix. This method enables to determine, for a given configuration of pallets, the critical pallet type.

I - INTRODUCTION

L'atelier flexible est un système de production complexe, ce qui rend sa mise en oeuvre particulièrement difficile. Il est donc intéressant de disposer de différentes méthodes qui pourront faciliter sa conception. En particulier, les problèmes de dimensionnement, d'optimisation, et de définition de la conduite d'un atelier flexible se posent et demandent souvent une approche nouvelle et spécifique à un tel système de production. Cela est particulièrement évident si on s'intéresse au problème de la répartition des palettes dans un atelier flexible. La palette est le support physique sur lequel la pièce est fixée en vue de permettre une manutention automatisée. Chaque palette est spécifique à un type de pièces ; de plus, si la pièce nécessite plusieurs posages, un type de palettes sera associé à chacun des différents posages de la pièce. Les palettes doivent donc être considérées comme une ressource limitée dont il faut déterminer la quantité. Un grand nombre de palettes entraîne d'une part un surcoût important et, d'autre part, augmente le niveau d'encours dans l'atelier. Un petit nombre de palettes risque de conduire à une sous-utilisation du potentiel de production (principalement des machines). Ce problème de dimensionnement du nombre de palettes est rendu encore beaucoup plus difficile par la spécificité de chaque type de palettes.

Le problème que nous nous posons ici est donc de déterminer la répartition optimale des palettes, c'est-à-dire le nombre de palettes de chaque type. Rappelons que l'objectif d'un atelier flexible est de fabriquer simultanèment différents types de pièces et ceci en respectant des ratios de production, c'est-à-dire des quantités relatives fixées entre les différents types de pièces. Le problème de l'optimisation de la répartition des palettes a déjà fait l'objet de travaux et mis en évidence deux approches. D'une part, Hildebrant [1] a développé un algorithme qui utilise la méthode de l'analyse de la valeur moyenne [2]. Cette méthode permet d'évaluer les performances d'un atelier flexible en supposant les files d'attente gérées avec la politique PAPS (premier arrivé, premier servi). L'algorithme d'Hildebrant permet par itération successive de répartir un nombre total de N palettes entre les différents types, de façon à ce que la répartition des palettes conduise à l'obtention des ratios de production désirés, et ceci avec la gestion PAPS. L'inconvénient de cette approche est, d'une part, qu'elle suppose l'atelier géré avec la règle PAPS et, surtout, d'autre part, qu'elle conduit à un nombre de palettes de chaque type non entier. La deuxième approche, quant à elle, suppose connue la gestion de l'atelier. Par exemple, Cohen [3] suppose un lancement périodique et un ordonnancement fixé au niveau de chaque machine. Il est alors possible de déterminer le type de palettes critique et d'augmenter d'une unité le nombre de palettes de ce type. On arrive ainsi, après incrémentations successives, à la répartition optimale des palettes. L'inconvénient majeur de cette approche est qu'elle est dépendante de la gestion imposée.

Nous proposons dans cet article, une troisième approche qui permet d'éviter les inconvénients majeurs des deux méthodes précédemment citées. Notre méthode est elle aussi basée sur une incrémentation successive de la palette critique ; mais nous utilisons, afin de déterminer celle-ci, une méthode d'évaluation de performances qui ne fait aucune hypothèse sur la gestion de l'atelier. Nous présentons au paragraphe II cette méthode d'évaluation des performances, et au paragraphe III l'algorithme d'optimisation de la répartition des palettes. Nous montrons au paragraphe IV un exemple d'application.

II - EVALUATION DES PERFORMANCES

Nous présentons ici la méthode d'évaluation des performances que nous avons développée. Pour plus de détail, le lecteur se reportera à [4 et 5]. Notre méthode modèlise l'atelier flexible par un réseau de files d'attente multiclasse fermé et utilise les résultats de l'analyse opérationnelle [6].

II.1. Position du problème

Soit un atelier flexible destiné à la fabrication simultanée de différents types de pièces. Le système de conduite a pour objectif de maximiser la production tout en respectant les ratios de production entre les différents types de pièces. Nous modélisons cet atelier flexible par un réseau de files d'attente. Les différents postes de travail (machines) seront les stations du réseau, et comme nous nous intéressons aux palettes, nous les considérerons comme les clients du réseau. Le modèle sera donc un réseau de files d'attente multiclasse fermé. Il est multiclasse car à chaque type de palettes est associé une classe de clients du réseau et fermé car les palettes circulent indéfiniment à travers l'atelier et sont en nombre constant, ceci pour chaque type de palettes. Les caractéristiques des clients (temps de service, cheminement dans le réseau) sont celles définies par les gammes de fabrication des pièces.

Les méthodes analytiques utilisant la théorie des files d'attente en vue d'évaluer les performances d'un réseau de files d'attente fermé ont été beaucoup étudiées [7]. Leur application au cas des ateliers flexibles, bien que très intéressante, est pourtant limitée ; en effet, ces méthodes supposent que toutes les files d'attente sont gérées avec la politique PAPS. Nous avons donc développé une nouvelle méthode analytique. Notre approche ne fait pas d'hypothèse sur le mode de gestion de chacune des files d'attente ; elle suppose seulement que le système de conduite de l'atelier flexible a comme objectif de maximiser la production tout en assurant des ratios de production entre les différents types de pièces.

II.2. Approche opérationnelle [5]

Les données du problème sont les suivantes :

M : nombre de postes de travail
R : nombre de types de palettes
N_r : nombre de palettes de type r
α_r : ratio de production relatif aux palettes de type r
$t_{r,m}$: temps de service pour une palette de type r à la station m
$V_{r,m}$: taux de visite d'une palette de type r à la station m (cette grandeur est définie par le cheminement des palettes dans l'atelier).

On peut déduire de ces quantités le débit relatif de palettes de type r à la station m, soit $\alpha_{r,m}$, par la formule :

$$\alpha_{r,m} = \frac{V_{r,m} \cdot \alpha_r}{\sum_{i=1}^{R} (V_{i,m} \cdot \alpha_i)}$$

Le principe de notre méthode est de décomposer le réseau multiclasse en R réseaux monoclasses fictifs relatifs chacun à un type de palettes. Si on sait évaluer les performances de chacun des réseaux, on pourra en déduire les performances du réseau multiclasse total. L'évaluation des performances de chacun des réseaux monoclasses est basée sur les résultats de l'Analyse Opérationnelle |6|. Cette méthode permet de calculer les paramètres de performances d'un réseau de files d'attente fermé monoclasse à partir des deux paramètres suivants :

N : nombre total de clients du réseau

Y_m : charge de travail moyenne apportée par un client à la station m (produit du temps moyen de service S_m par le taux de visite V_m).

L'analyse opérationnelle conduit aux mêmes algorithmes que ceux de l'approche classique de la théorie des files d'attente |8|, mais fait appel à des hypothèses moins restrictives. Nous ne détaillerons pas ici ces algorithmes |9|; il faut savoir qu'ils permettent de calculer les débits, les taux d'utilisation, les longueurs moyennes des files d'attente, etc.

Pour appliquer l'analyse opérationnelle, il nous faut donc, pour chacun des réseaux fictifs, définir les grandeurs $N(r)$ et $Y_m(r)$ (ce sont les grandeurs N et Y_m définies ci-dessus, l'indice r supplémentaire signifie qu'elles sont relatives au réseau fictif dans lequel ne circulent que les clients de type r). Nous avons évidemment $N(r) = N_r$, c'est-à-dire le nombre de palettes de type r ; il reste à définir la quantité $Y_m(r)$. Cette grandeur que nous noterons $Y_{r,m}$ est le produit du temps moyen de service d'un client de type r à la station m, dans le réseau fictif r, noté $S_{r,m}$, par le taux de visite $V_{r,m}$ défini précédemment. La quantité $S_{r,m}$, très importante pour la suite de notre exposé, doit ici être précisée. $S_{r,m}$ correspond au temps de service apparent d'un client de type r à la station m, et ne doit pas être confondu avec son temps de service réel $t_{r,m}$. En effet, chacun des réseaux monoclasses est un réseau fictif car il doit partager le potentiel de chacune des stations avec les autres réseaux. Ainsi, les clients de différents types vont se ralentir mutuellement et vont donc se comporter pour le réseau fictif comme si leur temps de service était $S_{r,m} \geq t_{r,m}$.

Les grandeurs N_r et $V_{r,m}$ sont des données au problème ; nous allons d'autre part, et dans un premier temps, supposer que les quantités $S_{r,m}$ sont connues. Nous pouvons donc, en appliquant pour chaque réseau fictif l'algorithme de calcul des performances |9|, déduire de ces grandeurs les paramètres de performances suivants :

$X_{r,m}$: débit de clients de type r à la station m

$P_{r,m}$: taux de présence des clients de type r à la station m

$$P_{r,m} = S_{r,m} \cdot X_{r,m}$$

$u_{r,m}$: taux d'utilisation de la station m par des clients de type r

$$u_{r,m} = t_{r,m} \cdot X_{r,m}$$

u_m : taux d'utilisation de la station m

$$u_m = \sum_{r=1}^{R} u_{r,m}$$

II.3. Méthode itérative d'évaluation des performances [5]

Nous avons supposé précédemment que les grandeurs $S_{r,m}$ étaient connues. Or, si on veut évaluer les performances d'un atelier flexible, nous ne connaîtrons pas ces valeurs, a priori. Notons tout d'abord que nous pouvons recalculer les paramètres $S_{r,m}$ au moyen de la relation :

$$S_{r,m} = \frac{P_{r,m}}{u_{r,m}} \cdot t_{r,m} \qquad \text{pour } r = 1,\ldots, R \text{ et } m = 1,\ldots, M$$

Si on appelle S la matrice de dimension R x M dont les coefficients sont les $S_{r,m}$ on a donc une relation du type :

$$S = F(S)$$

où F est une fonction non explicite.

La matrice S que nous cherchons doit être solution de cette équation. Remarquons, d'autre part, que nous n'avons jusqu'à présent pas utilisé la corrélation qui existe entre le comportement des différents réseaux, due à l'objectif de ratios de production à assurer. Ce dernier impose au niveau de chaque machine m une relation :

$$\frac{X_{r,m}}{\alpha_{r,m}} = \text{Constante} = C_m \quad \text{pour } r = 1,\ldots R$$

ou de manière équivalente :

$$\frac{1}{\alpha_{r,m}} \frac{u_{r,m}}{t_{r,m}} = \text{Constante} = C_m \quad \text{pour } r = 1\ldots, R \tag{1}$$

Il y a une infinité de matrices S qui, d'une part, vérifient S = F(S) et, d'autre part, conduisent à satisfaire la relation (1). La solution S* qui nous intéresse est l'une de ces matrices S. Le dernier facteur qui doit être pris en compte est le fait que l'objectif est de maximiser la production. La détermination de cette solution va être faite par une technique itérative de résolution de l'équation S = F(S) à laquelle vont s'ajouter les différentes contraintes dues à l'objectif de production (ratios et maximisation). Nous allons donc à chaque pas de la méthode itérative calculer le taux d'affectation du service de la station m aux clients de type r, c'est-à-dire calculer les paramètres $u_{r,m}$. Les contraintes sont de trois types :

1) les $u_{r,m}$ doivent vérifier la relation (1).
2) on doit avoir $u_{r,m} \leq P_{r,m}$ (on ne peut servir des clients que lorsqu'ils sont présents).
3) l'utilisation totale u_m doit vérifier $u_m \leq 1$.

L'objectif étant de maximiser la production, on cherche évidemment les $u_{r,m}$ maximum qui vérifient ces contraintes. L'initialisation de la matrice S est importante car les valeurs finales, après convergence de l'algorithme, en dépendent. Comme l'objectif du système de gestion est de maximiser la production, l'initialisation qui conduit à l'obtention de résultats correspondant à la réalisation de cet objectif est :

$$S^{\circ}_{r,m} = t_{r,m}$$

C'est en effet le cas idéal où les clients ne se ralentissent pas mutuellement. Nous présentons à la figure 1 l'algorithme de calcul des performances relatif à la méthode que nous venons de définir.

Des justifications théoriques complémentaires de notre méthode sont données dans |5|. Nous allons simplement ici résumer les points importants. Notre approche fait deux types d'hypothèses : d'une part, des hypothèses explicites (celles de l'analyse opérationnelle) et d'autre part, des hypothèses implicites (dans la détermination de la solution S*). Les premières conduisent à des paramètres de performances pessimistes ; les secondes à des paramètres de performances optimistes. Les conséquences de ces deux types d'hypothèses vont donc en partie se compenser. On peut donc penser que, dans un cas général, notre algorithme conduit à des résultats assez proches de la réalité ; on ne peut toutefois pas savoir si ils seront optimistes ou pessimistes.

Remarque : Si on s'intéresse à une borne supérieure des performances, une variante de l'approche proposée peut être utilisée [5]. Elle consiste à décomposer le réseau multiclasse en un nombre de réseaux fictifs égal au nombre total de palettes du réseau. Dans ce cas, les hypothèses explicites sont toujours vérifiées. C'est pourquoi on obtient une borne supérieure des performances.

III - OPTIMISATION DE LA REPARTITION DES PALETTES

Nous en arrivons maintenant au problème que nous avons posé dans l'introduction et que l'on peut résumer de la manière suivante : étant donné un atelier flexible destiné à produire des pièces de différents types en quantités relatives fixées, quel est le nombre de palettes optimal de chaque type nécessaire ? Les contraintes que nous nous imposons sont 1) le nombre de palettes total ne doit pas dépasser une quantité fixée, notée N_{max} et 2) le nombre de palettes de type r est limité à $N_{max}(r)$.

L'algorithme d'optimisation que nous proposons se décompose de la manière suivante. On initialise le nombre de palettes de chaque type à sa valeur minimum, c'est-à-dire $N_r = 1$. On évalue les performances de l'atelier flexible et on détermine le type de palettes critique au moyen d'un critère à définir ultérieurement. On rajoute une palette du type critique et ainsi de suite. L'arrêt de cet algorithme itératif

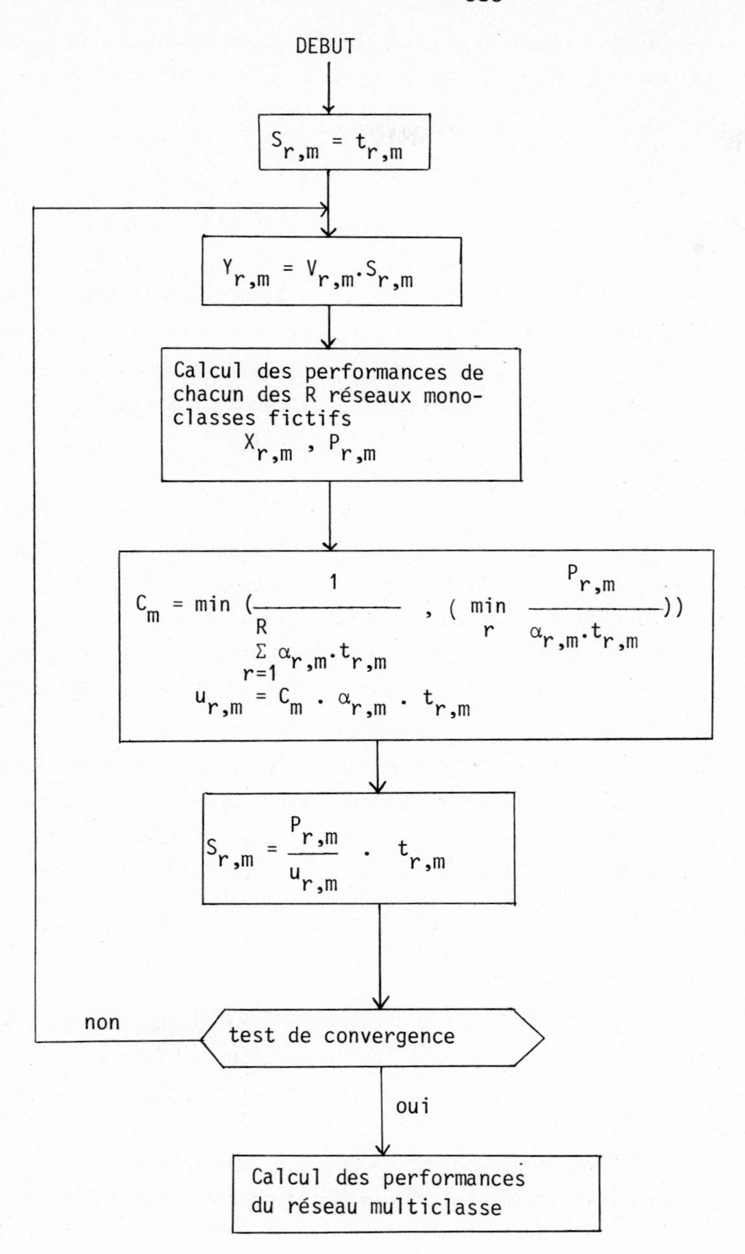

figure 1. Algorithme d'évaluation des performances

pourra se faire de deux manières : soit l'une des contraintes sur le nombre de palettes est atteinte, soit l'une des machines est saturée.

Nous abordons maintenant le problème de la détermination du type de palettes critique. Pour cela, nous disposons à chaque pas de l'évaluation des performances de l'atelier flexible relative à la distribution de palettes $(N_1, \ldots, N_R)$. En particulier, nous avons connaissance du paramètre $S_{r,m}$, temps de service apparent d'un client de type r à la station m. Il est intéressant de comparer cette grandeur au temps de service réel $t_{r,m}$; plus $S_{r,m}$ est proche de $t_{r,m}$ (et à la limite égal), plus les palettes de type r doivent être servies rapidement à la station m ; au contraire, si $S_{r,m}$ est grand devant $t_{r,m}$, les palettes de type r ne demandent pas à être servies rapidement. En étendant ce raisonnement à l'ensemble des stations, on est amené à définir le paramètre C_r, qui va nous servir de critère, par :

$$C_r = \frac{\sum_{m=1}^{M} V_{r,m} \cdot S_{r,m}}{\sum_{m=1}^{M} V_{r,m} \cdot t_{r,m}}$$

C_r définit le degré de liberté possible pour l'affectation des services des stations aux clients de type r. Le type de palettes critique sera donc celui qui minimise la quantité C_r.

En fait, tant que la saturation d'une des machines (la machine critique) n'est pas atteinte, la solution S* donnée par l'algorithme est telle que l'un des types de palettes, soit r_o, vérifie :

$$S^*_{r_o,m} = t_{r_o,m} \qquad \text{pour } m = 1, \ldots, M$$

C'est ce type de palettes qui est critique et qui sera mis en évidence lors de la détermination des critères C_r ; on aura $C_{r_o} = 1$

L'algorithme de détermination de la répartition optimale des palettes est donné à la figure 2.

IV - EXEMPLE D'APPLICATION

Nous allons appliquer notre algorithme sur un exemple proposé par Cohen et al. [3] et comparer avec leur résultat. L'atelier considéré comporte 8 machines et 6 types de palettes. Les gammes de fabrication sont données à la figure 3. L'objectif du système est de produire chacune des pièces en même quantité, c'est-à-dire que nous avons $\alpha_r = 1/6$. Les résultats obtenus par Cohen et al. [3] sont relatifs à un lancement périodique de produits et à un ordonnancement préfixé au niveau de chaque machine. Ils utilisent la théorie des dioïdes pour évaluer les performances en régime permanent. Ils peuvent de plus connaître le circuit critique et donc déterminer le type de palettes critique.

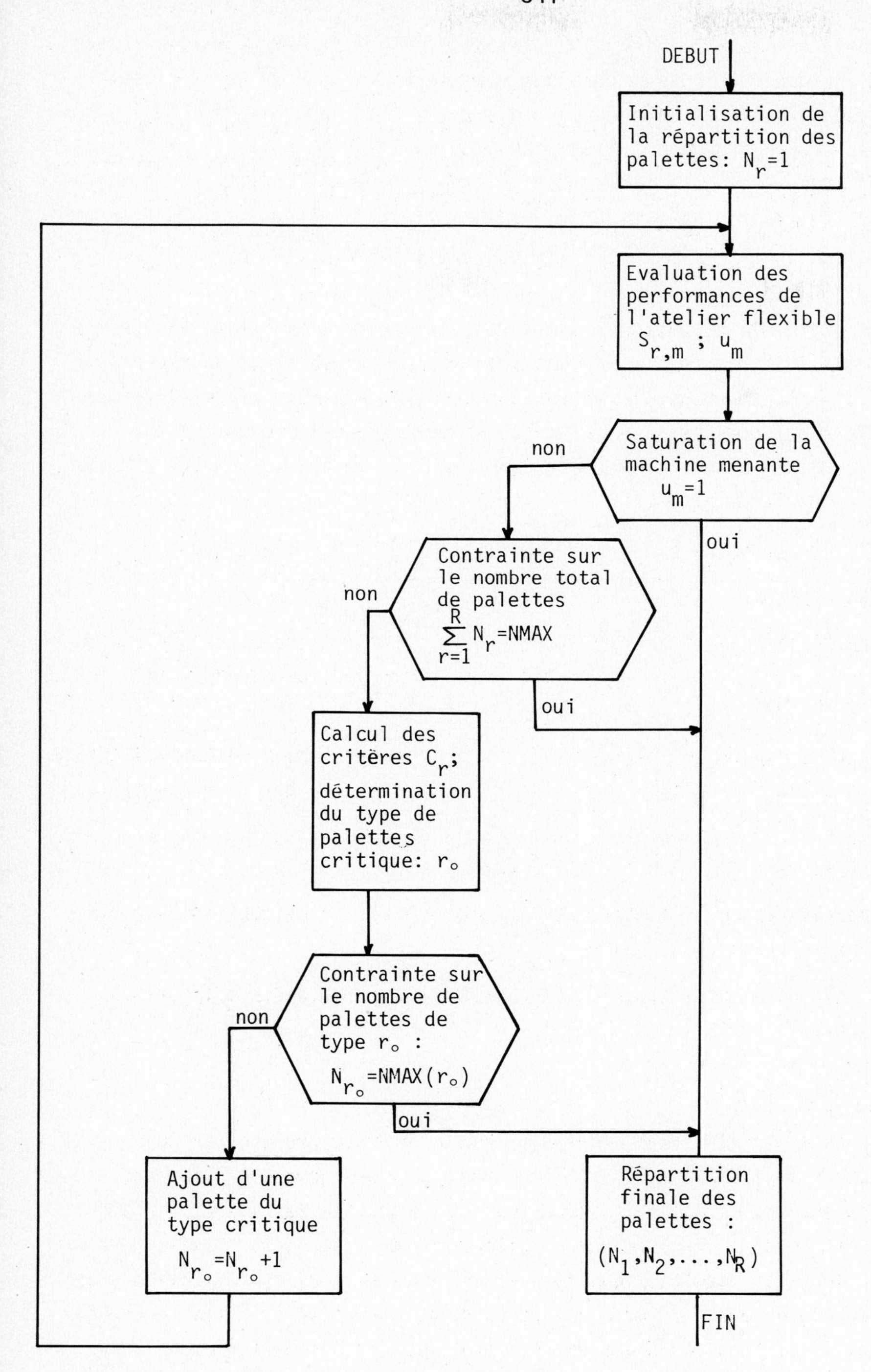

figure 2. Algorithme d'optimisation de la répartition des palettes.

Nous présentons la comparaison des résultats à la figure 4. D'une part, nous remarquons que les deux méthodes mettent toujours en évidence à chaque incrémentation le même type de palettes critique. D'autre part, notre algorithme d'évaluation des performances semble indiquer qu'à chaque répartition de palettes il peut exister une gestion meilleure que celle définie par l'ordonnancement fixé, et qui conduirait à une meilleure utilisation de la machine menante. Cette remarque ne s'applique pas pour la répartition initiale de palettes où l'ordonnancement fixé paraît conduire à la gestion optimale.

Notre algorithme d'optimisation conduit à la saturation de la machine menante avec un nombre total de palettes égal à 9, inférieur aux 12 palettes nécessaires dans le cas de l'ordonnancement périodique. Si on désire ajouter une palette supplémentaire, on utilise le critère C_r qui met en évidence comme palette critique le type 1, ce qui continue d'être similaire au résultat obtenu avec le cas de l'ordonnancement périodique fixé.

machine	1	2	3	4	5	6	7	8
type de palettes								
1	2.		3.7		1.7	0.5	1.	1.5
2	3.9				3.1	3.2	1.	1.5
3	0,95	2.	2.2	2.	3.	4.3	1.	1.5
4	1.1	1.2				1.9	1.	1.2
5	0.7		6.4	1.	1.3	1.6	1.	1.2
6	1.4	1.7		1.		0.4	1.	1.2

Le nombre en ligne r et colonne m est le temps de service $t_{r,m}$. Le taux de visite est $V_{r,m} = 1$ si $t_{r,m}$ est défini, $V_{r,m} = 0$ sinon.

Figure 3. Données de l'exemple

nb de palettes	6	7	8	9	10	11	12
ordonnancement fixe	3	5	2	1	6	4	
algorithme proposé	3	5	2	1			

- a - type de palettes critique

nb de palettes	6	7	8	9	10	11	12
ordonnancement fixé	0.726	0.814	0.823	0.856	0.859	0.869	1.
algorithme proposé	0.726	0.932	0.969	1.	1.		

- b - taux d'utilisation de la machine menance

Figure 4. Comparaison des résultats

V - CONCLUSIONS

Nous avons présenté ici un algorithme d'optimisation de la répartition des palettes dans un atelier flexible. Le principal intérêt de notre approche est qu'elle se base sur une méthode d'évaluation des performances assez proche de la réalité puisqu'elle prend en compte l'objectif de ratios de production. Cette méthode suppose une gestion adéquate de l'atelier flexible et ne fait donc aucune hypothèse sur la manière dont le système de conduite organisera la production en vue de réaliser son objectif.

L'algorithme d'optimisation de la répartition des palettes permet alors au moyen d'un critère fonction des paramètres de performances de déterminer à chaque pas le type de palettes critique et donc d'augmenter le nombre de palettes de ce type d'une unité. Le test d'arrêt de cet algorithme correspond à la condition de saturation de la machine menante.

La comparaison qualitative que l'on peut faire entre les deux approches possibles afin de résoudre le problème posé est la suivante : la première approche (gestion connue) a l'avantage de donner des résultats exacts et donc fiables. Son inconvénient majeur est qu'elle nécessite de prédéterminer tous les séquencements des tâches à l'intérieur de l'atelier flexible ; de plus, la répartition des palettes obtenue dépend de cette gestion. Notre approche a l'avantage d'être très facile à mettre en oeuvre ; son inconvénient est qu'elle ne conduit qu'à des résultats approximatifs (C.F. II-3) donc moins fiables.

On peut conclure en disant que si l'atelier flexible est simple (nombre de machines et de types de palettes modérés), on peut utiliser la premier approche. Si, par contre, l'atelier a une taille raisonnable, l'approche que nous proposons est nettement préférable ; ceci est d'autant plus vrai que plus le système est de taille importante, plus l'algorithme d'évaluation des performances donne de bons résultats.

REFERENCES

[1] - Hildebrant R.R. (1980) - Scheduling flexible manufacturing systems using mean value analysis-proceedings IEEE Conf. on Decision and Control, Albuquerque, 1980, 701-706

[2] - Reiser M., Lavenberg S. (1980) - Mean value analysis of closed multichain queuing networks - J. ACM, 27, 313-322.

[3] - Cohen G., Dubois D., Quadrat J.P., Viot M. (1983) - Analyse du comportement périodique de systèmes de production par la théorie des dioïdes - Rapport de recherche INRIA, n° 191.

[4] - Dallery Y., David R. (1983) - A new approach based on operational analysis for flexible manufacturing systems performance evaluation, IEEE Conf. on Decision and Control, San Antonio, Dec. 1983.

[5] - Dallery Y. (1984) - Une méthode analytique pour l'évaluation des performances d'un atelier flexible - Thèse de docteur-ingénieur, LAG, mai 1984.

[6] - Denning P.J., Buzen J.P. (1978) - The operational analysis of queuing network models - Computing surveys 10, 225-261.

[7] - Cavaillé J.B., Dubois D. (1982) - Intérêt de la théorie des réseaux de files d'attente pour l'évaluation des performances d'un atelier flexible. Conférence INRIA Analyse et Optimisation des systèmes, déc. 1982.

[8] - Baskett F., Chandy K.M., Muntz R.R., Palacios G.F. (1975) - Open closed and mixed networks of queues with different classes of customers - J. ACM, 22, 248-260.

[9] - Bruell S.C., Balbo G. (1980) - Computational algorithms for closed queueing networks - Operating and programming systems series, P.J. Denning Editor.

AN EFFICIENT DECOMPOSITION METHOD FOR THE APPROXIMATE EVALUATION OF PRODUCTION LINES WITH FINITE STORAGE SPACE

by Stanley B. Gershwin *

Laboratory for Information and Decision Systems
Massachusetts Institute of Technology

This paper presents an approximate decomposition method for the evaluation of performance measures for a class of tandem queuing systems with finite buffers in which blocking and starvation are important phenomena. These systems are difficult to evaluate because of their large state spaces and because they may not be decomposed exactly. This approach is based on such system characteristics as conservation of flow. It offers a dramatic reduction of computational effort. Comparison with exact and simulation results indicate that it is very accurate.

* 35-427
Laboratory for Information and Decision Systems
Massachusetts Institute of Technology
77 Massachusetts Avenue
Cambridge, Massachusetts 02139

This research has been supported by the U. S. Army Human Engineering Laboratory under contract DAAK11-82-K-0018.

1. INTRODUCTION

This paper presents a method for the analysis of a class of tandem queuing systems with finite buffers. Such systems are difficult to treat because of their large state spaces and because they may not be decomposed exactly. The method approximately decomposes a (k-1)-buffer system into k-1 single buffer systems. It has been developed around a specific class of models, but it is hoped that it may be extended to a wider class.

The tandem queuing system in Figure 1 consists of a series of k servers or machines (M_1, M_2,..., M_k) separated by queues or buffers (B_1, B_2,..., B_{k-1}). The buffers are each of finite capacity (N_1, N_2,..., N_{k-1}). Material flows from outside the system to M_1, then to B_1, then to M_2, and so forth until it reaches M_k, after which it leaves. The machines are assumed to spend a random amount of time with each item. (Here, randomness is due to the failures of machines. Operational machines spend a fixed amount of time processing the items.)

In all tandem queuing systems with finite buffer capacities (regardless of processing time distributions), if machine M_i spends a long time on a single item, buffer B_{i-1} tends to accumulate material and buffer B_i tends to lose material. If this condition persists, B_{i-1} may become full or B_i may become empty. Then machine M_{i-1} is blocked and prevented from working, or M_{i+1} is starved and also prevented from working.

The purpose of this paper is to present an approximation method for calculating the production rate and the average amounts of material in the buffers for a class of systems of this type. The class includes those in which the service process is deterministic but geometrically unreliable. That is, while a machine is operational and neither starved or blocked, a fixed amount of time is required to process a part. It is assumed that this time is the same for all machines and is taken as the time unit.

During a time unit when machine M_i is operational and neither starved nor blocked, it has probability p_i of failing. It may work on pieces only at such times, and it is only while it is working that it may fail. Its mean time between failures (MTBF) in working time is thus $1/p_i$. ("Working time" means time during which the machine is operational and neither starved nor blocked.) After a machine has failed, it is under repair and it has probability r_i of being repaired during a time unit (any time unit). Its mean time to repair (MTTR) is therefore $1/r_i$. This is actual elapsed time, not working time.

A detailed description of the mathematical model and a sur-

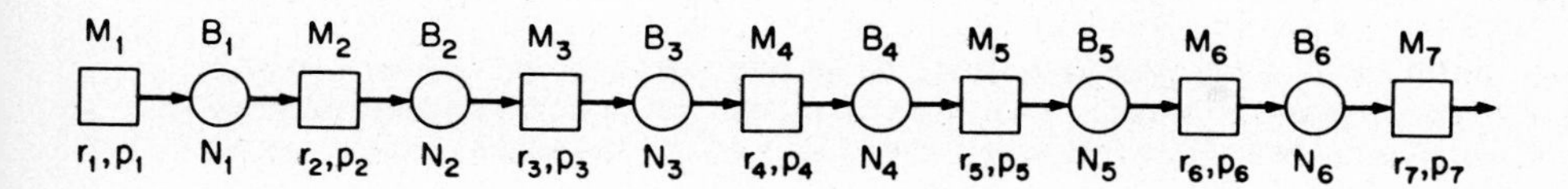

Figure 1. Upstream Portion of Transfer Line L.

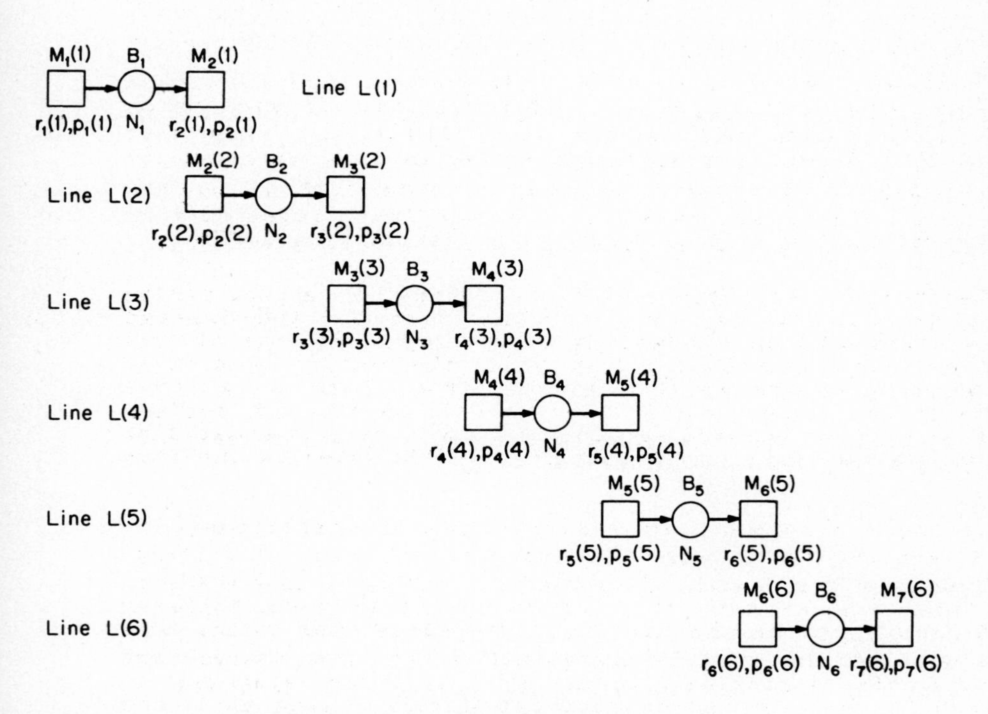

Figure 2. Decomposition of Upstream Portion of Transfer Line L.

vey of related work appear in Gershwin and Schick (1983). The model is based on that of Buzacott (1967). Approximate decomposition of tandem queuing models was discussed by Hillier and Boling (1966), Takahashi et al., (1980), Altiok (1982), and others. Simulation results for models of this type appear in Ho et al. (1979), Law (1981), and Vanderhenst et al. (1981).

The problem is difficult because of the great dimensionality of the state space. Each machine can be in two states: operational or under repair. Buffer B_i can be in N_{i+1} states: $n_i=0, 1,\dots,N_i$, where n_i is the amount of material in B_i. As a consequence, the Markov chain representation of a 20-machine line with 19 buffers each of capacity 10, for example, has over $6.41\text{X}10^{25}$ $(2^{20}11^{19})$ states.

2. TRANSFER LINE CHARACTERISTICS

Certain quantities are defined and relationships among them are described in this section. Approximations of the quantities and relationships are used in Section 3 to develop the decomposition method.

Two performance measures of great interest to designers of production lines are the production rate (throughput, flow rate, or line efficiency) E_i, and average buffer level (in-process inventory or work-in-process) $\bar{n}_i$.

The efficiency of machine M_i, in parts per time unit, is

$$E_i = \text{prob } (B_{i-1} \text{ not empty}$$

$$\text{and } M_i \text{ operational and } B_i \text{ not full}).$$

Conservation Of Flow

Because there is no mechanism for the creation or destruction of material, flow is conserved, or

$$E_1 = E_2 = \dots = E_k \tag{1}$$

The Flow Rate-Idle Time Relationship

Define e_i to be the isolated production rate of machine M_i. It is what the production rate of M_i would be if it were never impeded by other machines or buffers. It is given by (Buzacott, 1967)

$$e_i = r_i \;/\; (r_i + p_i)$$

and it represents the fraction of time that M_i is operational.

The actual production rate E_i of M_i is less because of blocking or starvation. In the Appendix, it is shown to be

$$E_i = e_i \text{ prob } (B_{i-1} \text{ not empty and } B_i \text{ not full})$$

This expression may be approximated by

$$E_i = e_i \left[1 - \text{prob } (n_{i-1}=0) - \text{prob } (n_i=N_i)\right]. \qquad (2)$$

3. DECOMPOSITION METHOD

The decomposition method is presented here. It is based on the equation of conservation of flow (1), the flow rate-idle time relationship (2), and a set of equations ((6) and (7)) developed below. The approach is to characterize the most important features of the behavior of the transfer line in a simple approximate way, and to find a solution to the resulting set of equations.

Decomposition

Consider Figure 2, a set of two-machine lines L(i), i = 1, in Figure 1. The object is to find the parameters (failure and repair rates $r_1(1)$, $p_1(1)$, $r_2(1)$, $p_2(1)$, $r_2(2)$, $p_2(2)$, etc.) of the machines so that the behavior of the material flow in the buffers of the two-machine lines closely matches that of the flow in the buffers of the long line. (All symbols with i in parentheses refer to the i'th two-machine line.)

That is, the rate of flow into and out of buffer B_i in line L(i) is found which approximates that of buffer B_i in line L. The probability of the buffer of L(i) being empty or full is close to that of B_i in L being empty or full. The probability of resumption of flow into (and out of) the buffer in line L(i) in a time unit after a period during which it was interrupted is close to the probability of the corresponding event in L. Finally, the average amount of material in the buffer of L(i) approximates the material level in buffer B_i in L. In order to find such parameter values, we use the relationships of the previous section as well as others described below.

Machine $M_i(i)$ models the part of the line upstream of B_i and $M_{i+1}(i)$ models the part of line downstream from B_i. There are four parameters per two-machine line (ie, per buffer in the long line): $r_i(i)$, $p_i(i)$, $r_{i+1}(i)$, $p_{i+1}(i)$. Consequently, four equations per buffer, or 4(k-1) conditions, are required to determine them.

Let E(i) be the efficiency or production rate of L(i). E(i) is a function of the four unknowns $r_i(i)$, $p_i(i)$, $r_{i+1}(i)$, $p_{i+1}(i)$ (and the buffer capacity N_i).

One set of conditions is related to conservation of flow:

$$E(i) = E(1), \quad i=2,\ldots,k-1 \tag{3}$$

There are k-2 equations here.

The second set of conditions follows from (2), the approximate flow rate-idle time relationship. Since we assume that the probability of B_i being empty or full in L is closely approximated by the probability of B_i being empty or full in L(i), we have

$$E(i) = e_i\left[1 - p_s(i-1) - p_b(i)\right], \quad i=2,\ldots k-1 \tag{4}$$

where $p_s(i-1)$ is the probability of the buffer in L(i-1) being empty and $p_b(i)$ is the probability of the buffer in L(i) being full. (The subscripts refer to starvation and blockage.) These quantities, like E(i), are functions of $r_i(i)$, $p_i(i)$, $r_{i+1}(i)$, $p_{i+1}(i)$, and N_i.

Equation (4), after some manipulation, can be written

$$\frac{p_i(i-1)}{r_i(i-1)} + \frac{p_i}{r_i} = \frac{1}{E(i)} + \frac{1}{e_i} - 2,$$

$$i=2,\ldots,k-1. \tag{5}$$

This is demonstrated in the Appendix. So far, 2(k-2) of the required 4(k-1) conditions have been determined.

Resumption of Flow

To characterize the repair rates of the two-machine lines, it is necessary to consider the meaning of failure and repair in those systems. Machine $M_i(i)$ in line L(i) represents, to buffer B_i, everything upstream of B_i in line L. Therefore, a failure of $M_i(i)$ represents either a failure of machine M_i or the emptying of buffer B_{i-1} (which, in turn, is due to a failure of M_{i-1} or the emptying of B_{i-2}, etc.).

The repair of $M_i(i)$ is thus the termination of whichever condition was in effect. The probability of repair of $M_i(i)$ in any cycle in which it is down is r_i if the actual failure is M_i and it is r_{i-1} or r_{i-2}, etc. if, instead, the "failure" is actually the emptying of B_{i-1}. It is r_{i-1} if B_{i-1} is empty because of the failure of M_{i-1}; it is r_{i-2} if B_{i-1} is empty because M_{i-2} has failed and B_{i-2} has emptied; and so forth.

We assume that the probability of B_{i-1} in L being empty, due to all causes, is the same as that of B_{i-1} being empty in line L(i-1). In L(i-1), however, B_{i-1} can be empty due only to one cause: the failure of $M_{i-1}(i-1)$. Consequently, the probability of repair of $M_i(i)$ is $r_{i-1}(i-1)$ if the cause of failure is the emptying of B_{i-1} and it is r_i otherwise.

Based on this, the probability of resumption of flow can be shown to be, approximately,

$$r_i(i) = \frac{r_{i-1}(i-1)p_s(i-1) + r_i\,\left[p_i(i)E(i)/r_i(i) - p_s(i-1)\right]}{\left[p_i(i)E(i)\ /\ r_i(i)\right]},$$

$$i=2,\ldots k-1. \qquad (6)$$

A similiar analysis yields the following equation for the second machine in the i-1'st line:

$$r_i(i-1) = \frac{r_{i+1}(i)p_b(i) + r_i\,\left[p_i(i-1)\ E(i-1)/r_i(i-1) - p_b(i)\right]}{\left[p_i(i-1)\ E(i-1)\ /\ r_i(i-1)\right]},$$

$$i=2,\ldots,k-1. \qquad (7)$$

Equations (6) and (7) contain 2(k-2) conditions each. A total of 4(k-2) conditions have thus been identified.

Finally, there are boundary conditions:

$$\begin{aligned} r_1(1) &= r_1 \\ r_k(k-1) &= r_k \\ p_1(1) &= p_1 \\ p_k(k-1) &= p_k \end{aligned} \qquad (8)$$

There are a total of 4(k-1) equations among (3), (5), (6), (7), and (8) in 4(k-1) unknowns: $r_i(i)$, $p_i(i)$, $r_{i+1}(i)$, $p_{i+1}(i)$, i=1,...,k-1. This is precisely what has been sought.

4. NUMERICAL TECHNIQUE AND RESULTS

These equations can be thought of as defining a two-point boundary value problem (TPBVP) of the form

$$f(x(i-1),\ x(i)) = 0,\quad i=2,\ldots,k,$$

$$x(1),\ x(k) \text{ partly specified}$$

where $x(i)$ is a 4-vector of the parameters of line $L(i)$; $x(i) = (r_i(i),\ p_i(i),\ r_{i+1}(i),\ p_{i+1}(i))$. The nonlinear function $f(\ )$ involves the evaluation of $E(i)$, $p_s(i)$, and $p_b(i)$.

Satisfactory results have been obtained with a modified shooting method consisting of three nested loops. It is described in detail in Gershwin (1983).

The production rate of the line is the common production rate of all the two-machine lines. The average buffer levels are approximated by those of the two-machine lines when convergence is reached.

Comparisons With Exact Results and Simulations

For a three-machine line, it is possible to compare the results of this algorithm with exact results by using the method of Gershwin and Schick (1983). A set of five cases are compared in Gershwin (1983). These cases represent a wide range of three-machine systems. Simulations (each run for 100,000 cycles) are also performed for comparison. The decomposition method produces results that are extremely close to the exact values obtained by solving the Markov chain exactly. The error is very small in the production rate E (less than 0.02%) and only a little larger (less than 2.6%) in the average buffer levels. The decomposition results are generally as close or closer to exact as the simulation results. No more than 86 evaluations of two-machine lines are required for these three-machine cases.

Exact methods are not available for systems of more than three machines and two buffers or for three-machine cases with very large buffers. Consequently, other techniques are required to assess the accuracy of the approximation. They include simulation and qualitative observations.

A large set of cases are considered in Gershwin (1983) which cover a wide range of failure probabilities, repair probabilities, and buffer sizes. The results also cover a wide range of production rates and average buffer levels.

There is close agreement between the decomposition and simulation results. In most cases, production rates and buffer levels agree to within a few percent. This remains true even for large buffer capacities (over 100) and long lines (20 machines.) There is no obvious trend indicating that the accuracy of the decomposition decreases as the line length increases.

The number of evaluations of the two-machine line increases with the length of the line. The number of evaluations appears to be less than approximately $2k^3$ where k is the number of machines. As a consequence, the computer time for the decomposition method is much less than that of simulation. For example, two 20-machine decomposition calculations took about 7 and 12 seconds while the corresponding simulations required 248 and 262 seconds. The computer time is that of the MIT Honeywell 68/DPS computer with the Multics operating system.

To assess the qualitative behavior of the method, closely related pairs of cases were treated. One case of each pair was longer than the other by one machine and one buffer. The additional machine was one that had a very small failure rate so that the unreliable part of the long system was identical to the whole short system. (When the reliable machine was internal to the long system, the adjacent buffers together had the same capacity as one of the buffers of the short system.) In each case, the production rates were very close, and the buffers levels displayed satisfactory behavior. The simulations behave very similarly.

Several authors have conjectured (Hillier and Boling, 1977) or shown (Dattatreya, 1978; Muth, 1979; Ammar, 1980; Ammar and Gershwin, 1981) that two tandem queueing systems which are the reverse of one another have the same production rates. In addition , the average levels of corresponding buffers are complementary (Ammar, 1980; Ammar and Gershwin, 1981). Symmetric lines are their own reverses. The results of the decomposition method agree with these observations exactly. Simulation, however, produces only approximate agreement.

Several cases were comparisons with simulations that appeared in the literature (Ho, Eyler, and Chien, 1979) and Law (1981). The decomposition and simulated results are in good agreement. However, we find that some of the statistical results are misleading in that the simulations seem to suggest that lines that are the reverses of one another have different production rates.

5. CONCLUSIONS AND FURTHER RESEARCH

A new decomposition method has been found for the analysis of tandem queuing systems with finite buffers in which blocking is important. Exact and simulation results indicate that the method, while approximate, is quite accurate. Current research is aimed at extending this work in two directions: other service processes, such as reliable and unreliable machines with exponential processing time; and assembly/disassembly networks. Future efforts will be devoted to systems such as Jackson-like networks with blocking.

REFERENCES

T. Altiok (1982), "Approximate Analysis of Exponential Tandem Queues with Blocking," European Journal of Operations Research, Vol. 11, 1982.

M. H. Ammar (1980), "Modelling and Analysis of Unreliable Manufacturing Assembly Networks with Finite Storages," MIT Laboratory for Information and Decision Systems Report LIDS-TH-1004.

M. H. Ammar and S. B. Gershwin (1981), "Equivalence Relations in Queuing Models of Manufacturing," Proceedings of the Nineteenth IEEE Conference on Decision and Control.

J. A. Buzacott (1967), "Automatic Transfer Lines with Buffer Stocks," International Journal of Production Research, Vol. 6.

E. S. Dattatreya (1978), "Tandem Queueing Systems with Blocking," Ph. D. thesis, Department of Industrial Engineering and Operations Research, University of California, Berkeley.

S. B. Gershwin (1983), "An Efficient Decomposition Method for the Approximate Evaluation of Tandem Queues with Finite Storage Space and Blocking," MIT Laboratory for Information and Decision Systems Report LIDS-P-1309.

S. B. Gershwin and O. Berman (1981), "Analysis of Transfer Lines Consisting of Two Unreliable Machines with Random Processing Times and Finite Storage Buffers," AIIE Transactions, Vol. 13, No. 1, March 1981.

S. B. Gershwin and I. C. Schick (1983), "Modeling and Analysis of Three-Stage Transfer Lines with Unreliable Machines and Finite Buffers," Operations Research, Vol. 31, No.2, pp 354-380, March-April 1983.

F. S. Hillier and R. W. Boling (1966), "The Effect of Some Design Factors on the Efficiency of Production Lines with Variable Operation Times," Journal of Industrial Engineering, Vol. 17, No. 12, December, 1966.

F. S. Hillier and R. Boling (1977), "Toward Characterizing the Optimal Allocation of Work in Production Lines with Variable Operation Times," in Advances in Operations Research, Proceedings of EURO II, Marc Reubens, Editor; North-Holland, Amsterdam.

Y. C. Ho, M. A. Eyler, and T. T. Chien (1979), "A Gradient Technique for General Buffer Storage Design in a Production Line," International Journal of Production Research, Vol. 17, No. 6, pp 557-580, 1979.

S. S. Law (1981), "A Statistical Analysis of System Parameters in Automatic Transfer Lines," International Journal of Production Research, Vol. 19, No. 6, pp 709-724, 1981.

E. J. Muth (1979), "The Reversibility Property of Production Lines," Management Science, Vol.25, No. 2.

I. C. Schick and S. B. Gershwin (1978), "Modelling and Analysis of Unreliable Transfer Lines with Finite Interstage Buffers," Massachusetts Institute of Technology Electronic Systems Laboratory Report ESL-FR-834-6, September, 1978.

Y. Takahashi, H. Miyahara, and T. Hasegawa (1980), "An Approximation Method for Open Restricted Queuing Networks," Operations Research, Vol. 28, No.3, Part I, May-June 1980.

P. Vanderhenst, F. V. Van Steelandt, and L. F. Gelders (1981), "Efficiency Improvement of a Transfer Line Via Simulation," Katholieke Universiteit Leuven, Faculteit Toegepaste Wetenschappen Afdeling Industrieel Beleid, Belgium; Report 81-04, January 1981.

APPENDIX

1. Proof of the Flow Rate-Idle Time Relationship

Proposition:

$$E_i = e_i \text{ prob } (B_{i-1} \text{ not empty and } B_i \text{ not full }) \tag{9}$$

Proof: This follows a similar proof by Gershwin and Berman (1981). Efficiency E_i has been defined verbally in the text. In symbols, it is given by

$$E_i = \text{prob } (\alpha_i = 1, n_{i-1} \neq 0, n_i \neq N_i).$$

Let

$$D_i = \text{prob } (\alpha_i = 0, n_{i-1} \neq 0, n_i \neq N_i).$$

Schick and Gershwin (1978) observe that

$$r_i D_i = p_i E_i \tag{10}$$

by noting that the left side is the probability of leaving the set of states

$$\{ (n_1, n_2, \ldots, n_{k-1}, \alpha_1, \ldots, \alpha_k) \mid \alpha_i = 0, n_{i-1} \neq 0, n_i \neq N_i \}$$

and the right side is the probability of entering that set.

By the definition of E_i and of conditional probability,

$$\text{prob } (\alpha_i = 1 \mid n_{i-1} \neq 0, n_i \neq N_i) = \frac{E_i}{\text{prob } (n_{i-1} \neq 0, n_i \neq N_i)} \tag{11}$$

or,

$$\text{prob } (\alpha_i = 1 \mid n_{i-1} \neq 0, n_i \neq N_i) = \frac{E_i}{E_i + D_i}.$$

Consequently,

$$\text{prob } (\alpha_i = 1 \mid n_{i-1} \neq 0, \; n_i \neq N_i) = r_i/(r_i+p_i) \tag{12}$$

and (11), (12), and the definition of e_i together imply

$$E_i = e_i \text{ prob } (n_{i-1} \neq 0, \; n_i \neq N_i) \tag{13}$$

which is equivalent to (9) and the proposition is proved.

This result is counter-intuitive because, as a reviewer pointed out, there is no reason to expect that the events of machine failure and adjacent buffers being empty or full are independent. However, failures may occur only while machines are not forced to be idle due to starvation or blockage. Furthermore, B_{i-1} can become empty and B_i can become full only when M_i is operational. Therefore, an idle period can be thought of as a hiatus in which a clock, measuring working time until the next machine state change event, is not running. The fraction of non-idle time during which M_i is operational is thus the same as the fraction of time it would be operational if it were not in a system with other machines and buffers.

While it is possible for n_{i-1} to be 0 and n_i to be N_i simultaneously, it is not very likely. The probability of this event is small because such states can only be reached from states in which $n_{i-1} = 1$ and $n_i = N_i - 1$ by means of a transition in which $\alpha_{i-1} = 0$, $\alpha_i = 1$, $\alpha_{i+1} = 0$. The production rate may therefore be approximated by

$$E_i = e_i \left[1 - \text{prob } (n_{i-1}=0) - \text{prob } (n_i=N_i) \right]. \tag{14}$$

2. Proof of Equation (5)

In the two-machine case, (4) reduces to

$$E(i) = e_i \; (1 - p_b(i)) \tag{15}$$

and

$$E(i-1) = e_i(i-1) \; (1 - p_s(i-1)) \tag{16}$$

in which $e_i(i) = r_i(i)/[r_i(i) + p_i(i)]$ is the isolated efficiency of machine $M_i(i)$ and $e_i(i-1)$ is the isolated efficiency of machine $M_i(i-1)$. Note that these equations are exact, not approximate. They can be written

$$p_b(i) = 1 - E(i) / e_i \qquad (17)$$

and

$$p_s(i-1) = 1 - E(i) / e_i(i-1) \qquad (18)$$

(since $E(i) = E(i-1)$). Substituting into equation (13),

$$E(i) = e_i \left[E(i)/e_i(i) + E(i)/e_i(i-1) - 1 \right].$$

Equation (14) follows after further manipulations using the expressions for the isolated efficiencies in terms of the parameters of the machines.

THE IMPULSE CONTROL PROBLEM WITH CONCAVE COSTS : ON THE SEARCH OF PLANNING HORIZONS.

J.M. PROTH

INRIA

Domaine de Voluceau -Rocquencourt

B.P. 105 -78153 LE CHESNAY CEDEX

FRANCE

Tél. : (3) 954 90 20

ABSTRACT

This paper is devoted to the continuous time problems with concave costs in the case of no backlogging and impulse control.

We first give some results concerning the finite horizon problem. We then prove that it may exist a planning horizon only if a forecast horizon holds. Some results are given in order to find a planning horizon knowing a forecast horizon.

I - INTRODUCTION.

Some papers have been devoted to the finite horizon problems in the case of continuous time and concave costs. Important results concerning the problems with no backlogging and impulse controls can be found in [1], [3] and [4]. Starting from the same hypotheses, we show that it never exists a planning horizon. We also give a sufficient condition in order that there exists a planning horizon knowing a forecast horizon.

The paper is organized as follows :

1. we first set the problem,
2. we then recall the main results concerning the finite horizon problem,
3. we finally consider the planning horizon problem.

II- SETTING OF THE PROBLEM.

Let $\xi(t) \geq 0$, $t \in [0,t]$, the instantaneous demand at time t.

A set $V = (\theta_i, v_i)_{i=1,2,\ldots,}$ where :

$$0 \leq \theta_1 < \theta_2 < \ldots < \theta_i < \ldots < T$$

and $v_i \geq 0$ for $i = 1,2,\ldots$

is called an impulse control. A couple (θ_i, v_i) is an impulse. v_i represents the amount ordered at time θ_i.

The corresponding trajectory for the inventory is given by :

$$\left.\begin{array}{l} dy(t) = -\xi(t)dt + \sum_i v_i\, \delta(t - \theta_i), \quad t \in [0,T] \\ y(0) = y_0 \text{ is known.} \end{array}\right\} \quad (1.1)$$

(1.1) leads to :

$$y(t) = y_0 + \sum_{i=1}^{n} v_i - \int_0^t \xi(s)ds \text{ for } t \in [\theta_n, \theta_{n+1}[\quad (1.2)$$

(θ_{n+1} must be replaced by $+\infty$ if θ_n is the last impulse time).

No backlogging is permitted, then :

$$y(t) \geq 0, \ \forall t \in [0,T] \quad (2)$$

We also introduce the ordering cost :

$$\left.\begin{array}{l} c : [0,T] \times R^+ \to R^+ \\ c(t,v) = K(t, \chi_{v>0}) + c^*(t,v) \end{array}\right\} \quad (3.1)$$

where

$$K(t,\chi_{v>0}) = \begin{cases} K(t,0) = 0 \text{ if } v = 0 \\ K(t,1) \geq K_0 > 0 \text{ if } v > 0 \end{cases} \quad (3.2)$$

c^* is continuous, concave and non decreasing in the second argument.

Moreover :

$$c^*(t,0) = 0, \quad \forall t \in [0,T] \tag{3.3}$$

The <u>inventory cost</u> f is defined as follows :

$$f : [0,T] \times R^+ \to R^+$$

f(t,y) is a Borel function, continuous, concave and non decreasing in the second argument. Moreover :

$$\int_0^T f(t,y)dt < +\infty, \quad \forall y \geq 0 \tag{3.4}$$

Let us consider an impulse control $V = (\theta_i, v_i)_{i=1,2,\dots}$ and the corresponding state function y (see (1.2)).

If y verifies (2), then V is said to be an <u>admissible impulse control</u>.

The corresponding cost is :

$$Q(y_0,V) = \int_0^T f[t,y(t)]\, dt + \sum_i c(\theta_i, v_i) \tag{4}$$

Let us denote <u>D the set of admissible impulse controls</u>.

V^* is an optimal impulse control if :

$$Q(y_0,V^*) = \min_{V \in D} Q(y_0,V) \tag{5}$$

The proof of the existence of V^* can be found in [3]. We first recall the most important results obtained in [1] and [3] concerning the above problem (also called below the T -horizon problem).

III - THE T -HORIZON PROBLEM.

The following results are already proven in [1]

<u>Theorem 1</u>

If $y_0 \geq \int_0^T \xi(t)dt$, then $V = \phi$ is optimal.

Theorem 2

If $y_0 < \int_0^T \xi(t)dt$, there exists an optimal control so that $y(T) = 0$, where y is the state function corresponding to the optimal control.

Theorem 3

An optimal impulse control consists in a finite number of impulses. In other words, if :

$$V = (\theta_i, v_i)_{i=1,2,\dots,n}$$

is an optimal impulse control, then $n < +\infty$.

Theorem 4

If $y_0 < \int_0^T \xi(t)dt$, then the optimal control $V = (\theta_i, v_i)_{i=1,2,\dots,n}$ verifies $y(\theta_i) = 0$ for $i = 2,3,\dots,n$ (y is the state function corresponding to V).

If the costs are both stationary, then $y(\theta_i) = 0$ for $i = 1,2,\dots,n$.

We now present some definitions and results concerning the planning horizons.

IV - THE PLANNING HORIZONS PROBLEM.

IV.1. - Définitions.

We suppose that :

a. $y_0 < \int_0^T \xi(t)\,dt$

b. the demand is known on $[0,T]$

c. the cost functions are defined on $R^+ \times R^+$

Then T is a planning horizon if :

1. $\forall\, T_1 > T$

2. $\forall$ the non negative demand on $]T,T_1]$ (6)

there exists an optimal impulse control on $[0,T_1]$ which is obtained by extending in an adequate manner the optimal impulse control on $[0,T]$ and leads to a state function equal to zero at time T. (Such an optimal impulse control exists, as we showed in theorem 2).

Let us now consider $T_0 \in]0,T]$

We suppose that :

a. $y_0 < \int_0^{T_0} \xi(t)dt$

b. the demand is known on $[0,T]$

c. the cost functions are defined on $R^+ \times R^+$

$T_0 \leq T$ is a planning horizon for the forecast horizon T if :

1. $\forall\, T_1 \geq T$

2. $\forall$ the non negative demand on $]T,T_1]$ (7)

there exists an optimal impulse control on $[0,T_1]$ which is obtained by extending in an adequate manner the optimal impulse control on $[0,T_0]$ and leads to a state function equal to zero at time T_0 (see theorem 2).

(2. vanishes if $T = T_1$)

Note that the following statments are equivalent :

T is a planning horizon

or

T is a planning horizon for the forecast horizon T.

IV.2 - Some results on the planning horizons.

We first give a necessary and sufficient condition in order that T be a planning horizon.

Using this result, we then show that it doesn't exist a planning horizon (in the sense of (6)) for the impulse control problem examined here.

Theorem 5

T is a planning horizon if and only if there exists, for every T_1 -horizon problem ($T_1 \geq T$), an optimal impulse control V_1 so that $y_1(T) = 0$ (y_1 is the state function corresponding to V_1).

Proof :

a. The condition of theorem 5 is obviously necessary (see definition (6) and theorem 2).

b. We now show that it is also sufficient.

Suppose that, whatever $T_1 \geq T$ may be, there exists an optimal impulse control $V_1 = (\theta_i^1, v_i^1)_{i=1,\ldots,n_1}$ so that $y_1(T) = 0$.

Let be θ_n^1 the largest impulse time less than T ($n \leq n_1$) and let us denote $V = (\theta_i^1, v_i^1)_{i=1,\ldots,n}$.

Suppose that V would not be an optimal impulse control for the T -horizon problem, and denote V^0 the optimal impulse control for the T -horizon problem so that the corresponding state function y^0 verifies $y^0(T) = 0$ (see theorem 2).

We consider :

$V^* = V^0 \circ (\theta_i^1, v_i^1)_{i=n+1,\ldots,n_1}$, where o represents the concatenation (if $n = n_1$, $V^* = V^0$).

V^* is an admissible impulse control for the T_1 -horizon problem (consequence of $y_1(T) = y^0(T) = 0$).

Moreover :

$$
\begin{aligned}
Q(y_0, V^*) &= Q(y_0, V^0) + Q(0, (\theta_i^1, v_i^1)_{i=n+1,\ldots,n_1}) \\
&< Q(y_0, V) + Q(0, (\theta_i^1, v_i^1)_{i=n+1,\ldots,n_1}) \\
&= Q(y_0, V_1)
\end{aligned}
$$

Finally :

$$Q(y_0, V^*) < Q(y_0, V_1)$$

and V_1 would not be optimal. It would be inconsistant with the hypothesis..

V is then optimal.

□

Theorem 6

Whatever be the T -horizon problem defined as in §II, T is not a planning horizon.

Proof

Let $V = (\theta_i,v_i)_{i=1,\dots,n}$ an optimal impulse control of a T -horizon problem. We suppose that $y(T) = 0$, where y is the state function corresponding to V.

We consider $T_1 > T$ and we extend the demand on $]T,T_1]$ as follows :

$$\xi(t) = \frac{\varepsilon}{T_1 - T}, \quad \forall t \in]T,T_1] \quad (\varepsilon > 0)$$

Let us now consider the impulse controls V_1^0 and V_1^R of the T_1- horizon problem defined as follows :

$$\text{and} \quad \begin{cases} V_1^0 = (\theta_i,v_i)_{i=1,\dots,n-1} \circ (\theta_n,v_n + \varepsilon) \\ V_1^R = V\circ(\theta_i, \dfrac{\varepsilon(\theta_{i+1} - \theta_i)}{T_1 - T})_{i=n+1,\dots,n+R}, \end{cases}$$

with $R \geq 1$, $\theta_{n+1} = T$ and $\theta_{n+R+1} = T_1$.

We can easily verify that these controls are both admissible. Let us compute the corresponding costs.

$$Q(y_0,V_1^0) = Q(y_0,V) + \int_{\theta_n}^{T} f[t,v_n - \int_{\theta_n}^{t} \xi(s)ds + \varepsilon]dt$$
$$+ \int_{T}^{T_1} f[t, \frac{T_1 - t}{T_1 - T}\varepsilon]\,dt - \int_{\theta_n}^{T} f[t,v_n - \int_{\theta_n}^{t} \xi(s)ds]dt \qquad (8)$$
$$+ c(\theta_n,v_n + \varepsilon) - c(\theta_n,v_n)$$

and

$$Q(y_0,V_1^R) = Q(y_0,V) + \sum_{i=n+1}^{n+R} \{\int_{\theta_i}^{\theta_{i+1}} f[t, \frac{\theta_{i+1} - t}{\theta_{i+1} - \theta_i}\varepsilon_i]dt + c(\theta_i,\varepsilon_i)\} \qquad (9)$$

with : $\varepsilon_i = \frac{\theta_{i+1} - \theta_i}{T_1 - T} \varepsilon$, $R \geq 1$

Equations (8) and (9) lead to :

$$Q(y_0,V_1^R) - Q(y_0,V_1^0) = \sum_{i=n+1}^{n+R} c(\theta_i,\varepsilon_i) - [c(\theta_n,v_n + \varepsilon) - c(\theta_n,v_n)]$$

$$+ \sum_{i=n+1}^{n+R} \int_{\theta_i}^{\theta_{i+1}} f[t, \frac{\theta_{i+1} - t}{T_1 - T} \varepsilon]dt - \int_T^{T_1} f[t, \frac{T_1 - t}{T_1 - T} \varepsilon] dt \tag{10}$$

$$- \int_{\theta_n}^{T} \{f[t,v_n - \int_{\theta_n}^{t} \xi(s)ds + \varepsilon] - f[t,v_n - \int_{\theta_n}^{t} \xi(s)ds]\} dt$$

We know (see (3.1), (3.2) and (3.3)) that :

$$c(\theta_i,\varepsilon_i) \geq K_0 > 0, \quad \forall \varepsilon_i > 0 \tag{11}$$

In addition, we can choose ε small enough in order that :

$$c(\theta_n,v_n + \varepsilon) - c(\theta_n,v_n) < \frac{K_0}{3} \tag{12}$$

$$\int_{\theta_n}^{T} \{f[t,v_n - \int_{\theta_n}^{t} \xi(s)ds + \varepsilon] - f[t,v_n - \int_{\theta_n}^{t} \xi(s)ds]\} dt < \frac{K_0}{3} \tag{13}$$

and :

$$\sum_{i=n+1}^{n+R} \int_{\theta_i}^{\theta_{i+1}} f[t, \frac{\theta_{i+1} - t}{T_1 - T} \varepsilon]dt - \int_T^{T_1} f[t, \frac{T_1 - t}{T_1 - T} \varepsilon] dt$$
$$\geq - \int_T^{T_1} f[t, \frac{T_1 - t}{T_1 - T} \varepsilon]dt > - \frac{K_0}{3} \tag{14}$$

Taking into account (11), (12), (13) and (14), equation (10) leads to :

$$Q(y_0,V_1^R) - Q(y_0,V_1^0) > (R - 1)K_0 \geq 0$$

Finally :

$$Q(y_0,V_1^R) > Q(y_0,V_1^0) \tag{15}$$

If T was a planning horizon, the T_1 -horizon problem defined above (with ε verifying (12), (13) and (14)) would have an optimal impulse control of the type V_1^R, and (15) shows that such an impulse control cannot be optimal. It completes the proof.

□

IV.3 - Some results on the planning horizon problem, knowing a forecast horizon.

In this paragraph, we propose a sufficient condition in order that τ be a planning horizon knowing the forecast horizon T. But this condition is not necessary.

IV.3.1. - The property $\mathcal{P}(\tau)$.

Let us consider the T -horizon problem defined in paragraph II and $\tau \in]0,T]$ so that :

$$y_0 < \int_0^\tau \xi(s)ds$$

We say that the T -horizon problem verifies the property $\mathcal{P}_T(\tau)$ if :

$$\forall Y \geq \int_\tau^T \xi(s)ds, \exists \theta_Y \in [\tau,T] \text{ so that :}$$

$$\int_\tau^{\theta_Y} f[t,\int_t^{\theta_Y} \xi(s)ds]\,dt + c[\theta_Y, Y - \int_\tau^{\theta_Y} \xi(s)ds] \leq \int_\tau^{\theta_Y} f[t,Y - \int_\tau^t \xi(s)ds]\,dt \qquad (16)$$

IV.3.2. - Theorem 7.

Suppose that $\mathcal{P}_T(\tau)$ holds.

We consider a T_1 -horizon problem defined as follows :

1. $T_1 \geq T$
2. y_0 is the initial state (state at time 0)
3. the demand is obtained by extending the demand of the T -horizon problem on $]T,T_1]$. Of course, this extension is non-negative.

Then the optimal impulse control of the T_1 -horizon problem has at least one impulse time which belong to $[\tau,T]$.

Proof :

a. We first prove that if $\tau_1 \in [0,\tau]$:

$$\forall\, Y_1 \geq \int_{\tau_1}^{T} \xi(s)ds, \exists\, \theta_Y \in [\tau,T] \text{ so that :}$$

$$\int_{\tau_1}^{\theta_Y} f[t, \int_t^{\theta_Y} \xi(s)ds]\, dt + c(\theta_Y, Y_1 - \int_{\tau_1}^{\theta_Y} \xi(s)ds)$$

$$\leq \int_{\tau_1}^{\theta_Y} f[t, Y_1 - \int_{\tau_1}^{t} \xi(s)ds]\, dt$$

$\mathcal{P}_T(\tau)$ is true, i.e. (see (16)) :

$$\forall\, Y \geq \int_{\tau}^{T} \xi(s)ds, \exists\, \theta_Y \in [\tau,T] \text{ so that :}$$

$$\int_{\tau}^{\theta_Y} f[t, \int_t^{\theta_Y} \xi(s)ds]dt + c[\theta_Y, Y - \int_{\tau}^{\theta_Y} \xi(s)ds] \leq \int_{\tau}^{\theta_Y} f[t, Y - \int_{\tau}^{t} \xi(s)ds]dt \tag{17}$$

Furthermore :

$$Y \geq \int_{\tau}^{T} \xi(s)ds \Rightarrow Y + \int_{t}^{\tau} \xi(s)ds \geq \int_{t}^{T} \xi(s)ds \geq \int_{t}^{\theta_Y} \xi(s)ds, \ \forall \theta_Y \in [\tau,T] \tag{18}$$

Consequently :

$$\int_{\tau_1}^{\tau} f(t, \int_t^{\theta_Y} \xi(s)ds)dt \leq \int_{\tau_1}^{\tau} f(t, Y + \int_t^{\tau} \xi(s)ds)dt \tag{19}$$

(see (18) and (3.4)).

Adding up (17) and (19) we obtain :

$$\forall Y \geq \int_{\tau}^{T} \xi(s)ds, \exists\, \theta_Y \in [\tau,T] \text{ so that :}$$

$$\int_{\tau_1}^{\theta_Y} f[t, \int_t^{\theta_Y} \xi(s)ds]dt + c[\theta_Y, Y - \int_{\tau}^{\theta_Y} \xi(s)ds] \leq \int_{\tau_1}^{\tau} f(t, Y + \int_t^{\tau} \xi(s)ds)dt \tag{20}$$

Let us introduce

$$Y_1 = Y + \int_{\tau_1}^{\tau} \xi(s)ds, \text{ which belongs to } [\int_{\tau_1}^{T} \xi(s)ds, +\infty[$$

if $Y \in [\int_{\tau}^{T} \xi(s)ds, +\infty[$.

(20) can then be rewritten :

$$\forall Y_1 \geq \int_{\tau_1}^{T} \xi(s)ds, \exists \theta_Y \in [\tau,T] \text{ so that :}$$

$$\int_{\tau_1}^{\theta_Y} f[t, \int_t^{\theta_Y} \xi(s)ds]dt + c(\theta_Y, Y_1 - \int_{\tau_1}^{\theta_Y} \xi(s)ds) \leq \int_{\tau_1}^{\theta_Y} f[t, Y_1 - \int_{\tau_1}^{t} \xi(s)ds]dt \tag{21}$$

b. We consider $\tau_1 \in [0,\tau]$.

$\forall t \in [\tau_1, T_1]$, we denote V_t^* the optimal impulse control on $[t,T_1]$ for the restriction of the T_1 -horizon problem to $[t,T_1]$ with an initial stock level equal to :

$$y_t = (y_0 - \int_0^t \xi(s)ds)^+$$

$Q(y_t, V_t^*)$ is the corresponding optimal cost.

Suppose that τ_1 is an impulse time for the optimal impulse control $V_{\tau_1}^*$.

$\theta_0 \in [0,\tau]$ verifies :

$$y_{\tau_1} = \int_{\tau_1}^{\theta_0} \xi(s)ds$$

Let us denote :

$$\theta^* = \text{Max } (\theta_0, \tau_1) \in [\tau_1, \tau[$$

Using the backward dynamic formulation, we obtain :

$$Q(y_{\tau_1}, V_{\tau_1}^*) = \inf_{\theta \in [\theta^*, T_1]} [c(\tau_1, \int_{\tau_1}^{\theta} \xi(s)ds - y_{\tau_1}) + \int_{\tau_1}^{\theta} f(t, \int_t^{\theta} \xi(s)ds)dt + Q(0, V_\theta^*)] \tag{22}$$

and, if θ_m is the optimal value for θ, then the optimal impulse at time τ_1 is :

$$v_{\tau_1} = \int_{\tau_1}^{\theta_m} \xi(s)ds - y_{\tau_1}.$$

We have to consider three cases :

<u>b_1. $\theta_m \in [\tau_1,\tau[$.</u>

Then, we rewrite (22) with θ_m instead of τ_1 and so on. Theorem 3 shows that θ_m will be greater than τ after a finite number of steps.

<u>b_2. $\theta_m \in [\tau,T]$.</u>

It means that there exists an impulse time on $[\tau,T]$.

<u>b_3. $\theta_m \in]T,T_1]$.</u>

In that case, (22) can be rewritten, using the optimal value θ_m :

$$Q(y_{\tau_1},V^*_{\tau_1}) = c(\tau_1,\int_{\tau_1}^{\theta_m} \xi(s)ds - y_{\tau_1}) + \int_{\tau_1}^{\theta_m} f(t,\int_{t}^{\theta_m} \xi(s)ds)dt + Q(0,V^*_{\theta_m}) \tag{23}$$

$$= c(\tau_1,\int_{\tau_1}^{\theta_m} \xi(s)ds - y_{\tau_1}) + \int_{\tau_1}^{\theta_m} f(t,\int_{\tau_1}^{\theta_m} \xi(s)ds - \int_{\tau_1}^{t} \xi(s)ds)dt + Q(0,V^*_{\theta_n})$$

Note that $\int_{\tau_1}^{\theta_m} \xi(s)ds \geq \int_{\tau_1}^{T} \xi(s)ds$

We then can find $\theta_y \in [\tau,T]$ so that (see (21)) :

$$\int_{\tau_1}^{\theta_m} f(t,\int_{\tau_1}^{\theta_m} \xi(s)ds - \int_{\tau_1}^{t} \xi(s)ds)dt$$

$$= \int_{\tau_1}^{\theta_y} f(t,\int_{\tau_1}^{\theta_m} \xi(s)ds - \int_{\tau_1}^{t} \xi(s)ds)dt + \int_{\theta_y}^{\theta_m} f(t,\int_{\tau_1}^{\theta_m} \xi(s)ds - \int_{\tau_1}^{t} \xi(s)ds)dt$$

$$\geq \int_{\tau_1}^{\theta_Y} f(t,\int_{t}^{\theta_Y} \xi(s)ds)dt + c(\theta_y,\int_{\theta_y}^{\theta_m} \xi(s)ds) \tag{24}$$

$$+ \int_{\theta_Y}^{\theta_m} f(t, \int_{\tau_1}^{\theta_m} \xi(s)ds - \int_{\tau_1}^{t} \xi(s)ds)\, dt$$

On the other hand (c is non decreasing in the second argument) :

$$c(\tau_1, \int_{\tau_1}^{\theta_m} \xi(s)ds - y_{\tau_1}) \geq c(\tau_1, \int_{\tau_1}^{\theta_Y} \xi(s)ds - y_{\tau_1}) \tag{25}$$

Finally :

$$Q(y_{\tau_1}, V^*_{\tau_1}) \geq c(\tau_1, \int_{\tau_1}^{\theta_Y} \xi(s)ds - y_{\tau_1}) + \int_{\tau_1}^{\theta_Y} f(t, \int_{t}^{\theta_Y} \xi(s)ds)\, dt$$
$$+ c(\theta_Y, \int_{\theta_Y}^{\theta_m} \xi(s)ds) + \int_{\theta_Y}^{\theta_m} f(t, \int_{t}^{\theta_m} \xi(s)ds)dt + Q(0, V^*_{\theta_m}) \tag{26}$$

Because $Q(y_{\tau_1}, V^*_{\tau_1})$ is optimal, (26) is an equality. Then $\theta_Y \in [\tau,T]$ is an impulse time for the optimal control.

These results are also true if $\tau_1 = 0$.

Finally, if $\mathcal{P}_T(\tau)$ holds, the optimal impulse control of every T_1 -horizon problem ($T_1 \geq T$) has an impulse time which belongs to $[\tau,T]$. □

IV.3.3. - The sufficient condition.

The following result allows to find a planning horizon knowing a forecast horizon by solving only a set of finite horizon problems. We outline that the condition is sufficient, but not necessary.

Corollary.

If $\mathcal{P}_T(\tau)$ holds and if $\theta^* \in [0,\tau]$ is an impulse time for the optimal impulse control of every π -horizon problem, where $\pi \in [\tau,T]$, then θ^* is a planning horizon knowing the forecast horizon T.

Proof

Obvious, starting from theorem 7 and from the definition of the planning horizon knowing a forecast horizon (see (7)).

V - CONCLUSION.

Using the backward dynamic formulation, we obtained a sufficient condition in order that there exists a planning horizon knowing a forecast horizon. Note that this condition requires to choose first the forecast horizon without criteria. If it is too small, it may be that the planning horizon cannot be found. If it is too large, the horizons of the finite horizon problems which have to be solved will also be large and leads to a large amount of computations.

To end, we outline that the sufficient condition presented above will certainly be used in some particular cases (for instance, small ordering cost on a given period, or linear costs on R^{*+}).

BIBLIOGRAPHY.

[1] A. BENSOUSSAN, M. CROUHY, J.M. PROTH, "Mathematical Theory of Production Planning", North Holland Publishing, 1983.

[2] A. BENSOUSSAN, J.L. LIONS, Contrôle impulsionnel et inéquations quasi variationnelles, Dunod, Paris, 1982.

[3] A. BENSSOUSSAN, J.M. PROTH, Inventory Planning in a deterministic environment. Concave set up in discrete and continuous time, Vienna, Nov. 1981.

[4] A. BENSOUSSAN, J.M. PROTH, On some impulse control problems with concave costs, C.D.C., Déc. 1982.

GESTION D'UN STOCK MULTI-PRODUITS AVEC COÛTS CONCAVES ET INCITATION AUX LANCEMENTS GROUPES : UNE HEURISTIQUE

DIAGNE S., LEOPOULOS V.I. et J.M. PROTH

INRIA

Domaine de Voluceau -Rocquencourt

B.P. 105 -78153 LE CHESNAY CEDEX

FRANCE

Tél. : (3) 954 90 20

ABSTRACT

This paper is devoted to the multi-product lot size model with concave costs, the production cost being joint. We give an heuristic which leads to a "good" solution, the amount of computation being only proportional to $N * M$, where N is the horizon of the problem and M the number of products involved.

RESUME

Ce papier est consacré aux problèmes de production multi-produits à coûts concaves, avec incitation aux lancements groupés. Nous proposons une heuristique qui conduit à une "bonne" solution avec un volume de calcul seulement proportionnel à $N * M$, où N est l'horizon du problème et M le nombre de produits considérés.

INTRODUCTION

Le problème déterministe de gestion de stocks en temps discret, sur un horizon fini et lorsque les coûts de lancement et de stockage sont concaves et non décroissants, a donné naissance à une abondante littérature dans le cas d'un produit unique (voir en particulier [6], [7] et [8]).

Plus récemment, nous avons repris ce problème sous les hypothèses les plus générales et en utilisant la technique de la programmation dynamique de type rétrograde (voir en particulier [1], [2], [3]). Cette approche a permis de mettre en évidence un ensemble de propriétés du coût optimal considéré comme fonction du stock initial. Ces propriétés ont été utilisées pour obtenir un théorème nouveau sur le problème des horizons de planification ([5]) et un algorithme basé sur ce théorème qui conduit à un horizon de planification pour un grand nombre de situations ([4]).

Nous avons envisagé le cas multi-produits à coûts concaves avec incitation aux lancements groupés.

Certains auteurs se sont déjà intéressés à ce problème (par exemple [9] et [10]).

Nous avons donné un algorithme qui utilise la programmation dynamique de type rétrograde ([11]). Nous avons montré que le nombre de calculs est proportionnel à N^{M+1} où N est l'horizon de planification et M le nombre de produits.

L'objet de cette publication est de proposer une heuristique qui conduit à une "bonne" solution, le nombre de calculs étant seulement proportionnel à N*M.

On se limitera à l'étude du problème à deux produits.

La publication est organisée de la façon suivante :

I Exposé du problème : notations, définitions, hypothèses et résultats

II L'heuristique

III Exemples d'application

I. NOTATIONS POUR L'APPLICATION DU CAS DE DEUX PRODUITS ET RESULTATS.

Afin que la présentation soit simple, on va se limiter au problème de deux produits. Les résultats peuvent être facilement généralisés au cas de plusieurs produits.

Les notations et relations à prendre en compte sont les suivantes :

$V = \{v_i^j\}_{i=0,1,\dots,N-1}^{j=1,2} = \{V_i\}_{i=0,\dots,N-1} = \{V^j\}^{j=1,2}$ est un contrôle.

v_i^j est le réapprovisionnement en produit j décidé à l'instant i et qui prend effet à l'instant i+1. De plus :

$$J\ (V_i) = \{j/j\in(1,2) \text{ et } v_i^j > 0\} \text{ pour } i = 0,1,\dots,N-1 \qquad (1)$$

Nous notons $Y_0 = (y_0^1, y_0^2)$ les niveaux des stocks à l'instant 0.

L'équation d'état s'écrit :

$$y_{i+1}^j = y_i^j + v_i^j - \xi_{i+1}^j,\ i=0,1,\dots,N-1 \text{ et } j=1,2 \qquad (2)$$

où y_i^j est le niveau du stock de produit j sur $[i,i+1]$

Les ruptures de stock ne sont pas admises. En d'autres termes :

$$y_i^j \geq 0 \text{ pour } i = 0,1,\dots,N \text{ et } j = 1,2 \qquad (3)$$

Le contrôle V est admissible si (3) est vérifié et si, bien entendu, $v_i^j \geq 0$ pour $i = 0,1,\dots,N-1$ et $j = 1,2$.

Nous notons $D\ (Y_0)$ l'ensemble des contrôles admissibles.

Pour $i = 0,1,\dots,N-1$, le coût de lancement s'écrit :

$$C_i^*\ (V_i) = k_i^{J(V_i)} + C_i^1\ (v_i^1) + C_i^2\ (v_i^2) \qquad (4)$$

où C_i^1 et C_i^2 sont concaves, non décroissantes, définies sur R^+ et à valeurs dans R^+. En outre :

$$\max\ (k_i^1, k_i^2) \leq k_i^{(1,2)} \leq k_i^1 + k_i^2, \text{ avec } k_i^j \geq 0 \qquad (5)$$

f_i^j (i=0,1,...,N-1 et j=1,2) est une fonction concave, non décroissante, défini sur R^+ et à valeurs dans R^+.

$f_i^j\ (y)$, $y \geq 0$ est le coût de stockage d'une quantité y de produit j sur $[i,i+1)$.

Si $V\in D\ (Y_0)$, ensemble des contrôles admissibles pour l'état initial Y_0, alors :

$$K\ (Y_0,V) = \sum_{i=0}^{N-1} \{C_i^*\ (V_i) + f_i^1\ (y_i^1) + f_i^2\ (y_i^2)\} \tag{6}$$

est le coût lié au contrôle V

($\{y_i^j\}_{i=1,\dots,N}^{j=1,2}$ est la suite des états qu'on obtient en appliquant le contrôle V si l'état initial est Y_0).

Nous recherchons toujours V^* (contrôle optimal) tel que :

$$K^*\ (Y_0,V^*) = \min_{V \in D(Y_0)} K\ (Y_0,V) \tag{7}$$

On appelle P le problème à deux produits.

Considérons maintenant $I \subset \{0,1,\dots,N-1\}$

Pour j = 1,2 on appelle P_I^j le problème à un produit défini comme suit :

a) P_I^j est un problème à horizon N (8)

b) $\{\xi_i^j\}_{i=1,2,\dots,N}$ est la demande de produit j à l'instant i

c) $\{f_i^j\}_{i=0,1,\dots,N-1}$ est le coût de stockage à l'instant i

d) Pour i = 0,1,...,N-1, le coût de lancement à l'instant i est défini comme suit :

$$q_{I,i}^j\ (v) = \begin{cases} k_i^{(1,2)} - k_i^{3-j} + C_i^j\ (v) & \text{si } i \in I \text{ et } v>0 \\ k_i^j + C_i^j\ (v) & \text{si } i \notin I \quad v>0 \\ C_i^j\ (0) & \text{si } v=0 \end{cases}$$

e) $y_0^j \geq 0$

Soit $V^j = \{v_i^j\}_{i=0,\dots,N-1}$ un contrôle admissible pour le problème P_I^j à horizon N.

Le coût correspondant est :

$$K_I^j\ (y_0^j,V^j) = \sum_{i=0}^{N-1} [q_{I,i}^j\ (v_i^j) + f_i^j\ (y_i^j)] \tag{9}$$

où y_i^j (i=0,1,...,N) sont les niveaux de stock qui correspondent à

v_i^j (i=0,1,...,N-1)

Pour $V = \{v_i^j\}_{i=0,1,\dots,N-1}^{j=1,2}$ on définit aussi

$$I_V^j = \{i / i \in [0,1,\dots,N-1) \text{ et } v_i^{3-j} > 0\} \text{ pour } j = 1,2 \tag{10}$$

La relation suivante est évidente ($\emptyset$ est l'ensemble vide)

$$K(Y_0,V) = K_{I_V^j}^j\ (y_0^j,V^j) + K_\emptyset^{3-j}\ (y_0^{3-j},V^{3-j}) \text{ pour } j = 1,2 \tag{11}$$

On va maintenant donner les principaux résultats de cette publication.

Nous allons démontrer le théorème suivant :

Théorème 1

Soit $V^* = \{v_i^{*j}\}_{i=0,1,\dots,N-1}^{j=1,2}$ un contrôle optimal pour P.

alors :

$V^{*j} = \{v_i^{*j}\}_{i=0,1,\dots,N-1}$ un contrôle optimal pour $P^j_{I^j_{V^*}}$ (j = 1,2)

Démonstration

Nous noterons encore :

$V = (V^1, V^2)$

Supposons pour le moment que V^{*1} ne soit pas un contrôle optimal pour $P^1_{I^1_{V^*}}$, et désignons par W^{*1} un contrôle optimal pour ce problème à un seul produit.

Si on pose : $W^* = (W^{*1}, V^{*2})$

alors (11) nous conduit à la formule :

$$K\,(Y_0, V^*) = K^1_{I^1_{V^*}}\,(y_0^1, V^{*1}) + K^2_{\emptyset}\,(y_0^2, V^{*2})$$

$$> K^1_{I^1_{V^*}}\,(y_0^1, W^{*1}) + K^2_{\emptyset}\,(y_0^2, V^{*2})$$

$$= K(Y_0, W^*)$$

et V^* ne peut pas être un controle optimal.

La même démonstration est valable pour le cas où V^{*2} n'est pas un contrôle optimal pour $P^2_{I^2_{V^*}}$.

Le théorème 1 nous donne une condition nécessaire pour que V^* soit un contrôle optimal. Nous présentons maintenant une condition suffisante dans quelques cas particuliers.

Théorème 2

Notons L = {0,1,...,N-1} et

$V^{*1} = \{v_i^{*1}\}_{i=0,1,\dots,N-1}$ un contrôle optimal pour P^1_L.

$V^{*2} = \{v_i^{*2}\}_{i=0,1,\dots,N-1}$ un contrôle optimal pour $P^2_{\emptyset}$

$V^* = (V^{*1}, V^{*2})$

$X^{*1} = \{x_i^{*1}\}_{i=0,1,\dots,N-1}$ un contrôle optimal pour $P^1_{\emptyset}$

$X^{*2} = \{x_i^{*2}\}_{i=0,1,\dots,N-1}$ un contrôle optimal pour P_L^2

$X^* = (X^{*1}, X^{*2})$

Si $I^1_{V^*} \supset I^2_{V^*}$, alors V^* est optimal pour P

Si $I^2_{X^*} \supset I^1_{X^*}$, alors X^* est optimal pour P

(voir (10) pour la définition de I^j_V)

Démonstration

a) Soit $L \supset I \supset I^2_{V^*}$

Supposons que V^{*1} soit optimal pour P^1_I

Nous allons d'abord montrer la proposition suivante :

Proposition :

Si $J \notin I^2_{V^*}$ et $j \in I$ alors V^{*1} est optimal pour le problème $P^1_{I-\{j\}}$ (12)

Démonstration :

Supposons que V^{*1} ne soit pas optimal pour $P^1_{I-\{j\}}$ et soit $W^{*1} = \{w_i^{*1}\}_{i=0,1,\dots,N-1}$ un contrôle optimal pour ce problème alors :

$$K^1_{I-\{j\}}(y^1_0, W^{*1}) < K^1_{I-\{j\}}(y^1_0, V^{*1}) \quad (13)$$

Il faut considérer deux cas différents

a_1. Si $w_j^{*1} > 0$

$$K^1_{I-\{j\}}(y^1_0, W^{*1}) = K^1_I(y^1_0, W^{*1}) + K^1_j + K^2_j - k_j^{(1,2)}$$

où (voir (5)) $k^1_j + k^2_j - k_j^{(1,2)} > 0$

alors :

$$K^1_{j-\{j\}}(y^1_0, W^{*1}) > K^1_I(y^1_0, W^{*1}) \quad (14)$$

en plus, $j \notin I^2_{V^*}$ et $j \in I$ nous conduit à :

$$K^1_{I-\{j\}}(y^1_0, V^{*1}) = K^1_I(y^1_0, V^{*1}) \quad (15)$$

Finalement on considère (13), (14) et (15) et on obtient la relation :

$$K^1_I(y^1_0, W^{*1}) < K^1_I(y^1_0, V^{*1}) \quad (16)$$

laquelle est en contradiction avec le fait que V^{*1} est un contrôle optimal pour P_I^1.

a_2. Si $w_j^{*1} = 0$

Alors la relation (14) devient :

$$K^1_{I-\{j\}}(y^1_0,W^{*1}) = K^1_I\ (y^1_0,W^{*1}) \tag{17}$$

et (15) reste vraie. Alors (16) est encore vraie et la conclusion est la même qu'auparavant.

La démonstration de (12) est maintenant complète.

b) Si on commence par I=L et si on prend en considération le fait que V^{*1} est optimal pour P^1_L (voir le théorème), la proposition (12) et le fait que $I^2_{V^*}$ est inclus dans $I^1_{V^*}$ ($I^1_{V^*} \supset I^2_{V^*}$) nous conduit pas à pas à la proposition suivante :

V^{*1} est optimal pour le problème $P^1_{I^1_{V^*}}$ (18)

c) Il reste que :

$$K\ (Y_0,V^*) = K^1_{I^1_{V^*}}(y^1_0,V^{*1}) + K^2_{\emptyset}\ (y^2_0,V^{*2}) \tag{19}$$

Mais on connaît (18) et le fait que V^{*2} est optimal pour $P^2_{\emptyset}$ alors (19) montre que V^* est optimal pour P. La deuxième partie de la démonstration est similaire.

Théorème 3

Considérons L = {0,1,...,N-1}, I⊂L et $V^j = \{v^j_i\}_{i=0,1,\dots,N-1}$ un contrôl optimal pour P^j_I (j=1,2)

Alors V^j est un contrôle optimal pour $P^j_{I\cap I^{3-j}_V}$

Démonstration

Supposons que V^j ne soit pas optimal pour $P^j_{I\cap I^{3-j}_V}$

On note $W^j = \{w^j_i\}_{i=0,1,\dots,N-1}$ un contrôle optimal pour ce problème.

On définit : $M = \{m/m\in(0,1,\dots,N-1),\ m\in I,\ m\notin I^{3-j}_V \text{ et } w^j_i > 0\}$

(M peut être vide)

Alors :

$$K^j_I\ (y^j_0,W^j) + \sum_{m\in M}(k^j_i + k^{3-j}_i - k^{(1,2)}_i) = K^j_{I\cap I^{3-j}_V}\ (y^j_0,W^j) \tag{20}$$

$$K^j_{I\cap I^{3-j}_V}\ (y^j_0,W^j) < K^j_{I\cap I^{3-j}_V}\ (y^j_0,V^j),\ (W^j \text{ est optimal pour } P^j_{I\cap I^{3-j}_V}) \tag{21}$$

$$K^j_{I\cap I^{3-j}_V}\ (y^j_0,V^j) = K^j_I\ (y^j_0,V^j) \tag{22}$$

Si on prend en considération (20), (21) et (22) on déduit la relation :

$$K_I^j\ (y_0^j, W^j) < K_I^j\ (y_0^j, V^j)$$

laquelle est en contradiction avec le fait que V^j est un contrôle optimal pour P_I^j. La démonstration et maintenant complète.

Théorème 4

On peut réécrire la relation (11) de la façon suivante :

$$K\ (Y_0, V) = K^j_{I_V^j \cap I_V^{3-j}}\ (y_0^j, V^j) + K_\emptyset^{3-j}\ (y_0^{3-j}, V^{3-j}) \quad j = 1,2 \tag{23}$$

Démonstration

Ce résultat est une conséquence immédiate du fait que si $i \in I_V^j$ et $i \notin I_V^{3-j}$ alors $v_i^j = 0$ et $C_i^*\ (V_i) = k_i^{3-j} + C_i^j\ (0) + C_i^{3-j}\ (V_i^{3-j})$ (voir (1) et (3)) quelque soit $v_i^{3-j} \geq 0$.

Théorème 5

Soit V_1^1 un contrôle admissible pour le problème $P_\emptyset^1$ et V_1^2 un contrôle optimal pour $P^2_{I^2_{V_1}}$ (voir (10))

Nous notons encore $V_1 = (V_1^1, V_2^2)$

On pose $V_2 = (V_2^1, V_2^2)$ où :

$$\begin{cases} V_2^1 \text{ est un contrôle optimal pour } P^1_{I^1_{V_1} \cap I^2_{V_1}} \\ V_2^2 = V_1^2 \end{cases} \tag{24}$$

alors :

$$K\ (Y_0, V_2) \leq K\ (Y_0, V_1)$$

Démonstration

$$K\ (Y_0, V_1) = K^1_{I^1_{V_1} \cap I^2_{V_1}}\ (y_0^1, V_1^1) + K_\emptyset^2\ (y_0^2, V_1^2) \quad \text{(voir théorème 4)}$$

$$\geq K^1_{I^1_{V_1} \cap I^2_{V_1}}\ (y_0^1, V_2^1) + K_\emptyset^2\ (y_0^2, V_2^2) \quad \text{(voir (24))} \tag{25}$$

Mais si on considère le théorème 3 :

$$I^1_{V_1} = I^1_{V_2} \quad \text{car } V^2_2 = V^2_1 \quad \text{(voir (24))}$$

Alors :

$$K^1_{I^1_{V_1}} \cap I^2_{V_1} (y^1_0, V^1_2) = K^1_{I^1_{V_2}} \cap I^2_{V_1} (y^1_0, V^1_2)$$

$$\geq K_{I^1_{V_2}} (y^1_0, V^1_2)$$

et la relation (25) devient :

$$K(Y_0, V_1) \geq K^1_{I^1_{V_2}} (Y^1_0, V^1_2) + K^2_{\emptyset} (y^2_0, V^2_2)$$

Ces résultats conduisent à la relation suivante (voir (11)) :

$$K\,(Y_0, V_1) \geq K\,(Y_0, V_2)$$

On va maintenant proposer une heuristique déduite des théorèmes 2 et 4

IV L'HEURISTIQUE

L'idée de l'algorithme est simple. Partant d'une solution admissible, elle consiste à fixer la partie du contrôle correspondant à l'un des produits, et à optimiser le problème mono-produit correspondant au produit restant. Les coûts de lancement considérés dans ce problème mono-produit tiennent compte, au niveau des coûts fixes, du contrôle fixé. On procède ainsi alternativement sur l'un et l'autre des produits. On démontre que l'on tend vers un contrôle global dont chaque composante est optimale pour le produit correspondant si le coût de lancement est établi en tenant compte de l'autre produit.

Dans le cas de plus de deux produits, on procède de la même manière en balayant l'ensemble des produits, le coût de lancement correspondant au produit considéré à un instant donné tenant compte de l'ensemble des autres produits.

Cet algorithme est divisé en deux parties. La première partie utilise le théorème 2. Si le théorème 2 ne conduit pas à un contrôle optimal, alors la deuxième partie de l'algorithme utilise le théorème 4 pour obtenir une "bonne" solution (i.e. une solution proche de la solution optimale).

Partie A (voir théorème 2)

1. Première étape

1.1 Calculer :

1.1.1. V^{*1}, contrôle optimal pour P_L^1

1.1.2. V^{*2}, contrôle optimal pour $P_\emptyset^2$

1.2 Poser :

$V^* = (V^{*1}, V^{*2})$

1.3 Tester

Si $I_{V^*}^1 \supset I_{V^*}^2$ alors V^* est optimal pour P et l'algorithme prend fin.

Sinon aller en 2.

2. Deuxième étape

2.1 Calculer :

2.1.1. X^{*1}, contrôle optimal pour $P_\emptyset^1$

2.1.2. X^{*2}, contrôle optimal pour P_L^2

2.2 Poser :

$X^* = (X^{*1}, X^{*2})$

2.3 Tester :

Si $I_{X^*}^2 \supset I_{X^*}^1$, alors X^* est optimal pour P et l'algorithme prend fin.

Sinon aller en 3.

Partie B (voir théorème 4)

3 z = 1

4 Engendrer $I \subset L$ au hasard et calculer V_1^1 contrôle optimal pour P_I^1.

5 Calculer V_1^2, contrôle optimal pour $P^2_{I^2_{V_1}}$, où $V_1 = \{V_1^1, V_1^2\}$ (voir (10)) pour la définition de $P^2_{I^2_{V_1}}$) et noter $V_1 = (V_1^1, V_1^2)$

6 Pour n = 2,3,4,...

6.1 Si n est pair

6.1.1. Calculer

V_n^1, contrôle optimal pour $P^1_{I^1_{V_{n-1}} \cap I^2_{V_{n-1}}}$

6.1.2. Poser

$V_n^2 = V_{n-1}^2$

6.1.3. Noter

$V_n = (V_n^1, V_n^2)$

6.1.4. Calculer le coût correspondant à V_n

$K\ (Y_0, V_n)$

6.1.5. Tester

Si $K\ (Y_0, V_n) = K\ (Y_0, V_{n-1})$, aller en 7

Sinon aller en 6 (la prochaine valeur de n)

6.2 Si n est impair

6.2.1. Poser

$V_n^1 = V_{n-1}^1$

6.2.2. Calculer

V_n^2, contrôle optimal pour $P^2_{I^1_{V_{n-1}} \cap I^2_{V_{n-1}}}$

6.2.3. Aller en 6.1.3.

7. Tester

7.1 Si z = 1 poser

7.1.1. $C = K\ (Y_0, V_n)$

7.1.2. $W = V_n$

7.1.3. Aller en 8

7.2 Si z > 1

Si $C' < K\ (Y_0, V_n)$ aller en 8

Sinon aller en 7.1.1.

8. Tester

8.1 Si $z < Z$ (voir plus bas la définition de Z)

8.1.1. Calculer

$z = z + 1$

8.1.2. Aller en 4

8.2 Si $z \geq Z$

8.2.1. Imprimer la "bonne" solution W et le coût C

8.2.2. Fin

Z est le nombre d'essais choisi par l'utilisateur.

V. EXEMPLES D'APPLICATION.

Dans tous les exemples que nous présentons ici :

1. l'horizon est égal à 25
2. les stocks initiaux sont nuls
3. les coûts de stockage attachés aux produits 1 et 2 sont respectivement :

$$\left.\begin{array}{l} f_i^1(y) = y/100 \\ f_i^2(y) = 0.3y \end{array}\right\} \begin{array}{l} \text{pour } i = 1,2,\ldots,25 \\ \text{et } y \geq 0 \end{array}$$

4. toujours avec les notations introduites dans le paragraphe I :

$$C_i^2(v) = \left\{\begin{array}{l} 0 \text{ si } v = 0 \\ 2 + v/10 \text{ si } v > 0 \end{array}\right\} \text{ pour } i = 1,2,\ldots, 25$$

et

$$C_i^1(v) = v/100 \quad \text{pour } i = 1,2,\ldots, 25$$

5. les coûts fixes de lancement sont générés au hasard de façon à répondre aux contraintes qui leurs sont imposées.
6. les demandes sont générées au hasard.
7. Z (c.f. l'algorithme), nombre d'essais effectués lors de l'application de l'heuristique, est égal à 20.

V.1. Exemples 1.

Nous avons porté, dans le tableau suivant, les coûts optimaux et les coûts obtenus à l'aide de l'heuristique. Les expériences pour lesquelles ces coûts sont différents sont marquées d'une étoile.

	Expérience	Coût optimal	Coût donné par l'heuristique
*	1	77.14	77.25
	2	69.10	69.10
	3	64.87	64.87
	4	65.99	65.99
	5	79.01	79.01
	6	66.73	66.73
*	7	71.59	71.65
	8	77.08	77.08
	9	71.27	71.27
	10	77.91	77.91
	11	71	71
	12	71.1	71.1
	13	67.63	67.63
*	14	73.96	74.20
	15	70.76	70.76

V.2. Exemples 2

Nous avons dit, plus haut, que 20 essais sont effectués lors de l'application de l'algorithme.

Nous donnons ici, pour chacune des expériences :

1. le coût optimal
2. le coût fourni par l'heuristique
3. la liste des coûts différents rencontrés au cours des vingt essais effectués

dans l'heuristique. (l'horizon ici est égal à 15).

Expérience 1

Coût optimal : 41.29 Coût donné par l'heuristique : 41.44

Coûts différents rencontrés : 42.14, 41.44, 42.55, 42.12, 42.39, 41.79, 42.33, 42.06.

Expérience 2

Coût optimal : 45.92 Coût donné par l'heuristique : 45.92

Coûts différents rencontrés : 46.93, 46.92, 46.56, 46.20, 47.01, 46.70, 46.52, 46.60, 46.32, 46.93, 46.42, 45.92, 46.79, 46.38, 46.90, 47.17.

Conclusion.

L'heuristique que nous venons de présenter se révèle très efficace et converge en probabilité vers la solution optimale. Elle permet de remplacer un problème dont la complexité est de l'ordre de N^{M+1}, où N est l'horizon et M le nombre de produits, par un problème dont la complexité évolue comme $N*M*\alpha$, où α est le nombre d'itérations pour chaque essai. Ce nombre n'a jamais dépassé 10 au cours des exécutions.

On peut espérer trouver des propriétés qui, lors de l'application de l'heuristique, permettront de partir de situations initiales plus favorables que celles qui sont actuellement générées au hasard. Le problème reste ouvert.

(1) A. BENSOUSSAN, M. CROUHY, J.M. PROTH, "Mathematical theory of Production Planning" Advanced Series in Management, North Holland.

(2) A. BENSOUSSAN and J.M. PROTH, "Inventory Planning in a Deterministic Environment : Concave Cost set up in Discrete and Continuous Time" Vienna, 81, November.

(3) A. BENSOUSSAN and J.M. PROTH, "Gestion de Stocks avec Coûts Concaves" RAIRO Automatique/Systems Analysis and Control Vol. 15, n°3 pp. 291-220.

(4) A. BENSOUSSAN and J.M. PROTH "Production Planning in a Deterministic Environment : Concave and Convex Cost-Planning Horizon", TIMS Congress, Bordeaux (FR), August 82.

(5) J.M. PROTH "Problèmes à Coûts Concaves : Notion d'Horizon de planification", Sciences de Gestion, n°3, December, 80.

(6) H.M. WAGNER and T.M. WITHIN, "Dynamic Version of the Economic lot Size Model," Man. Sc, 10, 1964, 465-471.

(7) R.A LUNDIN and T.E. MORTON "Planning Horizons for the dynamic lot size model : Zabel v.s. protective procedures and computational results", O.R., vol. 13, n°1, July-August 1975.

(8) L.A. JOHNSON and P.C. MONTGOMERY "Operations Research in Production Planning" Scheduling and Inventory Control, Wiley, NEW-YORK, 1974.

(9) A. EDWARD SILVER, "Coordinated Replenishments of items under Time-Varying demand : Dynamic Programming Formulation", Naval Research Logistics Quaterly, March, 79, vol. 26, n°1.

(10) P.C. EDWARD KAO, "A Multi-Product Dynamic lot Size model with individual and Joint Set-up Costs" O.R., vol. 27, n°2, Mach-April 1979.

(11) J.M. PROTH, "Gestion d'un Stock Multi-Produit avec Coût Concaves et incitation aux Lancements groupés" Analysis and Optimization of Systems, Proceedings of the Fifth International Conference on Analysis and Optimization of Systems, Versailles December 15-17, 1982.

A L L A N

Un préprocesseur pour faciliter l'utilisation du progiciel ASTEC 3 (CISI)

SUMMARY

Access to fast computing simulation codes, like ASTEC 3, and wise use of them become easy through use of the preprocessing code ALLAN. The aim of this code is to let technical research engineer free of coding tasks for the resolution of mathematical problems involved by simulation (set of differential equations) and to offer him some help in analysis of systems re-combining some elements used in previous problems.

RESUME

L'accès à des logiciels de simulation performants, tel ASTEC 3, et leur utilisation rationnelle, sont facilités par la mise en oeuvre du préprocesseur ALLAN. L'objectif de ce logiciel est de dégager l'ingénieur de recherche technique de tâches informatiques pour la résolution des problèmes mathématiques posés par la simulation (systèmes d'équations différentielles) et d'offrir à ce dernier une forme d'aide pour l'analyse de systèmes qui recombinent certains des éléments utilisés dans des problèmes déjà traités.

1 - INTRODUCTION

La simulation numérique sur ordinateur est aujourd'hui un des plus puissants outils à la disposition du chercheur; elle lui permet de mener à bien les études souvent fort complexes quant au volume de calculs nécessaire, et donc quasi inaccessibles aux méthodes traditionnelles (calcul analogique, expériences sur maquettes...).

La taille ou la difficulté des études ainsi conduites entraîne souvent le chercheur dans un domaine qui ne le concerne pas toujours directement : celui de l'analyse numérique et de l'informatique.

Le temps passé à comprendre et à analyser le problème risque de devenir inférieur au temps passé à écrire le programme de calcul pour la résolution numérique. La réflexion sur les phénomènes physiques ou même l'exploitation complète des résultats issus du code de calcul passe alors au second plan, ce qui est fort préjudiciable à la bonne santé de l'étude.

En réponse à ces objections de principe sur la méthodologie de l'organisation de la recherche, certains informaticiens ont d'ores et déjà mis au point des logiciels très performants qui permettent la simulation générale de larges classes de système physique (par exemple ceux que gouvernent des systèmes d'équations différentielles). Ces logiciels sont en principe orientés vers l'utilisateur, c'est-à-dire qu'ils sont utilisables sans lourdes connaissances mathématiques et informatiques, ou avec un apprentissage restreint.

Astec 3 est un de ces logiciels, offert par la Compagnie Internationale de Services en Informatique (CISI), filiale du Commissariat à l'Energie Atomique (CEA).

2 - PRESENTATION D'ASTEC 3

ASTEC 3 (3è version, mise en service en 1977) est un programme général de simulation des circuits électriques et de systèmes décrits par des équations algébro-différentielles. Jusqu'à ce jour les domaines d'application sont :

- l'électronique générale
- l'électricité de puissance
- l'étude par analogie électrique : thermique, hydraulique, biologie...

mais, comme nous le verrons plus loin dans cette note, nous pensons que ces domaines s'élargissent à tout problème pouvant s'étudier sous forme de réseaux par l'intermédiaire de boîtes noires.

Il y a trois étapes dans l'utilisation d'ASTEC 3 :

a - La description du circuit

En utilisant les modèles de base stockés dans une bibliothèque ou ceux que l'on écrit soit-même ; le circuit est décrit à partir de noms de noeuds .. associés à chaque élément.

La description d'un modèle peut faire intervenir des équations algébro-différentielles, qui doivent être exprimées en différentielles explicites, mais peuvent être non linéaires.

b - La description des simulations

On doit préciser le type de simulation désirée, en ayant la possibilité de modifier les valeurs d'un certain nombre de paramètres liés aux méthodes numériques, sinon données par défaut :

- régime statique ou continu pour le calcul d'un état d'équilibre ou le point de fonctionnement d'un circuit linéaire ou non,

- régime transitoire : solution dans le temps pour un réseau linéaire ou non soumis à des excitations spécifiées par l'utilisateur,

- régime alternatif : solution dans le domaine fréquentiel, fournissant la réponse fréquentielle petits signaux d'un circuit linéaire soumis à une excitation sinusoïdale. Les circuits non linéaires sont linéarisés autour du point de fonctionnement.

Les simulations peuvent être effectuées en mode nominal ou, pour une étude de dispersion de résultats liée à celles de valeurs de composants ou de conditions, en mode statistique.

c - Edition des résultats

Un ensemble de commandes est offert pour faciliter la présentation des résultats, soit sous forme de tableaux de valeurs, soit sous forme graphique.

Langage d'entrée :

Il est structuré en commandes (grandes fonctions du simulateur), contenant des séquences (sous-fonction dans chaque commande), elles-mêmes structurées en instructions élémentaires séparées par des points virgule.

Diverses commandes existent et correspondent à des mots-clés avec préfixe :

$ DESC	Description du circuit ou système à simuler
$ CONT	Simulation en continu, nominale ou statistique
$ TRAN	Simulation en transitoire, nominale ou statistique
$ EDIT	Edition des résultats (courbes, histogrammes)...
$ BIB	Gestion des bibliothèques de modèles.
$ SAUV	Sauvegarde de l'état d'un travail.
$ REST	Restauration d'un état sauvegardé.
$ FIN	Fin d'un jeu de données.

Ces commandes sont totalement disjointes, et ne communiquent entre elles que par des fichiers temporaires ou permanents, ce qui permet à un utilisateur de scinder une étude en plusieurs étapes, ou de ne reprendre qu'une partie de l'exécution du programme en cas de modification dans ses requêtes, ou en cas d'erreur.

Il est ainsi possible de modifier les ordres d'édition sans reprendre la simulation, de modifier des valeurs numériques sans reprendre l'exécution de la commande $DESC, et même de comparer entre eux les résultats de plusieurs simulations, puisque ASTEC a la possibilité de garder tous les résultats des simulations effectuées sur le même circuit.

L'utilisation d'ASTEC, implanté sur matériel IBM, peut se faire soit sous TSO, soit en Batch Processing, soit par soumission en Batch sous TSO.

Depuis 1983 ASTEC est disponible pour implantation sur VAX.

3 - UN PREPROCESSEUR EN LANGAGE NATUREL : ALLAN

C'est encore au bénéfice de l'utilisateur, c'est-à-dire au bénéfice d'un ingénieur plus concerné par l'analyse physique et l'étude approfondie d'un système que par la résolution mathématique et la programmation informatique, que diverses études sont en cours autour d'ASTEC.

On a vu que le langage d'ASTEC est très performant (puisqu'il permet presque, entre autres, l'introduction en clair des équations différentielles), mais son maniement aisé réclame un entraînement, dont, semble-t-il, on devrait pouvoir décharger l'utilisateur. En outre, ASTEC, sous la forme actuelle, ne se prête pas directement à l'analyse fonctionnelle d'un composant : si le composant souhaité ne figure pas dans la bibliothèque d'ASTEC (aujourd'hui riche seulement dans les secteurs de l'électronique et l'analogie thermique), l'utilisateur devra lui-même en faire l'analyse fonctionnelle, c'est-à-dire, à partir des connaissances dont il dispose à son sujet, en dégager les variables d'état, les paramètres de commande, et les équations associées.

L'élaboration d'un logiciel complémentaire, ALLAN (pour "Accès à des logiciels en langage Naturel"), répond aux objectifs suivants :

1 - offrir à l'utilisateur un accès totalement transparent et donc rendre ASTEC utilisable sans aucune connaissance préalable du produit. Après l'analyse fonctionnelle de son problème, l'utilisateur peut conditionner chacune des fonctions mises en évidence dans un module : les variables d'état du problème sont représentées par des bornes d'entrée et de sortie, et reliées par des équations algébro-différentielles.

A ce niveau, ALLAN permet la création directe de ces modules grâce à un jeu de questions/réponses en langage naturel. Ces modules sont conservés dans une bibliothèque personnelle de l'utilisateur.

De la même façon, la traduction du problème est obtenue en implantant à l'écran les modules nécessaires ; puis les liens qui établissent les égalités entre variables du problème sont créés entre les bornes des modules. Pour cela il a été fait appel aux techniques récentes de graphique interactif avec utilisation du "crayon lumineux".

La gestion des noeuds associés à ces liens est automatiquement assurée.

2 - permettre lorsque le problème l'impose, la création de macromodèles, qui résultent de l'association des fonctions de plusieurs modèles. Cette création est assurée par une démarche identique à la précédente.

En particulier, cette option est une aide précieuse dans le cas où il est nécessaire d'introduire une discrétisation d'espace.

3 - autoriser sur un problème isolé, qui ne mérite pas la conservation de modèles en bibliothèque, une description par les équations différentielles seulement, fournies successivement sans autres règles que celles de la syntaxe FORTRAN.

4 - générer automatiquement le texte ASTEC, libérant ainsi l'utilisateur des contraintes syntaxiques et le mettant à l'abri de tout risque d'erreur. Tout ce processus rend les simulations moins fastidieuses et permet aisément des modifications du problème.

4 - CONCLUSION

Il faut noter que la démarche est tout à fait générale et pourrait être appliquée à d'autres logiciels qu'ASTEC ; la seule part de ce travail qui soit spécifique au logiciel utilisé est la traduction du problème (analysé fonctionnellement) en langage ASTEC, selon les règles de syntaxe qui sont propres à ce logiciel. Pour utiliser d'autres logiciels du même genre, il suffirait donc de changer le traducteur de texte et de l'adapter au nouveau langage-cible, celui du nouveau logiciel envisagé. Cette remarque prend tout son poids quand on sait que les informaticiens proposent tous les jours de nouvelles méthodes de résolution et que l'évolution des logiciels de simulation, déjà rapide, ne fera dans l'avenir que s'accélérer. Citons à titre d'exemple le nouveau logiciel CISI, NEPTUNIX, qui, s'il est incapable d'analyser un réseau, possède une plus large gamme de problèmes mathématiques dans son domaine d'entrée.

ANALYSE SUR MICROORDINATEUR APPLE II
DU RYTHME VEILLE-SOMMEIL CHEZ LE RAT

LACOSTE G., RODI M., GANDOLFO G. et GOTTESMANN Cl.

Laboratoire de Psychophysiologie. Faculté des Sciences et Techniques.
Parc Valrose. 06034 NICE Cedex (93-51 91 00)

Summary : On-line analysis of sleep-waking behaviour in the rat was effected on microcomputer APPLE II. Because of some short-lasting behavioural stages (seven on the whole), the analysis is performed second by second. The percentage of stage occurrence and distribution (histogram) are computed by quarter, hour and 24 hours, the on-line analysis extending during several recording days. The interindividual differences don't ever permit to use standard calibration parameters : an off-line program allows an adjustment of determinative parameters from specific digitalized sequence in each rat.

Les méthodes d'analyse automatique de tracés électrophysiologiques sont devenues indispensables dans l'étude des comportements, notamment celui de veille et de sommeil, qui constitue un biorythme fondamental. Elles ont évolué parallèlement au marché des calculateurs. Un système automatique de quantification a d'abord été proposé sur calculateur IBM 1800 (1), puis sur miniordinateur TEXAS 980A (2). L'introduction des microcalculateurs a apporté des avantages indéniables et c'est sur APPLE II que nous présentons aujourd'hui une méthode d'analyse du rythme veille-sommeil chez le Rat. Ainsi au moyen d'un appareillage peu volumineux et de coût modique, d'importantes applications pourront être envisagées dans des domaines variés, depuis la chronobiologie fondamentale jusqu'à la pharmacologie appliquée.

Sept phases ont été retenues comme constituant le cycle veille-sommeil du rat : 1/ l'éveil attentif, caractérisé par la présence d'un rythme hippocampique de type thêta; 2/ l'éveil normal sans thêta; 3/ les ondes lentes lors de l'endormissement; 4/ la période des fuseaux avec l'approfondissement du sommeil lent; 5/ le stade intermédiaire, phase brève (1 à 3 sec.)

mais importante sur le plan fonctionnel (3); 6/ le sommeil paradoxal; 7/ les périodes de bouffées de mouvements oculaires du sommeil paradoxal. Ces états sont caractérisés à partir de quatre dérivations électrophysiologiques : le cortex frontal, le cortex occipital (activité thêta), les mouvements oculaires et l'électromyogramme des muscles dorsaux de la nuque. Du fait de la brièveté du cycle (12 min. environ) et de ses phases (surtout la cinquième), l'analyse se fait seconde par seconde.

Les programmes de détermination en temps réel des états sont écrits en Assembleur 6502 et BASIC sur APPLE II plus. Ils s'exécutent avec la configuration suivante :

- un APPLE II plus (ou IIe) équipé de 48 K Octets RAM,
- une extension mémoire de 16 K Octets,
- une horloge temps réel,
- une carte de conversion AD/DA 8 bits (temps de conversion : 9 µs),
- une ou deux disquettes APPLE DISK (144 K Octets),
- et une imprimante (Silentype).

L'organisation des traitements fait appel à trois programmes distincts :

1/ La détermination en temps réel des états assure les fonctions suivantes :

- l'initialisation du temps réel;
- l'acquisition des signaux (64 Hz, tranches de 1 sec.);
- le filtrage par FFT et le calcul des énergies;
- la détermination des états;
- l'application des logiques d'organisation;
- le calcul des statistiques;
- le stockage par enregistrement sur disque des énergies, des états et des statistiques.

2/ Le tracé en temps différé des statistiques comprend les opérations suivantes :

- la reprise sur disque des données stockées;

- la génération d'image en mémoire virtuelle sur disque;
- le tracé des histogrammes sur imprimante.

3/ Le programme de réglage en temps différé comprend :

- l'entrée et l'éventuelle modification des valeurs de réglage;
- le stockage des paramètres de fonctionnement en temps réel;
- la détermination en temps différé des états, à partir de données enregistrées;
- l'affichage des critères de détermination.

Le principe des traitements des états est le suivant :

1/ Pour la détermination des états :

Un état est caractérisé par une certaine répartition spectrale de l'énergie recueillie sur chacune des quatre dérivations traitées. La détermination se fait chaque seconde par évaluation d'un critère d'appartenance à chacun des sept états reconnus. Chaque dérivation permet le calcul des critères partiels qui sont combinés pour obtenir des critères globaux dont le maximum définit l'état le plus probable. Les calculs d'énergie nécessaire à l'évaluation des critères partiels sont faits en numérique après analyse spectrale par FFT (donnant un ΔF de 1/2 Hz) de manière à synthétiser des filtres très raides et facilement réglables. Les critères partiels sont des évaluations de la probabilité de réalisation de chacun des états, compte-tenu de l'énergie observée pour la dérivation par rapport au domaine de variation de l'énergie pour l'état considéré.

2/ Pour la logique d'organisation :

La détermination par tranche d'observation d'une seconde peut conduire en présence de signaux peu typés à la création d'états parasites brefs encadrés d'états stables bien typés. Une logique d'organisation interdit la présence de ces états incompatibles avec la réalité du phénomène physiologique traité. Pour cela, la logique d'organisation dispose de table préétablie donnant des patterns d'états incompatibles ainsi que l'état à forcer pour rétablir une séquence acceptable du point de vue du physiologiste. Les patterns ont des longueurs pouvant aller jusqu'à 9 secondes.

3/ Pour l'affichage et l'édition en temps réel :

Les voies acquises sont présentées en temps réel sur l'écran de la console du APPLE sous forme oscilloscopique à quatre traces. Les états bruts déterminés sont affichés. Les états issus de la logique d'organisation sont utilisés pour un calcul de statistique par tranche (1/4 d'heure; 6 heures; 24 heures) dont les résultats sont édités en temps réel sur imprimante.

4/ Pour l'enregistrement et le stockage des données :

Les informations en cours de traitement peuvent être enregistrées sur disquette pour constituer les données d'entrée du réglage en temps différé (durée stockée : jusqu'à 148 sec.). Les états et les statistiques sont systématiquement stockées sur disque pour utilisation par les programmes en temps différé.

Toutefois, les activités cérébrales spontanées ne révèlent qu'un aspect des processus centraux. L'excitabilité centrale, testée par la technique dite des potentiels évoqués, et l'activité psychomotrice de l'animal, dans sa répartition circadienne, fournissent également des informations précieuses au chercheur. Aussi est à l'étude un programme de quantification de ces données, qui pourra être couplé à la détermination automatique des états de veille et de sommeil. Un système analogue sera envisagé chez l'Homme, du fait de l'importance de ses applications cliniques.

1/ GOTTESMANN et al., Rev. EEG Neurophysiol., 1976, 6: 37.
2/ GOTTESMANN et al., Brain Res., 1977, 132: 562.
3/ GOTTESMANN et al., Waking and Sleeping, 1980, 4: 111.